BASIC HUMAN
GENETICS

BASIC HUMAN
GENETICS

SECOND EDITION

Elaine Johansen Mange
Arthur P. Mange

University of Massachusetts

Sinauer Associates, Inc.
Publishers
Sunderland, Massachusetts

The Cover

Red hair has always stirred interest, but how it is inherited has never been clear. One recently discovered gene encodes a hormone-binding receptor on pigment-forming cells. See Chapter 7.

Photograph © 1991 by Joel Meyerowitz. From *Redheads* by Joel Meyerowitz, 1991, Rizzoli International Publications, Inc. Used with the kind permission of the photographer.

For information address Sinauer Associates Inc., 23 Plumtree Road, Sunderland, MA 01375-0407 U.S.A.

FAX: 413-549-1118
Email: publish@sinauer.com
www.sinauer.com

Inset photograph, p. 3, © Jeff Greenberg / Visuals Unlimited.

Quotes from *The Demon-Haunted World* by Carl Sagan, pp. 6 and 8, © 1995 by Carl Sagan and reproduced with permission from the Estate of Carl Sagan.

Detail of Figure 10.10 on p. 191 appears with the permission of the Metropolitan Museum of Art, New York.

Library of Congress Cataloging-in-Publication Data

Mange, Elaine Johansen
    Basic human genetics / Elaine Johansen Mange, Arthur P. Mange -- 2nd ed.
        p.   cm.
    Includes bibliographical references and index.
    ISBN 0-87893-497-9
    1. Human genetics.   I. Mange, Arthur P.   II. Title.
QH431,M2785 199
599.93'5--dc21   98-40485
                                                           CIP

Printed in U.S.A.

5 4 3 2 1

Dedicated to the memory of
Laurence (Larry) M. Sandler, 1929–1987,
teacher, scholar, friend

Contents in Brief

Contents

PRACTICES AND PROSPECTS

Boxes

Preface

In this edition, as in our first edition (which won awards from the American Medical Writers Association and the Text and Academic Authors Assocation), we have tried to be interesting, accurate, up-to-date, and clear. The book is intended as a one-semester course in human genetics for students with diverse interests. We introduce the findings of genetics, show how they operate in humans, and discuss their implications for individuals and for society. Readers with a good background in high school biology and chemistry should have little difficulty with the material.

User-friendly features include new full-color illustrations with distinctive "balloon" captions, case histories of genetic disorders, sidelights of human interest, and an extensive glossary. In a few special cases, we provide the context of discovery, because science is more than just methods, facts, and concepts. Learning aids include chapter summaries, lists of key terms, questions (with full answers), and a detailed index. For readers who may need review, we have retained the appendix on basic chemical concepts. A new appendix cross-references human genetic disorders to the print and on-line versions of the encyclopedic *Mendelian Inheritance in Man* (McKusick 1997). Citations to Internet sources are given where appropriate.

The text is divided into six parts, with a total of 20 chapters:

Part	Topic	Number of chapters
One	Essential Ideas	5
Two	DNA	3
Three	Extensions	4
Four	Cytogenetics and Development	3
Five	Genetic Disease	3
Six	Practices and Prospects	2

Compared to the first edition, the chapters on the structure, function, and manipulation of DNA are closer to the beginning of the book. Molecular explanations of facets of human genetics appearing throughout the book have thus been made more useful. We describe, of course, dozens of new research results that have come to light in the last few years, including:

- The discovery, science, and politics of genes that influence the occurrence of breast cancer
- The role of pattern formation genes in early embryonic development, and the striking similarity of these genes among widely different animals

- The cloning of mammals from single embryonic and adult cells
- Progress of the Human Genome Project
- New discoveries about many genetic conditions, including Williams syndrome, hemochromatosis, albinism, facial and limb defects, and heart problems among well-conditioned athletes
- The position of Neandertal man in human evolution
- The proper forensic use of DNA fingerprinting and the reaction of the jury in the Simpson trial
- Maintaining genetic privacy in an age of increasing genetic testing

Despite all the new material, the text is the same length as before. The chapters on population genetics and evolution have been merged and condensed, the chapter on immunogenetics has been shortened, and nonessential details throughout have been omitted.

We hope that our readers will sharpen their ability to evaluate reports about scientific discoveries. However, the problem with virtually all science textbooks, including this one, is that they do not tell the whole truth. Instead, they give shortened and therefore unrealistic accounts of how science operates. But telling the whole truth would take more space than we have been allotted, and reading the whole truth would take more time than the average student can devote to a one-semester course.

Real science often wanders, sometimes without a map, and may wind up far astray on dead-end trails. But just as travelers occasionally stumble into fascinating out-of-the-way places that provide new experiences and insights, scientists may find some amazing things during their meanderings. For many of them, this uncertainty is what makes the trip worthwhile. We hope that our short account of basic human genetics has captured some of both the uncertainty and the wonder of science.

We have carefully chosen references to past and current scientific literature. The citations in the text, giving the author's name and year of publication, refer the reader to a listing of bibliographic data at the back of the book. Non-science students should be able to enrich their comprehension by looking up many of these references, which include articles in *The New York Times*, *Scientific American*, and the news sections of *Science* and *Nature*.

We are grateful for the help of many people in the library system at the University of Massachusetts at Amherst. We thank especially the staffs of the Biological Sciences Library and the Interlibrary Loan Office at the W. E. B. Du Bois Library. We are also grateful to the following persons for reviewing parts of the manuscript:

- Robert E. Braun (University of Washington, Seattle)
- Rita Ann Calvo (Cornell University)
- Richard W. Erbe (The Children's Hospital of Buffalo)
- James Fristrom (University of California, Berkeley)
- Stanton F. Hoegerman (The College of William and Mary)
- Carl A. Huether (University of Cincinnati)
- Mindaugas S. Kaulenas (University of Massachusetts, Amherst)
- Cran Lucas (Louisiana State University)
- Randall W. Phillis (University of Massachusetts, Amherst)
- Mary S. Tyler (University of Maine, Orono)

Their comments and suggestions for improvement were extremely helpful. We also owe a continuing debt of gratitude to the many colleagues and students who reviewed our previous books. The responsibility for any errors that remain, however, is ours.

We thank J/B Woolsey Associates for preparing the multitude of excellent illustrations. Finally, we are pleased to acknowledge the help of the entire staff of Sinauer Associates, especially Andrew Sinauer (president), Carol Wigg (editor), Christopher Small (production manager), Janet Greenblatt (copy editor), Wendy Beck (page layouts), Jeff Johnson (book and cover design), and Dean Scudder, Marie Scavotto, and Susan McGlew (marketing). Their expertise, attention to detail, creativeness, and endless patience have contributed greatly to the finished product.

Elaine Johansen Mange

Arthur P. Mange

PART 1

Essential Ideas

CHAPTER 1

A Frame for Genetics

The year was 1934 and the place was Norway. Mrs. E., an intelligent and persistent young mother, had consulted many doctors, hoping that somebody could explain why her two children were profoundly retarded and why they gave off such an odd, mousy odor. Both her 7-year-old daughter and 4-year-old son had seemed completely normal at birth, but within a few months had begun to show serious developmental problems. The girl did not walk until she was 22 months old. She could only speak in single words and was only partly toilet trained. She was hyperactive, always moving about aimlessly. She showed some muscle rigidity and a skin rash, but otherwise her physical health was relatively good. The boy could not walk, sit without support, speak, feed himself, be toilet trained, or focus his eyes. All he could do was cry, smile, and gurgle unintelligibly. Aside from some urinary problems, irregular muscle spasms, and rough skin, his physical health was also fairly good.

Finally, the mother brought her children to Professor Asbjørn Følling, a biochemist and physician. He noted the severe

3

retardation and peculiar odor and also a somewhat stooped posture and quite fair skin in both children. While analyzing their urine, he discovered that it turned green after he added a chemical called ferric chloride—a very unusual reaction. In two months he isolated and purified the unexpected compound that was responsible for the green color.

Følling then undertook a nationwide search for more cases of what he suspected was a newly discovered disease associated with severe retardation. After testing the urine from hundreds of patients in institutions for the retarded, mentally ill, or aged, he found a total of 34 cases from 22 families. With the help of a Norwegian geneticist, Følling also showed that the condition was inherited in a simple way. Although the mothers and fathers of affected individuals were themselves normal, they both carried the same undetected (*recessive*) disease-causing hereditary factor (*gene*) and passed it on to their children. The children, having two doses of the defective gene and lacking a normal counterpart of that gene, would thus be affected.

In 1938, Følling and a colleague reported that all of the patients had too much of a substance called *phenylalanine* in their urine and blood and hypothesized that these patients suffered from a defect in the way their bodies handled phenylalanine (Følling and Closs 1938). But the exact cause was not known until 1947, when the American scientist George Jervis reported that the culprit was a specific enzyme that in normal people converts phenylalanine to a substance called *tyrosine*:

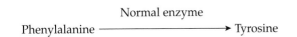

$$\text{Phenylalanine} \xrightarrow{\text{Normal enzyme}} \text{Tyrosine}$$

An **enzyme** is a protein molecule that is specialized to greatly increase the rate of a certain chemical reaction without itself being changed by that reaction. Enzymes are made by cells under the control of genes. The reaction described here (the conversion of phenylalanine to tyrosine) did not occur in people whose enzyme was either missing or defective. So they had too little tyrosine in their bodies and far too much accumulated phenylalanine—some of which was converted to another substance and excreted in their urine. There it was detected by the ferric chloride reaction. The excess of phenylalanine (and perhaps some derivatives) or the lack of tyrosine (and its derivatives), or some combination of these, must cause the brain damage in all these patients.

Følling and many others continued to study this disease, which later was named **phenylketonuria**, or **PKU**. (The name refers to the phenyl ketones—conversion products of excess phenylalanine—found in the urine.) It turns out that most of the phenylalanine present in our bodies comes from the breakdown of the proteins we eat. Thus, the scientists reasoned, maybe the disease could be treated by removing much of the protein and phenylalanine from patients' diets. This strategy was tried in the 1950s—and, amazingly, it worked. If PKU was detected within a few weeks of birth, babies who were fed a specially concocted low-phenylalanine diet escaped brain damage (Figure 1.1). If kept on the restrictive diet until their teen years, they usually developed intelligence and behavior patterns within the range shown by normal people.

But the special diet was ineffective if started later than one or two months after birth, so early identification of affected babies was the key to preventing brain damage. By 1963, a simple test had been devised that would detect excess phenylalanine in a few drops of blood. Soon after, entire populations of babies were being tested for PKU within a few days of birth. Clearly, there was much to rejoice about. For the first time, an inherited disease was being conquered: Its tragic effects could be prevented by a change in the environment, namely, dietary therapy. Those who were born with a double dose of the PKU gene variant (and no normal version of this gene) could be detected. Indeed, thousands of young people who previously would have been doomed to profound retardation by PKU have been rescued and allowed to live normal lives.

Unfortunately, the story does not end here. When the young women who were spared the ravages of this condition reached child-bearing age, it was found that their babies were often irreversibly retarded. The reason is this: The mothers had been allowed to stop their restrictive diets some years before, and with a regular protein-rich diet the levels of phenylalanine in their blood increased dramatically. This situation did not seem to seriously affect the young mothers, but it was devastating for the babies they bore. Now it is clear that unless these women return to the restrictive low-phenylalanine diet before they become pregnant, the medical triumph over PKU will be largely undone. So the struggle to conquer PKU continues.

Genetics and Social Issues

The case of phenylketonuria illustrates how far the science of human genetics has progressed in the past 60 years. It also shows how the solution to one medical/scientific/social problem can sometimes quite unexpectedly give rise to other medical/scientific/social problems that are equally complex and difficult to solve. Finally, it exemplifies the extremely important but often ignored principle that genetic conditions can be treatable:

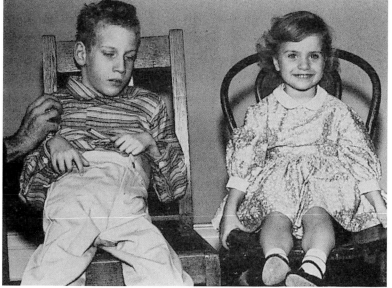

Figure 1.1 A photograph of siblings born with phenylketonuria (PKU). The untreated 11-year-old boy (left) is severely retarded. His 2½-year-old sister (right), who was treated from early infancy with a low-phenylalanine diet, has normal intelligence. (Photograph by Willard Centerwall, from Lyman 1963.)

> It would be well at an early stage in this book to dispel a common misconception. This is the concept that what is inherited is fixed and immutable, and nothing can be done about it. At the very least, this may become a convenient excuse for all one's shortcomings. Carried to an extreme, it leads to a philosophy of "biological determinism" which may stifle initiative and justify all manner of personal failures. That this concept is spurious is quite apparent from a brief consideration of everyday medical practices. We have already quoted several illustrations of the environmental modification of the expression of the [genetic makeup of a person]. Many more instances could be quoted from the field of medicine, involving the successful treatment of inherited disability. (Neel and Schull 1954)

The science of human genetics is now a thriving discipline with sophisticated techniques for the diagnosis and treatment of many inherited conditions. Indeed, these days we can hardly pick up a paper or magazine or watch the news on TV without hearing about some new genetic breakthrough. Along with these possibilities for reducing human suffering, however, have come many personal, social, political, and legal problems:

• What happens if a woman who was born with PKU (and rescued by dietary therapy) refuses to return to the low-phenylalanine diet before or just after she becomes pregnant? Does society have any right to interfere in this individual's decision-making process?

• It is now possible for the relatives of individuals suffering from certain devastating but incurable disorders (such as Huntington disease) to be tested for the presence or absence of the causative gene. How might the knowledge of their future fate affect their everyday lives? If results of such tests are not kept private, how might they affect employment and insurance opportunities?

• Should certain populations be screened to detect medically normal individuals who carry certain recessive detrimental genes, such as the sickle-cell gene in blacks or the cystic fibrosis gene in whites? Will such carriers feel tainted or defective?

• Should families be told when a child is born with a poorly understood genetic abnormality? Will the benefits of studying and perhaps later helping such children be offset by the possibility of stigmatizing them?

• Should society be asked to pay for expensive treatments for individuals with genetic disease? If so, should society have a say in whether such affected individuals ought to be born, perhaps even preventing their birth through prenatal diagnosis and selective abortion?

• Should parents be allowed to choose the sex of their children? Could such choices lead to significant changes in the sex ratio of some populations? (In China, where sons are favored and couples are limited to one child, male children now outnumber females by over 10%.)

• If it becomes possible to change our genetic destiny through techniques such as gene therapy, how should these methods be controlled? Might they lead to changed ideas of what constitutes a "normal" body and "normal" behavior? Might they cause parents (and society) to expect and demand ever more "perfect" children?

Tidy answers to such questions do not appear at the back of this book, or anywhere else for that matter. The facts and their interpretations often turn out to be quite slippery, changing as scientists ask new questions and develop new techniques for answering them. However unsatisfying (and perhaps risky) it may be, and however much we crave certainty, we usually have to make do with incomplete information. And because public policy decisions on complex social issues must be based on much more than the facts alone, they should involve the thoughtful participation of all citizens, and not just the scientists.

Science and Scientists

If the public is to have a say in developing guidelines for dealing with certain scientific problems, it should first have a clear sense of what science can and cannot be expected to do. This, in turn, depends on some understanding of how research is carried out. Ironically, great scientific achievements in many fields have led the public to expect more of science than it can possibly deliver: Too little food? Not enough fuel? Too many people? Not to worry—science will take care of everything! Then, when science fails to solve all our old problems and sometimes even creates new ones, disillusionment and hostility set in (LaFollette 1990). Yet withdrawing support for science can be as harmful to public welfare as ignoring the ethical problems that arise when science begins to impinge on certain human rights and values.

What is science, anyway, and how can we apply it to our daily lives? In an engaging and provocative book, Carl Sagan (1995) considered these questions at some length:

Science invites us to let the facts in, even when they don't conform to our preconceptions. It counsels us to carry alternative hypotheses in our heads and see which best fit the facts. It urges on us a delicate balance between no-holds-barred openness to new ideas, however heretical, and the most rigorous skeptical scrutiny of everything—new ideas and established wisdom. This kind of thinking is also an essential tool for a democracy in an age of change. …

Science is a way to call the bluff of those who only pretend to knowledge. It is a bulwark against mysticism, against superstition, against religion misapplied to where it has no business being. If we're true to its values, it can tell us when we're being lied to. It provides a mid-course correction to our mistakes. …

Finding the occasional straw of truth awash in a great ocean of confusion and bamboozle requires vigilance, dedication and courage. But if we don't practice these tough habits of thought, we cannot hope to solve the truly serious problems that face us—and we risk becoming a nation of suckers, a world of suckers, up for grabs by the next charlatan who saunters along.

It would help if scientists tried harder to explain their everyday work to the public. On the other hand,

---

BOX 1A   A Baloney Detection Kit

We know we can't believe everything we see or hear or read. And we are constantly bombarded with the results of studies about health risks, for example, that are often contradictory (Taubes 1995). So how do we figure out which stories to ignore and which ones to take seriously?

"In science we may start with experimental results, data, observations, measurements, "facts." We invent, if we can, a rich array of possible explanations and systematically confront each explanation with the facts. In the course of their training, scientists are equipped with a baloney detection kit. … What's in the kit? Tools for skeptical thinking. What skeptical thinking boils down to is the means to construct, and to understand, a reasoned argument and—especially important—to recognize a fallacious or fraudulent argument." (Sagan 1995)

Sagan offers several all-purpose tools for evaluating information or ideas of any kind. They include the following considerations:

Have the "facts" been independently confirmed by several researchers?
Have the "facts" been carefully quantified?
Does every link (not just most links) in a chain of argument work?

Have we chosen the simplest hypothesis that will explain the data?
Have other hypotheses and points of view been considered and tested and debated? Do any of them explain the data just as well?
Can the hypothesis be tested and thus potentially disproved?

To these general tools we add some more detailed questions to our handy baloney detection kit, for use especially in evaluating medical news:

Who did the study and where was it done? Was the work carried out at a university or a research institution? At a drug or biotechnology company?
Who paid for the research? A government agency? A not-for-profit foundation or organization? A drug company or other industry? Are desired results likely to be profitable to the sponsor or to the scientists? The quality of research is not necessarily dictated by the type of sponsor, but we should always consider the possibility of vested interests when we assess the latest scientific breakthroughs.
What and how large were the good effects of a drug or treatment? What and how large were the bad effects? Rather than accepting an unspecified "statistically significant" difference, which could be small and medically unim-

portant, look for a specific percentage of difference in effects between the treated group and the untreated controls. Also look for large benefits and small risks.
What was the organism studied? Humans? Rats? Fruit flies? Yeast? For practical, legal, and ethical reasons, many preliminary studies cannot be done on humans. And because biochemical processes in other organisms resemble those in humans—often to an astonishing degree—it makes sense to use other organisms first. But we should be skeptical about applying those results to ourselves until scientists have actually carried out the same studies in people.
Was the human research done on cells in test tubes or on whole people? Reports of test tube cures may raise our hopes, but we can't really be sure that a drug will work safely until it has actually been swallowed by or injected into or rubbed on intact bodies.
How large were the numbers? 20? 200? 2,000? *And how long did the tests last?* A month? A year? A decade? Sometimes the good effects are only temporary. Sometimes bad side effects occur in only a very small proportion of test subjects, or they develop only after some period of time has elapsed. Thus, the bigger the numbers and the

premature publicity—for example, divulging research results to the press before publishing them in scientific journals—can create problems too, especially if the results do not hold up to further scientific scrutiny (Maran 1991). If hyped-up media announcements of new discoveries were matched by lists of what we still do not know and judged against some standard rules of evidence, a more balanced picture would emerge (Box 1A). Even the most prominent medical journals may publish an occasional article of somewhat questionable scientific merit on "hot" topics guaranteed to generate publicity (and, incidentally, healthy advertising and subscription revenues). One recent example is an article, published by *The Journal of the American Medical Association* (often referred to familiarly as "JAMA"), suggesting that eating fish could reduce sudden cardiac death by 50%. Although the study was flawed, it received worldwide media attention and even led to a temporary increase in the sale of fish. Interviews with editors of this journal and *The New England Journal of Medicine*, reveal that the continued financial success of these highly competitive businesses may now depend on their appeal to the media and the lay public as well as to their professional audience (Shell 1998).

Nonetheless, science is usually viewed with awe as well as some fear or suspicion. It is perceived as a realm that most people cannot understand, let alone hope to enter. Such stereotypes, hopes, and fears can be dangerous if they affect public policies concerning the use of science. For these reasons, we wish to elaborate a bit on the popular images of scientists.

The people who "do" science are, like any other group, quite varied. Yet they tend to be stereotyped and are often misunderstood. In cartoons and comic strips, "mad scientists" abound (Figure 1.2). In movies and books, scientists are often pictured as brilliant but out of touch with reality, or as totally objective and dedicated to their work, or as coldly sinister (Tudor 1989). Seldom are they treated as just plain folks capable of pettiness or insecurity. Moreover, the social aspects of science—the daily collaboration with co-workers in the laboratory, the informal exchange of ideas at meetings and at parties, the personal interac-

BOX 1A   *(continued)*

longer the tests are, the more precise the results and the more reliable the benefit-to-risk estimate will be.

Was a simultaneous control group used for comparison? Control organisms must be very similar to the tested subjects in every way except that they did not undergo the experimental treatment in question. This is always important in biological research, but critical for testing new drugs or other medical treatments.

How were the test subjects and the controls chosen, and how carefully were they monitored? If humans were used, were they volunteers or nonvolunteers? Did they represent the general population or a special subgroup? Might they have been healthier or less healthy than average, more motivated or less motivated? Did supervised treatment take place in a hospital or clinic, or were subjects left alone and just assumed to have followed the prescribed regimen?

Were the studies double-blind? In a typical double-blind study, the subjects do not know whether they belong to the test group or the control group, and the researchers also do not know which group any subject belongs to until after the results are tallied. The *placebo effect*—an improvement reported by people who think they are taking medication but who are actually given fake pills or treatment—is well documented. Likewise, especially if differences are subtle, even the most honest investigators may unconsciously bias individual judgments if they know whether a subject is in the test group or the control group. (Even Mendel has been accused of generating data that were "too good"!)

Did the researchers look at the future or at the past? The better studies are *prospective* (looking forward and actually observing what happens to the subjects during an experimental period) as opposed to *retrospective* (looking backward and relying on possibly incomplete or faulty memories and past records of the subjects).

Has the research been published in a reliable scientific journal, or has it only been announced at a meeting or to the media? Be especially wary of "science by press conference," that is, announcing results to the media before they have been reviewed by peers and published in a scientific journal.

Are the new findings reported as part of a "big picture" survey and compared with other studies on that topic? When new research is presented as an isolated breakthrough rather than in the context of related research, we have trouble assessing its merit. Sometimes the media will give a big play to a small study that finds a weak association between some risk factor and a particular disease while ignoring a much larger study that shows no such linkage. It seems that scarier news may be preferable to more reliable news! Reporters point out, however, that some scientists and scientific journals may be partly to blame for publicizing their new studies without providing adequate background information.

If there are testimonials by medical authorities or celebrities, were they paid for by the research sponsor? If so, such opinions may be biased and unreliable.

How long will it take for the experimental advance to be parlayed into actual medical treatment? It usually takes years to repeat research results in test animals, develop a drug or treatment for humans, test the drug or treatment in humans, get government approval for the new drug or treatment, and produce it for public consumption. The premature announcement of new cures based on preliminary test results may raise false hopes for millions of people.

"Crack out the liquid nitrogen, dumplings ... we're on our way."

Figure 1.2 Misconceptions about genetics. This cartoon shows the imagined outcome of research involving the transfer of genes from one organism to another. Although whimsical and amusing, it has no basis in reality; yet many people believe that such results are possible. For information on this type of research, see Chapters 6 and 8. (Reprinted courtesy of *The Boston Globe*.)

It might be useful for scientists now and again to list some of their mistakes. It might play an instructive role in illuminating and demythologizing the process of science and in enlightening younger scientists. Even Johannes Kepler, Isaac Newton, Charles Darwin, Gregor Mendel, and Albert Einstein made serious mistakes. But the scientific enterprise arranges things so that teamwork prevails: What one of us, even the most brilliant among us, misses, another of us, even someone much less celebrated and capable, may detect and rectify.

Furthermore, even successful research involves a lot of tedium, and were it not for the hope of finding something exciting now and then, some scientists might at times wish they had chosen another line of work. So why do people do science? Some are strongly motivated by practical or humanitarian goals—such as finding new and better ways of solving old problems, or developing new commercial products, or expanding the space program, or conquering diseases, or feeding the hungry, or solving environmental problems.

Others, however, are simply curious about the world around them and get tremendous pleasure and satisfaction from finding out how things work and what basic laws govern our universe. If (as sometimes happens) their research ends up having some practical or humanitarian value, well and good; but that is not the main reason they do it. In an essay on science and scientific attitudes, astronomer S. Chandrasekhar (1990) points out that curiosity was the motivating force behind the ancient Greeks' fascination with geometric curves:

> Apollonius of Perga wrote eight monumental volumes devoted to these curves. ... But it did not occur to him, or to the other Greek mathematicians, that the curves, which they studied so earnestly for their intrinsic beauty, had any relevance to the real physical world.

Yet some eighteen centuries later, when Kepler was analyzing the orbits of the planets on the copernical system, he discovered that the very curves that the Greek mathematicians had studied for their intrinsic mathematical beauty were exactly those needed to represent the orbits of the planets.

A more recent example of a scientific project undertaken purely for the sake of curiosity was the 1990

tions that can speed up or slow down the solution to a problem—are rarely mentioned.

Textbooks (including this one) usually describe only the successes; rarely do readers learn about the intervening false starts, dead ends, misinterpretations, or outright failures. As Harvard scientist Stephen Jay Gould (1986) points out, the same criticism can be made of scientific journals, whose standard format (introduction, materials and methods, results, discussion, conclusions) makes scientific research look more orderly than it really is. Gould notes, "The false starts are in the wastebasket, not [in] the *Science Citation Index*." Francis Crick, codiscoverer of the DNA double helix, puts it this way (Horgan 1992): "Exploratory research is really like working in a fog. You're just groping. Then people learn about it afterwards and think how straightforward it was."

Truth be told, scientific research quite often yields confusing or negative or equivocal results, even at the hands of the world's greatest scientists. Sagan (1995) made the following suggestion:

space launch of the Hubble telescope. This instrument was designed to see ten times farther into the past than any earthbound telescope—perhaps even as far back as the Big Bang that may have produced our universe. As astronomer Timothy Ferris (1990) comments:

> It has no practical purpose at all. ... Hubble is an instrument of discovery; its paternity lies as close to the Niña, the Pinta and the Santa Maria as to Pasteur's test tubes or Einstein's notebooks; and it is intrinsic to discovery that one cannot accurately predict what will be discovered. ... The Hubble Space Telescope is a machine for subjecting our conception of the wider universe to an ordeal by fire.

The disastrous failure of its giant lenses to focus properly, which was due to faulty production and inspection procedures, also subjected its manufacturers to an ordeal by fire. But the error was corrected by astronauts in space in December 1993, and since that time, Hubble has performed more spectacularly than astronomers had even dared to hope.

The incredibly clear images it relays back to earth have included many amazing firsts: the collision of two distant galaxies, giving rise to the birth of new stars; the collision of the comet Shoemaker-Levy 9 with the planet Jupiter; the explosion of a giant supernova; a supermassive black hole; a new quasar surprisingly close to Earth; a volcanic eruption on Jupiter's moon Io; climate changes on Mars; several new planets outside our solar system, two of which might be able to sustain life; and a huge new star, the brightest ever seen in the Milky Way. Images and data from the Hubble space telescope have also shaken up some fundamental ideas about the universe, which now appears to contain 50 billion (rather than 10 billion) galaxies, each one with billions of stars. And its age may be a few billion years less than previously thought:

> The Hubble still has at least a decade of useful life, and astronomers are convinced that before it's mothballed the telescope will answer many of the most profound mysteries of the cosmos: How big and how old is the universe? What is it made of? How did the galaxies come to exist? Do other Earth-like planets orbit other sunlike stars? ... Indeed, before the Hubble is finished, all sorts of lectures, textbooks and astronomical theories will have to be rewritten. (Lemonick 1995)

In the course of solving practical problems and satisfying human curiosity, the pursuit of science has usually involved several recurring themes. One is the codiscovery of a principle by several researchers working independently. In some cases, codiscoveries are not surprising because the exchange of ideas in scientific journals and at meetings may draw many people to work on an interesting problem, especially in hot new areas of inquiry. In other cases, and particularly in recent times, the problems have become so complex that solutions require teamwork by many laboratories, each specialized in a different type of research.

A second theme in the history of science is the so-called *premature discovery*, which, although correct, is neglected for a long time by other scientists. There are several striking examples of this in genetics, starting with the work of Gregor Mendel. On the other hand, we also know of a few *postmature discoveries*, which are long past due and swiftly accepted by other scientists. One example is the 1946 discovery by Joshua Lederberg, then a 21-year-old medical student, that bacteria can reproduce by sexual union and not just by solitary fission—a finding that opened up the vast field of bacterial genetics (Lederberg 1986) and later earned him a Nobel prize. Another postmature discovery occurred in 1983, when Cetus Corporation scientist Kary B. Mullis "stumbled across a process that could make unlimited numbers of copies of genes, a process now known as the polymerase chain reaction (PCR)" (Mullis 1990). This procedure, which revolutionized the field of molecular biology, earned Mullis a Nobel prize in 1993.

A final major theme is the critical role that new technology plays in the development and testing of scientific ideas. The process is never-ending: Better microscopes, new ways of staining cells and cell parts, improvements in growing and handling cells, new techniques for isolating and analyzing genetic material, increasingly powerful and accessible computers, and many other technological advancements were all necessary for knowledge to progress to its current level. These research tools do not by themselves generate, prove, or disprove hypotheses. People do that, sometimes alone but rarely in complete isolation.

Of all the world's scientists who ever existed, most are alive today. A tour of their laboratories would reveal a wide range of intellects and personalities—from geniuses to plodders, from people with many talents to those with no outside interests, from the showy to the shy, from highly competitive workers to those with modest aspirations, from generous and caring individuals to selfish boors.

Very few of even the most intelligent and ambitious scientists will ever be mentioned in a textbook. Indeed, some of their published papers will never be cited anywhere, because scientific reports are now being turned out at such a rate that nobody can keep track of more than a tiny fraction of them. Browsing through scientific journals, we find a broad and amazing array of articles whose titles are so specific that they may sound silly to a layperson, especially one who feels that science should only solve practical problems and not bother with apparently trivial pursuits. But the gradual accumulation of unexciting bits of knowledge and experimental techniques by many workers often sets the stage for the great discoveries or insights of a few, usually in ways that could never have been predicted.

Genetics in Agriculture, Medicine, and Society

From feature stories in the popular media, and perhaps from personal experience, most of you are probably aware that genetics now plays an important role in medicine and in certain social issues. But its key role in agriculture goes much further back.

For thousands of years, people have been puzzled and intrigued by the similarities and differences seen within families of humans, other animals, and plants. Sometimes they were able to develop special breeds of crop plants or domesticated animals by selective mating. Yet their successes were hit-and-miss, depending partly on luck and partly on past observations, but with no real understanding of the biological principles behind them.

From anthropological findings, written records, and ancient art, we know that agriculture itself began at least 10,000 years ago. Although experts disagree on the details, it now appears that there were several centers of origin. These include the Near or Middle East, Asia, Africa, and Central America. The earliest herders and farmers selectively bred both animals and plants, using as parental stock those individuals or strains possessing the most desirable traits and culling out the offspring with the least desirable traits. In this way, they gradually improved on or developed animal breeds (such as goats, sheep, oxen, camels, dogs, cattle, and horses) and crops (such as wheat, rice, and maize). Figure 1.3 shows how the process of selection, carried out for thousands of years, can take advantage of the tremendous genetic diversity within a given species.* Trade and migration patterns also played important roles in the evolution of domesticated animals and plants.

Only during the past century have clear explanations for such improvements emerged. By experimenting with many plants, animals, and microorganisms,

Figure 1.3   Dog breeds. Archaeological evidence suggests that the domestic dog (*Canis familiaris*) was probably domesticated from the gray wolf (*Lupus*) about 14,000 years ago. But some recent (and controversial) DNA studies suggest that dogs might have been domesticated over 100,000 years ago. In any event, they now exist as more than 100 different breeds that range in size from Great Danes (30 inches shoulder height, 140 pounds) to tiny Chihuahuas (5 inches shoulder height, 4 pounds). All, however, still belong to one species and can interbreed. Perhaps the first wolf-to-dog changes that occurred were an upcurved tail and smaller teeth. Other traits that have been selected by humans include short legs, foreshortened face, drooping ears, and various coat colors and textures. Behavioral variations are seen, too, in dogs bred for different tasks, such as herding, hunting, guarding, pulling, guiding, and household companionship (Ostrander and Giniger 1997).

scientists have learned (1) how traits are passed on from one generation to another, (2) how genetic information is chemically "decoded" and expressed during each individual's development, and (3) how the genetic variability that arises among the members of groups can account for the gradual evolution of populations. These three major areas of inquiry are known, respectively, as transmission genetics, molecular (and biochemical) genetics, and population genetics.

Statistical techniques devised in the early 1900s also improved agricultural practices. By even more intensive selective breeding, interspecies crosses, and gene transfers, agronomists have developed crops that are higher yielding, more uniform, more vigorous, and more resistant to diseases, insect pests, and weed killers (Figure 1.4). In part, the results are seen around the world as the Green Revolution. Using statistical techniques coupled with technological advances such

---

*But there are limits to the selectability of some traits, including, for example, the speed of thoroughbred horses. Despite improved breeding, better training, and excellent health care, the finishing times of American and English racehorses today are no faster in the "classic races" (about 2.5 minutes) than they were 4 or 5 decades ago. Yet the speeds of American quarter horses running 20-second sprinting races have shown small increases. The limiting physiological factor may be the accumulation of lactic acid in the muscles, which occurs in the longer races but not in short sprints (Cunningham 1991). Humans, on the other hand, are still improving all their track records—not by selective breeding, but by better training, health care, and equipment. Whether we will reach performance limits is not known (Masood 1996). The metabolisms of glycogen and fat may be important factors.

as artificial insemination and embryo transfer, breeders have also developed livestock with more uniform and commercially desirable traits, such as higher yield and lower butterfat content of milk in cows and better quality and yield of wool in sheep.

The 1980s brought significant advances in tissue culture and in the new molecular techniques for cloning individual genes and implanting them into foreign species. These methods may revolutionize crop and livestock development, but close attention must be paid to the biological, environmental, and social effects of such programs.

Is it possible to significantly change or "improve" the human species through genetics? This is not a reasonable goal if human values and rights are to be preserved. Sir Francis Galton coined the term *eugenics* to describe the hoped-for improvement of our own genetic endowment. It reflects a naive optimism about applying methods of artificial selection, so successful in agriculture, to humans: Just encourage "superior" individuals to be fruitful and discourage or prevent "inferior" people from reproducing at all. Unfortunately, the eugenics movement attracted demagogues

along with idealists, and distortion of the meaning of Darwinian selection contributed to the genocidal Holocaust in Nazi Germany. With its emphasis on eliminating the "bad" genes of paupers, criminals, the feebleminded, and immigrants (in addition to its main targets—Jews, Gypsies, and homosexuals), the Nazi brand of eugenics fell into disrepute.

We now realize that even the most honorable attempts at improving the human gene pool create major problems. For example, the simplest equations of population genetics show that preventing the births

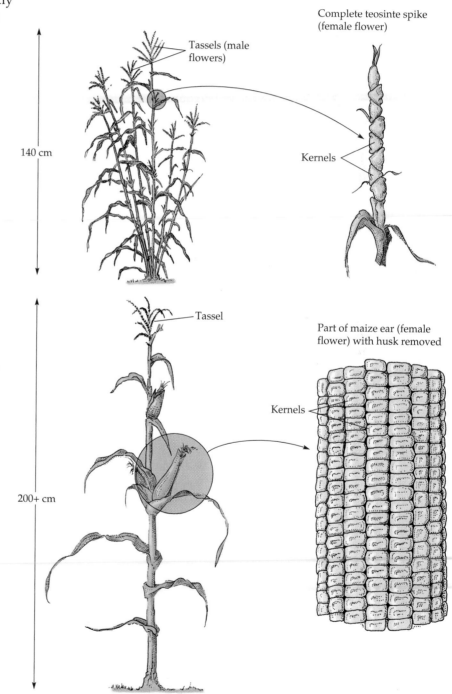

Figure 1.4 Corn evolution. Modern maize (*Zea mays*) probably evolved from a wild grass called teosinte (*Zea mexicana*). The male flowers (tassels) in the two species are similar, as is the general form of the stem and leaves. Teosinte, however, has many stems and numerous small female flowers (spikes), whereas maize has one stem and a few large female flowers (ears). The small spike of teosinte consists of a single row of six to ten kernels, each enclosed in a hard, triangular fruit case. Beadle (1980) and Galinat (1977) suggest that during domestication, the fruit case of ancient teosinte gradually evolved into the large cob structures of modern maize. Strong evidence for this hypothesis was recently obtained by other researchers. They found that just a few changes in a short stretch of teosinte DNA caused dramatic changes in flower development and seed structure. Individual kernels in the resultant teosinte plants were exposed and maizelike instead of being covered by a hard case (Roush 1996).

of people with rare recessive disorders will not significantly reduce the frequency of these detrimental genes. Preventing the births of people with dominant disorders would quickly eliminate these detrimental genes (except for new mutations), but such coercive programs are unthinkable in free societies. Furthermore, the criteria used to determine whether reproduction should be encouraged or discouraged are fuzzy at best and highly susceptible to social and political mischief.

But human genetics can be employed humanely at the personal level. Increasingly sophisticated prenatal testing and genetic counseling are able to avert family tragedy or reassure those individuals who turn out not to be at risk for certain hereditary disorders. Great caution, care, and sensitivity are required, however, when considering or carrying out the large-scale testing of populations that are known to be at increased risk for certain conditions. Even for those who oppose testing, there is still the problem of just maintaining our current genetic endowment, as when, for example, medical advances preserve detrimental genes that would otherwise be eliminated by the death of these individuals. The possible long-term genetic and social consequences of medical practices will be debated for years to come.

The discoveries reported in this book took place between the 1860s and the 1990s. Our goal, however, is not to give a highly detailed description of all these findings. Rather, we hope to define the broad scope of human genetics while stressing the important experimental role of other species in learning about our own. We try to show how answers to scientific questions rest on layers of knowledge accumulated by many investigators. These layers do not form at a constant rate. Indeed, for certain periods, there may have been no progress at all in some disciplines, perhaps because scientists did not ask the right questions or perhaps because they did not have the right technologies available to answer their questions. Each age sees itself as the modern age; but in science, final answers are rarely final (Box 1B).

Actually, human beings are not the best subjects for genetic study. Researchers are hampered by our long generation time, our small families, our highly varied genetic and environmental backgrounds, and the absence or incompleteness of our family records. Human geneticists also have to search the world's populations for those few families that provide some information on inheritance patterns, rather than relying on the experimental matings that are the stock in trade of all other geneticists. Yet human biochemistry is better understood than that of any other complex species, and research on humans has led to new discoveries—such as the first insight into the relationship between human genes and enzymes and the precise effect of gene mutations on protein structure in humans. And some people with rare genetic conditions do seek out physicians and geneticists. Obviously, the use of experimental organisms has been essential for discovering and testing fundamental genetic phenomena. But because we are so curious about ourselves, scientists have been willing to cope with the special problems of human genetics, or to find ways around them. What is difficult today may be less difficult tomorrow with new ideas and new technology.

Using the Literature

We encourage you to delve further into the literature of genetics, both of the traditional printed variety and on the computer. Pertinent articles and books are mentioned throughout the text and at the end of each chapter. Wherever possible, we cite nontechnical and general review articles, so most references should be understandable to most readers. All print citations are by author and date of publication. A single alphabetical listing of these works, with bibliographical information, appears at the end of the book. For keeping abreast of the latest genetic discoveries, we highly recommend *The New York Times*. Its articles, especially those in the Tuesday science section, are both readable and reliable.

New to this edition are references to Internet web sites, which have proliferated at a remarkable rate. A brief guide to Internet history, jargon, resources, and how-to-connect is presented in a supplement to the February 1996 issue of *Trends in Genetics* (Rashbass and Walsh 1996) and other *Trends* journals published by Elsevier. A short version of this guide is available online at http://www.elsevier.com/locate/trendsguide. A brief listing of some important human genetics web sites can also be found in an article by Lewitter (1996). Because you can easily become frustrated with delays and saturated with such information, step away from your computer occasionally and ask what you really want from it. In addition to specific web sites, various electronic referencing tools should also be available from your college library. We have made good use of one called InfoTrac, which references many newspapers and magazines.

Further Reading

The following readings are appropriate to Chapter 1. Sagan (1995) compares scientific thought with pseudoscience, giving many examples of both. Bauer (1992) very engagingly discusses the roles of (and misconceptions about) science and technology in modern life. Nelkin (1987) and Heussner and Salmon (1988) discuss how the media inform and misinform the public about scientific and medical news. Nelkin and Lindee (1995) describe the varied images and perceived powers of the gene in American culture. Weir et al. (1994) discuss the impact of modern genetics on human self-knowledge.

BOX 1B *Facts Aren't Forever: Changing Ideas about Genes*

Since the birth of the science of genetics, ideas about the nature of hereditary factors have undergone many changes as scientists have devised increasingly sophisticated ways of studying these factors. When he published the results of his experiments in 1865, Gregor Mendel, the father of genetics, didn't have a clue about the physical nature of his "hereditary factors" because major discoveries about cell structure and cell division had not yet been made. But after his publication was rediscovered in 1900, the striking parallels between the behavior of Mendel's "factors" and the behavior of chromosomes in cell division were almost immediately recognized. By 1902, it was accepted that the factors (now called *genes*) reside on chromosomes. Further knowledge about the nature of genes depended on the development of many new technologies and experimental methods. Here are some major examples of changing facts or ideas about the nature and behavior of genes.

Older Facts or Ideas	Newer Discoveries
Genes must be made of protein, because the 20 amino acids in proteins would provide a better code than the four bases of DNA.	Genes are made of deoxyribonucleic acid (DNA), in which a triplet code of three adjacent bases can specify all the amino acids.
One gene gives rise to a whole protein.	One gene gives rise to a string of amino acids (a polypeptide), which may be only part of a protein.
Humans have 48 chromosomes.	Humans have 46 chromosomes.
Within an organism's body, gene expression is the same in all the cells of a given tissue.	In each cell of a female mammal, nearly all of the genes on one of her X chromosomes are randomly inactivated, so gene expression may vary among the cells in a tissue.
A single chromosome may contain several molecules of DNA.	A single chromosome contains just one double-stranded molecule of DNA.
Genes always stay put at a given site on a chromosome.	In rare cases, some genes can move around from one spot to another
All the DNA in our chromosomes consists of genes that code for proteins.	Genes coding for proteins account for only 3% of the DNA in human chromosomes, and some code for RNA products. What the remaining DNA (over 90%) does is not clear.
In all organisms the transfer of genetic information goes in one direction only: from DNA to RNA to protein.	In certain viruses, where RNA is the genetic material, DNA can be made from an RNA template.
A gene consists of an unbroken coding sequence of bases.	In cells with nuclei, a gene consists of coding sequences (exons) that are interrupted by many noncoding sequences (introns) whose total length may greatly exceed that of the exons.
One gene makes only one polypeptide.	In some cases, one gene can give rise to different polypeptides if there are alternative ways of splicing together the RNA exons.
All genes are totally separate and do not overlap.	A few genes contain overlapping or even nested sequences.
A given allele of a gene will be expressed the same way, no matter which parent it is inherited from.	The alleles of some rare "imprinted" genes may behave differently, depending on the parent of origin.
The size of a gene stays constant as it is transmitted from one generation to the next.	A few diseases are associated with unstable genes that change in size from one generation to the next due to expanding (or contracting) triplet repeats.
The two alleles of a gene never interact or affect each other while present in the same cell.	A few cases are known in which one allele can alter or inactivate another allele.

Among the best histories of early genetics are those by Sturtevant (1965) and Judson (1996). Neel (1994) describes his life as a human geneticist. Articles on the history and development of human genetics include Dunn (1962), McKusick (1975), Motulsky (1978), Jacobs (1982), Caskey (1986), and Bodmer (1986). Kevles (1984a–d), in a very readable series of magazine articles, describes the development and evolution of the eugenics movement. And for an overview of some of the more recent discoveries in human genetics, see Epstein (1993, 1997).

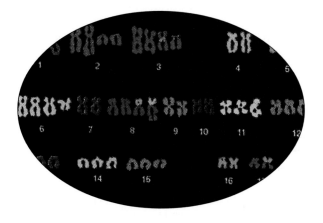

CHAPTER 2

Cells and Chromosomes

In August 1996, scientists reported some exciting new findings about a natural resistance to *AIDS* (*acquired immune deficiency syndrome*) that some people possess. AIDS is a usually fatal disease caused by *HIV* (*human immunodeficiency virus*). It is transmitted by sexual acts or blood exchange, or from mother to fetus or nursing baby. It is most prevalent among homosexual men, intravenous drug users, recipients of blood products, and the sexual partners of all these groups (Chapter 17). No cure exists, but some increasingly effective treatments are being developed. And for reasons that had baffled researchers for years, a few people seem to be immune to the disease.

Two homosexual men who remained free of infection, despite many high-risk exposures to HIV-positive sexual partners, were studied in great detail. Researchers finally discovered that they both had the same defect in a submicroscopic structure called a receptor, usually present in thousands of copies on the surface of each cell. These particular receptors, when normal, provide the docking points through which HIV attaches and enters the cells of people who become infected. But about 1% of the white population have nonfunctional and thus "resistant" receptors because they carry two doses of a previously unknown mutant gene and no normal variant of that gene (Altman 1996).

The scientists who made these discoveries could not have done so without the detailed knowledge of cell structure and the modern genetic technologies slowly and painstakingly developed by thousands of other scientists during the past century. In this chapter we present a brief overview of their most basic findings on cell structure and function.

With the invention of the compound microscope around 1600, a whole new world of beauty and science opened up. Between 1655 and 1879, researchers described the many kinds of *cells* seen in very thin slices of plant and animal tissues. They discovered the darkly staining threads (*chromosomes*) present in every cell and studied their remarkable dancelike behavior during cell division. They put forth a *cell theory*, stating that the cell is the underlying unit of structure in all living organisms. They also proposed that new cells come only from preexisting cells. With these ideas arose the conception of

> an unbroken series of cell-divisions that extends backwards from our own day throughout the entire past history of life. ... It is a *continuum*, a never-ending stream of protoplasm in the form of cells, maintained by assimilation, growth and division. The individual is but a passing eddy in the flow which vanishes and leaves no trace, while the general stream of life goes forwards. (Wilson 1925)

The kinds of cells that researchers usually study—whether taken from multicellular organisms (plants, animals, and fungi) or from single-celled organisms (such as protozoa, yeasts, and some algae)—have a very complex structure (Figure 2.1). Most of their hereditary material resides in **chromosomes** (Greek *chroma*, "color"; *soma*, "body") located in a **nucleus**. The nucleus is separated by a double membrane from the rest of the cell's contents, which make up the **cytoplasm**. Organisms with this kind of cell structure are called **eukaryotes**. Single-celled **prokaryotes** (bacteria and archaebacteria) lack nuclei and have a simpler type of chromosome.

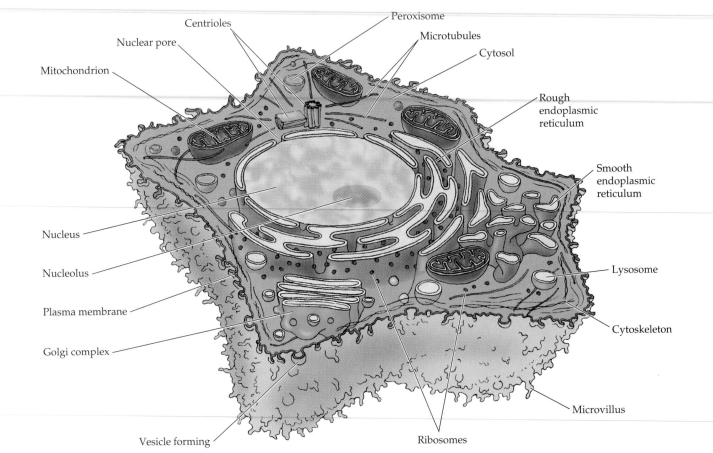

Figure 2.1 A three-dimensional representation of an idealized animal cell, sectioned to show internal structures as they appear in the electron microscope. The proportions of the various structures may differ from one type of cell to another. (Adapted from Cooper 1997.)

Multicellular eukaryotes contain two main kinds of cells. **Somatic cells** make up all the body parts: the skin, bones, muscles, brain, and internal organs of animals, as well as the leaves, stems, and roots of plants. **Germ cells** are reproductive cells—the eggs and sperm of animals, the egg cells and pollen grains of plants—that unite sexually to produce new individuals in the next generation. In this chapter we will focus on the structure of somatic cells only.

The human body contains about 10 trillion cells. All of them came from just one cell, the fertilized egg, but they are no longer identical to each other. Instead, up to 200 different types of somatic cells exist, each specialized to perform one or more unique functions.* Despite the differences in fine structure and function that exist among different cell types, however, there are certain basic structures and functions common to virtually all cells. We concentrate on these in the following overview, going from the inside to the outside of a cell.

Cell Structure

Keep in mind that the structures and substances inside a somatic cell do not just slosh around at random. They are distributed among several different compartments, separated from each other by at least one selectively permeable membrane. In almost every nondividing cell, a nucleus forms the large central compartment. Its **nuclear envelope** consists of two membranes that contain many *pores*, through which materials are exchanged with the surrounding cytoplasm. Within the nucleus are at least one dark-staining **nucleolus** and many finely dispersed chromosomes.

Another compartment is the **cytosol**, a jellylike cytoplasmic matrix in which organelles and intricate membrane systems are suspended. Each cell type has a somewhat different proportion and arrangement of these various components. The cytosol also contains thousands of **enzymes**, protein molecules that speed up chemical reactions. Proteins, the major end products of gene action, are synthesized on tiny particles called **ribosomes**, which number over a million per cell. Some ribosomes lie free in the cytosol, while others are bound to membranes. Supporting and giving shape to the cytosol is the **cytoskeleton**, a network of filaments and tubules made up of fibrous proteins. The tubules, called **microtubules**, act as an internal transport system on which cytoplasmic components can be shuttled to wherever they are needed in the cell. They also move chromosomes during cell division. Lying just outside the nucleus in animal cells is a

pair of tiny **centrioles**, which function as part of an organizing center for microtubules.

The remaining cytoplasmic compartments, which form narrow channels in some areas and large cavities or small vesicles in others, provide the huge surface area needed for biochemical processes. They are also involved in the distribution of substances to different regions of the cell. Attached to these internal membranes or to the cell's outer membrane and acting as "door openers" or "signal senders" are numerous **receptor proteins**. Each is specialized to recognize and admit (or bind to) only a certain type of signaling molecule. Each such complex of receptor plus a signaling molecule triggers a specific chemical reaction (or cascade of reactions) inside the cell.

Enzyme systems are everywhere in all cells. Some are embedded in or loosely bound to membranes, while others lie free in the cytosol. Processes of construction and destruction are constantly going on in cells—but in a highly ordered fashion, with each cellular structure specialized to carry on a certain activity. For example, all worn-out cell parts and invading organisms plus many imported materials end up in the **lysosomes**, membrane-bounded vacuoles that act as the cell's garbage disposal and recycling system. Inside the lysosomes, large molecules are broken down into their smaller components by powerful enzymes. Fortunately, these enzymes function only in the acidic innards of lysosomes; they might destroy the entire cell if released into the surrounding cytoplasm! Several dozen hereditary conditions called *lysosomal storage diseases* are known, each caused by the absence or malfunctioning of a specific lysosomal enzyme. Most of these disorders are very serious, involving skeletal and nervous system abnormalities, and some are lethal (Chapter 16). Other tiny membrane-bounded structures called **peroxisomes** contain enzymes that use oxygen to detoxify certain substances, including alcohol. The malfunctioning or absence of peroxisomes can also cause very serious illnesses (Chapter 16).

The digestion products released from the lysosomes, as well as many other substances coming into the cell, are gradually degraded, releasing the energy needed for doing the work of the cell. An oxygen-requiring stage, called *cellular respiration*, takes place inside **mitochondria**. These miniature power plants—between 300 and 600 per cell—are bounded by two membranes and contain hundreds of mitochondrial enzymes.† The released energy is used to form new molecules and to drive many metabolic reactions.

The **endoplasmic reticulum**, a compartment continuous with the outer membrane of the nuclear enve-

---

*Two or more groups of similar cells can associate to form *tissues* such as bone or muscle; different tissues organize themselves into *organs* (such as the brain or heart); and groups of related organs form *organ systems* (such as the digestive or skeletal system). All of these associations are specialized to perform certain functions.

†Mitochondria (and in plant cells, chloroplasts too) are self-replicating organelles. They are thought to have arisen from free-living, oxygen-requiring bacteria that invaded and were retained by primitive cells that could not use oxygen.

lope, is a large and ever-changing network of tubules and cavities of various shapes and sizes. It is involved in the production of all secretory and membrane proteins, nearly all lipids (fatty substances), and certain complex carbohydrates. The *rough endoplasmic reticulum* is studded with ribosomes, whereas the *smooth endoplasmic reticulum* has none.

The **Golgi complex** is the cell's "shipping agent"—receiving, sorting, tagging, and directing many newly made proteins and membrane parts to their final destinations inside or outside the cell. It is made up of saucerlike arrays of smooth membrane sacs that abut the endoplasmic reticulum on one face (often near the nucleus) and the nuclear membrane on the other. Enzymes in the Golgi complex modify or add sugar groups to the proteins and lipids passing through. The resulting molecules, surrounded by sections of membrane, form secretory or storage **vesicles** that later fuse with the plasma membrane and release their contents outside the cell.

The **plasma membrane** encloses the cytoplasm. In some kinds of cells, all or part of its surface is fuzzy with slender cytoplasmic extensions (*microvilli*) that protrude into the surrounding space. The plasma membrane contains many specialized proteins. It acts as the "gatekeeper" that controls which substances or bodies (such as viruses) will pass in and out of the cell. This function is accomplished by molecule-specific channels, which pass specific ions and molecules through the membrane, and also by the receptor proteins.* The absence or malfunctioning of receptors can have tragic consequences. One example is the genetic disorder *familial hypercholesterolemia* (Chapter 5), which accounts for 5% of all heart attacks in patients under the age of 60.

But defective receptors can in rare cases have a good effect, by blocking the entry of a damage-causing agent. As described at the beginning of this chapter, the absence of a recently discovered receptor prevents the AIDS-causing HIV virus from invading cells. It has no other known physiological effect. The gene that controls production of the normal receptor protein, present in most people, is called *CKR-5* (for *chemokine receptor* 5). The mutated variant of this gene produces an abnormally short and therefore nonfunctional receptor protein that renders these individuals resistant to HIV infection. This mutation is surprisingly common. Up to 20% of some populations carry it in a single dose, which might make them somewhat less likely to be infected with HIV or allow them to survive longer after infection. And 1% of white populations (fewer in other tested populations) carry it in double dose. Some scientists think that this mutation is common because it served some useful purpose in the past. They propose that it actually arose about 700 years ago, and greatly increased in frequency because it may have conferred resistance to the bacterium that caused a different plague 650 years ago. This mutation "is astonishingly common in people whose ancestors lived in areas of Europe that were ravaged by the Black Death, also known as the bubonic plague" (Kolata 1998).

The plasma membrane "swallows" certain outside materials (such as fluids, large molecules, bacteria, and parts of degenerating cells) by engulfing them and then pinching off internally to form vesicles of various sizes. The new vesicles may fuse with other intracellular structures (e.g., lysosomes) to be degraded, or they may be stored, or they may return to the plasma membrane and release their contents elsewhere. The release of ingested materials and the secretion of cell products occur through a reverse process whereby certain vesicles bud off from the Golgi complex and fuse with the plasma membrane. They then open to expel their contents from the cell surface.

The plasma membrane also contains a variety of specialized structures called **junctions** that act as connectors to adjacent cells. In some tissues, very tight junctions keep the cells closely packed together and impermeable to many substances. Other kinds of junctions allow certain substances to pass from cell to cell or communicate with the extracellular matrix.

The **extracellular matrix** is a collection of substances (mostly proteins and carbohydrates) that are secreted by certain cells into the spaces between cells. **Collagen**, the most abundant protein in the body, is a major component of the extracellular matrix. Together with other adhesive proteins, it binds cells together and gives shape to tissues. Over a dozen kinds of collagen are known. Mutations in collagen genes can cause very severe disorders that especially affect the skeletal system. One example is a highly variable type of crippling disease called *osteogenesis imperfecta* (brittle bone disease); some forms are relatively mild, involving few bone fractures, whereas others are lethal.

The extracellular matrix also provides a highway for certain substances (especially hormones and growth factors) to travel from their place of origin to their place of action. Various kinds of cells also migrate through the extracellular matrix. During early life, some cells move from one place to another in the developing embryo. And in later life, white blood cells travel through the extracellular matrix (as well as in blood vessels) on the way to performing their many functions. All in all, cells (and the spaces around them) have complex organizations and multiple roles. How they develop and carry on their work is controlled in large measure by the genetic material in their nuclei.

*On the cell surface, for example, there may be 500 to 100,000 receptors for a given substance.

Chromosomes

Over a century ago, researchers devised ways of staining cells so that their internal features could be seen clearly. In Germany, Walther Flemming (1879) described "a very delicately interconnected basket-work of winding threads of uniform thickness" inside stained nuclei. (A more sophisticated version is shown in Figure 2.2.) Flemming described cyclical changes in these threads, which were later named *chromosomes*, and recognized their central role in cell division.

A few years earlier, Friedrich Miescher had developed ways of isolating intact nuclei from cells and analyzing their chemical content. He extracted nuclear substances (now called nucleic acids) rich in phosphorus and nitrogen. These substances were different from any other chemical group then known. Miescher correctly predicted that someday they would be considered as important as the proteins. These two lines of research finally converged in the 1950s with the final determination that **deoxyribonucleic acid** (**DNA**) is the hereditary material. In this section we present a brief overview of what is now known about the physical and chemical nature of chromosomes.

With few exceptions, the number of chromosomes is the same for all somatic cells in an organism and indeed within a given species. Humans, for example, have 46 chromosomes; dogs, 78; carp, 104; red ants, 48; and fruit flies, 8. Among plants, potatoes have 48 chromosomes; broad beans, 12; and white oaks, 24. The relation, if any, between chromosome number and size or complexity of the organism is not clear.

In most species, the chromosomes of all cells (except eggs and sperm) occur in pairs. The two members of a pair look alike and carry the same genes; they are called **homologues**. Human cells have 23 pairs of chromosomes; carp cells, 52 pairs; and white oak cells, 12 pairs. The normal number of paired chromosomes in a nucleus is called the **diploid** (or **2n**) number. During the production of gametes, this number is halved in such a way that an egg or sperm contains only one member of each chromosome pair; such cells are said to contain the **haploid** (or *n*) chromosome number. This special reduction process prevents the doubling and redoubling of the chromosome number in successive generations. Instead, the union of two haploid (*n*) gametes restores the diploid (2*n*) chromosome number in each generation.

Sex Chromosomes

In many plants and animals, the sexes are separate, and the mode of inheritance of certain traits can differ according to the parent of origin. For such organisms, sex is nearly always an inherited trait. Specifically involved is either a pair of identical-looking **sex chromosomes**, the **X chromosomes**, or two different sex chromosomes, one **X** and one **Y chromosome**. All the remaining chromosomes are called **autosomes** (Greek *autos*, "same"). Human males, for example, have an XY pair of sex chromosomes and 22 pairs of autosomes.

How sex is determined varies from one organism to another, but a common pattern is the one found in humans. Individuals with two X chromosomes are female, and those with one X chromosome and one Y chromosome are male. As is the case with many organisms, human X and Y chromosomes are structurally and functionally quite different. The medium-sized X is about 2.5 times as long as the tiny Y chromosome. Because of differences in chromosomal organization, the two also look quite different when stained and viewed under a microscope. In later chapters we discuss the sex chromosomes, sex determination, inheritance patterns of the genes on sex chromosomes, and the syndromes associated with sex chromosome abnormalities.

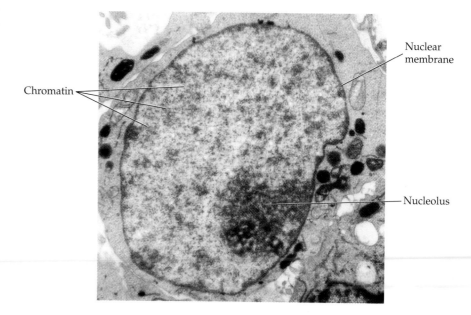

Chromatin

Nuclear membrane

Nucleolus

Figure 2.2 Chromatin in an ultrathin slice of a nondividing human white blood cell, as photographed through an electron microscope. The nucleus is outlined by a dark nuclear membrane. Inside the nucleus is a large, dark nucleolus surrounded by small dark clumps of chromosomal material called chromatin. (Courtesy of J. S. Heslop-Harrison.)

Gross Structure of Chromosomes

While cells are going about their metabolic business, their chromosomes exist as ultrafine threads of chromatin dispersed throughout the nucleus (Figure 2.2). But chromosomes duplicate themselves before the onset of cell division, and then each chromosome consists of two identical copies, known as **sister chromatids**, lying side by side. After cell division begins, these tremendously long and threadlike chromatids gradually shorten and thicken, becoming more rodlike (Figure 2.3).

During this period, every chromosome exhibits an indented region, the **centromere**, which connects the sister chromatids and is necessary for movement during cell division. It contains a special kind of DNA. The position of the centromere is constant for a given chromosome, dividing it into two **arms** of specific lengths. During their most condensed period (a brief stage called *metaphase*), the chromatids become more separated; consequently, each chromosome—depending on the relative lengths of its arms—takes on a somewhat X-like or V-like shape.

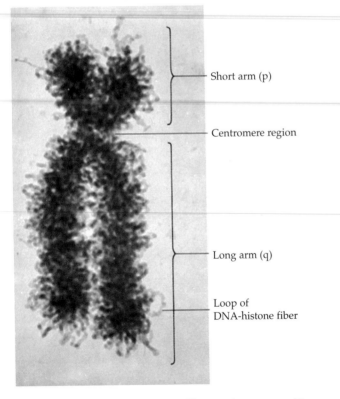

Short arm (p)

Centromere region

Long arm (q)

Loop of
DNA-histone fiber

Figure 2.3 Electron micrograph of human chromosome 12 during metaphase of cell division. The two identical chromatids are joined at the centromere (constricted region), which divides this chromosome into a short arm (p) and a long arm (q). Each chromatid consists of one long DNA-protein fiber that is coiled and folded. The thinnest fiber seen in this photo is 30 nm wide. Shown about 25,000 times actual size. (From DuPraw 1970.)

Certain treatment procedures reveal that each centromere is flanked by darkly staining blocks of highly condensed material called **heterochromatin**. Smaller regions of less condensed heterochromatin are dispersed throughout all the chromosomes. Heterochromatin contains few known genes, and its function is not well understood. Another, more lightly staining material, called **euchromatin**, contains most of the genes. Surprisingly, however, the genetically active regions of DNA make up only 5–10% of the total DNA in most eukaryotic cells. The chromosome tips, also heterochromatic, are known as **telomeres**. Their DNA caps have a unique chemical structure that normally keeps chromosomes from shortening during replication (Chapter 6). But with aging and with certain types of cancer there is a gradual accumulation of changes (mutations) in telomeres, as well as the gradual loss of some telomeres. For unknown reasons, the regions next to telomeres contain a high concentration of genes.

Besides the centromere, some chromosomes have additional pinched-in sites called **secondary constrictions**. When appropriately treated and magnified, for example, five pairs of human chromosomes (chromosomes 13, 14, 15, 21, and 22) exhibit secondary constrictions near the tips of the short arms. These **nucleolar organizer regions (NORs)** give rise to the *nucleoli*, which contain the special substances destined to form ribosomes in the cytoplasm. The sites and number of NORs are constant for a given species, but their nucleoli tend to fuse together (Figures 2.1 and 2.2). The five pairs of chromosomes with NORs also have tiny knobs of chromosomal material (called *satellites*) at the very tips of their short arms.

Special Features

Other kinds of constrictions, called **fragile sites**, were first discovered in the 1960s. Fragile sites are weak spots, stretched-out areas where chromosomes tend to break. Most of them can be seen only under certain treatment conditions and look like nonstaining gaps or constrictions in metaphase chromosomes. Unlike NORs, they do not occur in all individuals; nor are they detectable in all cells from a given person. But when one exists in an individual or family, it always occurs at the same spot on a particular chromosome, and it is transmitted from one generation to the next.

In humans, some fragile sites are common and some are rare. The best-known and medically most important example is a nonstaining constriction connected to a tiny knob at the tip of the long arm of the X chromosome (Figure 2.4). It is associated with a type of heritable mental retardation, called *fragile X syndrome*, that occurs mostly in males (Chapters 7 and 10). At least 16 rare heritable fragile sites have also been identified in the autosomes, but these do not seem to be associated

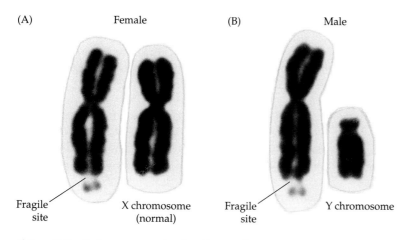

(A) Female (B) Male

Fragile site | X chromosome (normal) | Fragile site | Y chromosome

Figure 2.4 X chromosomes (each with two chromatids) containing a fragile site. (A) Sex chromosomes from a female with a normal X chromosome (right) and an X with a fragile site. (B) Sex chromosomes from a male with a normal Y chromosome (right) and an X with a fragile site. Also note the size difference between the X and Y chromosomes. (From Richards and Sutherland 1992.)

to form a **double helix**. This structure can be crudely pictured as a twisted ladder (Figure 2.5), the vertical support of each chain being a monotonous series of alternating sugar and phosphate groups—a **sugar-phosphate backbone**. Projecting inward from each sugar, to form the horizontal rungs, are pairs of **bases**.

Four different bases occur in DNA: two larger ones, called **purines**, and two smaller ones, called **pyrimidines**. The purines are *adenine* (A) and *guanine* (G); the pyrimidines are *thymine* (T) and *cytosine* (C). Each rung of the DNA ladder consists of a purine connected to a pyrimidine: either A with T (forming an AT or TA base pair) or G with C (forming a GC or CG base pair). Any other base pair combination leads to instability in the regular overall structure of the double helix. Thus, if you know the sequence of

with mental retardation or any other congenital abnormalities. Some analyses of chromosomal defects present in certain cancer cells suggest a possible connection between the breakage points and specific fragile sites, but this interpretation is controversial.

Normal individuals may harbor other chromosomal variations, such as slight differences in banding patterns between the two members of a homologous pair of autosomes. These observed variations were at first thought to be artifacts induced by the treatment procedures. But recent evidence indicates that some chromosome variations are inherited, unaccompanied by any known abnormalities. The best example is seen in the human Y chromosome, whose long arm is unusually long in some normal males and unusually short in others.

The Chemical Components of Genes and Chromosomes

The year 1953 marks a colossal advance in genetics. Two researchers at Cambridge University, James Watson (an American) and Francis Crick (an Englishman) proposed a structural model for the stuff that genes are made of: deoxyribonucleic acid (DNA). Two other workers, Rosalind Franklin and Maurice Wilkins at the University of London, had been analyzing the crystalline structures of DNA. Their X-ray photographs led Watson and Crick to deduce that the molecule consists of two chains wound about each other

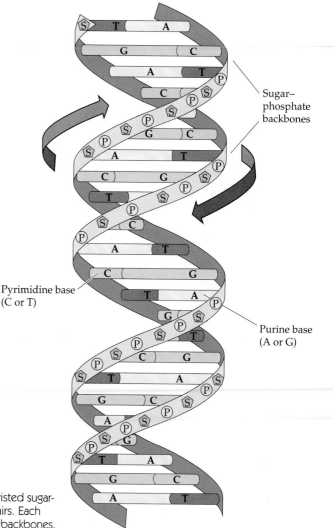

Sugar–phosphate backbones

Pyrimidine base (C or T)

Purine base (A or G)

Figure 2.5 A highly simplified view of the DNA double helix. Two twisted sugar-phosphate backbones are linked together by purine-pyrimidine base pairs. Each twist includes five base pair "rungs" attached to sugar molecules on the backbones.

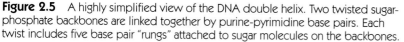

bases on one chain of the helix, you automatically know the sequence of bases on the opposite chain. The two chains are said to be *complementary*.

The particular sequence of bases along the length of the molecule can and does vary enormously. Indeed, it is the lengthwise base sequences that provide all the genetic information within DNA molecules. Like any written language, a sequence is a coded message. A *gene* is a segment of the DNA molecule with a unique sequence of thousands of bases that usually tells the cell how to make a particular protein. Chapters 6 and 7 provide details about the nature of genes and how they function.

After Watson and Crick's discovery, it became immediately obvious that the general properties of DNA could explain the behavior of genes. First of all, genes can scrupulously *reproduce* themselves. Although they usually copy themselves perfectly, genes are known to exhibit rare mistakes. These **mutations**, if not repaired, are perpetuated when copies are made. Thus arises new material for evolution. In addition to replicating and mutating, genes must also *store the information that is transmitted from mother to daughter cells or from parents to offspring*. The stored information in genes resides within the sequence of millions of bases along the DNA molecules—a developmental blueprint that is converted into each organism's unique set of characteristics through the process of protein synthesis. A gene is said to be *expressed* when its information is used by a cell to make a specific protein, this protein consisting of a special sequence of units called *amino acids*.

Fine Structure of Chromosomes

Each chromosome contains just one very long molecule of double-stranded DNA. Stretched out and laid end to end, the DNA molecules in a single human egg or sperm nucleus would measure about 1 meter (1 m = 3.28 feet).* Eukaryotic chromosomes consist of more than DNA, however. Their other main ingredients are collections of proteins known as histones and nonhistones. The best understood of these are the histone proteins.

Histones are small molecules that participate in the first level of compaction of the DNA double helix.† After being produced in the cytoplasm, histones migrate into the nucleus and become tightly bound to newly replicated DNA. This DNA-histone complex

forms the elementary structural unit of chromosomes. Histones also play a role in regulating genes—that is, turning them on and off as gene products are needed in different cells at different times (Grunstein 1992; Wolffe 1995).

During cell division, the chromosomal packing process involves repeated coiling and folding of the DNA-histone fiber. It is very hard to study because, as shown in Figure 2.3, a whole chromosome under the electron microscope looks like a pile of spaghetti. Although it is impossible to follow the tortuous path of a single DNA-histone fiber through its entire length, much has been learned from electron microscopic studies combined with biochemical and physical analyses.

The elementary chromatin fiber in all higher organisms is made up of repeating units called **nucleosomes** (Figure 2.6). Each nucleosome includes a total length of about 200 base pairs of somewhat stretched and twisted DNA wrapped around a spool-like core of eight histone molecules. Each histone molecule has a protruding stringlike tail that contacts neighboring nucleosomes, helping them to form stacked clusters (Wade 1997a). When organized into nucleosomes, a DNA double helix undergoes a sevenfold shortening, or *compaction*, to form a fiber 11 nanometers (nm) in diameter. This structure is further coiled and folded—now compacted about 50 times—to form a 30-nm-wide chromatin fiber. And during cell division, further levels of coiling and folding yield a human metaphase chromatid (see Figure 2.3) whose DNA double helix has been compacted 8,000- to 10,000-fold (Lewin 1997).

Nonhistone proteins (**NHPs**) are a mix of over 1,000 kinds of chromosomal proteins that are not histones. Unlike histones, NHPs seem to vary, in both amount and composition, with the degree of metabolic activity in different cell types, in the same cell at different times, or in differentially active regions of the same chromosome. The mix includes an enzyme called RNA polymerase, other DNA-binding proteins, gene regulators, and structural components of the chromosomes.

Certain nonhistone proteins also play an important structural role in chromosomal compaction. If all histones are removed from tightly condensed human chromosomes, the DNA becomes greatly unraveled but still keeps its identifiable chromosomal shapes. This is because each histone-free chromosome contains a central scaffold made of some nonhistone proteins. The DNA is attached to this scaffold in loops (Figure 2.7). When the DNA is also removed from these histone-depleted chromosomes, the scaffolds remain intact. How these structures arise is not known, but they do persist in some form throughout the entire cell cycle.

---

*See Appendix 1 for some common measurements.

†Histones have scarcely changed throughout evolution; indeed, some types taken from organisms as different as calves and pea plants are virtually identical! Such extreme conservation of sequence probably means that (1) histones have the same function in all organisms and (2) almost every amino acid in the molecule is functionally important.

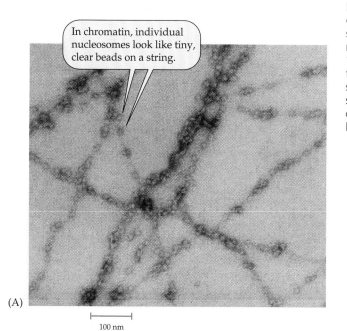

Figure 2.6 Nucleosome structure. (A) Electron micrograph of chromatin from chicken blood cells, showing individual nucleosomes (×150,000). (Courtesy of Christopher Woodcock.) (B) Two nucleosomes. A 146-base-pair length of DNA (color) is wrapped 1.8 times around a "spool" of eight histone molecules (two each of four kinds) to form each disk-shaped core particle. The remaining stretch of DNA, the linker, varies somewhat in length from one species to another. A fifth type of histone molecule, called H1, lies outside the histone core and "seals" the DNA as it enters and leaves the core particle.

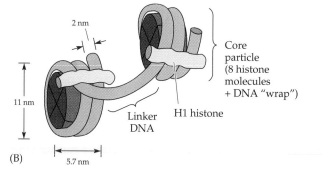

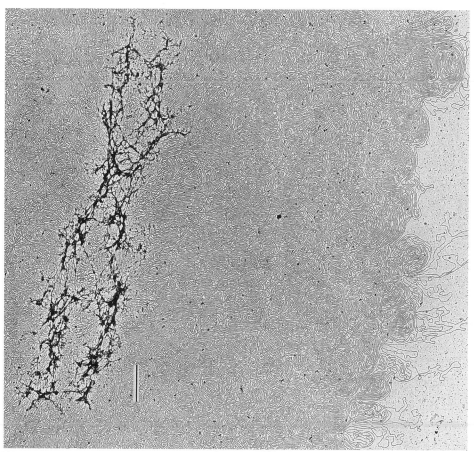

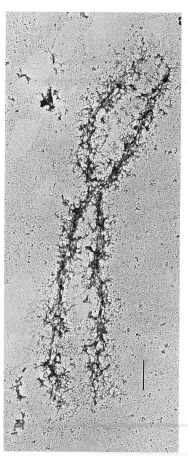

Figure 2.7 Human chromosomes, each with a dark-staining central scaffold of nonhistone protein outlining the shape of two chromatids joined at the centromere region. (A) After all histones are removed from this chromosome, an immense and tortuous maze of DNA loops out from the scaffold. (B) With both histones and DNA removed, all that remains is the central scaffold. Its major component is the protein topoisomerase II, an enzyme that cuts, untangles, and reseals knots of newly replicated DNA. The bar in each photo represents 1 μm. (Courtesy of Ulrich K. Laemmli.)

Human Chromosomes

During cell division, individual chromosomes can usually be identified by size, shape, and the patterns produced by various biochemical stains. These properties remain constant from cell generation to cell generation. Every species has a distinctive array of chromosomes known as its **karyotype** (Greek *karyon*, "nut" or "nucleus"). Deviations from the normal karyotype may be associated with abnormalities or even death of an organism (Chapters 13 and 14).

Each method for preparing chromosome spreads has unique advantages and disadvantages. Because the technique chosen depends on the type of analysis being undertaken, there is no single all-purpose protocol. Often, white blood cells are isolated from a blood sample and stimulated to divide in a culture medium. Special treatment stops cell divisions at the metaphase stage, when chromosomes are the most highly condensed. The cells are then killed, dropped onto a microscope slide, and dried. The slides are then stained and photographed. A karyotype is prepared by matching up the chromosome pairs from one nucleus according to their size, shape (i.e., X- or V-shaped, depending on the position of the centromere), and unique banding pattern. The slide can be de-stained and re-stained with a different dye so that the identical chromosomes can be compared with different stains.

Before modern karyotyping methods were developed (i.e., from the early 1920s until the mid-1950s), it was thought that humans had 48 chromosomes. But in 1956, two researchers working with the newer treatment techniques reported that they could find only 46 chromosomes in human cells, and within months this finding was verified by others. Such experiences teach us that objective reexamination and reevaluation of "established facts" are very important for scientific progress. Figure 2.8 shows that the chromosome number in our species is 46.

Recall that after replication, the replicas (chromatids) remain attached at the centromere region. Chromosomes with centromeres near the middle are X-shaped and called **metacentrics** (Greek *meta*, "among" or "after"). The V-shaped chromosomes, with centromeres near one end, are called **acrocentrics** (Greek *akros*, "topmost" or "extreme"). Most chromosomal shapes fall somewhere in between and are called **submetacentric**.

Chromosome Banding and Painting

The original techniques for staining chromosomes used dyes that produced uniform coloration throughout each human chromosome (Figure 2.8). Thus, the pairs of autosomes in a metaphase spread could be only crudely matched by size and shape. But since 1970, newer staining procedures have used pretreat-ments and dyes that produce dark and light cross-bands of varying widths. The exact relation of these bands to general chromosome structure is not well understood. But their patterns do seem to reflect local differences in (1) the timing of DNA replication during its synthesis period, (2) the relative content of GC versus AT base pairs, and (3) the relative length of the repeated gene sequences they contain. Because each chromosome in a haploid set has a unique banding pattern, every pair of homologues can now be distinguished from other chromosomes of similar size and shape.

The original banding methods could produce up to 320 metaphase bands in a haploid set of human autosomes. Such techniques allowed scientists to detect many new chromosomal anomalies in humans and to study the evolutionary relations among the chromosomes from various species. Newer staining tricks expanded the power of such analyses many times, up to 2,000 bands. They produced high-quality preparations—not only in metaphase, but in the preceding less-condensed stages (called prophase and prometaphase) as well (Figure 2.9).

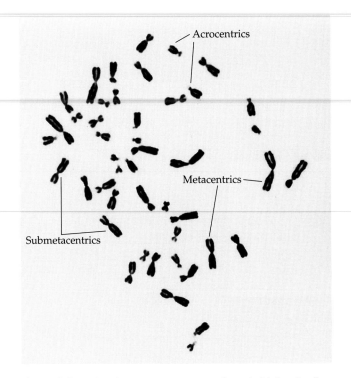

Figure 2.8   The chromosomes at metaphase (middle of cell division) from a white blood cell of a man (×2,000). The chromosomes are stained with Giemsa, which here colors all parts about the same. Each chromosome consists of two chromatids, and some chromosomes appear roughly X-shaped because the centromere is near the middle. A few chromosomes look more V-shaped, however, because the centromere is close to one end. (Courtesy of Irene Uchida.)

To describe a site on a particular banded chromosome, we first list the chromosome number, then the arm (**p** designating the short arm and **q** the long arm), then the region number within an arm, and finally the specific band within that region. For example, 1p13 refers to chromosome 1, short arm, region 1, band 3—a broad, light band. With this new **high-resolution banding** technique for examining prophase chromosomes, researchers were able to construct even more detailed karyotypes. Now they can detect tiny chromosome defects that were previously unidentified in some patients with congenital disorders. Similar analyses have allowed investigators to identify chromosomal abnormalities in the cancerous cells from many more patients.

Other methods of chromosome preparation, especially *fluorescent in situ hybridization (FISH)*, were developed in the late 1980s. They result in preparations with brilliantly colored whole chromosomes or chromosome parts (Figure 2.10). Even the centromeres or telomeres of individual chromosomes can be uniquely identified. Such **chromosome painting** requires complex mixtures of specially constructed DNA sequences, called *probes*, that are complementary to different chromosome regions (Chapter 8), together with a set of fluorescent dyes. Also needed are a special microscope with special optical filter sets, as well as computer software to analyze the images and convert them into a dazzling multicolored display. These techniques can be applied to both nondividing and dividing cells, and their interpretation does not require as much expertise as chromosome banding methods.

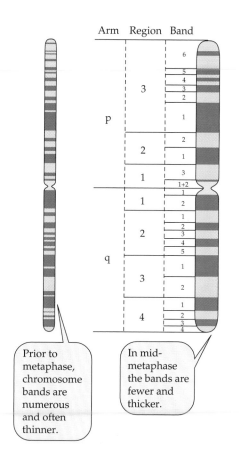

Figure 2.9 Diagram of banded chromatids of human chromosome 1 at different stages. The one on the left shows the banding pattern seen in late prophase; the one on the right shows the mid-metaphase banding pattern. The dark bands are those that fluoresce brightly with quinacrine or stain darkly with Giemsa. (From Yunis 1976.)

> Prior to metaphase, chromosome bands are numerous and often thinner.

> In mid-metaphase the bands are fewer and thicker.

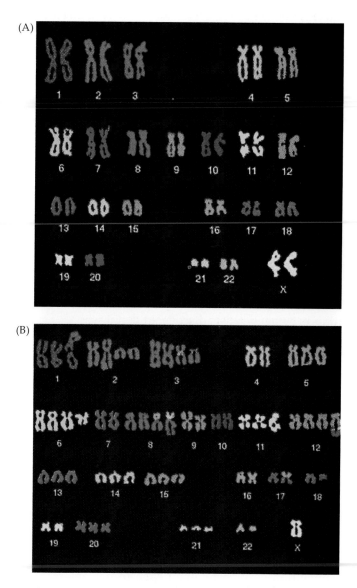

Figure 2.10 Human karyotypes prepared by a 24-color FISH (fluorescent in situ hybridization) method that uniquely identifies each chromosome. (A) Metaphase spread of a cell from a normal female, showing 46 chromosomes. (B) Metaphase spread of a cell from an ovarian cancer cell line. Note that it contains 63 chromosomes; at least 10 of them are obviously abnormal, each being made up of pieces from two or more chromosomes. (Courtesy of David C. Ward, Yale University; from Le Beau 1996.)

Chromosome painting is particularly useful for quickly detecting cells with too many or too few chromosomes, or with other types of chromosomal abnormalities, especially in the cancer cells of solid tumors (Figure 2.10B). It also allows scientists to more easily find genetically identical chromosomal segments in a wide range of organisms. Le Beau (1996) lists many additional applications of this exciting and visually stunning new technology.

The Normal Human Karyotype

Recall that 22 of the 23 pairs of human chromosomes occur in the cells of both sexes. These *autosomes* are numbered from 1 to 22 according to length (with chromosome 1 being the longest). The two remaining chromosomes—the sex chromosomes—are not numbered. The larger one is the **X** chromosome and the smaller one is the **Y** chromosome. Male cells normally contain one X and one Y chromosome, whereas female cells contain two X chromosomes.

The **haploid autosomal length (HAL)** is the total amount of chromosomal material present in a haploid set of autosomes. The *relative lengths* of human autosomes decrease with increasing chromosome number. For example, chromosome 1 (the longest) is 8.4% of HAL, while chromosome 21 (the shortest) is 1.9% of HAL. Why is the shortest chromosome called 21 rather than 22? This one exception is due to a historical fluke involving the extra chromosome present in Down syndrome individuals; the chromosome was originally named 21 in the mistaken belief that it was the next-to-shortest autosome. Later, with improved techniques, cytogeneticists discovered that it was really the shortest autosome and should be called 22. But by this time, the error was so entrenched in the scientific literature that it was easier to accept the inconsistency in relative length than to change the number.

Before human autosomes were distinguishable by banding methods, they could only be arranged according to length and centromere position into seven groups (A–G). Within each of these groups, the centromeric position is quite similar. (For example, groups A and F contain the four most metacentric chromosomes; groups D and G consist of acrocentric chromosomes; and the chromosomes in groups B, C, and E are submetacentric.) This convention is often retained even in newer banded karyotypes (Figure 2.11).

It is often convenient to describe an individual's karyotype with cytogenetic shorthand rather than by a picture. At the simplest level, this description is merely the total number of chromosomes and the sex chromosomal complement; thus, a normal female is designated 46,XX and a normal male 46,XY. The fact that the bands appear much fuzzier in a photograph (Figure 2.11) than in a diagram (Figure 2.12) points up the need for great technical expertise in analyzing karyotypes.

To describe abnormalities in chromosome structure and number, cytogeneticists use standardized symbols. For example, a boy with typical Down syndrome is designated 47,XY,+21. He has 47 chromosomes, including normal X and Y sex chromosomes, with the forty-seventh chromosome being an extra chromosome 21—that is, three 21s altogether. The karyotype of a female with Turner syndrome is 45,X—which means that she lacks one sex chromosome. Other symbols and some interspecies comparisons will be presented in later chapters.

Artificial Chromosomes

In the early 1980s, scientists first produced tiny artificial chromosomes in yeast. These *yeast artificial chromosomes* (*YACs*) contain the minimum requirements for survival and replication: a centromere, telomeres, and a few other bits of yeast DNA. Because they can accept and replicate large pieces of foreign DNA, including whole eukaryotic genes, YACs quickly became an invaluable tool for analyzing and mapping the chromosomes of other species (Chapter 9).

But there is a limit to the amount of DNA that YACs can accept, so geneticists also dreamed of creating artificial human chromosomes that could carry vastly larger DNA inserts—and perhaps even provide a reli-

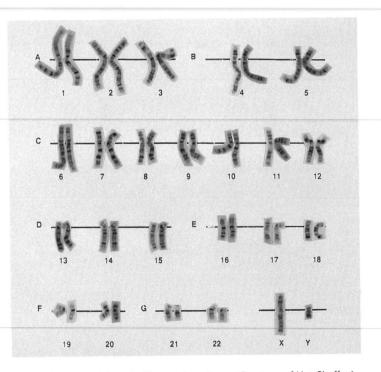

Figure 2.11   A G-banded human karyotype. (Courtesy of Lisa Shaffer.)

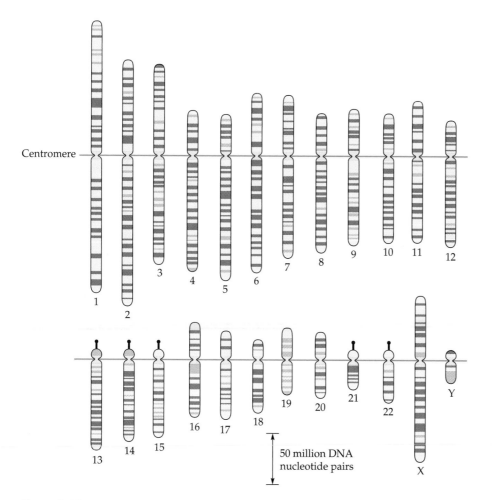

Figure 2.12 Diagram of a human karyotype at the prometaphase stage of mitosis, showing a total of 850 G-bands. (Adapted by Alberts et al. 1994, from Francke 1981.)

Summary

1. The cell is the underlying unit of structure in all living organisms. Eukaryotic cells have a nucleus, which contains the chromosomes. The surrounding cytoplasm contains many structural components specialized to carry out the varied activities of the cell.

2. Chromosomes, which carry hereditary determiners called genes, are the genetic link between generations. They occur in homologous pairs in body cells, one member of each pair derived from the female parent and the other from the male parent.

3. Chromosomes are best seen during cell division, when they are tightly coiled. Each replicated chromosome at first consists of two identical chromatids joined by an undivided centromere.

4. DNA is a long molecule composed of two complementary chains wound about each other to form a double helix. Genetic information is encoded in the sequence of bases in DNA. This information is decoded in the cytoplasm to form specific proteins. When rare mutations occur, they are copied during DNA replication.

5. Chromosomes of eukaryotes contain DNA, histone proteins, and nonhistone proteins. The basic chromosomal fiber is a DNA-histone complex, and chromosomal structure is provided by some nonhistone proteins.

6. Each chromosome can be identified in metaphase or late prophase by its size, shape, and banding pattern (or FISH color). The human karyotype consists of 44 autosomes (22 pairs) plus 2 sex chromosomes (1 pair). Males are designated 46,XY and females 46,XX. In addition, the karyotypes of normal individuals sometimes have special features, such as secondary constrictions or variations in Y chromosome length.

able vehicle for human gene therapy (Chapter 20). Human chromosomes are about 100 times larger and much more complex than yeast chromosomes, however, and their centromeres are not well understood. But in 1997, Huntington Willard and colleagues at Case Western Reserve University announced that they had constructed human artificial **microchromosomes** from synthesized centromeric and telomeric DNA plus some genomic DNA (Wade 1997b). These tiny chromosomes, one-fifth to one-tenth the size of normal human chromosomes but about 100 times the size of YACs, were taken up by laboratory cultured cells and stably replicated along with their normal chromosomes for six months (about 250 cell divisions). Researchers feel confident that these human microchromosomes will accept and express human genes, an important first step toward the possibility of treating certain genetic disorders by inserting normal genes into abnormal somatic cells. Meanwhile, they will provide another valuable tool for the ongoing study of human chromosome structure.

Key Terms

acrocentric
autosome
base
centriole
centromere
chromatin
chromosome
chromosome arm
chromosome band
chromosome painting
collagen
cytoplasm
cytoskeleton
cytosol
deoxyribonucleic acid (DNA)
diploid (2n)
double helix
endoplasmic reticulum
enzyme
euchromatin
eukaryote

Key Terms (continued)

extracellular matrix	nucleolus
fragile site	nucleosome
germ cell	nucleus
Golgi complex	peroxisome
haploid (*n*)	plasma membrane
haploid autosomal length	prokaryote
(HAL)	purine
heterochromatin	pyrimidine
high-resolution banding	receptor protein
histone	ribosome
homologue	scaffold
karyotype	secondary constriction
lysosome	sex chromosome
metacentric	sister chromatid
microchromosome	somatic cell
microtubule	submetacentric
mitochondrion	sugar-phosphate backbone
mutation	telomere
nonhistone protein (NHP)	vesicle
nuclear envelope	X chromosome
nucleolar organizer region	Y chromosome
(NOR)	

Questions

1. Distinguish between the terms in each of the following sets:

   somatic cell, germ cell
   cytoplasm, cytoskeleton, cytosol
   nucleus, nucleolus, nucleosome
   plasma membrane, nuclear membrane
   mitochondrion, endoplasmic reticulum, Golgi complex
   lysosome, ribosome, peroxisome
   autosome, sex chromosome
   haploid, diploid
   chromosome, chromatid, chromatin
   centromere, telomere
   euchromatin, heterochromatin
   acrocentric, metacentric, submetacentric
   X chromosome, Y chromosome
   p, q
   receptor protein, enzyme
   DNA, histone protein, nonhistone protein

2. Suppose that along one strand of a double helix, a tiny segment of DNA includes the following sequence of bases (where A = adenine, C = cytosine, G = guanine, and T = thymine): CGGATGTAACCCT. What is the order of bases at the same nucleotide positions on the complementary strand of this double helix?

3. Using "shorthand" symbols, give karyotypes for the following syndromes, which will be described in Chapter 13. (a) Males with Klinefelter syndrome have an extra X chromosome. (b) Females with Turner syndrome are missing one X chromosome.

4. Although the presence of an extra autosome is usually lethal before birth, some cases survive to birth. (a) In Down syndrome, there is an extra chromosome 21. (b) Trisomy 13, with an extra chromosome 13, is rare; it leads to gross malformations and early death. (c) Also rare and characterized by serious deformities and a very short life span is trisomy 18, caused by an extra chromosome 18. Describe these karyotypes using the standard shorthand symbols and assuming that (a) is female, (b) is male, and (c) is female.

5. If two breaks occur in a chromosome, the intervening section may be turned around 180° before the broken ends heal, a process producing an *inversion*. Within the inversion, gene order is reversed and the location of the centromere may be altered (e.g., *a b c ♦ d* can change to *a ♦ c b d*). For each of the following hypothetical inversions (the portion of the chromosome between the arrows), indicate (1) the new gene order, centromere (♦) position, and banding pattern; (2) whether the inversion would be detectable in unbanded spreads; and (3) whether it would be detectable in banded spreads.

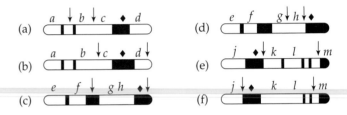

Further Reading

Much of the information on cell and chromosome structure is gleaned from three superbly written and illustrated books: Lodish et al. (1995), Alberts et al. (1994), and Lewin (1997). A shorter and very readable introduction to cell biology is Cooper (1997). Books on chromosomes include Therman and Susman (1993) and Wagner et al. (1993). For very readable accounts of the early history of human chromosome studies, see Jacobs (1982) and Hsu (1979).

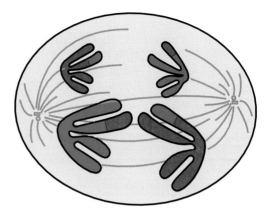

CHAPTER 3

Cell Division

The development of a single cell, the zygote, into an exquisitely complex organism—whether it be an orchid or a giant sequoia, a butterfly or a human being—is an awesome and fascinating process. The development of a single somatic cell into a malignant tumor is awesome and fascinating, but fearsome as well. Both processes result from cell division, but one is an orderly progression of events, while the other is an out-of-control situation. Normal development and cancer will be discussed in detail in later chapters, but the underlying principles of the cell cycle and cell division are presented here.

Like most plants and animals, humans reproduce sexually. This **sexual reproduction** involves the alternation of a *diploid (2n)* generation of cells (the body cells) with a *haploid (n)* generation of cells (the eggs and sperm). With the union of one haploid egg and one haploid sperm, another diploid life begins. This new cell, the **zygote**, immediately divides (as do its daughter cells) by a process called *mitosis.*

Following successful implantation in the human uterus, this tiny ball of cells continues to replicate—about 50 times in about 38 weeks—forming a marvelously complex individual of over a trillion cells. Continued divisions of many (but not all) types of cells occur throughout life, replacing the millions of body cells that are always dying for one reason or another. Within the sex organs, certain cells also undergo the special process of meiosis to produce the eggs or sperm needed to complete the human sexual cycle.

The Cell Cycle and Mitosis

Adult humans, like all multicellular organisms, are formed by the proliferation of a single cell first into 2 cells, then into 4, 8, 16, and so on, followed by gradual differentiation of the resulting cellular mass into tissues, organs, and organ systems. Whatever their developmental fate, most of these cells exhibit the same basic behavioral pattern: a long period of growth and metabolism, known as **interphase**, alternating with a short period of cell division called **mitosis**. These two delicately integrated functions constitute the **cell cycle**, whose orderly course includes four distinct phases (Figure 3.1). The first three are called **G_1** (gap 1), **S** (DNA synthesis), and **G_2** (gap 2). Together they make up interphase, during which the cell doubles its volume and replicates its DNA. The final phase (**M**) is the mitotic division, during which the duplicated

chromosomes segregate into two identical groups and the whole cell splits in two.

Most cultured cells, including those of mammals, complete the entire cycle in 16 to 24 hours, of which no more than 1 hour is spent in division. Each cell type seems to have its own precise schedule, which is not easily altered by experimental means. But the continued progress of any eukaryotic cell through the cycle is controlled at one or more **checkpoints**:

> The cell contains a number of systems that are responsible for monitoring the proper completion of events. When such a system detects the failure of some event it signals the inhibition of downstream events. The surveillance systems have been called *checkpoints* because they define stages of the cell cycle at which the cell checks to see if it is okay to pass on. (Hartwell 1995)

In vertebrate somatic cells, the main checkpoint, called the restriction point, occurs during late G_1.

Interphase

The first phase, **G_1**, is the initial time gap between cell "birth" at division and the beginning of DNA synthesis. All the metabolic activities described in Chapter 2 are carried on during this period. By far the most variable in duration among different cell types, this phase may be too brief to measure in rapidly dividing cells,* or it may last for weeks, months, or years in cells that seldom or never divide. G_1 in the "average" dividing cell takes 6 to 12 hours. Late in this period is the **restriction (R) point**. Prior to this time a cell may enter a *resting state* with no growth and no cell division, or may even self-destruct if conditions are unsuitable (e.g., if growth factors aren't available, if the cell isn't big enough, if it doesn't have adequate food reserves or oxygen, if its DNA is damaged, if it has been invaded by certain viruses; or if a cancer-causing gene has been activated). R is a "point of no return," because cells that pass it will almost always proceed through S, G_2, and M, regardless of local conditions. Also, at the end of G_1, the cell's centriole pair (see p. 000) begins to replicate. (Whether centrioles contain nucleic acid and exactly how they duplicate are not clear.)

Then molecular signals at the end of G_1 trigger the onset of the **S** (synthesis) **period**, during which nuclear DNA replicates (Chapter 6). Meanwhile, histones are being synthesized in the cytoplasm and transported through pores of the nuclear envelope into the nucleus. There they combine with replicating DNA to form new nucleosomes and chromatin fibers (Figure 3.2). After the DNA wraps around histone proteins to form nucleosomes, the nucleosome string is com-

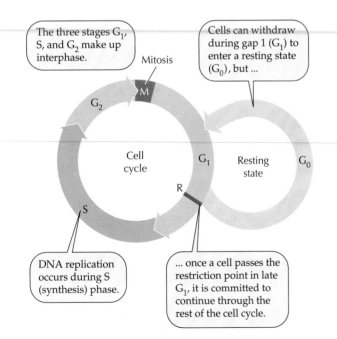

The three stages G_1, S, and G_2 make up interphase.

Mitosis

Cells can withdraw during gap 1 (G_1) to enter a resting state (G_0), but ...

G_2

M

G_1

Resting state

G_0

Cell cycle

R

S

DNA replication occurs during S (synthesis) phase.

... once a cell passes the restriction point in late G_1, it is committed to continue through the rest of the cell cycle.

Figure 3.1   A diagrammatic representation of stages of the cell cycle in rapidly dividing mammalian cells. Strictly speaking, mitosis (M) is nuclear division, but it is usually followed by cytoplasmic division. Interphase (stages G_1, S, and G_2) is a period of cell growth. (Adapted from Cooper 1997.)

*The cell cycles in most early embryos are extremely short—just alternating S and M phases, without intervening G^1 and G^2 phases.

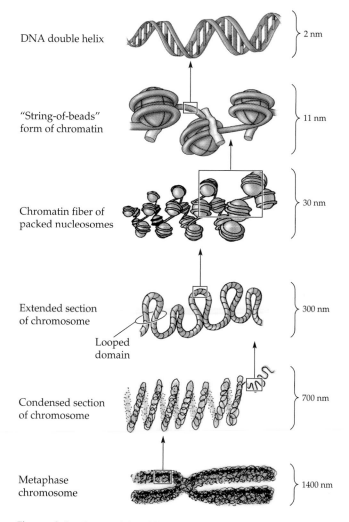

DNA double helix — 2 nm

"String-of-beads" form of chromatin — 11 nm

Chromatin fiber of packed nucleosomes — 30 nm

Extended section of chromosome — 300 nm

Looped domain

Condensed section of chromosome — 700 nm

Metaphase chromosome — 1400 nm

Figure 3.2 Some of the different orders of chromatin packing assumed to give rise to a mitotic (or meiotic) chromosome. By metaphase, each DNA helix has undergone about a 10,000-fold compaction. (Adapted from Alberts et al. 1994.)

pacted into a 30-nm-wide chromatin fiber. This fiber gets folded into **looped domains** that are attached to the chromosomal scaffold and stabilized by a network of nonhistone proteins. Besides being important in chromatin compaction, looped domains are thought to have an important genetic function. Each one contains a specific group of genes regulated as an independent unit and protected by the domain boundaries from regulation by surrounding domains.

How does DNA in a chromatin fiber manage to replicate? Rather than beginning at just one point on each chromosome, the unwinding of nucleosomes and replication of DNA starts at one specific spot (the **origin of replication**) in each looped domain and continues in both directions at a rate of about 50 base pairs per second. It takes 15 minutes to replicate the average domain. The mammalian diploid nucleus contains about 30,000 domains, all of which must replicate once

(and only once) within an S period. The initiations of replication are highly ordered; usually, euchromatin replicates early and heterochromatin replicates late in the S period. But certain regions may replicate early or late, depending on whether their tissue-specific genes are active or inactive. The S phase in mammals lasts about 6 hours. It is not well understood how the timing of replication is controlled or how the cell manages the amazing feat of replicating every segment of its DNA once and only once during a single cell cycle.

The two copies of each chromosome, when they eventually become visible during mitosis, are called *sister chromatids*. The centrioles also complete their replication during the S period, after which a shorter gap of 3 to 4 hours, G_2, occurs before the nucleus is stimulated to undergo mitosis—during **M**. Before entering mitosis, however, the cell's molecular sensors need to determine that its DNA is fully replicated, unmutated, and unbroken.

Some cells, especially those of blood-forming and epithelial tissues, continue to divide regularly throughout the lifetime of the individual. Other cells, including some nerve cells and all skeletal muscle fibers, do not normally divide at all after birth. Such *noncycling*, or G_0, cells, when removed from the body, can often be induced to divide in tissue culture; thus, the capacity for cycling has not been lost. Conversely, cancerous cells are those whose cycling controls have run amok; as a result, they divide uncontrollably. If scientists can learn how the division cycle is regulated in normal cells, they may be able to use this knowledge to prevent the growth of cancer cells and to promote wound and tissue healing. Thus, in addition to its scientific value, the study of the cell cycle has immense practical value.

Mitosis

In the 1870s, with the development of special dyes and improved microscopes, European scientists first discovered how cells divide. Since then, the process has been studied in ever more exquisite detail by focusing with more and more sophisticated microscopes on all kinds of cells and cell parts that have been treated with increasingly complicated physical and biochemical techniques. Yet, strange as it may seem to beginning students, we still do not fully understand the most basic molecular events of cell division: exactly how chromosomes replicate and condense; what holds the sister chromatids together for awhile; how football-shaped structures called *spindles* form and operate; how chromosomes attach to the spindle; how homologous chromosomes pair with each other during meiosis; how daughter chromosomes separate and move to opposite poles of the spindle; how the cytoplasm divides; what limits the total number of cell divisions in the descendants of normal cells (as opposed to transformed or cancerous cells); and how all these processes are controlled and integrated.

All higher plants and animals follow, with some variations, the same general routine during nuclear replication and cell division. Although it is a continuous process, cytologists have for the sake of convenience separated it into several stages: prophase, prometaphase, metaphase, anaphase, telophase, and cytokinesis (Figure 3.3).

Prophase begins as the nucleus swells and its chromatin fibers, previously stretched out to their ultimate interphase length, start to

condense (Figure 3.3A). Heavy phosphorylation of H1 histones is associated with this process. The looped domains that are attached to nuclear scaffold proteins also gradually shorten or coalesce. Finally, each chromatin fiber is densely packed into a short, thick, coiled chromosome whose looped domains radiate out in all directions from a central scaffold. Its two identical sister chromatids, replicated in the preceding S phase, be-

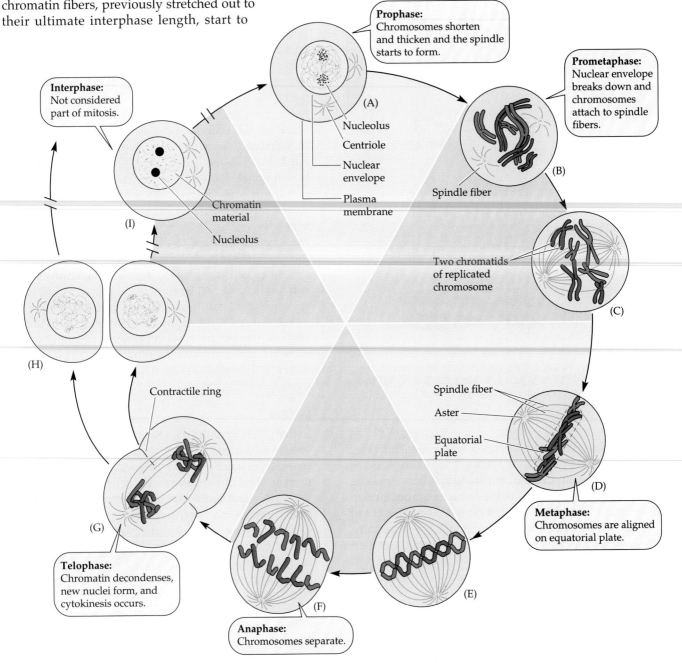

Figure 3.3 Chromosome behavior during mitosis. Only six chromosomes (three homologous pairs) are shown, with centromeres drawn as constrictions. One member of each pair came from the female parent and one came from the male parent. Two nucleoli are shown; they and the nuclear envelope disappear in prophase and reappear in telophase. *Note:* The partitioning of this diagram does not reflect the relative amount of time needed for each mitotic stage, or for the much longer interphase period.

come visible by late prophase. The nucleoli (containing ribosomal RNA and proteins) are dispersed by late prophase.

In the cytoplasm, the two pairs of **centrioles**—replicated during the previous interphase and lying just outside the nuclear envelope—separate and start to migrate in opposite directions around the nucleus. As they do this, **spindle microtubules** arise from the diffuse material surrounding the centrioles. (A centriole pair plus this surrounding material is called a **centrosome**.) Most of these fibers will form the **spindle**. The remaining microtubules make up the two *asters*, whose role in cell division is not yet clear.

The next division stage, **prometaphase**, begins when the nuclear envelope starts to undulate in the polar regions (Figure 3.3B). Ruptures then appear, and the whole envelope rapidly breaks down into small vesicles that remain scattered just outside the developing spindle. Meanwhile, the centriole pairs have become stationed at opposite poles of the spindle, which has taken its final shape. The centromeres of replicated chromosomes, which have started to converge toward the midregion of the spindle, develop platelike **kinetochores**. These proteinaceous structures, which bind on one side to special sequences of centromeric DNA, become attached on their free surface to bundles of *kinetochore microtubules* (Figure 3.4). Each chromatid pair gets violently jerked back and forth as one chromatid's kinetochore attaches to a bundle that goes to one pole and its counterpart finally gets connected to the opposite pole (Figure 3.3C). The key rule governing proper chromatid segregation is that the two members of a chromatid pair must never be connected to the same pole at prometaphase.

In **metaphase**, the chromosomes, now short and thick, have collected midway across the spindle in a region called the **equatorial plate** (Figure 3.3D). Each chromatid pair is suspended between the two poles by the equal tension exerted on each centromere through the kinetochore fibers. Metaphase is the longest stage of mitosis, lasting until all chromosomes are correctly aligned on the equatorial plate. In fact, the metaphase-to-anaphase transition is a checkpoint. Anaphase will not start, for example, if the spindle is defective or if the kinetochores are not properly attached to spindle microtubules.

Anaphase begins when the centromeres of each sister chromatid pair suddenly jump apart. Then the kinetochore fibers appear to tow the attached centromeres, with the two chromosome arms trailing behind them, toward opposite poles (Figure 3.3E and F). At this point, each sister chromatid is now called a chromosome. Exactly how chromosomes get moved on the spindle is not well understood, but cytologists have identified two clearly separable components. First the spindle elongates. Then the kinetochore mi-

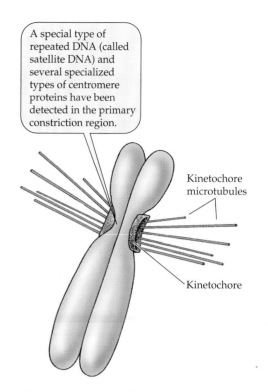

A special type of repeated DNA (called satellite DNA) and several specialized types of centromere proteins have been detected in the primary constriction region.

Kinetochore microtubules

Kinetochore

Figure 3.4 Diagram of the mammalian centromere, showing the primary constriction, the platelike kinetochore associated with the outer edges of the centromere, and kinetochore microtubules attached to the kinetochore. (Redrawn from Willard 1990.)

crotubules shorten, thereby moving the chromosomes to the separating poles. In addition, the kinetochores contain "motor proteins" that help propel chromosomes along the microtubules. The net result is that each side of the elongated dividing cell gets one copy of each chromosome.

The events of **telophase** are essentially the reverse of what happens during prophase and prometaphase. Chromosomes begin to decondense, and nuclear envelopes reaggregate around the two identical groups of chromosomes. Chromosomes continue to decondense, becoming individually indistinguishable. Nucleoli, produced by special ribosomal DNA (rDNA) sites on nucleolar organizer chromosomes, appear inside the two daughter nuclei (Figure 3.3G and H).

Meanwhile (actually starting in late anaphase), a **contractile ring** of actin and myosin fibrils forms just under the plasma membrane, in the equatorial plane midway across the long axis of the cell. Like a tightening belt, it gradually pinches in until the cytoplasm and spindle are split into two. This process of bisection, called **cytokinesis**, is independent of nuclear division. The cytoplasm in cells whose nuclei have been experimentally removed can still divide; conversely, some nuclei can undergo mitosis without any accompanying cytoplasmic division, thereby yielding multinucleate cells.

The end result of mitosis is the production, from one original cell, of two daughter cells with identical nuclei. *One interphase chromosome doubling (S period) has been followed by one division (M); so chromosome number and composition remain constant from one cell generation to the next.* The cytoplasms of daughter cells are usually about the same, too; but because there is no mechanism for the precise distribution of cytoplasmic components, differences may arise between the two cells from a mitotic division. This variation is particularly true of the early divisions of the fertilized egg (**zygote**), whose organelles and inclusions, such as yolk, may be present at different concentrations in various regions. In some organisms (such as frogs), such gradients result in the formation of daughter cells with greatly disparate cytoplasms.

All the cells descended from one zygote generally have the same nuclear genotype, yet they somehow manage to differentiate into many different types of cells, each with a specialized function. Among these specialized cells are the **germ cells**, which will divide by a special process called meiosis to give rise to eggs and sperm.

Cell Cycle Mutants

The cell cycle progresses by a series of ordered steps, each of which is usually required before the next step can proceed and be catalyzed by one or more specialized proteins. Thus, the cell cycle itself must be under genetic control, and *cell cycle genes* must be subject to occasional mutation. (Geneticists learn how normal genes work only by finding mutations that lead to errors in the functioning of normal genes. Without mutations, geneticists would be out of business.) Work on cell cycle mutants is difficult and slow; nevertheless, since the 1960s, researchers have managed to isolate and study several hundred cell cycle genes present in the cells of yeast, frogs, fruit flies, and mammalian tissue cultures. In yeast alone, over 400 genes are known to affect the cell cycle.

What can we say about cell cycle mutants in humans? Not as much as we know about those in yeast. But recent high-tech cloning experiments have shown that cell cycle genes isolated from human cells are actually comparable to those found in yeast. Indeed, normal cell cycle genes from a yeast species can function in human cells, and vice versa. From these and other experiments it has become clear that

the cell cycles of all eukaryotes … are governed by the same principles. The key proteins that regulate the cell cycle have been so well conserved during evolution that human proteins can successfully regulate the cell cycle of simple yeasts even though our ancestors diverged more than a billion years ago. (Murray and Hunt 1993)

These various cell cycle proteins have many different functions. They monitor conditions outside and inside cells, control cell size, signal cells to divide or not to divide, replicate DNA, produce RNA from DNA, translate RNA into proteins, condense chromosomes, assemble the spindle, move chromosomes along the spindle fibers, split cells into two, and so on. Of course, behind every cell cycle protein is at least one gene. The genes isolated so far include suppressors as well as activators of the cell cycle. Some details about these functions and about the myriad genes involved in their control are discussed in Chapter 17.

Meiosis and Gametogenesis

Every egg that a human female will ever produce is already present at birth, although immature and in a state of "suspended animation." Thus, any egg is as old as its host is at the time of the egg's release from the ovary. Although ovulated at the rate of one (or perhaps a few) per month for the 30 to 40 years between puberty and menopause, these eggs will not fully complete their meiotic divisions unless fertilized. It has been suggested that the increased risk of offspring with birth defects among older mothers may be somehow related to the longer storage time of their eggs.

Although estimates vary widely, a female produces perhaps several million immature eggs, of which only a few hundred are matured and ovulated during her lifetime. A male, on the other hand, probably makes and releases several trillion sperm in his lifetime. Sperm production does not begin until puberty, however, and usually continues for at least 50 years. From start to finish, a human sperm takes only about 3 months to mature, and it is released soon thereafter.

Egg formation (**oogenesis**) and sperm formation (**spermatogenesis**) occur by a special kind of cell division known as **meiosis**, during which diploid ($2n$) cells give rise to haploid (n) cells. (Figure 3.5) In humans, this means that cells containing 46 chromosomes (23 pairs) produce gametes with just $n = 23$ chromosomes. Unlike the mitotic cycle, in which one chromosome replication is followed by one division,

Figure 3.5   Chromosome behavior during the meiotic divisions of gametogenesis. In females, the smaller cells represent polar bodies; in males (nomenclature in parentheses), the four products of meiosis are the same in size and function. Only two sets of homologous chromosomes are shown, one member of each being in color and the other black. It is convenient to assume that the colored chromosomes were received from one parent and their black homologues from the other. Only one crossover in each chromosome pair is indicated, and the very brief prophase II is omitted, as is telophase II. The nuclear envelope and nucleoli are also omitted. Note that (1) the segregation of maternal and paternal chromosomes at anaphase I could equally well have been both black chromosomes to one pole and both colored ones to the other, (2) the secondary oocyte will not progress past metaphase II unless it is fertilized, and (3) each of the four end products contains one of the four chromatids that make up each tetrad.   ▶

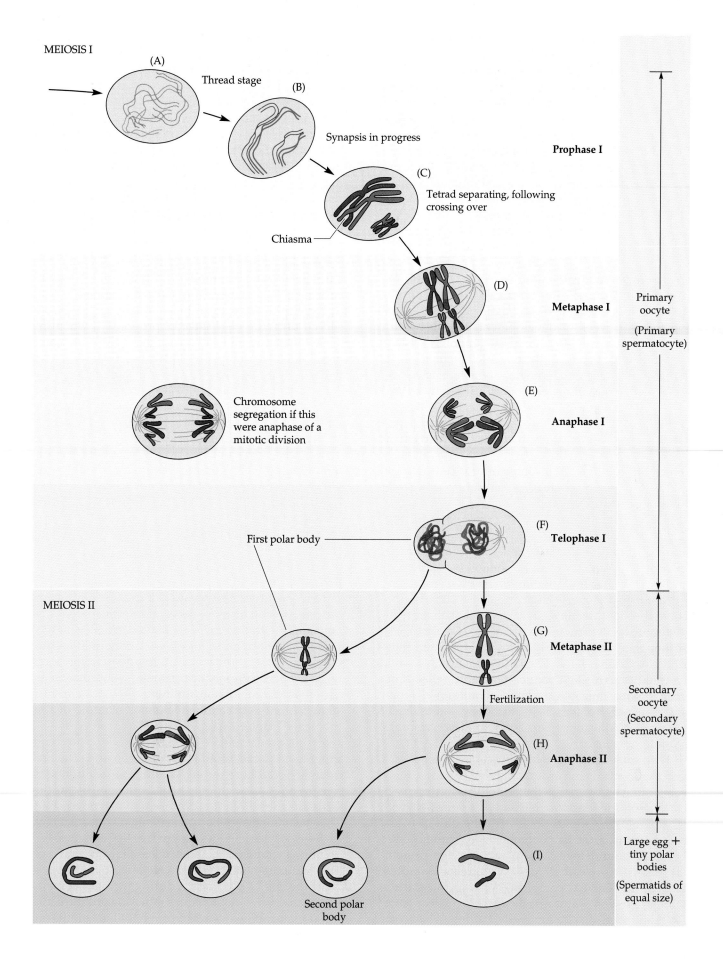

MEIOSIS I

(A) Thread stage

(B) Synapsis in progress

Prophase I

(C) Tetrad separating, following crossing over

Chiasma

(D) **Metaphase I**

Primary oocyte

(Primary spermatocyte)

Chromosome segregation if this were anaphase of a mitotic division

(E) **Anaphase I**

(F) **Telophase I**

First polar body

MEIOSIS II

Metaphase II (G)

Fertilization

Secondary oocyte

(Secondary spermatocyte)

(H) **Anaphase II**

Second polar body

(I)

Large egg + tiny polar bodies

(Spermatids of equal size)

meiosis involves a single replication (*premeiotic S period*) followed by *two* divisions, called **meiosis I** and **meiosis II**. Along the way, the two members of each pair of chromosomes become separated from each other. This event results from a unique feature of meiosis, the pairing of homologous chromosomes, and from the exchange of homologous parts during prophase of the first meiotic division. These processes, called, respectively, *synapsis* and *crossing over*, ensure that the resultant daughter cells contain one member of each chromosome pair. Both meiotic divisions include the stages observed in mitosis, but by far the longest and most complex of these—occupying up to 90% of the total time—is prophase of the first division, or *prophase I*.

First let us follow the developmental fate of some primordial germ cells in a female. These cells appear in the undifferentiated gonadal tissue of the fetus during the fifth week of development. By the twelfth week, they have undergone repeated mitotic divisions to form daughter cells called **oogonia** (singular, oogonium) in the developing ovaries of the females. Some of these oogonia continue dividing mitotically to produce more oogonia, but others begin to divide by meiosis. The behavior of the latter cells is illustrated in Figure 3.5. The first sign of change is the enlargement of an oogonium, which is then called a **primary oocyte**. Like the chromosomes of any cell in interphase, the 46 chromosomes appear diffuse and nearly invisible in stained cells.

Meiosis I

To understand meiosis, we must keep in mind that maternal and paternal chromosomes are *associated as homologous pairs during the first division.*

As a primary oocyte enters prophase I, its chromosomes become visible as delicate threads that are indistinguishable from one another (Figure 3.5A). Although already doubled during the previous S period of interphase, they are not visibly double when viewed through a light microscope. As in mitosis, the progressive shortening and thickening of chromosomes during prophase is due to folding and coiling of the fine DNA-histone fibers and to coalescence of the scaffold proteins.

Then something occurs that is unique to prophase I and that controls the orderly halving of the chromosome number. By means that are still a mystery, each chromosome manages to find and pair intimately with its homologue (Figure 3.5B). During this process, called **synapsis**, the pair of homologous chromosomes becomes bound together by a ribbonlike proteinaceous structure called the **synaptonemal complex**. Each synaptonemal complex begins to form at the telomere regions (now attached to the nuclear envelope) and then zips up gradually. Pairing, when complete, is so specific that if the order of genes on one chromosome should be different from that on its homologue, as sometimes occurs (Chapter 14), the two homologues will contort themselves like pretzels into a shape that aligns them region by region. Because sister chromatids are not individually recognizable until later in prophase I, the synapsed homologues appear to be a pair of threads, and the term **bivalent** describes their association from initial pairing to the end of metaphase I.

When synapsis is complete, chromatin condensation continues. Then another meiotic event takes place. **Crossing over**, a process involving the breakage and exchange of parts between two of the four homologous chromatids, results in some shifting of genes from each chromosome to its homologue. (In Chapter 9, we discuss the genetic consequence of such swapping.)

Physical evidence of these exchanges is not observed until the next stage of prophase I, when the synaptonemal complexes are shed and the doubleness of individual chromosomes finally becomes visible in a light microscope (Figure 3.5C). Sister chromatids, however, remain together until the onset of anaphase I. Because each chromosome consists of two sister chromatids, the bivalent has four strands and is called a **tetrad**. Along the length of each bivalent are found one or more X-shaped connections between nonsister chromatids; these **chiasmata** (singular, chiasma) are the places where crossovers have occurred. When—as if mutually repelled—the pairs of homologues within each bivalent start to move apart, the chiasmata prevent the homologues from separating completely. Depending on the number and positions of these connections, the chromosome pairs may become shaped like an X, a figure 8, or a more complex form.

Cytogeneticists have known for decades that synapsis and crossing over are necessary for the proper disjunction (separation) of homologues. Whenever chromosomal abnormalities interfere with normal pairing or crossing over, an increase in **nondisjunction** leads to gametes with too many or too few chromosomes. Down syndrome, associated with an extra chromosome 21, is the preeminent human example of this problem (Chapter 13).

By the time a female is born, all her oocytes have progressed to the tetrad stage of prophase I. They remain arrested here for anywhere from about one to five decades, their chromosomes still in bivalents but with the chromatin in a diffuse and extended state. Then beginning at puberty, every month one (or a few) of these oocytes enlarges, progresses through meiosis I and into metaphase II, and is released from the ovary. This maturation process has the following stages.

At the end of prophase I, during meiotic prometaphase, the paired homologues move toward the equatorial plate. Some events in prophase I are similar to

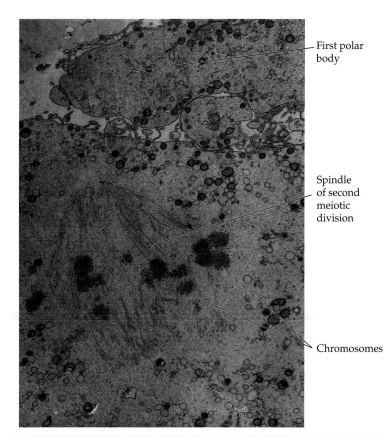

First polar
body

Spindle
of second
meiotic
division

Chromosomes

Figure 3.6 An unfertilized human egg at metaphase II. The egg has just been released from the ovary. Its chromosomes are lined up on the equatorial plate of the meiotic spindle and will remain at this stage until the egg is fertilized. The first polar body (just above and to the right of the spindle) is a product of the first meiotic division. (From Baca and Zamboni 1967.)

rate), moving to opposite poles. Unlike mitotic anaphase—during which the sister centromeres separate, thus bringing about the separation of their two chromatids—we see here the *disjunction of whole chromosomes*, each still composed of two chromatids that are held together at their centromere regions. At *telophase I* (Figure 3.5F), the chromosomes at one pole, together with a small amount of cytoplasm, are extruded and pinched off to form the tiny **first polar body**. The remaining chromosomes plus the bulk of the cytoplasm constitute a **secondary oocyte**, which begins to divide again immediately.

Meiosis II

Meiosis II involves the separation of sister centromeres and thus the separation of the two chromatids of each chromosome. It is like a mitotic division, but with 23 rather than 46 chromosomes. During the very brief *prophase II*, a new spindle arises just under the surface of the egg. In *metaphase II*, the stage at which mature human eggs are expelled from the ovary, chromosomes are lined up in the equatorial region (Figures 3.5G and 3.6) and remain so until the penetration of a sperm stimulates them to continue to *anaphase II* (Figure 3.5H). At the onset of anaphase II, the cohesion of sister chromatids finally lapses, and they move to opposite poles. At *telophase II* (not shown in Figure 3.5, but seen in Figure 3.7), one of the haploid chromosome groups is expelled, again with only a small glob of cytoplasm, to form the **second polar body**. By this time, the first polar body might also have

those in mitotic prophase: The nucleoli disappear and the nuclear envelope starts to break down while spindle microtubules arise.

During *metaphase I* (Figure 3.5D), the spindle is completed. Lined up on it are 23 bivalents. Their homologous chromosomes are now held together by chiasmata, and their centromeres are ready for departure in opposite directions. During *anaphase I* (Figure 3.5E), the two members of each chromosome pair disjoin (sepa-

Figure 3.7 Telophase II in a just-fertilized secondary oocyte, with the second polar body being formed. The two groups of separated chromosomes lie at opposite poles of this second meiotic spindle. Cytokinesis has begun, and the dense material in the equatorial region of the spindle determines where it will be pinched in two. (From Sathananthan et al. 1986.)

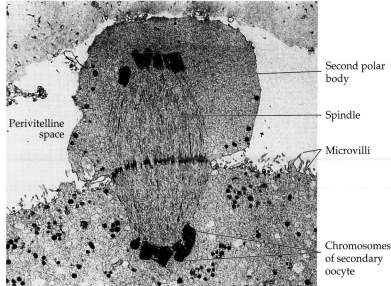

Second polar
body

Spindle

Microvilli

Perivitelline
space

Chromosomes
of secondary
oocyte

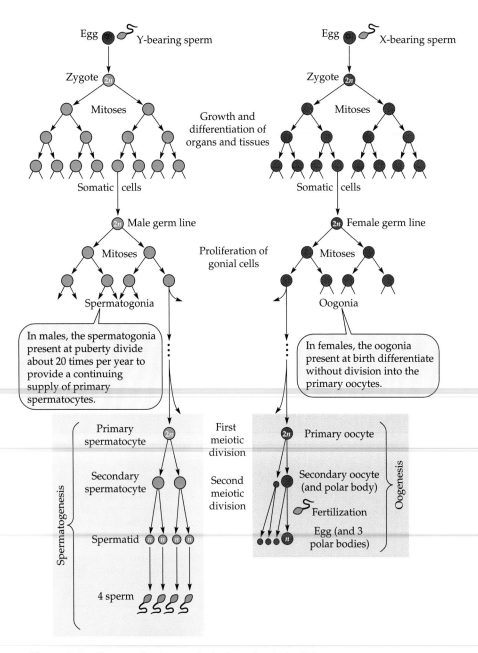

Figure 3.8 The overall scheme of mitotic and meiotic divisions in the two sexes, from zygotes to gametes and back to zygotes. The haploid number of chromosomes found in the gametes is designated *n*.

chromosome pairs on the metaphase I spindle is the result of a random process (i.e., it is equally likely that a maternal or a paternal centromere points toward a given pole), the chromosomes found at telophase I poles are mixtures of those originally derived from both the mother and the father. In each oocyte or spermatocyte undergoing meiosis, different combinations of maternal and paternal chromosomes will be formed. In addition to this **independent assortment** of chromosome pairs, the process of crossing over causes a further reshuffling of genes: Superimposed on the number of different whole chromosome combinations possible from a single individual (2^{23}, or over 8 million) is the additional variability produced by exchanges of chromosome parts between homologues. Hence, no two eggs (or sperm) produced by an individual are ever expected to be identical.

The events of meiosis in the male are the same as those in the female with regard to the nucleus, but they begin at puberty (rather than during fetal development) and proceed without interruption once started. After puberty, diploid **spermatogonia** (descendants of primordial germ cells) in the walls of the seminiferous tubules in the testes continually enter meiosis as **primary spermatocytes** (Figures 3.5 and 3.8). From one primary spermatocyte, meiosis I produces two **secondary spermatocytes** of equal size. Each of these in turn forms two **spermatids** after meiosis II. Division of the cytoplasm at telophase I and telophase II is equal, however, so that four haploid cells of equivalent size and functional ability are formed (see Figure 3.8). There follows a complex developmental process whereby each spermatid becomes an extremely specialized **spermatozoon (sperm)**. Much cytoplasm is lost. Histones are replaced by simpler proteins called protamines, which tightly pack the chromosomes into the sperm head.

divided into two tiny cells. Thus, meiosis in the female leads to the formation of four haploid cells: one egg and three very small, nonfunctional polar bodies (Figures 3.5I and 3.8).

Overview of Meiosis

The whole process from oogonium to egg takes from 12 to 50 years (Figure 3.9). Its most important feature is the accurate reduction of chromosome number from diploid to haploid, but the genetic shake-up that occurs may also be important. Because the orientation of

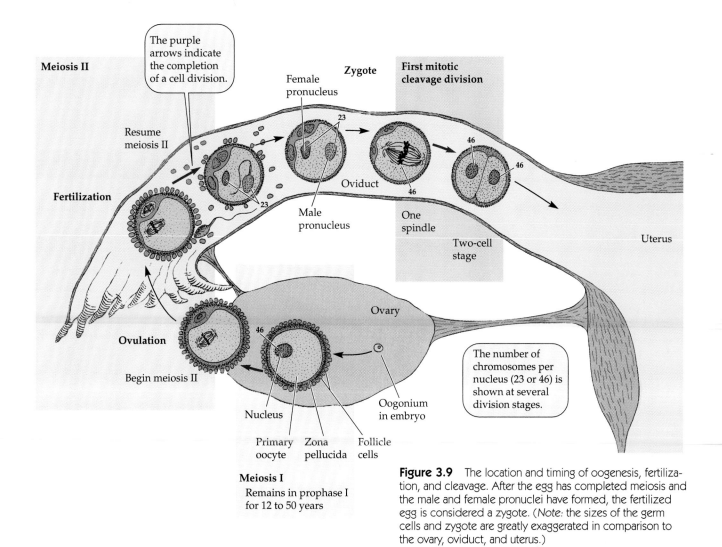

Figure 3.9 The location and timing of oogenesis, fertilization, and cleavage. After the egg has completed meiosis and the male and female pronuclei have formed, the fertilized egg is considered a zygote. (*Note:* the sizes of the germ cells and zygote are greatly exaggerated in comparison to the ovary, oviduct, and uterus.)

Human spermatogonia develop into mature sperm in roughly 9 weeks, of which about 3 weeks are spent in meiosis. Some men may continue to produce sperm throughout their lifetime, but in most males, the fertilizing ability of sperm is greatly reduced by the time they are in their 60s. As depicted in Figure 3.9, fertilization restores the diploid chromosome number to the newly created zygote.

Mature Gametes and Fertilization

Let us now follow the fate of a mature egg after its fertilization by a mature sperm.

The Mature Egg

The volume of a mature **egg** is vastly greater than that of the average human somatic cell (Figure 3.10). Because it is suspended partway through division, in metaphase II of meiosis, it lacks a well-defined nucleus. At the stage shown in Figure 3.10, the egg's nuclear envelope has disintegrated into bits and pieces

that will later be reused. The chromosomes, having already participated in one meiotic division and part of the next, are lined up on a spindle, awaiting the signal to complete the second cell division.

In the space just outside the egg membrane lies the tiny *first polar body*, a product of the first meiotic division. It carries the same number of chromosomes found on the egg spindle, but very little cytoplasm. Enveloping both the first polar body and the egg is the **zona pellucida**, a dense, jellylike layer secreted by the egg and whose surface is thickly peppered with species-specific sperm receptors. This layer protects the egg and early embryo and prevents the entry of foreign sperm. Surrounding the zona pellucida are *follicle cells*, which by the time of fertilization have begun to decompose and to separate from each other.

The Mature Sperm and Fertilization

The sperm, extremely tiny compared with the egg, is a cell stripped down to bare essentials—basically a *nucleus* enclosing the chromosomes, plus *mitochondria* for

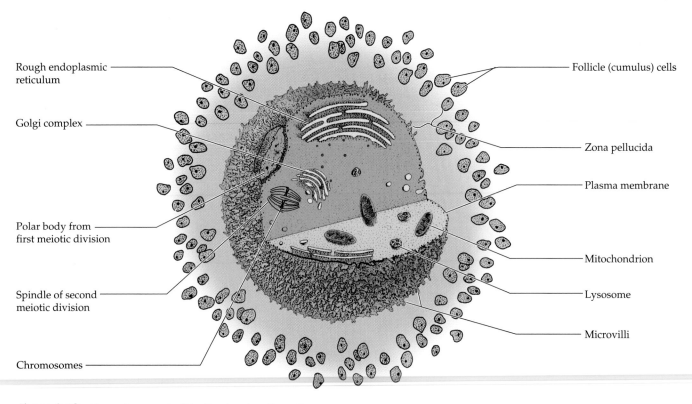

Rough endoplasmic reticulum

Golgi complex

Polar body from first meiotic division

Spindle of second meiotic division

Chromosomes

Follicle (cumulus) cells

Zona pellucida

Plasma membrane

Mitochondrion

Lysosome

Microvilli

Figure 3.10   Secondary oocyte (idealized and sectioned) ready for fertilization. The various inclusions shown here are suspended in the cytosol. At this stage of development, the nuclear membrane has broken down. Drawn about 500 times actual size.

energy and a relatively long *tail* (Figure 3.11). The head contains the nucleus with very densely packed DNA. Hugging the upper part of the nucleus is a thin **acrosomal cap** produced by the Golgi complex and filled with several kinds of enzymes. A short neck, composed of a centriole surrounded by segmented columns, attaches to the long tail. The core of the tail (see the cross section in Figure 3.11) arose from a second centriole. Mitochondria are spirally wrapped around the core of the upper part of the tail, providing the energy needed for movement; and a tough, fibrous sheath covers most of the remaining length. A plasma membrane encloses the entire sperm.

One ejaculate from a fertile male contains at least 250 million sperm, of which up to 100 may find and attach to the material surrounding the egg (Box 3A). Sperm are swept to the fertilization site mainly by rhythmic contractions of the female reproductive tract, and they need to swim only during their final approach. Whether they are attracted to the egg or simply collide with it is not clear. In any event, some will attach to receptors on the zona pellucida (Figure 3.12) and penetrate the egg's coatings, thereby reaching the plasma membrane. But normally, only one sperm enters the egg and forms a zygote.

The binding of the first sperm to the egg's plasma membrane triggers two events. First, electrical changes in the egg membrane and cytosol render it relatively impervious to additional sperm for a short while. As a second line of defense, the underlying *cortical granules* expel their contents into the surrounding space. This reaction alters the receptors in the zona pellucida and permanently blocks any more sperm from adhering to the egg.

Following fusion of its plasma membrane with the egg's plasma membrane, the entire sperm is engulfed by microvilli and pulled into the egg's cytoplasm. The mitotic apparatus of the egg is suddenly activated and completes the second meiotic division. The egg chromosomes separate into two groups, one of which is pinched off to form the *second polar body*. The remaining egg chromosomes decondense, and a newly constituted nuclear envelope encloses them to form the **female pronucleus**.

Meanwhile, the sperm's tail, midpiece, and nuclear envelope disintegrate. Its centriole becomes the zygote's centrosome, located next to the sperm nucleus. Its chromatin material becomes more swollen and dispersed, and the special proteins that were present in the sperm nucleus are replaced by histones produced by the

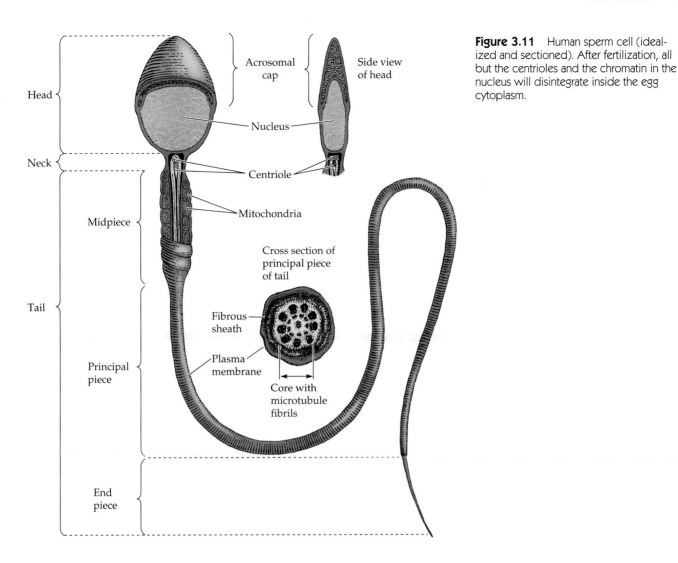

Figure 3.11 Human sperm cell (idealized and sectioned). After fertilization, all but the centrioles and the chromatin in the nucleus will disintegrate inside the egg cytoplasm.

Sperm Abnormalities

For unknown reasons, a surprisingly high proportion of human sperm show obvious defects.

"Since the examination of ejaculates by light microscopy became a technically simple matter, the disconcerting conclusion has emerged that man is the species that has the greatest percentage of malformed, anomalous and immotile spermatozoa, even in samples from healthy individuals of proven fertility. A possible explanation for this is that the temperature of the scrotum in which spermatogenesis occurs is raised above its natural physiological level as [a] result of wearing clothing … but there are also a significant number of known genetic defects that affect the entire sperm population of an individual." (Bacetti 1984)

Frequently observed abnormalities include an immature nucleus with uncompacted DNA, a malformed or absent acrosome, an abnormally shaped mitochondrial helix, extra cytoplasm surrounding and immobilizing the tail, an abnormal tail attachment, a short or missing tail, extra tails, and completely double sperm. Although many abnormal sperm are unable to progress through a female's reproductive tract, some not only pass through but even manage to fertilize an egg. Several abnormalities of human sperm are shown here. The figure is redrawn from Fujita 1975, which is included (along with many scanning electron micrographs) in Hafez and Kenemans 1982.

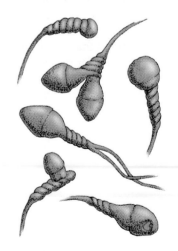

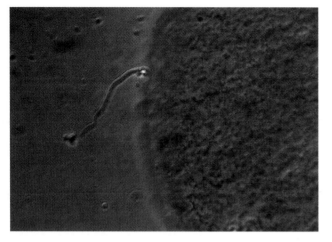

Figure 3.12 Human sperm touching the surface of an oocyte. (Copyright © Richard G. Rawlins/Custom Medical Stock.)

oocyte. Formation of a nuclear envelope around this chromatin completes the **male pronucleus**. The two pronuclei approach each other but do not fuse (Figures 3.13 and 3.9). Within each pronucleus, the chromosomes duplicate, condense, and (following breakdown of the two nuclear envelopes) arrange themselves on the first cleavage spindle. Maternal and paternal chromosomes are not enclosed within the same nuclear envelope until the two-cell stage (Figure 3.14). In humans, division of the zygote (the first cleavage division) occurs 24 to 36 hours after fertilization.

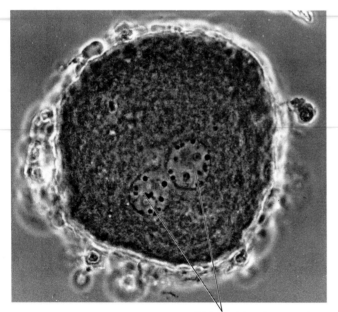

Male and female pronuclei

Figure 3.13 The male and female pronuclei lying side by side, shortly after penetration of the sperm has triggered completion of the second meiotic division in the egg. The first mitotic division of this human zygote will follow. (From Dickmann et al. 1965.)

Summary

1. The cell cycle consists of a long interphase, during which protein synthesis and DNA replication occur, alternating with a short mitotic division, during which the replicas are distributed to daughter cells.

2. The dimensions of a chromosome vary during the cell cycle, but its basic structure remains constant. One DNA double helix combines with histones to make a long string of nucleosomes, which then coils into a chromatin fiber. This fiber folds into a series of looped domains, which attach to scaffold proteins. During prophase of cell division, the fiber becomes highly condensed.

3. Mitosis has five main stages. Chromosomes shorten and thicken in prophase; in prometaphase, the nucleus breaks down and chromosomes attach to microtubules; chromosomes line up on the spindle in metaphase; at anaphase, the centromeres separate and pull identical sister chromatids (now called chromosomes) to opposite poles; in telophase, chromosomes uncoil, nuclear envelopes re-form, and the cytoplasm splits. Thus, two genetically identical cells are formed from one. No pairing of homologous chromosomes occurs during mitosis.

4. Each step of the cell cycle is genetically regulated. Many of the genes that control the cell cycle are the same or similar in all eukaryotic species.

5. The formation of haploid eggs and sperm by meiosis involves one replication and two successive cell divisions. Meiosis I results in the separation of pairs of homologous chromosomes, each individual chromosome consisting of sister chromatids held together in their centromere regions. Meiosis II results in the separation of sister chromatids.

6. Two key meiotic processes—the independent assortment of nonhomologous chromosomes and crossing over between homologous chromosomes—result in a tremendous reshuffling of gene combinations.

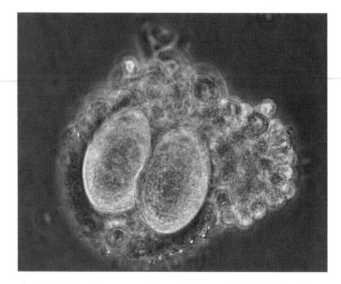

Figure 3.14 A human embryo 24 hours after fertilization. The first mitotic division has taken place, creating two cells. Each cell's nucleus will now contain identical genetic information, derived from both parents. (Copyright © Richard G. Rawlins/Custom Medical Stock.)

7. A primary oocyte forms one egg and up to three nonfunctional polar bodies; a primary spermatocyte yields four functional sperm. Oocytes are in prophase I by the time of birth and in metaphase II at ovulation and will complete the second meiotic division only if fertilized. Spermatocytes do not begin to develop until after puberty.

8. In the human life cycle, a haploid sperm fertilizes a secondary oocyte that is in metaphase II of meiosis, stimulating it to complete meiosis. Then the egg and sperm pronuclei, each with 23 chromosomes, unite to make a diploid zygote with 46 chromosomes. Repeated mitotic divisions form diploid body cells that differentiate into all tissues and organs. Diploid oocytes and spermatocytes undergo meiosis to yield haploid eggs and sperm, thus completing the life cycle.

Key Terms

acrosomal cap	oogenesis
anaphase	oogonium
bivalent	origin of replication
cell cycle	primary oocyte
centriole	primary spermatocyte
centrosome	prometaphase
checkpoint	pronucleus
chiasma	prophase
contractile ring	restriction (R) point
crossing over	S period
cytokinesis	secondary oocyte
egg	secondary spermatocyte
equatorial plate	second polar body
first polar body	sexual reproduction
G_1	sperm
G_2	spermatid
germ cell	spermatogenesis
independent assortment	spermatogonium
interphase	spindle
kinetochore	spindle microtubule
looped domain	synapsis
meiosis	synaptonemal complex
meiosis I	telophase
meiosis II	tetrad
metaphase	zona pellucida
mitosis (M)	zygote
nondisjunction	

Questions

1. Review oogenesis in human females by listing the meiotic stages that occur during each of the following periods: embryo/fetus until birth; birth until start of oocyte maturation; oocyte maturation until ovulation; fertilization until first cleavage division. Give the amount of time that elapses for each period.

2. Explain why a metacentric chromosome has an X shape at mitotic metaphase and a V shape at mitotic anaphase. Explain why an acrocentric chromosome has a nearly V shape at mitotic metaphase and a sort of J shape (or nearly rod shape) at mitotic anaphase.

3. Consider these metaphase homologues:

(a) After one division, the two daughter cells had the following chromosomes. Was the division mitosis or meiosis?

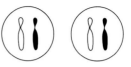

(b) After one division, the two daughter cells had the following chromosomes. Was the division mitosis or meiosis?

(c) What would the chromosomes in (b) look like, diagrammatically, before and after the second meiotic division?

4. Consider a spermatogonium with three sets of homologues: 1, 1'; 2, 2'; and X, Y. List, in a systematic fashion, the eight different kinds of sperm that can be formed, considering the three pairs together and assuming no crossing over within the intervals we are following. Choose one of the possibilities and diagram the meiotic divisions that produced it.

5. Consider two sets of homologues: 1, 1' and 2, 2'. (a) If a secondary oocyte contains the centromeres of 1 and 2, what does the first polar body contain? (b) If an egg cell contains the centromeres of 1 and 2, what does the second polar body contain?

6. During interphase and before DNA replication, a diploid human cell contains 46 chromosomes, and the term *chromatid* is inapplicable. How many chromosomes and chromatids should be visible in this cell at metaphase? (Recall that at anaphase, chromatids become the chromosomes of incipient daughter cells.)

7. How many human chromosomes and chromatids (if applicable) are present at these stages of meiosis: (a) a spermatogonium before the S period? (b) a primary spermatocyte at metaphase I? (c) a secondary spermatocyte at metaphase? (d) a spermatid? (e) a sperm?

8. When miscarriages occur, many of the fetuses are triploid (3*n*), and some are tetraploid (4*n*). How many chromosomes do these fetuses have in their somatic cells?

9. How many chromosomes has a mule whose mother (a horse) had 64 and whose father (a donkey) had 62?

Further Reading

The literature on chromosomes, cells, and cell division is vast. Three excellent but massive cell biology books are by Lodish et al. (1995), Alberts et al. (1994), and Lewin (1997). A shorter and more accessible one is by Cooper (1997). The cell cycle is described in books by Murray and Hunt (1993) and Baserga (1985) and in articles by Murray and Kirschner (1991), Chandley (1988), Stewart (1990), Manuelidis (1990), Strange (1992), Baserga (1981), and Mazia (1987). On gametes and fertilization, see articles by Wassarman (1988), Grobstein (1979), and Epel (1977, 1980) and a book by Longo (1987).

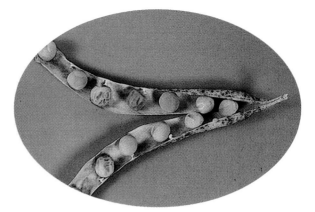

CHAPTER 4

The Rules of Inheritance

The Austrian monk Gregor Mendel experimented for nine years to develop and confirm his remarkable new theories of inheritance. Far from being a shy recluse who labored in obscurity, Mendel was a broadly trained, energetic, and highly respected scientist and teacher. He was also widely involved in local affairs: a chairman of the Moravian Mortgage Bank, a founder of the Austrian Meteorological Society, and very active in the agricultural societies organized by local sheep, apple, and grape growers to further their interests in animal and plant breeding and to keep the Czechoslovakian town of Brno (Brünn) a thriving commercial center (Orel 1996). He was noted, too, for his kindness to the poor.

Mendel tackled the problem of heredity at the suggestion of his superior, Abbot Napp, a brilliant administrator who fostered scientific research at the monastery in Brno (Box 4A). Yet despite their confidence in Mendel and their interest in the hybridization problem, his contemporaries failed to grasp the significance and generality of his findings.

BOX 4A *Gregor Johann Mendel: Some Highlights from His Life*

Johann Mendel was born in July 1822 in a village in the eastern part of a region now known as the Czech Republic. His parents, who were of German and Czech ancestry, tended a small farm that included many fruit trees and beehives. The villagers valued education, and a great uncle of Mendel's had once taught in a local school.

From early childhood, Mendel excelled as a student. Despite his family's modest means, he was able to attend school until age 21, but from the age of 16 had to support himself as a private tutor. In 1843, he entered the Augustinian monastery at Brno, taking the name Gregor. This move gave him some financial security as well as the opportunity to study natural science. The monastery had a mandate from the government to train its members to teach in local schools. Its extraordinary abbot, F. C. Napp, interpreted these orders to mean that the monks should do original scientific research while also keeping up with and publicizing the latest findings in their various fields of expertise. Special attention was given to practical applications for agricultural use. In this highly stimulating environment, young Mendel thrived.

But he also encountered many difficulties. After finishing his theological studies at age 26, he became a parish priest—a role so stressful for him that after six months, he suffered an emotional breakdown. (He had also experienced a serious mental crisis at age 18, while a lonely and impoverished student, and spent a year recovering at home.) So in 1849, Abbot Napp transferred him to a teaching job in a nearby town, where with great distinction he taught mathematics, Latin, and Greek. Before he could teach physics and natural history, however, he needed to pass a teacher's certification test in Vienna. At age 28, after studying natural history on his own while teaching full-time, he took the test and —to everyone's surprise—failed.

Abbot Napp (whose confidence in Mendel never flagged) then sent him to the University of Vienna for two years to prepare for a second try. There Mendel concentrated mostly on physics but also studied mathematics (including probability theory), chemistry, biology, and paleontology. Experimentation and inductive reasoning were stressed in all of these fields; indeed, one of Mendel's distinguished professors was accused of "spreading agnostic and socialist views" and came close to being fired (Orel 1996). It was a time of exciting scientific developments, and Mendel soaked it all up.

After completing his studies in Vienna, Mendel returned to the monastery. At age 32, he became a substitute teacher at a new type of secondary school in Brno where the science faculty was expected to do research. At the suggestion of Abbot Napp, he tackled the problem of heredity. Using a monastery garden plot and a greenhouse built especially for him, Mendel began to experiment with garden peas. He also established himself as a knowledgeable, gifted, and popular teacher. But to advance to a regular position with full pay, he had to retake the certification exam, which he did the following year. Again he failed, for reasons that remain a mystery. Despite this stunning setback, he was allowed to continue with his teaching and research.

About that time, Abbot Napp ran into serious trouble with the Catholic Church because his view of his monastery's mission was not always shared by his superiors. After a formal investigation in 1854, the local bishop "saw the scientific and teaching activities of the monks as being in contradiction to their spiritual calling" (Orel 1996) and recommended that the monastery be closed down. But Abbot Napp managed to stave off this potential disaster just at the time when Mendel was starting his experiments with pea plants.

Before setting up any actual crosses with pea plants, Mendel first had to find some simple, sharply defined pairs of traits to study. Starting with 34 fairly distinct varieties, he spent about two years determining whether or not they bred true. From these he selected seven pairs of traits, with which he performed many experimental crosses from about 1856 to 1863. In 1865, at age 43, he described his research in two lectures to the Natural Science Society in Brno and then published these findings in a German scientific journal. Reprints of his paper were sent to scientists and libraries of major universities throughout Europe. Although his paper was cited now and then, nobody grasped the fundamental importance of Mendel's work.

Even Carl Nägeli, a leading botanist of the day who was deeply interested in plant hybridization and evolution, didn't get it. Between 1867 and 1873, he received and responded to ten letters from Mendel, in which they mostly discussed experiments Mendel did with other kinds of plants. Mendel even sent Nägeli pea seeds with which to repeat his experiments, but Nägeli never did. Mendel, however, made many crosses with the hawkweed (*Hieracium*) seeds and plants received from Nägeli, hoping that they would confirm his results with peas. Alas, unbeknownst to both of them, hawkweed often reproduces without undergoing meiosis, so that the offspring are just clones of the female parent! Thus, Mendel was never able to verify his original results with this plant.

But Mendel had experimented with about 15 other kinds of plants, several of which yielded results like those found in peas. He did not publish these studies in scientific journals, however. He was also very successful in breeding new fruit trees and was an active member of the Pomological, Viticultural and Horticultural Society. In 1862, he traveled abroad with a local delegation to see the Great Exhibition in London, stopping along the way in Paris and six other cities.

Mendel carefully read translations of Darwin's books on evolution. Although he never mentioned these works in his own publication, he did mention them in his correspondence with Nägeli. He accepted the idea that new species could evolve by natural selection, but (correctly) disagreed with Darwin's hypothesis about how germ cells were formed.

In 1870, seeking to study inheritance in animals, Mendel attended a congress of beekeepers in Kiel, Germany. Then he joined the Association of Moravian Beekeepers, had a large beehouse built at the monastery, and started describing and trying to cross many strains of bees. Although unsuccessful at producing specific crosses, Mendel greatly improved beekeeping techniques and made many important observations about the behavior and natural history of bees. This highly regarded work continued until around 1880. None of it was published.

Meanwhile, Mendel was also making significant contributions in meteorology. Between 1857 and 1883, he

BOX 4A (continued)

recorded various weather-related observations and was among the first to analyze them statistically. He was keenly interested in weather forecasting by means of the most modern instruments and scientific analyses and even investigated sunspots as a possible influence on weather. His most remarkable meteorological publication was a detailed description and analysis of a tornado that passed by the monastery in 1870.

But by age 46, the focus of Mendel's other responsibilities had changed drastically. In 1868, following the death of Abbot Napp, he was elected to that lifelong administrative position. It afforded him financial security (to pay for his nephews' education and contribute to various causes) but otherwise brought him nothing but grief. The new Austro-Hungarian monarchy had decided to impose taxes on all monasteries for the years 1830–1864. At great length, Mendel explained why his monastery could not possibly pay this huge debt, but all his waiver applications were denied. Rather than trying to work out

some kind of compromise, he stubbornly continued to fight.

By 1876, Mendel's health had begun to fail, and he was becoming isolated from the world. Gradually, he stopped attending agricultural and scientific meetings and was visited by fewer and fewer people. He continued some experimental observations, however, and—ever searching for patterns in apparent randomness—even collected lists of names and tried to find some mathematical order in their frequencies!

By June 1883, when his last official protest was denied, Mendel was so ill that his duties had been assumed by another monk. In January 1884, at age 61, Mendel died from kidney and heart disease. Hundreds of dignitaries, teachers, agriculturalists, and ordinary citizens attended his funeral. They knew they had lost a fine teacher, an outstanding agricultural and meteorological expert, and a modest and kindly friend. But they had no idea that a great scientist had lived and worked in their midst. Virtually none of Mendel's personal or scientific papers

remain, having been destroyed either by Mendel himself or by the abbot who succeeded him.

Gregor Johann Mendel (1822–1884)

Mendel's First Law: Segregation of a Pair of Alleles

In this chapter, we answer the first and most basic question in transmission genetics: What are the rules for predicting how simple traits are passed on to succeeding generations? Gregor Mendel solved the basics of this problem of heredity by 1865, but nobody realized it until 1900. His work, which marked the beginning of quantitative biology, still stands as a model of careful experimental design. Drawing on his training in mathematics, physics, and chemistry, Mendel applied mathematical principles to the analysis of his biological studies. He noted the different classes of offspring that resulted from specific matings, as well as how many individuals occurred in each class. He calculated the ratios among these classes and showed that they fit the terms of simple algebraic expansions. He recognized that "the greater the numbers, the more are merely chance effects eliminated" (Mendel 1865).

Mendel's experiments were designed to test a specific hypothesis about inheritance. Rather than worrying about how *all* traits are inherited in *all* organisms, he focused on the inheritance of easily distinguishable forms of a *few* traits in a *single* organism, the garden pea. This plant was chosen for three reasons: (1) It has

a number of clearly defined variations, such as tall versus short plants, round versus wrinkled seeds, yellow versus green seeds, inflated versus constricted pods, and white versus gray seed coats. (2) Its flower structure, tightly surrounding both male and female parts, normally guarantees *self-fertilization*. But *cross-fertilization* can be arranged by removing the male parts from each flower before the pollen matures and later dusting the female parts with pollen from a different line. (3) The hybrids between two lines are as fertile as the parental lines, so large numbers of progeny can be studied in every generation.

Mendel started with **pure-breeding** lines, in which all the progeny from self-fertilization resemble the parents, generation after generation. A pure-breeding line has little or no genetic variation within it. When making crosses between two different lines, Mendel always set up **reciprocal matings**: females of line A mated with males of line B, as well as males of line A mated with females of line B. In this way, he could see whether the inheritance of a trait was influenced by the sex of the parents. Mendel was careful to keep the plant generations separated from each other and to study the descendants of each hybrid line through at least four generations of self-fertilization. He seemed to recognize the importance of keeping environmental conditions as constant as possible. The need for every one of these precautions is obvious today, but apparently was not understood in Mendel's time.

Mendel's Experiments

With each of seven pairs of traits, Mendel did the same series of experiments and got the same kinds of results. Here we present data for only one pair: *tall* plants (about 200 centimeters, or 6.5 feet) versus *short* plants (about 30 centimeters, or 1 foot). (We now know that this difference in height is due to the presence or

absence of a gene-controlled growth hormone.) All the crosses are diagrammed in Figure 4.1. Here each generation represents one year's work.

First, Mendel made many reciprocal matings between a pure-breeding tall line and a pure-breeding short line. These made up the **parental generation (P)**. Because the parents differed from each other in just one characteristic, the mating is called a **monohybrid cross**. The progeny of these matings, called **F_1 hybrids**, were all tall. Thus, Mendel reasoned, the tall plant character must be **dominant** in expression to the short

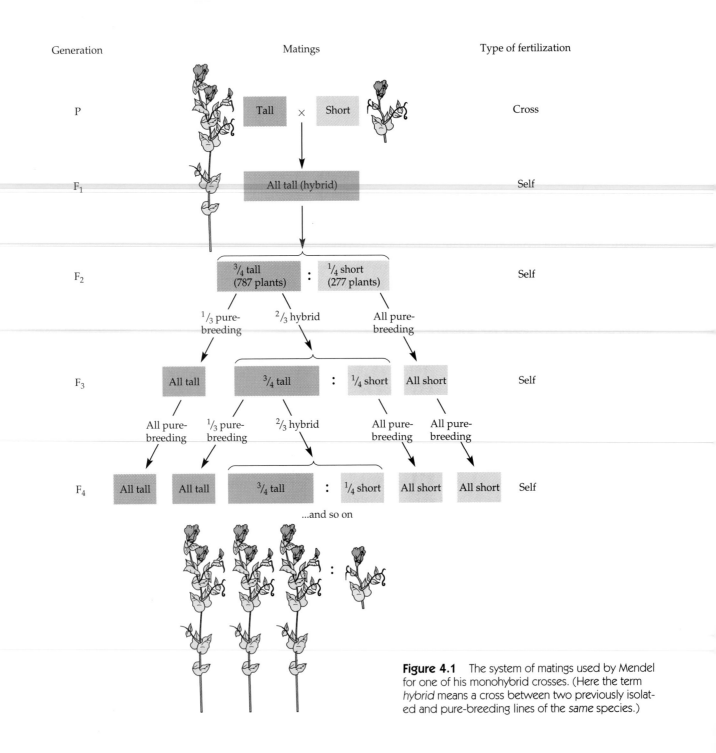

Figure 4.1 The system of matings used by Mendel for one of his monohybrid crosses. (Here the term *hybrid* means a cross between two previously isolated and pure-breeding lines of the *same* species.)

plant character, which was designated **recessive**. For the other six pairs of traits, likewise, only one of the two alternative forms was expressed in the F_1 hybrid.

Next, Mendel allowed the F_1 hybrid plants to self-fertilize, collected the resulting seeds, and sowed them the following year. Among this **F_2 generation**, 787 plants were tall and 277 were short, a ratio of about 3 to 1 (3:1). Comparable F_2 ratios were obtained for the other six pairs of traits. No plants of medium height, or with any characteristic intermediate between the two parental lines, ever appeared.

The third step was to again allow self-fertilization, sow the seeds from individual F_2 plants, and look at the F_3 generation. The plants with the recessive trait were always found to be pure-breeding: short plants yielded only short progeny. But the progeny of the F_2 plants with the dominant trait were of two types. About ⅓ of the F_2 tall plants were pure-breeding, giving rise to only tall plants after self-fertilization. But the remaining ⅔ of the tall F_2 plants, after self-fertilization, yielded both tall and short progeny in a ratio of 3:1. Thus, Mendel concluded, they must be hybrids like the F_1. As shown in Figure 4.1, he continued this procedure for several generations and found that this pattern repeated in all cases.

These experimental results did not fit prevailing ideas about heredity, which held that the progeny of two different lines would all be intermediate in appearance as a result of a permanent blending of their characteristics. According to this hypothesis, the F_1 plants should have been all of medium height, rather than all tall. From the blending hypothesis, it also followed that the original parental traits (in this case, tall and short) should never be recovered in future generations, having contaminated each other in the hybrids. This outcome was not observed either. Mendel found that each F_1 hybrid line gave rise to F_2 progeny resembling one or the other of the two original parents (i.e., tall or short) but never intermediate in appearance. Thus, *no blending or contamination of the two alternative traits occurred while they were combined in the hybrids.*

Mendel's First Law

Mendel's interpretation of these data is called the **law of segregation**. From the ratios described, he reasoned that each observable trait (e.g., plant height) is determined by a pair of cellular factors. During the formation of gametes, the *two members of each pair of factors are separated (segregated)*, so an egg cell or a pollen grain contains one or the other, but not both, of the original factors.*

---

*Although chromosomes had not yet been discovered, the cell theory propounded by Schleiden and Schwann in 1838–1839 was fully accepted in Mendel's time, as were reports that eggs and sperm were single cells. Cytological proof that an egg is fertilized by just one sperm did not come until 1879, however.

Because reciprocal crosses gave the same results, Mendel assumed that the male and female parents contribute equally to the formation of offspring. Thus, as shown in Figure 4.2, each F_1 hybrid inherits one "tall" factor (A) from its tall parent plus one "short" factor (a) from its short parent, and its factor composition can be designated as A/a. Although the recessive a factor is not expressed in the A/a hybrid, it remains unchanged and may be expressed in future generations.

When the hybrid A/a plants form gametes, half of them will contain the A factor and the other half will carry the a factor.

> It remains, therefore, purely a matter of chance which of the two sorts of pollen will become united with each separate egg cell. According, however, to the law of probability, it will always happen, on the average of many cases, that each pollen form, A and a, will unite equally often with each egg cell form, A and a. (Mendel 1865)

The four equally likely fertilization combinations are as follows:

Type of egg cell		Type of pollen grain		Type of seed and plant		
¼	A	×	A	→	A/A	¾ tall
¼	A	×	a	→	A/a — ½ A/a	
¼	a	×	A	→	a/A	
¼	a	×	a	→	a/a	¼ short

Hence the recovery, in the F_2 generation, of ¼ A/A : ½ A/a : ¼ a/a plants. Because the tall trait is dominant, however, it is impossible to distinguish the A/A plants from the A/a plants just by looking at them. What one sees is ¾ tall plants and ¼ short plants.

Modern Nomenclature

Several terms were coined after Mendel's time to describe these situations. The factors A and a are **alleles** of each other; they are alternative forms of a **gene** that influences plant height. We can thus refer to A as the dominant allele of the gene determining plant height and to a as its recessive allele. The terms *gene* and *allele* are sometimes used interchangeably, but in most contexts only one term is correct. It is proper to say that we are discussing the gene (not the allele) for plant height and that this gene exists chemically in different allelic forms, one form determining taller growth than the other. (Most genes, by the way, exist in more than two allelic forms, but a given individual carries only two alleles of a particular gene.)

Genotype refers to the precise allelic composition of a cell, such as A/A or A/a or a/a. **Phenotype** refers to what is actually seen in an organism, that is, tall plants or short plants. The phenotype *short* is due to the

Generation	Matings	Type of fertilization

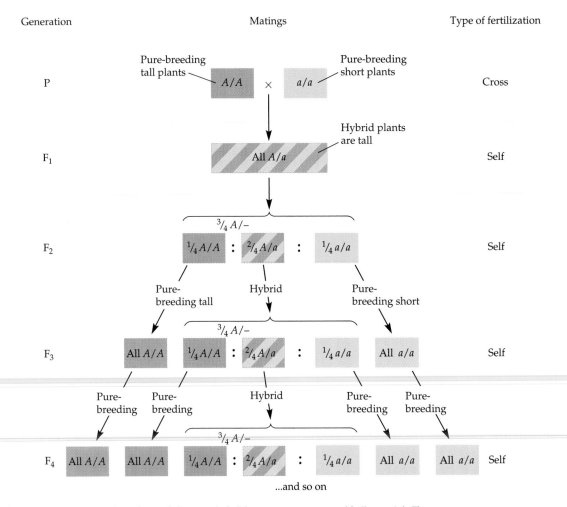

Figure 4.2   Genetic interpretation of Mendel's monohybrid crosses; compare with Figure 4.1. The pure-breeding tall plants (A/A) are represented by boxes of solid gray, the pure-breeding short plants (a/a) by boxes of solid color, and the hybrid tall plants (A/a) by boxes with stripes. In the F₂, F₃, and F₄ generations, the ¼ A/A and ²⁄₄ A/a are summed to give ¾ A/–. The latter notation indicates that at least one dominant factor (in this case A, representing tall) is present.

genotype a/a. The phenotype *tall* is due to either the genotype A/A or the genotype A/a. A **homozygous** genotype is one in which the two alleles are the same (A/A or a/a); in a **heterozygous** genotype, the two alleles are different (A/a). When we are not sure whether a phenotypically tall plant is genotypically homozygous (A/A) or heterozygous (A/a), we write the genotype as $A/–$, the dash representing an allele that could be either A or a. The designation $A/–$ can also be used when we are not interested in distinguishing between the dominant homozygote and the heterozygote.

Genes are present on **chromosomes** inside the cell nucleus of higher organisms (Chapters 2 and 3). Mendel was unaware of the existence of chromosomes. But just as Mendelian factors (alleles) are paired, so, too, are chromosomes. In fact, the two alleles of the same gene occupy the same position on **homologous chromosomes**. The particular site or location where a gene is found along the length of a

chromosome is the **locus** of that gene. For example, in a heterozygous tall plant, A/a, the allele A occupies a particular locus on a particular chromosome. At a corresponding position on the homologous chromosome is found its allele, a:

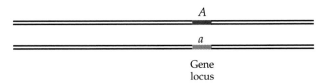

Gene
locus

A Checkerboard Method for Predicting Offspring

The time-honored *Punnett square** helps us visualize simple crosses involving just one or two gene loci. Our checkerboard includes fractions as well as the geno-

*Reginald Punnett is mentioned in Box 12A.

types of gametes and offspring. The steps, pictured in Figure 4.3, are as follows:

1. Label the rows of the checkerboard with the various kinds of eggs that the female parent can make. There should be one row for each *different* kind of egg. The labeling of rows should also include the fraction represented by each egg type—what we call the *gamete fraction*—and these egg gamete fractions should together add up to exactly 1.

2. Label the columns of the checkerboard in a similar fashion with the different kinds of male gametes. Include the gamete fractions of each different kind. If the number of kinds of pollen or sperm is different from the number of kinds of eggs, you will end up with a rectangle rather than a square.

3. Fill in the body of the checkerboard with the *offspring genotypes* by combining the headings of the corresponding row and column. The fraction of each offspring genotype is the *product* of the fractions heading the corresponding row and column (see also Chapter 9). In Figure 4.3, for example, $\frac{1}{2} A \times \frac{1}{2} A$ gives $\frac{1}{4} A/A$ for the upper left box.

Summing the two identical genotypes on the ascending diagonal boxes (lower left + upper right) gives a total of $\frac{2}{4} A/a$. Phenotypically, A/A and A/a are tall ($\frac{3}{4}$), and a/a is short ($\frac{1}{4}$). In this checkerboard, the egg and pollen types are the same because the cross was by self-fertilization, but this will not always be the case. Also, the fractions heading the rows and columns are all $\frac{1}{2}$, reflecting Mendel's law of segregation; the alleles separate randomly into the gametes. But we will now be coming to situations in which the different kinds of gametes are not formed in equal proportions (see also Chapter 9).

Progeny Testing and Matings

Mendel figured out the genotypes of the tall F_2 plants by **progeny testing**. He took ten seeds from each self-fertilized F_2 plant, grew them, and observed the F_3

phenotypes (Figure 4.2). Homozygous tall F_2 plants (A/A) can produce only A gametes; thus, when self-fertilized, they yield only tall plants. But heterozygous tall F_2 plants (A/a) produce equal numbers of both A and a gametes ($\frac{1}{2} A$ and $\frac{1}{2} a$). Like their hybrid F_1 parents, they yield both tall and short progeny after self-fertilization. What Mendel found among the tall F_2 plants, by these breeding tests, was a 1:2 ratio of homozygotes to heterozygotes. This he expressed as $\frac{1}{3}$ pure-breeding to $\frac{2}{3}$ hybrid. (Recall that a ratio is converted into fractions by dividing each member of the ratio by the sum of the members.)

Actually, allowing a tall plant to self-fertilize is not the best way of determining whether it is homozygous or heterozygous. To maximize the probability of detecting the recessive allele in a heterozygote, we should cross each tall plant with a short plant ($A/- \times a/a$). As shown in Figure 4.4A, heterozygous tall plants, when crossed with short plants, will yield a ratio of 1 tall to 1 short progeny, rather than the ratio of 3 tall to 1 short plant recovered after self-fertilization (Figure 4.3). Note in Figure 4.4B and C that homozygous tall plants produce all tall progeny, whether mated with short plants or self-fertilized. Note also in

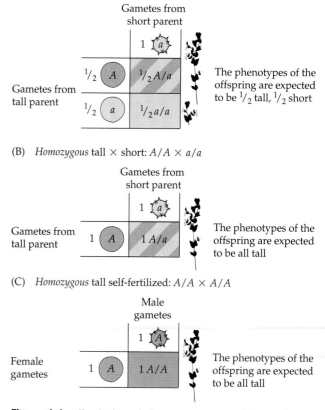

(A) *Heterozygous* tall × short: $A/a \times a/a$

Gametes from short parent

The phenotypes of the offspring are expected to be $\frac{1}{2}$ tall, $\frac{1}{2}$ short

(B) *Homozygous* tall × short: $A/A \times a/a$

Gametes from short parent

The phenotypes of the offspring are expected to be all tall

(C) *Homozygous* tall self-fertilized: $A/A \times A/A$

Male gametes

Female gametes

The phenotypes of the offspring are expected to be all tall

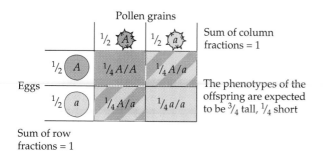

Figure 4.3 Checkerboard depicting Mendel's F_1 cross of an A/a female with an A/a male. In the boxes within the body of the checkerboard appear the genotypes of the F_2 progeny generation and the proportion of each genotype (the gamete fraction).

Figure 4.4 Checkerboards for one gene, two alleles, with A dominant to a. The phenotype of $A/-$ is tall; the phenotype of a/a is short. (The phenotypes of progeny arising from the self-fertilization of a heterozygous tall plant are shown in Figure 4.3.)

Figure 4.4 that when all the gametes from a parent are the same, they are represented by the fraction 1, and by just one row or one column in the checkerboard.

Consider that Mendel planted only ten seeds from each self-fertilized F_2 plant. From heterozygous tall plants ($A/a \times A/a$), he could expect to recover two or three short plants among the progeny ($1/4 \times 10 = 2.5$). Of course, as a result of random variation, he observed sometimes more and sometimes fewer than this number of short plants. The occurrence of even one short plant among the ten progeny would be enough to identify its parent as a heterozygote. But what if, by chance, $A/a \times A/a$ yielded ten tall and no short progeny, instead of the expected seven or eight tall and two or three short plants? In such a case, the parent plant would *mistakenly* be called a homozygote rather than a heterozygote. We can calculate how often this might happen, and it is likely that Mendel missed a few heterozygotes this way.

Now, if ten seeds from the cross $A/a \times a/a$ are planted, we would expect to find five short plants among the progeny ($10 \times 1/2 = 5$). Here, too, we would observe sometimes more and sometimes fewer than five short plants because of random fluctuation. But the occurrence of ten tall and no short plants from this cross is much less likely than in the case of self-fertilization. For this reason, geneticists will, whenever possible, cross a dominant phenotype to a recessive phenotype to determine whether the former is homozygous or heterozygous. Such a mating is called a **testcross**. A testcross is one of the six matings possible for a single pair of alleles, A and a. That is, there are six different ways that the three genotypes (A/A, A/a, and a/a) can be combined in pairs, as shown in Table 4.1. In the course of describing Mendel's experiments, all but one of these has been considered. We suggest that you verify for yourself the outcome of each of these matings, using checkerboards if necessary, and then simply commit them to memory.

Mendel did many additional experiments to see what happens when more than one pair of characters are considered simultaneously, for example, the cross $A/a \ B/b \times A/a \ B/b$. Before discussing these results, however, we turn our attention to multiple alleles and some rules of probability.

Multiple Alleles

So far, we have only considered genes with two alleles. But there is nothing magical about this number. The various forms of a gene—the alleles—are generated by the process of **mutation**. This is the chemical change of one allele—say, the "standard," or "wild-type," allele—into another—a "variant," or "mutant," allele—or vice versa. Mutations can theoretically generate as many different alleles as there are chemical changes possible in a gene. Indeed, some genes have hundreds of known alleles.

To describe the various allelic forms of a gene, called **multiple alleles**, geneticists may use sequential superscripts (or subscripts): A^1, A^2, A^3, ..., A^N, where N is the highest number in the allelic series. The important thing to remember about multiple alleles is that any one individual still has only two alleles (or *one*, for X-linked genes in a human male). These can be selected from a larger number, N, existing in the population as a whole.

A good example of a multiple allelic series is the gene that codes for the beta (β) polypeptide chain of the hemoglobin molecule (Chapter 7). About 400 alleles of this gene are known; among these, the most common are Hb^A, Hb^S, Hb^E, and Hb^C. The superscript A refers to the normal allele, and S, E, and C refer to allelic forms of the gene that lead to various abnormal β polypeptide chains. Phenotypically, homozygotes for Hb^S have sickle-cell disease and are often quite ill. Homozygotes for Hb^E or Hb^C have only a mild or moderate anemia. Individuals who are heterozygous for the normal allele combined with either Hb^S, Hb^E, or Hb^C are unaffected, or nearly so.

Consider a gene with *three* alleles: B^1, B^2, and B^3. For each allele there is a homozygous genotype: B^1/B^1, B^2/B^2, and B^3/B^3. For each pairing of different alleles selected from the three available, there is a heterozygous genotype: B^1/B^2, B^1/B^3, and B^2/B^3.

With *four* alleles of a gene—C^1, C^2, C^3, and C^4—there are ten genotypes: four homozygous and six heterozygous. One methodical way for writ-

---

TABLE 4.1 Matings for one pair of alleles, A and a

Parental genotypes[a]	Genotypic ratios among progeny	Phenotypic ratios[b]
1. $A/A \times A/A$	All A/A	
2. $A/a \times A/a$	$1/4 \ A/A : 1/2 \ A/a : 1/4 \ a/a$	$3/4 \ A/- : 1/4 \ a/a$
3. $a/a \times a/a$	All a/a	
4. $A/A \times A/a$	$1/2 \ A/A : 1/2 \ A/a$	All $A/-$
5. $a/a \times A/a$	$1/2 \ a/a : 1/2 \ A/a$	
6. $A/A \times a/a$	All A/a	

[a]In the first three, the genotypes of the two parents are the same. In the last three, the genotypes of the two parents are different, but here we do not care to specify which sex has which genotype. If reciprocal matings were considered separately, then there would be a total of nine different matings, rather than the six listed here.

[b]Where different from genotypic ratios and assuming that A is dominant to a.

ing down all the genotypes is to first list all the alleles, as we have done for these four. Then combine the first allele (C^1) with itself and sequentially with each allele to the right of it in the list, here yielding a subtotal of four genotypes (C^1/C^1, C^1/C^2, C^1/C^3, C^1/C^4). Then combine the second allele (C^2) with itself and sequentially with each one to the right of it (C^2/C^2, C^2/C^3, C^2/C^4), here a subtotal of three genotypes. Continue this process until you have run out of combinations, in this case as far as C^4/C^4, a subtotal of one genotype.

In general, for a gene with N alleles—A^1, A^2, A^3, ... , A^N—this listing and combining will yield all possible genotypes, albeit in a tedious fashion. Readers proficient in algebra can derive a formula for the number of possible different genotypes for a gene with N alleles:

$$\frac{N(N+1)}{2}$$

Of this number, N are homozygous and the rest are heterozygous. Using this formula, confirm the number of possible different genotypes given earlier for N = 2, 3, and 4 alleles. As the number of alleles of a gene increases, the number of possible different genotypes increases dramatically: With 10 alleles, there are 55 different genotypes; with 20 alleles, 210 different genotypes. Although all the possible genotypes may occur in some populations, some genotypes may be missing—that is, may not be present in any individual now living or even once living. In any event, we emphasize again that any individual genotype is made up of only two alleles for each gene.

Some Probability

Up to this point, we have used checkerboards to predict the offspring of crosses involving the alleles of one gene. This method could be expanded to deal with several genes, but there are quicker, less tedious ways of arriving at the same predictions: using the rules of probability. Although some aspects of this kind of thinking might be new to you, much of the following material will seem familiar.

When we flip a coin, we can ask, What is the *chance*, or *likelihood*, of a head? When we flip the coin 100 times, and then repeat the series of flips several more times, we might ask how often the *proportion*, or *frequency*, of heads will be greater than 80 (or some other number). These terms are roughly synonymous with **probability**. More specifically, however, probability is always written as a number between 0 and 1, expressed as a simple fraction (like $\frac{1}{3}$) or a decimal fraction (like 0.333).

Definition

More rigorously, we can think of doing an experiment like picking a card, rolling a die (plural, dice), or se-

lecting a gamete. In each case, there is a certain number of equally likely outcomes: 52 for a deck of cards, 6 for a die, and 2 for the case of a person heterozygous at a given locus. Before doing the experiment, we ask about the likelihood some event A occurring; its probability can be written

$$P(A) = \frac{\text{number of outcomes "favorable" to A}}{\text{total number of equally likely outcomes}}$$

The notation $P(A)$ should be read "the probability of event A occurring," and its definition is a common-sense one. In the case of drawing a card randomly from a deck, there are 52 equally likely outcomes (the denominator). Of these, 13 are in a suit "favorable," say, to the event "drawing a heart" (the numerator). Therefore, the probability of drawing a heart is

$$P(\text{heart}) = {}^{13}\!/_{52} = \frac{1}{4} = 0.250$$

Using the definition again, note that for a die, fairly balanced and thrown,

$$P(\text{even number}) = {}^{3}\!/_{6} = \frac{1}{2} = 0.500$$

$$P(\text{at least a 3}) = {}^{4}\!/_{6} = {}^{2}\!/_{3} = 0.667$$

The phrase "at least a 3" means throwing a 3 or a higher number: a 3 or 4 or 5 or 6, a total of four favorable outcomes. For a B/b heterozygote,

$$P(B \text{ gamete}) = \frac{1}{2} = 0.500$$

$$P(b \text{ gamete}) = \frac{1}{2} = 0.500$$

If an event is *impossible*, there will be *no* outcomes favorable to it, so the numerator is 0. On the other hand, if an event is *inevitable*, then *every* outcome is favorable to it, and the numerator is the same as the denominator. Thus,

$$P(\text{impossible event}) = 0$$

$$P(\text{inevitable event}) = 1$$

For example, the probability of rolling a 7 with one die is 0, and the probability of rolling at least a 1 is 1.

If the probability of event A occurring is some number x, what is the probability of A *not* occurring? Something must happen; it is inevitable that either A occurs or it does not occur, and so the probabilities of events A and not-A must add to 1. Therefore, if

$$P(A) = x$$

then

$$P(\text{not-A}) = 1 - x$$

For example, if the probability of snow tomorrow is $\frac{1}{10}$, then the probability of no snow is $\frac{9}{10}$.

We can generalize the foregoing a bit by saying that the probabilities of a set of events must add to 1 if every possible outcome is included. Thus, the sum of probabilities of a set of exhaustive events equals 1. This statement provides a useful check on arithmetic. For example, from the mating $B/b \times B/b$, the probability of heterozygous offspring plus the probability of homozygous offspring must equal 1.

Two Rules for Combining Probabilities

Sometimes the probability of interest is not a simple one that can be gotten directly from the definition. For example, the probability of a straight in poker (five cards in sequence regardless of suit) is far from obvious.* What we need are some rules for combining several simple probabilities to give us the probability of a more complex event. Here we present, in a fairly intuitive way, two such rules.

The Addition Rule ("Or" Rule). Consider an experiment that can result in any of the events A, B, C, D, … x. Our interest is in not just one of these events, but in, say, A or B occurring. Event A could be "picking a heart"; event B, "picking a spade." In shorthand notation, we are interested in $P(A\ or\ B)$. Intuitively,

$$P(A\ or\ B) = P(A) + P(B)$$

or, for more than two events,

$$P(A\ or\ B\ or\ C\ or…) = P(A) + P(B) + P(C) + …$$

In words, if we want to know the probability of one event or another or yet another, and so on, we add their individual probabilities together. This is the **addition rule**.

Whenever the conjunction or is used in everyday language, the addition rule is usually appropriate. For example, for a gene with two alleles, B dominant to b, what is the probability of a person having the dominant phenotype? Note that the dominant phenotype arises if a person is either B/B or B/b. Therefore,

$$P(\text{dominant phenotype}) = P(B/–) = P(B/B) + P(B/b)$$

There is a restriction in the use of the addition rule: *The events in question must be mutually exclusive.* This means that the occurrence of one event excludes the other(s) from happening. You will discover that in genetics, this is usually the case. If an organism has the genotype B/B, it does not have the genotype B/b or any other genotype at that particular locus. If a card is a heart, it is not a spade or any other suit; thus,

$$P(\text{heart } or \text{ spade}) = \frac{1}{4} + \frac{1}{4} = \frac{1}{2}$$

$$P(\text{heart } or \text{ spade } or \text{ diamond } or \text{ club})$$
$$= 4 \times \frac{1}{4} = 1 \text{ (inevitably)}$$

*The probability is $\frac{768}{216,580}$, or about 0.35%.

To illustrate events that are not mutually exclusive, let's say that A is "picking a heart" and B is "picking a queen." Because it is entirely possible for a card to be both a heart and a queen at the same time,

$$P(\text{heart } or \text{ queen}) \neq \frac{13}{52} + \frac{4}{52}$$

The sign ≠ means "does not equal." Since we cannot use the addition rule here, the probability P(heart or queen) must be obtained some other way. It can be calculated directly from the definition of probability as $\frac{16}{52}$ (i.e., 13 hearts plus 3 other queens out of the 52 cards in the deck).

The Multiplication Rule ("And" Rule). Consider an experiment that has as possible results the events A, B, C, D, … x. If the experiment is repeated several times, we might be interested in the joint outcome of successive trials. For example, the joint probability of A on the first trial and B on the second is

$$P(A\ and\ B) = P(A) \times P(B)$$

or, for more repetitions,

$$P(A\ and\ B\ and\ C\ and … x)$$
$$= P(A) \times P(B) \times P(C) \times … P(x)$$

In words, if we want the joint probability of one event and another and another, and so on, we just multiply their individual probabilities together. This is the **multiplication rule**.

With two decks of cards before you, what is the probability of drawing an ace from the first and another ace from the second? By the multiplication rule, we obtain

$$P(\text{ace } and \text{ ace, two decks}) = (\tfrac{1}{13})(\tfrac{1}{13}) = \tfrac{1}{169}$$

Note that each factor of $\frac{1}{13}$ is itself obtained either from the addition rule ($\frac{1}{52} + \frac{1}{52} + \frac{1}{52} + \frac{1}{52}$) or directly from the definition of probability: 4 favorable outcomes (aces) out of 52 (cards).

The restriction in the use of the multiplication rule is that *the events must be independent.* This means that the occurrence of one event does not affect in any way the probability of other events occurring. Selections from separate decks are independent events because handling one deck of cards cannot affect the other.

If we draw two cards from the same deck, however, the outcomes are *not* independent. Having drawn one card, ace or not, the probabilities applying to the second draw are changed. To compute the probability of picking two aces from the same deck *without replacing the first card*, we need to take note of the altered probability:

$$\tfrac{4}{52} \text{ (ace on first draw)} \times \tfrac{3}{51} \text{ (ace on second draw)} = \tfrac{1}{221}$$

Questions of independent events arise in very different situations than do questions of mutually exclusive events, a condition of the addition rule. Mutually exclusive events can never be independent, because excluding one event affects drastically the probability of another event's occurrence.

Whenever the conjunction *and* is used in everyday language, the multiplication rule is usually appropriate. For example, what is the probability that four siblings will all be girls? This can be interpreted as

$$P(\text{4 girls}) = P(\text{1st born a girl } and \text{ 2nd a girl } and \text{ ... })$$

$$= P(\text{girl}) \times P(\text{girl}) \times P(\text{girl}) \times P(\text{girl})$$

$$= \text{approximately } (\tfrac{1}{2})^4$$

$$= \tfrac{1}{16} = 0.0625$$

Based on actual birth records, the likelihood of a live-born girl is slightly, but significantly, less than $\tfrac{1}{2}$, about 0.48 or 0.49, depending on which real population is sampled. Thus, a slightly more accurate answer might be

$$P(\text{4 girls}) = (0.49)^4 = 0.0576$$

As an example of using both the addition and multiplication rules, consider the offspring from $A/a \times A/a$. From a checkerboard, the genotypic results are

$$P(A/A) = \tfrac{1}{4} \qquad P(A/a) = \tfrac{1}{2} \qquad P(a/a) = \tfrac{1}{4}$$

Here are the rules of probability that lead to these same results:

$$P(A/A) = P(A \text{ egg } and \text{ } A \text{ sperm}) = \tfrac{1}{2} \times \tfrac{1}{2} = \tfrac{1}{4}$$

$$P(A/a) = P(A \text{ egg } and \text{ } a \text{ sperm, } or \text{ } a \text{ egg } and \text{ } A \text{ sperm})$$
$$= (\tfrac{1}{2} \times \tfrac{1}{2}) + (\tfrac{1}{2} \times \tfrac{1}{2}) = \tfrac{1}{4} + \tfrac{1}{4} = \tfrac{1}{2}$$

$$P(a/a) = P(a \text{ egg } and \text{ } a \text{ sperm}) = \tfrac{1}{2} \times \tfrac{1}{2} = \tfrac{1}{4}$$

$$P(\text{dominant phenotype}) = P(A/A) + P(A/a) = \tfrac{1}{4} + \tfrac{1}{2} = \tfrac{3}{4}$$

Other basic rules of probability answer questions such as, What is the probability of getting two *or more* girls in a family of five? These rules are based on expansion of the binomial equation, however, and will not be presented here. Interested readers can find them in Mange and Mange (1990) or in any textbook on general genetics.

Mendel's Second Law: Independent Assortment of Two Pairs

Mendel developed his second law by studying the joint inheritance of two or more independent traits in garden peas. Among these were seed shape (*smooth* versus *wrinkled*) and seed color (*yellow* versus *green*). For the parental generation, he crossed two pure-breeding lines, one with smooth yellow seeds and the other with wrinkled green seeds. Because the parents differed from each other in just the two characteristics, this type of mating is called a **dihybrid cross**. The F_1 progeny of these matings all had smooth yellow seeds. Thus, smooth seed is dominant to wrinkled seed (Box 4B), and yellow seed is dominant to green seed.

Calling the smooth allele S, its wrinkled allele s, the yellow allele (at a different locus) Y and its green allele y, we can summarize the parental cross yielding F_1 hybrid offspring:

$$S/S \; Y/Y \times s/s \; y/y \rightarrow S/s \; Y/y$$

After sowing the F_1 hybrid seeds and allowing the resulting plants to self-fertilize, Mendel collected a total of 556 F_2 seeds. They fell into four phenotypic categories:

- 315 smooth yellow ($S/-\; Y/-$)
- 101 wrinkled yellow ($s/s \; Y/-$)
- 108 smooth green ($S/- \; y/y$)
- 32 wrinkled green ($s/s \; y/y$)

What Mendel observed, roughly, was a *9/16:3/16: 3/16:1/16 phenotypic ratio* among these F_2 offspring. From these and additional experiments, Mendel determined that the joint inheritance of two or more different traits results in progeny whose ratios mirror the laws of probability for independent events. More specifically, they obey the multiplication ("and") rule. Let us now verify Mendel's findings for his dihybrid crosses involving smooth or wrinkled seeds and yellow or green seeds. First we must determine the types and proportions of gametes formed by the F_1 parents. Then we will determine the offspring ratios by means of (1) the checkerboard method and (2) the gene × gene method.

The Proportions of the Gametes in a Dihybrid Cross

The F_1 hybrid parents with smooth and yellow seeds were derived from two different pure-breeding parental lines. Thus, they had to be doubly heterozygous, $S/s \; Y/y$, and produced four kinds of gametes: SY, Sy, sY, and sy.

In what proportions were these gametes formed? Mendel's law of segregation states that $\tfrac{1}{2}$ the gametes are expected to carry S and $\tfrac{1}{2}$ to carry s. The same law states that $\tfrac{1}{2}$ the gametes are expected to carry Y and $\tfrac{1}{2}$ to carry y. The question is this: How are these two pairs of alleles distributed into gametes with respect to each other? Mendel was able to show that alleles of different genes are assorted into gametes independently: *The segregation of one pair of alleles in no way alters the segregation of the other pair of alleles.* Each gene pair is acting alone, not interacting with the other, so

BOX 4B   *Why Are Some Pea Seeds Wrinkled*

For decades, it has been known that a difference in starch metabolism accounts for these two alternative phenotypes. Compared to smooth seeds, wrinkled seeds have less starch and smaller starch granules, but a higher lipid and sugar content. The latter then leads to a higher water content. As they mature, these seeds also lose more water; but because the seed coat does not shrink at all, they end up being wrinkled.

In 1990, English researchers reported that wrinkled seeds lack (and smooth seeds contain) a certain starch-branching enzyme I (SBEI). After isolating the *SBEI* gene from both phenotypes, they were surprised to find that the mutant allele in wrinkled seeds is larger than the normal allele present in smooth seeds. Permanently inserted into the mutant allele is an extra piece of DNA, a type of "jumping gene" first discovered in corn but now known to occur in many organisms (Chapter 10). This discovery adds an unexpectedly modern twist to a classic story.

The pea pod shown here contains nine smooth (also called round) peas and three wrinkled peas, the expected result from a cross of two F₁ heterozygotes.

---

the various kinds of gametes have random frequencies. Thus, we calculate the probability of any kind of gamete by multiplying together the probabilities for all the alleles it contains.

A *tree diagram* can be useful to determine the different kinds of gametes from a double heterozygote. Recalling the "and" rule for probabilities, here is the diagram for our example:

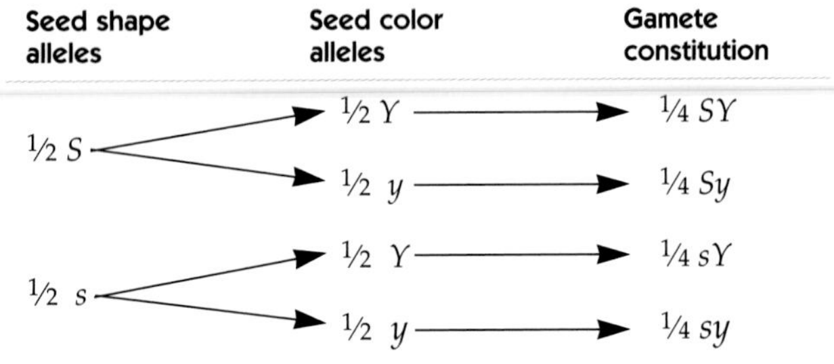

Seed shape alleles	Seed color alleles	Gamete constitution
½ S	½ Y →	¼ SY
	½ y →	¼ Sy
½ s	½ Y →	¼ sY
	½ y →	¼ sy

This diagram is the essence of Mendel's second law, the **law of independent assortment**.

Offspring Types for Two Genes

Let us now determine the offspring phenotypes from the mating

$$S/s\ Y/y\ \text{male} \times S/s\ Y/y\ \text{female}$$

In a checkerboard, rows and columns are labeled with the various kinds of gametes and their probabilities. The four different kinds of pollen and egg cells have already been given in the tree diagram, and Figure 4.5 shows the offspring checkerboard.

Each box in the checkerboard corresponds to a probability of ¹⁄₁₆, which is the product of ¼ and ¼, the gamete fractions heading the rows and columns. Summing the fractions and genotypes in these boxes,

we get the following offspring phenotypic expectations:

$$\left.\begin{array}{l}\text{⁹⁄₁₆ smooth yellow seeds}\\\text{³⁄₁₆ wrinkled yellow seeds}\\\text{³⁄₁₆ smooth green seeds}\\\text{¹⁄₁₆ wrinkled green seeds}\end{array}\right\} \text{Sum}=1$$

The first three phenotypic probabilities are tallied by combining probabilities according to the "or" rule. For example, the probability for ³⁄₁₆ smooth green offspring comes from the following rows and columns in Figure 4.5:

Second row, second column: ¹⁄₁₆ *S/S y/y*

Second row, fourth column: ¹⁄₁₆ *S/s y/y*

Fourth row, second column: ¹⁄₁₆ *S/s y/y*

Notice that among these smooth green offspring, there is a 1:2 ratio of *S/S* to *S/s*. Thus, ⅓ of them are homozygous smooth and ⅔ are heterozygous smooth. The wrinkled yellow offspring are constituted in the same way. You can verify the following fractions among the ⁹⁄₁₆ smooth yellow offspring: ⅑ are homozygous at both loci, ²⁄₉ are homozygous only at the *S* locus, ²⁄₉ are homozygous only at the *Y* locus, and ⁴⁄₉ are heterozygous at both loci.

Gene × Gene Method

The checkerboard can be called the **gamete × gamete** method for solving genetics problems, because first the various types of gametes are determined and then their proportions are multiplied together. Another pro-

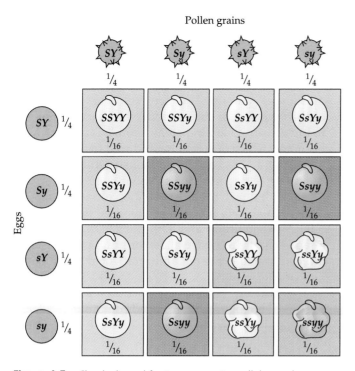

Pollen grains

Figure 4.5 Checkerboard for two genes, two alleles each, with S dominant to s and Y dominant to y. The phenotype of S/– is smooth seeds, and the phenotype of s/s is wrinkled seeds. The phenotype of Y/– is yellow seeds, and the phenotype of y/y is green seeds.

cedure, called the **gene × gene method**, is useful when two or more loci are considered simultaneously and just one or a few of the possible offspring types are of interest. We illustrate the gene × gene method using the same mating: *S/s Y/y × S/s Y/y*. Considering one pair of alleles at a time, we know that Mendel's first law predicts the following:

S/s × S/s   Genotypes: ¼ *S/S*, ½ *S/s*, ¼ *s/s*

Phenotypes: ¾ smooth, ¼ wrinkled

Y/y × Y/y   Genotypes: ¼ *Y/Y*, ½ *Y/y*, ¼ *y/y*

Phenotypes: ¾ yellow, ¼ green

Now we use the fact that the two genes are inherited independently and so multiply the individual expectations. From the phenotypic results, we obtain

$P(\text{smooth yellow}) = P(\text{smooth}) \times P(\text{yellow}) = (\tfrac{3}{4})(\tfrac{3}{4}) = \tfrac{9}{16}$

$P(\text{smooth green}) = P(\text{smooth}) \times P(\text{green}) = (\tfrac{3}{4})(\tfrac{1}{4}) = \tfrac{3}{16}$

$P(\text{wrinkled yellow}) = P(\text{wrinkled}) \times P(\text{yellow})$

$= (\tfrac{1}{4})(\tfrac{3}{4}) = \tfrac{3}{16}$

$P(\text{wrinkled green}) = P(\text{wrinkled}) \times P(\text{green}) = (\tfrac{1}{4})(\tfrac{1}{4}) = \tfrac{1}{16}$

These gene × gene results are, of course, the same as the gamete × gamete results, and in either case we might have considered genotypes rather than phenotypes.

Suppose that we want to know only the proportion of offspring expected to have the genotype *S/S Y/y*. Here we need only multiply $P(S/S) \times P(Y/y) = \tfrac{1}{4} \times \tfrac{1}{2} = \tfrac{1}{8}$.

Now consider a situation involving more than two genes: What proportion of the offspring from the following cross is expected to show the dominant phenotype for all genes? (Assume that each uppercase allele is dominant to the corresponding lowercase allele.)

D/d E/e F/F G/g × D/d e/e F/f G/g

For this, the checkerboard is quite tedious, but the gene × gene analysis is easy:

Mating by gene	P(dominant offspring)
D/d × D/d	¾
E/e × e/e	½
F/F × F/f	1
G/g × G/g	¾

Multiplying the fractions together, we get ⁹⁄₃₂, the proportion of offspring who have the dominant phenotype for all genes. Solving this problem by means of a checkerboard would require an 8 × 8 diagram. Should you care to try it, the different kinds of gametes from each parent can be obtained by a tree diagram that branches for each heterozygous gene.

In summary, we can always set up a checkerboard or a tree diagram, but the gene × gene method is usually much less work.

Mendel's Laws and Meiosis

By the time Mendel's work was uncovered, the basic facts of cell reproduction were already known. Thus, the striking parallels between the behavior of his "factors" and the behavior of chromosomes in cell division were almost immediately recognized. The parallels between Mendel's "factors" and nuclear chromosomes were synthesized as the **chromosome theory of inheritance.**

Direct microscopic proof for the chromosome theory of inheritance came about a decade later. Of course, it cannot be shown that one pair of chromosomes segregates independently of another if the two members of a homologous pair look exactly alike—as they usually do. Eleanor Carothers (1913) found in the grasshopper not only a sex chromosome difference but also an additional set of homologous chromosomes in which one member of the pair had an extra piece of chromosomal material stuck to it. Thus, she was able to see differences between the homologues of two sets. In microscopic examinations of testes, she found the four types of secondary spermatocytes in roughly equal numbers. This showed conclusively that chromosome behavior underlies both Mendelian rules of transmission. Shortly thereafter, another scientist was able to correlate the unorthodox inheritance of an eye color gene with the presence of an extra chromosome in a female fruit fly (Chapter 13).

Some Examples in Humans

Occasionally, a human chromosome carries a distinctive cytological feature that is found in several generations of a family. Such chromosomes are known as **marker chromosomes**. If this family is also segregating for a trait whose genetic locus is present on the unusual chromosome, it should be possible to observe their joint inheritance. This is in fact how the first assignment of an autosomal gene to a specific human chromosome came about. While karyotyping members of his own family, Roger Donahue (1968), a graduate student at Johns Hopkins University, discovered a peculiar "uncoiled" region in the centromeric heterochromatin of the long arm of chromosome 1 (Figure 4.6). Chromosomes with this property are considerably longer than their normal homologues, but carriers of this variant chromosome are phenotypically normal. Donahue noted that his family also showed segregation patterns for the Duffy blood group (*Fy*) that coincided with the inheritance of the extra-long chromosome. Specifically, family members who carried the marker chromosome also had the *Fy^a* allele for the Duffy blood group, while those without the marker chromosome had the *Fy^b* allele. Statistical analyses of several such pedigrees in which *uncoiled* and *Fy* were segregating, as well as other studies, placed the Duffy locus very close to the heterochromatin on the long arm of chromosome 1. Other kinds of cytologically detectable chromosomal variants and their use in genetic studies will be discussed in later chapters.

Now let us consider the joint inheritance of two human traits, pituitary dwarfism and the ABO blood group, whose genes are known to be borne on different chromosomes. (The gene causing pituitary dwarfism lies near the middle of the long arm of chromosome 17, and the ABO gene is located near the tip of the long arm of chromosome 9.) Short stature has many causes. One kind stems from a lack of growth hormone, which is normally produced by the pituitary gland (a tiny but very important structure inside the brain). Affected individuals exhibit *pituitary dwarfism*. Those with type I, the most common form (Figure 4.7), have normal body proportions, pinched facial features with high foreheads, and high-pitched voices. Letting *d* represent the recessive allele for pituitary dwarfism, affected persons are *d/d* and unaffected persons are *D/D* or *D/d*.

The ABO blood type, expressed as antigens on the surface of red blood cells, depends on multiple alleles, *I^A*, *I^B*, and *I^O*, of a gene: The alleles *I^A* and *I^B* are

Figure 4.7 Charles S. Stratton, popularly known as General Tom Thumb, shown here with the circus showman P. T. Barnum. Stratton's short stature (3 feet, 2 inches) apparently resulted from homozygosity for recessive genes that interfered with the production of pituitary growth hormone. His parents were first cousins. (Photograph gift of Ben Kelley, from the Collection of the John and Mable Ringling Museum of Art, the State Art Museum of Florida.)

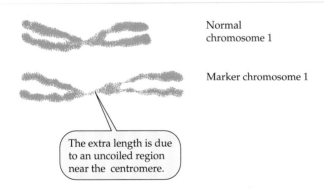

Normal chromosome 1

Marker chromosome 1

The extra length is due to an uncoiled region near the centromere.

Figure 4.6 An "uncoiled" region in the centromeric heterochromatin of human chromosome 1. Joint inheritance of this chromosome 1 marker and Duffy allele *a* in one family suggested that they were physically linked together. This 1968 finding led to the first assignment of a human autosomal gene to a specific human autosome.

codominant to each other and cause antigens A and B, respectively, to be expressed. These two alleles are both dominant to I^O. (A discussion of the gene and antigens is presented in Chapter 18.) We can summarize the genotype/phenotype correspondences as follows:

Full genotype	Shorthand genotype	Phenotype (blood type)
I^A/I^A or I^A/I^O	A/A or A/O	A
I^B/I^B or I^B/I^O	B/B or B/O	B
I^A/I^B	A/B	AB
I^O/I^O	O/O	O

The shorthand genotypes, which we use throughout this chapter, omit the basic locus designation, I, leaving just the superscripts.

Consider a doubly heterozygous male, $D/d\ A/B$, whose genotype and karyotype are roughly diagrammed as follows:

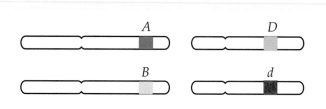

Because D and d are alleles of each other, they occupy corresponding places on homologous chromosomes. The same is true for alleles A and B. This male inherited one chromosome 9 and one chromosome 17 from his mother, and one each from his father. When he makes sperm, the maternal and paternal chromosomes 9 segregate at the first meiotic division independently of the segregation of maternal and paternal chromosomes 17 (Figure 4.8).

If this man marries a woman of genotype $D/d\ O/O$, what offspring and in what proportions might be expected? The four different sperm types are given in Figure 4.8. But the mother, being heterozygous at the D locus but homozygous for the ABO gene, will produce only two types of eggs: ½ DO and ½ dO. We could diagram this cross by means of a checkerboard, but instead let us simply use the gene × gene method. Considering one pair of alleles at a time, we know that Mendel's first law predicts the following:

$D/d \times D/d$ Genotypes: ¼ D/D, ½ D/d, ¼ d/d

              Phenotypes: ¾ unaffected, ¼ affected

$A/B \times O/O$ Genotypes: ½ A/O, ½ B/O

              Phenotypes: ½ type A, ½ type B

Knowing that the two genes are inherited independently, we can multiply together the individual expectations. From the phenotypic results, we obtain:

$$P(\text{unaffected, type A}) = P(\text{unaffected}) \times P(\text{type A})$$
$$= (\tfrac{3}{4})(\tfrac{1}{2}) = \tfrac{3}{8}$$
$$P(\text{unaffected, type B}) = P(\text{unaffected}) \times P(\text{type B})$$
$$= (\tfrac{3}{4})(\tfrac{1}{2}) = \tfrac{3}{8}$$
$$P(\text{affected, type A}) = P(\text{affected}) \times P(\text{type A})$$
$$= (\tfrac{1}{4})(\tfrac{1}{2}) = \tfrac{1}{8}$$
$$P(\text{affected, type B}) = P(\text{affected}) \times P(\text{type B})$$
$$= (\tfrac{1}{4})(\tfrac{1}{2}) = \tfrac{1}{8}$$

With this result, you can see that the offspring ratios from crosses of simply inherited autosomal traits may differ from the "standard" 3:1 or 9:3:3:1 ratios obtained in earlier examples. Thus, it is important to analyze each case very carefully.

Sex Determination and Sex Linkage

Mendel found no differences among the progeny from reciprocal matings. The same phenotypic classes occurred, and in the same proportions, whether a given trait was inherited through the male or through the female parent. Pea plants are not distinguishable on the basis of sex; both male and female structures appear in the same individual, a situation that characterizes some animals as well. In other plant and animal species, however, the sexes are separate, and the mode of inheritance of some traits does differ with reciprocal matings. For such organisms, sex is (with few exceptions) an inherited trait.

Sex Determination

Recall that among the sperm produced by a male, half carry an X chromosome and the other half carry a Y chromosome. All the eggs produced by a female, on the other hand, contain an X chromosome. So it is the father, rather than the mother, whose gamete determines the sex of their offspring. If an X-bearing sperm fertilizes the egg, a female is produced; the union of a Y-bearing sperm with an egg gives rise to a male. This system, diagrammed in Figure 4.9, perpetuates a presumed 1:1 sex ratio at the time of fertilization, generation after generation. Actually, the sex ratio at birth differs slightly from 1:1, perhaps as a result of a differential mortality of the sexes during intrauterine life. We should emphasize that sexual development is an immensely complex process controlled by many genes (Y-linked, X-linked, and autosomal) and susceptible to alteration by environmental conditions at any stage from before birth through adulthood. In many species (including humans), some sexual characteristics can be modified, or perhaps even reversed, by hormonal treatments, injury, surgery, and many other factors. So chromosomal sex and phenotypic sex are usually, but not always, the same (Chapter 15). Anyway, what concerns us here is the behavior of the sex chromosomes and the genes they carry rather than the development of sex *per se*.

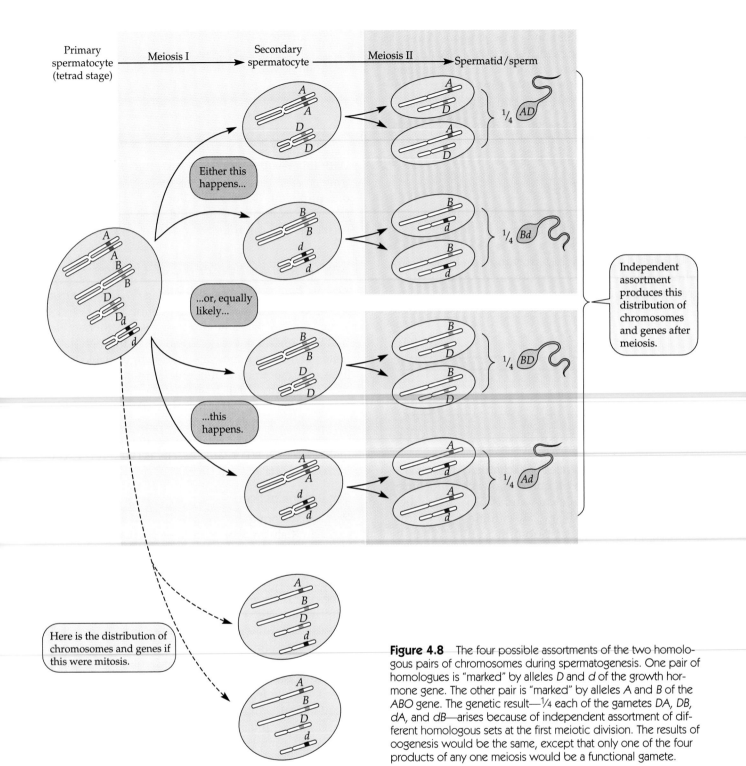

Primary spermatocyte (tetrad stage) — Meiosis I — Secondary spermatocyte — Meiosis II — Spermatid/sperm

Either this happens...

...or, equally likely...

...this happens.

Independent assortment produces this distribution of chromosomes and genes after meiosis.

Here is the distribution of chromosomes and genes if this were mitosis.

Figure 4.8 The four possible assortments of the two homologous pairs of chromosomes during spermatogenesis. One pair of homologues is "marked" by alleles *D* and *d* of the growth hormone gene. The other pair is "marked" by alleles *A* and *B* of the *ABO* gene. The genetic result—¼ each of the gametes *DA*, *DB*, *dA*, and *dB*—arises because of independent assortment of different homologous sets at the first meiotic division. The results of oogenesis would be the same, except that only one of the four products of any one meiosis would be a functional gamete.

Sex Linkage

The X chromosomes of most species bear roughly as many genes as do autosomes of similar size. The human X chromosome is medium-sized; it carries over 220 definitely identified loci, with many other loci possible but not yet fully validated. These genes are said to be **X-linked** or *sex-linked*, because their inheritance is coupled with that of the X chromosome. Keep in mind, however, that X-linked genes are no different in kind from those found on autosomes. They affect characteristics and systems as diverse as blood groups, vision, hearing, the nervous system, the muscular system, teeth, skin, and glucose metabolism. Indeed, most X-linked genes have nothing to do with sexual differ-

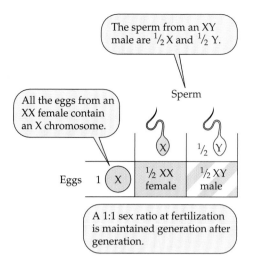

Figure 4.9 Checkerboard for the chromosomal sex-determining system that operates in humans and many other organisms.

entiation. Conversely, there are a number of autosomal genes known to play a role in sexual development. Thus, *X linkage is a matter of gene geography, not gene function.*

The human Y, on the other hand, is one of the smallest chromosomes and carries very few genes. Nevertheless, the X and Y chromosomes do possess a tiny homologous region at the tips of their short and long arms, and during the cell division that leads to the production of gametes, they form a homologous pair that obeys Mendel's first law.

X Linkage in Humans

A familiar example of X-linked inheritance in humans is **color blindness**. Actually, most "color-blind" people see a range of colors—especially among the blues and yellows. What we really mean is *partial* color blindness or *defective color perception*, which is mainly a problem of discriminating between reds and greens.* Although pedigree studies as early as 1911 indicated that the relevant gene resides on the X chromosome, complete understanding of its inheritance had to await molecular studies carried out in the mid-1980s. Researchers showed that there are actually *two* color vision loci on the X chromosome; they lie close to each other and differ by only 2% of their DNA base sequences. Indeed, one locus probably arose as a duplication of and subsequent divergence from the other locus (Nathans 1989).

Each of the two loci controls the production of a light-absorbing protein, one making up part of the *red-sensitive* color vision pigment and the other making up part of the *green-sensitive* pigment. These two pigment types occur separately in color-sensitive cells (cones) within the retina of the eye, as does a third, blue-sensitive pigment (autosomally inherited and not under consideration here). Four types of X-linked color vision defects are known: At each of the two loci, one mutant allele is associated with an anomalous (abnormal) pigment, and the other mutant allele is associated with a missing pigment (Rushton 1975). At each locus, the normal allele is dominant to either mutant, and the mutant with abnormal pigment is dominant to that with missing pigment.

Altogether, about 8% of males and up to 1% of females of western European extraction have defective red-green vision. One method for diagnosing color vision defects is shown in Figure 4.10. Here we will only consider the most common defect, *deuteranomaly* (which we call *green shift*), occurring in about 5% of white males. (For reasons that will become clear later, all these conditions are much less frequent among females.) Affected individuals have normal red and blue pigments, but their green-sensitive pigment is weak due to a defect of green color vision pigment in cone cells. In certain ranges, they see fewer hues than normal: Reds are perceived as reddish browns, brighter greens as tans, and olive greens as indistinguishable from browns. Individuals with the more serious condition of *deuteranopia* (*green blindness*)—which occurs with a frequency of 1% in white males—cannot tell reds from greens at all, nor can they distinguish either of these from yellow. Basically, what they see are yellows, blues, blacks, whites, and grays (Collins 1925).

The two green defects are controlled by alleles at the same locus.* A female inherits two sets of X-linked genes, one from her mother and one from her father. Since the green-shifted allele (call it *g*) is recessive to the normal *G* allele, a female must be homozygous *g/g* to express the trait. Heterozygous *G/g* females, with one normal allele and one green-shifted allele, are called **carriers**, and their own color vision is usually normal or near normal. Homozygous *G/G* females have normal color vision.

A male's single X chromosome can only be inherited from his mother. Whatever genes are carried on his X chromosome will be expressed, whether dominant or recessive, because his other sex chromosome is a Y received from his father. Thus, we cannot use the terms *homozygous* or *heterozygous* with respect to X-linked traits in a male. Rather, he is said to be **hemizygous** (Greek *hemi*, "half") for *G* (i.e., phenotypically normal) or hemizygous for *g* (i.e., phenotypically

---

*There are a few truly color blind people who see only in black and white (and very poorly during the day), due to the absence or non-functioning of *all* color-sensitive cells in the retina. But these disorders are extremely rare and are inherited as autosomal recessives.

*The parallel situation for the defectiveness or absence of red visual pigment in cone cells (see Figure 4.10) is controlled by alleles at an adjacent locus on the X chromosome.

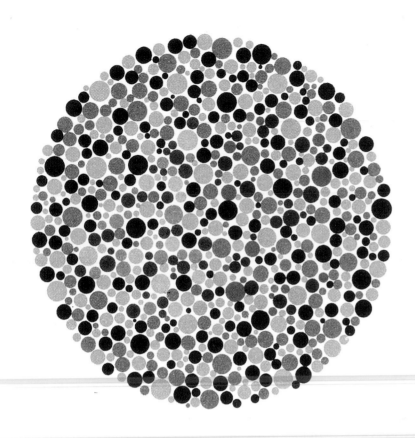

Number(s) Perceived by the Viewer	Phenotype
42	Unaffected
4 (and, less clearly, 2)	Green shift (deuteranomaly)
4 only	Green blindness (deuteranopia)
2 (and, less clearly, 4)	Red shift (protoanomaly)
2 only	Red blindness (protanopia)

Figure 4.10 The numbers the viewer perceives in the pattern of dots is a fairly good diagnostic criterion of the red-green color defects listed in the table. *Note:* This illustration is not intended to be used for the clinical diagnosis of color blindness. Such a diagnosis must be based on the original plates, which appear in Ishihara 1968.

green-shifted). The corresponding genotypic description is $G/(Y)$ or $g/(Y)$.*

Compare the results of reciprocal crosses between normal and green-shifted individuals. First consider the most common case, a homozygous normal mother and a green-shifted father (Figure 4.11A). All the mother's eggs have a normal G gene, so all her sons are normal. The father produces two kinds of sperm; one kind carries the Y chromosome, which is passed on to sons, and the other kind carries the X chromosome with the g gene. Thus, all daughters will be carriers.

*Alternatively, one could use superscripts to the X chromosome as gene notations—i.e., X^G and X^g. We use the less cumbersome notation, G and g.

What can be expected from matings between heterozygous normal mothers and green-shifted fathers? As shown in Figure 4.11B, the eggs are of two types, G and g. The sperm are of two types, g and Y. The progeny will include four possible genotypic combinations, which are also phenotypically distinct: ¼ carrier females, ¼ green-shifted females, ¼ normal males, and ¼ green-shifted males. Among the offspring of either sex, half will be normal and half will be green-shifted.

Finally, consider the reverse case, in which the father is normal but the mother is green-shifted (Figure 4.11C). The father must be $G/(Y)$ and the mother must be g/g. The mother's eggs all carry g; of the father's sperm, half carry G and half carry the Y chromosome. Thus, all sons will be green-shifted and all daughters will be carriers. Note that in this case, the pattern of color vision in the male and female parents is "crisscrossed" in their offspring.

Now you can see why deuteranomaly occurs much more frequently among males than among females. A male needs only one mutant gene, inherited from either a carrier or a green-shifted mother, to express the trait; the genotype of the father, who contributes only a Y chromosome, is irrelevant. A green-shifted female, however, must have not only a carrier or a green-shifted mother, but a green-shifted father as well. The low probability of such parentage leads to about a 20:1 ratio of green-shifted males to green-shifted females in the general population.

Because one mutant gene guarantees the expression of *X-linked recessive* traits in males, whereas two are needed for the expression of *autosomal recessive* traits in either sex, it is much easier to detect the former than the latter in pedigree studies. This is not true for *dominant* mutations, which are expressed in single dose whether X-linked or autosomal. Since many mutations are recessive, however, there used to be a striking difference between the number of genes assigned to the X chromosome and the number known for any one autosome. It is unlikely that more genes reside on the X chromosome than on autosomes of comparable size. Rather, X-linked mutations are easier to detect and verify. But with the advent of molecular genetic techniques, which do not favor X-linked over autosomal DNA, this distinction is disappearing.

(A) *Homozygous* normal mother × green-shifted father : $G/G \times g/(Y)$

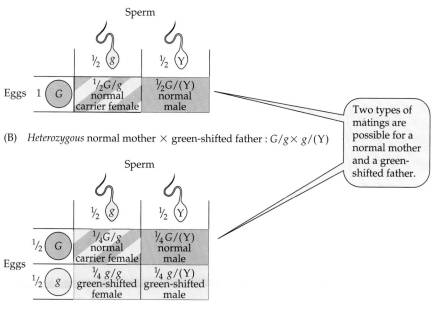

(B) *Heterozygous* normal mother × green-shifted father : $G/g \times g/(Y)$

Two types of matings are possible for a normal mother and a green-shifted father.

(C) *Homozygous* green-shifted mother × normal father : $g/g \times G/(Y)$

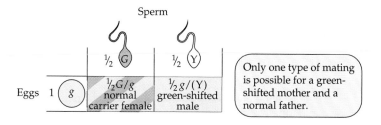

Only one type of mating is possible for a green-shifted mother and a normal father.

Figure 4.11 Result of reciprocal matings between normal and green-shifted individuals.

X-Linked Lethal Alleles and Shifts in Sex Ratio

So far we have discussed alleles that produce normal phenotypes and alleles that may cause conditions detrimental to the organism. There is another class of alleles, called **lethal alleles**, that disrupt processes absolutely essential to life. Like other alleles, lethals can be dominant or recessive and can exist on any chromosome. They can also take effect at various stages of the life cycle, from the earliest embryonic stage through adulthood. Some miscarriages, for example, are due to lethal alleles acting during fetal development.

Suppose that a recessive lethal mutation causing fetal death (call it *l* for lethal) occurs on the X chromosome. Any males that inherit it would die before birth. Females who inherit one lethal-bearing X chromosome are usually unaffected, but as shown in Figure 4.12, half of their sons are not expected to survive. Thus, the sex ratio among their live-born offspring will be 2 females to 1 male, rather than the usual 1 female to 1 male. Of course, to distinguish an abnormal sex ratio from random fluctuations about a normal sex ratio, large numbers of progeny from extended human families must be observed. X-linked lethals are rare in human populations, however, and so do not appreciably affect the overall sex ratio.

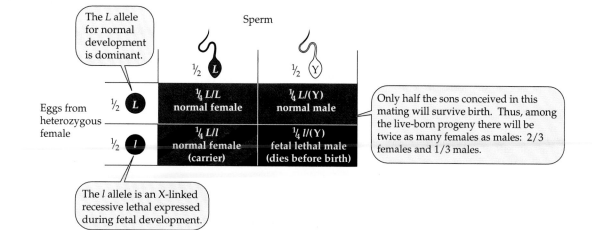

The *L* allele for normal development is dominant.

The *l* allele is an X-linked recessive lethal expressed during fetal development.

Only half the sons conceived in this mating will survive birth. Thus, among the live-born progeny there will be twice as many females as males: 2/3 females and 1/3 males.

Figure 4.12 Proportion of male and female progeny expected from a mating between a normal male and a female heterozygous for an X-linked recessive allele (*l*) that is lethal during fetal development.

Summary

1. Gregor Mendel was a gifted teacher and researcher (and later the abbot) at a monastery noted for its independence and scientific excellence.

2. In modern terminology, Mendel's law of segregation states that the two alleles of a gene separate during the formation of gametes, so each gamete receives one of the two alleles at random.

3. The law of segregation and the randomness of fertilization allow predictions of the genotypic and phenotypic ratios among the offspring of any mating. A convenient way to set up these predictions is to construct a checkerboard whose rows and columns are labeled with the various kinds of eggs and sperm and with gamete fractions representing their relative abundance.

4. The testcross is the most informative mating for determining whether an organism with the dominant phenotype is homozygous or heterozygous.

5. More than two (i.e., multiple) alleles may exist for one gene. The number is determined by the chemical changes (mutations) that alter one or more alleles of the gene. Although one individual has only two alleles for each gene, multiple alleles are the basis of a large number of possible genotypes in the general population.

6. The notion of probability is similar to intuitive ideas of chance or frequency. Probabilities are expressed as numbers between 0 and 1 inclusive.

7. The way probabilities are combined depends on the kind of question and the nature of the events. When the word *or* is used in describing mutually exclusive events, the combined probability is obtained by addition. When the word *and* is used for independent events, the joint probability is obtained by multiplication.

8. Mendel's second law, called the law of independent assortment, deals with the inheritance of two or more pairs of traits. It states that the segregation of one pair of alleles in no way alters the segregation of the other pair of alleles. Outcomes of crosses involving two or more genes can be predicted by the gamete × gamete (checkerboard) method or the gene × gene method.

9. The X chromosome carries many genes, most of which have nothing to do with sexual development. Very few genes are present on the tiny Y chromosome.

10. A father transmits his X chromosome to daughters and his Y chromosome to sons; a mother transmits an X chromosome to both sexes. A son receives a Y chromosome from his father and his single X chromosome from his mother; being hemizygous, he expresses all his X-linked genes. A daughter receives an X chromosome from both parents.

11. X-linked recessive lethal alleles that act during fetal development can alter the sex ratio to favor females.

Key Terms

addition rule	law of segregation
allele	lethal allele
carrier	marker chromosome
codominant	monohybrid cross
dominant	multiple alleles
F_1 hybrid	multiplication rule
gene	mutation
gene × gene method	parental generation (P)
genotype	phenotype
hemizygous	probability
heterozygous	recessive
homozygous	reciprocal mating
law of independent assortment	testcross
	X-linked

Questions

Classic albinism is caused by the lack of an enzyme necessary for the synthesis of melanin pigments. Enzyme production requires the dominant allele, *C*, so *C/C* or *C/c* (which can be abbreviated as *C/–*) represents normal pigmentation, and *c/c* represents albino.

1. What progeny are expected, and in what proportions, from a normally pigmented woman who has an albino husband and an albino father?

2. A normally pigmented woman and an albino man have nine normally pigmented children and one albino child. What is the best guess for the woman's genotype? Is any other genotype possible?

3. A normally pigmented woman and an albino man have ten normally pigmented children and no albinos. What is the best guess for the woman's genotype? Is any other genotype possible?

4. Two parents are heterozygous for albinism. Assume that they will have five children.

   (a) What is the probability that the children will have the following phenotypes in the order stated: firstborn = unaffected, second = albino, third = unaffected, fourth = albino, fifth = unaffected?

   (b) What is the probability that all five will be unaffected?

   (c) What is the probability that *at least* one child will be albino?

   (d) Suppose that the first three children are albino. What is the probability that the fourth child will be albino?

5. Consider a gene with two alleles, *B* and *b*. (a) List all the matings (the parental genotypes) that could produce a heterozygous child. (b) Which mating in your list gives the greatest proportion of heterozygous offspring?

6. Again considering alleles *B* and *b*:

(a) List all the matings whose offspring can be of only one genotype.

(b) List all the matings whose offspring can be of two, and only two, genotypes.

(c) What one mating is not on either of these lists?

7. For some rare autosomal dominant diseases, it is possible that the dominant homozygote has never been seen. One reason is that the fetus with this genotype may die early in development and be spontaneously aborted. Consider a situation in which *Q/Q* dies as an early fetus, *Q/q* lives to reproduce but is affected, and *q/q* lives and is unaffected. What genotypes are possible among the live-born children of the mating *Q/q × Q/q*? Among these live-born children, what is the proportion of each genotype?

8. What kinds of gametes, and in what proportions, can be made by a person who develops from an egg carrying alleles *A* and *B* and a sperm carrying *a* and *b* if the two loci are on different (nonhomologous) chromosomes?

9. List all the kinds of eggs that can be formed by a woman who is heterozygous for the three independently inherited genes for brachydactyly (short fingers and toes), pituitary dwarfism, and sickle-cell disease: *B/b D/d Hb^A/Hb^S*. What is the proportion of each kind of gamete?

10. If the woman in question 9 were heterozygous for a fourth independent gene, *Xg^a/Xg*, how many different kinds of gametes could she make?

11. What phenotypes, and in what proportions, are possible from the following mating involving the growth hormone and *ABO* genes?

$$D/d \ B/O \times D/d \ B/O$$

12. Consider these three independent genes and their alleles:

B = brachydactyly (short fingers); *b* = its normal recessive allele

P = polydactyly (extra fingers); *p* = its normal recessive allele

S = syndactyly (joined fingers); *s* = its normal recessive allele

Consider the mating *B/b P/p S/s × B/b P/p S/s*.

(a) What is the one most likely genotype among the offspring? What is its probability?

(b) What is the one most likely phenotype among the offspring and its probability?

(c) What is the probability of a child with completely normal fingers?

13. A good example of a multiple allelic series is the gene that codes for the β polypeptide chain of the hemoglobin molecule. About 400 alleles of this gene are known; among these, the most common are *Hb^A*, *Hb^S*, *Hb^E*, and *Hb^C*. List methodically all the genotypes possible when considering these four alleles of the hemoglobin β chain gene. (Your work will be easier if you omit the basic gene designation, *Hb*, and just use the superscripts.) How many genotypes are homozygous? How many are heterozygous?

14. Give the genotypes and probabilities for the offspring of the following matings. The basic gene designation, *Hb*, has been omitted.

(a) *A/A × A/S*

(b) *A/C × A/S*

(c) *A/E × C/S*

15. Consider the three-allele, X-linked locus that controls the green-sensitive pigment for color vision. Recall that the normal allele *G* is dominant to both mutant alleles; the deuteranomaly (green-shifted) allele, which we will call *g^1*, is dominant to the deuteranopia (missing pigment) allele, which we will call *g^2*. List the offspring genotypes, phenotypes (including sex and carrier status), and proportions of each, expected from the following matings:

(a) *g^1/g^1* female × *G/(Y)* male

(b) *G/g^2* female × *G/(Y)* male

(c) *G/g^1* female × *g^2/(Y)* male

(d) female heterozygous for deuteranomaly and deuteranopia × an unaffected male

Further Reading

For biographies of Mendel, see Orel (1996) and Iltis (1924). One translation of Mendel's paper appears in Stern and Sherwood (1966), which also reprints an interesting statistical analysis of his experimental results. Internet address http://www.netspace.org/MendelWeb/ also provides Mendel's paper in German or English, as well as commentaries and pictures. Corwin and Jenkins (1976), Srb, Owen, and Edgar (1970), Peters (1959), and L. Levine (1971) reprint research papers important to the early development of general genetics. For comments on Mendel's contribution to human genetics, see Stern (1965). Textbooks by Crow (1983), Strickberger (1985), and Levitan (1988) contain good, short accounts of probability. A short and very readable book by Rosenthal and Phillips (1996) describes the many everyday problems encountered by people who are color-blind. Collins (1925) is a fascinating study of ten color-blind students at Edinburgh University.

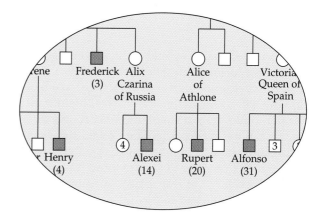

CHAPTER 5

Human Pedigrees

Stormie Jones was born in May 1977, in Cumbry, Texas. The blond, brown-eyed little girl seemed normal except for one thing. Beginning at age 3 months, yellow bumps appeared on her buttocks, knees, elbows, knuckles, and toes. But until she was 6 years old, no doctor recognized these fatty deposits as a sign of life-threatening illness.

Stormie had **familial hypercholesterolemia (FH)**, a defect of fat metabolism characterized by excess cholesterol in the blood, yellow fatty deposits in the skin and tendons, and coronary heart disease. This condition, inherited as an autosomal dominant, is now known to be one of the most common Mendelian disorders in humans. Most affected individuals are heterozygotes. But Stormie, having inherited one defective allele from each of her heterozygous parents, was a rare homozygote for FH. When finally diagnosed, she had a blood cholesterol level about nine times that expected for a normal child, and her arteries were already severely blocked. By the age of 6½ years, she had suffered two heart attacks and undergone two coronary artery bypass operations.

Soon it became clear that only a liver transplant could save Stormie's life,* but her own heart was too weak to withstand such a trauma. So at age 6 years and 9 months, Stormie made medical history as the first person to get a transplanted heart and liver in one operation (Figure 5.1). The procedure was successful, and she was able to return to a fairly normal home and school life. Six years later, however, she needed a second liver transplant. Then in November 1990, at age 13, she suddenly took ill and died from a totally unexpected event: rejection of the heart she had received almost seven years before.

During Stormie's short lifetime, researchers learned a tremendous amount about the physiology and genetics of FH. A few details will be presented later in this chapter, when we focus on the inheritance pattern of FH. But first we need to explain how geneticists collect and organize pedigree data.

Pedigree Construction

One might think that the recent explosion of molecular techniques would make old-fashioned pedigree analy-

---

*Normally, liver cells have the major responsibility for clearing cholesterol from the bloodstream; but being homozygous for FH, Stormie's liver cells were incapable of performing this function.

ses unnecessary. Not so. Even "high-tech" geneticists often need to know the exact interrelationships among the sources of DNA molecules they analyze, while clinicians and genetic counselors need to use all possible genetic tools, both new and old. Thus, they still summarize, in diagrammatic form, information from family studies. These **pedigrees** are drawn up in a fairly (but not completely) standardized way.* Figure 5.2, for example, represents a hypothetical pedigree in which affected persons possess a dominant autosomal gene.

Separate generations occupy separate horizontal lines, with the most ancestral at the top, and are numbered from top to bottom by Roman numerals. The individuals within a given generation are numbered from left to right by Arabic numerals. Note that consecutive individuals (such as II-2 and II-3 in Figure 5.2) may be genetically unrelated. Within each sibship (group of brothers and sisters), the birth order, from oldest to youngest, is arranged from left to right. An arrow points to the **propositus** (or **proband**), the individual through whom the pedigree was discovered (IV-15) and who must be accounted for in statistical analyses. Huge pedigrees are sometimes drawn in circular or spiral form.

Two parents are joined by a horizontal line, from which drops an inverted T, to which their offspring are attached by short vertical lines. Individuals III-4 through III-7, for example, are sibs, the children of II-3 and II-4. Note that III-2 had two mates, III-1 and III-3; thus, IV-1 and IV-2 are half sibs. A single child (such as III-8) is attached by a long vertical line directly to its parents' horizontal "mating line." Parents who are unaffected and unrelated, such as the mate of III-5, may be omitted from the diagram.

For dominant disorders, affected individuals may be shown as "half solid," because they are usually heterozygotes. Alternatively, carriers of autosomal genes, including those showing nonpenetrance, may also be represented by half-solid symbols. (Individual III-13, for example, did not express her dominant allele.) Partial expression of a trait may be indicated by lighter shading. When two or more traits are being studied simultaneously, more complex symbolism must be used.

Consanguineous matings (matings between related individuals) are very important. Individuals IV-8 and IV-9 are related as second cousins, having one set of great-grandparents (I-1 and I-2) in common. The mating line between IV-8 and IV-9 (doubled to show consanguinity) closes a genetic loop that also passes

---

*Some of the currently used symbols (e.g., for twins) may vary a bit among the experts. You may also notice variations in symbolism for some of the pedigrees reproduced from original sources. The legends, however, should make their interpretations clear. Also, not included in this discussion are symbols depicting adoption or the various methods of assisted reproduction that may include egg and sperm donation or surrogate motherhood.

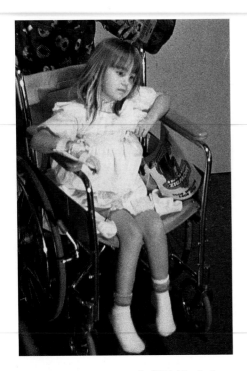

Figure 5.1 Stormie Jones at age 6. (U.P.I./The Bettman Archive.)

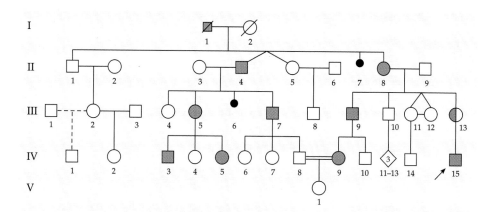

Figure 5.2 Hypothetical pedigree of a rare autosomal dominant trait, illustrating the symbols most commonly used by American geneticists. The gene is not fully penetrant in individual III-13, who therefore does not express the trait. Symbols present in the key are adapted from Kingston (1997).

Unaffected male and female

Affected male and female

Sex unspecified

Two unaffected individuals

Deceased male and female

Marriage of unrelated individuals

Divorce or severed relationship

Illegitimate offspring

No offspring

Consanguineous mating

Heterozygote (autosomal trait)

Heterozygote (X-linked trait)

Multiple traits

Miscarriage (either sex)

Stillbirth

Termination of pregnancy, affected male

Non-identical twins

Identical twins

Twins, unknown type

Propositus

deceased relatives. Recollections may be hazy or exaggerated, especially about medical histories; also, adoptions, abortions, artificial inseminations, and illegitimate births may go unmentioned. (One large study, reported by Ashton in 1980, indicated that 2.3% of tested children were apparently the result of infidelity, concealed adoption, or some other such event. Estimates from some other studies have ranged much higher.) Another problem is the reluctance to discuss family members who are in some way abnormal. Indeed, close relatives may be unaware of such individuals if they died young or were institutionalized. Stillbirths and miscarriages are even less likely to be remembered. Given all these difficulties, looking carefully at all the members of just a few generations is usually better than relying on incomplete or faulty information about generations further back.

Also complicating most pedigree analyses is the variation in phenotypic expression of a trait among individuals (Chapter 10). This variation can be due to differences in mutant alleles at a given locus, differences in background genotypes (including modifying genes), and differences in environmental conditions. The few populations (such as the Finns, the Amish, the Hutterites, and the Mormons) that produce large families, live in well-defined areas, marry almost exclusively among themselves, maintain a uniform lifestyle for generations, and also keep good records provide unique opportunities for genetic study.*

By analyzing all kinds of pedigrees and medical data, geneticists have, over many decades, determined the mode of inheritance for many phenotypic traits (Box 5A). At first they could only separate these phenotypes into two groups, X-linked and autosomal, without knowing which of the 22 autosomes harbored any given trait. But starting in 1968 with the assign-

through ancestral individuals III-7, II-4, II-8, and III-9. The daughter of IV-8 and IV-9, although inbred, is normal with respect to the trait considered in the figure.

Data Collection

The collection of pedigree information is not a simple task. Living family members may be widely scattered, and because few people keep careful written records, it is often difficult to get reliable information about

---

*A recently developed study source (called CEPH, for Centre d'etudes du polymorphisme humain) is the DNA from some large French families, in which many genetic markers have been identified.

BOX 5A   *Inheritance of Some "Regular" Human Traits*

What are the chances that our child will have brown eyes? Red hair? Do you think it might be left-handed like me? Tall like my husband? Will its nose be large or small? Will its skin color be light or dark? Is it likely to inherit my father's baldness or his bushy eyebrows? My mother's big ears? Sometimes the questions come *after* the fact: "My son and his wife both have blue eyes, but their child has brown eyes. How could this happen?"

In studying the inheritance of human traits, geneticists have quite understandably concentrated on diseases and physical abnormalities rather than on the ordinary phenotypic variants that may interest us. Still, it often comes as a surprise that we know so little about the inheritance of hair color, height, and other run-of-the-mill physical characteristics. Furthermore, the data in reports on such traits are not as extensive or reliable as we would like, so they need to be taken with a grain of salt.

Consider, for example, one genetic "fact" that everybody seems to remember from their biology course: Brown eyes are always dominant to blue eyes. Right? Wrong! It turns out that eye color is a complex rather than a simply inherited trait, and—although alleles for darker colors tend to be dominant over alleles for lighter colors—several loci are probably involved, and it is possible for two blue-eyed parents to produce a brown-eyed child. Indeed, the author of the "bible" of human genetics (McKusick 1997) writes, "My monozygotic twin brother and I, brown-eyed, had blue-eyed parents and blue-eyed sibs."

Also probably not inherited as single-gene traits are hair color, skin color, height, nose shape, earlobe type, tongue curling, handedness, hand-clasping pattern, and production of

Name of trait	Description
Achoo syndrome	Sneezing in response to bright light
Amylase enzyme variants	Variations in the breakdown of starch and glycogen
Camptodactyly	Palmward bending of a finger, usually fifth
Cleft chin	Chin dimple
Clinodactyly	Shortened and radially curved fifth finger
Distichiasis	Two rows of eyelashes
Ear malformation	Tightly curled, cup-shaped external ear
Ear pits	Pit in front of ear opening
Earwax, wet type	Sticky (as opposed to dry) earwax
Epicanthus	Vertical fold of upper eyelid skin covering inner corner of eye, as present in Asians
Extra nipples	Extra nipples/breasts in males or females
Hairy palms/soles	Hairy areas near wrist and on arch of foot
Male pattern baldness	Gradual loss of hair from top of scalp
"Michelin Tire Baby" syndrome	Multiple deep skin creases on limbs
Midphalangeal hair	Hair on top of middle segment of fingers
Nevus flammeus, nape of neck	Port wine stains on nape of neck
Odd-shaped teeth	Peg-shaped teeth
Piebald trait	White forelock, and lack of pigment in midline extending from face to abdomen
Prognathism	Hapsburg jaw; prominent lower jaw, often with thickened lower lip and flat cheekbones
PTC tasting	Ability to taste phenylthiocarbamide
Trembling chin	Trembling triggered by anxiety or upset
Tune deafness	Inability to distinguish different musical notes
Uncombable hair	Longitudinal groove present in each hair

excess intestinal gas. Another common misconception about simply inherited traits is that the dominance or recessiveness of a trait determines its frequency in a population. This is not true. As will be explained in Chapter 12, dominant traits can be rare and recessive traits can be common.

The table above lists some normal or benign morphological and behavioral traits from McKusick's 1997 listing of autosomal dominant genes. For these confirmed traits, the evidence for autosomal dominant inheritance is strong and the phenotype is known to be determined by a distinct locus.

---

ment of the Duffy blood group trait to chromosome 1, new cytogenetic and molecular techniques have allowed geneticists to map many genes to their exact locations. McKusick's on-line catalog of simply inherited human phenotypes (OMIM 1997) includes about 9,200 phenotypic entries, of which over 6,100 are established genes. Of the latter, 94% are autosomal, 5% are X-linked, 0.4% are Y-linked, and 0.6% are mitochondrial loci. Also, about 72% of the established autosomal loci have been assigned to specific chromosomes, and

most of the X-linked loci have been mapped. Nearly all single-gene disorders are rare, but together they account for over 5% of all pediatric hospital admissions and have a frequency of about 1% in the general population.

Gene Function

For most of the traits listed in McKusick's catalog, the gene products are unknown. Yet from what is known, an interesting rule of thumb has emerged: *Almost all disorders caused by enzyme deficiencies are recessive,* whereas those *disorders caused by defects in nonenzymatic (i.e., structural) proteins are often inherited as dominants.* The reason given for this distinction is as follows: Although heterozygotes make only half as much of the normal gene product as homozygous unaffected individuals do, this is usually still enough to get the job done in the case of enzyme reactions, where the catalyst is needed in only small amounts. But in the case of structural proteins, half of the gene product may not be enough to do the job—as is the case, for example, with LDL receptors in people who are heterozygotes for familial hypercholesterolemia. Or a mutant structural protein (of collagen, for example) may interfere with the functioning of the normal protein.

To a surprising extent, gene structure and function are *conserved* in closely related and even between distantly related species.* Thus, researchers can often make interspecies and interclass comparisons of phenotypes, genotypes, and genes. What they find are astonishing similarities between humans, mice, flies, worms, yeast, and bacteria. For inherited human disorders in particular, animal models (nonhuman) are both instructive and practical, allowing many kinds of crosses and experiments that—for ethical and legal reasons—could never be undertaken on people.

And now that traditional pedigree studies can be combined with powerful new techniques for analyzing differences in the structure of DNA itself, a new brand of genetics is emerging. From Mendel's time until the mid-1970s, allelic differences could be detected and studied only if they gave rise to phenotypic variation. But with the ability to isolate, sequence, and even insert specific mutations into small segments of DNA (Chapter 8) comes the possibility of studying and using genes whose allelic variants may show no obvious phenotypic differences whatsoever or whose phenotypes and gene products are unknown.

We begin, however, with the standard approach. Summarized in this chapter are the three types of single-gene inheritance most often found by traditional pedigree studies: autosomal dominant, autosomal recessive, and X-linked recessive inheritance. We will also look briefly at X-linked dominant, Y-linked, and mitochondrial inheritance. (The kinds of traits determined by the *interaction* of numerous genes will be discussed in later chapters.)

Some Warnings

In trying to determine how a given disorder is inherited, genetic sleuths must keep several things in mind.

Although a single pedigree may be used to rule out certain possibilities, it may not be enough to prove a specific genetic hypothesis. For example, some pedigrees for X-linked recessive inheritance would also fit an autosomal gene acting as a dominant in males and a recessive in females. (The gene for male pattern baldness, discussed in Chapter 10, may behave this way.) Thus, it may be necessary to pool together many studies and analyze the data by more sophisticated mathematical techniques.

On the other hand, because mutations at different loci may occasionally give rise to very similar phenotypes, some inherited disorders do have multiple modes of transmission. McKusick's 1997 catalog includes several dozen phenotypes with more than one pattern of inheritance (Table 5.1). Some of these phenotypes can be inherited in three different ways: as an autosomal dominant, an autosomal recessive, or an X-linked trait! In such cases, the initial diagnosis of a given disease does not always predict its gene location or its mode of inheritance, and a very careful clinical and genealogical follow-up is needed to determine which form it really is and to provide accurate genetic information to the family.

For these reasons, geneticists who are armchair physicians or physicians who are armchair geneticists should be very cautious about giving advice to families with suspected genetic conditions. It is better to refer such patients to specialists who are trained in both medicine and genetics. These genetic counselors are more likely to be familiar with new developments in diagnosis and treatment and with the various problems—psychosocial as well as medical and genetic—encountered by such families.

Autosomal Dominant Inheritance

Stormie Jones died because her body's cells lacked the ability to remove cholesterol from her bloodstream. Instead, much of this fatty substance remained trapped there, while some got deposited as lumps in her tendons and under her skin (Figure 5.3) and as plaques that clogged her arteries and caused her heart attacks. Before discussing the exact cellular defect involved, however, we should describe the phenotypes and present a pedigree of familial hypercholesterolemia.

Unaffected newborn babies have serum cholesterol levels of about 30 milligrams per deciliter (mg/dl). In unaffected adults raised on a low-fat diet, serum cholesterol levels are 50–80 mg/dl. But in adults raised on high-fat diets in Western societies, "ideal" serum cholesterol levels are 130–190 mg/dl.

FH heterozygosity is relatively common, with a worldwide frequency of 1 in 500, accounting for 5% of all heart attacks in patients under the age of 60.

---

*One of the most unusual examples is the gene for *hemoglobin*, the oxygen-carrying molecule of animal blood cells. A very similar gene is found in some plants, prompting the suggestion that one might be able to get blood from a turnip after all.

TABLE 5.1	Some human disorders associated with two or more gene loci and multiple modes of inheritance		
Name and inheritance modes[a]	**Brief description**	**Chromosome loci**[b]
All Three Modes: AD, AR, and X		
Cataracts	Clouding of the eye lens or of its capsule	1, 2, 16, X
Charcot-Marie-Tooth disease	Progressive weakness and atrophy of leg and arm muscles	1, 17, X
Cleft lip/palate[c]	Opening in midline of upper lip and/or palate	1, 6, X
Craniosynostoses	Premature fusion of skull bones	7, X
Ehlers-Danlos syndrome	Loose joints; stretchy and fragile skin; skin scarring	2, 7, 17, X
Hypoparathyroidism	Low or no parathyroid hormone, leading to low calcium and to muscle spasms	11, X
Two Modes		
Diabetes insipidus (AD, X)	Excess thirst and urination due to disorder of a pituitary hormone	20, X
Diabetes mellitus (AD, AR)	Faulty sugar metabolism due to lack of pancreatic insulin	6, 11, 19
Epidermolysis bullosa (AD, AR)	Occurrence of large blisters, spontaneously or with mild trauma	8, 11
Glycogen storage disease (AR, X)	Accumulation of glycogen in cells due to errors in metabolism	1, X
Hypothyroidism (AD, AR)	Deficient thyroid hormone, causing cretinism or mental and physical slowing	1, 8, 22
Osteogenesis imperfecta (AD, AR)	Brittle, easily broken bones	7, 17
Severe combined immuno-deficiency disease (SCID) (AR, X)	Impaired or absent immune response	20, X
Spherocytosis (AD, AR)	Abnormal spheroid red blood cells, causing chronic anemia	1, 8, 14
Von Willebrand disease (AD, X)	Deficiency of clotting factor VIII, causing gastrointestinal, urinary, and uterine bleeding	12, X
One Mode, Two or More Loci		
Adrenal hyperplasia (AR)	Ambiguous sex organs due to steroid hormone imbalances	1, 6, 8, 10
Elliptocytosis (AD)	Elliptical red blood cells; mild anemia due to red blood cell destruction	1, 14
Hyperphenylalaninemia (AR)	Excess phenylalanine in urine; other PKU and PKU-like symptoms, some mild	4, 12
Porphyria (AD)	Abnormal porphyrin metabolism, affecting skin, liver, and brain	1, 9, 11, 14
Tuberous sclerosis (AD)	Epilepsy; mental retardation; skin tumors and depigmented spots; bone cysts	9, 11
Ventricular hypertrophy, hereditary (AD)	Overgrowth of tissue between heart ventricles, obstructing blood flow	1, 14, 15
X-linked mental retardation (X)	Several phenotypically distinct types (all with X-linked loci)	X

[a]AD = autosomal dominant, AR = autosomal recessive, and X = X-linked.

[b]Relatively few human gene loci have been mapped to their exact locations. Thus, for any given disorder, the chromosomal sites listed here may not be exhaustive.

[c]Cleft lip and/or palate is a complex trait, often not inherited in a simple, predictable fashion; also, over 200 syndromes include cleft lip and/or palate as a feature. Thus, extreme care must be taken in diagnosis and in counseling families about the chances of recurrence.

Heterozygotes are born with cholesterol levels at least twice the normal adult levels. Heart attacks are 25 times more frequent than in unaffected relatives and begin to occur at about age 35; 50% of males and 15% of females die by age 60. Surprisingly, FH patients usually have a slim body build and normal blood

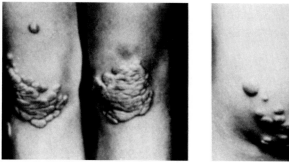

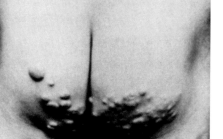

Figure 5.3 Lumpy deposits of cholesterol found in people (especially homozygotes) with hypercholesterolemia. (From Goldstein, Hobbs, and Brown 1995.)

also the raw material for bile acids and steroid hormones. But here's the rub:

> The very property that makes it useful in cell membranes, namely its absolute insolubility in water, also makes it lethal. For when cholesterol accumulates in the wrong place, for example within the wall of an artery, … its presence eventually leads to the development of an atherosclerotic plaque. … If cholesterol is to be transported safely in blood, its concentration must be kept low. (Brown and Goldstein 1986)

Our bodies get their cholesterol in two ways: from what we eat and from what our cells (mostly liver cells) make from scratch. These two pathways are interconnected by means that were discovered by Michael S. Brown and Joseph L. Goldstein of the University of Texas Health Science Center at Dallas—research that won them a Nobel prize in 1985.

Within our intestines, fats are broken down into cholesterol and fatty substances called triglycerides, which then move into the bloodstream and travel throughout the body. But being insoluble in water, they cannot be transported "as is"; instead, they are packaged into *lipoprotein particles*. The major carrier of cholesterol is the low-density lipoprotein (LDL) fraction. Each LDL particle has a cholesterol core protected by an outer coat and topped by a special protein molecule.

How is LDL-packaged cholesterol removed from the bloodstream? Brown and Goldstein (1984) found that cell surfaces are studded with binding sites for the "topping" protein. These **LDL receptors** are made of protein with some sugar chains attached, and their numbers increase or decrease according to a cell's need for cholesterol. After latching onto the LDL receptors, LDL particles get pulled into the cytoplasm and processed in various ways (Figure 5.5). But these regulatory mechanisms cannot control cholesterol levels *outside* cells when large amounts of fats and cholesterol are present in the bloodstream. Blocked from entering the liver, the excess cholesterol remains trapped in the bloodstream and ultimately gets deposited inside artery walls, causing atherosclerotic plaque formation and heart disease.

In genetically normal individuals with low-fat diets, the LDL pathway works very effectively. But the high-fat diets common in Western societies can swamp the LDL transport and metabolic pathways, leaving dangerously high levels of cholesterol circulating in the bloodstream.

In the 1980s, researchers purified the LDL receptor molecule and determined the sequence of its 839

pressure. Rare *FH homozygotes* such as Stormie occur only about once in a million births. They are born with serum cholesterol values about four to six times the normal levels; this excess is limited to the **low-density lipoprotein (LDL)** fraction of cholesterol (discussed shortly). Heart attacks are common from age 5 on and often cause death before age 20. Thus, homozygotes rarely reproduce.

Pedigree Analysis: Familial Hypercholesterolemia

The pedigree in Figure 5.4 shows all the characteristics of **autosomal dominant inheritance** when a gene is fully, or almost fully, penetrant—that is, expressed phenotypically wherever it occurs (Chapter 10). Note the following characteristics:

1. *All affected individuals have at least one affected parent.* Also, affected individuals are usually found in every generation of a sizable pedigree.

2. *Both males and females inherit and transmit the trait.* Male-to-male transmission rules out X linkage.

3. *Males and females are affected in about equal numbers,* which makes X linkage less likely.

4. *Matings of heterozygous affected with unaffected individuals produce about ½ affected and ½ unaffected offspring.*

5. *The trait does not appear in the descendants of two unaffected parents* unless the dominant allele is reintroduced by matings with unrelated affected individuals.

6. *The trait may be more severe or extreme in homozygotes than in heterozygotes, or it may even be lethal.* In such cases, the dominant allele is not completely dominant in heterozygotes.

Metabolic and Genetic Control of Cholesterol

Despite its bad press, cholesterol is needed in our bodies. It is a major component of cell membranes and

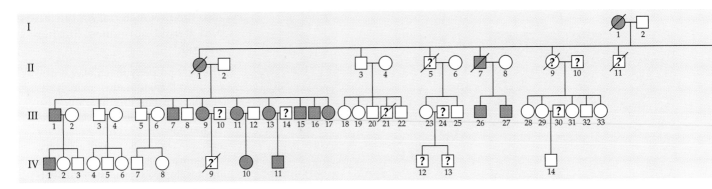

Figure 5.4 Pedigree of familial hypercholesterolemia, an autosomal dominant trait, in a large Aleutian kindred. According to the researchers, all affected persons are heterozygotes. (Redrawn from Schrott et al. 1972.)

amino acids. They also isolated the LDL receptor gene, a medium-sized locus residing on the short arm of chromosome 19. Worldwide, over 300 mutations of this gene have been found. Some produce no LDL receptors at all, while others produce LDL receptors that are nonfunctional for one reason or another. These different mutations are not randomly scattered throughout the peoples of the world; instead, particular de-

fects tend to be found in high frequency among particular populations.

In certain self-contained populations—especially Lebanese, South African Afrikaners, and French Canadians—FH is relatively frequent. Among South African Afrikaners, for example, FH heterozygotes are 1 in 100 and homozygotes are 1 in 30,000, and two particular mutations account for nearly all affected individuals. This situation is most likely due to the **founder effect**—the rapid multiplication in a small population of just one or two mutant *FH* alleles (Chapter 10). Similarly, there is good evidence for a founder effect among the ancestors of present-day

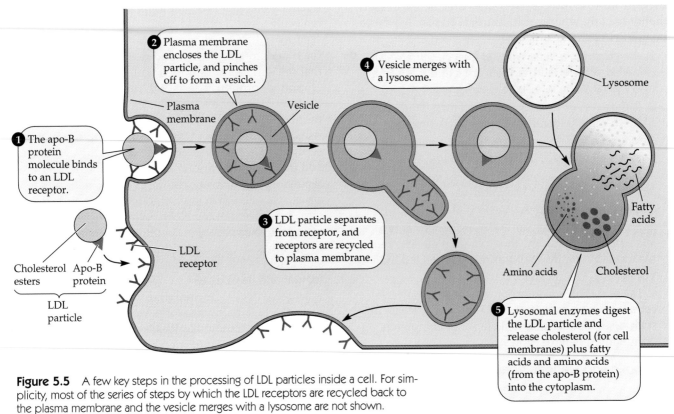

Figure 5.5 A few key steps in the processing of LDL particles inside a cell. For simplicity, most of the series of steps by which the LDL receptors are recycled back to the plasma membrane and the vesicle merges with a lysosome are not shown. (Adapted from Lodish et al. 1995.)

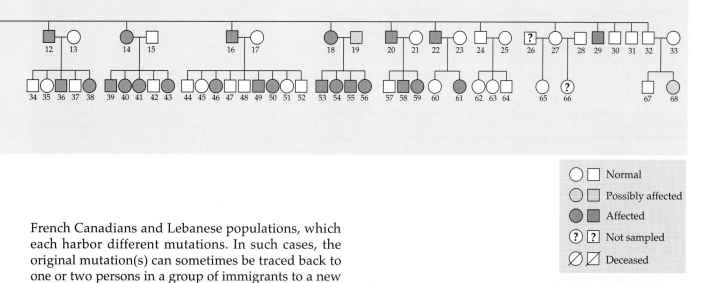

French Canadians and Lebanese populations, which each harbor different mutations. In such cases, the original mutation(s) can sometimes be traced back to one or two persons in a group of immigrants to a new territory.

From detailed molecular studies, it has also become clear that *many affected individuals, previously thought to be homozygotes, are actually heterozygous for two different LDL receptor mutant alleles*. This finding is true not only of FH, but of nearly every important genetic disorder that has been carefully analyzed in recent years. It helps to explain why the phenotypes of these disorders can be quite variable.

Appropriate drugs and low-fat diets can reduce cholesterol levels in both FH heterozygotes and genetically unaffected individuals. With proper treatment,

FH heterozygotes should be able to lead reasonably normal lives, even though their one normal allele of the LDL receptor gene produces only about half as many LDL receptors as do unaffected individuals. But FH homozygotes, having no normal alleles, cannot make receptors under *any* conditions. Thus, their only treatment possibilities are repeated plasma exchanges and/or surgery, including liver-heart transplants from organ donors with normally functioning LDL receptor genes.

TABLE 5.2 Some human disorders inherited as autosomal dominants and their chromosomal locations		
Name	**Brief description**	**Site**
Angioneurotic edema, hereditary	Allergy-caused itching and swelling, often involving the airways	11p
Dentinogenesis imperfecta	Teeth abnormally shaped, blue-gray or brown, and opalescent	4q
Hyperthermia of anesthesia	Abnormal/lethal response to anesthesia; high fever, muscular rigidity	19q
Li-Fraumeni syndrome	Cancers of the bladder, blood, bone, brain, breast, colon, liver, lung, and ovary	17p
Mandibulofacial dysostosis (Treacher Collins syndrome)	Abnormal jaw/facial bones, ears; down-slanting eyes, other defects	5q
Marfan syndrome	Tall, loose-jointed; long, thin limbs, hands, feet; eye and heart/artery defects	15q
Myotonic dystrophy	Difficulty in relaxing contracted muscles; muscle wasting, cataracts, defects in gonads, heart; mental retardation	19q
Nail-patella syndrome	Poorly developed or absent kneecaps and nails; defects of joints, kidneys, eyes	9q
Neurofibromatosis	Defects in nerves, muscles, bones, skin, and eyes; pigmented spots and soft tumors all over body	17q
Polyposis, adenomatous intestinal	Precancerous growths in lining of large intestine (usually colon)	5q
Retinoblastoma	Tumor(s) of retina of the eye(s), with predisposition to bone cancer	13q
von Hippel-Lindau syndrome	Blood-vessel-forming tumors in the retina and cerebellum	3p
Waardenburg syndrome	Wide-set inner corners of eyes, wide nose, white forelock, hearing loss	2q

BOX 5B   *Early Death of an Olympic Champion*

Ekaterina Gordeeva and Sergei Grinkov were young Russian figure skaters who dazzled the world with the strength and elegance of their pairs routines. They had won two Olympic gold medals and were expected to continue their brilliant careers for many years. But in November 1995, during a practice session, 28-year-old Grinkov suddenly collapsed and died of a heart attack (Gordeeva 1996). Although he was a nonsmoker, slim, and physically fit, and although there had been no warning signs, an autopsy showed that the arteries in his heart were very badly clogged. Also, his father had died of a sudden heart attack at age 52.

Analysis of a blood sample revealed that he carried a mutant gene (presumably inherited from his father) that is associated with heart attacks in people under age 60. The mutation is a single-base change, from a T (thymine) to a C (cytosine) at one spot in this gene, which causes the substitution of the amino acid proline for leucine at one spot in the resultant protein. The gene codes for a fibrinogen receptor on the surface of tiny blood cells called platelets, whose function is to stop the bleeding after any damage to blood vessels.

Clot formation (blood coagulation) is a complicated series of steps involving platelets plus many enzymes and a dozen coagulation factors. One of the latter is a soluble protein called *fibrinogen*, which gets converted to insoluble threads called fibrin. Near the end of the clotting process, 50,000 fibrinogen receptors on the surface of each activated platelet grab onto these and other adhesive molecules that act as a kind of molecular glue. Together they form a meshwork of threads in which platelets and other blood cells are trapped (Becker 1996). The resultant blood clot will plug up the injury so that healing can begin.

Without the benefit of blood clotting, we could bleed to death from even the smallest cuts. But there is a downside to this process too: Clots that form under the wrong circumstances in the wrong places can actually clog up our blood vessels and cut off the blood supply to vital organs, including the heart. The result can be heart attacks and death. Unfortunately, the mutant form of Grinkov's platelet fibrinogen receptor gene, the *PL(A2)* allele possessed by about 2% of the U.S. population, can play a role in the formation of "bad clots." It somehow binds fibrinogen more avidly than does the receptor encoded by the *PL(A1)* allele and makes an abnormally large clot.

Why do bad clots form? The inside surfaces of arteries in people with *atherosclerosis* are coated with fatty substances (cholesterol and triglycerides) that form sticky plaques. The space available for blood flow is decreased, blood pressure is increased, and the cells lining these blood vessels become damaged. If a plaque breaks loose (say, in a vessel that supplies the heart itself), a tiny hole may form in that artery wall and set off the blood-clotting mechanism. Then the clot itself might further block the blood flow in that region and trigger a heart attack.

Scientists expect that new insights into the *PL(A2)* mutation will allow them to develop drugs to neutralize the potentially lethal action of the abnormal platelet receptor it encodes.

Some additional examples of human disorders inherited as autosomal dominants are listed in Table 5.2. Two other autosomal dominant traits with tragic consequences are described in Boxes 5B and 5C. Box 5B deals with another receptor disorder that causes heart attacks. Box 5C describes Huntington disease, in which (for heterozygotes) the lethal effects usually occur later in life.

Autosomal Recessive Inheritance

Baby Alexandra—the first girl born into the Deford family since 1904—arrived in October 1971 in Redding, Connecticut. Although declared healthy at birth, she was from the very beginning "sick all the time":

> She ate voraciously, but food went right through her and she couldn't put on any weight. Besides, she always seemed to have colds and ear infections. The pediatrician finally said it was "failure to thrive." (Deford 1986)

Every day Alex got weaker. At age 4 months, scrawny and extremely ill with pneumonia, she was given a "sweat test" to measure the amount of salt in her sweat. The results showed that she had cystic fibrosis (CF)—such a serious case, in fact, that she could die at any time. Her parents, themselves healthy and having no known relatives with CF, were shocked and devastated by the news.

But Alex survived, becoming a bright, lively, and extraordinarily plucky child who charmed and amazed all those who knew her (Figure 5.6). She attended nursery school and then elementary school, for a few years managing to live a fairly normal life except for occasional hospital stays. But she also had to endure a daily routine of extensive treatments: taking enzyme preparations to help with digestion and antibiotics to prevent infections; inhaling decongestants to loosen the thick mucus that clogged her lungs; and having her chest pounded and pinched to force the mucus out. Nowadays, CF patients may also inhale an enzyme called DNase, which digests the long fibrils of DNA spilled out from dead infection-fighting cells. These and other components of pus clog up and greatly increase the viscosity of CF mucus. Ibuprofen and other anti-inflammatory drugs are also being tested.

BOX 5C *Huntington Disease*

Huntington disease (HD) is an autosomal dominant neurological disorder that killed folksinger Woody (Woodrow Wilson) Guthrie (1912–1967). His mother Nora, an intense and very musical person, was affected, and two of his eight children have also died from the disease.

Among the first to clearly describe this tragic and incurable condition was Johan Christian Lund, a physician in rural Norway. His 1860 report (translated by Ørbeck 1959) includes four-generation pedigrees of two families (both males and females affected) in which "chorea St. Vitus," his name for the condition, recurred as a hereditary disease:

"It is commonly known as the "twitches." … It usually occurs between the ages of 50 and 60, generally starting with less obvious symptoms, which at times only progress slowly … but more often after a few years they increase to a considerable degree, so that any form of work becomes impossible and even eating becomes difficult and circuitous. The entire body, though chiefly the head, arms, and trunk, is in constant jerking and flinging motion. … A couple of the severely affected patients have become demented."

Lund's obscure report was rediscovered in 1914, but attracted very little attention. Indeed, a few American physicians had written even earlier descriptions that were likewise ignored.

What did attract attention was a report (still considered a classic) by George Huntington, an American physician whose father and grandfather were also family doctors on Long Island, New York. His first encounter with this disorder—and the beginning of his medical education—took place when he was a young boy accompanying his father on professional rounds. In 1872, drawing on the 78 combined years of the three physicians' experiences in that area, the 22-year-old Dr. Huntington described a

"hereditary chorea … confined to certain and fortunately a *few* families transmitted to them, an heirloom from generations away back in the dim past. … It is attended generally by all the symptoms of common chorea, only in an aggravated degree, hardly ever manifesting itself until *adult* or *middle* life, and then coming on gradually but surely, increasing by degrees, and often occupying years in its development, until the hapless sufferer is but a quivering wreck of his former self. …

Of its hereditary nature. When either or both the parents have shown manifestations of the disease,. . . one or more of the offspring almost invariably suffer from the disease, if they live to adult age. But if by any chance these children go through life *without* it, the thread is broken and the grandchildren and great-grandchildren of the original shakers may rest assured that they are free from the disease. … Unstable and whimsical as the disease may be in *other* respects, in *this* it is firm, it never skips a generation to again manifest itself in another; once having yielded its claims, it never regains them." (Huntington 1872)

Note that this description of autosomal dominant inheritance in humans followed the publication of Mendel's laws by just seven years! Although the Huntington gene has now been isolated and cloned (Wexler 1995), the function of its protein product remains unclear. However, the progressive degeneration of certain groups of brain cells is well documented. Age of onset is highly variable, the average being about 38 years. There is no effective treatment, and death usually comes 10 to 25 years after the onset of symptoms. The frequency of

HD is roughly 1 in 20,000 in most Western countries, but is much lower among Asians and blacks.

The cloning of the *HD* gene makes it possible to determine with virtual certainty whether at-risk relatives possess the *HD* allele (Morell 1993; Pollen 1993; Revkin 1993). The decision about whether to take this test is an agonizing one, however. The possible relief of knowing that one is free of the disease must be balanced against the grief and loss of hope that might overwhelm those who learn that they have the mutant allele. Indeed, genetic counselors worry about the potential for suicides in the latter group. For a variety of reasons, most members of HD families have thus far chosen not to take the test.

This 1943 photo shows Woody Guthrie at age 31, about a decade before he began to show the symptoms of Huntington disease. (Courtesy of Woody Guthrie Publications, Inc.) For more information about Woody Guthrie and his family, see Klein 1980.

Cystic fibrosis, Alex's parents learned, is the most common lethal autosomal recessive disorder among Caucasians and a leading cause of childhood death. Frequency estimates vary considerably among different white populations; but overall, about 1 in 2,500 newborns are affected. About 1 in 20 to 25 whites are heterozygous for this gene, but most carriers never know they harbor a lethal allele unless they produce

an affected child. CF is rare among black Africans, Asians, and their American descendants.

The disease is variable, but its most common and serious form has three major manifestations: serious respiratory and digestive problems and extremely salty sweat. (An old German folk saying states that the baby who is kissed at birth and tastes salty will have a short life.) These and other effects are caused by the defec-

Figure 5.6   Alex Deford. (Courtesy of Frank Deford.)

tive transport of chloride, sodium, and water in epithelial cells lining the ducts of glands that release their products into tubules (rather than into blood or lymph vessels). As a result, thick, sticky secretions clog up all these tubules, causing irreversible damage and malfunctioning of the lungs, pancreas, intestines, liver, sweat glands, and (especially in males) the reproductive organs. In 85% of CF patients (called *pancreatic-insufficient*), the pancreatic enzymes fail to reach the intestines; thus, digestion and absorption are incomplete and intestinal complications are common. Growth may be retarded in children with CF. And given the lung problems, they all have a greatly increased (and often fatal) susceptibility to certain bacterial infections, especially pneumonia.

In 1950, few patients lived past infancy; but now the median age of survival is about 30 years for current CF patients and perhaps about 40 years for CF newborns (Harris and Super 1995). Many CF adults marry, and most lead active, productive lives. But most CF males are sterile because of damaged ducts (the vas deferens), and fertility is greatly reduced in females. Despite improvements in diagnosis and treatment over the last few decades (Ramsey 1996), however, many CF patients die within a year or two of birth. There is still no cure for this chronic, relentless disease, and the suffering extends to the entire family.

Alex was lucky to have a strong, loving family. But despite their best efforts, at age 5 her health began to deteriorate. She coughed and gagged, struggling to breathe—a result of severe mucus blockage. Her appetite waned and she lost weight. She tired easily, growing too weak to play at recess time. Her fingertips became pale and clubbed. She spit up blood. Her

heart weakened. She had liver problems, a regular high fever, and arthritis. Bouts with pneumonia grew more serious, and her hospital stays increased in length and frequency.

By age 8, Alex was in constant pain; her breathing was labored, and she often needed oxygen. A lung collapsed three times, requiring the insertion of tubes through her chest wall. Alex knew she was dying and talked freely about it. Because nothing more could be done for her, she returned home from the hospital just before Christmas; a few weeks later, she died in her parents' arms.

Since Alex's death in 1979, scientists have unraveled many mysteries of cystic fibrosis: determining the location of the CF gene on the long arm of chromosome 7; isolating it and identifying many of its mutations; analyzing phenotypic variants; and deducing the CF protein's structure and studying its functions in the cell membrane (Welsh and Smith 1995). These and other findings have greatly improved the accuracy of prenatal diagnosis and made possible, for the first time, carrier detection in families with cystic fibrosis and, to a certain extent, in the general population.

One major puzzle is how such a lethal gene can be so common, especially when so few affected individuals reproduce. Perhaps the mutation rate of this gene is unusually high. Or perhaps the CF mutations persist because of some small beneficial effect in unaffected carriers—such as an increase in fertility or a resistance to asthma, typhoid fever, or diarrhea caused by bacterial toxins (Glausiusz 1995). Another possibility, in some cases, is the founder effect (Angier 1994).

Pedigree Analysis: Cystic Fibrosis

The pedigree shown in Figure 5.7 is taken from a genealogical survey of a small area in northwest Brittany where cystic fibrosis is common (1 in 377 live births) and the population is quite isolated and stable. It exhibits some, but not all, of the features of **autosomal recessive inheritance:**

1. *Most affected individuals have two unaffected parents, so that the trait tends to show up erratically and unpredictably.* In fact, such a gene is often passed on invisibly for many generations, through unaffected heterozygotes. Only when two carriers produce an affected child is the condition manifested.

2. *The trait is expressed in both sexes and transmitted by either sex to both male and female offspring in roughly equal numbers.*

3. *With rare traits, the unaffected but heterozygous parents of an affected individual may be related to each other.* When this occurs, the recessive allele that each parent carries has usually originated in an ancestor that they have in common (Chapter 12).

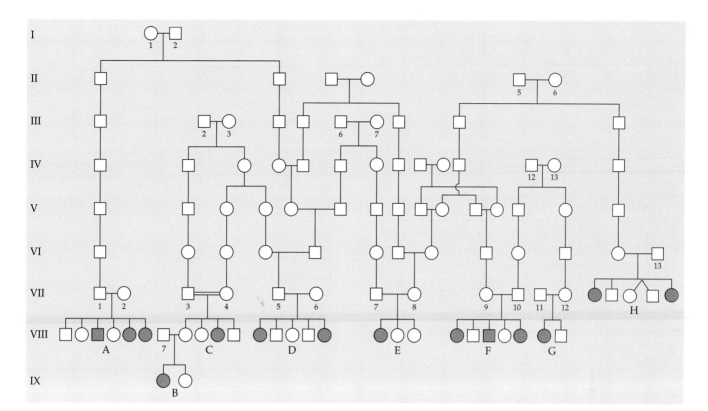

Figure 5.7 Autosomal recessive inheritance of cystic fibrosis in eight related families (A–H) from the department of Finistere in northwestern Brittany. By interviewing family members and examining birth, death, and marriage registers, geneticists were able to trace these families back six to eight generations, within which there are at least five ancestral couples: I-1 and I-2, II-5 and II-6, III-2 and III-3, III-6 and III-7, and IV-12 and IV-13. Note that there is only one known consanguineous marriage, in generation VII. One feature of an autosomal recessive trait is that it can remain "hidden" for generations. On the other hand, the apparent absence of individuals affected with CF in earlier generations may be related to the following factors: (1) CF was not widely recognized as a disease until about the 1930s. (2) Before antibiotics became available in the 1940s, most affected individuals died during infancy. (3) In this pedigree, the absence of childless individuals in generations I through VI suggests that no "singles" of any age were recorded. (4) Most or all (possibly) affected individuals in generations I through VI probably died young; but if any survived to adulthood, the males would be sterile and the females would have reduced fertility. (Redrawn from Bois et al. 1978.)

4. *Matings between unaffected heterozygotes yield about ¾ unaffected and ¼ affected offspring.*

5. *Matings between an affected and an unaffected parent will usually yield all unaffected offspring;* only if the unaffected parent is heterozygous can affected children be produced.

6. *When both parents are affected, all their children are affected.*

Cystic Fibrosis and Genetic Testing

Attempts to map the CF gene involved years of effort and numerous detours down blind alleys. But in 1989, a research team led by Lap-Chee Tsui at the Hospital for Sick Children in Toronto finally found this large gene near the middle of the long arm of chromosome 7.

In 1990, researchers isolated the CF protein, **cystic fibrosis transmembrane regulator (CFTR)**, which con-

tains 1,480 amino acids. Its structure is similar to that of several proteins known to transport small molecules across cell membranes. How it regulates the passage of chloride through epithelial cells is not completely understood. The normal CFTR molecule actually has two separate functions: to provide a chloride channel (a pore in the plasma membrane through which chloride ions pass) and to act as a regulator, opening and closing both chloride and sodium channels—i.e., to be both a gate and a gatekeeper.

About 600 *CFTR* mutations are known in combinations that give rise to tremendous variation in disease states (Estivill 1996), ranging from extremely severe (classic CF) to mild to almost normal. These errors of CFTR protein production fall into five general classes: (I) no protein produced at al; (II) improper processing of the protein, so that it never reaches the cell membrane; (III) abnormal regulation of the CFTR chloride channel; (IV) defective conduction of the chloride current; and (V) reduced synthesis of normal CFTR. Class I and II errors are the most severe, and class V errors are the least severe.

The most common mutation, a class II type, is found on 65–70% of all *CFTR* chromosomes worldwide. It is a deletion of three bases that results in the loss of the amino acid phenylalanine (abbreviated F) at position 508 in the CFTR protein. Thus, this mutation is called *ΔF508*. The ΔF508 protein is improperly folded and gets stuck in the endoplasmic reticulum rather than migrating to the cell surface. About 30 other mutations occur with some frequency (say, 1–3% or more) in certain populations, but the remaining *CFTR* mutations are individually very rare.

With regard to pancreatic function, there is considerable correlation between *CFTR* mutational genotypes and clinical phenotypes. For example, patients homozygous for the *ΔF508* mutation are often diagnosed at an earlier age and have a much greater frequency of pancreatic insufficiency (99%) than do patients with any other *CFTR* genotype. But the pancreatic-sufficient patients who are not homozygous for *ΔF508* may show milder and more heterogeneous symptoms. And with regard to the severity and course of lung disease or survival rates, there seems to be little correlation between genotype and phenotype (Hamosh and Corey 1993). Indeed, some cases are known in which patients with exactly the same mutational genotype nevertheless exhibit quite different phenotypes.

Identification of the *CFTR* gene and its CFTR protein have greatly improved prospects for the development of drugs to treat the disease. In addition, some researchers are now considering the possibility of *gene therapy* for CF—that is, getting normal copies of the gene into those CF epithelial cells that line the lungs and airways. Indeed, a few experimental steps in this direction have already been taken (Palca 1994). But it is a long way to go from curing CF cells in a laboratory dish or inserting human genes into rat lung cells to curing CF in people. In Chapter 20, we discuss gene therapy in detail.

Genetic Testing. In families known to be at risk, there are several ways to detect CF prenatally. The most reliable indicators involve DNA analysis together with testing the levels of certain intestinal enzymes in amniotic fluid samples. Carriers can also be directly detected in families where DNA is available from a CF patient. But what about CF testing in the general population?

One promising approach to mass screening of newborns is a test for blood levels of an antibody to the pancreatic enzyme trypsin, which tends to be higher in babies with CF. But even this test misses some cases.

Another goal is to find a simple and reliable DNA test for detecting CF newborns, as well as adult carriers before they produce an affected offspring. About 85% of CF carriers possess either the most common mutant allele, *ΔF508*, or one of about a dozen other mutants. But many of the remaining 15% of carriers have various rare mutant alleles for which good tests are not yet available. For this and other reasons (Chapter 19), there are currently no large-scale plans for mass screening to detect the 7 to 10 million unsuspecting CF carriers in the United States. But in 1997, an advisory panel of the National Institutes of Health did suggest that a CF carrier test be offered to all couples of childbearing age.

Hereditary Hemochromatosis: The Rusty Organ Syndrome

Once considered rare, *hereditary hemochromatosis (HH)* is an autosomal recessive disorder of iron overload—sometimes mild, sometimes serious—now known to be surprisingly common. Although not as well recognized as CF, its frequency is much higher: Overall, 1 in 200 to 400 people are affected, and 1 in 8 to 10 people are carriers. Both sexes are equally represented in young patients, but among adult patients, HH is seven to ten times more frequent in males than in females, probably because women lose some iron by menstruating.

Like cholesterol, iron is an essential but potentially toxic substance in our bodies. For unknown reasons, HH homozygotes absorb, through their small intestines, two to four times the normal amount of dietary iron. It travels in the bloodstream and gradually accumulates in the liver (causing liver enlargement, cirrhosis, and cancer), in the heart (causing arrhythmias and heart failure), in the pancreas (causing diabetes), in the skin (causing darkening), in the joints (causing arthritis), and in endocrine glands (decreasing the sex drive and stopping menstruation). Weakness, fatigue, abdominal pain, and weight loss are also common (Barton and Bertoli 1996; Brody 1997; Fackelmann 1997).

Because of its multiple effects and because until very recently the condition was thought to be rare, hemochromatosis as a specific diagnosis often goes unrecognized. More severely affected homozygous children may be diagnosed because of heart problems or failure to menstruate and other sexual problems, or they may be found (perhaps without symptoms) because they have an affected relative. Typically, disease symptoms may remain undiagnosed until males reach their 40s and females reach their 50s. (And some homozygotes, especially females, never develop the worst symptoms associated with iron overload.) Cardiac failure is usually the cause of death among untreated, severely affected patients. Yet the serious symptoms and premature death can be avoided if HH is diagnosed before the excess iron has damaged vulnerable organs, especially the liver. The surprisingly simple treatment—frequent bloodletting (a method similar to donating blood)—harks back to ancient medical practices, but works effectively.

Since the 1970s, it was known that the causative alleles lie on the short arm of chromosome 6, close to a large gene complex (called *HLA* for *h*uman *l*eukocyte *a*ntigen) that is important in the body's immune response and tissue transplantation genetics (Chapter 18). In 1996, researchers reported the discovery of a particular mutation in the related *HLA-H* region that is homozygous in 83% and heterozygous with a second mutation in 4% of the 178 HH patients in their study (Feder et al. 1996). (The remaining patients may have mutations in a different, unidentified gene.) Only 3% of chromosomes in the CEPH control subjects carried this mutation. Two other studies found even higher percentages of homozygosity for this mutant allele among hemochromatosis patients, strongly suggesting that it is indeed associated with the disease. But much more work needs to be done to learn how its defective protein product might lead to iron overload and to determine the function of the normal *HLA-H* allele(s).

Why is the detrimental mutation so common? One hypothesis is that it might help carrier females to avoid iron deficiency during pregnancy. Another hypothesis is that over the centuries, it allowed normal carriers (who absorb more iron than do noncarriers) to survive periods of famine, when dietary iron is scarce.

But nobody really knows the answer to this intriguing question.

Scientists expect to develop a blood test for direct detection of the allele(s) responsible for HH. But a simple, inexpensive, and underutilized screening protocol already exists (Edwards and Kushner 1993). *Transferrin* and *serum ferritin* are blood components that bind, transport, and store iron. The measured percentages of these substances in the blood indicate how much iron is present in an individual's body. People who show signs of iron overload from both tests can get a liver biopsy (removal and analysis of a tiny tissue sample) to detect any damage due to "rusting" of that vital organ. Because hemochromatosis is a common and potentially serious yet easily treatable condition, some researchers have suggested that whole populations should be screened (e.g., by testing all blood donors) to identify homozygotes before they become ill.

Sickle-cell disease, an autosomal recessive disorder that is relatively frequent among people of African and Mediterranean descent, is discussed in Box 5D. For some additional examples of human disorders inherited as autosomal recessives, see Table 5.3.

X-Linked Recessive Inheritance

All muscular dystrophies are characterized by gradual weakening and wasting of muscle tissue (*dys*, "abnormal"; *trophy*, "growth"), and all are inherited; but the symptoms, metabolic defects, and modes of inheritance vary considerably.

BOX 5D *Sickle-Cell Disease*

Another example of autosomal recessive inheritance is sickle-cell (or hemoglobin S) disease, which is most common among West African blacks (up to 1 in 50). Several centuries ago, tribes there gave it specific names—*ahotutuo* (Twi tribe), *chwech-weechwe* (Ga tribe), *nwiiwii* (Fante tribe), and *nuidudui* (Ewe tribe)—and recognized that it ran in families. They described two clinical types, "severe" and "not-so-severe," the latter type being essentially normal and vastly more frequent than the severe type. They even realized that matings between "severe" and "not-so-severe" types (which we would call homozygote × heterozygote) gave rise to a much greater proportion of severely affected children than did matings between two "not-so-severe" parents (Konotey-Ahulu 1991).

In the first published report in Western medical literature, Chicago physician J. B. Herrick (1910) described "peculiar elongated and sickle-shaped red blood corpuscles in a case of severe anemia" in a black dental student from Grenada, West Indies. Twelve years later, another case report and a review of previously published cases finally aroused widespread interest in this condition. Mason (1922) comments:

"The patients all complained of weakness and poor health since early childhood. All had swelling of the ankles associated with leg ulcers. ... Two patients had had repeated attacks of abdominal pain associated with fever and jaundice. ... The blood picture does not resemble that seen in any of the more common anemias, and it is possible that the disease represents a clinical entity. If that is true, it is of particular interest that up to the present the malady has been seen only in the negro."

Later studies established that sickle-cell disease also occurs in India, the Middle East, the Mediterranean area, and the Caribbean. Individuals affected with *sickle-cell disease* are homozygous for a defective hemoglobin *S* allele. Heterozygotes are unaffected under ordinary circumstances, but in rare situations of oxygen deprivation, they may show some clinical symptoms. Although phenotypically normal, heterozygotes are said to have the *sickle-cell trait* because their red blood cells become sickle-shaped on a microscope slide under very low oxygen conditions. Further discussions of sickle-cell disease appear in Chapters 7, 12, 19, and 20.

TABLE 5.3   Some human disorders inherited as autosomal recessives and their chromosomal sites

Name	Brief description	Site
Adenosine deaminase (ADA) deficiency	Severely impaired immunity; skeletal and neurological abnormalities	20q
Alkaptonuria	Darkening of urine and joints; arthritis	3q
α-1 antitrypsin deficiency	Lack of protease inhibitor (PI), causing severe lung and liver disease	14q
Ataxia telangiectasia	Poor muscle coordination; dilation of capillaries in eyes and skin, immune defects; high predisposition to cancer	11q
Albinism, "classic"	Absence of melanin, affecting eyes, hair, skin, and hearing	11q
Disaccharide intolerance I	Diarrhea due to inability to absorb sucrose that has not been split into glucose and fructose	3q
Friedreich ataxia	Degeneration of cerebellum and spinal cord; defects in limb coordination, tendon reflexes, speech, cardiac function	9q
Galactosemia	Inability to digest milk sugar; liver enlargement and damage; cataracts; mental retardation	9p
Gaucher disease	Blood abnormalities; enlarged spleen; bone and neurological defects; skin pigmentation	1q
Homocystinuria	Displaced eye lens; defects of skeleton and blood vessels; mental retardation	21q
Hurler syndrome	Mental retardation; enlarged liver and spleen; bone defects; coarse features	22q
Pyruvate kinase-1 deficiency	Hemolytic anemia (loss of red blood cells); liver damage	1q
Tay-Sachs disease	Developmental retardation; paralysis; dementia; blindness; red spot on retina	15q
Thalassemias, α (inactivation of one to four α-globin alleles at the two loci)	Varies from no symptoms to mild or moderate anemia to stillbirth (from lethal anemia, heart failure, fluid accumulation)	16p
Thalassemias, β (defects or inactivation of one or two β-globin alleles)	More than 70 different mutations known; varies from no symptoms to thalassemia major (severe anemia; defects of bone, liver, spleen; ultimately fatal)	11p
Wilson disease	Defects in copper metabolism; damage to liver and central nervous system; corneal changes, tremors, emotional changes	3q
Zellweger syndrome	Absence or defects of peroxisomes, leading to defects of skull, face, ears, eyes, hands, feet, liver, and kidneys	7q

The most common and deleterious of these syndromes is **Duchenne muscular dystrophy (DMD)**, which was described in 1861 and 1868 by French neurologist Guillaume Benjamin Amand Duchenne.* In 1879, 21 years before the rediscovery of Mendel's Laws, it was reported that affected males inherited this condition through unaffected mothers. DMD is now known to be a fully penetrant, X-linked recessive condition occurring in about 1 in 3,500 newborn males. Like CF and HH, it too remains at a higher-than-expected frequency in the population. But about one-third of all cases appear to be new mutations; that is, the mothers are not carriers.

*Actually, it was first described in great detail in 1852 by Edward Meryon, an English physician. Perhaps because he was less of a self-promoter than Duchenne, his work remained relatively unknown and the condition is not called Meryon disease (Emery and Emery 1995).

"Usually the onset ... occurs before age 6 years, and the victim is chairridden by age 12 and dead by age 20" (McKusick 1997). Most patients show normal intelligence, although some are mildly retarded and about 20% are seriously retarded. The earliest signs are developmental delays in walking and talking. Then come weakness and gradual wasting of the thigh and pelvic muscles, leading to unsteadiness, difficulty in walking stairs or rising from chairs, and a cautious, waddling gait. Although weak, the calf muscles are noticeably thickened. As the shoulder, trunk, and back muscles gradually weaken and degenerate, the child develops a "swayback" posture (Figure 5.8), has trouble maintaining balance, and falls a lot. With gradual contraction of the Achilles tendons (heel cords), the child also begins to walk on his toes. Later, tendons connected to the hips, knees, and elbows will also shorten, and curvature of the spine (scoliosis) will

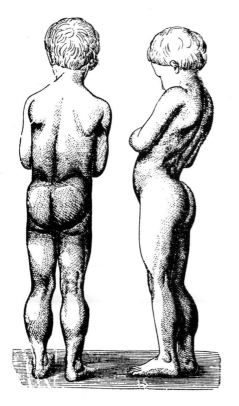

Figure 5.8 Duchenne muscular dystrophy, showing the thickened calves and swayback posture seen in young patients. (From Duchenne 1868.)

squeeze and interfere with the functioning of internal organs (notably the lungs). Heart problems are the rule, sometimes leading to sudden death. Usually, however, patients die of respiratory infections or of respiratory failure when diaphragm muscles become affected (Emery 1994).

Under the microscope, Duchenne dystrophic muscles show a significant loss of muscle fibers and variable sizes of those that remain, along with considerable infiltration of fat and connective tissue. Muscle physiology goes awry, and a muscle enzyme called *creatine kinase* is found in the serum at levels 50 to 100 times higher than normal. Because *serum creatine kinase* levels may be normal before birth, however, this diagnostic test is not reliable enough for prenatal use.

A similar but rarer X-linked condition is **Becker muscular dystrophy (BMD)**. Although Becker muscular dystrophy seems identical to Duchenne muscular dystrophy with regard to muscle tissue abnormalities, its overall phenotype is more benign and variable, with a tendency to later onset, slower progression, little or no intellectual impairment, and a much longer life span. The genetic relation between the DMD and BMD mutants—whether they are alleles of the same gene or found at independent loci on the X chromosome—was not determined until 1986 (discussed shortly).

There is no cure for either DMD or BMD, and no particularly effective drugs are available. All that can be done is to try to slow the progressive deformity of joints and the loss of muscle strength, thereby prolonging the patient's ability to sit and walk. Standard treatment includes the use of lightweight splints and braces, physical therapy, pulmonary therapy, and perhaps orthopedic surgery.

A recent and controversial experimental treatment, *myoblast transfer*, involves the injection of cultured muscle fiber cells (obtained from normal donors) into the muscles of DMD patients. The hope is that they might fuse with DMD muscle cells and replace the missing gene product—but clinical trials have not shown any significant improvement. A potentially more promising approach is *gene therapy* (Chapter 20), in which the defective allele is replaced by a normal one in affected somatic tissues. This experiment has been attempted with some success in mice, but many technical problems must be overcome before it can be tried in humans.

Pedigree Analysis: Duchenne Muscular Dystrophy

Here are the characteristics of **X-linked recessive inheritance**, many of which are illustrated in Figure 5.9:

1. *If the gene is not common, many more males than females are affected.* These males are hemizygous for the abnormal allele.

2. *Father-to-son inheritance is never seen. Instead, affected males always get their abnormal allele from their mothers,* who are usually unaffected but may have affected fathers, brothers, or uncles. Affected males produce only unaffected sons and carrier daughters.

3. *Among the sons of carrier mothers, about half are affected and half are unaffected.*

4. *All the daughters from the mating of a carrier female with an unaffected male are unaffected, but half will be carriers.*

5. *Occasionally, a carrier female may manifest some symptoms of a recessive X-linked condition.* Indeed, serum creatine kinase is elevated in about 70% of known female carriers, and about 20% of carriers show some muscle weakness. These effects are due to a phenomenon called **lyonization** (Chapter 10), a normal process whereby one or the other X chromosome in every female cell is rendered inactive. If by chance a female ends up with many cells whose active X chromosome carries the mutant allele, she may be partly or (in very rare cases) fully affected by a recessive condition that usually strikes only males. In fact, all females heterozygous for X-linked traits are **mosaics**; their somatic tissues contain various proportions of cells in which one or the other of the two alleles is expressed.

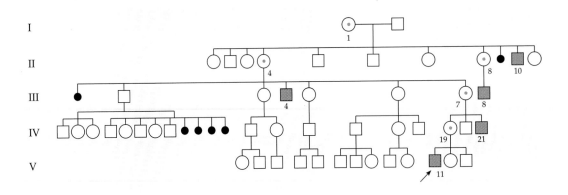

Figure 5.9 One of 33 Duchenne muscular dystrophy kindreds discovered in Utah. It illustrates X-linked recessive inheritance, which in this case results in the sterility of affected males. Note that carrier females (circles with dots inside) are the sole transmitters. (Redrawn from Stephens and Tyler 1951.)

Identification of the DMD Gene and Protein

Detection of the *DMD* gene itself became a possibility in the early 1980s. Investigators took several different approaches. In analyzing a few rare cases of females with full-blown Duchenne muscular dystrophy, geneticists found that all of them carried a chromosomal abnormality involving the breakage and fusion of parts of two nonhomologous chromosomes. Each female had a different autosome joined to an X chromosome, but in all cases the X chromosome had broken at the same spot, the so-called Xp21 band of the short arm. Along the way, an amazing case also turned up: a male patient who had Duchenne muscular dystrophy plus four other X-linked conditions and a big X chromosome deletion at Xp21 (Chapter 14). These discoveries pinpointed the *DMD* gene to the Xp21 region, which makes up about 20% of the short arm.

Zeroing in on the new reference point and using many clever tricks of molecular biology, scientists set out to identify the genes in that region. By sharing their DNA samples and information about their patients, an international group of 25 research teams amassed and analyzed data on 1,346 males with Duchenne or Becker muscular dystrophy. This study, published by Louis Kunkel, of Harvard University, and 76 coauthors (Kunkel et al. 1986), showed for the first time that (1) the DMD and BMD mutations are allelic, and (2) the normal locus is gigantic—over ten times larger than any other known human gene!

By 1988, Kunkel's team had found the product of the normal allele of the *DMD/BMD* gene. This previously unknown protein, dubbed **dystrophin**, exists at very low levels in the muscle cell membranes of unaffected individuals. It connects at one end to actin fibers of the cytoskeleton and at the other end to some *glycoproteins* (proteins with attached carbohydrates) that bridge the membrane and connect to the extracel-

lular matrix. The exact function of this complex is unknown. Perhaps it acts mainly as a strengthener, helping to protect the fibers (especially of skeletal muscle cells) from damage due to repeated contraction. Dystrophin is present at even lower levels in a dozen other normal tissues (including brain, lung, and kidney tissues), where its function is unknown. But it is missing from the muscles of nearly all DMD patients (even before birth) and present in altered form or amounts in BMD patients (Figure 5.10).

A mutant strain of mice carries a mutation called *mdx*. These mice exhibit some traits comparable to those of DMD: They have deletions in the same X-linked gene; they lack dystrophin in their muscle cells; and they show the same kind of tissue damage seen in human dystrophic muscle cells. Furthermore, there is great similarity between the DNA sequence of the normal alleles of the *DMD* and *mdx* genes, confirming this mutation's status as a genetic counterpart to DMD. Yet, for unknown reasons, these mice do not express a severe DMD-like phenotype; indeed, they are fairly normal and seem to manage quite well without dystrophin. But a few cases of full-blown DMD have been found in dogs, which provide the best (nonhuman) animal model now known.

Detection of Carriers and Hemizygotes

Until 1982, the only way of identifying DMD carriers was by measuring the levels of serum creatine kinase present in their blood: Levels exceeding the normal 95th percentile appear in about two-thirds of definite carriers. In addition, up to 80% of definite carriers show thickening of the calf muscles, about 8% exhibit some muscle weakness, and some show other serum and muscle abnormalities.

Unfortunately, elevated serum creatine kinase levels do not show up reliably in affected males before birth. Thus, in families known to be at risk for DMD,

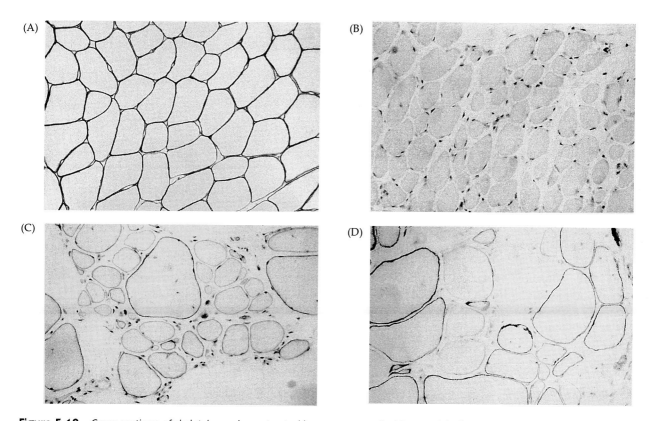

Figure 5.10 Cross sections of skeletal muscle are treated by a process called immunolabeling to detect the presence of dystrophin. (A) In a normal individual, dark-staining dystrophin is uniformly present in the membrane (sarcolemma) surrounding every muscle fiber. (B) In a person with Duchenne muscular dystrophy, dystrophin is absent from the muscle fibers. (C) In a person with Becker muscular dystrophy, dystrophin is nonuniformly present in some fibers and absent or nearly absent in others. (D) In a carrier female who shows some symptoms of DMD, dystrophin is present in some fibers and absent in others. (Reproduced by kind permission of Drs. Louise V. B. Anderson and Margaret A. Johnson.)

about the only option before 1982 was to test for sex and then abort all male fetuses—even though 50% would be unaffected. Now it is possible, by using molecular techniques described in Chapter 8, to detect carrier mothers and affected offspring with 95–100% accuracy in most known DMD families.

But it will never be possible to detect all cases of DMD in advance through prenatal testing of known carriers' offspring, because about one-third of all cases represent new mutations. That is, the mothers of these individuals do not carry the abnormal *DMD* allele. Why such a high mutation rate? Nobody knows for sure, but one factor might be the colossal size of the *DMD* gene, which presents a big target for mutation and in addition allows for a great deal of intragene recombination (exchange of homologous parts) during the formation of gametes in carrier mothers. There is also evidence that two areas within the *DMD* locus are hot spots, especially prone to breakage and deletions.

Some mothers who are apparent noncarriers but produce an affected offspring may actually be **germline mosaics**. That is, although their somatic cells are homozygous for the normal allele, an ovary may contain a cluster of germ cells with the mutant allele. Thus, they run the risk of producing more than one affected offspring. Perhaps as many as 12–20% of DMD cases can be attributed to germline mosaicism in mothers.

Because of the high mutation rate, the testing for DMD in all newborn infants (rather than just those in known high-risk families) might be worthwhile if reliable and cost-effective tests were available. It is possible, for example, to detect serum creatine kinase in the tiny amounts of dried blood (on filter paper) that are routinely obtained from all newborns to test for certain other inherited conditions. Although no effective treatment yet exists for the affected babies that might be found through such screening programs, their families could be offered genetic counseling, giving them the information and the choice of what to do about future pregnancies.

A pedigree for yet another X-linked recessive disease is presented in Box 5E. For examples of additional human disorders inherited as X-linked recessives, see Table 5.4.

BOX 5E *Hemophilia*

Another example of X-linked recessive inheritance is hemophilia, a failure of blood clotting. Although hemophilia is rare, its mother-to-son inheritance was recognized over 2,500 years ago. The Talmud states that if a woman produces two "bleeders," any additional sons shall be excused from circumcision. If as many as three sisters bear sons who are bleeders, the sons of any other sisters are not to be circumcised; but no such prohibition applies to the sons of their brothers.

There are two nonallelic X-linked recessive forms of this disorder: hemophilia A (the "classic" type, or factor VIII deficiency) and hemophilia B (Christmas disease, or factor IX deficiency). Hemophilia A appeared in several interrelated royal families of Europe, apparently arising from a mutation in one parent of Queen Victoria of England (1819–1901).

Probably the most famous victim of hemophilia was Victoria's great-grandson Tsarevitch Alexei, heir to the throne of Russia (Massie 1985). His illness caused his parents to put their faith, and considerable political power, into the hands of the "mystic" and charlatan Rasputin. Rasputin's evil doings may have hastened the onset of the Russian Revolution, to be followed by the murder of 14-year-old Alexei and his family by Bolshevik revolutionaries in 1918. The remains, of all except Alexei and Marie, have only recently been found and identified by DNA analyses (Anonymous 1996; Massie 1995).

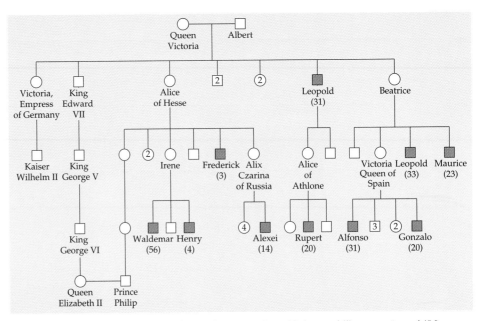

The pedigree shown here presents the ten known males with hemophilia—genotype *h*/(Y)—who are descended from Queen Victoria. McKusick (1965) gives brief sketches of their lives; most died from uncontrolled bleeding at the ages indicated in parentheses. The pedigree is incomplete in many respects, and children indicated by the multiple-person symbols are not necessarily in the birth orders indicated. But the pedigree includes all sibs of the affected males.

This 1914 photograph of the Romanovs includes (from left to right) Olga, Marie, Tsar Nicholas II, Alexandra (Alix), Anastasia, Alexei, and Tatiana. (Photograph © The Bettmann Archive.)

Other Modes of Inheritance

Here we describe three much less common ways by which genes can be transmitted: X-linked dominant inheritance, Y-linked inheritance, and mitochondrial inheritance.

X-Linked Dominant Inheritance: Hypophosphatemia

X-linked dominant inheritance is relatively rare—perhaps in part because when families are small, it may be impossible to distinguish X-linked dominant from autosomal dominant inheritance.

TABLE 5.4 Some human disorders inherited as X-linked recessives and their chromosomal locations

Name	Brief description	Site
Fabry disease	α-galactosidase A deficiency; kidney damage; corneal changes; vascular skin growths	Xq
G6PD deficiency (favism, primaquine sensitivity)	Acute hemolytic anemia brought on by certain substances, such as fava beans and more than 340 drugs (primaquine, aspirin, sulfas, etc.)	Xq
Hunter syndrome	Heparitin sulfate in urine; milder skeletal and mental symptoms, but more deafness, than seen with Hurler syndrome (see Table 5.3)	Xq
Lesch-Nyhan syndrome	Defect in purine metabolism; mental retardation; spastic cerebral palsy; involuntary movements; uric acid urinary stones; self-mutilation (biting fingers and lips)	Xq
Ornithine transcarbamy-lase (OTC) deficiency	Excess ammonia in blood; mental deterioration; extreme irritability; vomiting	Xp
Testicular feminization (androgen insensitivity)	XY individuals with testes but cells unresponsive to androgen, so develop female external genitals, breasts, blind vagina	Xq
X-linked mental retard-ation with fragile site	Mental retardation; large testes, ears, and jaw; high-pitched voice; jocular speech	Xq

Hypophosphatemia (Greek *hypo*, "less than"; Greek *emia*, "in the blood") is characterized by abnormally low levels of phosphorus in the blood due to defective reabsorption of phosphate in the kidneys. Deficient reabsorption of calcium in the intestines also occurs, leading to softening of the bones and *rickets* (bowlegs) in virtually all affected males and about 50% of affected females. Vitamin D metabolism is also abnormal, and affected individuals may be bowlegged despite the intake of normal amounts of this vitamin. Thus, this disorder may also be called *vitamin D–resistant rickets* or *X-linked hypophosphatemic rickets*. Its frequency is roughly 1 in 60,000.

Those individuals with bone disease have abnormally short legs and short stature, along with certain other skeletal abnormalities that can vary considerably within a family. Pain due to softening or overgrowth of the bones is common, as are tooth abnormalities. Onset is early, usually between 6 and 12 months of age, and the disease persists through adulthood. There is no cure, but treatment with four or five daily doses of oral phosphate plus vitamin D can lead to improvement. The use of growth hormone and thiazide diuretics is also being explored.

Much has been learned about the hypophosphatemic phenotype from a comparable mutation, also inherited as an X-linked dominant, in the laboratory mouse. In humans, the gene (called *PEX* for *p*hospate-regulating gene with homologies to *e*ndopeptidases, on the X chromosome) has been mapped to the distal part of the X chromosome short arm, but its exact structure and function remain unknown. Thus far, it is not possible to diagnose hypophosphatemia prenatally.

Figure 5.11 is a pedigree for hypophosphatemia. Among the following features of this inheritance pattern, note that affected males and their offspring and parents provide the best clues:

1. There is no male-to-male transmission of the trait.

2. Affected males (when mated to unaffected females) produce all unaffected sons and all affected daughters.

3. Affected females can produce affected sons and daughters as well as unaffected sons and daughters. Among these sons and daughters, the ratio of affected to unaffected is about 1:1.

4. Every affected individual has one affected parent; but for affected males, the affected parent is always the mother.

5. Males, being hemizygous, are usually more severely affected than females, who are heterozygous.

6. Owing to the phenomenon of X chromosome inactivation (Chapter 10), expression of the trait is also more variable among females.

7. If affected males and affected females reproduce at the same rate, in the population as a whole there will be approximately twice as many affected females as males. This is because females have two X chromosomes, whereas males have only one.

Box 5F presents an unusual case of excessive hairiness inherited as an X-linked dominant.

Y-Linked Genes

Very few genes have been detected on the Y chromosome of most species. In humans, the Y is less than half the length of the X chromosome (which contains

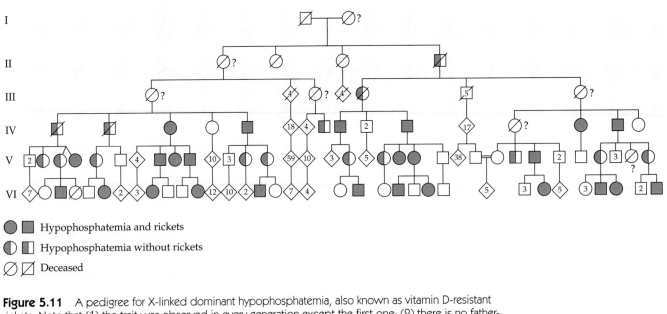

Ⓞ ▣ Hypophosphatemia and rickets

◑ ▥ Hypophosphatemia without rickets

⊘ ▨ Deceased

Figure 5.11 A pedigree for X-linked dominant hypophosphatemia, also known as vitamin D-resistant rickets. Note that (1) the trait was observed in every generation except the first one; (2) there is no father-to-son transmission; and (3) affected males produce daughters who are all affected. There are ambiguities among some deceased females, however. If the trait is indeed an X-linked dominant, then—because they have affected sons or are in the line of descent of the mutant gene—females I-2, II-1, III-1, III-6, III-17, IV-50, and V-161 (indicated with question marks) should have been affected. Perhaps they were only mildly affected and thus eluded detection. These kinds of problems are not uncommon in analyses of human pedigrees. (Redrawn from Winters et al. 1957.)

over 280 confirmed loci) and slightly longer than the smallest autosome (which contains about 60 confirmed loci).

Only about 20 loci have been definitively assigned to the Y chromosome (OMIM 1997). These include some genes associated with traits that have nothing to do with sexual development. The Y is critical to the development of maleness and the prevention of femaleness, however, because certain rare individuals possessing only one X chromosome but no Y are females (Chapter 13). More precisely, a particular region of the Y is needed for male development, and femaleness is what develops in its absence.

The tip of the Y chromosome short arm is actually homologous to a tiny region at the tip of the X chromosome short arm, and this region contains nine genes (Wade 1997). But what concerns us here is the rest of the Y chromosome, the *nonhomologous* part. Any gene located on this part of the Y chromosome has a very striking mode of inheritance. Such a **Y-linked gene** is received, and usually expressed, by all the male descendants of an affected male, generation after generation. (And in the other direction, it should extend from any male through his father, paternal grandfather, paternal great-grandfather, and so on, for countless generations.) Female descendants, lacking the nonhomologous Y region, could neither receive, express, nor transmit such a trait. In the past half century, claims of Y linkage have been advanced for over

a dozen traits. On closer examination of the family histories, however, exceptions to the given requirements were nearly always discovered.

Mitochondrial Genes

In the 1960s, scientists discovered that aerobic (oxygen-using) cells of animals contain two genetic systems: the main one (about 80,000 genes in humans) in the nucleus and a very tiny one (exactly 37 genes in humans) in the mitochondria, the "powerhouses" of the cell (Chapter 2).* Recall that each cell contains hundreds of mitochondria, and each mitochondrion contains several tiny circles of DNA. Thus, every cell may contain thousands of mitochondrial chromosomes.

All the mitochondria present in our cells are derived exclusively from our mothers. This is because egg cells contain many mitochondria, but sperm cells contain very few—and none of the sperm mitochondria survive inside the fertilized egg. So any trait associated with a **mitochondrial gene** must be transmitted by the mother to all of her children, both male and female. In the other direction, this lineage should extend from any individual through the mother, maternal

---

*The stripped-down mitochondrial genome (encoding just 13 polypeptides, 22 transfer RNAs, and 2 ribosomal RNAs) could not possibly make a mitochondrion by itself. Instead, at least 66 additional polypeptides are actually produced by *nuclear* genes and then transported from the cytoplasm into the mitochondria.

BOX 5F *Hypertrichosis*

Excessive hairiness (*hypertrichosis*) is usually caused by hormonal imbalance, but very rarely it results from genetic disorders. Some genetically affected individuals have been exhibited in circuses and called such things as "ape men" or "dog men." Two known conditions (with generalized congenital hypertrichosis as the main feature) are inherited as autosomal dominants, and half a dozen other conditions (with additional abnormalities) show various patterns of inheritance. But a large, five-generation Mexican family exhibits X-linked dominant inheritance. At birth, affected members have abundant hair on their face and upper body, males being more seriously affected than females.

Shown here is a severely affected 6-year-old male. Figuera et al. (1995) determined that the locus for this condition is on the long arm of the X chromosome.

A 6-year-old male severely affected with hypertrichosis. ▶

The pedigree depicts five generations of a large family from Mexico that exhibits the X-linked dominant inheritance of hypertrichosis. Note in Generation III that three of the affected females each had more than one husband. Indeed, III-2 had four and III-9 had six. ▼

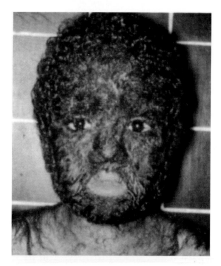

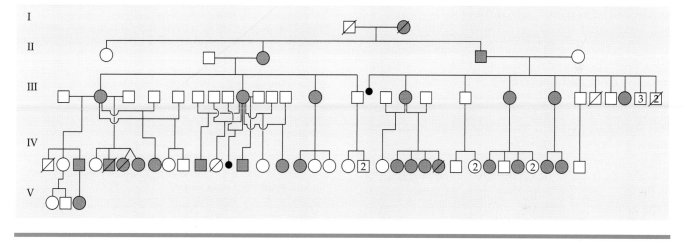

grandmother, the maternal great-grandmother, and so on, as far back as humans existed.

Clearly, this situation is an exception to Mendel's rules of inheritance. Mitochondrial genes have other unique features too. The mutation rate of mtDNA is at least ten times greater than that of nuclear DNA.[*] A

given cell may contain all normal or all mutant mtDNA molecules or a mixture of mutant and nonmutant molecules. When a cell containing a mixed population divides, the daughter cells may randomly receive differing proportions of normal and mutant mitochondrial chromosomes. For deleterious mutations, these cell-to-cell or tissue-to-tissue fluctuations will alter energy-generating capacity and consequently their phenotypic effects.[*] Additional (sporadic, noninherited) mutations also accumulate in our somatic cells with age, further eroding tissue function. Thus, the analysis of mitochondrial genetics is very complex, and genetic counseling becomes immensely difficult.

[*]This property has made mitochondrial DNA very useful in forensic and anthropological studies. For example, mtDNA recently recovered from some long-lost remains proved to belong to the murdered Romanovs of Russia (see Box 5E), because the bones contained a unique mutation shared by some living relatives, including England's Prince Philip (Anonymous 1996). Furthermore, mtDNA samples from a woman claiming to be Anastasia, a daughter of Tsar Nicholas II, proved to be unrelated to the Romanovs (Massie 1995; Stoneking et al. 1995). And starting in the mid-1980s, mtDNA has been used to identify (and reunite with their grandparents) many children who were illegally adopted following the disappearance and death of their parents at the hands of a brutal Argentinian dictatorship (Davies and White 1996).

[*]Although mitochondria normally generate about 90% of the energy needs of our cells, some cells with impaired mitochondria may still survive with only a fraction of their normal energy capacity.

Mitochondrial mutations that cause diseases can sometimes be detected by their maternal pattern of inheritance. They are also associated mostly with those tissues that have high energy requirements. These include the central nervous system (especially the brain), heart, skeletal muscles, kidneys, hormone-producing tissues, liver, eyes, inner ears, colon, and blood. Dozens of mitochondrial mutations have been described in humans; they include point mutations (base substitutions) and rearrangements (deletions and duplications) occurring in genes for polypeptides, transfer RNA (tRNA) molecules, and ribosomal RNA (rRNA) molecules. The resultant illnesses are quite variable and often have a delayed onset (Wallace 1997, 1995; Johns 1995).

One group of vision disorders is *Leber's hereditary optic neuropathy* (*LHON*). Many more males than females are affected, and the use of tobacco and alcohol may increase susceptibility. The probability of recovery varies with different mutations. Sometimes the blindness of LHON is accompanied by neurodegenerative disease (including movement disorders or multiple-sclerosis-like symptoms). *Leigh syndrome*, a fatal disorder of early childhood, affects the brain, muscles, eyes, and lungs. Each of these illnesses has been traced to a specific enzyme complex involved in the production of the high-energy compound ATP and is caused by the mutation of single nucleotide bases.

Mitochondrial mutations affecting protein synthesis—that is, those of tRNA or rRNA genes—cause defects in *all* the respiratory enzyme complexes (rather than just one) and thus give rise to more complex problems affecting multiple tissues. *Mitochondrial myopathy* (Greek *myo*, "muscle"; English *pathy*, "disease") is frequently associated with these disorders. It is caused by the degeneration of muscle fibers and leads to serious muscular weakness or wasting or both. (In 1995, bicyclist Greg LeMond—three-time Tour de France winner and two-time world championship winner—was diagnosed with possible mitochondrial myopathy and forced to retire at age 33. Changes in his skeletal muscle cells made them unable to work at peak capacity.) Other symptoms occurring with different tRNA mutations include myoclonic epilepsy (uncontrolled periodic jerking), heart defects, strokes, seizures, failure of muscle coordination, type II diabetes mellitus, deafness, retinal abnormalities, droopy eyelids, paralysis of eye muscles, kidney failure, and dementia (including Alzheimer disease). Only since 1988 have scientists linked these various human disorders to mitochondrial mutations; more such discoveries will no doubt be found in coming years.

Summary

1. Gathering reliable family data is not easy. Interpreting genetic data may be difficult, too, especially when dealing with rare traits.

2. Some inherited conditions show unique patterns of transmission that are recognizable by means of standard pedigrees. In humans, the most common types of simple inheritance are autosomal dominant, autosomal recessive, and X-linked recessive.

3. Autosomal dominant traits that are fully penetrant do not skip generations. They are expressed equally in both sexes and are transmitted by both sexes (including father-to-son transmission).

4. Autosomal recessive traits often do not appear in every generation, and affected individuals usually have unaffected parents. In cases of very rare traits, the unaffected parents of an affected child are usually related. The trait is expressed equally in both sexes and transmitted by both sexes (including father-to-son inheritance).

5. X-linked recessive traits skip generations, are more frequent in males than females, and are transmitted differently by the two sexes. Father-to-son inheritance is never observed; instead, males inherit the trait from carrier mothers, who may have affected male relatives and carrier female relatives. Because of the lyonization effect, cultured cells taken from unaffected carriers may consist of two phenotypic populations, and an occasional carrier female may be partly or fully affected.

6. Few X-linked dominant traits are known. Like autosomal dominants, they are expressed in every generation and are transmitted by both sexes. Unlike autosomal dominants, however, they show no male-to-male transmission, and the offspring of affected males consist of all affected daughters and all unaffected sons.

7. Disorders caused by mutations in mitochondrial genes mainly affect high-energy tissues, may have a delayed onset, and are highly variable. They are inherited only through the mother.

8. Some genetic disorders have multiple modes of inheritance or exhibit differences in the molecular basis of their mutations. For this and other reasons, genetic counseling should be done only by those trained in both clinical medicine and genetics.

9. Familial hypercholesterolemia is a common disorder, leading to increased risk of heart attacks. It is caused by autosomal dominant alleles on the short arm of chromosome 19. The normal gene product, the LDL receptor, removes excess cholesterol from the bloodstream.

10. Cystic fibrosis is a relatively common lethal condition caused by autosomal recessive alleles on the long arm of chromosome 7. The gene and its normal product, cystic fibrosis transmembrane regulator, have been isolated. The defect involves the regulation of chloride ion transport through the cells of affected organs.

11. Duchenne muscular dystrophy is caused by a recessive lethal gene located on the short arm of the X chromosome. About 60% of patients with Duchenne muscular dystrophy have deletions in this huge gene. The normal gene product, called dystrophin, is absent from the membranes of dystrophic muscle cells; its function is not well understood.

Key Terms

autosomal dominant
  inheritance
autosomal recessive
  inheritance
consanguineous mating
founder effect
germline mosaic
mitochondrial gene
mosaic

pedigree
proband
propositus
X-linked dominant
  inheritance
X-linked recessive
  inheritance
Y-linked gene

Questions

1. Why are the terms *homozygous* and *heterozygous* not applicable to X-linked genes in males?

2. (a) Construct, as completely as possible, a pedigree of your own family. (b) Ask as many relatives as possible to eat some asparagus and determine whether or not they smell a distinctive odor in their urine afterward. If feasible, suggest that they also try to detect this odor in the urine of other family members (without discussing their observations among themselves) and then report all the results back to you. Then see if you can figure out how this simple and harmless trait may be inherited.

3. What mode(s) of inheritance could explain the pedigree shown here?

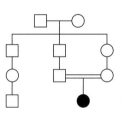

4. Galactosemia is an autosomal recessive disorder (genotype *gl/gl*) caused by an inability to metabolize the milk sugar galactose. The symptoms seen in untreated infants are lethargy, failure to thrive, severe vomiting, liver disease, cataracts, and mental retardation; early death may occur. But if newborns with galactosemia are placed on a diet free of lactose and galactose, symptoms regress, and development may even fall within the normal range. About 1 person in 150 is heterozygous (*Gl/gl*) and phenotypically unaffected.

(a) Two unaffected parents have a baby with galactosemia. What are the genotypes of the parents and baby?
(b) Two unaffected children are also born to this couple. What is the probability that their fourth child will have galactosemia? Would the probability be different if their first three children had been affected?
(c) If a treated individual reaches adulthood and is fertile, what are his or her chances of marrying a heterozygote and having children with galactosemia?

5. Refer to the pedigree of hemophilia in Box 5E. (a) Identify all females who must have been heterozygous (*H/h*). (b) What is the ratio of affected sons to unaffected sons among the children of known heterozygotes? (c) Is it possible that Queen Elizabeth II is a carrier?

6. What is the most likely mode of inheritance for the single condition illustrated in the following pedigrees? (From Mabuchi et al. 1978.)

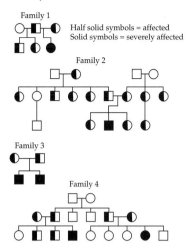

7. With regard to the X-linked gene for green-shift color blindness, females can have three different genotypes—*G/G*, *G/g*, or *g/g*; but males can have only two—*G/*(Y) and *g/*(Y). The normal allele *G* is dominant to its green-shifted allele *g*.

(a) Systematically list the six possible matings for this X-linked gene, identifying the sex of each parent. For each mating, give the expected offspring. List daughters and sons separately.
(b) Which of these matings can produce a carrier daughter? In which of these are all daughters carriers?
(c) Which of these matings is expected to produce some sons with green shift and some sons with normal color vision? What is the genotype of the mothers in these matings?
(d) What offspring phenotypes are possible from a green-shifted mother and a normal father? Identify offspring by sex. (Note that in this particular family, there is a crisscross type of inheritance.)
(e) A couple has a green-shifted daughter and a son with normal color vision. What are the genotypes of the parents?

8. Around 1930, several published reports described males who lacked some teeth and sweat glands and were partly bald. This condition had been noted much earlier by Charles Darwin (1875):

> I may give an analogous case … of a Hindoo family in Scinde, in which ten men, in the course of four generations, were furnished … with only four small and weak incisor teeth and with eight posterior molars. The men thus affected have very little hair on the body, and become bald early in life. They also suffer much during hot weather from excessive dryness of skin. It is remarkable that no instance has occurred of a daughter being affected …. [T]hough the daughters in the above family are never affected, they transmit the tendency to their sons; and no case has occurred of a son transmitting it to his sons. The affectation thus appears only in alternate generations, or after long intervals.

How is this trait, which is now called ectodermal dysplasia, inherited?

9. If a male desires to maximize his contribution of genes to future generations, is he better off having daughters or sons? Why?

10. You could not have gotten a sex chromosome from one of your four grandparents. Which grandparent could not have transmitted, via your parents, a sex chromosome to you? (*Note:* The answer depends on your sex.)

11. Describe the expected pattern of inheritance for a rare, fully penetrant X-linked dominant trait that is lethal (before birth) in hemizygous males. How can it be distinguished from an autosomal dominant trait?

12. (a) What mode(s) of inheritance could explain the pedigree shown here, assuming that the trait is *rare?* (b) What mode(s) of inheritance could explain the same pedigree if the trait is *common?*

13. In the pedigree shown here, assume that the trait is fully penetrant and that none of the matings are consanguineous. Also consider whether the causative allele is rare or common in the general population. Why could the trait represented by solid symbols *not* be due to (a) a rare autosomal recessive, (b) a rare X-linked recessive, (c) a rare autosomal dominant, (d) an X-linked dominant, or (e) a common Y-linked gene? (f) What type(s) of inheritance could explain this pedigree?

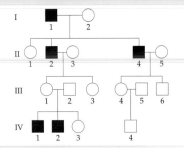

14. Is it possible to have defects in mitochondria that are not caused by mutations of mitochondrial DNA? Explain.

15. These pedigrees, published in 1997, all describe the inheritance of a particular trait. Blackened symbols represent affected individuals and question marks refer to untested phenotypes. Also, the first-generation affected female in family F had two (unaffected) husbands, and the one in fam-

ily J had three (unaffected) husbands. What is the most likely mode of inheritance of this trait? Why? Are there other possible modes of inheritance?

16. In our discussion of familial hypercholesterolemia, we give the frequencies of FH heterozygotes (1 in 500) and homozygotes (1 in 1 million). Assuming that homozygotes do not reproduce, how can the frequency of homozygotes be calculated from that of heterozygotes?

Further Reading

Kingston (1997) provides a clear description of genetic assessment and pedigree analysis. Milunsky (1991), written for the lay public, includes many interesting case descriptions and historical sidelights. Rimoin et al. (1997) and Scriver et al. (1995) are multivolume reference works that present detailed clinical and genetic information on a vast number of human disorders. For simply inherited human traits, the "bible" is *Mendelian Inheritance in Man*, an encyclopedic catalog compiled and regularly updated by McKusick (12th ed., 1997); the on-line version, called OMIM, is http://www3.ncbi.nlm.nih.gov/omim/. In addition to its detailed descriptions of about 5,000 phenotypes, the foreword and appendices are packed with clearly written information on human genetics. Krush and Evans (1984), an interesting nuts-and-bolts guide on how to design and conduct family studies, also discusses personal reactions and interactions between researchers and families.

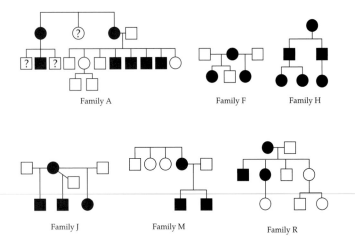

Family A

Family F

Family H

Family J

Family M

Family R

PART 2

DNA

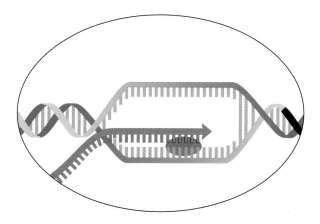

CHAPTER 6

DNA in the Organism

Our genes, segments of long DNA molecules, are one of the hottest subjects for modern biological research. Because genes provide the blueprints for development and growth, they affect the entire range of human differences from the obvious to the subtle, from the trivial to the life-threatening. Thus, unraveling the mysteries of DNA structure and function promises not only to refine the practice of medicine, but also to profoundly change how we see ourselves and others.

The more we learn about our genes, and the more deftly we are able to manipulate them, the less clearly defined become the boundaries between normal and abnormal conditions. For example, because human growth hormone can be produced by culturing genetically engineered cells, it is now available in substantial amounts. The question then arises as to who should receive growth hormone treatment to alter their growth patterns. In other words, "How short is too short?" (Werth 1991). As public debate heats up on more and more genetic fronts, many people worry about unwarranted uses of biotechnology and fear the possibility of unfair discrimination based on information about an individual's DNA.

But some situations are not subject to doubt. Scientists know, for example, that extraordinarily small alterations in DNA may lead to grave genetic diseases. In 1957, biochemist Vernon Ingram of Cambridge University showed that the abnormal hemoglobin in the red blood cells of patients with *sickle-cell disease* differs from normal hemoglobin by just one amino acid, the basic building block of proteins. A few years later, researchers learned that this amino acid change results from the alteration of just one base among the many thousands making up the genes that code for hemoglobin. Unfortunately, new scientific understanding does not always lead to quick cures, and we still have no effective long-term treatment for what one patient described as the "pain, strokes, seizures, and leg ulcers, the ridicule from peers, low self-esteem, desire to die, and diminishing hope for the future associated with sickle cell anemia" (Huntley 1984).

In Chapter 2, we presented a short introduction to **DNA, deoxyribonucleic acid,** pointing out that its basic structure is a double helix. This discovery, made in 1953 by Nobel prize winners James Watson and Francis Crick, was perhaps the most important biological advance of the century. In this chapter, we describe in greater detail the activity of DNA, showing that genes are like the 3 × 5 cards in a recipe box. Just as a recipe tells a cook how to put together a particular dish, the DNA segment that forms a gene tells the cells of the body how to assemble a particular RNA molecule or protein. The ingredients for RNA and protein synthesis are there in the cells, derived from the things we eat.

In Chapter 7, we will look at the variations (alleles) of a gene, whose simple inheritance patterns were first determined by Mendel (Chapters 4 and 5). We will be concerned with how the bases of DNA can change, or *mutate*, in response to chemicals and radiation in our environments. We have already seen some of the phenotypic consequences of these mutations, and we will further develop the idea of genetic disorders throughout the book.

In Chapter 8, we will look at the many clever manipulations of DNA that have provided both basic knowledge and practical applications. The first breakthrough in DNA laboratory technology occurred in 1973, when Paul Berg of Stanford University and his colleagues invented methods for hooking together pieces of DNA from different species. The product of such splicing operations is called *recombinant DNA*. Biochemical expertise has now been so sharpened that DNA molecules can be copied, tagged, sized, base sequenced, altered, sliced, and spliced in many ways. These techniques suggest the power of applied genetic research and point up the need for an informed public

to help guide and regulate the biological revolution in which we are now immersed.

The Double Helix

In his best-seller *The Double Helix* (1968), James Watson recounts how, while playing with cutouts of the parts of DNA, he suddenly became aware of regularities of structure that made a lot of sense. He and Crick then excitedly worked and reworked a three-dimensional wire model of the possible DNA double helix. Less than two months later, their historic one-page report announcing the structure of the genetic material appeared in print (Watson and Crick 1953).

Watson, a virus geneticist, and Crick, a physicist, had begun their collaboration at Cambridge University in 1951 (Figure 6.1). They realized that determining the structure of DNA would be a grand coup, because evidence was accumulating that DNA was the genetic material. For example, the amount of DNA was known to be constant for all cells of a species (except for gametes, which of course contained half as much), a pattern that was not observed for other cell components. In 1944, it had been shown by Oswald Avery and his colleagues at Rockefeller Institute that free DNA, but not other cell components, could enter and change the genotype of bacteria (Bearn 1996). These changes were then passed on to subsequent, untreated generations; thus genes (of bacteria, at least)

Figure 6.1 James Watson (left) and Francis Crick at Cambridge University in 1953 with their original wire model of the DNA double helix. Watson was 25 years old at the time and Crick was 37. (From Crick 1988.)

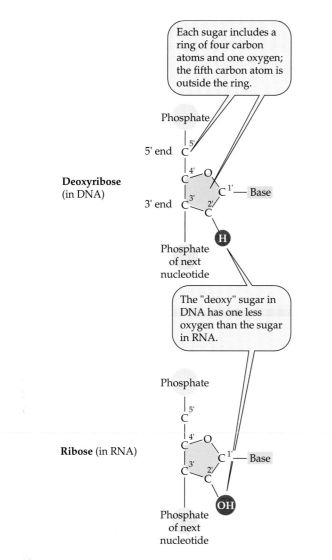

Figure 6.2 Partial structures of the five-carbon sugars, deoxyribose and ribose, that are present in nucleic acids. The carbons are numbered 1' to 5'. As indicated, the 3' and 5' carbons are bonded to phosphate groups, making a long alternating sugar and phosphate backbone structure. At the 1' position, each sugar binds one of four possible bases, whose structures are shown in Figure 6.3.

Furthermore, Erwin Chargaff at Columbia University had discovered that in DNA, the amount of A equals the amount of T, and the amount of G equals the amount of C; these two significant facts are now known as Chargaff's rules.

Watson and Crick also knew about the X-ray studies of DNA fibers done by Rosalind Franklin and Maurice Wilkins at King's College of the University of London.* They deduced that DNA had a helical structure, but it was unclear how many strands were present, and whether the bases stuck out from a narrow helical core or in from a wider helix. The fit of the DNA parts had to satisfy the laws of structural chemistry, as well as Chargaff's rules and some dimensions of the helix revealed by the X-ray photographs. The structure also had to make some biological sense (Box 6A).

---

*Along with Watson and Crick in 1962, Wilkins also won a Nobel prize for this work. Unfortunately, Franklin died of cancer before the prize (which is awarded only to living persons) was announced. With regard to the 1944 discovery that bacterial genes consisted of DNA, it is a matter of some embarrassment to the scientific community that Oswald Avery never won a Nobel prize.

appeared to be DNA. By 1952, even skeptics were convinced by these and other findings that heredity was controlled by nucleic acids (DNA or, in a few viruses, similar molecules of RNA).

We present here enough about DNA so that you can appreciate the nature of the genetic material and how genes replicate, mutate, store information, and have this information direct the structure of proteins. Watson and Crick knew that DNA contained:

• the sugar (**S**) deoxyribose (Figure 6.2)

• phosphoric acid (**P**)

• four bases: adenine (**A**), thymine (**T**), guanine (**G**), and cytosine (**C**) (Figure 6.3)

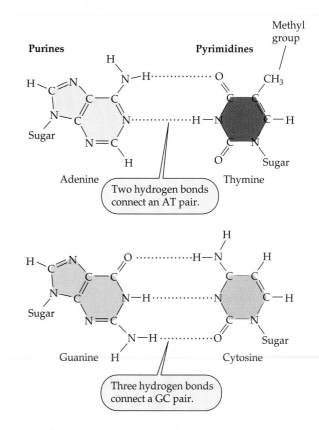

Figure 6.3 Structures of the four bases present in DNA. The bases are shown in their base-pairing configurations—that is, as if one base were on one strand of DNA and its pairing partner were on the other. Each pair consists of a larger purine (A or G) and a smaller pyrimidine (T or C), and could be flipped left for right.

BOX 6A *Insight from Cardboard Cutouts*

Before Watson and Crick assembled the components of DNA in the right way, they invented and discarded a long series of wrong structural choices. In this excerpt from *The Double Helix*, *Watson wrote about the remarkably simple events of February 1953 that followed these months of fruitless fiddling.*

When I got to our still empty office the following morning, I quickly cleared away the papers from my desk top so that I would have a large, flat surface on which to form pairs of bases held together by hydrogen bonds. Though I initially went back to my like-with-like prejudices, I saw all too well that they led nowhere. When Jerry [Donohue, an American crystallographer] came in I looked up, saw that it was not Francis, and began shifting the bases in and out of various other pairing possibilities. Suddenly I became aware that an adenine-thymine pair held together by two hydrogen bonds was identical in shape to a guanine-cytosine pair held together by at least two hydrogen bonds. All the hydrogen bonds seemed to form naturally; no fudging was required to make the two types of base pairs identical in shape. Quickly I called Jerry over to ask him whether this time he had any objections to my new base pairs.

When he said no, my morale skyrocketed, for I suspected that we now had the answer to the riddle of why the number of purine residues exactly equaled the number of pyrimidine residues. … Furthermore, the hydrogen bonding requirement meant that adenine would always pair with thymine, while guanine could pair only with cytosine. Chargaff's rules then suddenly stood out as a consequence of a double-helical structure for DNA. Even more exciting, this type of double helix suggested a replication scheme much more satisfactory than my briefly-considered like-with-like pairing. …

Upon his arrival Francis did not get more than halfway through the door before I let loose that the answer to everything was in our hands. Though as a matter of principle he maintained skepticism for a few moments, the similarly shaped A-T and G-C pairs had their expected impact. His quickly pushing the bases together in a number of different ways did not reveal any other way to satisfy Chargaff's rules.

The predominant form of DNA in living cells is a *right-handed* helix. (Almost all screws and bolts are also right-handed; they advance when twisted to the right.) The DNA molecule has two smoothly spiraling sugar-phosphate backbones outlining the helix (see Figure 2.4). The bases extend toward each other from the two backbones like the rungs of a twisted ladder, albeit a flexible, threadlike ladder. The width of the helix is constant at about 2 nm, and its length can be millions of times more than its width. (See the labyrinthine traces of DNA in Figure 2.7A.)

Complementary Bases

The point we emphasize here is not the three-dimensional measurements of the molecule but *the pairing of the bases*. The uniform width throughout the helix means that each rung of the ladder is the same size, a larger purine (A or G) being paired with a smaller pyrimidine (T or C). Chemical restraints require, further, that purine A pair only with pyrimidine T, and purine G only with pyrimidine C (Figure 6.4). Thus the sequence of bases in one strand is always **complementary** to the bases on the other strand according to the pairing rules: A opposite T (or vice versa) and G opposite C (or vice versa). The percentage of pairs that are AT or GC depends on the sequence along the length of a strand, however, and this varies from region to region, from molecule to molecule, and from organism to organism.

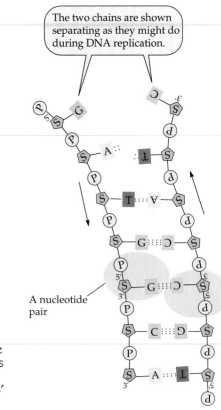

The two chains are shown separating as they might do during DNA replication.

A nucleotide pair

Figure 6.4 A portion of a double-stranded molecule of DNA showing AT and GC base-pairing joining the two strands. (The helical form of the molecule is not shown.) The order of bases on one chain is irregular, but given the bases on one strand, those on the other are determined. Note that the two strands have opposite directionality, represented here by arrows that point from the 5' to the 3' end of a given molecule of deoxyribose sugar.

The basic building block of DNA is called a **nucleotide*** and it consists of a deoxyribose sugar, a phosphate attached to one end of the sugar (the so-called 5′ carbon atom), and a base attached to the middle of the sugar (the 1′ carbon). (The other end of the sugar, the 3′ carbon, is attached to the *next* nucleotide.) The nucleotides that are incorporated into DNA are synthesized from simpler compounds in the cell. Because the four nucleotides present in DNA differ from one another only in which base they carry, the terms *nucleotide pairing* and *base pairing* express the same complementarity.

Another interesting point of DNA structure is that the two sugar-phosphate backbones of the double helix run in "opposite" directions, like two-way traffic. Specifically, the orientations of the sugars, marked by labeling the 3′ and 5′ carbons, are reversed on the two strands. Thus, at the very tip of a DNA molecule (refer to Figure 6.4), the sugar on one strand has its 5′ carbon unattached to any more nucleotides, whereas the sugar on the other strand has its 3′ carbon so exposed.

Replication

Complementary base pairing suggested to Watson and Crick a mechanism for the replication of DNA. Replication could occur by separation of the two strands, with each single (old) strand acting as a pattern (or guide or mold) for the formation of a new strand. Individual nucleotides present in the cell are captured and strung together successively, one at a time, opposite their complements in the two original (old) strands, a process resulting in two identical double helices where there was once just one.

That DNA actually replicates in the way Watson and Crick suggested was shown by ingenious experiments with the common bacterium *Escherichia coli* (more simply, *E. coli*; all of us play host to *E. coli* in our intestines) by Matthew Meselson and Franklin Stahl (1958) at the California Institute of Technology. The bacterial cells were grown for several genera-

**More precisely, the nucleotides in DNA are called* deoxyribo*nucleotides because the sugar is* deoxyribose. *Those in RNA are called* ribo*nucleotides because the sugar component is* ribose.

tions in medium with nutrients containing the heavy isotope nitrogen-15, rather than the common, lighter isotope nitrogen-14.* The purines and pyrimidines synthesized by the bacteria incorporated nitrogen-15 into their structure, and therefore the bacterial DNA became slightly heavier than if the cells had been grown in the common isotope of nitrogen.

At the start of the experiment, then, both strands of DNA in the double helix contained nitrogen-15, and the DNA was considered *heavy* (left test tube at the bottom of Figure 6.5). Next, the *E. coli* cells were grown for one generation in a medium containing only nutrients with nitrogen-14. If DNA replicates as outlined here, then each new double helix should contain one heavy strand and one light strand, thus being *intermediate* in density (middle tube). This was indeed the observed result. As further predicted, when the *E. coli* cells were allowed to undergo *two* generations of growth in nitrogen-14, half the DNA double helices were of *intermediate* density and half were *light* (right tube; see also question 3).

Bubbles and Forks

In higher organisms, the single DNA molecule that makes up each chromosome begins to replicate at

*Isotopes are variants of the same chemical element that may differ in some physical properties. For example, because it possesses an additional proton, nitrogen-15 (a rare isotope) is slightly heavier than nitrogen-14 (the common form). Some other isotopes are radioactive; for example, radioactive phosphorus-32 is sometimes used to label molecules of DNA, and sulfur-35 is sometimes used to label protein molecules.

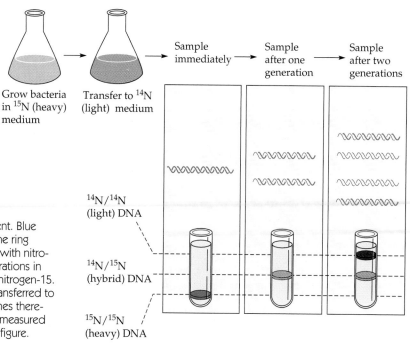

Figure 6.5 Results of the Meselson-Stahl experiment. Blue refers to a strand of DNA containing nitrogen-15 in the ring structures of the bases. Red indicatets a DNA strand with nitrogen-14. *E. coli* bacteria were grown for several generations in nutrient medium containing only the heavy isotope, nitrogen-15. At the start of the experiment, bacterial cells were transferred to medium with nitrogen-14 and sampled at various times thereafter. The densities of the DNA double helices were measured by centrifugation, as suggested at the bottom of the figure.

thousands of sites (rather than at just one site, as in *E. coli*). First, local untwistings of the two strands occur at places where replication will start (Figure 6.6). A special enzyme called **DNA helicase** helps open up the double helix, exposing the single-stranded sequences to be copied. Under an electron microscope, each local separation appears as a bubble in the otherwise still closely paired double helix.

The replication of DNA by complementary base pairing requires the presence of an enzyme called **DNA polymerase**, which travels along the old DNA strands and aids in the selection, placement, and bonding together of the new nucleotides. One of several interesting facts of biochemical life is that the new nucleotides are always added to the 3′ end of the sugar of the nucleotide already in place in the growing new strand of DNA. The two strands of the original double helix run in opposite directions; consequently, the two new strands forming at one of the ends of a replicating bubble cannot both be synthesized in the same direction. Consider one of the Y-shaped replication forks at the right side of a bubble (Figure 6.7). The upper branch of the replication fork can continuously

add nucleotides to the 3′ end, but replication on the lower branch must wait until a length of the old strand has been exposed and then proceed "backward." The new fragments are about 150 nucleotides long in mammalian cells. Eventually they are joined together by another enzyme called **DNA ligase**, thus making this new strand continuous too.

In higher organisms, DNA is replicated in preparation for each mitotic cell division and prior to the first meiotic division of gametogenesis. Accurate replication is needed for the transmission of exact genetic information from cell to daughter cells and from one generation to the next. High-fidelity reproduction is accomplished by several mechanisms, including what is called the *proofreading mechanism* of DNA polymerase. The last-attached nucleotide at the growing tip of the new strand must "settle in" and become firmly base-paired to its complementary nucleotide on the old strand before another nucleotide can be added. DNA polymerase will remove any improperly matched nucleotide and try again for an exact AT or GC fit before continuing down the line. It is estimated that only one replication mistake remains for every 10^{10} base pairings—an average of one mistake for each replication of a human cell with about 6×10^9 nucleotide pairs. For comparison, just one typographical error in this book of about 3×10^6 characters would constitute an error rate 2,000 times greater than that occurring in DNA replication.

Gene Expression: DNA ⇒ RNA ⇒ Protein

Within cells, information encoded in lengths of DNA—a gene—helps to direct the development and maintenance of the body. Decoding of the information proceeds in two steps. First, as shown in Figure 6.8, base sequences in DNA are **transcribed** into complementary sequences of **RNA, ribonucleic acid**. Note that DNA is not turned into RNA; rather, DNA is like a recipe: It provides information that the

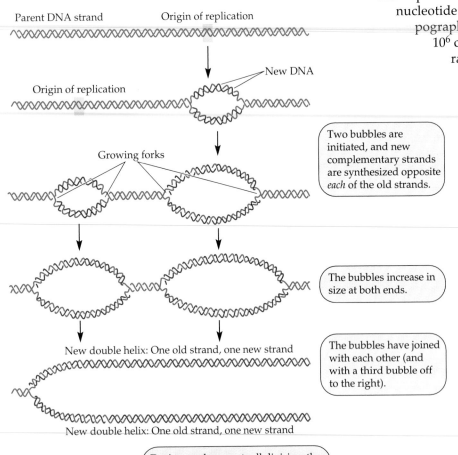

Parent DNA strand   Origin of replication

New DNA

Origin of replication

Two bubbles are initiated, and new complementary strands are synthesized opposite *each* of the old strands.

Growing forks

The bubbles increase in size at both ends.

The bubbles have joined with each other (and with a third bubble off to the right).

New double helix: One old strand, one new strand

New double helix: One old strand, one new strand

During a subsequent cell division, the newly formed double helices segregate to the two daughter cells.

Figure 6.6   DNA replication arising through local separations of the double helix and progressive synthesis. Newly synthesized DNA strands are shown in red.

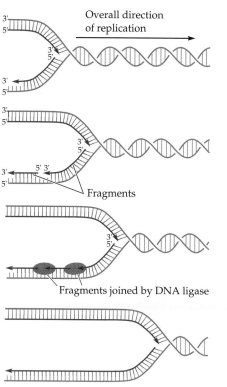

Overall direction of replication

Fragments

Fragments joined by DNA ligase

Figure 6.7 Events at a replicating fork. Newly synthesized DNA strands are shown in red. DNA polymerase adds new nucleotides on only the 3' end of the sugar already in place. Because the two strands run in opposite directions (see Figure 6.4), replication of one strand—here, the lower one—occurs in discontinuous steps. The double helix must open for a length before synthesis can start at the point of the fork and proceed "backward" in the proper 5' to 3' direction. These segments butt up against each other, and their sugar-phosphate backbones are joined into a continuous strand by DNA ligase.

The terms *transcription* and *translation*, the two steps in **gene expression**, are well chosen. The former term means a transformation within the *same* language, such as from oral to written English, or from a string of deoxyribonucleotides to one of ribonucleotides (but still a language of nucleotides). The latter term describes a more profound change: from one language to another (say, from English to French), or from a string of nucleotides to a string of amino acids.

The process of transcription is similar to DNA replication, except that the complementary molecule that is formed is RNA, containing the sugar ribose (rather than deoxyribose; see Figure 6.2) and the base uracil (**U**) instead of thymine (Figure 6.9). Transcription is controlled by the enzyme **RNA polymerase**, which links together ribonucleotides one after another into a

cell uses to put together RNA. These molecules of RNA are made at many chromosomal sites (i.e., at the genes) and then migrate across the nuclear membrane into the cytoplasm. In the second step, occurring at ribosomes in the cytoplasm, the information now encoded in the base sequence of RNA molecules is **translated** into a string of amino acids called a **polypeptide**. Often, one polypeptide by itself constitutes a protein, but sometimes several polypeptide chains are linked together to make a functional protein. Again we emphasize that an RNA molecule is not turned into a polypeptide; rather, RNA provides information that the cell uses to synthesize a polypeptide. Many of the resultant proteins are the enzymes required for cell metabolism; other proteins are hormones, antibodies, structural elements in bones, nails, and hair, or other vital components of our bodies.

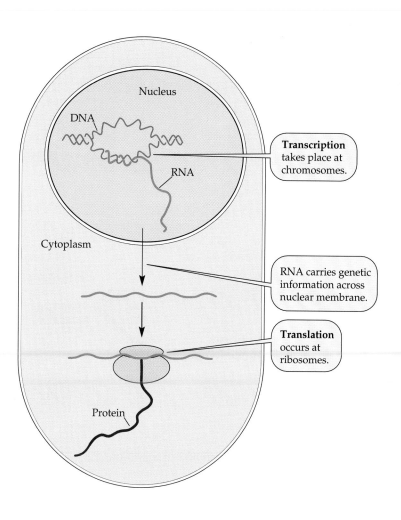

Transcription takes place at chromosomes.

RNA carries genetic information across nuclear membrane.

Translation occurs at ribosomes.

Figure 6.8 Decoding a gene. In a two-step process, a particular string of nucleotides in DNA directs a cell to assemble a particular string of amino acids into a polypeptide. The RNA molecules are intermediaries, carrying copies of genetic information from the nucleus to the cytoplasm.

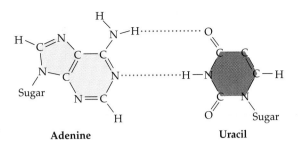

Adenine   **Uracil**

Figure 6.9 The pyrimidine base uracil, present in RNA. During transcription and translation, uracil pairs with adenine, as indicated. Thus, uracil plays the same role in RNA as the chemically similar thymine does in DNA. Comparison with Figure 6.3 shows that uracil lacks only the methyl group (CH_3) present in thymine.

sequence complementary to the sequence of bases along one of the strands of DNA.

The process is illustrated in Figure 6.10. The place where RNA polymerase first attaches is a specific sequence of bases called the **promoter**. Usually, each gene has its own promoter. Although promoter sequences vary from gene to gene and from organism to organism, they all contain certain sequences in common: In eukaryotes, a seven-base sequence of mostly T's and A's (the so-called TATA box) is part of all promoters. Transcription begins at a spot after the promoter, with new ribonucleotides being added in turn to the 3' end of the preceding ribonucleotides (following the same direction of synthesis used in DNA replication). The RNA polymerase moves off the DNA mol-

ecule at a certain **transcription termination sequence**, completing the transcription of a gene into a complementary RNA copy. Each human chromosome includes a single, very long molecule of DNA with thousands of genes along its length, usually separated from each other by sections of DNA that are not expressed.

The question of which strand of a double helix acts as the guide for the formation of complementary RNA is crucial, since one strand is obviously different from the other. For a particular gene, the strand that is actually transcribed is called the **template strand**; the other—the **nontemplate strand**—is not used in transcription (although both strands are used equally during DNA replication). The strand of the DNA double helix that is actually transcribed changes from place to place along one chromosome; that is, one of the strands of DNA (followed end to end along a chromosome) is the template for some genes and the nontemplate for others.

Not all of our 80,000 genes* are transcribed all the time. Regulation of gene expression, ensuring that the right proteins are made at the right time in the right place in the right amount, is essential to proper growth and maintenance of the body. Control is often mediated by regulatory proteins called **transcription factors**; they bind to DNA at promoter sequences and other checkpoints to determine which genes are "turned on" and which are "turned off." The control-

*The number 80,000 is a guess, as discussed in Box 9C.

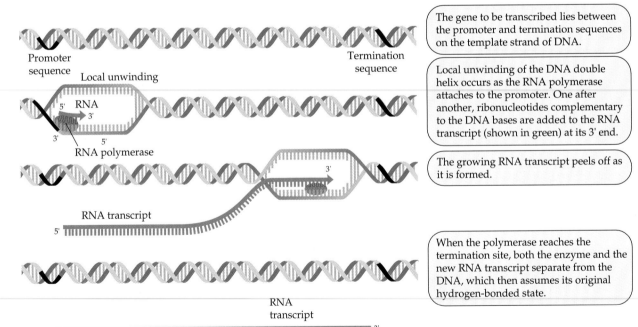

The gene to be transcribed lies between the promoter and termination sequences on the template strand of DNA.

Local unwinding of the DNA double helix occurs as the RNA polymerase attaches to the promoter. One after another, ribonucleotides complementary to the DNA bases are added to the RNA transcript (shown in green) at its 3' end.

The growing RNA transcript peels off as it is formed.

When the polymerase reaches the termination site, both the enzyme and the new RNA transcript separate from the DNA, which then assumes its original hydrogen-bonded state.

Figure 6.10 The process of transcription.

ling mechanisms are intricate and interrelated (and we omit the many details here). Some genes are transcribed only during a brief period in the life cycle of an organism; others, such as those needed in energy production, are operating all the time. Animals that undergo dramatic metamorphosis illustrate the important role of gene regulation: A caterpillar uses many genes the resulting butterfly does not need, and vice versa; yet exactly the same genes are present in both.

Types of RNA

There are three main types of RNA: *messenger* RNA (mRNA), *transfer* RNA (tRNA), and *ribosomal* RNA (rRNA). All of the RNAs are transcribed from genes and then enter the cytoplasm, where they play specific roles during translation (Figure 6.11).

Messenger RNA (mRNA) plays the key informational role in protein synthesis. Only mRNA is actually translated into protein chains, so mRNA is the RNA referred to in the shorthand scheme DNA ⇒ RNA ⇒ protein. Indeed, a gene is often defined as the sequence of nucleotides that dictates the order of amino acids in a polypeptide or protein via an mRNA intermediate. Any one mRNA molecule may be translated many times, although it is soon degraded.

We point out here (and expand upon the idea later) that genetic information is expressed in a *triplet* code; that is, three adjacent bases on the template strand of DNA, via three adjacent bases on the mRNA intermediate, stand for one of the 20 amino acids found in proteins. During translation, each successive group of three nucleotides in mRNA, called a **codon,** represents the code for one amino acid. But neither DNA nor the intermediary RNA are actually marked off into threes by any special chemical structure. Rather, *successive groups of three bases* are interpreted by the cellular chemistry as units of information.

There are as many codons strung together in a particular mRNA as there are amino acids in the corresponding protein. Because most polypeptides are at least 100 amino acids long, most mRNA molecules are at least 300 nucleotides long. (Actually, as we will see later, these statements are oversimplified.) Because there are tens of thousands of different polypeptides, there are tens of thousands of different mRNA molecules.

Transfer RNA (tRNA) molecules act as adapters, binding to specific amino acids in the cytoplasm and bringing them to appropriate codons along the length of an mRNA molecule. They accomplish this task through a double specificity. At the end of a given tRNA molecule, a specific enzyme attaches one and only one of the 20 amino acids. The other end of this tRNA contains a sequence of three bases called the **anticodon,** which recognizes a specific codon of mRNA by base pairing. For example, the tRNA anticodon CCC recognizes and binds to the mRNA codon GGG by the usual rules of base pairing. The structure of the tRNA molecule with the anticodon CCC allows this tRNA to attach to only one specific amino acid (proline, in this case). Thus, the anticodon of a given tRNA molecule and the amino acid that it picks up are "matched" to each other (Mlot 1989).

There are several dozen different tRNA molecules, one or a few different types for each amino acid. Each tRNA molecule is about 80 nucleotides long and is coiled and folded into an intricate three-dimensional structure that helps it carry out its function. Although they help in translation, the tRNA molecules are not themselves translated. Thus, unlike the genes that encode mRNA, the genes that encode tRNA have no protein product.

Ribosomal RNA (rRNA), a third type of RNA, helps form the ribosomes, the "workbenches" on which protein synthesis occurs. The ribosomes in eukaryotes are composed of four types of rRNA and about 80 types of protein molecules arranged in a complex three-dimensional structure (Lake 1981). Some of the rRNA molecules are transcribed off special chromosomal regions called *nucleolus organizers* (briefly noted in Chapter 2). There are ten of these regions in the human genome, one on each short arm of the five pairs of acrocentric chromosomes. These regions have multiple copies—perhaps hundreds—of rRNA genes. Like the tRNA genes, the rRNA genes are transcribed but not translated.

The ribosomes come apart into two subunits—one small and one large—enabling

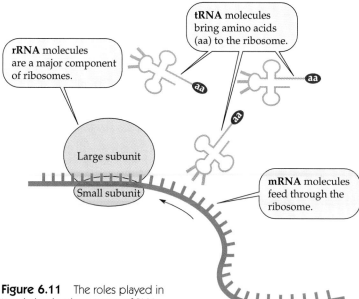

Figure 6.11 The roles played in translation by three types of RNA.

rRNA molecules are a major component of ribosomes.

tRNA molecules bring amino acids (aa) to the ribosome.

Large subunit

Small subunit

mRNA molecules feed through the ribosome.

them to clamp onto a molecule of mRNA, which then feeds through the ribosome during protein synthesis. The amino acids needed for incorporation into the protein chains are transported to the ribosomes by the tRNA molecules. Thus, all three types of RNA are brought together at the ribosomes, as shown in Figure 6.11.

Proteins

During the process of translation, anywhere from a dozen to a thousand or more amino acids are strung together to form unbranched polypeptide chains. One or more of these chains—coiled, folded, intertwined, cross-linked—constitute a protein. The marvelous diversity of proteins is suggested by the following rather random list: muscles, spider webs, tiger claws, snake venoms, eye lenses, blood clots, and catalysts for DNA replication (Table 6.1). The properties of a cell are often determined by the proteins it makes. For example, red blood cells manufacture large amounts of hemoglobin; certain white blood cells make antibodies; liver cells make an array of enzymes; and hair follicle cells make keratin. The human body contains more than 10,000 different proteins, each with a characteristic structure and function and each assembled from the same set of 20 amino acids.

Amino acids all have the same general structure: a central carbon atom (C) bonded to four groupings, two of which—the basic positively charged *amino group* (NH_3^+) and the acid negatively charged *carboxyl group* (COO^-)—give the class its name. A third group is a single hydrogen atom (H). The fourth group, the *side chain* (Figure 6.12), accounts for the differences in the properties of the various amino acids. The simplest amino acid is *glycine*, in which the side chain is another hydrogen atom. In *alanine*, the side chain is a methyl group (CH_3). In *phenylalanine*, a six-membered ring of carbons (a phenyl or benzene group) substitutes for one of the three hydrogens of the methyl group of alanine. A few more structures are shown in Figure 6.12, and all 20 amino acids are listed along with their common abbreviations in Table 6.2.

Amino acids in polypeptide chains are joined to one another by the so-called *peptide bond* between the carboxyl group of one amino acid and the amino group of the next. The successive bonding of many amino acids into a polypeptide chain leaves an amino group free on one end of the molecule and a carboxyl group free on the other.

The complete sequence of amino acids in a small two-chain protein, human insulin, is illustrated in Figure 6.13. Note that bonding between the sulfur atoms of some of the cysteine side chains (called disulfide bonds) holds the two chains together. (In addition, there is a disulfide bond between the cysteines at positions 6 and 11 of the A chain.) Not all proteins

---

TABLE 6.1 Representative proteins

Name of protein	Type of protein[a]	Specific function	Polypeptide chains
DNA polymerase	Intracellular enzyme	Replicates and repairs DNA	One chain of about 1,000 amino acids
Trypsin	Digestive enzyme	Cleaves proteins into smaller polypeptides in small intestine	One chain of 223 amino acids (bovine)
Insulin	Hormone	Aids in moving glucose from the bloodstream into cells	Two chains of 21 and 30 amino acids
Hemoglobin	Transport protein	Carries oxygen in red blood cells from lungs to tissues	Four chains: two with 141 and two with 146 amino acids, plus heme (nonprotein part)
Gamma globulin	Protective protein	Antibody molecules, which react with and incapacitate foreign substances	Four chains: two with 214 and two with 446 amino acids
Myosin	Contractile protein	Sliding arrays of myosin and actin molecules cause muscles to contract	Six chains: two fiberlike chains of about 1,800 amino acids each, and four small globular chains
Collagens	Structural element	Connective tissue in cartilage, tendons, bones, arteries, etc. (one-third of body protein)	Three chains of about 1,000 amino acids each woven into various triple helices
Keratin	Structural element	Main fibrous component of hair, skin, nails (and wool, silk, horns, hooves, feathers, turtle shells)	Complex helical or zig-zag structures of thousands of amino acids each

[a]The first type, intracellular enzymes, is the largest class of proteins. The first five types are largely globular in shape and water-soluble. The last three types are largely fiberlike and water-insoluble.

(A) Uncharged side groups

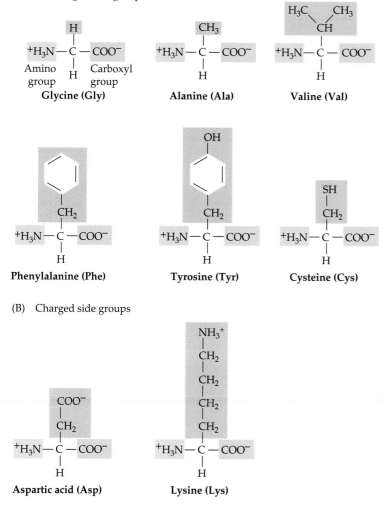

Figure 6.12 Representative amino acids found in proteins. The amino acids are shown in their charged state as they would be in cells. The portions of the molecule that are not in purple are common to all the amino acids.

Translation

The three-dimensional structure of a protein ultimately comes down to its amino acid order, which is determined by the order of codons in mRNA (which in turn is determined by the order of corresponding DNA triplets). As noted earlier, translation occurs on the ribosomes with the aid of tRNAs, each of which has been loaded with a specific amino acid. In Figure 6.14, we see how the sixth amino acid of a growing polypeptide chain becomes bonded to the first five, which are already linked together. The large subunit of the ribosome has two pockets for positioning tRNA

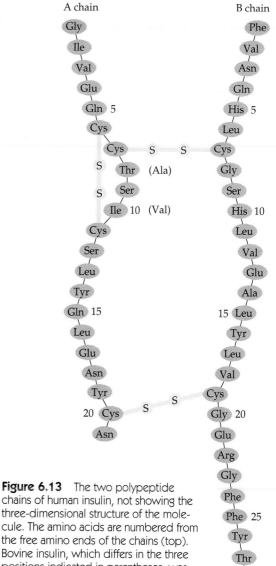

Figure 6.13 The two polypeptide chains of human insulin, not showing the three-dimensional structure of the molecule. The amino acids are numbered from the free amino ends of the chains (top). Bovine insulin, which differs in the three positions indicated in parentheses, was the first protein to be sequenced; this was accomplished by Cambridge University chemist Frederick Sanger, a Nobel laureate in 1958.

with multiple chains are held together this way, however. The single insulin gene actually encodes one chain of 109 amino acids. This polypeptide is subsequently cut into four pieces, with two pieces assembled into the form shown in Figure 6.13 and two pieces discarded.

The three-dimensional shape of a protein is crucial to its biological activity. Most proteins are globular, in part or in whole, because the polypeptide chains are compactly folded and looped, often in complex ways. In enzyme molecules, some groups of amino acids fold in from the surface to form a precisely shaped cavity called the *active site*, into which fits the substance or substances that the enzyme acts on. Critical sites of function are also present in other proteins, such as the oxygen-binding sites of hemoglobin or the antigen-binding sites of antibody molecules.

TABLE 6.2 The common abbreviations for the amino acids found in proteins	
Amino acid	**Abbreviation**
Alanine	Ala
Arginine*	Arg
Asparagine	Asn
Aspartic acid	Asp
Cysteine	Cys
Glutamine	Gln
Glutamic acid	Glu
Glycine	Gly
Histidine*	His
Isoleucine	Ile
Leucine*	Leu
Lysine*	Lys
Methionine*	Met
Phenylalanine*	Phe
Proline	Pro
Serine	Ser
Threonine*	Thr
Tryptophan*	Trp
Tyrosine	Tyr
Valine*	Val

*The so-called essential amino acids. We cannot synthesize these nine amino acids for ourselves but must obtain them from the proteins in the food we eat.

position (with its amino group facing to the left and its carboxyl to the right).

In Figure 6.14B, the tRNA-leucine unit is stationed in the right pocket. The leucine (i.e., the sixth amino acid) becomes attached to the fifth amino acid, which was released from its tRNA in the left pocket. Thus, the length of the polypeptide is being increased by one amino acid.

In Figure 6.14C, the mRNA molecule has moved one codon to the left, and the tRNA-leucine complex has also moved left to the next pocket. We are back to a diagram like that in Figure 6.14A, but now one amino acid has been added to the polypeptide. The right pocket is free to accept the seventh tRNA and its attached amino acid. The fifth tRNA molecule, having

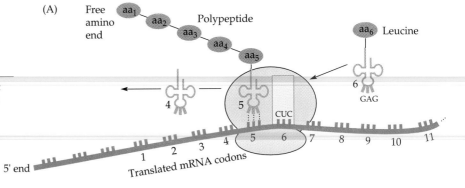

molecules. A pocket will fit only the tRNA whose anticodon complements the codon of mRNA that forms the lower boundary of the pocket. For example, if the sixth mRNA codon is CUC, then by the rules of base pairing, only that tRNA with the anticodon GAG will fit the pocket (Figure 6.14A). The tRNA molecule that has the anticodon GAG picks up the amino acid leucine. In this way, whenever the codon CUC passes through the ribosome, leucine is brought into a nearby

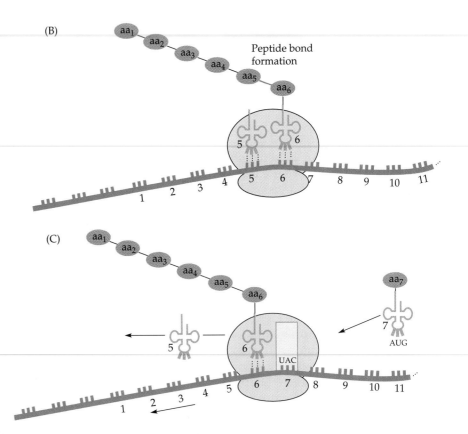

Figure 6.14 The process of translation. For clarity, the bases in mRNA are shown crowded together into three-base codons; in reality, the bases are equally spaced along the molecule. The 5' end of the messenger feeds through the ribosome first, and the mRNA codon, labeled 1, leads to the placement of the amino acid that ends up with a free amino group.

contributed its amino acid to the polypeptide, is free to bind another molecule of the same amino acid to be incorporated later (if called for) as the mRNA feeds through the ribosome.

The mRNA molecule contains not only the codons for the polypeptide but also certain head (5′) and tail (3′) sequences that are not translated into amino acids. In particular, the first translated codon (some distance in from the 5′ tip of the molecule) always calls for the amino acid methionine. (Later, this first amino acid and sometimes additional amino acids may be cut off from the polypeptide.) Following the last translated codon is a so-called *stop codon*, discussed shortly. Thus, the amino acids that are hooked together to make a polypeptide start with the first codon for methionine and continue up to (but not including) the first stop codon. The 3′ tip of the mRNA molecule is some distance beyond the stop codon.

Note that the processes of transcription and translation both depend on base pairing of T with A (or U with A) and G with C (Figure 6.15). To say that a gene "codes for" a protein means that successive DNA triplets provide the information for the assembly of successive amino acids through two rounds of base pairing. In bacteria, in which there is no separation of nucleus from cytoplasm, one end of the mRNA molecule (the 3′ end) is still being transcribed while the other end (5′) is already being translated! Miller (1973) has published some remarkable electron microscope photographs of bacterial genes being transcribed as the mRNA transcripts are being translated.

Three-Letter Words

By 1960, molecular biologists had established the importance of base pairing in the reactions of DNA and RNA. But they were pessimistic about quickly cracking the code—that is, discovering *which* specific sequences of nucleotides caused the assembly of *which* specific amino acids. Yet by 1965, several groups of investigators had ingeniously worked out the complete coding dictionary.

Because proteins are assembled from 20 amino acids, a string of three nucleotides was considered the most reasonable size for a coding unit: *One* nucleotide can specify only 4 items (Table 6.3A); *two* nucleotides in sequence can specify only 16 items (Table 6.3B); *three* nucleotides, however, carry more than enough information, 64 items in all (Table 6.3C), to code for all 20 amino acids. (Note that these numbers are successive powers of 4: 4^1, 4^2, and 4^3.) Using a type of muta-

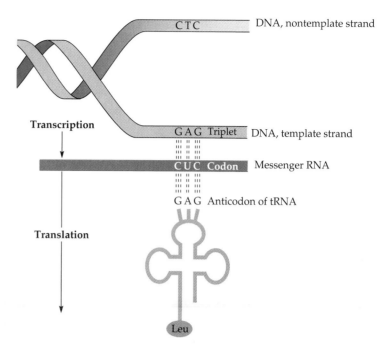

Figure 6.15 Base pairing during transcription and translation. The DNA triplet GAG in a polynucleotide provides the information that finally leads to the incorporation of the amino acid leucine into a polypeptide.

tion in which a base is added to (or deleted from) a gene, Crick and his associates were able to show that indeed *three* was the correct coding unit.

Cracking the Code

Early experimenters trying to determine which specific group of three bases coded for which amino acid made their own *synthetic messenger RNA*. These informational molecules were added to a test tube containing all the ingredients that a cell uses to make proteins:

- All the individual amino acids
- All the transfer RNA molecules
- Ribosomes extracted from *E. coli*
- ATP and other energy-rich compounds
- A batch of appropriate enzymes

Remarkably, these in vitro systems synthesized polypeptides. For example, when the artificial mRNA was polyuracil, which has a monotonous sequence of one base,

. . . UUUUUUUUUUUU . . .

the test tube was found after a time to contain a polypeptide containing only phenylalanine:

. . . Phe–Phe–Phe–Phe . . .

TABLE 6.3 One-, two-, and three-letter code words that can be formed from an alphabet with the four letters, A, G, T, and C

(A) The 4 different one-letter words

A	G	T	C

(B) The 16 different two-letter words

AA	GA	TA	CA
AG	GG	TG	CG
AT	GT	TT	CT
AC	GC	TC	CC

(C) The 64 different three-letter words

AAA	AGA	ATA	ACA
AAG	AGG	ATG	ACG
AAT	AGT	ATT	ACT
AAC	AGC	ATC	ACC
GAA	GGA	GTA	GCA
GAG	GGG	GTG	GCG
GAT	GGT	GTT	GCT
GAC	GGC	GTC	GCC
TAA	TGA	TTA	TCA
TAG	TGG	TTG	TCG
TAT	TGT	TTT	TCT
TAC	TGC	TTC	TCC
CAA	CGA	CTA	CCA
CAG	CGG	CTG	CCG
CAT	CGT	CTT	CCT
CAC	CGC	CTC	CCC

The conclusion was clear: UUU in messenger mRNA (or, by inference, AAA in DNA) was the genetic word for phenylalanine.

As biochemists developed more sophisticated means for hooking together different ribonucleotides in specific orders, the nucleic acid vocabulary was extended. Here are two examples (see also question 11).

Example 1. When synthetic mRNA with a repeating set of two nucleotides,

... UGUGUGUGUGUG ...

was added to the protein-synthesizing system, the polypeptide

... Val–Cys–Val–Cys ...

was made. Thus, valine is coded for by UGU and cysteine by GUG, or vice versa. Because the in vitro system lacks a proper signal for initiating the translation process, it could start at the first or second or third base. These different starting places are said to produce different **reading frames** for the mRNA. In this particular example, however, the different reading frames give the same polypeptide product, an alter-

nating string of valines and cysteines (although the first amino acid in the string does depend on the reading frame).

Example 2. The synthetic mRNA

... UUCUUCUUCUUC ...

(a repeating trinucleotide) made *three different* polypeptides:

... Phe–Phe–Phe–Phe ...

... Ser–Ser–Ser–Ser ...

... Leu–Leu–Leu–Leu ...

The three reading frames for the mRNA lead to repeated codons UUC or UCU or CUU; thus, these three codons stand for the three amino acids, although which stands for which is not clear from this experiment by itself. (You can confirm that a coding unit of two bases or four bases would have generated molecules consisting of three different amino acids in one polypeptide.)

On the basis of similar experiments, especially those using RNA molecules only three nucleotides long, researchers gave all 64 possible codons in mRNA the unambiguous assignments shown in Table 6.4. Crick (1988) provides an engaging account of this exciting period of biochemical sleuthing.

Properties of the Code

Examination of Table 6.4 reveals some interesting properties of the genetic code.

1. *Several codons often stand for the same amino acid.* This so-called **code degeneracy** might be expected when 64 codons are available to specify just 20 amino acids. The most striking examples are the six codons for leucine, the six for serine, and the six for arginine. Fourfold degeneracy is seen for several other amino acids (e.g., valine); here the third base of the codons can be ignored, effectively reducing the genetic dictionary to two-letter words. Twofold degeneracy is seen for many others (e.g., phenylalanine); here the third RNA base is either of the two pyrimidines (U or C) or either of the two purines (A or G). Thus, code degeneracy usually involves the third base. The number of synonymous codons for an amino acid (one, two, three, four, or six) is roughly proportional to the actual frequency of that amino acid in proteins. For example, tryptophan, an infrequent amino acid, has a single codon; serine and leucine, very frequent amino acids, have six codons.

Although the code is degenerate, the genetic language is not imprecise: A specific DNA triplet, through its mRNA codon, will direct the assembly of a specific amino acid.

2. *Three of the 64 codons do not specify any amino acid.* While deciphering the code, investigators found that

the RNA codons UAA, UAG, and UGA failed to incorporate any amino acid. These sequences are sometimes called *nonsense codons* because they do not code for amino acids; however, the term **stop codon** (or *chain termination codon*) is more accurate, because UAA, UAG, and UGA stop the translation process, releasing the now-completed polypeptide from the ribosome. The laboratory evidence that translation ends this way came from mutant strains of microorganisms in which the synthesis of a polypeptide stopped prematurely. This could happen, for example, if the third base in a tyrosine mRNA codon near, say, the middle of a gene transcript changed to A or G. (Verify this logic by examining Table 6.4.) Also note that a stop codon only affects translation, not transcription.

3. *The code is nearly universal.* Although the code was first deciphered using synthetic RNA molecules and translation components from the bacterium *E. coli*, further tests showed that almost all organisms use the same array of codons. Biochemists can obtain specific genes or messenger RNA molecules in large enough quantities to determine the ordering of the bases; matching the codons with the amino acids in the resultant proteins confirms that the genes of viruses, bacteria, yeast, maple trees, earthworms, butterflies, dogs, humans, and most other organisms rely on the same genetic blueprint.

Furthermore, protein synthesis can run smoothly even with mixed components. Toad oocytes injected with the messenger RNA for rabbit hemoglobin synthesize genuine rabbit hemoglobin; when injected with bee venom mRNA, they synthesize bee venom. Researchers have even made a remarkable strain of tobacco that glows in the dark by transferring to the tobacco the firefly gene that encodes the luminescence enzyme *luciferase* (Root 1988). Pigs injected as embryos with human genes can sometimes make human hemoglobin and transmit this property to their offspring (Hilts 1991). As described in Chapter 8, bacterial cells can also replicate, transcribe, and translate human genes that are incorporated into their DNA. At the biochemical level, basic processes are often the same in many different species.

Scientists have found several exceptions to the codon assignments in Table 6.4. For example, in all mitochondrial DNAs that have been studied, the mRNA codon UGA encodes

tryptophan rather than chain termination. A few other exceptions are also known. The near universality of the code, however, suggests that it evolved just once in ancient life-forms, perhaps 3 billion years ago. Possibly the code began as a sequence of two nucleotides, because some proteins in very primitive bacterial species have fewer than 16 amino acids. But how each of the 20 amino acids in the current coding dictionary came to be specified by a particular nucleotide triplet is not known.

Gene Organization

The preceding sketch gave a short account of DNA construction, replication, and expression. In this section, we add a few details about the way base sequences are arranged along the DNA molecule, including some mysteries about the nature and operation of the genetic material. (The way DNA is packaged into chromosomes was presented in Chapter 2.)

Exons and Introns

In the years following the 1953 discovery of DNA structure by Watson and Crick, geneticists developed the view that the DNA from a chromosome consists of a succession of thousands of genes. Each gene was seen as an unbroken string of nucleotides separated, perhaps, by "spacer" DNA or control regions that allowed polymerases to gain a foothold on the molecule. For most of the genes of higher organisms, however, this portrait turns out *not* to be entirely true.

Observations leading to a more complex view included electron micrographs of the template strand of

TABLE 6.4   The genetic code displayed as mRNA codons

FIRST BASE[a]	SECOND BASE				THIRD BASE
	U	**C**	**A**	**G**	
U	UUU ⎫ Phe UUC ⎭ UUA ⎫ Leu UUG ⎭	UCU ⎫ UCC ⎪ Ser UCA ⎪ UCG ⎭	UAU ⎫ Tyr UAC ⎭ UAA ⎫ Stop UAG ⎭	UGU ⎫ Cys UGC ⎭ UGA Stop UGG Trp	U C A G
C	CUU ⎫ CUC ⎪ Leu CUA ⎪ CUG ⎭	CCU ⎫ CCC ⎪ Pro CCA ⎪ CCG ⎭	CAU ⎫ His CAC ⎭ CAA ⎫ Gln CAG ⎭	CGU ⎫ CGC ⎪ Arg CGA ⎪ CGG ⎭	U C A G
A	AUU ⎫ AUC ⎪ Ile AUA ⎭ AUG Met	ACU ⎫ ACC ⎪ Thr ACA ⎪ ACG ⎭	AAU ⎫ Asn AAC ⎭ AAA ⎫ Lys AAG ⎭	AGU ⎫ Ser AGC ⎭ AGA ⎫ Arg AGG ⎭	U C A G
G	GUU ⎫ GUC ⎪ Val GUA ⎪ GUG ⎭	GCU ⎫ GCC ⎪ Ala GCA ⎪ GCG ⎭	GAU ⎫ Asp GAC ⎭ GAA ⎫ Glu GAG ⎭	GGU ⎫ GGC ⎪ Gly GGA ⎪ GGG ⎭	U C A G

[a]By convention, the first base is the one on the 5' side of the mRNA codon.

a gene (isolated from the nucleus) joined with its complementary messenger RNA (isolated from the cytoplasm). These threads are hybrid molecules in which one strand is DNA and the other RNA. The expected picture would have been a simple double-stranded structure throughout its length (Figure 6.16, top), because DNA, base after base, would have paired exactly with the complementary mRNA, base after base.

In fact, what is often seen are short double-stranded DNA-RNA regions separated from one another by single-stranded loops of unpaired DNA (Figure 6.16, bottom). As expected, the paired regions are found to consist of DNA segments hybridized with their mRNA product. These are the parts of the gene, called **exons** or **expressed sequences**, that actually code for a polypeptide. The unpaired loops of DNA, called **introns** or **intervening sequences**, consist of segments of a gene that appear to have no corresponding mRNA product. The introns are often many and long: In the gene seen at the bottom of Figure 6.16, eight exons (labeled 1–8) coding for successive segments of the polypeptide are separated from each other by seven introns.

What an extraordinary surprise to discover that some nucleotide sections within the boundaries of a gene have no complement in mature mRNA and therefore code for no amino acids in the polypeptide product. An analogy would be for readers of this book to suddenly confront scattered discourses on music or sports, or perhaps just pages of nonsense. The discovery of introns was made in 1977 by Phillip Sharp of the Massachusetts Institute of Technology and Richard Roberts, now of New England Biolabs (a biotechnology company). For this work they were awarded Nobel prizes in 1993 (Altman 1993).

Investigators have discovered that most genes of vertebrates are interrupted by one or more introns—segments that may be much longer than the coding regions of the gene they interrupt. Among human genes, the one that codes for clotting factor VIII (and is mutated in persons with hemophilia) is an extreme (but not the most extreme) example: There are 26 exons coding for about 2,000 amino acids but constituting only 4% of the total length of the gene. Ninety-six percent of the length of the gene is not involved in coding for a polypeptide!

The genes coding for the polypeptides of hemoglobin provide other well-studied examples. The protein part of the molecule includes two α polypeptide chains and two β polypeptide chains. Each type of chain is encoded by a different gene, and each gene has three exons separated by two introns. The gene that codes for

the β polypeptide chain, for example, contains 1,606 bases. The initial RNA transcript of the human β-globin gene (Figure 6.17) also contains 1,606 nucleotides, an exact complement of the entire gene. But before it leaves the nucleus, the 130 bases complementary to intron 1 and the 850 bases complementary to intron 2 are *cut out* of this RNA molecule. The splicing together of the cut ends of successive *exons*, called **RNA splicing**, must be done with pinpoint precision to preserve the step-by-step progression of three-base codons in the different segments. Specific base sequences at the two ends of an intron (GU at the beginning and AG at the end) and other signals mark the region to be cut out. Mutations within introns sometimes cause RNA splicing to be inaccurate, leading to genetic disease (Chapter 7).

Such alterations to RNA molecules—after transcription but before translation—are collectively called **mRNA processing.** The removal of introns and the joining together of adjacent exons constitute just one

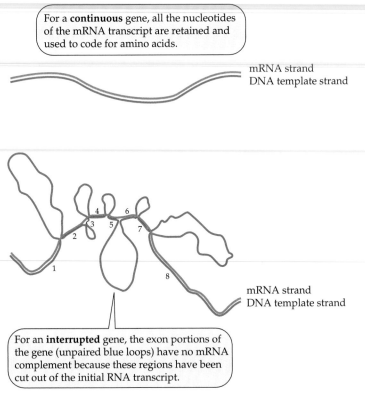

For a **continuous** gene, all the nucleotides of the mRNA transcript are retained and used to code for amino acids.

mRNA strand
DNA template strand

mRNA strand
DNA template strand

For an **interrupted** gene, the exon portions of the gene (unpaired blue loops) have no mRNA complement because these regions have been cut out of the initial RNA transcript.

Figure 6.16 The results of hybridizing the DNA template strand of a gene (blue) with its mature messenger RNA (green). Continuous genes (typical of prokaryotes) are compared with interrupted genes (typical of eukaryotes). Double-stranded regions (blue/green) result from complementary base pairing between DNA and RNA. Single-stranded segments (blue strand only) result whenever DNA has no mRNA complement. The bottom drawing, based on the chicken gene that codes for the egg-white protein ovalbumin (Chambon 1981), shows eight exons (numbered) separated by seven looping introns.

part of the process, however. Prospective mRNA is modified in two other ways before it leaves the nucleus (Figure 6.18). After transcription, some nucleotides are cut off from the 3′ end of the molecule, and a string of 50 to 200 adenine nucleotides is added; this *poly(A) tail* may help in stabilizing the RNA molecule, but its function is not well understood. At the 5′ end of the molecule, a *cap* consisting of one special nucleotide (a modified G nucleotide in reverse orientation) is added; its function is to attach the mRNA to the ribosome. The term *messenger RNA* is usually reserved for the finished product that attaches to ribosomes, that is, the sequence of codons specifying amino acids and some leader and trailing sequences.

Gene Families

Within an individual, there are several different but similar genes that code for various hemoglobin chains. Apparently, a gene present in a human ancestor millions of years ago was duplicated, so that it was present twice rather than once. Over successive generations, the duplicated loci evolved independently of each other to produce genes that code for variants of a single polypeptide. Further replications of similar loci can lead to a family of many related genes. The members of such a **gene family** may be clustered on one chromosome, or (if pieces of chromosomes break apart and rejoin) they may be scattered on different chromosomes. Known gene families include those that code for histones, antibodies, and ribosomal RNA components, but one of the best understood is the family of genes that code for the β and β-like polypeptides of hemoglobin.

Hemoglobin is the red pigment of blood that binds oxygen and transports it from the lungs to all parts of the body. The oxygen that is delivered to tissues is es-

sential for cell respiration, the process by which the chemical energy of food is converted to the high-energy compound ATP (Chapter 2). Each oxygen molecule is associated with an iron atom contained within a nonprotein *heme* group of hemoglobin. Four heme groups are pocketed in folds of the four polypeptides, the *globin* part of the molecule (Figure 6.19).

Humans produce several kinds of hemoglobin. In adults, about 98% of the hemoglobin is characterized as $\alpha_2\beta_2$: Two of the four polypeptides are identical *alpha* (α) *chains* with 141 amino acids each, and two are identical *beta* (β) *chains* with 146 amino acids each. The α and β polypeptides are encoded by genes on different chromosomes.

About 2% of hemoglobin in adults, however, consists of two α chains and two chains similar to, but not the same as, the β chains. These are the *delta* (δ) *chains*, in which just 10 of the 146 amino acids differ from those in β chains. This hemoglobin is therefore characterized as $\alpha_2\delta_2$. A very small portion of adult hemoglobin (under 1%) has two α chains and two *gamma* (γ) *chains*, in which 39 of 146 amino acids differ from those of β. There are two different types of γ chains, one with the amino acid glycine in position 136 (called $^G\gamma$ or G-gamma chains) and one with alanine (called $^A\gamma$ or A-gamma chains). During most of fetal life, hemoglobin with two γ chains is the predominant form of hemoglobin. *Early* embryos, however, possess yet another β-like polypeptide, the *epsilon* (ε) *chain*, which differs from the β chain in 35 amino acids.

To summarize, all hemoglobin has four polypeptide chains, two α chains and two β or β-like chains. The β and β-like chains of hemoglobin are of *five* types—beta, delta, two types of gamma, and epsilon—whose quantities depend on the stage of development. Each polypeptide is coded for by a different gene, and the

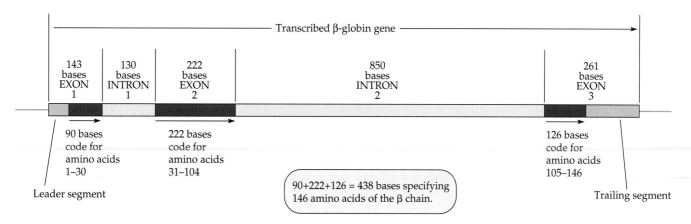

| 143 bases EXON 1 | 130 bases INTRON 1 | 222 bases EXON 2 | 850 bases INTRON 2 | 261 bases EXON 3 |

Transcribed β-globin gene

90 bases code for amino acids 1–30

222 bases code for amino acids 31–104

126 bases code for amino acids 105–146

90+222+126 = 438 bases specifying 146 amino acids of the β chain.

Leader segment

Trailing segment

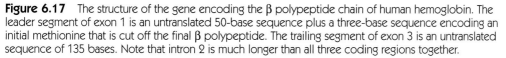

Figure 6.17 The structure of the gene encoding the β polypeptide chain of human hemoglobin. The leader segment of exon 1 is an untranslated 50-base sequence plus a three-base sequence encoding an initial methionine that is cut off the final β polypeptide. The trailing segment of exon 3 is an untranslated sequence of 135 bases. Note that intron 2 is much longer than all three coding regions together.

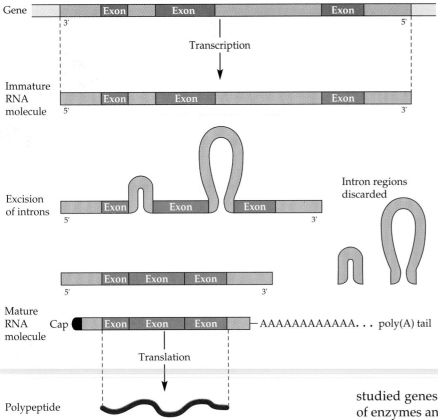

Gene

Transcription

Immature RNA molecule

Excision of introns

Intron regions discarded

Mature RNA molecule

Cap — AAAAAAAAAAAA . . . poly(A) tail

Translation

Polypeptide

Figure 6.18 Messenger RNA processing. Between transcription and translation, mRNA is modified in three ways: The regions corresponding to introns are cut out, and all the exons are joined into an uninterrupted sequence; a special nucleotide is put on the 5′ end (the head cap); and a string of 50 to 200 nucleotides, all adenine, is put on the 3′ end, forming the poly(A) tail.

muddle the reading of the triplet code and the intron-exon junctions. Globinlike pseudogenes have been found elsewhere; several are present in a clustered family of α-globin genes.

A second curious aspect of the intergenic DNA in the β-globin cluster are four very similar sequences of about 310 bases that serve no useful function. These repetitious sequences are called *Alu* **repeats**, and there are many other *Alu* repeats in nearby regions. In fact, *Alu* sequences are a strikingly common feature of our genome and are discussed further in the next section. Figure 6.20 shows how the various elements of the α-globin and β-globin clusters are arranged.

Repetitive DNA

Geneticists have traditionally studied genes that are responsible for the production of enzymes and other protein molecules via an mRNA intermediate. These genes are called the *structural genes*, and each is present once per gamete and therefore (for autosomal genes) twice in every diploid cell of the body. Taken collectively, the structural genes are known as **single-copy DNA**, and the bases exhibit no particular repeating patterns. This state of affairs is expected, since the structural genes are coding for proteins with no particular repeating patterns of amino acids.

five genes form a cluster on the short arm of chromosome 11. Each gene has three exons and two introns similar to those of the β-globin gene depicted in Figure 6.17. Although no other genes are in this region, the five β and β-like genes occupy only a small fraction of the total length of this stretch of DNA. The nucleotides that constitute the noncoding "spaces" between the genes are not transcribed and translated, but they have many intriguing aspects, including two interesting types of sequences.

One is a gene-sized length of DNA called *psi beta* (ψβ), which in structure and nucleotide sequence is similar to the β-like genes. Because it has no polypeptide product, however, it is called a **pseudogene**. Pseudogenes represent evolutionary dead ends, having accumulated one or more critical mutations that inactivate the gene. Usually the gene cannot be transcribed, or if transcribed, its mRNA cannot be translated into a functional polypeptide. The human ψβ pseudogene, for example, has small insertions that

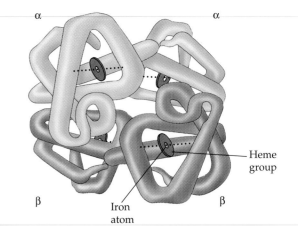

Heme group

Iron atom

Figure 6.19 The folding of the four polypeptide chains in normal adult hemoglobin, $\alpha_2\beta_2$. Each chain envelops a heme group containing an iron atom that reversibly attaches a molecule of oxygen.

Only about 3% of human DNA, however, is transcribed into mRNA, rRNA, or tRNA molecules. The question of what the remaining 97% of DNA does—within introns and between genes—is largely unanswered. Some of this nongenic DNA consists of regulatory sequences (such as promoters, enhancers, and silencers) that control the rate at which nearby genes are transcribed. Some of it consists of the centromere regions and special sequences at chromosome tips. Much of the nongenic DNA, however, may simply be nonfunctional "junk" (Angier 1994; Nowak 1994). In the 1960s, researchers found that about half of this DNA exists as multiples of relatively short sequences of nucleotides called **repetitive DNA**. Because most repetitive DNA never gives rise to protein products, repeated sequences are generally not associated with any detectable phenotypes. Repetitive DNA is classified as highly repetitive or moderately repetitive.

The *highly repetitive* portion consists of short lengths of DNA, up to a few hundred bases that are repeated hundreds of thousands of times. Highly repetitive sequences include the *Alu* repeats such as those in the globin region noted previously. About half a million copies of *Alu* per haploid genome are generously spread between genes and within introns throughout the chromosomes. This amount of DNA represents about 5% of human DNA, the rough equivalent of a whole average-sized chromosome. On the average, an *Alu* sequence occurs every four kilobases! *Alu* repeats have also been identified in DNA extracted from ancient archaeological specimens (see Box 8B). The function of *Alu* sequences is something of a mystery (Novick et al. 1996).

The *Alu* repeats are an example of fairly short repetitive elements (hundreds of nucleotides long) that are interspersed throughout the genome and known collectively as *SINES* (for *s*hort *in*terspersed *e*lements). Longer repetitive elements (thousands of nucleotides long) known as *LINES* (for *l*ong *in*terspersed *e*lements) are also present. Both SINES and LINES have nucleotide sequences that allow them to move around in the genome.

Additional highly repetitive DNA with no known function consists of very short sequences of two to a dozen bases repeated in tandem many times (Sutherland and Richards 1994). The number of copies of, say, the dinucleotide AC (with TG on the other strand) at a particular locus in the genome may vary widely from person to person and between the two "alleles" in one person.* Each repeat region has a few to a few dozen repeats, and there are thousands of different regions. For that reason, these repeats are useful in mapping and in personal identification (see later chapters). Some of this type of highly repetitive DNA is concentrated in the centromere regions of chromosomes; it also accounts for over half the DNA in the human Y chromosome.

Sequences with very short repeat units are referred to as di-, tri-, tetra-, or pentanucleotide repeats, as appropriate. Alternatively, these repeat regions are often referred to as **microsatellites** and somewhat longer repeated units as **minisatellites**. The use of *satellite* in these terms does not refer to chromosome knobs (see Chapter 2). Rather it derives from work in which solu-

---

*A site or locus along a chromosome with repeated sequences in tandem is not a gene in the sense that it codes for a protein. Nevertheless, the variations in copy number lead to localized differences in length that can be distinguished from each other by simple laboratory techniques (Chapter 8). Thus, even though apparently nonfunctional, sites with repeated segments of DNA have different variants (alleles).

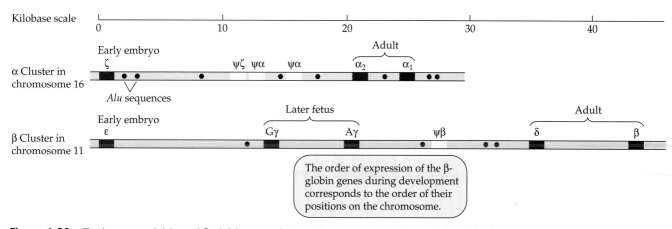

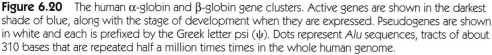

Figure 6.20 The human α-globin and β-globin gene clusters. Active genes are shown in the darkest shade of blue, along with the stage of development when they are expressed. Pseudogenes are shown in white and each is prefixed by the Greek letter psi (ψ). Dots represent *Alu* sequences, tracts of about 310 bases that are repeated half a million times times in the whole human genome.

tions of DNA are subjected to high-speed centrifugation (not considered in this book).

The *moderately repetitive* portion of DNA consists of varied lengths of DNA that are repeated tens to thousands of times in the genome. Some genes with well-known functions are moderately repetitive, a redundancy perhaps reflecting the need for large amounts of the gene product. For example, the genes that code for ribosomal and transfer RNA molecules are present in multiple copies. Other sequences that have repetitive elements are the gene families coding for histones and for antibodies. (The extraordinary nature of the antibody genes is discussed in Chapter 17.) Also, the genes encoded in mitochondrial DNA represent a type of repetitive DNA, since a mitochondrion has about ten identical molecules of DNA, and a typical human cell has hundreds of these organelles.

Although never transcribed into any product, the sequence TTAGGG is repeated an average of 2,000 times at the tips of all chromosomes. These special structures, known as **telomeres** (Greek *telos*, "end"), are vital to the stability of chromosomes (Box 6B). Conserved through many millions of years of evolutionary time, exactly the same sequence makes up telomeres in the chromosomes of organisms as widely separated as yeast and humans.

Mobile Genes

Similar genetic structures in related species suggest that the organization of DNA is *stable* over long periods of time. Yet there are now known to be genetic elements that actually move around within the genome. This phenomenon was first demonstrated in the 1940s by Barbara McClintock of Cold Spring Harbor Laboratory, New York. By examining color patterns on kernels of corn, she was able to show that certain genetic elements within the maize genome modified the expression of other genes at adjacent sites. From time to time these controlling elements seemed to disappear from their old locations and reappear at new locations!

McClintock's interpretations were considered innovative and unusual at the time, but did not lead to studies on other organisms until the 1960s, when somewhat similar DNA segments called **transposable elements** or **transposons** were discovered in bacteria. A transposon is a DNA element that excises itself from one chromosomal location and inserts itself into another without making additional copies of itself. Various types of transposons with a variety of effects have been found in many species, including maize, bacteria, yeast, fruit flies, and mice. In humans, some repetitive SINES and LINES resemble transposons.

In addition, some LINES have sequences that code for an enzyme called *reverse transcriptase*. Such an element is called a *retrotransposon*, and it can make mobile copies without itself moving away from its origi-

nal site. In the process, the retrotransposon is first transcribed into an RNA copy, which is then *reverse transcribed* into a DNA copy by reverse transcriptase. The DNA copy becomes double-stranded and can insert elsewhere in the genome (Novick et al. 1996). The terms *reverse transcriptase* and *retrotransposon* reflect the reversal of information flow—from RNA to DNA, rather than the usual route, DNA to RNA. (We return to this idea in Chapter 16.)

Although *Alu* repeats and other SINES are not long enough to encode reverse transcriptase, they can sometimes make use of reverse transcriptase from elsewhere and so be successively copied from DNA to RNA to DNA and then be inserted at other sites. Insertion of *Alu* or any other segment of DNA into a functioning gene constitutes a kind of mutation. Examples of half a dozen different genetic disorders, including hemophilia, have been shown to have such a causation (next chapter).

Summary

1. DNA, the genetic material, consists of two nucleotide chains wound in a helix. In the backbone of each chain, the sugar deoxyribose alternates with phosphate groups. Attached to the sugars of both strands are the paired bases: A opposite T, and G opposite C.

2. DNA is replicated by separation of the helix, each old strand becoming a template for the enzymatic synthesis of a new complementary strand. DNA polymerase, the replicating enzyme, proceeds in only one direction along the helix.

3. DNA is transcribed by the enzyme RNA polymerase. The process is turned on and off by transcription factors. Transcription forms complementary single-stranded copies of RNA of three types:

(a) Messenger RNA molecules are the transcripts of genes that will be translated to produce polypeptides. Successive groups of three nucleotides make up coding units, called codons, for amino acids.

(b) Transfer RNA molecules act like adapters during translation. One end binds a particular amino acid; the other end has a three-base sequence, the anticodon, that pairs with a codon on messenger RNA.

(c) Ribosomal RNA molecules are structural elements of ribosomes. Neither ribosomal nor transfer RNA is translated into a polypeptide.

4. As it feeds through a ribosome, the mRNA is translated into a polypeptide. The complex process links amino acids together into a chain. The particular order of amino acids causes the polypeptide molecule to fold into a unique three-dimensional shape that determines its function. Proteins consist of one or several polypeptide chains with a total of dozens to thousands of amino acids.

5. Specific codons were deciphered by adding artificial RNA molecules of known composition to an in vitro protein-synthesizing system. The same genetic code, which includes several codons for most amino acids and signals to stop translation, applies to almost all forms of life.

BOX 6B   *Does Telomere Shortening Contribute to Aging?*

Think about these two facts:

1. The tips of chromosomes—the telomeres—have hundreds to thousands of repetitions of the simple nucleotide sequence TTAGGG.

2. The progress of DNA replication proceeds in opposite directions on the two strands of the double helix (Figure 6.7). The strand that is being replicated in the "backward" direction (the discontinuous strand) needs a length of DNA to keep restarting the copying process. As a quirky result, the very end of this strand (the 5' end) does not get replicated at all. Consequently, the telomere sequences at the tips of all chromosomes get shorter by about 100 base pairs with each cell replication.

The loss of some TTAGGG repeats is not important. Like the plastic tips on shoelaces, the presence of the repetitive telomere units protects significant genes located near the chromosome tips, at least for a number of cell generations. Eventually, the tips may get too short, however, and some researchers think that the degenerative processes of aging may stem in part from this chiseling away at the tips of chromosomes, perhaps even beyond the telomeres and into structural gene territory. Several observations support this idea. First of all, researchers have shown that the telomeres from cells of older people are shorter than telomeres from younger people. And individuals with progeria, a premature aging disease, have significantly shorter telomeres.

When cells are removed from the body and grown in tissue culture, they reproduce for about 50 generations and then stop, a point called the Hayflick limit, after its discoverer. (Cells divide somewhat more than 50 times if taken from a younger person, somewhat fewer if taken from an older person.) There are important exceptions to this rule. For example, cells from persons with Werner syndrome (Chapter 7), another premature aging syndrome, stop after about 20 divisions.

But the most important exception to the Hayflick limit is in the opposite direction. Malignant cancer cells growing in tissue culture divide forever. The prime example is a line called HeLa cells, which has been important to researchers all over the world.

HeLa cells were taken from a cervical cancer of a woman named Henrietta Lacks in 1951 and they continue to be used in thousands of cell culture experiments (including the telomere labeling pictured here). Malignant cells can continue to divide because they produce a special enzyme called telomerase, which is able to extend the TTAGGG tips that would otherwise be lost each cell generation (Haber 1995; Hilts 1997; Travis 1995). Although telomerase is also present in germ line cells, the enzyme is not present in normal adult human tissues. Presumably, the gene is turned off. But when the telomerase gene is active, as in cancer cells, the enzyme product stabilizes the telomeres and confers immortality. Without telomerase, the cells would eventually die. Indeed, when researchers inactivate the telomerase of HeLa cells growing in culture, the cells stop dividing after about 20 cell divisions.

Researchers from the Geron Corporation of California (from *gerontology*, the medical speciality concerned with changes in the mind and body that accompany aging) and the University of Texas have recently been able to reactivate the telomerase gene in normal human cells growing in tissue culture (Wade 1998). As of this writing, these cells with active telomerase are growing and dividing vigorously 40 generations beyond untreated control cells, which stopped dividing at the 50-generation Hayflick limit. You can see why researchers are interested in how to turn telomerase on in vivo as an anti-aging treatment. On the other hand, they might want to turn telomerase off in tumors growing in vivo as a possible anticancer treatment. Because the latter purpose clearly conflicts with the former, much more needs to be learned to even begin to consider telomerase as a healthful prescription (see, for example, question 13).

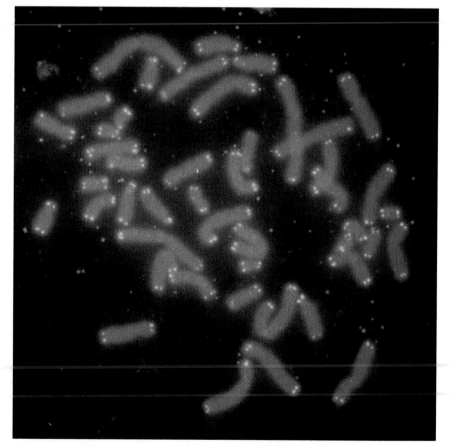

In this micrograph, the TTAGGG sequences at the tips of the chromosomes have been tagged with a fluorescent dye. (Courtesy of R. K. Moyzis.)

6. Genes are often interrupted by noncoding regions called introns. Although the introns are transcribed, they are cut out of RNA molecules before translation.

7. The β-globin genes represent a gene family. The cluster consists of five β-type chains of hemoglobin and a pseudogene that has no polypeptide product.

8. A great deal of nongenic DNA exists as repetitive multiples of relatively short nucleotide sequences.

(a) Moderately repetitive DNA sequences are present up to thousands of times per genome. Examples include the genes that encode ribosomal and transfer RNA molecules.

(b) Highly repetitive DNA sequences may occur as many as a million times per genome. Some short tandem sequences are concentrated at certain chromosomal locations, but longer sequences, like SINES (including *Alu* repeats) and LINES, may be interspersed throughout the genome.

9. Some segments of DNA, called transposons and retrotransposons, can move around in the genome.

Key Terms

Alu repeat	pseudogene
amino acid	reading frame
anticodon	repetitive DNA
code degeneracy	ribosomal RNA (rRNA)
codon	RNA polymerase
complementary	RNA splicing
DNA helicase	single-copy DNA
DNA ligase	stop (chain termination)
DNA polymerase	codon
exon	telomerase
gene expression	telomere
gene family	template strand
intron	transcription
messenger RNA (mRNA)	transcription factor
microsatellite	transcription termination
minisatellite	sequence
mRNA processing	transfer RNA (tRNA)
nucleotide	translation
polypeptide	transposon
promoter	

Questions

1. The elements C, H, N, O, P, and S are the common constituents of organic matter. Which is present in DNA but not in protein? Which is present in protein but not in DNA?

2. Chargaff's rules (A = T; G = C) do *not* hold for the DNA in the virus φX174 and in some other small viruses, nor do comparable rules (A = U; G = C) hold for the RNA of retroviruses and most other molecules of RNA. Explain these exceptions.

3. If the cells in a Meselson-Stahl experiment (Figure 6.5) were grown for three generations in nitrogen-14, what frac-

tion of the DNA double helices would be intermediate in density? What fraction would be light?

4. A messenger RNA molecule contains the following base composition: 21% A, 33% U, 28% G, and 18% C. What is the base composition of (a) the template strand of DNA from which it was transcribed; (b) the nontemplate strand; and (c) both strands of DNA considered together? Assume there are no introns in this gene.

5. A tRNA molecule that binds the amino acid glycine has the anticodon CCC; that binding phenylalanine has AAA. Give the sequences of bases in mRNA and in the template strand of DNA that code for the dipeptide glycine–phenylalanine.

6. The digestive enzyme trypsin specifically cleaves a protein molecule on the carboxyl side of lysines and arginines. What peptide pieces result from the trypsin digestion of the B chain of human insulin (Figure 6.13)?

7. What fraction of the bases transcribed from the β-globin gene actually codes for amino acids in the β polypeptide chain of hemoglobin?

8. The β-globin cluster of genes extends over about 50 kilobases. It includes the five genes that code for polypeptide products (β, δ, Aγ, Gγ, and ε chains of hemoglobin). Each of these chains is 146 amino acids long. What fraction of that segment of DNA actually codes for amino acids in polypeptides?

9. Let + stand for a one-nucleotide addition and – for a one-nucleotide deletion. Would the following mutant genes (with *combinations* of nearby additions and deletions) change the reading frame beyond the last mutational site?

(a) +–   (d) +– –+
(b) +–+   (e) +++–
(c) +++

10. The human brain hormone somatostatin was the first product of recombinant DNA technology. Bacteria synthesized the hormone after biochemists made and inserted into the cells DNA with the following base sequence (template strand):

CGA CCA ACA TTC TTG AAG AAA ACC TTC TGA AAG TGA AGC ACA

What is the amino acid sequence of somatostatin?

11. What polypeptides could be made by each of the following artificial mRNA molecules?

(a) polyadenine, or poly(A)
(b) ... UCAUCAUCAUCA ...
(c) ... UUCCUUCCUUCC ...

12. List the nine possible DNA triplets and corresponding amino acids that would be formed by changing a single base in the CCC triplet for glycine. (As explained in the next chapter, these mutations are called base substitutions.) Which do not produce an amino acid change?

13. Some cell lines in our bodies, such as those that produce red blood cells, continue to divide throughout life. With this in mind, comment on trying to inactivate telomerase as an anti-cancer treatment (see Box 6B).

Further Reading

For more detailed information on all aspects of gene structure and function, we recommend the large, beautiful books by Lodish et al. (1996), Alberts et al. (1994), Lewin (1997), and Watson et al. (1987). Scriver et al. (1995), in three encyclopedic volumes of 5,600 pages, provide individual chapters on the molecular basis of human diseases as well as several general chapters relevant to the discussions here. Among the best of several histories of the events leading to the Watson-Crick double helix is Judson (1979). See Watson (1968) and Crick (1988) for lively personal accounts, and Sayre (1975) for another view on the role of Franklin in the discovery of the DNA double helix.

Among the wealth of fairly recent *Scientific American* articles on genes and proteins, we recommend Nomura (1984) on ribosome structure and function, Gallo and Montagnier (1988) on the AIDS virus, Radman and Wagner (1988) on DNA replication, Steitz (1988) on intron splicing, Richards (1991) on protein structure, Moyzis (1991) and Greider and Blackburn (1996) on telomeres, and Tjian (1995) on control of transcription. In addition, the October 1985 issue is devoted to the molecules of life and includes articles on DNA by Felsenfeld, on RNA by Darnell, and on proteins by Doolittle.

CHAPTER 7

The Nature of Mutations

A head of red hair singles out a person as few other physical characteristics do, and everyone seems interested in its inheritance (Figure 7.1). But professional geneticists have not rushed to study red-headed people, because pedigrees have never consistently shown a standard dominant or recessive pattern, and reddish hair has limited medical significance (McKusick 1997). Interest in red hair and in the failure of red-headed and other light-skinned persons to tan, however, may be rekindled as the result of recent studies at the cellular and molecular level.

The most noticeable pigments in hair and skin are the *melanins*. There are two kinds: brown-black and reddish. The brown-black melanins protect the body against the damaging effects of ultraviolet light by screening out these rays. Reddish melanins are not protective, and may even contribute to skin damage. The melanins are produced in specialized cells called *melanocytes* that are present in hair follicles, the skin, and the eyes (Figure 7.2).

Figure 7.1 This strikingly redheaded boy posed for his portrait in about 1520. But his drawing, which he proudly shows off, looks remarkably modern. The painter Francesco Caroto and his brother Giovanni were prominent in the art world of Verona, Italy, during the early 16th century. (Courtesy of Museo di Castelvecchio, Verona.)

The production of brown-black melanins is increased by a pituitary hormone called MSH (melanocyte-stimulating hormone). To do the job, MSH binds to protein receptors on the surface of melanocytes. This **MSH receptor protein** and its encoding gene are the focus of the recent studies. In brief, people with red hair often have mutations in the gene that encodes the MSH receptor protein. A **mutation** is simply a change in DNA of one sort or another. In the case of red hair, the mutant allele produces altered MSH receptor proteins that cannot bind the hormone, effectively preventing the production of brown-black melanins.

In an initial study of British and Irish individuals, mutations of the MSH receptor gene were found in 21 of 30 redheads, but in none of 30 control individuals with brown or black hair (Valverde et al. 1995). In most of the redheads, only *one* allele was mutated. In some people with the reddest hair, however, mutations were present in *both* alleles. Calling the normal allele R (for receptor) and the mutant allele R', the phenotype/genotype relationships would be:

Phenotype	Brown/black	Red	Very red
Genotype	R/R	R/R'	R'/R'

Thus, the red hair allele acts in an intermediate fashion. Further study of more individuals pointed in the same general direction, but mutations of the MSH receptor gene turned out to be neither necessary nor sufficient for the expression of red hair. Thus, variation of the MSH receptor gene appears to be a major player, but not the only player, in the production of red hair, whose several modes of inheritance remain unclear. It is interesting that mutations of the MSH receptor gene in other mammals (such as dogs, pigs, foxes, and cows) result in reddish coat colors comparable to red hair in people.

Kinds of Mutations

The effect a mutation has on the phenotype of an organism ranges from trivial to lethal. It all depends on the properties of the protein encoded by the altered gene and how that alteration affects development and maintenance of the body. Many mutations produce phenotypic effects even less consequential than red hair. Examples include clockwise versus counterclockwise cow-

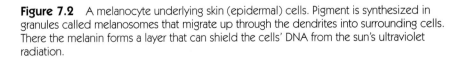

Skin surface

Melanin

Epidermal cells

Melanosomes

Dendrite of melanocyte

Nucleus of melanocyte

Interior of body

Figure 7.2 A melanocyte underlying skin (epidermal) cells. Pigment is synthesized in granules called melanosomes that migrate up through the dendrites into surrounding cells. There the melanin forms a layer that can shield the cells' DNA from the sun's ultraviolet radiation.

lick patterns on top of the head, cleft versus noncleft chin, and soft versus dry earwax. In contrast are the genes that cause death during infancy (such as Tay-Sachs disease), young adulthood (Duchenne muscular dystrophy and cystic fibrosis), or later years (Huntington disease). In between are genetic conditions that are inconvenient or somewhat burdensome (pattern baldness, color blindness, alkaptonuria, and dwarfism) as well as more serious conditions (familial hypercholesterolemia, hemophilia, sickle-cell anemia, and phenylketonuria).

Here we will be primarily concerned with **germinal mutations,** which occur in the *germ line*—the cells that are destined to become eggs or sperm. A germinal mutation does not affect in any obvious way the person in whom it occurs, but it can be transmitted to and harm future generations. **Somatic mutations,** on the other hand, occur in body cells (liver, lung, intestine, etc.) not ancestral to gametes. A somatic mutation can affect the phenotype of its carrier, but it will not be transmitted to offspring. Somatic mutations are involved in one or more steps that lead to the development of cancer (Chapter 17).

The occurrence of an altered phenotype may be far removed in time from the germinal mutation itself, especially if the mutated allele behaves as a recessive, as many do. In these cases, it must be combined in a zygote with another allelic recessive mutation for its effect to be revealed during the course of development. Thus, the mutational event itself and the mutant phenotype that it produces may be separated by generations.

Mutations that seem to occur haphazardly, irrespective of known environmental conditions, are called **spontaneous mutations.** To a large extent, they result from random thermal motions of atoms and molecules in and near DNA, often at the time of DNA replication, but at other times as well. This motion is characteristic of all matter. In addition, it is likely that some spontaneous mutations are due to foreign chemicals or radiation that get inside cells and near DNA. Because we cannot observe the mutational events, it is not possible to pinpoint the precise cause (heat, chemicals, radiation) of any spontaneous mutation, especially since the observable phenotypic effects may not be evident for a long time.

Induced mutations refer to gene changes following exposure of an organism to an agent, called a **mutagen,** known to produce mutations. Mutagens of concern include components of cigarette smoke, some industrial chemicals, pesticides, various substances in waste landfills, food additives, drugs, medical and dental X-rays, ultraviolet light, and radiation from past atomic weapons tests, atomic warfare, and the nuclear power industry—a list that reflects our technological society, attitudes toward the environment, and

modern life-styles. The list of mutagens overlaps to a large degree the list of **carcinogens,** substances that cause cancer, since gene mutations are critical in the development of malignancies. In addition to synthetic substances, some largely inescapable *natural* products are known to be mutagens, including the radioactivity found in soil, rocks, and our bodies. Many common raw or cooked foods also contain mutagenic chemicals; some of these are manufactured by plants as natural pesticides, toxic to the species that would otherwise eat them (National Research Council 1996).*

Substitutions and Frameshifts

At the molecular level, the actual genetic change may be a *replacement* of one base by another, called a **base substitution mutation.** Because one base can change to any of the other three bases, a DNA triplet can mutate in nine different ways by a single base substitution. For example, one of the red hair mutations in the MSH receptor gene changes the first C in a CAC triplet to a T. This mutation occurs in the unit that codes for the ninety-second amino acid of the MSH receptor, and you should confirm (from Table 6.4) that the altered protein contains a methionine in place of a valine. (Why this change inhibits the synthesis of brown-black melanins is unknown.) A base substitution like this one that replaces one amino acid with another is often called a **missense mutation.** About 73% of random base substitutions in the coding regions of genes are missense mutations.

A mutation in coding DNA that changes a triplet to a synonym for the same amino acid would be innocuous. For example, a DNA base substitution in an exon that changes AAA to AAG would still lead to the incorporation of phenylalanine; such a mutation would probably be undetectable except by sequence analysis of DNA or mRNA molecules. Although it would usually have no phenotypic effect, it is still a mutation by the usual definition, and is often called a **silent mutation.** Silent mutations constitute about 23% of random base substitutions. Also innocuous, of course, would be most substitution mutations that change a base in the noncoding spacer DNA between genes.

Sickle-Cell Disease. The classic example of a missense base substitution is the one that causes *sickle-cell disease.* This illness was the first to be dubbed a *molecular disease* (by the American chemist and Nobel laureate

*Not wishing to add to scary stories that bombard the public daily, we hasten to point out that the levels of both natural and synthetic mutagens to which we are exposed are, in general, low and pose no significant biological risk (see Box 1A, "Judging the Media"). In particular, the food supply in the United States is varied and abundant; it is regulated to a large degree and widely recognized as safe. What you choose to eat in large amounts, however (fruit, vegetables, cereals, meat, desserts, etc.), can certainly affect your well-being.

Linus Pauling), one brought on by an altered protein encoded by a mutated gene. The mutation occurs in the gene coding for the β chain of hemoglobin. The middle base of a DNA triplet coding for the amino acid *glutamic acid* is mutated so that the triplet now codes for *valine* instead (Figure 7.3). The resulting amino acid switch changes the shape of hemoglobin so that the altered molecules stack into narrow crystals, distorting the smooth, rounded surface of red blood cells (Figure 7.4). Distorted cells have a life span of only a few weeks rather than the normal life span of about four months, leading to anemia (i.e., too few red blood cells). In addition, the sickled cells become trapped in small vessels throughout the body, leading to local oxygen depletion, tissue damage, infections, and periodic episodes of excruciating pain that occur throughout life. These various effects of the sickle-cell mutation markedly reduce the quality of life, reproductive ability, and life span.

Affected persons are homozygous (Hb^S/Hb^S, or more simply S/S) for the sickling allele; heterozygotes (A/S; A is used here for the normal β-chain allele) are said to have the sickle-cell *trait*; they are essentially normal and rarely come to medical attention. The sickle-cell allele is prevalent in African populations and in their descendants worldwide because of a curious interaction with malaria (see Chapter 12).

Although clinical work has continued for many years, there has been no good basic treatment for the disease until very recently. Now, the drug *hydroxyurea* offers some relief from the painful crises and associated problems, at least for adults (Fackelmann 1995; Schechter and Rodgers 1995). Daily treatment with hydroxyurea reduces by about half the number of sickle-cell episodes and also the need for blood transfusions.

A/A
Normal
phenotype

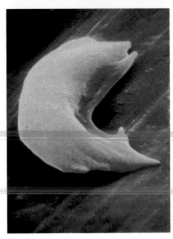

S/S
Sickle-cell
phenotype

Figure 7.4   A red blood cell from a normal homozygote (top), and one from an individual who is homozygous for the sickle-cell allele (bottom). Although red blood cells in heterozygotes (*A/S*; not shown) generally look rounded and function normally (as the top cell), they can be made to sickle when the amount of oxygen is reduced—for example in a sealed film of blood on a microscope slide. (Copyright © Stanley Flegler/Visuals Unlimited.)

The drug apparently works by turning on the gene for the gamma (γ) chain of hemoglobin, which, while active in fetal life (Chapter 6), is normally turned off soon after birth. The presence of some normal fetal hemoglobin ($\alpha_2\gamma_2$) dilutes out the adult sickle-cell hemoglobin ($\alpha_2\beta_2$) within red blood cells and inhibits their sickling. Although fetal hemoglobin is not usually present in adults except in tiny amounts, it seems to work in adults as well as their "regular" hemoglobin does. The safety of using hydroxyurea in children and the drug's long-term effects in adults are still being evaluated.

β-Thalassemia.   When the α or β polypeptides of hemoglobin are synthesized in reduced amounts or not at all, the resulting disease is *thalassemia* (Greek *thalass-emia*, "sea-blood"), a seriously debilitating or

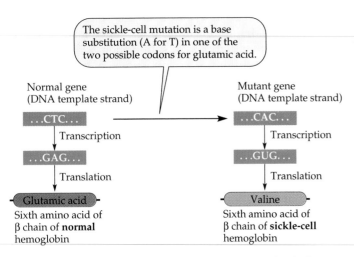

The sickle-cell mutation is a base substitution (A for T) in one of the two possible codons for glutamic acid.

Normal gene
(DNA template strand)

...CTC...

↓ Transcription

...GAG...

↓ Translation

Glutamic acid

Sixth amino acid of
β chain of **normal**
hemoglobin

Mutant gene
(DNA template strand)

...CAC...

↓ Transcription

...GUG...

↓ Translation

Valine

Sixth amino acid of
β chain of **sickle-cell**
hemoglobin

Figure 7.3   The sickle-cell mutation. The normal gene for the β polypeptide chain of hemoglobin incorporates glutamic acid as the sixth amino acid from the free amino end (left). Verify the amino acid codons by referring to Table 6.4.

lethal condition common in some Mediterranean peoples and in their worldwide descendants. Affected children who are homozygous cannot usually be treated very satisfactorily, but the disease can be diagnosed prenatally.

Different forms are designated by the globin chain that is deficient. The *β-thalassemias* constitute one of the most serious health problems worldwide, accounting for hundreds of thousands of childhood deaths per year. The major symptom is anemia, a reduction in the hemoglobin needed to carry oxygen. The disease (in homozygotes) is accompanied by fatigue, shortness of breath, recurrent infections, stunted growth, abnormalities of bones and bone marrow, and enlargement of the spleen. Although the mutations are in the same β-globin gene as the mutation leading to sickle-cell disease, the red blood cells do not sickle. This situation—different phenotypic traits from different mutations of the same gene—is not at all unusual, and we will meet other examples later. Heterozygotes have *β-thalassemia minor* and usually only a mild anemia, but homozygotes—afflicted with *β-thalassemia major*—often die before 10 years of age.

About 100 different mutations have been shown to produce β-thalassemia. They affect not only the actual amino acid coding regions of the β-globin exons, but also sequences surrounding the gene and even within the noncoding introns. The severity of the resulting disease varies with the particular mutation, and sometimes from patient to patient with the same mutation (Figure 7.5).

Mutation 1 in Figure 7.5 alters the promoter region of the gene, thereby markedly reducing the efficiency of transcription. Although the globin chains are normal, very few of them are synthesized in homozygotes, and a severe anemia results.

Mutation 2 changes codon 39 from a triplet specifying glutamine to a stop codon. (What was the base substitution? See Table 6.4.) Mutations of this type—from a codon specifying an amino acid to a stop codon—are sometimes called *nonsense mutations*. A better name is **chain termination mutations**, and they constitute about 4% of random base substitutions. They have the effect of stopping translation prematurely, producing a too-short polypeptide. In this case, the polypeptide is only 38 amino acids long, no functional β chains are present, and homozygotes are severely anemic. This particular mutation accounts for about 30% of β-thalassemia alleles in the Mediterranean region and for almost all of the disease on the island of Sardinia.

Mutations 3 and 4 perturb RNA splicing so that the intron regions are not cut out of the molecule properly. In fact, about 15% of all mutations that result in human disease affect RNA splicing. Another example is given later in this chapter (see Figure 7.14) in which an exon is mistakenly excluded from the mature RNA transcript, rather than an intron mistakenly included.

Mutation 5 causes the poly(A) tail on the messenger RNA to be attached in the wrong place. As a consequence, the mRNA is unstable and a mild thalassemia results.

Mutations 6, 7, and 8 are **frameshift mutations**, which are defined as the addition or deletion of one or more bases. Frameshift mutations not only alter the codon in which they occur, but may shift the *reading frame*, the division of the DNA message into *successive* three-letter words. If one extra base is *inserted* (as in mutation 8), all the coding units after the insertion are read incorrectly. The result is an altered polypeptide in which virtually all the amino acids are wrong (Figure 7.6). You will easily see that a single-base *deletion* has a similar result. A two-base addition or a two-base deletion also shifts the reading frame. Often when the reading frame is shifted, a stop codon will soon be encountered, terminating the polypeptide after a string of all wrong amino acids. The term **point mutation** is often used for both base substitutions and small frameshift mutations, since neither are microscopically visible.

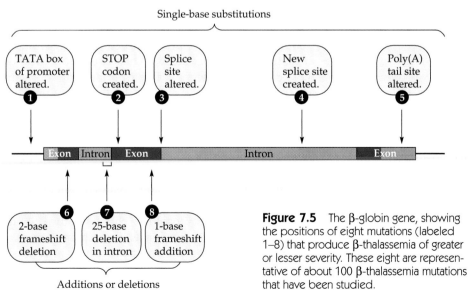

Figure 7.5 The β-globin gene, showing the positions of eight mutations (labeled 1–8) that produce β-thalassemia of greater or lesser severity. These eight are representative of about 100 β-thalassemia mutations that have been studied.

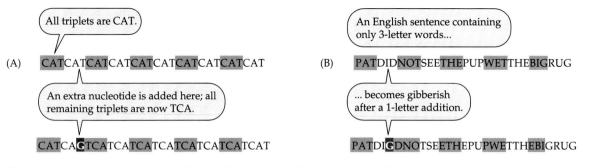

Figure 7.6 The consequences of a frameshift mutation resulting from the addition of a single base. (A) A hypothetical gene. Adding a base causes a shift in the reading frame. (B) An English language analogy. It is assumed that the sentence—like DNA—continues to be read in successive three-letter words.

Unstable Trinucleotide Repeats

In 1991, molecular geneticists discovered a totally unexpected type of mutation. It was a DNA triplet that was repeated many times, one after another, and thus called a **trinucleotide repeat**. Such a repetition had never been seen in any of the well-studied species such as bacteria, fruit flies, yeast, maize, or mice. But about a dozen human diseases are now known to have this causation (Rosenberg 1996; Sutherland and Richards 1994).

A triplet repeat in the coding region of a gene will, of course, produce an amino acid repeat in the encoded polypeptide. The number of nucleotide and amino acid repeats can increase in successive generations and cause disease symptoms. The longer the repeat, the more severe the symptoms tend to be, and the earlier in life the symptoms appear.

Huntington Disease. *Huntington disease* is a dominantly inherited, slowly progressive brain disorder that results in mental disturbances, physical degeneration, and eventually death within 15 to 20 years after onset (see Box 5B). Early symptoms include involuntary jerky motions of the body (chorea), mood disorders, and personality changes. Speech becomes slurred, and walking resembles that of a drunk person. Although symptoms often begin in the 30s or 40s, the age of onset may be earlier or later. Often the symptoms first appear after reproductive age, so the mutant allele will have already been passed to half the children (on average). No satisfactory treatments are available, and long-term home care or institutionalization can devastate family finances and family stability. The overall frequency in the United States is about 1 in 25,000 individuals.

By 1983, researchers had located the approximate site of the Huntington disease gene near the tip of the short arm of chromosome 4. It took another ten years of intense international effort by dozens of scientists, led by James Gusella of Harvard University Medical School, before the gene was finally isolated, cloned, and sequenced.

The gene is large, with 67 exons spanning 170,000 base pairs. It codes for a 3,144-amino acid polypeptide, dubbed *huntingtin*, an unusual protein whose normal function is still unknown. Figure 7.7 shows the first 180 coding nucleotides and the resulting 60 amino acids from a *normal* individual. Note that near the beginning of the gene, starting at nucleotide 52, there is a string of CAG repeats that code for the amino acid glutamine (Q in standard one-letter designations). In this particular normal allele, you can count 23 CAG repetitions. (Actually the next to last one is CAA, a synonymous triplet, but for simplicity we will call them all CAG.) Normal individuals have alleles with 6 to 39 CAG repetitions, while individuals with Huntington disease possess an allele with 36 to 180 CAG repetitions (almost always heterozygous with a normal allele).

Therein lie many puzzles. How do the expanded, disease-causing stretches of DNA come about? Why is the repeat unit three rather than some other number? What does normal huntingtin protein do in the body, and what does the abnormal protein with extra glutamines not do, or do wrong? Can anything be done for people who carry the mutant allele, either before or after the disease develops? These questions do not have complete answers yet. Furthermore, if you were at 50:50 risk (because a parent had Huntington disease), would you want to be tested to see if you received an expanded allele?

What is known about Huntington disease is that brain cells die during the course of the disease, especially deep within the brain (Figure 7.8). As judged by mice that carry the abnormal human *HD* allele (*transgenic* mice; see Chapter 8), huntingtin protein with the extra glutamines form insoluble clumps in the cell nuclei, similar to protein fibrils seen in Alzheimer disease (Blakeslee 1997). Yet huntingtin is found in cells all over the body, so it is unclear why the disastrous clumping is apparently restricted to the brain. A possi-

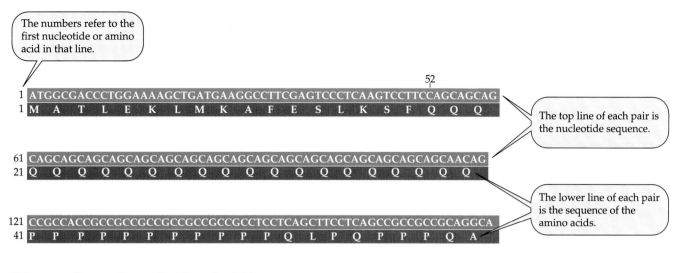

The numbers refer to the first nucleotide or amino acid in that line.

52

```
  1  ATGGCGACCCTGGAAAAGCTGATGAAGGCCTTCGAGTCCCTCAAGTCCTTCCAGCAGCAG
  1  M  A  T  L  E  K  L  M  K  A  F  E  S  L  K  S  F  Q  Q  Q
```

The top line of each pair is the nucleotide sequence.

```
 61  CAGCAGCAGCAGCAGCAGCAGCAGCAGCAGCAGCAGCAGCAGCAGCAGCAGCAGCAACAG
 21  Q  Q  Q  Q  Q  Q  Q  Q  Q  Q  Q  Q  Q  Q  Q  Q  Q  Q  Q  Q
```

The lower line of each pair is the sequence of the amino acids.

```
121  CCGCCACCGCCGCCGCCGCCGCCGCCTCCTCAGCTTCCTCAGCCGCCGCCGCAGGCA
 41  P  P  P  P  P  P  P  P  P  P  P  P  Q  L  P  Q  P  P  P  Q  A
```

181 . . . extending to coding nucleotide number 9,432

 61 . . . extending to amino acid number 3,144

One-letter abbreviations for amino acids:

A = alanine   M = methionine
E = glutamic acid P = proline
F = phenylalanine Q = glutamine
K = lysine   S = serine
L = leucine   T = threonine

Figure 7.7 The first 180 coding nucleotides and the resulting 60 amino acids of a normal allele for the Huntington disease gene. The sequence starts with ATG, the initiating triplet coding for methionine. The nucleotide sequence (top line) is that for the nontemplate strand. This practice is standard in the research literature, and it has the effect of making the displayed DNA sequence the same as the messenger RNA sequence (except that T replaces U). (Data from the Huntington's Disease Collaborative Research Group 1993.)

ble answer is that huntingtin binds to another protein that *is* mainly expressed in the brain. One such brain protein is abbreviated *HAP*, for huntingtin-associated protein; its function is also unknown. HAP binding is particularly strong for huntingtin molecules that have extra glutamines (Blakeslee 1995). Another protein that binds polyglutamine segments is an important enzyme abbreviated *GAPDH*. This enzyme is involved in glycolysis, the first step in energy production, on which brain tissue is particularly dependent because of its large ATP requirements (Barinaga 1996). The importance of these observations is currently being evaluated.

One conclusion stands out, however: The disease-causing Huntington allele seems to be doing something wrong, rather than just failing to do something right. That is, the altered protein with its extra glutamines is not merely innocuous, but is actively harmful. Such mutations are often called **gain-of-function mutations**, and are typical for dominantly acting mutations. (On the other hand, most recessively acting mutations involve a loss of function; the mutated allele simply fails to make a functioning enzyme, for example. In these cases, one good allele makes enough effective enzyme for a normal phenotype.)

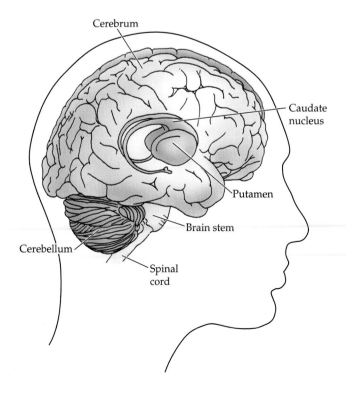

Figure 7.8 The human brain, showing the position of the caudate nucleus and the putamen, which are adversely affected in Huntington disease. The caudate nucleus and putamen are embedded within each of the two hemispheres; their positions are shown as if projected to the surface.

The age of onset of Huntington disease is strongly correlated with the number of CAG repeats. Note in Figure 7.9 that people carrying an allele with about 40 CAG repeats have an average age of onset of about 60 years, while people with about 80 repeats typically show signs of the disease as 10-year-old children.

But where do the expanded alleles come from in the first place? This is not entirely clear, but it appears that the gene is unstable and may increase in length during the process of gamete formation in either parent. When an affected person has no family history of the disease, one of the parents is likely to have a Huntington allele with a repeat number in the 30s, that is, just below the abnormal range. Such an allele is sometimes called a **premutation** because it is near the upper limit of normality and is sensitive to further change. Then, during the course of meiosis, the premutation lengthens a bit more to a **full mutation,** sufficient for disease expression when transmitted to a child.

A surprising finding is that the expansion of Huntington premutations to full mutations is much more likely to happen in a male than in a female. The preferential involvement of the father is especially strong when Huntington disease appears in a young child. Although an affected child is equally likely to be male or female, 90% of the time the Huntington allele was inherited from the father. Among affected adolescents (age 11 to 20), 75% of the time the Huntington allele came from the father.

In summary, the CAG repeat in the Huntington disease allele is unstable and may increase in size in successive generations. It is much more likely to expand when transmitted through males.

Other Trinucleotide Repeat Diseases. Table 7.1 lists a number of diseases that are due to the repetition of a trinucleotide. The first five are all due to repetition of CAG within the coding region of the gene. As a consequence, the amino acid glutamine is repeated in the encoded polypeptide, and for all five the dividing line between normal and abnormal alleles is in the neighborhood of 40 to 50 repetitions. Because of this similarity and somewhat similar clinical phenotypes, medical scientists think that the mechanism of pathology may be the same. Indeed, all five diseases involve progressive loss of neurons in the brain when the gene is expanded. The areas that are affected, in addition to the caudate nucleus and the putamen (Huntington disease), include the cerebrum, the cerebellum, and the brain stem (see Figure 7.8).

The last three entries in Table 7.1 involve different repeat units (CGG, CTG, and GAA), which are not even within the coding region of the corresponding genes. Instead, the repeated region is just "upstream" or "downstream" of coding triplets, or in an intron—that is, in regions that are transcribed into messenger RNA but not translated into a polypeptide. How expansions of these regions cause disease symptoms is not entirely clear, but the processes of transcription or translation are certainly not normal.

The *fragile X syndrome* was in 1991 the first disease to be shown to be due to a DNA sector with an expanded trinucleotide repeat. It is the most frequent cause of inherited mental impairment (appearing in about 1 in 4,000 to 5,000 males) and was so named because the site of the gene (at band q27.3 of the X chromosome) that is constricted or broken when a chromosome karyotype is prepared in particular ways (see Figure 2.4).

Males who exhibit the fragile site on their single X chromosome are moderately to severely retarded (IQs from 20 to 60) and may have distinctive physical features, including protruding ears and very large testes. Their speech is often high-pitched and repetitive. Although some of them may be hyperactive or autistic as children, their later behavior tends to be shy but friendly.

Females who exhibit the fragile site on one of their X chromosomes may show mild mental retardation or some kind of learning disability, but about half of them have IQs within normal ranges. Thus, there are more hemizygous, seriously affected males than heterozygous, mildly affected females. This situation partly accounts for

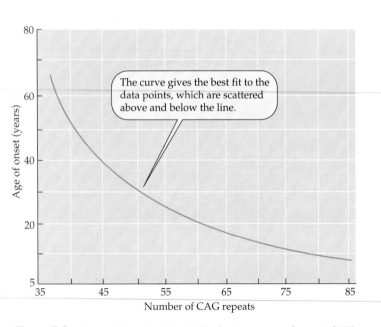

Figure 7.9 A graph showing the relation between age of onset of 479 cases of Huntington disease and the CAG trinucleotide repeat length in each patient's disease-causing gene. (After Hayden and Kremer 1995.)

TABLE 7.1 Unstable trinucleotide repeat diseases.

| Disease | Repeat sequence | Location of repeat within gene | Number of repeats in | | | Chromosomal location |
			Normal alleles	Premutations	Full mutations	
Huntington disease	CAG	In coding region	6–30	30–39	36–180	4p
Kennedy disease	CAG	In coding region	11–33	?	40–62	Xq
Spino-cerebellar ataxia type 1	CAG	In coding region	6–44	?	40–82	6p
Dentatorubral-pallidoluysian atrophy	CAG	In coding region	7–23	?	49–75	12p
Machado-Joseph disease	CAG	In coding region	13–40	?	68–82	14q
Fragile X syndrome	CGG	Before Met initiation codon (untranslated)	6–55	52–200	200–2,000	Xq
Myotonic dystrophy	CTG	After stop codon (untranslated)	5–35	50–100	100–2,000	19q
Friedreich ataxia (see Box 7A)	GAA	Within the first intron	8–22	?	120–1,500	9q

the often observed excess of males in institutions for the mentally retarded.

As noted in Table 7.1, normal alleles of the fragile X gene have between 6 and 55 CGG repetitions, which are usually interrupted by a few AGG triplets. The AGG triplets seem to stabilize the normal allele because their loss often leads to expansion of the gene in subsequent generations. Expansion directly from a normal allele to a full mutation allele apparently never occurs; the intermediate premutation level, 52 to 200 repeats, is always required. Individuals with the premutation number continue to show normal phenotypes.

Further expansion of the allele shows a parental sex effect opposite to that seen in Huntington disease. For the fragile X syndrome, expansion from the premutation level to full mutation, with 200 or more repeats, occurs only during egg formation. Males, even those with a large premutation, do not have daughters with the fragile X syndrome.

One clue to understanding the adverse effects of fragile X expanded alleles is that the C nucleotides in the repeat region and in neighboring promoter DNA are found to be chemically modified by the addition of methyl (—CH₃) groups. Methylation of the cytosines within these control regions prevents transcription of the fragile X gene. Thus, the protein product is absent. Although it interacts with RNA molecules, the function of the normal protein is unknown. But its absence is undoubtedly the primary event in the development of the fragile X

symptoms. Chapter 10 has more information on this syndrome.

Myotonic dystrophy, (Figure 7.10), is the most common form of muscular dystrophy. It occurs as an autosomal dominant in about 1 in 8,000 births. *Myotonia* is a defect of muscle fibers that results in abnormally prolonged contractions. Patients have trouble relaxing a movement after any vigorous effort, such as letting go their grip after a handshake. *Dystrophy* refers to inadequate growth of some tissue, usually muscle. Here it involves progressive muscle wasting and weakness, especially in the face, jaw, neck, and limbs. In addition, central nervous system defects—mental retardation in children and unusual apathy and drowsiness in adults—are common. Expansion of the gene from nor-

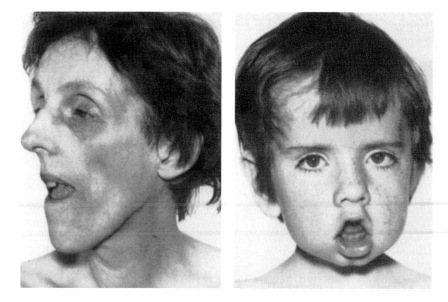

Figure 7.10 Facial features of myotonic dystrophy in two severely affected patients. (A) Drooping eyelids, facial weakness, and wasting of the muscles of the jaw and neck. (B) "Tent mouth" and jaw weakness. (From Harper 1990.)

mal lengths through premutations to full mutations occurs about equally in both sexes. Furthermore, the expanded alleles continue to expand in somatic cells throughout an individual's life. Analyses of various tissues from older adults show a greater number of repeats.

All the unstable trinucleotide repeat diseases discussed so far, but especially myotonic dystrophy, show a phenomenon, called **anticipation**. It was first observed by early human geneticists, who could not explain it. Anticipation means that a disease appears in a family at younger and younger ages in successive generations, and its expression may become more and more severe. In one of the first textbooks on human genetics, Curt Stern (1949) wrote:

> We have seen that the age of onset of Huntington's chorea is variable from one affected individual to another. Such variability is typical for many inherited diseases whose symptoms appear late in life. It is a widely held opinion among medical men, and some statistics seem to support it, that the age of onset of these diseases becomes earlier and earlier in successive generations.

At first, medical personnel thought that they were more likely to look for and find the disease in the children and grandchildren of affected individuals. The phenomenon was thus thought to be a bias of observation—an artifact of looking harder—rather than a condition with a genetic explanation. But now we can see that at least some cases of anticipation are real and due to continuing expansion of trinucleotide repeats in successive generations.

In Box 7A, we note briefly a somewhat atypical example of a trinucleotide repeat disease.

Changes Due to Unequal Crossing Over

A heterogeneous group of mutational changes consists of deletions, duplications, insertions, and fusions involving blocks of DNA. These blocks are generally larger than those discussed above, but smaller than chromosomal abnormalities that are visible under the microscope (Chapters 13 and 14). Although deletions and duplications, for example, can involve just one or a few nucleotides (frameshift mutations), they sometimes involve hundreds or thousands of nucleotides, or even the loss or gain of a whole gene. One mechanism for generating these alterations of DNA is a curious phenomenon called **unequal crossing over**.

As noted in Chapter 3, homologous chromosomes undergo a very close pairing called synapsis during the first meiotic division. Synapsis is thought to involve the yoking together of two double helices that have very similar nucleotide sequences, although how this trick is accomplished is still a big mystery. After synapsis, the strands undergo crossing over at random places along the chromosomes. In a formal way, crossing over can be explained as breakage of two different strands at exactly the same point and their reunion in opposite positions (Figure 7.11A); the actual molecular mechanisms of crossing over, however, are much more complicated than this. The end result is that alleles of genes on corresponding parts of homologous chromosomes are interchanged with respect to each other. We emphasize that crossing over is a perfectly normal phenomenon; virtually every chromosome pair has multiple crossover sites during every meiosis. (We ignore here the fact that a crossover involves chromatids rather than whole chromosomes, but that refinement need not concern us in the present discussion.)

When the same DNA sequence occurs more than once, either nearby or even at widely separated places on a chromosome, the proper alignment for synapsis can be fooled. In Figure 7.11B, we show synapsis of three adjacent regions sufficiently similar to each other that synapsis can still occur when one chromosome has slipped a unit passed the other. Such misalignments are apparently quite common, and have no particular consequence except when crossing over occurs in the misaligned region, in which case the products of

---

BOX 7A   *An Atypical Trinucleotide Repeat Disease*

Friedreich ataxia is fairly rare, having a frequency of about 1 in 50,000. (*Ataxia* means a failure of muscular coordination.) It is characterized by degeneration of the cerebellum and spinal cord, leading to progressive defects in limb coordination, reflexes, and speech. Patients are confined to a wheelchair. In addition, heart abnormalities are common and often fatal. Onset of Friedreich ataxia is usually between 8 and 20 years, but can happen at age 5 or even younger. Two unusual aspects of the mutant allele are the location of the GAA repeat (in an intron) and its mode of inheritance (recessive rather than dominant). Affected individuals are almost always homozygous for alleles with the expanded repeat length (but not usually of the same length). In a few cases, affected persons possess one allele with an expanded repeat and a second allele of normal length but with a base substitution mutation. Perhaps because of its recessive inheritance, there are no obvious signs of anticipation. Thus, the recent discovery that Friedreich ataxia is associated with an unstable trinucleotide repeat destroys some of the generalizations that are possible on the basis of all the other examples (Warren 1996).

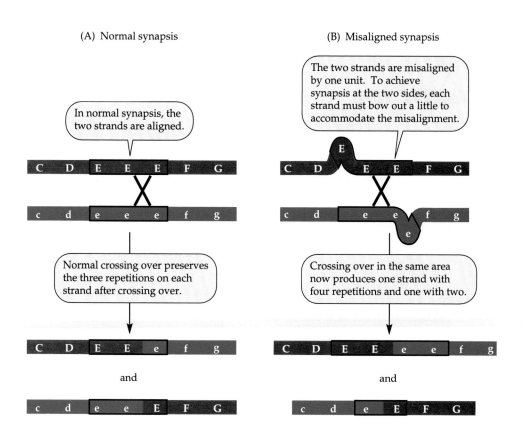

Figure 7.11 A comparison of (A) normal synapsis and (B) misaligned synapsis and subsequent unequal crossing over. The chromosome region has a sequence that is repeated three times (labeled E on one strand and e on the other).

the meiosis (eggs or sperm) will contain chromosomes with gains and losses of material. We will consider two examples of the kinds of mutant alleles that can result from unequal crossing over.

Lepore Hemoglobin. The first example is a type of β-thalassemia that occurs in persons with so-called *Lepore hemoglobin* (first found in a family with that name). The β-like chain in Lepore hemoglobin has the usual 146 amino acids, but the first part is like a normal δ chain and the remainder is like a normal β chain. Recall from Chapter 6 that most adult hemoglobin has the formula $\alpha_2\beta_2$, but a little adult hemoglobin is $\alpha_2\delta_2$. The δ and β chains are very similar, differing in only ten amino acids. The δ and β genes, therefore, are expected to be very similar. As seen in Figure 6.20, the δ and β genes are close to each other on chromosome 11, and it has been suggested that during meiosis they might sometimes mistakenly synapse with each other rather than with their own alleles (Figure 7.12A). If a crossover should then occur along the length of the misaligned DNA sequences, the result would be a mixed or fused gene. One crossover product, the Lepore chromosome, having neither an intact δ gene

nor an intact β gene, leads to a thalassemia phenotype. The other crossover product, the "anti-Lepore" chromosome, produces a near-normal phenotype.

The second example involves a duplicated gene that leads to a syndrome called *Charcot-Marie-Tooth (CMT) disease*, named after the three physicians who first described it in 1886 (and having nothing to do with teeth). With a frequency of about 1 in 2,500, CMT is the most common inherited neurological disorder. Although at least three different genes can lead to very nearly the same phenotype, we are concerned here with the most common form, type 1A, inherited as an autosomal dominant with a locus near the centromere of chromosome 17. Beginning in their early teens, affected individuals show weakness and wasting of the muscles of the legs, arms, feet, and hands. The cause is gradual degeneration of the myelin sheath of nerves that supply these regions. (Myelin is a fatty insulating substance that encases many nerves and is needed for normal nerve impulses.) Without normal nervous stimulation, the muscles become progressively weakened. Although treatments (physical therapy, for example) are not very effective, life expectancy is near normal.

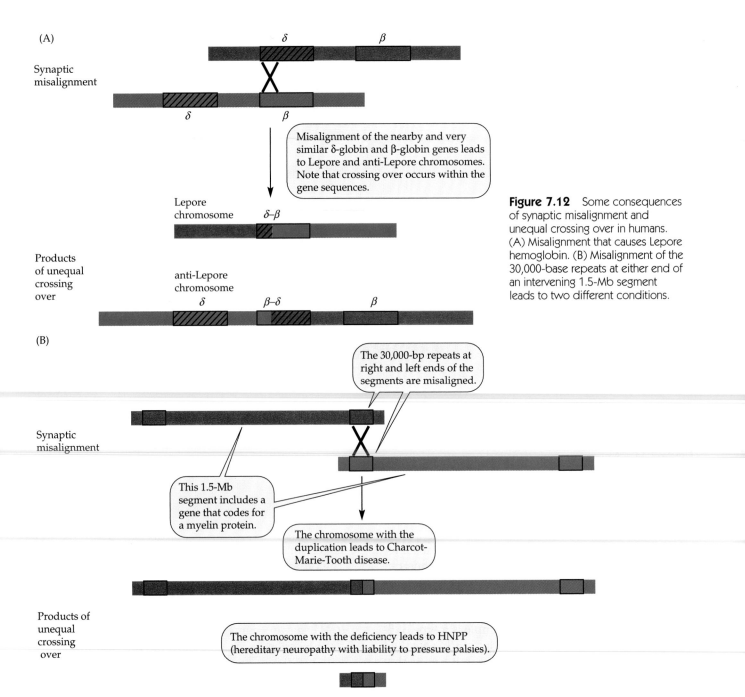

(A)

Synaptic misalignment

δ β

δ β

Misalignment of the nearby and very similar δ-globin and β-globin genes leads to Lepore and anti-Lepore chromosomes. Note that crossing over occurs within the gene sequences.

Lepore chromosome δ–β

Products of unequal crossing over

anti-Lepore chromosome
δ β–δ β

(B)

The 30,000-bp repeats at right and left ends of the segments are misaligned.

Synaptic misalignment

This 1.5-Mb segment includes a gene that codes for a myelin protein.

The chromosome with the duplication leads to Charcot-Marie-Tooth disease.

Products of unequal crossing over

The chromosome with the deficiency leads to HNPP (hereditary neuropathy with liability to pressure palsies).

Figure 7.12 Some consequences of synaptic misalignment and unequal crossing over in humans. (A) Misalignment that causes Lepore hemoglobin. (B) Misalignment of the 30,000-base repeats at either end of an intervening 1.5-Mb segment leads to two different conditions.

CMT is associated with a duplication of 1,500,000 bases, or 1.5 megabases (Mb), of DNA. The duplication includes a gene that encodes a protein found in myelin, and the disease symptoms are probably due to the double dose of this gene on one chromosome. It is remarkable that in completely unrelated families with CMT, exactly the same duplication is seen. Thus, the event that leads to the sizable chromosome change must occur over and over.

James Lupski and other researchers at Baylor College of Medicine in Houston have recently sequenced the relevant DNA. They found that at either end of the single 1.5-Mb region in normal people there is the same 30,000-base sequence. By searching DNA databases, they showed that a portion of this sequence is similar to a transposon named *mariner* that is found in insects! Transposons like mariner have the ability to release themselves from a chromosome and insert themselves elsewhere. To do this, they break the chromosomal DNA in ways perhaps similar to what happens in crossing over. Indeed, the researchers ultimately found a "hot spot" for crossing over near the transposon-like DNA, especially when the end sequences are misaligned, as seen in Figure 7.12B. The researchers called this segment of human DNA *MITE*, for *m*ariner *i*nsect *t*ransposon-like *e*lement. MITE is not an active transposon, but researchers believe that it nevertheless invites crossing over nearby. Still mysterious is whether and how MITE got into human DNA from insects (Grady 1996). Mariner-like elements are found elsewhere in the human genome and in the genomes of other mammals.

One other fact makes this example very similar to the Lepore/anti-Lepore example. The two products of crossing over are a too-long chromosome producing CMT, and a complementary too-short chromosome that also produces a disease—but not the same one (Figure 7.12B). The disorder corresponding to the deletion of the 1.5-Mb segment is called HNPP, for *hereditary neuropathy with liability to pressure palsies*.

Insertions

The transposable element implicated in CMT disease is not *within* the CMT gene itself. However, there are genes that become mutant through the insertion of DNA segments that don't belong there. We will consider two examples of such **insertion mutations**.

Hemophilia A. Hemophilia A (classic hemophilia) is due to a recessive mutation in the X-linked gene that codes for plasma factor VIII, one of the many proteins needed for proper blood clotting (see Box 5D). The disease affects about 1 in 10,000 males, and its effects are variable with mild, moderate, and severe cases. Typically, boys with severe hemophilia are diagnosed because of bleeding into joints when they first begin to walk. Untreated, the condition leads to inflammation, pain, restricted motion, and permanent disability (Hoyer 1994).

The disorder can be treated with injections of factor VIII, purified and concentrated from pooled human plasma (usually from paid donors). Side effects, however, include immunological problems and liver damage. In addition, during the late 1970s and early 1980s, before methods to inactivate viruses were perfected, patients with hemophilia were at risk for receiving the human immunodeficiency virus (HIV). Tragically, about two thirds of several thousand people with hemophilia who contracted AIDS from their factor VIII injections have died (Meier 1996). The synthesis of factor VIII by recombinant DNA methodologies was spurred by these events (Chapter 8). Since 1992, virus-free recombinant factor VIII has been available in addition to factor VIII from pooled human plasma that is also virtually free of disease-causing viruses.*

In 1988, Hugh Kazazian and colleagues at Johns Hopkins University analyzed the DNA sequences of

two mutant alleles of the factor VIII gene and found that a piece of DNA from another chromosome was inserted into exon 14 of the gene. The inserted piece was a *LINE* transposable element, about 3.8 kilobases (kb) long in one case, and 2.3 kb in the other. Researchers were able to show that the 3.8-kb insert was a copy of DNA that was near the centromere of chromosome 22. The insertions completely disrupted normal gene function: No factor VIII was detectable in either patient, and their hemophilia was severe. In both cases, neither parent had any abnormality in their factor VIII genes, so both mutations were newly arising during either egg or sperm formation.

We should add that most cases of hemophilia A are *not* due to an insertion of errant DNA. The gene itself, located near the tip of the long arm of the X chromosome, is big—186 kb comprising 26 exons. The protein it codes for has 2,351 amino acids separated (after transcription and translation) into two polypeptides. Many different types of mutations of the factor VIII gene have been found, including a wide variety of base substitutions and frameshifts. The single most common mutation—constituting about a third of all new mutations—is a complex disruption in which part of the gene is turned around end for end, a so-called *inversion* (Chapter 14).

Neurofibromatosis Type I. The second example of a disease-causing insertion involves the gene for *neurofibromatosis type 1*, or *von Recklinghausen disease*, a fairly frequent (about 1 in about 3,500 births) autosomal dominant disease of the nervous system. Although expressed to some degree in anyone who possesses the mutant allele, the phenotype is extremely variable—between families, within families, and even in the progression of the disease within a single individual. Mild cases may involve only a few lightly pigmented areas called *café-au-lait spots* (which are the color of coffee with milk), freckles in the armpits, and a few *skin neurofibromas* (benign but disfiguring tumors that arise from the fibrous coverings of nerves) (Figure 7.13A). Also present are so-called *Lisch nodules* (small lumps) in the iris of the eye. In more serious cases, however, patients may develop thousands of neurofibromas of various sizes (Figure 7.13B) and sometimes a wide range of additional symptoms, including learning disabilities, curvature of the spine, and other bone deformities. Physicians cannot treat the major symptoms very effectively, although cosmetic surgery may lessen severe disfigurement and improve social interaction.

Like the hemophilia gene, the neurofibromatosis gene is large, about 350 kb long with at least 59 exons. It encodes a protein called *neurofibromin* over 2,800 amino acids long. The normal protein helps to control cell division by blocking a series of chemical growth signals that pass from the surface of cells to the genes.

---

*Both sources of factor VIII are expensive. The recombinant DNA product costs about $40,000 to treat a person with hemophilia for a year. The plasma concentrate costs $15,000 to $35,000, depending upon the degree of purity.

(A) (B) (C)

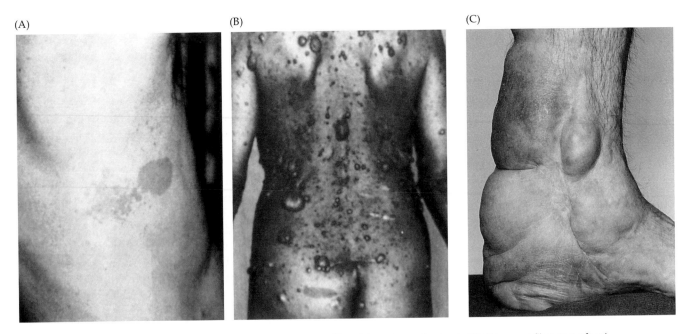

Figure 7.13 Some features of neurofibromatosis type 1. (A) Café-au-lait spots on the trunk. (B) Skin neurofibromas of various sizes on the back. (C) Neurofibroma with tissue overgrowth on the lower leg. (A and B from Riccardi 1990. C courtesy Susan M. Huson.)

It is one of a class of agents called *tumor suppressors*, which we consider further in Chapter 17 in our discussion of cancer. But mutant alleles produce altered neurofibromin that fails to block these signals. Mitosis then proceeds unchecked, leading to excessive cell growth and the formation of tumors.

Many kinds of mutations are known to produce neurofibromatosis type 1. A mutation in one patient was found to be due to the insertion of an *Alu* repeat into the intron between exons 5 and 6, fairly close to the beginning of exon 6. You might expect such an insert to have no effect, but this transposon-like element interferes with the splicing of the mRNA transcript such that exon 6 is deleted; that is, exon 5 is directly connected with exon 7 (Figure 7.14). The incorrect splicing also alters the reading frame for exon 7 and a stop codon is soon encountered, leading to a truncated protein product. The parents of this patient did not have neurofibromatosis, so the transposition mutation must have occurred during gamete formation in either the mother or father.

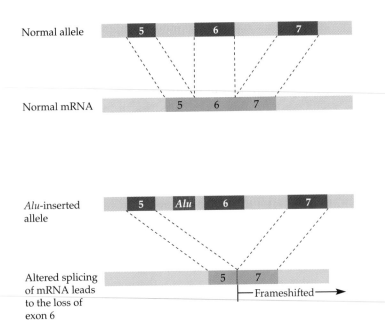

Figure 7.14 The *Alu* insertion between exons 5 and 6 of the neurofibromatosis type 1 gene and its effect on splicing of the messenger RNA. (After Wallace et al. 1991.)

Agents of Mutation

As noted earlier, agents that can induce mutations are called mutagens. They fall into two broad categories: chemicals and radiation. Direct information on how these agents affect the DNA of human beings is scanty, but geneticists who work with experimental organisms are able to obtain some fairly precise data that may or may not apply to humans. They can, for example, deliberately subject laboratory animals to known mutagens to produce phenotypic variations in the descendants of the treated animals.

Chemical Mutagenesis

The following three chemical mutagens are often used in a laboratory setting:

1. *Proflavin*, a brownish dye used as an antiseptic in veterinary medicine, causes the addition or deletion of single bases during the synthesis of new DNA strands. Proflavin's mutagenic action is caused by its shape: It is a very flat molecule that wedges itself between the base pair rungs of the DNA helix.

2. *Nitrous acid* is a powerful mutagen that replaces amino groups ($-NH_2$) with keto groups ($=O$). By this action, the base cytosine is converted to uracil, and adenine is converted to a base called hypoxanthine. Because the pairing properties of changed bases are modified, DNA treated with nitrous acid produces errors during subsequent replications.

3. *5-Bromouracil* is one of many *base analogues*, substances that so closely resemble the natural bases that they can be mistakenly incorporated into DNA at the time of replication. A molecule of 5-bromouracil is similar to thymine. Whereas thymine always base-pairs with adenine, 5-bromouracil occasionally base-pairs with guanine, which leads to mistakes during subsequent replications.

Although these chemicals are not part of the general environment, a variety of substances that may be mutagenic impinge daily on the general human population. Furthermore, it is possible that genetic damage is occurring without our ever being able to link a particular mutagen with its eventual deleterious phenotype. We are unable to match a mutagen to its effect because induced mutations are generally not different in their phenotypic effects from spontaneous mutations, and their germinal effects may not appear for generations. But one clearly dangerous material over which each of us has some direct control is tobacco smoke, which contains about 50 substances known to be mutagenic or carcinogenic. The somatic consequences of cigarette smoking are well documented: sharply increased rates of cardiovascular disease and cancer, especially lung cancer (Bartecchi et al. 1994, 1995; MacKenzie et al. 1994). Whether smoking induces germinal mutations in humans is unknown.

Because it is neither ethical nor feasible to deliberately expose humans to a chemical in order to detect possible mutations, no reliable data have been obtained directly on human populations. Instead, test systems have been developed that use bacteria, fungi, plants, insects, mammals, or mammalian cells in tissue culture. In principle, because DNA is DNA, mutagenicity in one organism should mean mutagenicity in another. But in practice, differences exist between species, both in the degree of protection afforded germinal tissue and in the type of metabolism developed by an organism over evolutionary time. For example, in a bacterium a single cell membrane and cell wall separate its DNA from its surroundings. But in humans an ingested potential mutagen is first exposed to the harsh environments of the mouth, stomach, and intestines before it is absorbed into the blood via the capillaries of the digestive organs. This blood is then delivered to the liver, where absorbed materials may be processed by myriad enzyme systems. Only then are ingested substances or their by-products transported to other sites throughout the body, including the germinal cells.

Extensive testing of substances for their mutagenic or carcinogenic potential stems in part from the 1958 Delaney amendment to the Food, Drug, and Cosmetic Act, which barred from the market any food additive found to cause cancer in experimental animals, *whatever* the dosages tested. As analytical methods improved, traces of pesticides could be detected in foods (in parts per trillion) far below levels considered to be possible health risks, so this law was modified in 1996; pesticide concentrations in the food supply must now be shown to be harmless with "reasonable certainty" (Hoyle 1996).

Because animal tests must be conducted within a reasonable time on a finite budget, and because damage to any given cell is a rare event, experimental animals must be exposed to very high concentrations of the test agent. Only in this way is the investigation practical. Although extrapolating from rodent bioassays based on high doses to humans exposed to low doses is far from direct, and in some cases is known to be invalid, animal tests are to some extent unavoidable. In short, mutagen testing is not straightforward, as illustrated by the case of caffeine (Box 7B).

Because it is costly and time-consuming to test small mammals directly with compounds that may be mutagenic or carcinogenic, assays on lower organisms are used to prescreen a wide variety of chemicals. Those chemicals that give positive results can be investigated further by using more difficult but more pertinent test systems. Of the many screening tests for chemical mutagenicity, perhaps the most rapid, simple, sensitive, and economical test was developed by Bruce Ames of the University of California at Berkeley. The **Ames test** can suggest whether a compound is mutagenic or possibly carcinogenic in animals by determining whether it is mutagenic in bacteria.

The Ames test starts with strains of *Salmonella* bacteria that are unable to make the amino acid histidine because they carry a mutant allele *his⁻* (coding for a defective enzyme). Histidine is a vital amino acid and the bacteria need it to grow. Bacterial cells carrying *his⁻* are spread over the surface of nutrient medium lack-

BOX 7B   *Is Caffeine a Mutagen?*

Caffeine is present naturally or artificially in coffee, tea, chocolate, many soft drinks, and a wide range of common medicines. The plants that synthesize caffeine apparently use it as a natural pesticide. Caffeine is a purine, very similar in structure to adenine and guanine, but different enough (with three methyl groups) that it does not substitute for the normal purines in DNA (see figure). When tested on most bacteria and fungi, caffeine produces mutations. It is thought that caffeine may act not by being a mutagen itself, but by interfering with the ability of cells to repair damage to DNA from other causes.

In plant, hamster, and human cell cultures, high concentrations of caffeine have been shown to cause chromosome breaks, but the relationship between breakage and point mutations

is unclear. With fruit flies, the evidence is contradictory. Extensive experiments with mice yield no indication of caffeine-induced mutations. In humans, ingested caffeine is rapidly excreted, although it does reach all body fluids (including milk) and tissues (including the gonads), and it does cross the placental barrier. Fetuses appear to be unaffected by moderate maternal coffee consumption, however. Thus,

while pondering all the data over a cup of coffee, we probably need not worry too much about the caffeine in it. (But note that the International Olympic Committee considers caffeine in urine as a disqualifying factor for athletes.) Risk taking is personal, and you may wish to investigate whether caffeine has other detrimental or beneficial effects (Apple 1994; Eskenazi 1993; Garattini 1993).

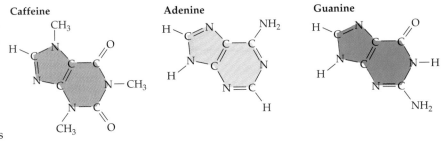

ing histidine. Therefore, at the start of the test, the cells cannot grow; but they *will* grow if a random mutation changes the allele coding for the defective enzyme to an allele coding for a functioning enzyme (a so-called back mutation, symbolized *his⁻ → his⁺*). Individual bacteria with back mutations will eventually be visible as discrete bacterial colonies (Figure 7.15).

Several different histidine-requiring strains of *Salmonella* can be used in different tests of the same potential mutagen. Some tester strains are unable to synthesize their own histidine because of a base substitution mutation, and others because of a frameshift mutation. Thus, a mutagen with a particular mode of action that might be missed using one tester strain may be detected using another. For example, a frameshift mutation in a gene due to a single-base *deletion* can be most simply "corrected" by a nearby single-base *addition*. No base substitution can correct a frameshift mutation.

The tester strains of *Salmonella* also carry other heritable traits that make the test very sensitive. A modification of the cell wall allows easy entry of the test chemical into the bacteria. In addition, the bacterial enzymes that are normally used to repair DNA damage are inactivated. Sometimes added to the growth medium are rat liver enzymes that are capable of modifying some test chemicals from nonmutagenic to mutagenic form, or vice versa. In this way, the bacterial test is made to resemble, to some extent, a mammalian system in which nutrients absorbed into the bloodstream first travel to the liver.

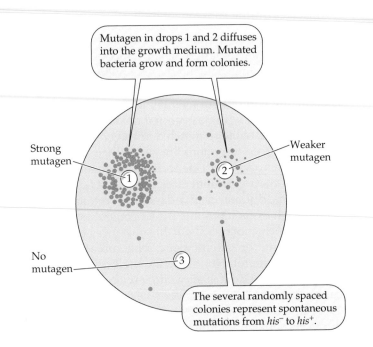

Figure 7.15   An Ames test. In a petri dish, a thin, nearly invisible lawn of bacteria is spread on a nutrient medium that lacks the vital amino acid histidine. Because the bacteria are *his⁻* and thus unable to synthesize histidine, they fail to grow. A mutation from *his⁻* to *his⁺*, however, permits a cell to make its own histidine and thus to reproduce. Three drops of test substances are spotted on the lawn. Mutated cells grow and divide, forming visible colonies—the small, dark spots. (Many other types of mutations not revealed by this test presumably also occurred in the bacteria.)

Radiation Mutagenesis

All living things are exposed to natural, and largely unavoidable, radiation. It emanates from the sun, from cosmic rays, and from the radioactivity of uranium, radium, and other unstable elements in rocks, soils, food, and air. Very soon after the discovery of X-rays and radioactivity in the 1890s, it became clear that the body could be damaged by these agents that could not be seen, felt, or smelled. In the 1920s, Herman J. Muller, then at the University of Texas, reported that X-rays induce mutations in fruit flies; at about the same time, Lewis J. Stadler of the University of Missouri got the same effects in barley plants. Public awareness of the biological consequences of radiation began in the 1950s as a result of the atomic bombing of two Japanese cities at the end of World War II and the continued atmospheric testing of nuclear weapons. Current medical uses and the presence of about 90 nuclear power plants in the United States (about 400 worldwide) continue to focus public attention on the implications of radiation. Interest has been heightened by aging nuclear power facilities and by accidents at Three Mile Island, Pennsylvania, in 1979 and at Chernobyl in the Ukraine, in 1986.

To be mutagenic, radiant energy must usually reach DNA. When it does, so-called **ionizing radiation** may knock an electron out of an atom in DNA; this event leaves the atom as a charged ion and in a very reactive state, subject to further chemical changes. A direct hit on DNA is not necessary, however; nearby ionized chemicals can subsequently affect DNA. With our emphasis on germinal mutations, we are not concerned here with *ultraviolet* light, since it does not penetrate below the skin (although ultraviolet radiation is implicated in the induction of skin cancer). Nor are we concerned here with the *electromagnetic fields* (EMFs) associated with power lines. This radiation is nonionizing (not able to produce ions) and generally very weak; no evidence connects EMFs with germinal mutations, and the evidence that it causes cancer is inconclusive and contradictory (Bennett 1995; Broad 1995; Campion 1997).

The amount of radiation absorbed is called the *radiation dose*. It can be calculated from physical principles and measured by various devises (e.g., Geiger counters or badges containing sensitive film) that determine the ions produced or the energy absorbed. Radiation can be received at different rates. A dose received from a high-intensity source is called *acute* radiation, whereas a dose received at low intensity is called *chronic* or low-level radiation. The public is concerned primarily with chronic doses, because these characterize many natural and artificial sources to which the general population is exposed.

A typical American receives 10–11 rem (a unit of radiation dose) of low-level ionizing radiation from all sources over the average reproductive cycle of 30 years (Box 7C). Roughly 80% of this is from natural sources, including radioactive isotopes in our environment and within our bodies. The major artificial source of radiation is from medical procedures. Only a fraction of this radiation reaches the gonads, where it could induce germinal mutations. For abdominal or pelvic X-rays, a greater fraction is genetically significant because the gonads (especially the ovaries) cannot always be shielded effectively.

Radiation has many different kinds of effects. These include chromosome breaks and point mutations in germinal and somatic tissues, the induction of cancer (particularly leukemia), localized radiation burns and tissue scarring, and at higher doses, generalized radiation sickness (nausea, vomiting, diarrhea, internal bleeding, general weakness) and death. One important and much-debated question is whether there is a "safe dose," or **threshold,** an amount of radiation below which there is no risk. To answer this question, investigators irradiate experimental animals with smaller and smaller doses (necessitating greater costs and the use of more and more test organisms) and then measure a particular effect—for example, point mutations. The plotted result is a *dose-response curve*, which unfortunately will have no experimental points below a certain dose. The form of the curve, however, gives theoretical information on the way radiation induces damage, as well as practical information to help formulate medical and public health policies.

The *somatic* effect of major concern with respect to low-level radiation is the possible induction of cancer. Using Japanese data on atomic bomb survivors and patients treated with radiation, as well as other data, researchers have estimated that 1 additional rem of X-radiation per person (above that normally received) leads very roughly to 2 cancer deaths per 1,000 people (Davis and Bruwer 1991). For comparison, note that roughly 150 of 1,000 births currently end in a cancer death. No single cancer fatality can be unequivocally identified as radiation-induced because (1) the time elapsing between irradiation and the detection of the malignancy is long—years or even decades; and (2) the cancers induced by excess radiation are not different in kind from those occurring from other causes (although leukemia seems to be induced at a greater rate by radiation than by other agents). Thus, it is unclear whether there is a radiation threshold for the induction of cancer (Goldman 1996).

The *germinal* effect of major concern is the possible induction of point mutations or chromosomal aberrations that may cause harm to the immediate offspring or later descendants. Reliable estimates of the germinal effect of radiation are especially difficult to obtain

BOX 7C *Exposure to Low-Level Ionizing Radiation*

Every day we are bombarded with radiation from both natural and artificial sources. Natural radiation, which is largely unavoidable, accounts collectively for about 82% of our total exposure. Only recently has it been recognized that a sizable proportion (55% of the total) comes from radioactive radon gas escaping from the earth. Some of this radon seeps into houses from soil and groundwater. When it does, the tighter and more energy-efficient the house, the higher the concentration of radon. Although the matter is still uncertain, radon appears to pose a risk much greater than that from other radiation sources, especially in certain geographical areas (Horgan 1994). Here are some other sources of natural radiation:

Internal emitters (11%) refer to radioactive elements that become part of our body structure through eating and breathing. A major source is radioactive potassium in many common foods.

Cosmic rays (8%) reach the earth from the sun and interstellar space. Progressively filtered by layers of the atmosphere, the dose is several times greater in mountains than at sea level.

Terrestrial radiation (8%) is due largely to the radioactive elements uranium and thorium in rocks and soil. Individual doses vary considerably from one geographical area to another.

The sources of artificial ionizing radiation include the following:

X-rays for medical diagnosis (11%) and other health uses of radiation differ from those previously listed in that they are administered deliberately, rather than being incidental to the activities of living. Exposures include dental and chest X-rays, CAT scans, and mammography. Fluoroscopic examinations produce much higher doses than most other types of X-ray diagnoses. Pregnant women should be irradiated only in unusual circumstances, because the developing fetus is particularly sensitive to radiation damage.

Nuclear medicine (4%) includes many diagnostic and therapeutic techniques. Radioactive isotopes are often used to trace the path of chemicals as they are metabolized in the body or to make an image of an organ or tissue, such as the heart, liver, lung, bone, or thyroid gland.

Consumer products (3%) include a variety of sources, such as radioactive materials in tobacco smoke (polonium), building materials (uranium), well water (radon), and lawn and garden fertilizers. Smaller exposures come from color televisions, computer screens, fire detectors, and luminous dials.

The "other" category (less than 1%) includes occupational risks to miners, medical personnel such as X-ray technicians, and flight crews (because they work at high altitudes). (In the diagram, these exposures are assumed to be spread over the total U.S. population.) Also included here are sources that generate considerable political interest: nuclear and coal-fired power production and fallout from testing nuclear weapons.

The pie diagram shown here gives rough estimates of the amount of low-level ionizing radiation to which an average American is exposed (BEIR V 1990). The total amount is 10–11 rem per reproductive generation (30 years). For each source, the numbers give the 30-year dose in rem and the corresponding percentage of the total radiation exposure.

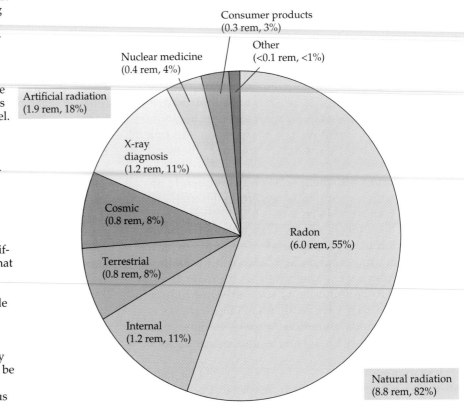

directly from humans because many new mutations are recessive. Therefore, the appearance of the mutant phenotype may be several generations removed from the mutational event (if it is ever expressed at all). The direct studies from Hiroshima and Nagasaki of the germinal effects of radiation were also plagued by great uncertainties about the doses received by the survivors.

On the basis of animal (especially mouse) and human data, researchers have estimated that 1 rem per generation (over successive generations) produces roughly one case of genetic disease (Mendelian or chromosomal) per 7,000 births. For comparison, note that approximately 3% of all newborns are currently affected with a moderate to severe Mendelian or chromosomal disorder. On theoretical grounds and from the experiments with flies and mice, most geneticists think that any amount of radiation carries with it a proportional risk of mutation. *There is probably no threshold, no safe dose, for germinal effects.*

On the basis of all information available, we conclude that the genetic hazards of radiation for the general human population are small. The most reliable data come from extensive mouse experimentation and from 50 years of studying survivors of the atomic bombs dropped on Japan in 1945. Researchers have estimated that about 400 rem of chronic radiation would be required to double the rate at which human mutations occur spontaneously (Schull 1995, pp. 264–265). This amount is called the **doubling dose.** Over a reproductive lifetime, however, a typical person receives only 10–11 rem of radiation; so it appears that the radiation we normally receive cannot account for most of the mutations that occur spontaneously.

Augmenting that temperate view of the mutagenic role of radiation are recent studies of the DNA of offspring of persons who received acute radiation from the bombings of Hiroshima and Nagasaki. Japanese researchers found no significant differences in the number of nucleotide changes between the children of parents exposed to substantial radiation and control children born to lightly exposed or unexposed parents (see commentary by Neel 1995).

Meanwhile, the continuing studies of human populations in Ukraine, Belarus, and western Russia exposed to fallout from the Chernobyl accident have had diverse results. Some firefighters and other heroic personnel who were the first on the scene after the explosion received massive radiation doses, and several dozen of them died from burns or radiation sickness. Among children exposed to fallout that was heavy with radioactive iodine, there has been a sharp increase in thyroid cancer—more than 700 cases (Balter 1996).* This tragedy has been one the most devastating health-related results of the accident. But other physical health consequences of the widespread radiation, such as the induction of leukemia and other cancers, have been difficult to document.

Studies of risk-taking have shown that people's perceptions of the hazards of life sometimes correlate poorly with real dangers. For example, actuarial esti-

mates of the causes of death in the United States show that cigarettes, alcohol, motor vehicles, and handguns are far more dangerous than police work, plane travel, nuclear power, and pesticides (Anonymous 1996; Upton 1982). Yet we often tolerate the more serious risks while concentrating on the lesser ones, perhaps because some risks are voluntary and others are imposed on us. With regard to radiation, we also tend to ignore medical and dental X-rays as a significant source of radiation over which we have some personal control. Although the benefits from diagnostic and therapeutic X-rays are often substantial, their risks should be evaluated and minimized whenever possible, and alternative methods should be considered.

Thwarting Mutations

That DNA is a *double-stranded* and *complementary* structure means that the sequence on one strand can be determined on the basis of the other. Thus, if an environmental agent alters one strand, the cell has not totally lost the correct genetic information; it could rebuild the original double helix if the right tools were available. Indeed, cells possess repair kits in the form of enzymes that act on the genetic material, so that a DNA change need not be permanent.

It is estimated that most injuries that occur from the action of toxic chemicals, radiation, random thermal motion, or replication mistakes are quickly mended, so that lesions are not transmitted to subsequent cell generations as mutations. The importance of DNA repair in the life of cells is suggested by the multitude of systems involved: In yeast, more than 50 genes have been identified that code for proteins used in the complex and interrelated processes of DNA repair, replication, recombination, and transcription. The enzymes that help in DNA repair were designated "molecules of the year" by *Science* magazine in 1994 (Koshland 1994).

Typically, enzymes in the cell are able to recognize damage to one strand of DNA, remove the single-stranded segment that includes the damaged site, and then fill in the correct nucleotides complementary to those on the undamaged strand. The process has been well studied in bacteria after ultraviolet light has induced the formation of so-called *thymine dimers,* which are adjacent thymines in one strand that become abnormally bonded to each other. This versatile cut-and-patch process, called **excision repair**, is illustrated in Figure 7.16. Excision repair is able to rectify many different kinds of insults to one strand of DNA.

Abnormal Repair Syndromes

In humans, a number of rare but interesting autosomal recessive diseases are caused by faulty DNA repair systems or faulty DNA replication systems. The two systems are related because part of the repair process

---

*Iodine is concentrated into the thyroid gland, where it is incorporated into thyroid hormones that control metabolic rates, growth, and development. Radioactive iodine is no longer a threat in the Chernobyl area because radioactive iodine-131 has a half-life of 8 days—meaning that half of its radioactivity is dissipated in 8 days. Other radioactive isotopes of iodine have shorter half-lives.

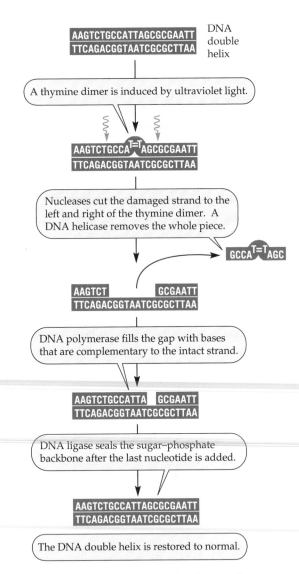

DNA double helix

A thymine dimer is induced by ultraviolet light.

Nucleases cut the damaged strand to the left and right of the thymine dimer. A DNA helicase removes the whole piece.

DNA polymerase fills the gap with bases that are complementary to the intact strand.

DNA ligase seals the sugar–phosphate backbone after the last nucleotide is added.

The DNA double helix is restored to normal.

Figure 7.16 Repair of a DNA-distorting thymine dimer by the various enzymes (nuclease, helicase, polymerase, ligase) of excision repair. Details of this and several other DNA repair systems are based primarily on bacterial experiments.

is the synthesis of a new section of DNA. The phenotypes of affected individuals often include a tendency to cancer and chromosomal aberrations of various kinds. We will consider several of these **DNA repair syndromes** (Table 7.2).

Xeroderma Pigmentosum. One abnormal repair syndrome is called xeroderma pigmentosum (XP) (Greek *xeros*, "dry"; *derma*, "skin"), a group of rare diseases with a total frequency of about 1 in 250,000. In areas exposed to sunlight, patients have very heavy freckling, pigment splotches, open sores on the face and elsewhere, and several types of cancer (Figure 7.17). Some forms of XP involve neurological abnormalities, including microcephaly (small head) and progressive

mental deterioration. Death usually occurs before adulthood, often after metastases (spreading) of the skin cancers (Williams 1997).

The several different recessive genes leading to XP phenotypes code for abnormal enzymes that are unable to repair those DNA defects that are induced by ultraviolet light, such as the thymine dimers noted earlier. Nine somewhat similar but distinguishable forms of XP suggest that the repair of these defects requires the correct functioning of nine different gene-encoded enzymes involved in excision repair. Several of these genes have been cloned—the first in 1990 by Japanese researchers at Osaka University, who are investigating the functions of the encoded proteins.

Ataxia-Telangiectasia Ataxia-telangiectasia, or AT, occurs at a frequency of about 1 in 40,000 children. Although normal at birth, affected infants begin to display diverse symptoms. *Ataxia* refers to loss of muscle control (often beginning with an unsteady gait) due to progressive damage in the cerebellum, the portion of the brain concerned with posture, balance, and coordination (see Figure 7.8). Patients may eventually be confined to wheelchairs and have difficulty speaking. *Telangiectasia* refers to skin redness (especially on the eyes, ears, and neck) resulting from tiny dilations of capillaries. About 15% of patients develop malignancies, especially leukemias and lymphomas. Immunological deficiencies are common and contribute to a

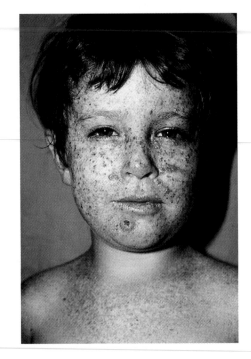

Figure 7.17 A patient exhibiting the skin symptoms characteristic of xeroderma pigmentosum. A squamous cell carcinoma (skin cancer) is visible on the boy's chin. (Copyright © Kenneth Greer/Visuals Unlimited.)

TABLE 7.2 Some abnormal DNA repair syndromes inherited as autosomal recessive disorders

Syndrome	DNA sensitive to	Clinical appearance	Molecular defect
Xeroderma pigmentosum (XP)	UV	Sun sensitivity; multiple metastasizing skin cancers; mental deterioration in some forms	Impaired excision repair (some forms with defective DNA helicase)
Ataxia-telangiectasia (AT)	Ionizing radiation (e.g., X-rays)	Loss of muscle control; skin redness; immunological problems; predisposition to cancers, especially leukemias and lymphomas; chromosome aberrations	Abnormal DNA synthesis following radiation (exact defect unknown)
Bloom syndrome	UV; chemotherapeutic agents	Short stature; disfiguring facial rashes made worse by sunlight; immunological problems; predisposition to all cancers; chromosome aberrations	Defective DNA replication (defective DNA helicase)
Werner syndrome	?	Premature aging, including early graying and loss of hair, tough and thickened skin, arteriosclerosis, several kinds of cancer, diabetes, osteoporosis, cataracts. Cells in culture exhibit chromosome rearrangements	Defective DNA helicase
Fanconi anemia	UV; some chemical mutagens	Decreased blood cells; short stature; limb, heart, and kidney malformations; patchy skin pigments; mental retardation; cancers, especially leukemias; chromosome aberrations	Impaired excision repair
Cockayne syndrome	UV	Sun sensitivity; short stature; mental retardation; deafness; precociously senile appearance	Defect in transcription

high frequency of lung infections, a major cause of death, which usually occurs in young adulthood.

A distinctive feature of AT is extreme sensitivity to ionizing radiation but not to ultraviolet light. For example, X-rays used for therapeutic treatment of the cancers in patients with AT often produce devastating death of normal tissues. X-irradiation also easily kills cultured cells from these patients. In the cells that survive, X-ray-induced DNA breaks in one or both strands of the double helix may go unrepaired, leading to a variety of chromosome breaks and rearrangements.

The gene for AT has recently been identified and cloned by researchers at Sackler School of Medicine in Tel Aviv, but it is still unclear how the encoded protein (deciphered from knowledge of the genetic code) works. It is certainly involved in the repair of DNA that is injured by ionizing radiation. The AT protein is similar to large proteins in yeast and fruit flies that are involved in the detection of DNA damage and the control of cell cycle progression. In particular, it has been suggested that the mutant AT protein fails to act properly at so-called **cell cycle checkpoints.** At each of several of these checkpoints, cells normally delay progress through mitosis to allow repair of damage to DNA, which would otherwise be perpetuated.

Heterozygotes for AT are generally healthy, but they too are more susceptible to cancer, especially breast cancer, than people in the general population, and they are particularly vulnerable to the induction of cancer by radiation. This finding raises questions about the advisability of routine mammography for all women, since about 1% of women are heterozygous for AT. Although the X-ray dose during mammography is very low, the risk-benefit ratio for AT heterozygotes is uncertain (Kastan 1995).

Bloom Syndrome.   Bloom syndrome is also very rare. Fewer than 200 cases worldwide have been examined since the disorder was described. Patients are much smaller than normal from the time of birth; they have a narrow face, a prominent nose, and a receding chin. Mild to extremely disfiguring telangiectases appear on the face, although protection from the sun helps reduce the severity of the skin lesions. Life-threatening infections and malignant tumors of all kinds are common, often leading to death in teenage or early adult years.

Cultured cells from Bloom syndrome patients grow slowly, and metaphase spreads show many broken and abnormal chromosomes. These processes are also presumed to occur in vivo. One common type of chromosome irregularity is called **sister chromatid exchange**, in which crossing over occurs in somatic cells between the two chromatids of one chromosome (so-called sisters of each other). This process is abnormal on two accounts. First, crossing over usually occurs in germ cells (at the first meiotic division) rather than in somatic cells. Second, crossing over usually occurs between two chromatids of homologous but separate chromosomes (nonsister strands) rather than between the two chromatids of one chromosome.

Long-standing and inventive research by James German and his colleagues at the New York Blood Center has resulted in the recent identification and cloning of the Bloom syndrome gene on chromosome 15. Sequence analysis of the gene suggests that it codes for a DNA helicase, an enzyme that is involved in separating the double helix into single strands during DNA replication and other processes (Chapter 6). In addition, abnormalities occur in the action of DNA ligase, an enzyme that connects sections of the sugar-phosphate backbone of DNA, but this defect seems to be a secondary effect. DNA repair processes, on the other hand, are normal. The serious and diverse phenotypic effects in Bloom syndrome thus relate to chromosome instability during the cell divisions of development, but in ways that are not yet clear (Seachrist 1995).

Werner Syndrome.   Another recently cloned gene that encodes a DNA helicase has mutant alleles that lead to Werner syndrome, an odd disease with a frequency of about 1 in 100,000 births. The symptoms are features that usually occur only as people get old, but in Werner syndrome they start much earlier in life. Graying and loss of hair begin in affected teenagers. As young adults, these patients develop rough and wrinkled skin, cataracts, hardening of arteries, brittle bones, diabetes, and certain cancers. Typically, they die of cardiovascular disease or cancer in their 40s.

The gene for Werner syndrome was cloned and characterized in 1996 by researchers at the University of Washington in Seattle and their worldwide collaborators (Pennisi 1996). That it codes for a DNA-unwinding enzyme is known primarily by the resemblance of its protein product to other helicases in humans, yeast cells, roundworms, and bacteria. Humans possess a variety of helicases, and additional research will have to sort out their specific functions during replication, transcription, DNA repair, and crossing over. The connection of Werner syndrome with signs of growing old may in fact be coincidental, because normal aging is related to much more varied and complex factors than just a single mutation in a helicase gene.

Summary

1. Mutations are unrepaired changes in DNA occurring in germinal or somatic tissue. Their effects are varied, and they may occur spontaneously or be induced by a mutagen.

2. Mutations are often base substitutions or frameshifts. The serious symptoms of sickle-cell disease stem from a single base substitution, a missense mutation that leads to a changed amino acid.

3. The thalassemias result from the underproduction of functional α or β chains of hemoglobin. Studies of the mutational sites in the globin genes have shown many different ways by which transcription and translation can be modified.

4. Some neurological diseases are caused by unstable trinucleotide repeat regions that grow in length in successive generations, first from a normal to a premutation length and then to a full mutation. In Huntington disease, the repeated region is within the coding region of the gene and is especially prone to expansion when passed through males. The dominant pattern of inheritance suggests that the expansion is a gain-of-function mutation.

5. In the fragile X syndrome, the expanded region is not in the gene exons and is not translated. The full mutation has hundreds or thousands of repeats, which arise more often when the gene is transmitted through females. Anticipation is commonly observed for trinucleotide repeat diseases.

6. Misalignment of synapsis can occur when separated DNA sequences are similar. Then unequal crossing over can produce duplications and deficiencies. Examples include β-thalassemia, due to a deficiency in globin genes, and Charcot-Marie-Tooth disease, due to a duplication that affects a myelin gene.

7. Some gene mutations are due to insertions of DNA that do not belong in the gene. Examples include a type of hemophilia and a type of neurofibromatosis. The latter represents a mutation in a tumor suppressor gene.

8. Chemical mutagens in our environment can be identified through screening tests using microorganisms, cells grown in tissue culture, or small mammals. The Ames test is a simple system for determining mutagenesis in bacteria.

9. The human population is exposed to both natural and artificial sources of radiation, which can cause cancer via somatic mutations and hereditary defects via germinal mutations. Most spontaneously occurring mutations, however, are not caused by radiation, and the rate of induction of mutations by atomic bombs and nuclear accidents seems to be less than once feared.

10. Changes in DNA are usually fixed by enzyme repair systems before they are transmitted as mutations. Mutations in the genes that encode repair enzymes lead to a number of diseases often characterized by extra sensitivity to radiation, chromosome aberrations, and cancer.

Key Terms

Ames test	ionizing radiation
anticipation	missense mutation
base substitution mutation	MSH receptor protein
carcinogen	mutagen
cell cycle checkpoint	mutation
chain termination mutation	point mutation
DNA repair syndrome	premutation
doubling dose	silent mutation
excision repair	sister chromatid exchange
frameshift mutation	somatic mutation
full mutation	spontaneous mutation
gain-of-function mutation	threshold
germinal mutation	trinucleotide repeat
induced mutation	tumor suppressor
insertion mutation	unequal crossing over

Questions

1. Explain why red hair might be expected as the result of a mutation in the gene that codes for the MSH (melanocyte-stimulating hormone) receptor protein.

2. In the study of red hair, Valverde et al. (1995) found that the most common mutation in the gene for the MSH receptor changed an aspartic acid (Asp) to a histidine (His). Assuming a single-base substitution mutation, which DNA base (A, T, G, or C) was changed to which other base? See Table 6.4.

3. Distinguish between somatic and germinal mutations. Would it be possible for a single mutation to be both?

4. For several different reasons, a base substitution mutation can result in no phenotypic effect whatsoever. Explain.

5. What would be the effect of a mutation that:
   (a) changed an amino-acid-coding triplet into a stop triplet?
   (b) changed a stop triplet into an amino-acid-coding triplet?

6. The successive triplets beyond the point of a frameshift mutation become essentially random as far as the genetic code is concerned. How often, on average, would you expect to find a stop triplet in the DNA sequence beyond a frameshift mutation?

7. Over 200 different mutations have been detected in the gene that (when mutated) leads to the recessive disease phenylketonuria. In Northern Ireland, the most common mutation is designated R408W, in which arginine (R) at

amino acid position 408 is replaced by tryptophan (W). What single-base substitution mutation produces this amino acid change? See Table 6.4.

8. Over 600 different mutations have been detected in the gene that (when mutated) leads to the recessive disease cystic fibrosis. The most frequently occurring mutation is designated ΔF508, in which the 508th amino acid, F (= phenylalanine), is deleted from the encoded protein. What is the gene mutation?

9. How might expansions of trinucleotide repeat regions bring about disease symptoms?

10. Could misalignment and unequal crossing over account for the expansion of alleles in successive generations that leads to the trinucleotide repeat diseases?

11. In what way is the Ames test for chemical mutagenicity in bacteria made to resemble a mammalian test system?

12. A man who worked in a radiation laboratory sued his company because his son was born with hemophilia, which is due to a recessive X-linked allele. He claimed that his working environment was responsible for his son's disease, inasmuch as hemophilia was not present in him, his wife, or any other relative. You are asked to testify in a pretrial hearing as an expert witness in genetics. What would you say?

13. What is the argument for the statement that most spontaneous mutations in humans are not caused by radiation?

14. Newborns homozygous for β-thalassemia are often healthy, the severe disease symptoms developing during the first several months *after* birth. On the other hand, newborns for α-thalassemia are often severely affected *at* birth or are spontaneously aborted. Explain.

15. Following are some of the enzymes involved in DNA repair and replication. What is the function of each? (See also Figure 7.16.)
   (a) DNA nuclease
   (b) DNA helicase
   (c) DNA polymerase
   (d) DNA ligase

16. This chapter discusses some relatively frequent genetic disorders. Identify the following diseases, their frequencies, and the type of mutation that causes each.
   (a) the most common inherited mental impairment
   (b) the most common inherited neurological disorder
   (c) the most common muscular dystrophy

Further Reading

For more information on all aspects of gene structure and function, consult the recommendations for Chapter 6. We particularly point out Scriver et al. (1995) for detailed information on the molecular basis of human diseases. (But have a medical dictionary handy if you attempt to study these volumes.) In addition, Schull (1995) has written a nontechnical but detailed book about the biological effects of the atomic bombs. Among timely topics, see Campion (1997) for the effects of electromagnetic fields, Bartecchi et al. (1994, 1995) and MacKenzie et al. (1994) on the human costs of tobacco use, and the National Research Council (1996) on dietary factors contributing to human health and disease.

CHAPTER 8

DNA in the Laboratory

Starting with a bit of minced animal tissue, students in introductory biology courses can isolate nearly pure DNA during one laboratory period. They first add a detergent to dissolve the cell membranes and let the insides spill out. Then the proteins in the resultant cellular soup are removed by gentle shaking with an organic solvent. This treatment produces a separate water layer containing the DNA, which is very gooey because the long, thin, stretchy threads are intertwined with each other. Adding alcohol precipitates the DNA molecules, which can be twirled around the tip of a glass rod and lifted from the liquid as a glistening globule.

Obtained this way or by other methods, DNA is used for the ever-expanding areas of basic research and applied science called *molecular biology* and *biotechnology* (the latter also called *genetic engineering*). In these fields, researchers identify specific sequences of nucleotides, multiply given stretches of DNA many times, and join together DNA molecules from different sources. A major feature is the speed with which scientists report new findings—important discoveries that deepen our knowledge of basic life processes. More than a decade ago in *Molecular Biology of Homo sapiens* (Cold Spring Harbor Symposia 1986), James Watson wrote, "The scientific advances … amaze, stimulate, and increasingly often overwhelm us. Facts that until recently were virtually unobtainable now flow forth almost effortlessly." The same brisk pace of DNA research exists today. The practical benefits—as well as ethical and legal problems created by the new information—have touched medicine, agriculture, environmental management, and courts of law. Biotechnology will continue to permeate many human endeavors in important ways that cannot be fully foreseen (Box 8A).

In this chapter, we describe a few of the many techniques for manipulating DNA, as well as some applications to human genetics. We start with one of the newest and cleverest techniques, a remarkably simple chemical method for **cloning** selected short pieces of DNA. (The term *cloning* refers in general to making multiple copies of something. Identical twins, for example, are whole-body clones of each other. In this chapter, however, we are dealing only with molecular clones.) We continue with three other innovative but well-established processes: cloning larger pieces of DNA in unrelated species, determining the order of bases in DNA, and using DNA for personal identification. Other manipulations of DNA are described in subsequent chapters.

The Polymerase Chain Reaction

To study any chemical process, we need adequate amounts of the starting materials, often in relatively pure form. To prepare an adequate amount of a particular segment of DNA, researchers can now begin with extraordinarily tiny scraps of DNA, scraps that may be impure, fragmented, or ancient. Indeed, scientists in diverse fields are learning how to deal with DNA extracted from pathology collections, old and dusty museum exhibits, and archaeological specimens such as Egyptian mummies, ice-age dung, and extinct animals—"all sorts of dead things" (Sykes 1991). They may also start with the DNA present in the cells at the base of a single human hair or a trace of saliva on a cigarette butt—perhaps the only evidence found at the scene of a crime.

The process begins with the **polymerase chain reaction (PCR).** PCR singles out a stretch of about 50 to 4,000 nucleotides from all the DNA that may be present in a sample. The system then multiplies this short *target DNA* a millionfold or more in a few hours in a way that mimics repeated DNA replication. Just as an office machine can quickly photocopy a piece of a long document for unlimited users, investigators can rapidly amplify one copy of the target DNA pulled from a complex background to do whatever analyses are called for. This technique was conceived and developed in 1985 by Kary Mullis and other researchers of the Cetus Corporation, a California biotechnology

BOX 8A   *Big-Time Biology*

There are more than a thousand biotechnology companies in the United States alone. Many include a form of the word *gene* in their name: Amgen, Biogen, Calgene, Genentech, Genetics Institute, Genzyme, and Neurogen, to name a few. In addition, many giant pharmaceutical and agricultural companies worldwide have major stakes in biotechnology research; these include Eli Lilly, Hoffmann-La Roche, Merck, Monsanto, Novartis, Schering-Plough and SmithKline Beecham.

The smaller companies pursue many different research and economic strategies. They have attracted venture capital, formed alliances, been merged and split, been bought and sold, sued and countersued, gone public, and gone bankrupt, sometimes in a frenzy of entrepreneurial activity. On October 14, 1980, the first biotechnology company to go public, Genentech, saw its

initial stock offering soar from $35 to $89 per share in 20 minutes (Metz 1980). Investors saw promise in a company that could insert a gene for human insulin into bacterial DNA, transforming trillions of cells into tiny factories for the production of a commercially valuable drug. Yet only a handful of biotech companies have turned a profit in two decades. Still, the largest of them, Amgen, had revenues of $2 billion in 1995 (Thayer 1996).

The term *biotechnology* has been used since early in this century to mean the large-scale production of various chemicals (e.g., penicillin) by microbial fermentation. A dramatic change in biotechnology came with the discovery of methods for making recombinant DNA—that is, for combining the genetic material of the fermenting microbe with that of a different species. Proven and possible

applications of recombinant DNA processes are the source of the frenetic, highly competitive, and speculative economic activity revolving around the science of biology. Will such activities lead to a deeper understanding of human biochemistry? More potent drugs for infectious and genetic diseases? Improvements in waste management? Sustained increases in crop and livestock yields? We shall see. As the harvests from biotechnology industries are gathered and evaluated, we must also be concerned with possible side effects: damage to the environment, disruptions in farming practices, increased costs for medical care, restrictions on the use of patented procedures, reins on the publication of scientific information, and invasion of personal privacy. The implications of commercial biotechnology go far beyond the annual reports of biotechnology companies.

firm (Mullis 1990). PCR quickly became one of the most convenient tools of molecular biology with widespread and growing applications, and Mullis was awarded a Nobel prize in 1993 for his innovation.

To use it, however, researchers must already know the order of at least 20 nucleotides at each end of the target region. Lack of this sequence information usually prevents the use of PCR, requiring that other methods of cloning DNA be used instead. (These alternatives, technically more difficult but perhaps more useful in certain situations, predate the discovery of PCR. They are described in the next section.)

The first step in PCR is to synthesize a pair of **oligonucleotide primers** (Greek *oligo*, "few"). These are single-stranded DNA molecules whose base sequences are complementary to the known sequences flanking the target DNA. Each primer contains about 20 nucleotides. (Oligonucleotides of any sequence up to about 50 bases long are routinely synthesized using automated devices and ingredients such as purines, pyrimidines, and deoxyribose.)

The next step in PCR is to separate the double-stranded DNA (including the region of interest) into its single strands, a process called **denaturation.** Then the key reagents, the two oligonucleotide primers, are added to the denatured DNA, along with the ingredients for DNA replication: DNA polymerase and the four nucleotides. Because of their complementary sequences, the primers will anneal to one or the other end of the short target region present in the long, original strands of denatured DNA. Figure 8.1 shows in detail how the primers bind to *opposite strands* at *opposite ends* of the region to be duplicated.

After binding to the denatured DNA, each primer acts as a starting point for replication. Using the original DNA strands as templates, replication proceeds under the influence of the enzyme *DNA polymerase* in the usual 5' to 3' direction. That is, DNA polymerase adds new complementary nucleotides only to the 3' end of DNA segments already in place (in this case, one end of the primers). In the first cycle of replication shown in Figure 8.2, note that synthesis of new strands proceeds from each primer in the direction of the other primer. The joining of nucleotides into a new strand of DNA encom-

passes the region of interest and continues past the place where the other primer can anneal. Thus, in the second cycle of copying, both old and new strands provide sites for the binding of primers.

After the third cycle of replication, some of the double-stranded DNA is in short segments, containing only the target region with its flanking primer sites (Figure 8.3). After many cycles of replication, the short segments, doubling (approximately) in each cycle, will outnumber by far the longer segments that contain extraneous sequences to the left and right of the target region. The numbers of copies are shown in Table 8.1 (see also question 1). The entire PCR process occurs automatically and continuously in a single small reaction tube with the initial set of reagents. No further additions to the tube are required, although the starting reagents must include a great deal of the primers and four nucleotides, since more and more of these materials are needed in successive cycles of replication.

Changes in the temperature of the reaction mixture control the phases of each cycle: 95°C to denature, 40°C to anneal the primers, and 70°C to extend the complementary copies with polymerase, each cycle taking 4–5 minutes. The temperatures are programmed into and rapidly changed by a small machine called a *thermal cycler*. After about 2 hours, the target DNA region—that spanned by the two primers—is amplified a million times or more in the original test tube.

The process makes use of a special replicating enzyme called **Taq DNA polymerase**, isolated from *Thermus aquaticus*, bacteria that live in the hot springs and geysers of Yellowstone National Park.* Having adapted to its environment, the enzyme can survive extended incubation at 95°C (just under the boiling point of water) during the denaturation phases.

---

*Because of its importance, thermally stable *Taq* DNA polymerase was chosen by *Science* magazine as its 1989 "Molecule of the Year" (Koshland 1989).

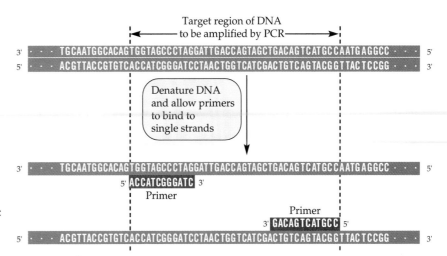

Figure 8.1 The primers (red) used in the polymerase chain reaction are chosen so that they bind to opposite ends of opposite strands of the DNA section to be amplified. In this oversimplified diagram, the primers are only 12 bases long, rather than about 20. In the next step, the primers will be elongated at their 3' ends (see Figure 8.2).

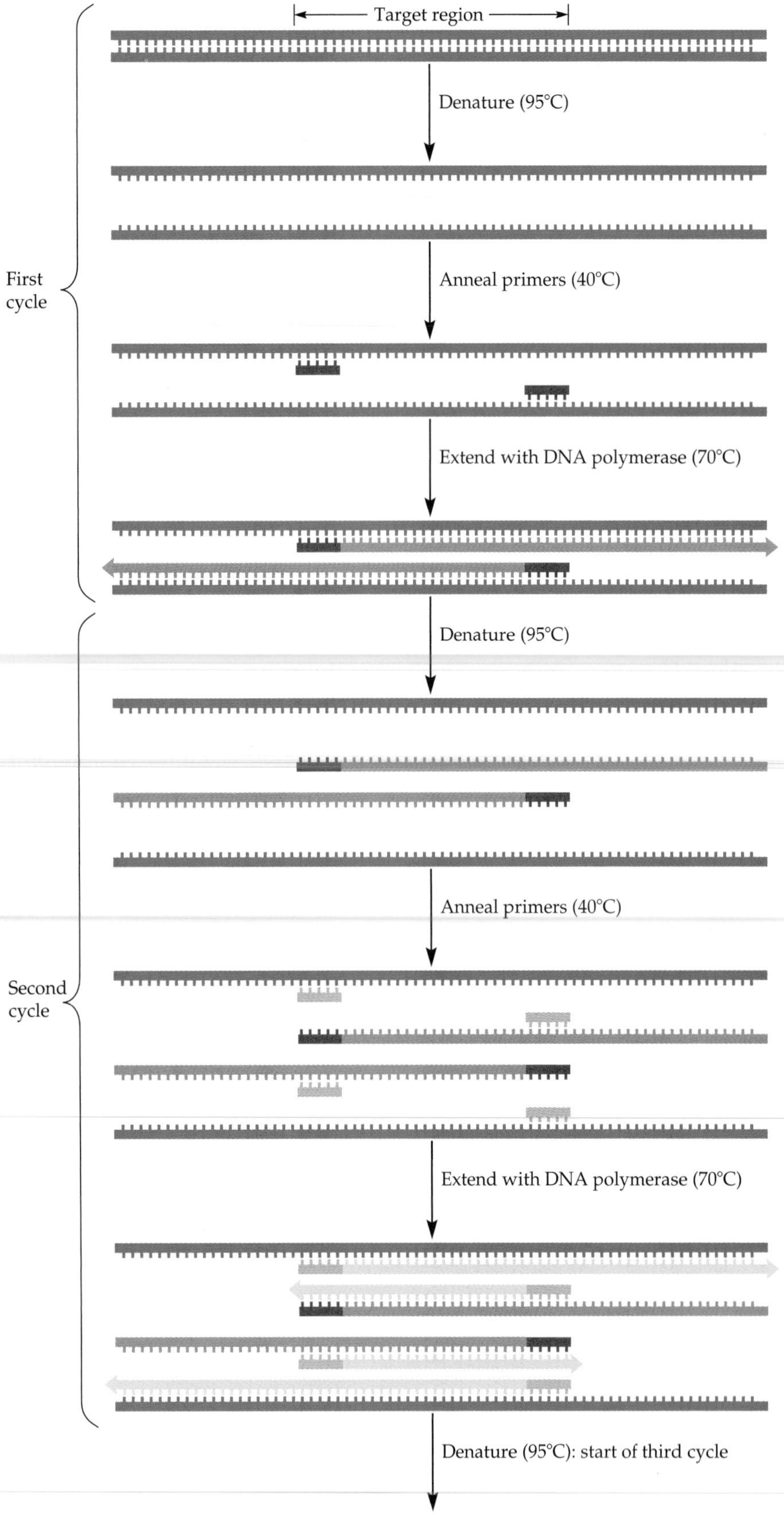

Figure 8.2 The first two cycles of the polymerase chain reaction. Each cycle denatures the DNA at a high temperature (95°C), then anneals the primers (shown in shades of red) by base pairing at a low temperature (40°C), and finally extends the primers at their 3′ ends by DNA polymerase at an intermediate temperature (70°C). Confirm that the next cycle will produce short segments of DNA that contain only the region of interest, that portion bounded by the 5′ ends of the primers (see Figure 8.3).

Virtually all ordinary enzymes are inactivated by such high temperatures.

You may wonder why the primers do not anneal to complementary 20-base sequences elsewhere in the DNA sample, leading to the amplification of unwanted segments. We can answer with a probability calculation: What is the likelihood of finding a particular sequence of 20 bases, assuming a random order of nucleotides? Because each base can be any one of four items (A, T, G, or C), the answer is 1 in 4^{20}. This number is about 1 in 10^{12}, or 1 in a trillion. Since the total bases in a person's DNA is less than 1% of a trillion, a specific sequence of 20 nucleotides (of nonrepetitive DNA) is not likely to be repeated by chance. Although a one- or two-base mismatch might still allow some annealing of the primer, the PCR conditions tend to minimize such spurious binding. Furthermore, for the PCR amplification to be spoiled to a significant degree, *both* 20-base primer-binding regions would have to be present outside the real target DNA, on opposite strands, relatively close together, and with their 3′ ends facing each other. These conditions are very unlikely, but only a single copy of mistaken target DNA is needed to start PCR. Thus laboratory workers take unusual care to guard against the amplification of a minute amount of an unwanted stretch of DNA.

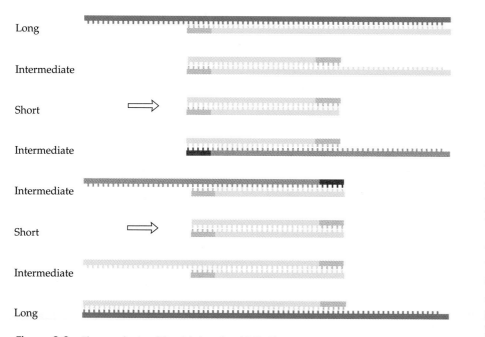

Long

Intermediate

Short

Intermediate

Intermediate

Short

Intermediate

Long

Figure 8.3 The products of the third cycle of PCR. The arrows indicate short segments that increase exponentially in subsequent cycles (see Table 8.1).

The polymerase chain reaction has been applied to an ever increasing number of questions. Major uses of PCR include prenatal diagnosis of genetic disease, screening for mutations, gene therapy, detection and identification of infectious microorganisms (including the AIDS virus), crime investigation where cells are left behind (e.g., in rape), analysis of archaeological samples, and evolutionary investigations.

A controversial use of PCR is to help verify the sex of female athletes at Olympic Games. While karyotyping was used in the past, PCR was used at the 1992 Barcelona Games and the 1996 Atlanta Games. The first step in this process (in Barcelona) was screening for a segment of repetitive DNA, called *DYZ1*, that is only present on the long arm of the Y chromosome—not on the X or on any autosome. This segment contains a tandem array of about 700 pentanucleotides, mostly TTCCA or variants thereof. The testers used oligonucleotide primers specific for opposite ends of this segment (see Figure 8.1 again). Scraping cells from inside the cheek to get some DNA, they found that 11 of 2,406 female competitors had a positive result. That is, their DNA contained a segment that was amplified by PCR when using these primers. (How this amplified DNA can be visualized is explained in following sections.) The conclusion was that these athletes contained at least a snippet of Y chromosome.

In a follow-up screening, these 11 DNA samples were further analyzed by PCR for another Y-specific sequence, the *SRY* gene (sex-determining region of the Y chromosome). The product of this gene initiates testicular development during early embryology, and in the absence of *SRY* the fetus develops into a female (Chapter 3). The five Barcelona athletes who were positive for both *DYZ1* and *SRY* were invited to have a medical examination. One refused, and the other four were found to have some morphological abnormalities. The detailed findings are confidential, and we do not know if these athletes were allowed to compete (Serrat and de Herreros 1993). It is important to realize, however, that genetic sex as determined by possessing DNA sequences specific to the Y chromosome does not always correspond to phenotypic sex, much less to how a person regards his or her own physiological and psychological sexual status. What criteria are the "correct" ones to use in defining sex? The question is not easy to answer in the context of athletic competition or in certain other situations.

Cloning Human DNA in Other Species

To a large extent, the same basic chemicals and metabolic reactions, especially those needed for energy production, occur in all cells, from bacterial to human.

TABLE 8.1 The amplification of DNA segments in the polymerase chain reaction, starting with one double-stranded molecule

Segment[a]	Number of segments after cycle number n										
	1	2	3	4	5	6	7	8	9	10	11
Short	0	0	2	8	22	52	114	240	494	1,004	2,026
Intermediate	0	2	4	6	8	10	12	14	16	18	20
Long	2	2	2	2	2	2	2	2	2	2	2
Total (2^n)	2	4	8	16	32	64	128	256	512	1,024	2,048

[a]The short, intermediate, and long segments are identified in Figure 8.3.

This biochemical unity of life includes the genetic code as well as the processes of DNA replication, transcription, translation, and mutation. Given the high degree of biochemical similarity, it was inevitable that scientists would try to mix and match the genes themselves.

By 1973, the technology for splicing together DNA segments from different sources had been developed by Stanley Cohen at Stanford University and Herbert Boyer at the University of California at San Francisco. Using complementary base pairing as a kind of glue, they (and others) were able, for example, to insert DNA from the primitive frog *Xenopus* into the DNA of the bacterium *E. coli*, where the frog DNA subsequently reproduced itself as if it were "at home." Such a hybrid molecule (in this case, a frog-bacterial molecule) is called **recombinant DNA.** Its formation is called **gene splicing**, referring to the attachment of a piece of DNA from one species to that of a second species, followed by insertion of the hybrid molecule into a host organism (often a bacterium).

The high-precision biological tool for making recombinant DNA is a special class of enzymes that cut across the double helix in an interesting way. Called **restriction enzymes,** these special cutting enzymes are widespread among bacteria. Their properties were first investigated by Werner Arber, a Swiss biochemist at the University of Geneva who later received a Nobel prize for this work. Acting like scissors that cut the DNA of invading viruses, the restriction enzymes in bacteria protect their owners from viral infection. Thus, they *restrict* the range of viruses that can successfully attack. Hundreds of different restriction enzymes have been isolated from many species of bacteria; each recognizes a different target sequence of usually four to seven bases.

For example, one restriction enzyme called *Eco*RI* recognizes the following sequence of six base pairs wherever it occurs in long DNA molecules:

$$\cdots \text{G A A T T C} \cdots$$
$$\cdots \text{C T T A A G} \cdots$$

The enzyme causes breaks in the backbones between the G and A nucleotides on *both* strands. Then the weak hydrogen bonds joining the four intervening base pairs also break, and a staggered cut results:

$$\cdots \text{G} | \text{A A T T C} \cdots \quad \rightarrow \quad \cdots \text{G} \qquad \text{A A T T C} \cdots$$
$$\cdots \text{C T T A A} | \text{G} \cdots \qquad\qquad \cdots \text{C T T A A} \quad + \quad \text{G} \cdots$$

Double-stranded DNA molecules with protruding single-stranded tips like these are said to have **sticky** or **cohesive ends.** This is because each broken end has four unpaired bases, AATT (A being the terminal base), which tend to spontaneously join with any complementary end that is available. The complementary bases can be the single-stranded tip from which it has just been separated or with the tip of any other DNA molecule cut with the *same* restriction enzyme. A key point is that the DNA molecules that join together need not be from the same organism or even from the same species.

The hosts for much recombinant DNA work are special strains of *E. coli* carrying deleterious mutant genes that prevent them from surviving outside a laboratory. Each *E. coli* cell has a circular chromosome, a DNA double helix coding for about 4,000 polypeptides. In addition to chromosomal genes, some strains of *E. coli* possess extra genes that confer attributes that the cells can usually do without. These dispensable genes are present not on the large bacterial chromosome, but on small, circular pieces of DNA called **plasmids,** which often contain just a few genes (Figure 8.4). For example, some plasmids possess genes that make the *E. coli* resistant to common antibiotics.* Anywhere from one to a few hundred copies of a plasmid may be present in a cell, and they replicate and are inherited by daughter cells the same as chromosomal DNA. Because researchers can easily purify the small plasmids, operate on them, and cause them to be taken up again by other strains of *E. coli*, the plasmids are good tools for manipulating DNA.

One particular plasmid, dubbed pBR322, is of special interest to recombinant DNA workers. Along its length of about 4,000 base pairs, it has exactly one GAATTC sequence recognized by the restriction enzyme *Eco*RI. A researcher can isolate pBR322 plasmids in a test tube and treat them with *Eco*RI to open up each DNA circle at that one point, exposing two sticky, single-stranded ends. *The crux of gene splicing is to mix the opened plasmids with foreign DNA that has been treated with the same restriction enzyme* (Figure 8.5).

When the plasmid DNA is enlarged by the addition of a piece of foreign DNA, it becomes a vehicle, or **vector,** for transporting the foreign DNA into other *E. coli* cells. The recipient bacteria are treated with calcium chloride, which alters the cell wall and permits an occasional plasmid to enter. Because any bacterium carrying the new plasmid has thousands of its own genes but only a few genes from the foreign source (equivalent, say, to adding just one word to six pages of this

---

**Eco*RI stands for *E. coli*, strain R, first (Roman numeral I) restriction enzyme isolated.

* Such plasmids, called R or resistance plasmids, are often found in hospital populations of pathogenic bacteria. Because these germs may possess resistance to more than one antibiotic, they are difficult to control.

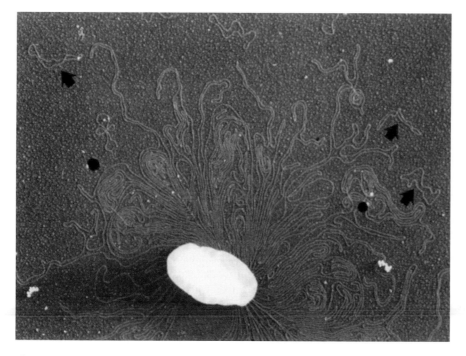

Figure 8.4 An electron micrograph showing DNA extruded from a ruptured E. coli cell (the white area). The long, tortuous path of chromosomal DNA is easily seen, as well as at least three small plasmids (arrows). (Courtesy of David Dressler and Huntington Potter, Harvard Medical School.)

Manufacturing Useful Proteins

Because the bacteria can be grown in prodigious quantities (a 10,000-gallon industrial fermenter has roughly the volume of a 12 × 14-foot living room), the foreign gene can be gotten in substantial amounts for investigative or practical purposes—all this from as little as one copy of one gene inserted into one plasmid that was taken up by one bacterial cell! The foreign DNA has been cloned, just as in the polymerase chain reaction. But this type of molecular cloning goes beyond the capabilities of PCR. Because bacteria are living cells, they not only replicate their genetic material, but also transcribe it into messenger RNA and translate it into polypeptide products. The bacteria can do the same with foreign genes.

book), we cannot really call it a new species. The bacterium is essentially its old self—except that it harbors a bit of foreign DNA that replicates along with the plasmid DNA.

A California research team was the first to report on a human polypeptide "engineered" this way, having tricked E. coli cells into making a small hormone called *somatostatin*. Just 14 amino acids long, somatostatin is normally made in the brain and acts to limit the production of other hormones. Although no inherited disease is known to involve this polypeptide directly, a deficiency of somatostatin can lead to the overproduc-

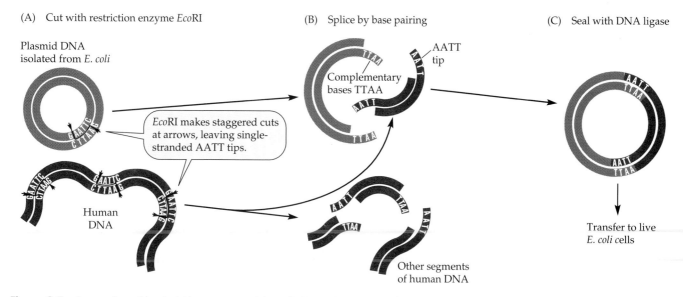

Figure 8.5 Preparation of bacterial-human recombinant DNA. (A) In one test tube, plasmids isolated from E. coli are treated with the restriction enzyme EcoRI. In another test tube, human DNA is treated with EcoRI. (B) When the cut-up DNAs are combined, some human DNA is spliced into the E. coli plasmids by complementary base pairing. (C) The chemical bonds in the sugar-phosphate backbones are rejoined by another enzyme, DNA ligase. The enlarged plasmids can then be transferred into live E. coli cells.

tion of growth hormone and a form of gigantism called *acromegaly*.

Knowing the amino acid sequence of somatostatin and the genetic code, these researchers first deduced a possible base sequence for the gene. Then they constructed a double-stranded helix, both template and nontemplate strands, by synthetically linking together the appropriate nucleotides. They also tacked the DNA triplet for methionine onto the beginning and *two* stop triplets (just to be sure) onto the end (Figure 8.6A). Affixing appropriate single-stranded tips to this double-stranded structure of 17 coding units, they used two different restriction enzymes to splice it into a section of an *E. coli* plasmid. When the plasmid was incorporated into *E. coli*, the live cells transcribed and translated the foreign insert into somatostatin as if it were an *E. coli* gene.

Actually, the somatostatin molecule that the researchers isolated from the *E. coli* cells was attached to the end of another polypeptide (Figure 8.6B) because the somatostatin gene had been deliberately spliced into the middle of a bacterial gene. Thus, the *E. coli* could use the promoter of the bacterial gene to start transcription of this section of the plasmid DNA. The resultant polypeptide began as a bacterial protein and ended as a human protein! But investigators were able to detach somatostatin from the rest by treating the hybrid protein molecule with a reagent (called cyanogen bromide) that specifically cleaves a polypeptide after any methionine residue (Figure 8.6C). This clever biochemical trick worked only because somatostatin itself contains *no* methionine. Thus, the cyanogen bromide stratagem cannot be used to isolate the more typical proteins of a hundred or more amino acids that usually contain methionine. Altogether, the in vivo bacterial synthesis of human somatostatin was a biochemical tour de force.

Many valuable pharmaceuticals have been produced by recombinant DNA technology. More than a dozen human recombinant DNA products, having gone through years of clinical testing, have been approved by the U.S. Food and Drug Administration for commercial use. These include several hormones, three different forms of interferon, and various enzymes (Table 8.2). Hundreds of other recombinant DNA products for medicine, agriculture, or industry are in various stages of development. Whether the basic discoveries, testing, and development are carried out in universities or private industry, the general aim is the same: to have a constant and reliable source of pure, concentrated polypeptides (which may or may not be available from other sources). Biotechnology companies are, of course, interested in turning a profit too.

The recombinant DNA production of human insulin in 1982 (the first recombinant product approved for commercial use) was very similar to that of somatostatin. Researchers synthesized the gene for the A polypeptide chain from scratch (see Figure 6.13) and inserted it into a plasmid in one strain of *E. coli*. Then they made the gene for the B chain from scratch and inserted it into a second strain of *E. coli*. Subsequently, they combined the

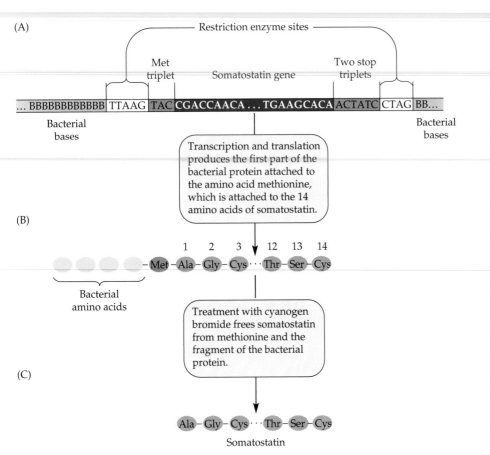

(A)

Restriction enzyme sites

Met triplet   Somatostatin gene   Two stop triplets

… BBBBBBBBBBBB TTAAG TAC CGACCAACA … TGAAGCACA ACTATC CTAG BB…

Bacterial bases   Bacterial bases

Transcription and translation produces the first part of the bacterial protein attached to the amino acid methionine, which is attached to the 14 amino acids of somatostatin.

(B)

Bacterial amino acids

1   2   3   12   13   14
– Met – Ala – Gly – Cys ··· Thr – Ser – Cys

Treatment with cyanogen bromide frees somatostatin from methionine and the fragment of the bacterial protein.

(C)

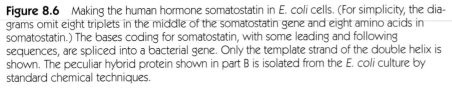

Ala – Gly – Cys ··· Thr – Ser – Cys

Somatostatin

Figure 8.6 Making the human hormone somatostatin in *E. coli* cells. (For simplicity, the diagrams omit eight triplets in the middle of the somatostatin gene and eight amino acids in somatostatin.) The bases coding for somatostatin, with some leading and following sequences, are spliced into a bacterial gene. Only the template strand of the double helix is shown. The peculiar hybrid protein shown in part B is isolated from the *E. coli* culture by standard chemical techniques.

TABLE 8.2 Human proteins made by recombinant DNA methods and approved by the FDA for clinical use

Protein	Year approved	Approved for	Cloned in
Insulin	1982	Children with insulin-dependent diabetes, to replace missing insulin	*E. coli*
Growth hormone	1985	Children with pituitary dwarfism, to replace missing growth hormone	*E. coli*
α-Interferon	1986	Patients with hairy cell leukemia, AIDS-related Kaposi's sarcoma; also for genital warts and hepatitis C	*E. coli*
Hepatitis B vaccine	1986	Health care workers exposed to blood, also drug users and homosexuals; to ward off liver infection	Yeast
Tissue plasminogen activator (tPA)	1987	People with heart attacks, to dissolve blood clots blocking coronary arteries	Hamster cells
Erythropoietin	1989	Patients with kidney failure, to treat anemia due to deficiency of erythropoietin (made in kidney); also for AIDS-related anemia	Hamster cells
γ-Interferon	1990	Patients with chronic granulomatous disease, a rare genetic defect of the immune system, to prevent severe infections	*E. coli*
Granulocyte colony-stimulating factor	1991	Patients with cancer who are undergoing chemotherapy, to prevent infections by stimulating growth of white blood cells	*E. coli*
Granulocyte/macrophage colony-stimulating factor	1991	Patients with leukemia or Hodgkin disease, to speed the growth of white blood cells in transplanted bone marrow	Yeast
Interleukin-2	1992	Patients with kidney cancer, although side effects may be serious	*E. coli*
Clotting factor VIII	1992	Patients with hemophilia A (classic hemophilia), to replace missing protein	Hamster cells
β-Interferon	1993	Patients with multiple sclerosis, to regulate immune reactions	*E. coli*
Glucocerebrosidase	1994	Patients with Gaucher disease, to replace missing lysosomal enzyme	Hamster cells
Deoxyribonuclease	1994	Patients with cystic fibrosis, to help break up viscous secretions in lungs	Hamster cells

two chains. The strictly chemical formation of the disulfide bonds between and within the A and B chains was inefficient, however, so researchers tried an alternative method of making insulin, which allowed the disulfide bonds to form more naturally. To explain what "naturally" means, we point out that a *single* human insulin gene codes for a *single* polypeptide about twice the length of insulin. The longer molecule, called *proinsulin*, folds up spontaneously; then the disulfide bonds are formed, and a section between the A and B chains is removed, producing active insulin. In the alternative method for the bacterial production of human insulin, investigators inserted into a single *E. coli* plasmid the nucleotides for the proinsulin polypeptide.

The recombinant DNA production of human growth hormone (HGH) differed from that of somatostatin and insulin in that the gene was not primarily synthesized from individual nucleotides. Instead, the gene was copied by the enzyme reverse transcriptase from the HGH messenger RNA that had been isolated from pituitary glands. Recombinant HGH was approved by the FDA just a few months after they had

banned the only other source of HGH—pituitary glands from human cadavers. The ban was instituted because several people who had taken cadaver-derived HGH had died of Creutzfeldt-Jakob disease, a very rare and mysterious neurodegenerative ailment that is sometimes inherited and sometimes communicable.* If the HGH from recombinant DNA processes had not existed, thousands of children and adolescents with HGH deficiencies would have been without effective treatment.

Other Ways to Clone

Vectors other than plasmids can be inserted into *E. coli* cells. Viruses that infect bacteria, called *bacteriophages* (literally, "bacteria-eaters") or simply *phages*, consist of a molecule of DNA inside a protein coat. *Lambda* (λ)

*The occurrence of ten unusual cases of Creutzfeldt-Jakob disease in England in 1993–1995 was the impetus for destroying thousands of cattle suspected of having the related brain disease bovine spongiform encephalopathy (BSE, or mad cow disease). The export of British beef was banned, with dire economic repercussions for British herdsmen. But it is still unclear whether the threat to people from BSE is real or trivial (Boffey 1996).

phage that attacks *E. coli* can have a 20-kb length of its own DNA replaced with foreign DNA without altering its infectivity. Another type of vector, called a *cosmid*, consists of a plasmid joined to very short end pieces (so-called *cos* sites) of λ DNA and housed in a λ protein coat. Both phages and cosmids are convenient vectors in many situations, in part because they accommodate somewhat larger pieces of foreign DNA than plasmids.

Organisms other than *E. coli* can also be used for cloning. An increasing favorite (used also for many other types of genetic studies) is baker's yeast, *Saccharomyces cerevisiae*. This one-celled organism is easily grown and manipulated in laboratory glassware and can be maintained in either a haploid or a diploid state. Yeast is, moreover, a *eukaryotic* organism. Although we have stressed the biochemical unity of all organisms, there are clear distinctions in cell architecture and function between prokaryotes like bacteria (with no distinct nucleus enclosing the chromosomes) and eukaryotes like yeast and humans (with a distinct nucleus). In particular, eukaryotes have mechanisms, lacking in prokaryotes, for modifying proteins *after* translation. These systems help form disulfide bonds, cut out nonfunctional segments of amino acids, and add sugars or other groups to some amino acids.

Yeast has 16 chromosomes, about 6,000 genes, and 12 million nucleotides. It is the first eukaryote to have its DNA 100% deciphered—a remarkable achievement of international (largely European) cooperation in sequencing (Kolata 1996a, and later discussions). Researchers have developed some very useful vectors that they can shuttle back and forth between yeast and *E. coli*. For example, a **yeast artificial chromosome (YAC)** has been made to order by combining DNA from several different sources. Specific restriction enzyme sites are positioned along the artificial chromosome for inserting foreign DNA. A major advantage of this vector is its ability to incorporate very large inserts—100 to 1,000 kb, as opposed to inserts of only 10 to 40 kb of foreign DNA in plasmids, phages, and cosmids.

YACs, as well as phages and cosmids, are useful vectors for making *libraries* of human DNA. Such libraries are assembled by cloning random pieces of human DNA into the vectors and selecting so many of them that the probability is high that every piece of human DNA is present in one or another clone. For a **genomic library**, the starting material is total human DNA isolated from any cell type—for example, sperm or liver tissue. A genomic library contains all DNA, including coding regions, introns, control sequences, pseudogenes, and repetitive sequences.

For a so-called **cDNA library**, the starting material is the DNA that is obtained by copying from the messenger RNA present in a particular cell type. The "c" in cDNA stands for *copy* or *complementary*, and it is obtained from RNA via the enzyme reverse transcriptase. A cDNA library contains the coding regions of genes, as well as the leading and trailing sequences present in mRNA, but no introns or intergenic DNA. Not all genes are represented, however—just those genes that are expressed in the cells from which the mRNA is obtained. For example, consider the cDNA derived from the mRNA of immature red blood cells. This cDNA has large quantities of the α-globin and β-globin genes that encode the polypeptides of hemoglobin, since these genes are transcribed at high levels and not many other genes are. On the other hand, a cDNA library derived from the mRNA in brain tissue has about 40,000 different genes. This large number of active genes reflects the complexity of brain metabolism, which requires the participation of perhaps half of all genes, including those controlling energy production, gene expression, cell-to-cell communication, cell maintenance, and cell repair.

There are other cloning systems too. The human factor VIII gene, for example, was cloned in hamster cells grown in laboratory culture dishes. Many different methods have been developed for getting foreign genes into cultured mammalian cells, where they may become stably incorporated into the host's DNA. One method is to treat the foreign DNA with calcium phosphate, making tiny clumps of DNA that a cell readily engulfs with its cell membrane. Still another method is to insert the foreign gene into a virus that naturally infects the cells. Another is to microinject the foreign DNA into the nucleus of the cells with a very fine needle. Each method has advantages and disadvantages.

It is also possible to inject foreign DNA into the fertilized eggs of mammals. When such eggs are then implanted in surrogate mothers, they can lead with low frequency to the development of newborns with the foreign gene—the **transgene**—in every cell, creating so-called **transgenic animals**. The injected gene seems to be incorporated into the host DNA at random places (if it is incorporated at all), with consequences that have not been fully explored. Research teams in the United States, Scotland, and the Netherlands have produced some unusual transgenic goats, sheep, cows, and pigs that secrete small to moderate amounts (grams per liter) of the protein product of a *human* gene into the female animal's milk (Figure 8.7)! To accomplish this feat, researchers manipulate the DNA segment to be injected into the eggs so that the human gene is joined to the promoter of a host gene that encodes a milk protein, casein. By this maneuver, the human gene is expressed, and the resultant protein is secreted in the mammary gland during lactation (Maga and Murray 1995; Velander et al. 1997). Although the actual embryonic development to term of such a transgenic animal is rare and unpredictable, just *one* transgenic animal could transmit its new capability to many descendants. To improve the odds of success, Scottish researchers have first added a clotting factor–milk promoter gene complex to fetal sheep cells growing in culture ("readily" done, they say). Then they used the techniques by which they cloned the sheep Dolly (see Chapter 20), fusing one of the cultured cells with an enucleated sheep egg. The eggs, when transferred to foster mothers, produced many lambs carrying the human clotting factor IX gene (Schnieke et al. 1997.)

Figure 8.7 A transgenic goat that secretes the human enzyme tissue plasminogen activator (tPA) in her milk. This goat developed from eggs that were injected with human DNA encoding tPA. She has passed the human gene on to two of her offspring (also shown here). During the process of normal wound healing, tPA converts the serum protein plasminogen to an active form that dissolves blood clots. For this reason, tPA is also useful in treating heart attack patients and is available now as a recombinant DNA product (see Table 8.2). The drug however is, very expensive—over $2,000 per treatment. Opinion is divided over the value of tPA versus other clot-dissolving products that cost far less. (Courtesy Karl M. Ebert, Tufts University School of Veterinary Medicine, and Genzyme Transgenic Corporation.)

The secretion of a useful pharmaceutical in milk may give a convenient and generous source, since an average cow makes about 10,000 liters of milk per year. Purifying the drug from the complex mix of components in milk may be difficult, however, and the issue of safety still needs to be addressed. In any case, research in such barnyard biotechnology—called "pharming" in news reports—continues, although it will be some time before drugs produced this way will be available for general human use.

Public Concerns

Many people believe that the techniques just outlined promise great benefits to human welfare; they are confident that accompanying risks can be evaluated and managed. But others claim that the risks to public health and to the environment are unacceptably large or that it is morally wrong to tinker with nature in this manner.

In fact, some molecular biologists were the first to voice safety concerns. In the summer of 1973, a small group attending a conference on nucleic acid research requested that the National Academy of Sciences (NAS) study the potential hazards of recombinant DNA research. One year later, a distinguished NAS committee headed by Paul Berg (a 1980 Nobel laureate in chemistry) recommended that scientists throughout the world refrain from certain types of recombinant DNA experiments pending further evaluation. This call for a moratorium on research by the researchers themselves was unprecedented in the history of science.

As one consequence, an international panel of scientists, lawyers, members of the press, and government officials met in 1975 at the Asilomar Conference Center near Monterey, California. After days of spirited discussion, the conferees decided that research with recombinant DNA should be regulated by a set of safety rules to protect laboratory personnel, the general public, and the environment. By the time the National Institutes of Health (NIH) set forth these rules in July 1976, considerable public interest had been generated by eminent scientists who spoke out for or against such government regulation (Berg and Singer 1995; Watson and Tooze 1981).

The original NIH guidelines for recombinant DNA research provided successively more rigorous standards for experiments that were perceived to be increasingly dangerous to public health and safety. (Strictly speaking, the guidelines applied only to research funded by NIH, but it was hoped that other groups—including pharmaceutical companies—would also observe them.) In addition to the successive levels of *physical containment*, increasingly risky experiments required increasingly enfeebled strains of *E. coli*. This type of safeguard is referred to as *biological containment* and seeks to ensure that the engineered organisms do not survive if released into the general environment. Experience to date suggests that the original NIH guidelines for laboratory research were too stringent. Indeed, the guidelines were relaxed several times without untoward effects, and most recombinant DNA experiments today are not regulated at all. In thousands of laboratories performing millions of experiments around the world, no known harm has yet been documented as a consequence of recombinant DNA research. The settings for such experiments are now sometimes student laboratories in college or even in high school.

But controversy continues. Recent debates have focused on the effects (on cows, farmers, and consumers) of treating cows with bovine growth hormone (from recombinant DNA) to increase milk yields, on disruptions to natural fish populations from transgenic, faster-growing fish that escape from net-enclosed pens, and on the environmental safety of recombinant agricultural products that are tested in open fields. Examples of the latter include transgenic crops that resist pests and herbicides, tolerate drought or soil contaminants, provide better nutrition, or can be more easily transported and preserved (Rouhi 1996). Microorganisms are being created that attack weeds and insects, clean up toxic chemicals, enhance oil recovery, and leach minerals from their ore beds.

Some people believe it would be morally wrong *not* to vigorously pursue these technologies in the face of increased environmental degradation from a burgeoning human population. For others, genetic engineering is the route to ecological disaster, spearheaded by nightmarish blights from mutated pathogens or uncontrollable weeds. Small farmers are concerned about their livelihood, because biotechnological advances are seen as benefiting only large-scale operations. Still others worry that wrong decisions will be made because the public is not objective in balancing benefits and risks and lacks a basic understanding of probability. Academic and industrial personnel, environmental activists (including Greenpeace and the Sierra Club), biotechnology critics and proponents, the Environmental Protection Agency, the Department of Agriculture, organic farm associations, and residents near proposed test sites have all become involved in spirited give-and-take of fact and opinion (Hileman 1995; Kling 1996).

DNA Sequencing

DNA sequencing means determining the exact order of bases along a length of single-stranded DNA. Researchers may sequence both strands of the double helix, matching them up A to T and G to C to ensure that no mistakes were made. Practically all extended studies of genes and gene products involve sequence determination somewhere along the way. For example, knowledge of the nucleotide sequence of a human disease gene often precedes knowledge of the amino acid sequence of the encoded protein. With the DNA sequence in hand, researchers can usually quickly develop ways to diagnose the disease before symptoms appear. But management and treatment of the disease are often difficult and delayed, if possible at all. Such sequencing studies also include analysis of the different mutations—sometimes hundreds of them—that cause the genetic disease. Another increasingly used application of sequencing is the study of evolutionary relationships among a group of related species and among races or varieties within a species.

The details of sequencing methodologies are largely beyond the scope of this book (but see Mange and Mange 1990; Rosenthal 1995). The inventor of the most frequently used method is Frederick Sanger* of the Medical Research Council in Cambridge, England. His method relies on working with the original piece of DNA (amplified by PCR, if necessary) in four separate reaction tubes, labeled G, C, A, and T. Within the G tube, for example, a chemical reaction produces many fragments of DNA of different lengths, all ending in G (Figure 8.8). Through this approach, the G tube eventually reveals the positions of all the G's along the length of the original piece of DNA. Combining the information from the four reaction tubes gives the exact positions of every G, C, A, and T.

The key process in DNA sequencing is **electrophoresis** (Greek *phoresis*, "to be carried"), one of the most widely used methods in chemistry. Electrophoresis is the movement of electrically charged molecules, such as proteins or nucleic acids, in a solution to which an electric field is applied. Negatively charged molecules move toward the positive pole, and positively charged molecules move toward the negative pole. Electrophoresis was invented in 1937 by the Swedish Nobel laureate Arne Tiselius, who was a chemistry graduate student at the time. Although Tiselius carried out electrophoresis with proteins free in solution, more modern methods utilize a supporting medium through which the charged molecules move. The support can be simply a sheet of filter paper or a gel-like substance such as starch, agar, or often a polymer called *polyacrylamide*.

In gel electrophoresis of DNA, for example, the DNA molecules migrate through a thin slab of gel material supported horizontally (or sometimes vertically between glass plates) in an electric field. Because

---

*Having invented first a method for determining the sequence of amino acids in proteins and then a method for determining the sequence of bases in DNA, Sanger is one of a very few individuals to win *two* Nobel prizes (Ruthen 1993). The first person to win two Nobel prizes was Marie Curie.

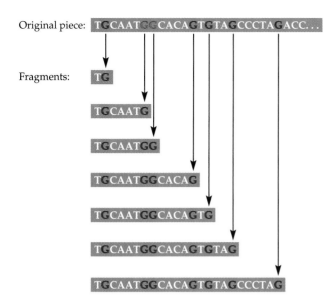

Original piece:

Fragments:

Figure 8.8 Fragments of DNA that would be present in the G reaction tube used to sequence a hypothetical piece of DNA. Included are all possible fragments that end at the G's present in the original piece, which in practice would be about 500 bases long.

phosphate groups are negatively charged, DNA molecules move toward the positive pole. And because the supporting medium impedes the migration of bigger DNA molecules, the smaller pieces of DNA migrate faster through the pores of the gel. Under good conditions, just *one* base more or less makes a noticeable difference in speed.

An apparatus for gel electrophoresis of DNA is illustrated in Figure 8.9. The wells at the left of the gel hold the solutions in the starting gates. We could, for example, pipet into the four wells the contents of the four reaction tubes, G, C, A, and T, used in DNA sequencing. When the electricity is turned on, the fragments in the G lane are separated from each other according to size, parallel to the lanes for the other three reactions. Since each fragment ends in a G, the resulting bands in the G lane represent the positions of all the

Figure 8.9 An electrophoresis apparatus for separating lengths of DNA by size. Hot liquid agarose is poured into the plastic gel support, which is initially taped shut at both ends and which holds a comb with four rectangular teeth to make the wells. After the gel cools and sets, the tape and comb are removed. Test solutions are pipetted into the wells of the slightly submerged gel, and voltage (20–200 volts) is applied to the two ends.

G's in the original DNA segment. At this stage of the procedure, however, the DNA fragments are not visible, just as you cannot see the sugar or salt molecules dissolved in water.

The DNA fragments in the gel can be made visible by incorporation of radioactive phosphorus into the DNA samples. After researchers complete electrophoresis, they lay a sheet of X-ray film over the gel. The high-energy particles striking the film from the radioactive DNA develop the silver grains in the film and produce a high-resolution image, called an **autoradiograph** (self-image). The fastest-migrating band of DNA represents position 1, the next fastest position 2, and so on (Figure 8.10). Thus, the sequence of bases is read directly off an X-ray film exposed to the gel, from the bottom up. A length of several hundred nucleotides can be read off one gel. Longer sequences can be obtained by combining the information of several cloned DNA segments that overlap one another.

Automated DNA sequencing devices based on the Sanger method are being perfected. The prototype for these expensive machines was developed by Leroy Hood, now in the Department of Molecular Biotechnology at the University of Washington in Seattle, and was partially funded by computer guru Bill Gates, chairman of Microsoft Corporation. Such corporate support reflects the required input of engineers and computer experts for the automation of sequencing and especially for the handling of the resultant massive amounts of sequence data.

Some of these automated machines use large, ultrathin gels and differently colored *fluorescent* tags on each of the four kinds of fragments. Avoiding radioactive tags and the bothersome safety precautions needed for using them is a big plus. In addition, since the G, C, A, and T fragments are tagged with different colors, the four lanes shown in Figure 8.10 can be run together in one lane. The four different fluorescent labels are identified one after another and the bases they represent are automatically computerized. Although

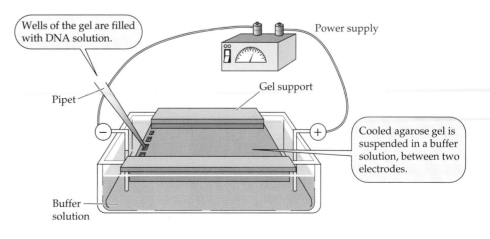

Wells of the gel are filled with DNA solution.

Power supply

Pipet

Gel support

Cooled agarose gel is suspended in a buffer solution, between two electrodes.

Buffer solution

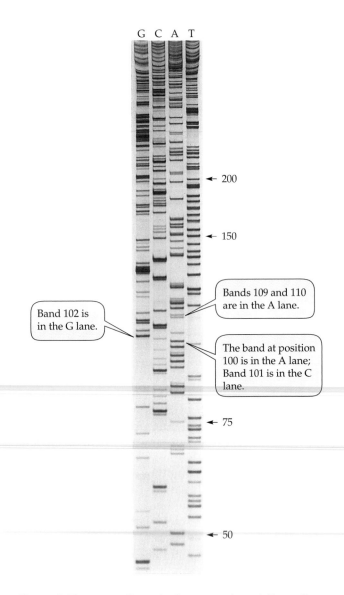

Figure 8.10 Autoradiograph of a sequencing gel. The wells are above the top of the picture, and the faster-moving (shorter) DNA fragments are toward the bottom. The bands in the four parallel lanes represent the positions of different-sized segments of DNA ending in either G, C, A, or T. The bands appear dark because the radioactivity in the DNA sensitized and developed the X-ray film. Reading upward and switching back and forth between lanes as required, successive bands give the sequence of bases. Confirm that the sequence of bases from position 100 to position 110 is ACGAGCCGGAA. (From Williams et al. 1986.)

(Labels on figure:)
G C A T

Band 102 is in the G lane.

Bands 109 and 110 are in the A lane.

The band at position 100 is in the A lane; Band 101 is in the C lane.

← 200
← 150
← 75
← 50

The DNA sequencing procedure is now so good that the amino acid order of a *protein* is often determined by sequencing the *gene* that encodes the protein. Researchers then infer the amino acids of the protein from knowledge of the genetic code. This approach was taken in 1987 for X-linked Duchenne muscular dystrophy. The massive gene extends over 2.3 million bases, by far the largest gene known. The expressed portion of the gene is divided into about 75 exons that account for only about 1% of the total sequence. That is, the gene is about 99% noncoding introns. After excision of the corresponding intron regions of the RNA transcript, the final messenger RNA encodes a protein with 3,685 amino acids. Dubbed *dystrophin*, the previously unrecognized protein is present in muscle tissue but in barely detectable amounts. It is absent in boys with Duchenne muscular dystrophy, as discussed in more detail in Chapter 5.

A new field called **bioinformatics** deals with the storage, annotation, retrieval, and comparison of DNA sequences within human DNA and between species (Boguski 1995; Kolata 1996b). Many extended sections of human DNA have been sequenced, including start and stop signals, exons, introns, repetitive DNA, and other genetic features. Many such sequences from a wide variety of organisms have been brought together in computerized databases on the Internet, accessible by a variety of software packages. One of these, called GenBank and sponsored by the National Library of Medicine, now contains information on millions of segments of DNA from thousands of biological species. In the neighborhood of 100 million bases of human DNA have been sequenced. Thus, a small start has been made in knowing the finest details of the human genome. We say *small*, because 100 million is only about 3% of 3 billion, the approximate number of total bases in a haploid set of human chromosomes (Figure 8.11). We return to other aspects of DNA sequencing in the next chapter, where we discuss the Human Genome Project.

DNA Fingerprinting

Fingerprinting using ink and paper is the standard method for personal identification in police work. Constant throughout life, fingerprint ridge patterns can be used to distinguish one person from any other, even identical twins, although their patterns are very similar. The eugenicist Francis Galton (see Box 11A), began fingerprint classification in the 1890s, making practical the classification, filing, and retrieval of records.

DNA fingerprinting has a similar aim, but it is based on certain nucleotide sequences that differ greatly among people. (Actually, the method is better called *DNA profiling* or *DNA typing*, since it has nothing to do with fingers, but we will stick to the original

capable of sequencing tens of thousands of bases a day, such a machine does not avoid the need for generating a set of DNA fragments differing in length by single bases and separating them by size via electrophoresis. Totally different approaches to sequencing are being considered, but these fundamentally new systems are many years away from practical application (Hodgson 1995; Olson 1995).

name.) Although the DNA fingerprinting procedure has been contentious in legal matters over the last decade, it is now accepted as valid by virtually all courts of law. Even in the bitterly contested 1995 murder trial of football star O. J. Simpson, the defense lawyers did not challenge the scientific basis of DNA fingerprinting. Rather, the defense attacked the way the police obtained and handled the blood samples from which the DNA was extracted.

DNA fingerprinting was first used in England in 1986 to prosecute a rape-murder trial. DNA from sperm found on the victims' genitals was compared with DNA from white blood cells of possible suspects (Box 8B). The same basic procedure has been used increasingly in the United States, in many thousands of criminal and paternity cases each year. Because DNA evidence may be very convincing, a proportion of these investigations never go to trial, especially cases involving rape, in which the location of the incriminating DNA-yielding semen leaves no room for argument.

Periodically, we read in the newspaper that a man, convicted of rape long ago and jailed for years, is released on the basis of newly performed DNA tests that prove his innocence (Connors et al. 1996). Currently, 53 convicts have been exonerated after DNA fingerprinting was used on their behalf (Goldberg 1998).

The uses of DNA analysis for personal identification are increasing. For example, the U.S. military routinely collects and stores DNA samples from members of the armed forces. Remains of some Vietnam War veterans were identified by DNA methods two decades after they died in combat. Some of the severely burned victims of the 1993 government raid on the Branch Davidian compound in Waco, Texas were also identified in this way. And the similarly gruesome task of identifying body fragments from TWA Flight 800 and other air disasters was made possible by DNA technology (Ballantyne 1997).

Animal geneticists also make good use of DNA fingerprinting, tracking the movement of cheetahs,

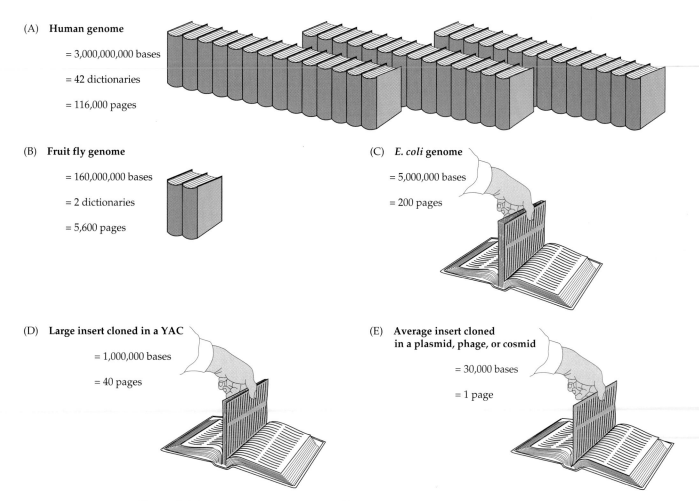

(A) **Human genome**

= 3,000,000,000 bases

= 42 dictionaries

= 116,000 pages

(B) **Fruit fly genome**

= 160,000,000 bases

= 2 dictionaries

= 5,600 pages

(C) ***E. coli* genome**

= 5,000,000 bases

= 200 pages

(D) **Large insert cloned in a YAC**

= 1,000,000 bases

= 40 pages

(E) **Average insert cloned in a plasmid, phage, or cosmid**

= 30,000 bases

= 1 page

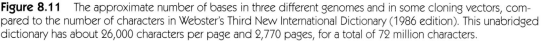

Figure 8.11 The approximate number of bases in three different genomes and in some cloning vectors, compared to the number of characters in Webster's Third New International Dictionary (1986 edition). This unabridged dictionary has about 26,000 characters per page and 2,770 pages, for a total of 72 million characters.

whales, and members of other endangered species. Moreover, DNA information helps zoos and wildlife preserves establish relationships among animals as an aid to captive breeding programs. Using DNA extracted from bone has helped investigators probe illegal trading in ivory and other valuable horns and tusks.

How To Do DNA Fingerprinting

Here we describe the most discriminating method for doing DNA fingerprinting, requiring a dime-sized splotch of fresh or dried blood or an equivalent source of DNA. The method includes the use of restriction enzymes, electrophoresis, and a few additional laboratory procedures.

Recall from Chapter 6 that there are many places in human DNA where sequences of nucleotides are repeated over and over. In some cases, a relatively short sequence (15 to 35 bases) is repeated in tandem many times within a defined DNA region, and the same pattern with perhaps a different number of repeats may occur at other places in the genome. For example, a particular sequence that we all have in our DNA is 16 nucleotides long:

AGAGGTGGGCAGGTGG

This unit is repeated in tandem hundreds of times at dozens of different regions scattered over many chromosomes. The repetitive regions are not in genes and do not code for polypeptides. They are just there.

DNA fingerprinting—that is, individual-specific DNA identification—is made possible because no two people are likely to have the same number of copies of these repetitive DNA sequences. Regions where such differences occur are called **VNTRs,** which stands for **variable number of tandem repeats**. VNTRs are equivalent to standard genetic loci with multiple alleles, except that they have no gene product. It is not known why people differ so much in the number of repetitive DNA units anymore than it is known why people have different fingerprints. But the variation may result from *unequal crossing over* occurring in past generations. As we saw in Figure 7.11, this process can generate DNA segments with different numbers of repeated units.

The variation in copy number leads directly to variation in the lengths of the repetitive regions—lengths that can be measured by procedures that we have in part described in previous sections. To begin, let us assume that two blood samples are at hand, one from blood spots found at the scene of a crime and another from a suspect. As outlined in Figure 8.12, the analysis proceeds as follows:

1. We separately treat the two DNA samples with a restriction enzyme such as *Eco*RI, which cuts DNA at all GAATTC sites. It is important that this recog-

---

BOX 8B   *The First Use of DNA Fingerprinting in Criminal Investigation*

In 1983, on a narrow footpath that passed by a psychiatric hospital in the central English village of Narborough, 15-year-old Lynda Mann was raped and strangled. Investigating detectives found few substantive clues. Conventional forensic tests on recovered semen, for example, could only narrow the identity of the killer to about 10% of the population. Despite a year of wide-ranging police work, the case was eventually dropped when hundreds of leads all proved futile.

Three years later, 15-year-old Dawn Ashworth was similarly sexually attacked and slain within a few hundred yards of the first crime, and semen was again recovered from a vaginal swab. Richard Buckland, a 17-year-old kitchen worker in the psychiatric hospital, had known the victim and had a history of sexual misbehavior. He was arrested, and after extensive interrogation, he confessed to Ashworth's rape and murder.

The unsolved Lynda Mann case was then reopened by sending the semen samples taken from Mann, Ashworth, and Buckland to Professor Alec Jeffreys at nearby Leicester University. Jeffreys had used his newly developed technique of DNA fingerprinting in a highly publicized immigration dispute. On the basis of DNA band matching using a multilocus probe, Jeffreys concluded that both girls had been raped by the same man, but the man was not Buckland! In fact, Buckland—mentally a bit dull and easily confused—did not even fit into the 10% class of possible suspects.

Police persevered in their investigation of the two frightful crimes. Men and teenage boys in Narborough and nearby villages were asked to voluntarily submit to blood drawing, and any blood that met the right criteria by traditional tests was sent to Jeffreys for DNA analysis. The police hoped that this massive undertaking would somehow flush out the killer. After thousands of samples, however, representing about 98% of possible males, nothing turned up.

Colin Pitchfork, a 28-year-old bakery worker who was well known for creative cake decoration, had not given blood when sent his notice. Worried about prior convictions for flashing (not in the Narborough community), he had used a stand-in with falsified identification: Pitchfork's passport with the stand-in's picture. The break in the case came in a local pub over a few pints of ale, when the stand-in casually spoke of his trickery to fellow workers, a disclosure that was reported to the police. Pitchfork, who had managed to conceal his sociopathic personality, was eventually arrested and convicted of the crimes against both girls, as well as two earlier unreported sexual assaults. Pitchfork's blood was the last of 4,583 to be drawn and tested. His DNA proved a perfect match to the genetic fingerprint obtained from the semen samples (Wambaugh 1989).

nition site *not* occur in the repetitive unit to be analyzed, say, the 16-base sequence already noted. *Eco*RI cuts total human DNA into about 100,000 pieces of varying sizes, a few of which will include the VNTR regions (i.e., the tandem repetitions of the 16-base sequence in question). For example, letting r represent one of the 16-base units and x represent the sequence GAATTC where *Eco*RI cuts, we see that one person's DNA might have the following pattern at a particular place on one chromosome (12 repeats of r):

——x——rrrrrrrrrrrr————————x——

At the corresponding place, another person's DNA might have 16 repeats of r and the same *Eco*RI sites to the left and right:

——x——rrrrrrrrrrrrrrrr——————x——

The distance between cutting sites is greater in the second case, thus yielding a bigger fragment.

2. We use electrophoresis to spread out the 100,000 fragments, or bands, of DNA according to size. The two samples are run in side-by-side lanes along with control lanes containing DNA segments of known size. Although there are so many bands of somewhat similar lengths within a lane that they overlap and smear into each other, somewhere along the length of the gel, invisible at this time, are the several bands that contain the VNTR regions in question. We want to know their exact positions in the gel, which are indicative of their lengths. The method for revealing the positions is a bit involved, but proceeds as follows (Housman 1995; McElfresh et al. 1993; Zurer 1994).

3. Since gels are difficult to work with, we transfer the pattern of DNA bands in the gel to a nylon membrane. As shown in Figure 8.13, this is accomplished by simply laying the membrane on top of the gel and letting the materials, including DNA, diffuse up from the gel. This technique was invented by Scottish biochemist E. M. Southern in 1975 and is appropriately called **Southern blotting**. The liquid that moves up through the gel and membrane also denatures the DNA, so that what sticks to the membrane are lengths of *single-stranded* DNA corresponding to the bands in the gel (the bands of interest and lots of other bands).

4. We remove the membrane and bathe it with a so-called **radioactive probe**, which is single-stranded DNA complementary to the repetitive DNA. In this case, we can use a probe developed by English geneticist Alec Jeffreys, who pioneered DNA finger-

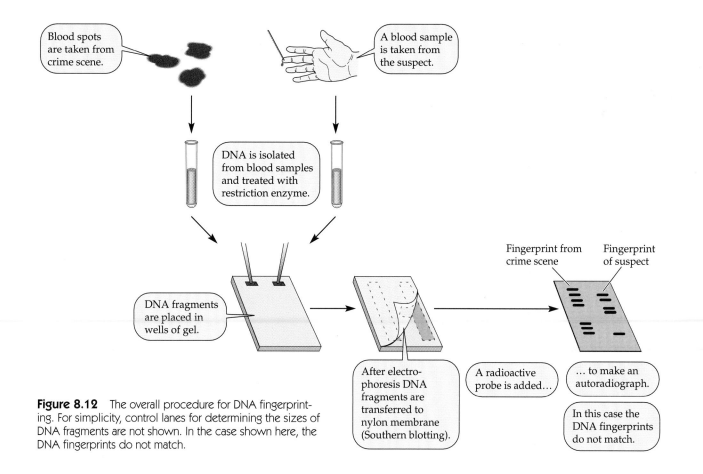

Figure 8.12 The overall procedure for DNA fingerprinting. For simplicity, control lanes for determining the sizes of DNA fragments are not shown. In the case shown here, the DNA fingerprints do not match.

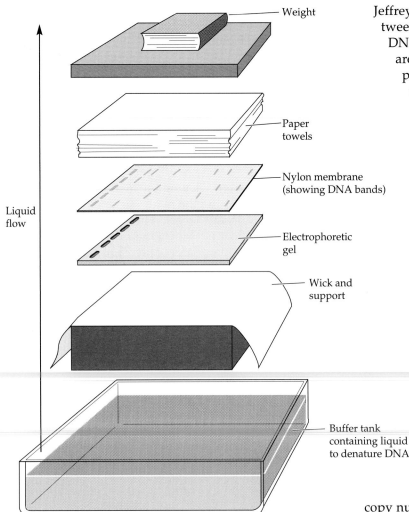

- Weight
- Paper towels
- Nylon membrane (showing DNA bands)
- Electrophoretic gel
- Wick and support
- Buffer tank containing liquid to denature DNA

Liquid flow

Figure 8.13 The setup for Southern blotting, one of the most frequently used procedures in molecular biology. (For clarity this drawing includes space between the several layers.) A stack of absorbent paper towels draws up the liquid from the buffer tank through the wick, through the gel, and through the nylon membrane. DNA molecules from the gel are denatured in the upwardly moving liquid and stick to the membrane (where they can be easily treated one way or another). Although the diagram shows bands of DNA, they are actually invisible at this stage

Jeffreys's probe detects about 15 distinct bands between 4 and 20 kb long. (Some shorter repetitive DNA segments migrate off the gel, and other bands are faint or smeared.) The film looks like an expanded supermarket bar code. Some of the detected bands represent DNA fragments that are alleles of each other; that is, they represent the same region on the two homologous chromosomes, but they have different lengths because of different numbers of copies of the repeat unit. Other bands are nonalleles, deriving from different places in the genome. This type of probe, which detects many different regions, is called a **multilocus probe** (Figure 8.14A). The degree of variation in bands from one person to another is so large that the theoretical probability that the bands seen in one individual are present in another unrelated individual may be one in many millions or billions. Essentially, each person has a unique DNA fingerprint.

A **single-locus probe** will hybridize to just one VNTR region in an individual's DNA and is easier to analyze mathematically. Only one or two bands can be present in such an autoradiograph, corresponding to the two alleles (homozygous or heterozygous). But because there is so much variation in DNA fragment sizes (owing to the variation in copy number of the repetitive unit), these alleles will often be different in different people and lead to distinguishable bands. If several different single-locus probes are examined, the bands from one person will almost always be different in some way from the bands in another individual (Figure 8.14B).

Calculating the exact probability that random people will match is a bit complicated, and we omit the details here. The probability depends on how **polymorphic** the VNTR locus is, that is, on how many different alleles there are in the population from which the samples are drawn and on the frequencies of those alleles. Some DNA fingerprinting sites may have more than 10, or even more than 50, alleles distinguishable from each other by the electrophoretic process (Figure 8.15). Because a gene with 50 alleles can produce 1,275 different genotypes,* even a single-locus probe can do

printing (and who coined the phrase). This probe, consisting of repetitions of the 16-base sequence, hybridizes by base pairing and therefore sticks to the corresponding repetitive DNA sequences wherever they occur on the nylon membrane. No other DNA becomes tagged in this way with radioactivity.

5. The final step is *autoradiography*, which was described earlier with regard to DNA sequencing. We lay a sheet of X-ray film on the membrane, and bands appear wherever the radioactive rays from the probe sensitize the film. That is, the film shows dark bands representing the various lengths of DNA "discovered" by the probe.

Polymorphic means "many forms." A gene with two or more alleles with appreciable frequencies is said to be polymorphic. The type of investigation we are dealing with here is often called *restriction fragment length polymorphism*, or *RFLP*, analysis. (We return to RFLPs in the next chapter.) The formula for the number of different genotypes is $[n(n + 1)]/2$, where n is the number of alleles. Try out the formula for $n = 1, 2, 3$, etc. Mathematically minded readers should try to derive the formula.

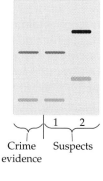

(A) Multilocus probe: 15 bands

Each band represents a DNA fragment of a particular length, dependent on the number of tandem repeats of the short nucleotide that binds the probe. Some bands represent alleles and some nonalleles, but it is not possible to say which are which.

(B) One single-locus probe: 2 bands   Another single-locus probe: 2 bands

Crime evidence | Suspects 1 2

Crime evidence | Suspects 1 2

Figure 8.14 Hypothetical (but typical) DNA fingerprinting autoradiographs, which implicate suspect 1 and exonerate suspect 2. (A) Results using a *multilocus* probe that "discovers" about 15 bands. (B) Results using two different *single-locus* probes, each of which binds to just two bands representing allelic DNA fragments.

Crime evidence | Suspects 1 2

a good job of distinguishing unrelated persons. Combining the information in several different single-locus probes can provide each person with an individualized DNA fingerprint. Thus, when there is complete matching between DNA evidence from a crime scene and DNA from a suspect, it is highly likely that the two samples represent the same person. On the other hand, even a single band mismatch means that the suspect is excluded from consideration.

PCR-Based Methods

Many variations on the fingerprinting theme are available to investigators (Inman and Rudin 1997). The most sensitive methods utilize the *polymerase chain reaction*. In one PCR-based method, investigators amplify a "real" polymorphic gene such as that for a histocompatibility antigen (Chapter 18). The various alleles that are present in a given biological sample are revealed by using a set of specific probes, each of which binds to

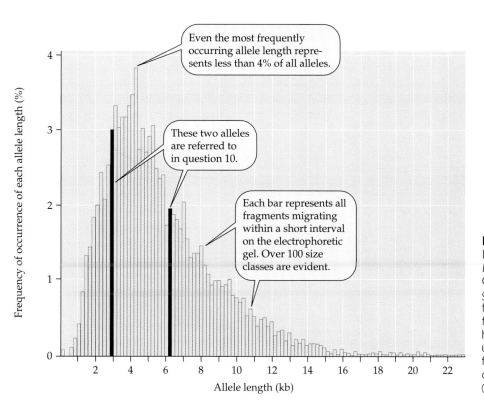

Even the most frequently occurring allele length represents less than 4% of all alleles.

These two alleles are referred to in question 10.

Each bar represents all fragments migrating within a short interval on the electrophoretic gel. Over 100 size classes are evident.

Figure 8.15 A polymorphic VNTR locus on chromosome 1 (known as *D1S7*), one that was analyzed for the O. J. Simpson trial. The horizontal axis gives the lengths (in kilobases) of DNA fragments or alleles obtained by DNA fingerprinting of 22,000 Caucasians. (A histogram for an African American population would have the same general form.) The vertical axis shows the frequencies of the various allele lengths. (After Pena and Chakraborty 1994.)

only one allele. A half dozen *single-stranded* probes complementary to corresponding alleles are spotted at known places on a nylon membrane. The membrane is then bathed with the PCR-amplified, *single-stranded* gene from evidence or a suspect. The hybridization of alleles (whichever are present) to their probes is revealed by a dye that binds only to double-stranded DNA (Figure 8.16). Electrophoresis is not needed; radioactivity is not needed—a great simplification. But the major advantage of PCR-based typing is that the tests can be performed quickly (in a day or two) with minuscule original amounts of DNA—less than 1 nanogram. The disadvantage is that the genes available for this type of analysis are less polymorphic—and therefore less discriminating—than the VNTR sites analyzed by the more complex methodology.

The polymerase chain reaction has also been used more recently to amplify sites called **STR loci** for **short tandem repeats**. These sites are similar to the VNTR loci except that the repetitive unit is only 3 to 5 bases long. Because STR loci are relatively short, they are amenable to amplification by PCR, whereas the longer VNTR loci are not. Since PCR provides millions of copies of the pertinent alleles, the electrophoretic analysis is much simpler. In particular, radioactive probes are not needed to make the relevant bands stand out on a background of irrelevant DNA. Essentially only relevant DNA is on the electrophoretic gel, and *any* DNA stain (often a fluorescent stain) can be used. STR loci are therefore ideal: very polymorphic and easily analyzed with tiny amounts of sample, even if the DNA is degraded to some degree. Usable DNA can sometimes be obtained from a trace of blood, a licked envelope, a sweaty hatband, or a plucked hair. The methodology is being automated, and the Federal Bureau of Investigation and police laboratories nationwide are beginning to standardize their procedures and pool their computerized records (Goldberg 1998).

Court Proceedings

There are several difficulties in using DNA fingerprinting in forensic matters, that is, in court proceedings. One problem is deciding what constitutes a match between bands—bands that may be faint or fuzzy. Forensic laboratories try to minimize such problems by including DNA fragments of known sizes as controls, by refining the electrophoretic system (using longer gels, for example), and by developing objective statistical criteria for comparing bands. Quality controls on DNA fingerprinting include proficiency testing, during which laboratories are expected to reach the right conclusions 100% of the time from samples provided by an accrediting agency. Critics have pointed out that it makes no sense at all to compute the chance of a random match as one in millions if the chance of a laboratory error is one in hundreds.

A second problem is the meaning of a match once it is declared. For example, the prosecution may display a complete match with five different single-locus probes between crime evidence and the defendant. The prosecution then says that the likelihood of somebody other than the defendant having this DNA fingerprint is 1 in 10 million (or some even more amazingly low likelihood). The accuracy of this statement depends on the method of calculation and on the nature of the polymorphism—how many alleles and how frequent each allele is for each of the single-locus probes. The specifics of the polymorphism may differ, however, from one ethnic group to another.

In practice, the method of calculation and the matter of race do not matter all that much, a conclusion that has been worked out laboriously by proponents and critics of DNA fingerprinting. Even the most conservative method of calculation usually leads to a sufficiently low value of a random match to be convincing (Kolata 1994; Lander and Budowle 1994). A recent report by the National Research Council also argues that greater and greater accumulated knowledge of ethnic variation in VNTRs and STRs shows that racial identity is less significant than once believed. From extensive FBI and other databases, it is evident that a wide range of allelic variation is present in *all* ethnic groups—African American, Caucasian, Hispanic, and in subgroups within these broader classifications (Marshall 1996; National Research Council 1996).

How a Jury Reacts. There is the further question of what a jury might make of all this. In the

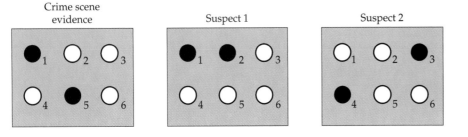

Figure 8.16 PCR-based typing in a hypothetical situation. Single-stranded probes complementary to six different alleles of a gene are spotted at corresponding places (1 through 6) on three nylon membranes. The membranes are bathed with the single-stranded PCR-amplified genes from the crime scene and from two different suspects. A dye is added that binds only to double-stranded DNA (black). In this case, both suspects are excluded. As with traditional blood group evidence, a match at spots 1 and 5 would not have definitely incriminated a suspect, only put him or her in a class of possible suspects.

Simpson trial, for example, many different blood samples were analyzed independently by three different forensic laboratories using dozens of different single-locus VNTR fingerprinting probes and PCR-based genes. Prosecutor Marcia Clark was able to state that the chance that random persons would have the incriminating DNA fingerprints found in blood at the scene of the crime, in Simpson's car, and at Simpson's home would be 1 in many billions, a base number far greater than the population of the world.

The jury members essentially ignored the DNA evidence, probably because defense lawyers were able to argue credibly that police and laboratory workers were sloppy—perhaps even dishonest—in collecting and handling the samples (Levy 1996). Perhaps the jury members were also uncomfortable with complicated laboratory protocols and with concepts of probability that were beyond their everyday experiences. Rather than clear documentation, DNA evidence was perceived as a mystery, deepened by the adversarial nature of the televised courtroom drama, with its argumentative lawyers and rival expert witnesses "dry as sand and about as digestible." (Aranella of the *Los Angeles Times*, quoted in Weir 1995). As noted by the National Research Council (1996), it need not be that way. Better means are needed to present DNA evidence to juries, and behavioral research can be directed toward that end. Also, judges need to be more knowledgeable. In an educational seminar on Cape Cod, for example, 35 federal judges under the tutelage of 20 scientists learned how to sequence their own DNA, make DNA fingerprints, and use DNA databases (Blakeslee 1996).

To summarize, DNA fingerprinting can detect variation in the lengths of DNA fragments based on restriction digests followed by electrophoresis, Southern blotting, and autoradiography. PCR-based methods make for quick analyses from tiny samples. As a method for the unique identification of human beings, the scientific base of DNA fingerprinting is firmly rooted. It is much more discriminating than blood typing (for ABO, Rh, etc.) and can be more useful in crime investigation than traditional fingerprints. Provided appropriate precautions are taken in handling DNA evidence, in executing laboratory procedures, and in explaining the results, DNA fingerprinting is a powerful tool to convict the guilty and exonerate the innocent.

Summary

1. The polymerase chain reaction (PCR) is one of the basic techniques of biotechnology. It quickly replicates a million-fold or more a small target section of DNA that has been selected out from a complex background.

2. PCR requires short primer sequences complementary to the two ends of the target DNA sequence, as well as a heat-resistant DNA polymerase. The process takes place automatically in a thermal cycler. PCR has many applications involving the identification and selective amplification of a short DNA sequence.

3. DNA segments from different species can be spliced together to make recombinant DNA molecules. The first step in this procedure uses restriction enzymes, precision cutting tools of molecular biologists. Once inserted into a plasmid of an *E. coli* host cell, a foreign gene can be multiplied to any desired amount.

4. Genes for insulin, growth hormone, clotting factors, interferons, enzymes, and other products can be cloned and expressed in bacterial or cell culture systems. Biotechnology companies make these valuable medical products commercially.

5. Transgenic animals or plants have a foreign gene incorporated into their cells. Scientists have even constructed transgenic farm animals that secrete useful amounts of a human protein in their milk.

6. Recombinant DNA laboratory procedures have proved to be safe. Debate continues, nevertheless, on the effect of recombinant methods and products on people, animals, and the environment.

7. The sequence of bases along a segment of DNA can be readily determined. Electrophoresis of four separate preparations of a DNA sample allows researchers to determine the positions of all four bases, read sequentially from an autoradiograph. Sequencing using automated devices is now so advanced that a human protein is often identified and characterized by first cloning and sequencing the gene that encodes it.

8. Personal identification is possible by DNA fingerprinting. One method is based on regions of DNA with variable numbers of tandem repeats (VNTRs). Small amounts of DNA, such as that isolated from semen samples or dime-sized spots of blood, have made this type of DNA fingerprinting useful in forensic science.

9. DNA fingerprinting using VNTRs makes use of restriction enzyme digests, electrophoresis, Southern blotting, hybridization with a radioactive probe, and autoradiography. A single VNTR locus can be very polymorphic, that is, have 50 or more distinguishable alleles spread liberally throughout a population.

10. When the results from many different VNTRs or PCR-based methods are combined, the probability that two random people will have the same DNA fingerprint is extraordinarily small. Judges and juries are coming to grips with DNA fingerprinting technology and its interpretation.

Key Terms

autoradiograph	*Eco*RI
bioinformatics	electrophoresis
cDNA library	gene splicing
cloning	genomic library
denaturation	multilocus probe
DNA fingerprinting	oligonucleotide primer

Key Terms (continued)

plasmid
polymerase chain reaction (PCR)
polymorphic
radioactive probe
recombinant DNA
restriction enzyme
short tandem repeats (STRs)
single-locus probe
Southern blotting

sticky (cohesive) end
Taq DNA polymerase
transgene
transgenic animal
variable number of tandem repeats (VNTRs)
vector
yeast artificial chromosome (YAC)

Questions

1. Using the last line of Table 8.1, indicate how many PCR cycles are required to produce a million copies of a target sequence, starting with one copy. How long will this take at, say, 5 minutes per cycle?

2. A *palindrome* is a sequence of letters that reads the same forward and backward: WASITACATISAW, for example. In what way is the sequence recognized by *Eco*RI (GAATTC on one strand) palindromic?

3. Assume that you have cloned a 1-kb human gene in a plasmid using restriction enzyme *Eco*RI as outlined in Figure 8.5. You have subsequently isolated from a culture of *E. coli* a high concentration of the recombinant plasmid shown in (C) of Figure 8.5. What is the first step in getting the human gene separated from the plasmid DNA ?

4. Assume that YAC vectors hold, on the average, 300-kb inserts of foreign DNA. Ideally, how many different YACs would be needed in a library to accommodate all 3 billion bases of human DNA?

5. The restriction enzyme *Eco*RI cuts DNA at the sequence GAATTC, and another restriction enzyme *Hind*III cuts at the sequence AAGCTT. Assume restriction sites along a piece of DNA as follows, letting E = *Eco*RI site and H = *Hind*III site:

3 kb	6 kb	6 kb	7 kb
——E——	——H——	——E——	——

What bands will be present in an electrophoretic gel when this DNA is digested with *Eco*RI alone? with *Hind*III alone? with both enzymes?

6. Animal rights activists have objected to several aspects of recombinant DNA technology and more generally to all areas of medical research using animals. Discuss.

7. From the original piece of DNA shown in Figure 8.8, what fragments would be found in the A reaction tube?

8. From Figure 8.10, determine the sequence of the ten bases from position 50 to position 59.

9. When 300 million bases of human DNA have been sequenced, what percentage of the human genome will remain to be sequenced?

10. In Chapter 12, we will describe the Hardy-Weinberg law, which says that in many populations the frequency of a heterozygous genotype is $2pq$, where p and q are the frequencies of the alleles. For the VNTR locus shown in Figure 8.15, what would be the expected frequency of the genotype composed of the two alleles represented by solid bars?

11. In a criminal trial, the prosecution shows a complete match at five single-locus DNA fingerprinting sites between crime scene evidence and the DNA of the defendant. The testing laboratories calculate the probability of a random match to the crime scene evidence at each of the loci as 0.003, 0.008, 0.02, 0.07, and 0.12. If these loci are inherited independently of each other (an often-used assumption), what is the probability that the sample at the crime scene and that from the defendant represent different individuals?

12. There has been much discussion of whether or not it is possible to say that each person (other than an identical twin) has a *unique* DNA fingerprint. Discuss.

Further Reading

Designed for general college students, the nicely written book by Drlica (1992) covers much of the material in this chapter at an understandable level. Of the many books on the societal impact of biotechnology, we like one by the same author, Drlica (1994). In addition, Hileman (1995) and Anonymous (1995) are thoughtful analyses of the environmental, social, and commercial aspects of genetic engineering. On the polymerase chain reaction, see Eisenstein (1990) for easy how-to-do-it guidelines, and the collection edited by Mullis et al. (1994), which is mostly but not completely technical. Good sources on DNA fingerprinting include the National Research Council (1992, 1996), Lander and Budowle (1994), Zurer (1994), McElfresh et al. (1993), and Inman and Rudin (1997).

Scientific American articles related to biotechnology include Velander et al. (1997) on transgenic pigs, Kessler and Feiden (1995) on drug testing, Beardsley (1994) on the interaction between university-based and commercial research, Neufeld and Colman (1990) on DNA fingerprinting, and Mullis (1990) on the discovery of the polymerase chain reaction.

PART 3

Extensions

CHAPTER 9

Chromosome Mapping and the Human Genome Project

Who owns the *BRCA1* gene, whose mutant alleles are implicated in some cases of breast cancer? The question may seem silly, since in some sense we all own our own genes. Nevertheless, Myriad Genetics (a biotechnology company in Salt Lake City) and researchers at the University of Utah and the National Institutes of Health have filed an application with the U.S. Patent and Trademark Office to obtain exclusive rights to the gene. Opposing their bid are a coalition of hundreds of women's and cancer advocacy groups, and Jeremy Rifkin, a longtime opponent of biotechnology. They say that genes are part of nature and cannot be "invented," and that patenting *BRCA1* will impede cancer research (Lewin 1996).

Your genes direct the production of your polypeptides—your hormones, muscle proteins, enzymes, structural elements—and certainly they cannot be assigned to anybody else. What Myriad Genetics did, however, was pinpoint the gene locus after others had established its approximate position near the middle of the long arm of chromosome 17. Then they isolated and cloned it. They generated quantities of the gene and characterized the control regions and the intron/exon organization. They sequenced its bases. Along with other research groups, they determined the specific mutations in hundreds of different disease-causing alleles and devised diagnostic tests. They further hope to investigate the protein product of *BRCA1* (extrapolated from knowledge of the base sequence) and to devise therapeutic approaches to breast cancer.

In the laboratories of Myriad Genetics, the gene is not in its natural state. Through the work of their scientists, *BRCA1* now exists in a form that is new, nonobvious, and economically useful—the three basic criteria for protecting an invention through a patent (Glick and Pasternak 1994). Thus, by traditional criteria, *BRCA1* is justifiably patentable. Although you could argue that genes are not traditional inventions and therefore should not be judged by traditional criteria, biotechnology companies have already sought and received patent protection for many cloned genes and other DNA sequences.*

Biotechnology enterprises say that patents encourage investment that can lead to both basic discoveries and products with medical applications. Although they retain control of commercial deals, biotechnology companies assert that they promote open exchange of scientific information. Others have argued that patenting restricts competition, produces higher prices, cramps the pace of research, favors established corporations over small entrepreneurs, promotes self-interest over the public good, and affronts human dignity. In short, they fear that business pressures will distort the paths of true science. In any event, the phrase "gene patenting" is just a handle for the real issue: safeguarding for the patent holder (for 20 years) the marketable uses to which the sequence information might be put.

Although the functioning of *BRCA1* in your body is not in question, we might better understand the patentability or nonpatentability of a base sequence if we examine how a gene like *BRCA1* is cloned. The first step is often to find out where the gene is. The encoding regions of a typical gene are 3,000 nucleotides long (expressed as a 1,000-amino-acid polypeptide). Since the haploid complement contains about 3×10^9 nucleotides, a single gene represents about one-millionth of the total DNA in a cell. So how do you get your hands on this proverbial needle in a haystack? A first step is often **gene mapping**, that is, locating the position of a particular gene among other genes along a chromosome.

Genetic Linkage Maps

Gene mapping often begins with a statistical analysis of the consequences of crossing over during meiosis in successive generations in a pedigree. Started in 1913, this kind of "old-fashioned" mapping is a long-standing activity of geneticists (Box 9A). Although mapping by crossing over may one day be obsolete, it is currently performed in hundreds of modern laboratories to help locate specific disease-causing genes and to form a framework for the Human Genome Project (discussed later). The analysis of crossing over leads to the construction of **genetic linkage maps**. We describe the method briefly before considering more recent developments.

Consider for a moment two genes that lie next to each other on the same chromosome. Assume that the dominant alleles of two genes (*D* and *E*) are on the

---

*To keep up with demand, during 1996 and 1997 the Patent and Trademark Office hired scores of new biotechnology patent examiners, most of whom have advanced degrees in genetics, immunology, or molecular biology.

BOX 9A *Dr. Julia Bell*

Julia Bell was a 21-year-old graduate student in mathematics at Cambridge University when Mendel's work was rediscovered in 1900. At age 41 she earned a medical degree. She worked as a statistical assistant at University College London, was awarded a prize from Oxford University for her contribution to biometric science, and was elected Fellow of the Royal College of Physicians—all this at a time when few women entertained scientific careers. A pioneer in human genetics, she retired from her research position at age 86 and died in 1979 at age 100.

A perfectionist in gathering family data and a skilled statistician in analyzing the assembled pedigrees, she was well known for work on muscular dystrophy and hereditary diseases of the eye. She was one of the first to investigate inbreeding rates in different populations. Moreover, she and J. B. S. Haldane were the first to establish the chromosomal positions of two human genes with respect to each other by an analysis of crossing over. They determined that the X-linked gene for color blindness was several "map units" from the gene for hemophilia (shown in Figure 9.2). They said that it was important "to demonstrate that the principles of linkage which have been worked out for other animals also hold good for man" (Bell and Haldane 1937).

same chromosome (inherited from one parent), and the recessive alleles of both genes (*d* and *e*) are on the homologue (from the other parent). The genotype of the individual can thus be written:

$$\frac{DE}{de}$$

The lines represent the two homologous chromosomes with the genes they carry shown above and below the lines. This information is often written simply *DE*/*de*. Since the two homologous chromosomes segregate from each other during the meiotic divisions, half the gametes contain the upper chromosome carrying *DE* and half contain the lower chromosome carrying *de*.

Now let us consider what would happen if the genes were moved farther and farther apart along the chromosome:

$$\frac{D\ E}{d\ e} \quad \text{or} \quad \frac{D \qquad E}{d \qquad e} \quad \text{or} \quad \frac{D \qquad\qquad E}{d \qquad\qquad e}$$

The more separated the genes are, the greater is the opportunity for crossing over to occur somewhere along the length of the chromosome between them. (Crossovers are depicted in Figures 3.8 and 7.11A; we add a few more details about the process here.) Whenever crossing over occurs, some of the gametes will contain chromosomes with new assortments of genes, either *De* or *dE*. The gametes that are formed are either **parental gametes** or **recombinant gametes**:

Parental gametes: *DE* and *de*

Recombinant gametes: *De* and *dE*

The parental gametes are those formed by the individual that are the *same* as the gametes that formed the individual. The recombinant gametes are those with a *new assortment* of genes. From matings involving a *DE*/*de* parent, it is often possible to determine the percentage of progeny that come from parental or recombinant gametes.

The essence of traditional genetic mapping techniques is that the farther apart two genes are along the length of a chromosome the greater is the percentage of recombinant-type gametes arising from crossing over. Early geneticists were able to show that under some conditions the relationship between the percentage recombination and the distance apart was strictly proportional. Thus, they defined a **map unit**:

$$x\% \text{ recombination} = x \text{ map units}$$

In essence, the observable percentage recombination becomes a unit of distance along a chromosome. A 1% recombination is interpreted as 1 map unit, 5% recombination as 5 map units, and so on. Two genes that are shown to be within a small recombinational, or crossover, distance from each other are said to be **linked genes**. And since many genes can be shown to be linked to each other in a line—like towns along a highway—the resulting map is called a *genetic linkage map*.

The occurrence of crossing over can be seen through a microscope when homologous chromosomes synapse with each other during the prophase of the first meiotic division. The visible exchanges of chromosome segments are called *chiasmata* (singular, chiasma). They appear as X-shaped—or more aptly, chi-shaped (χ)—configurations (Figure 9.1), and there can be several chiasmata between one pair of synapsed chromosomes. Unfortunately, chiasmata are difficult to see in the meiotic cells of most mammals, including humans; but they are starkly evident in the cells of some plants, insects, and amphibians.

The formal representation of crossing over in Figure 9.2 shows the two X-linked genes mapped by Bell and Haldane in 1937:

G = dominant allele for normal color vision

g = recessive allele for green-shift (deuteranomaly)

H = dominant allele for normal blood clotting

h = recessive allele for hemophilia A (classic type)

Note that a crossover involves just *one* of the chromatids of each of the homologous chromosomes. The net result is that the egg that develops from this meiosis could be either parental (from a chromatid not involved in the crossover) or recombinant (from a chromatid involved in the crossover). In essence, crossing over occurs when two *nonsister* chromatids appear to break and exchange pieces. At the molecular level the process is more complicated than this formal description. However, we emphasize that crossing over is a normal, relatively frequent event.

We present the following three extensions of the idea of mapping.

1. *The phase can be either coupling or repulsion.* Mapping two genes with respect to each other requires that they both be heterozygous in the test parent. Otherwise, there is no way to make the distinction between a crossover chromatid and a noncrossover chromatid. (Confirm this statement for yourself by trying to tell a crossover from a noncrossover gamete when one, or the other, or both of the genes are homozygous.) But it does *not* matter whether the genotype of the doubly heterozygous parent is:

$$\frac{g\ h}{G\ H} \quad \text{or} \quad \frac{g\ H}{G\ h}$$

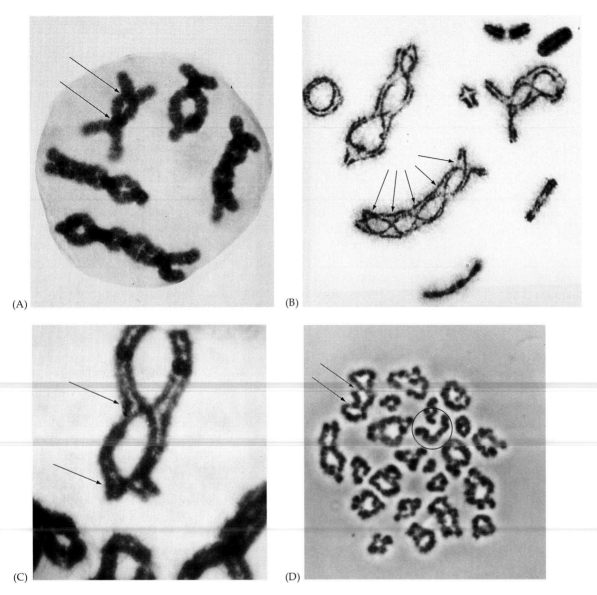

Figure 9.1 Micrographs of chiasmata. These X-shaped configurations (some indicated by arrows) are seen in late prophase of the first meiotic division, when homologous chromosomes start to move apart. (A) During pollen formation in the trillium plant. (B) During sperm formation in a grasshopper; five chiasmata can be seen in one tetrad. (C) During sperm formation in a salamander; the centromeres and interchange points show particularly clearly. (D) During sperm formation in a man; the circled X and Y chromosomes appear to be synapsed end to end. (A, photograph by R. F. Smith; courtesy of A. H. Sparrow, Brookhaven National Laboratory. B, courtesy of Bernard John, The Australian National University. C, courtesy of James Kezer, University of Oregon. D, courtesy of Paul Polani, Guy's Hospital, London.)

This distinction is called the **phase** of the double heterozygote. **Coupling** is shown on the left, **repulsion** on the right. The only consequence of different phases is that the gametes that are called parental in one case are recombinant in the other, and vice versa. The measurement of genetic linkage distances by percent recombination is the same whether the genes are in coupling or repulsion, but we do need to know the phase of the parent.

2. *Geneticists measure only short distances.* For accuracy, the mapping of two genes with respect to each other is restricted to relatively short distances— about 20 map units or less. The reason is that more than one crossover along the chromosome spanning two genes becomes increasingly likely when two genes are farther and farther apart. (Note the multiple chiasmata in Figure 9.1.) Such multiple crossovers cause errors in measurement. There is no easy way to correct for these errors, and the solution has been to simply avoid the problem. It is, of course, possible to build up chromosome maps extending for hundreds of map units by successively linking many genes, each one within 20 map units of its immediate neighbors.

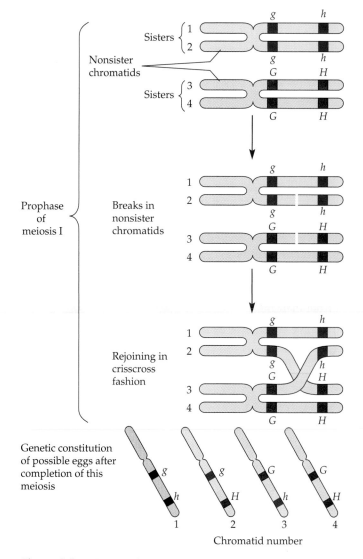

Figure 9.2 Diagram of the occurrence and consequence of a single crossover along the length of the X chromosome that spans the loci for green-shift (*G,g*) and hemophilia (*H,h*). We assume that the woman is heterozygous for both genes, with the recessive alleles of each gene (*g* and *h*) on the upper chromosome and the dominant alleles (*G* and *H*) on the lower one. The four chromatids are numbered at their left ends. Although the crossover is shown as occurring between chromatids 2 and 3, the same results would come from a crossover between strands 1 and 3, or between 1 and 4, or between 2 and 4—that is, between any two nonsister chromatids. (Question 1 refers to this figure.)

3. *The maximum recombination is 50%*. A consequence of multiple crossing over is this: If you try to map two genes far apart along a chromosome, you always find 50% recombination. Converted to map units, this result can be a gross error, of course. The 50% maximum comes about partly because crossing over involves only two of the four chromatids that exist at the tetrad stage of meiosis.

It is interesting that two genes on different (nonhomologous) chromosomes also show 50% recombination, this being the natural result of Mendel's law of independent assortment: With *D* and *E* on different chromosomes, an individual with the genotype *D/d E/e* makes four kinds of gametes: ¼ *DE*, ¼ *De*, ¼ *dE*, and ¼ *de*. Two of these (50%) are always recombinant and two (50%) parental. Thus, mapping by crossover percentages does not reveal whether two genes are far apart on one chromosome or on different chromosomes. The corollary is that mapping is easier and more accurate the closer the genes. The limit is two genes that are adjacent (zero map distance), in which case their alleles are transmitted as one element with no recombinants.

But if you have no prior knowledge of the relative position of two genes with respect to each other, mapping by genetic linkage is often very tedious. The problem is threefold: (1) You must find parents who are heterozygous for the two genes to be mapped. (2) You must know the phase or make a good guess at it, since what is recombinant in one case is parental in the other. (3) Even if the two previous conditions are met, two randomly chosen genes are simply not likely to be within mappable distance of each other, since the total genetic length of all the human chromosomes is about 3,700 map units. (That two genes are *not* linked is information, of course, but it is not usually very useful.)

With many experimental organisms, you can regularly generate double heterozygotes of known phase for whatever genes you want to map. In fact, the principles of mapping that we have described were all worked out using fruit flies (Box 9B). But with people, the statistically-minded geneticist must ferret out informative matings from what is available in the medical literature, hospital records, or population data. Far-ranging searches may be needed if the two genes both relate to rare recessive diseases. Virtually all of the early linkage results in humans involved a blood group locus or other polymorphic gene, since heterozygotes for such genes are not rare. (Recall that a polymorphic gene is one with several common alleles.) Of course, if no other gene of interest happened to be close to a blood group locus, linkage researchers were out of luck, and such was usually the case throughout the 1970s.

DNA Marker Maps

Fortunately for gene mappers, cell and molecular biologists came to their rescue in the 1980s with the identification of so-called **DNA markers**. The first DNA markers that were extensively investigated are called *RFLPs*. Others are *microsatellites*, very short units repeated 20 to 30 times (Chapter 6), and *VNTRs*, somewhat longer repeat units that are the basis of some DNA fingerprinting (Chapter 8). When examined in a

BOX 9B   *Morgan's Fly Room*

Thomas Hunt Morgan was born in Kentucky of an aristocratic family that included Confederate general John Hunt Morgan (his uncle), composer Francis Scott Key (his great-grandfather), and financier J. P. Morgan (a remote cousin). From 1904 to 1928, as Professor of Experimental Zoology at Columbia University, he assembled a remarkable group of scientists who crowded together in a small space that was affectionately dubbed the Fly Room.

Early on, Morgan invited two undergraduates, Calvin Bridges and Alfred Sturtevant, to join the Fly Room; later, a graduate student named H. J. Muller joined the group. The lab atmosphere was remarkably relaxed, with students treated as equals. The four of them (and visiting colleagues) discovered many of the details of basic genetics: multiple alleles, sex-linked genes, crossing over and mapping, recessive lethal genes, sex determination, chromosomal aberrations, and nondisjunction.

Crow (1988) provides a fascinating account of how Sturtevant in 1913 (as an undergraduate student) constructed the first genetic linkage map of five X-linked loci from "seemingly irrelevant counts of the number of different kinds of offspring from various matings." Members of the Morgan group established the fruit fly *Drosophila*, as

the premier organism for genetic research and wrote the first authoritative textbook of genetics, *The Mechanism of Mendelian Heredity* (Morgan et al. 1915).

The physical setting of the laboratory, with its haphazard and make-do simplicity, was unconventional. The culture medium, which was poured into a miscellaneous assortment of glass milk bottles, contained mashed bananas.

"The resulting smell was fierce and drew constant complaints from the rest of the biology department. … Near the entrance of the room a stalk of bananas hung conspicuously, serving as a center of attraction for the numerous fruit flies that had escaped from their milk bottles. … Morgan was by nature a careless and sloppy man, but he also delighted in shocking others, in playing the imp. … More than once Morgan was mistaken for the janitor." (Shine and Wrobel 1976)

Describing the work in the Fly Room, Sturtevant wrote:

"Each carried on his own experiments, but each knew exactly what the others were doing, and each new result was freely discussed. There was little attention paid to priority or to the source of new ideas or new interpretations. What mattered was to get ahead with the work. … There can have been few

times and places in scientific laboratories with such an atmosphere of excitement. … This was due in large part to Morgan's own attitude, compounded of enthusiasm combined with a strong critical sense, generosity, open-mindedness, and a remarkable sense of humor." (Sturtevant 1959)

Morgan was awarded the 1933 Nobel prize in physiology or medicine, but he did not attend the award ceremony in Stockholm. Perhaps he was too busy, or perhaps he just disliked fancy-dress occasions.

Thomas Hunt Morgan (1866–1945)

---

laboratory, all three types have distinct phenotypes. This property puts them into the same class of inherited characteristics as one's ABO blood type and many other biochemical traits. Whether the laboratory test relies on a clumping reaction, as in blood typing, or on electrophoretic bands in a gel, as with DNA markers, the heterozygous condition is clear-cut and frequent. DNA markers have proved very useful in linkage studies, and over 10,000 such sites have been positioned on the human chromosomes (White and Lalouel 1988).

RFLPs, short for **restriction fragment length polymorphisms**, are based on base substitution mutations that (by chance) alter the recognition sites of restriction enzymes (Chapter 8). Thus, at those particular spots, the enzyme cuts one person's DNA but not another's. For example, suppose that a certain section of John's DNA has four *Eco*RI sites (GAATTC, shown by ×'s) spaced as follows:

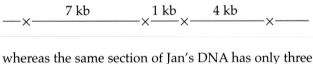

whereas the same section of Jan's DNA has only three *Eco*RI sites:

A mutation in any one of the bases that make up John's second *Eco*RI recognition site could have produced the difference between John and Jan. This base change, occurring perhaps many generations back, produced two alleles—one allowing the restriction enzyme to cut (GAATTC in John) and one preventing the cut (say, AAATTC in Jan). Most of these mutations in recognition sites are not in protein-encoding genes. The variation they produce, however, can be detected by the techniques described for VNTRs in Chapter 8. The experimenter cuts a DNA sample with *Eco*RI, runs the digests in an electrophoretic gel, denatures and blots the migrated DNA onto a nylon membrane, hybridizes the fragments with a radioactive, single-stranded probe complementary to this region, and autoradiographs. Remembering that the shorter the fragment the farther it migrates, the result looks like this (assuming that both John and Jan are homozygous):

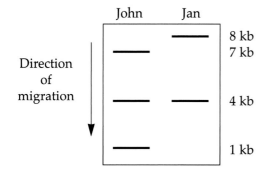

Some people resemble John and some resemble Jan, some may be heterozygous, and some may have completely different banding patterns, based on mutations of the other *Eco*RI sites. (You should be able to predict the banding pattern for a heterozygote having one allele like John's and one like Jan's.) Here we are studying a polymorphic gene that leads to different patterns of migration of restriction fragments—thus the term *restriction fragment length polymorphism*. The DNA sites that result in RFLPs occur at thousands of random places throughout the genome, and the heterozygous state is common and easy to detect.

DNA markers—RFLPs, microsatellites, or VNTRs— provide background reference sites (comparable to mile markers along a highway) for plotting "regular," protein-encoding genes. This framework approach is the specialty of Généthon and the Centre d'Etudes du Polymorphisme Humaine (CEPH) in Paris.* Over the years, CEPH has played a coordinating role in the Human Genome Project, especially by distributing, free throughout the world, cultured cell samples from reference families for DNA analysis (Dausset and Cann 1994). One of their latest reports is entitled "A Comprehensive Genetic Map of the Human Genome Based on 5,264 Microsatellites" (Dib et al. 1996). This map is based entirely on sites where the dinucleotide AC (TG on the other strand) is repeated, and where the repeat number is polymorphic. Some of the sites are represented several times in their library of microsatellites. The actual number of different positions on the map is 2,335, spread somewhat unevenly

among the total of 3,700 map units for the 22 autosomes and the X chromosome (Figure 9.3). (Confirm that the average distance between these markers is 1.6 map units.) An additional advantage of this monumental work is the ability of any laboratory in the world to analyze any particular microsatellite region in their own DNA samples. They can do this because CEPH has determined and published DNA sequences on either side of each microsatellite. Scientists anywhere can synthesize (or, more usually, buy) primers that recognize these sequences and thus amplify by the polymerase chain reaction the microsatellites of any individual.

Researchers use DNA markers in linkage studies the same way they use ordinary phenotypic characteristics. Huntington disease was the first disease-producing gene to be mapped against DNA markers (in 1983). Researchers, led by James Gusella of Harvard University, examined members of Huntington pedigrees and identified people heterozygous for the disease and also for a nearby RFLP site (genotype $\underline{HR_1}/\underline{hR_2}$, say). By looking for recombinants among the children—those receiving an $\underline{HR_2}$ or $\underline{hR_1}$ gamete—they were able to link the Huntington disease gene to an RFLP site about 4 map units away near the tip of the short arm of chromosome 4. Later they were able to find DNA markers within 1 or 2 map units on *both* sides of the gene. These are then called **flanking markers.** Delimiting the section containing the disease gene aids in eventually identifying and cloning the gene.

In the search for the *BRCA1* gene, the first mapping results were reported in 1990 by a group at the University of California at Berkeley headed by Mary-Claire King. The mapping was difficult. About 90% of breast cancer cases are sporadic—that is, they are not inherited through generations at all. Also, the disease is common enough that multiple cases within a family do not necessarily mark it as having an inherited form. Furthermore, because women can sometimes have the gene without getting breast cancer, families with an inherited form might only exhibit a single, seemingly sporadic case. Another difficulty is that *BRCA1* accounts for only about half of inherited, early-onset breast cancer. Despite these problems, the King group was eventually able to show that a gene predisposing to breast cancer was within about 10 map units of a VNTR with multiple alleles in band q21 of chromosome 17 (Figure 9.4). Later, other researchers mapped the *BRCA1* gene between ever-closer flanking markers, first to a 4-map-unit interval, then 2, and finally down to a 0.6-map-unit interval.

We should note that the crossover distances may not strictly represent actual physical distances along a chromosome measured in, say, nanometers or base pairs (1 base pair = 0.34 nm). Nonproportionality can exist because crossing over does not occur evenly all along a

---

*Following a strong tradition in molecular biology, French scientists consider genome research a national priority. Généthon, near Paris, is a high-technology genetic center supported largely by the French muscular dystrophy organization, Association Française Contre les Myopathies. CEPH is also supported largely by private funding; its founder is Jean Dausset, a 1980 Nobel prize-winning immunologist.

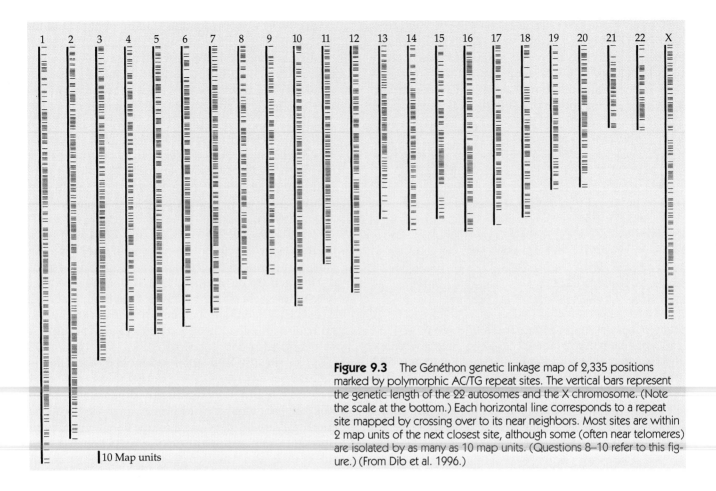

Figure 9.3 The Généthon genetic linkage map of 2,335 positions marked by polymorphic AC/TG repeat sites. The vertical bars represent the genetic length of the 22 autosomes and the X chromosome. (Note the scale at the bottom.) Each horizontal line corresponds to a repeat site mapped by crossing over to its near neighbors. Most sites are within 2 map units of the next closest site, although some (often near telomeres) are isolated by as many as 10 map units. (Questions 8–10 refer to this figure.) (From Dib et al. 1996.)

10 Map units

chromosome. You can see this in Figure 9.3, where the density of DNA markers is clearly less near many of the telomeres. See, for example, the upper ends of chromosomes 1, 9, 21, and 22. The best guess for this unevenness is that the frequency of crossing over is greater near the ends of chromosomes. On the average, however, the approximately 3×10^9 base pairs of a haploid human genome are divided into 3,700 map units for all human chromosomes. Thus, $(3 \times 10^9)/3,700$, or approximately 800,000 base pairs, correspond to 1 map unit. So if a disease gene is flanked by known DNA markers, each 1.25 map units away, the gene resides in a DNA interval of 2 million base pairs.

A DNA marker map with over 7,000 dinucleotide loci has been constructed for the mouse genome by researchers at the Whitehead Institute for Biomedical Research in Cambridge, Massachusetts. Comparing the maps of different species is an active and exciting area of research, revealing surprising similarities between species. Over 2,000 genes have been mapped in both humans and mice.

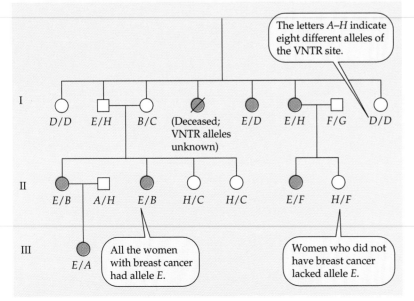

The letters *A–H* indicate eight different alleles of the VNTR site.

All the women with breast cancer had allele *E*.

Women who did not have breast cancer lacked allele *E*.

Figure 9.4 One of the 23 families in the study that initially linked *BRCA1* to a VNTR site (called *D17S74*) on chromosome 17. All women with breast cancer (solid circles) were diagnosed before age 46; the average age at diagnosis was 33. The pedigree indicates a close linkage of allele *E* with the occurrence of breast cancer. (After Hall et al. 1990.)

Physical Maps

To nongeneticists, maps based on crossing over and gene recombination probably seem esoteric. The behavior of chromosomes during meiosis is just not the kind of thing one easily associates with measured lengths. Yet despite their abstractness, genetic linkage maps are curiously timeless. If microscopes had not been devised and if subcellular structures were unknown, we would still have Figure 9.3 with its thousands of microsatellite loci spaced along dozens of separate cords of varying lengths. In no way do these data depend upon our having *seen* the chromosomes. If Figure 9.3 had somehow preceded the microscope, the marker-rich human chromosomes or equivalent stringlike structures would have had to be invented to explain the known facts.*

Other kinds of chromosomal representations are **physical maps** based on actual nucleoprotein material. One fairly coarse type relies on chromosome bands or additional cytological landmarks (e.g., duplications and rearrangements; see Chapter 14) that can be seen. At an intermediate level are physical maps based on segments of cloned DNA that can be arranged in a unique order. At this intermediate level, physical mapping often ties together gene-length pieces of DNA. Table 9.1 shows, however, that gene lengths differ greatly, from the small tRNA genes (a few hundred base pairs long) to the large factor VIII or cystic fibrosis genes (hundreds of thousands of base pairs long) or even longer genes. But by and large, the intermediate physical maps deal with DNA segments much smaller than the intervals between genetically mapped DNA markers. At the finest level, physical mapping deals with sequencing lengths of DNA by ordering

the individual bases—A, T, G, and C. (The "in" joke among sequencers is that all the billions of human bases were ultimately determined; however, the computer then alphabetized them!)

Cytological Studies

Physical mapping began when geneticists were able to assign genes to specific autosomes. The human X chromosome genes, of course, were identifiable as such on the basis of pedigree information soon after Mendel's laws became known. Until 1967, however, not a single autosomal gene could be assigned to a specific chromosome. Now about 3,000 autosomal genes have been located, not just to a particular chromosome, but usually to a small recognizable portion

TABLE 9.1   **The size of some genes**[a]

Size class	Gene or gene product (disease)	Length of whole gene (kb)	Length of exons in kb (% of entire gene)	Number of introns
Small				
	Transfer RNA gene	0.2	0.2 (100)	0
	Histone H4	0.5	0.5 (100)	0
	β-globin (β-thalassemia; sickle-cell disease)	1.5	0.6 (38)	2
	Insulin (diabetes)	1.7	0.4 (33)	2
	Apolipoprotein E (Alzheimer disease)	3.6	1.2 (33)	3
Medium				
	Collagen type 1, α-1 chain (osteogenesis imperfecta)	18	5.0 (28)	50
	Albumin	25	2.1 (12)	14
	Adenosine deaminase (ADA deficiency)	32	1.5 (5)	11
	Clotting factor IX (Christmas disease)	34	2.8 (8)	7
	LDL receptor (hypercholesterolemia)	45	5.5 (17)	17
Large				
	Phenylalanine hydroxylase (PKU)	90	2.4 (3)	12
	*BRCA*1 (breast cancer)	100	5.6 (6)	24
	Factor VIII (hemophilia)	186	9.0 (3)	26
	RB1 (retinoblastoma)	200	2.8 (1)	27
	Cystic fibrosis transmembrane regulator (cystic fibrosis)	250	6.5 (2)	26
Giant				
	Dystrophin (Duchenne muscular dystrophy)	>2,000	16.0 (1)	>60

Sources: McKusick (1997), Strachan and Read (1996), with additions.

[a]The average gene is perhaps 10–20 kb long. The longer genes tend to have more and longer introns.

*Similarly, if microscopes were unknown, meiosis or something like meiosis would have had to be invented to explain Mendel's laws.

of that chromosome. This stunning turnabout in putting human genes in their places started when Mary Weiss and Howard Green, at New York University School of Medicine, assigned to chromosome 17 the gene for thymidine kinase, an enzyme needed in the synthesis of DNA. They used a curious technique called **somatic cell fusion**, in which laboratory cultures of human cells and mouse cells are joined together to form hybrid cells (Ephrussi and Weiss 1969). (We hasten to add that these hybrid cells remain in their petri dishes and never develop into any kind of critter.) Now used in conjunction with molecular techniques, somatic cell fusion is a powerful tool for human genetic analysis.

Somatic cells of different species (say, human and mouse) are placed together in a culture dish with nutrient medium and left for hours or days to grow and divide. In the mixed population, several human-mouse hybrid cells will form spontaneously, but the frequency of fused cells can be enhanced up to a thousandfold by adding to the culture medium either a chemical (polyethylene glycol) or an inactivated virus (the Sendai virus). These agents cause cells to clump together, aiding in their intimate fusion. The chromosomes of both species eventually collect into one nucleus in the hybrid cells, which divide and redivide by mitosis. Furthermore, the genes of both parental types may be expressed in the hybrid cells.

The cardinal property of fused cells that enables researchers to localize genes is the loss of chromosomes over the course of cell generations (Figure 9.5). Interestingly, it turns out that in human–mouse hybrid cell lines, only *human* chromosomes are lost. The reason for a preferential loss is not clear, but it may hinge upon the slower replication rate of human chromosomes. Which particular human chromosomes are lost and which ones are retained seems to be fairly random. The process of elimination continues until only one or a few human chromosomes are left in a hybrid nucleus. At this point, the various subclones (shown at the bottom of Figure 9.5) become stabilized.

This loss of chromosomes is accompanied by a loss of the proteins encoded by the genes present on those chromosomes. By monitoring the loss of particular chromosomes by standard karyotyping and also the loss of proteins, researchers gather information on which functions are associated with which chromosomes (Ruddle and Kucherlapati 1974). Although all mammalian cells contain mostly the same proteins to regulate their metabolism, the amino acids usually differ somewhat from species to species. Thus, two species may possess different molecular forms of an enzyme, which can often be detected by gel electrophoresis. Like the separation of DNA segments by migration in an electric field, the separation of protein molecules depends on charge and size variations. In the case of similarly sized proteins, their differential migration rates come about primarily because of charge differences in their constituent amino acids.

For example, in the early 1970s, Frank Ruddle and his colleagues at Yale University analyzed 26 hybrid clones for the presence or absence of a human enzyme called peptidase C

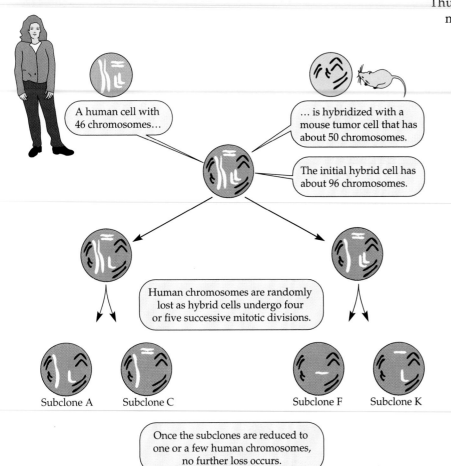

A human cell with 46 chromosomes…

… is hybridized with a mouse tumor cell that has about 50 chromosomes.

The initial hybrid cell has about 96 chromosomes.

Human chromosomes are randomly lost as hybrid cells undergo four or five successive mitotic divisions.

Subclone A   Subclone C   Subclone F   Subclone K

Once the subclones are reduced to one or a few human chromosomes, no further loss occurs.

Figure 9.5 The derivation of stable cells containing one or a few human chromosomes. All the chromosomes of the two somatic cell parents are present in the initial hybrid cell. However, human chromosomes (white) are successively lost during divisions of the hybrid cells. Because the loss is random, each eventual subclone usually ends up with a different array of human chromosomes (but a full complement of mouse chromosomes).

(which migrates a little faster than the corresponding mouse peptidase C). With respect to the corresponding presence or absence of chromosome 1, their results were as follows:

		Chromosome 1	
		Present	Absent
Peptidase C	Present	14	0
	Absent	0	12

Thus, in all hybrid cell lines that retained a chromosome 1, they found the human enzyme; in all lines that had lost chromosome 1 (both copies), they failed to find the human enzyme. Thus, the human gene for peptidase C must reside on human chromosome 1.

The strategy we have outlined for assigning a locus to a specific chromosome is limited to genes that are *expressed*—genes that are transcribed and translated into detectable amounts of the protein product. Hence, this approach is suitable for only some of the genes present in the cell. A newer approach can locate *any* gene or other DNA sequence, provided researchers have a corresponding radioactive probe.

The experimental procedure starts, as before, with the isolation of human-mouse hybrid cell lines containing a small number of human chromosomes whose identities are established by standard cytogenetic techniques. A researcher will likely have handy dozens of hybrid cell lines, each with a different array of human chromosomes. From each hybrid cell line, DNA is extracted, cut with a restriction enzyme, spread out by size on an electrophoretic gel, denatured, and transferred to a nylon membrane by Southern blotting. The membrane is hybridized with a radioactive probe complementary to the DNA sequence in question and autoradiographed. A band will appear in a particular lane of the gel only if the original hybrid cell contained a chromosome with the DNA sequence in question.

Gusella and his colleagues localized the Huntington disease gene to chromosome 4 by this method. Recall that they located an RFLP site about 4 map units from the disease gene. The probe that hybridized to this RFLP region was called G8. They examined 18 different human-mouse hybrid lines for the presence or absence of DNA sequences complementary to the G8 probe. The data for six of the lines are given in Table 9.2. The radioactive G8 probe bound only to those lanes on the Southern blot membrane that included DNA from human-mouse hybrid cells containing chromosome 4.

A rather different method for gene localization is called **fluorescence in situ hybridization (FISH)**, introduced in Chapter 2. FISH evolved from techniques devised in 1969 by Joseph Gall and Mary Lou Pardue of Yale University. These techniques use radioactivity (rather than fluorescence) as a tag. With FISH, researchers first prepare human metaphase chromosomes on a microscope slide in the standard way. Next, the chromosomes are treated with enzymes to get rid of contaminating RNA and proteins. Without disturbing the structural integrity of the chromosomes, the remaining DNA is then denatured in place (i.e., "in situ") to form single strands. Researchers bathe the slide with a probe (also single-stranded DNA) specific for a given gene or other DNA feature. This probe is chemically linked to a *fluorochrome*, a molecule that emits visible light of a specific wavelength when illuminated with ultraviolet light. Various fluorescent chemicals shine with different colors. For example, researchers can use different fluorescent color tags joined to probes rich in repetitive DNA sequences to outline *all* chromosomal contours. Washing off excess probe, they finally examine their preparation microscopically under ultraviolet light.

In Figure 9.6, the probe of interest was specific for repetitive DNA sequences found only near the centromere of the X chromosome. (The centromeric regions of most chromosomes have unique repetitive DNA sequences.) The photomicrograph shows that this unusual individual had five X chromosomes (see Chapter 13). Other examples of FISH are shown in Figure 2.10, where the chromosome are "painted" different colors, and in Box 6B, where the telomeres are labeled with a repeated TTAGGG probe.

Another application of FISH is the mapping of unique DNA sequences to specific chromosomal features on metaphase chromosomes. For example, by combining FISH and standard banding techniques, researchers can locate the physical position of a gene or other DNA sequence to a particular chromosome band. The DNA sequence in question is inserted into a

TABLE 9.2 Assignment of the Huntington disease gene to chromosome 4 by analysis of human–mouse hybrid cells

Hybrid cell line	Presence (+) or absence (–) of G8[a]	Human chromosome present
W-5	+	4 17 18 21 X
J-22	+	4 6 10 11 14 17 18 20 21
N-16	+	3 4 5 7 12 17 18 21 X
W-2	–	8 12 17 21 X
N-5	–	12 14 15 16 18 20
R-11	–	11 13 16 20 21 X

Source: Gusella et al. 1983.
[a]G8 is an RFLP marker located 4 map units from the Huntington disease gene.

Figure 9.6 Fluorescence in situ hybridization (FISH). In this micrograph, metaphase chromosomes were hybridized with a yellow fluorescent DNA probe specific for sequences in the centromere region of the X chromosome. The five yellow regions indicate an 49,XXXXX karyotype, or person with five X chromosomes. A red fluorescent counterstain (called propidium iodide) was used to color all chromosomes about equally. (Courtesy Oncor Incorporated, Gaithersburg, MD.)

A major problem is the sheer size of the whole human genome (3,000 Mb), or the smaller but still unmanageable size of an average human chromosome (120 Mb), or even the size of the smallest human chromosome (chromosome 21, at 39 Mb). The largest segment of DNA that can be sequenced in one operation is at most 500 bases. You can calculate that it would take a minimum of 240,000 such pieces placed end to end to span the average human chromosome. And even that is not enough sequencing work, because you need a method for putting the 240,000 segments in their proper order. This could be done if all the pieces were long enough to overlap one another. Then researchers (at their computers) could simply match up the overlaps to provide continuity.

The only solution is to use a hierarchy of steps based on ever smaller sets. That is, a chromosome must be broken into large pieces, each of which is then broken into medium-sized pieces, each of which is broken into small pieces. This process is continued until the researcher gets segments suitable for sequencing. Although there are different ways of going about it, one hierarchy of steps is presented in Figure 9.8.

Suppose the researchers are interested in the 1- to 2-Mb section of DNA between linkage markers A and B. They can make use of a library of segments cloned in **yeast artificial chromosomes** (**YACs**) that includes this region (see also Chapter 8). Libraries of YAC clones covering large portions of human DNA are available from Généthon and other sources. YACs contain the four elements of a proper chromosome: a telomere at each end,

cloning vehicle, such as a *yeast artificial chromosome*, to amplify it (see the next section). The actual order of two or more separately cloned human DNA sequences along a chromosome can often be determined by tagging the different DNA sequences with different fluorochromes. This procedure can be refined by using the stretched-out chromosomes of prometaphase or even the diffuse chromosomes of interphase (Figure 9.7).

In summary, FISH has become one of the most powerful techniques for visualizing chromosomal abnormalities and for locating DNA sequences to particular chromosomal regions. Although assignment of a particular gene to a chromosomal band with its 5 to 10 million base pairs—100 to 200 genes—is coarser than the best genetic linkage maps, the FISH technique brings to light a tangible spatial organization of genes. They are really there.

Ordered DNA Segments

Finer physical mapping techniques involve cloning segments of DNA and arranging the segments in order. The collection of cloned segments can span a region that is cut out of human DNA with a restriction enzyme or a region between two DNA markers—markers whose linkage distances have been determined or whose physical positions have been established by FISH. Such a collection of fragments is called a **contig**, short for "contiguous set of cloned segments."

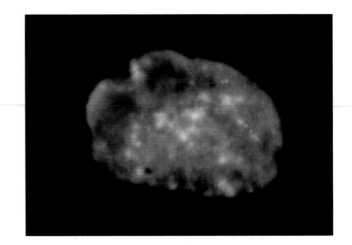

Figure 9.7 The positions of three DNA sequences within band 3p14 using the FISH technique on an interphase nucleus. Each specific sequence is cloned in a different YAC and tagged with a different fluorochrome (red, orange, green). The three sequences are lined up red-orange-green on each of the two presumed chromosomes 3. Because the extended chromatin in interphase could easily fold back upon itself, this same pattern would have to be observed consistently in many cells to establish the order of genes. (From Wilke et al. 1994.)

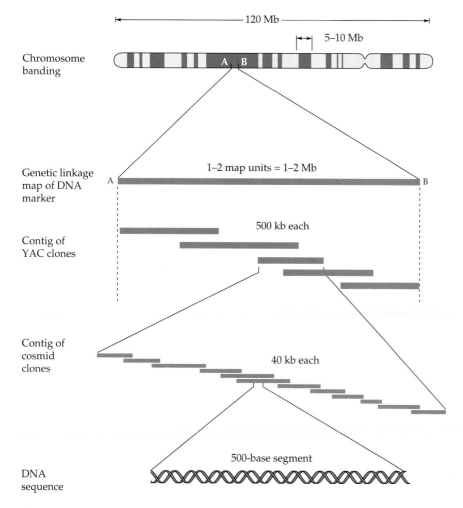

Chromosome banding

Genetic linkage map of DNA marker

Contig of YAC clones

Contig of cosmid clones

DNA sequence

Figure 9.8 The relationship between chromosome banding and genetic linkage maps (top), contig construction (middle), and DNA sequencing (bottom).

a centromere for attachment to mitotic spindle fibers, and a region that initiates DNA replication. In addition, chromosomes need to be above a certain size, and this can be achieved by inserts of human DNA about 500 kb long. **Bacterial artificial chromosomes (BACs)**, which are more stable than YACs and can contain inserts of about 150 kb, are preferred for some applications. **Phage artificial chromosomes (PACs)**, which contain inserts of about 100 kb, and **cosmids**, which contain inserts of about 40 kb, are used for contig sets over smaller distances, as indicated in Figure 9.8. Construction of a contig of overlapping clones sometimes utilizes the partial digestion of DNA with a restriction enzyme so as to get overlapping pieces (Figure 9.9).

Sequencing all the bases of all segments of a contig is prohibitively tedious as an initial step. (Of course, the eventual aim of the Human Genome Project is to do just that, but a framework must be established to avoid willy-nilly sequencing and duplication of effort.) The method of choice to align and order a set of YACs, BACs, PACs, or cosmids involves sequencing just short randomly selected segments of the cloned human DNA. These sites are called **sequence**

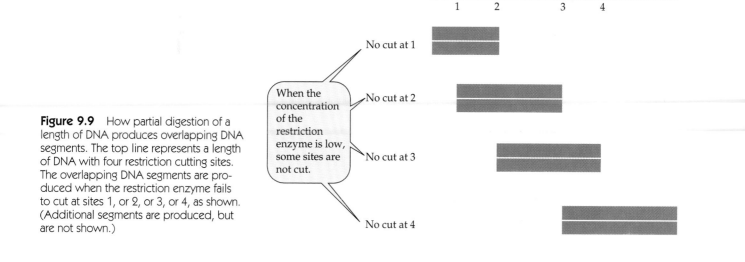

Figure 9.9 How partial digestion of a length of DNA produces overlapping DNA segments. The top line represents a length of DNA with four restriction cutting sites. The overlapping DNA segments are produced when the restriction enzyme fails to cut at sites 1, or 2, or 3, or 4, as shown. (Additional segments are produced, but are not shown.)

tagged sites (**STSs**), and each is about 200 nucleotides long. By publishing 20-base sequences toward the ends of STSs, researchers make it possible for any laboratory in the world to reproduce the STSs in their *own* DNA samples by the polymerase chain reaction, using the published sequences as primers. Figure 9.10 shows in a simplified way how STSs can be used to order a set of clones into a contig. Figure 9.11 shows a real example of a YAC contig that spans the 2.5-Mb region around the cystic fibrosis gene on the long arm of chromosome 7.

An STS-based map of almost the entire human genome has been assembled by researchers at the Whitehead Institute for Biomedical Research and Massachusetts Institute of Technology. This map includes about 15,000 STSs whose positions are anchored by known genetic linkages or known physical linkages involving YACs or other DNA segments. You can calculate that the average distance between these STS sites spaced along the 3 billion bases of the human genome is about 200,000 bases. A major interim goal of the Human Genome Project is to establish an STS map of 30,000 sites spaced, on average, 100,000 kb apart. Mappers think that once this finely detailed and ordered blueprint is established (perhaps by the time you read this chapter), the work of sequencing every chromosome can begin in earnest. Thus, the Whitehead-MIT map goes a long way toward setting up the last and major stage of the Human Genome Project (Cox and Myers 1996). In addition, the STS map is a key tool for identifying disease genes.

Base Sequence Data

The ultimate physical map of human DNA is the specification of the base sequence. In Chapter 8 we described the method of base-by-base determination using the common Sanger method. Researchers sequence the human DNA insert (10,000 to 40,000 bases) in a plasmid, phage, or artificial chromosome by first breaking up the DNA as necessary into overlapping 300 to 500-base pieces. They then sequence the pieces and order them by the overlapping regions. Some improvements in sequencing technology have been developed, but more are needed to better automate the process (Rowen et al. 1997). Advances in computer-based methods for collecting, storing, distributing, and analyzing the data are also anticipated. As noted in Figure 8.11, just to print 3 billion nucleotides would fill the space of 42 unabridged dictionaries. For this reason, most sequence information will exist only as unprinted electronic databases. Analysis of variation in nucleotides among people will necessitate still greater computer-related capabilities.

Then there arises the crucial question: What have we got? Perhaps 97% of human DNA is noncoding—the often long intron

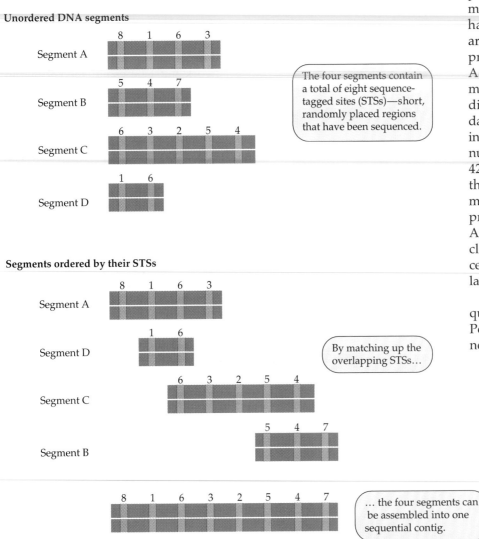

Unordered DNA segments

Segment A   8   1   6   3

Segment B   5   4   7

Segment C   6   3   2   5   4

Segment D   1   6

The four segments contain a total of eight sequence-tagged sites (STSs)—short, randomly placed regions that have been sequenced.

Segments ordered by their STSs

Segment A   8   1   6   3

Segment D   1   6

Segment C   6   3   2   5   4

Segment B   5   4   7

By matching up the overlapping STSs...

8   1   6   3   2   5   4   7

... the four segments can be assembled into one sequential contig.

Figure 9.10 Ordering a set of four DNA segments (cosmids or YACs labelled A, B, C, D) into a contig by matching up eight sequence-tagged sites (STSs). Although the order of STSs within a segment is given, that information is not actually necessary to assemble this contig. Note that sites 5 and 4 could be reversed.

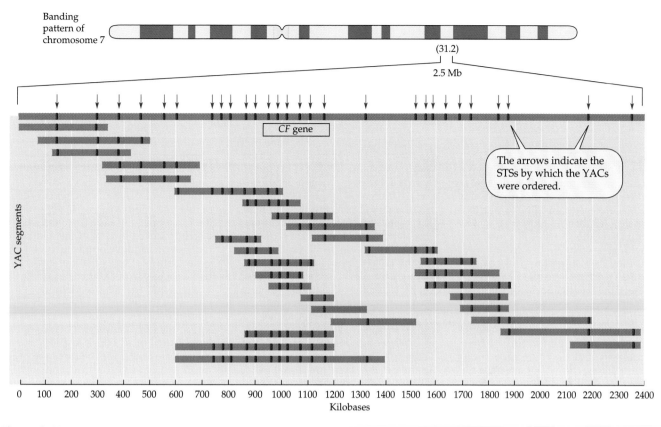

Figure 9.11 The YAC contig spanning the 2.5-Mb section of chromosome 7 that contains the cystic fibrosis gene. The cystic fibrosis gene is roughly 250 kb long. It has 26 introns that constitute over 97% of the 250 kb. (After Green et al. 1995.)

regions and the even longer regions between genes. *How do you know when the bases read off a sequencing gel constitute a gene?* We address this question in the next section with respect to DNA isolated directly from cells. But first we describe a sure way of knowing that *genes* are being sequenced.

Recall that cDNA is transcribed from messenger RNA using reverse transcriptase. It therefore contains only the exon base sequences for structural genes, that is, the coding triplets for the amino acids of proteins, along with some untranslated sequences at the beginning and end. (The end section of an mRNA molecule contains the poly(A) tail by which a researcher can distinguish messenger RNA from other forms of RNA in a cell.) In cDNA there are no introns or regulatory sequences or any intergenic DNA with its repetitive sequences (e.g., the 500,000 copies of the *Alu* repeat). Furthermore, a particular library of cDNA molecules includes only the genes that are expressed in the tissue from which the mRNA is isolated.

Researchers at The Institute for Genomic Research (TIGR) in Gaithersburg, Maryland, directed by J. Craig Venter, have extensively analyzed the cDNA molecules obtained from the mRNAs isolated from dozens of different human organs and tissues (Goodfellow

1995). But instead of sequencing the several kilobases of each cDNA (which would be prohibitively tedious at this stage), TIGR researchers sequenced only a few hundred bases from each cDNA molecule. These segments are called **expressed sequence tags** (**ESTs**), that is, markers for genes that are expressed in a particular tissue. Altogether, the TIGR researchers tagged (with one or more EST) close to 30,000 different genes that were expressed in 37 different organs and tissues.

The EST procedure does not specifically identify genes that cause disease or those that are polymorphic. But ESTs that appear useful are being mapped, or the genes that they mark are being sequenced in their entirety. Analysis of the ESTs is also being used to give an estimate of the total number of human genes (Box 9C). Another interesting component of the TIGR studies is a comparison of their EST-tagged genes with established gene and protein databases, giving rough estimates of the genes involved in different roles in the cell (Table 9.3). In the bottom row of Table 9.3, note that the largest role of genes, 21.9%, deals with basic gene and protein expression. Another 16.4% deals with metabolism, which includes energy production and the synthesis of cell metabolites, including amino acids, nucleotides, sugars, and fats. The brain (first row)

BOX 9C *So How Many Genes Do I Have?*

The short answer is, nobody knows—yet. Most estimates of the haploid number of genes range from 50,000 to 100,000, but some estimates are lower or higher (Cohen 1997).

One problem is how to define a gene. A good definition is a nucleotide sequence of DNA that is transcribed into RNA. Almost all these RNA molecules are messenger elements that are then translated into proteins. (Some RNA molecules are tRNA and rRNA molecules that function as such without translation.) There is a problem with genes that are present in multiple copies. Do we want to count all genes or only different genes? There is also the problem of genes with alternative splice sites, so that several different proteins are formed from one DNA sequence. Is such a sequence counted once or several times? There is the problem of overlapping genes and nested genes, in which one gene is found within an intron of another gene. And there is also the problem of the antibody genes, which are generated by somatic recombination from sev-

eral adjacent sets of germline "mini-genes" (see Chapter 18). But all told, these are relatively minor considerations given that we are dealing with such a large total number.

A rough estimate of gene number is based on how many genes have been found in several extended lengths of sequenced human DNA. This method yields about one gene per 25 kilobases, or 120,000 genes in the 3,000 megabases of human DNA. But the still relatively meager DNA regions that have been chosen for large-scale sequencing are probably comparatively gene-rich, so the 120,000 estimate may be high. Other gene workers, especially those in biotechnology companies (where a big number might be more profitable), also put the gene number in the 100,000 range.

Based on an analysis of ESTs from cDNA, J. Craig Venter, head of TIGR, estimates that there are 60,000 to 70,000 human genes (Fields et al. 1994). Venter's group counted the number of matches between ESTs and selected nucleotide sequences from

GenBank, the major human DNA database. One problem with this method is estimating how many genes in the cDNA collection are represented by more than one EST.

An interesting estimate toward the lower end is based on the Japanese puffer fish, a gourmet delicacy, which does without most "junk" DNA. With much reduced intron sizes and very little spacer DNA between genes, 92% of its DNA codes for proteins (versus 3% for people). Sydney Brenner (who also introduced the use of the nematode worm for developmental and genetic studies) finds that the puffer fish has close to 60,000 genes, and he believes that any other vertebrate, including us, would have about the same number.

For the purposes of this book, we have settled on 80,000, as a guess in the middle range. Maybe there are fewer genes, maybe more. Generally, "people like to have a lot of genes," says Brenner. "It makes them feel more comfortable."

---

expresses the largest number of genes, and more than in other tissues, these active genes are responsible for cell signaling and communication. Note the differences between white blood cells, which are vigorously dividing and participating in immune functions, and red blood cells, which do little more than transport oxygen.

Finding Disease Genes

Prior to the full flowering of physical mapping techniques, some genes were identified and cloned by a process called **functional cloning** that used knowledge of the protein to eventually get back to the gene. Examples include the globin genes and the color vision genes. In the former case, it had been known since the 1950s that people suffering from sickle-cell disease had an amino acid substitution in the β-globin chain. Later, it became clear that messenger RNA from immature red blood cells contained little else except globin mRNA, from which relatively pure globin cDNA could be obtained. Furthermore, patients with either α- or β-thalassemia due to deletions of the corresponding globin genes allowed researchers to construct cDNA probes that would hybridize to either the α- or β-globin genes, but not both. These probes eventually led to the isolation and cloning of the globin

genes, the α cluster on chromosome 16 and the β cluster on chromosome 11.

In the case of the color vision genes, the initial information base was five adjacent amino acids of the protein rhodopsin isolated from the rod cells of retinas of cattle. As it turned out, rhodopsin has about 40% homology to the red, green, and blue colorvision pigments in cone cells. Moreover (as expected), the cattle proteins and genes are very similar to our own. Functional cloning of a disease gene starts in this way with some knowledge of the encoded protein.

But when they know nothing about the protein product of a gene involved in a particular disease, researchers must employ an approach called **positional cloning**. That is, they use the techniques previously described in this chapter to identify and clone the gene's position in the genome. First they establish the approximate location of a gene by genetic linkage to flanking markers or by fluorescence in situ hybridization. They then construct contigs that span the region where the gene is likely to be. Researchers "walk" along overlapping segments of DNA in the contig set. Walking goes step-by-step, using the DNA sequences in one segment to identify an overlapping segment. The process is tedious and can be blocked easily by a stretch of repetitive DNA, which does not easily lend

TABLE 9.3 Approximate percent of genes playing selected cellular roles in cDNA samples from different tissues

Tissue	Number of expressed genes identified	Percent of genes involved in			
		Cell signaling and communication	Gene and protein expression	Cell division and DNA synthesis	Metabolism
Brain	3,195	17.3	17.5	4.7	18.1
White blood cell	2,164	12.8	20.7	5.7	16.7
Testis	1,232	10.8	23.7	5.3	16.4
Heart	1,195	11.8	19.3	4.8	15.6
Ovary	504	7.9	30.5	4.2	13.2
Small intestine	297	9.8	18.9	2.9	20.2
Red blood cell	8	12.5	12.5	0	0
Total of 37 tissues	30,000	12.4	21.9	4.4	16.4

Source: Adams et al. 1995.

itself to contig construction. To overcome this obstacle, techniques are available to "jump" from one clone to another that is not contiguous. At each stage of this process, researchers test to see whether the base sequences they are obtaining are likely to be the gene.

In isolating the gene for cystic fibrosis, for example, researchers did not know that the normal allele codes for a protein that regulates the passage of chloride ions in and out of cells. In fact, only after researchers cloned the gene were they able to predict for the first time the protein product from knowledge of the genetic code. Indeed, for some of the genes that have been identified, isolated, cloned, and sequenced recently, the function of the encoded protein is not yet known. For example, as of this writing, no one knows the basic function of huntingtin, the protein encoded by the normal allele of the gene involved in Huntington disease. The protein was not even named until its amino acid sequence was inferred from the gene. In summary, positional cloning of a disease gene starts with locating the gene without prior knowledge of how it functions in the disease process (Table 9.4).

A shortcut on the positional cloning approach is available if the disease phenotype maps to a chromosomal region where genes for previously known proteins also reside. Researchers then ask if any of these so-called **candidate genes** coding for the known proteins could plausibly be involved in the disease phenotype. They test to see if persons affected with the disease have mutations for the selected candidate gene. Cloning of the gene for Marfan syndrome, for example, fell into this pattern. By family linkage studies, the gene for Marfan syndrome was located at a particular band on the long arm of chromosome 15. Separate from that study, the gene for the protein fibrillin was mapped to the same region by FISH. Now, fibrillin is an elastic protein found in tissues affected in Marfan syndrome (namely, the aorta), support for the

eye lens, and bone coverings. The coincidence was obvious, and it was shown in short order that persons affected with Marfan syndrome had mutations in the fibrillin gene. Other genes that have been identified by this positional-candidate approach include genes involved in amyotrophic lateral sclerosis (Lou Gehrig's disease), Alzheimer's disease, and Charcot-Marie-Tooth disease (Chapter 8).

How Do You Know It's a Gene?

Positional cloning typically generates cloned segments of DNA near and including, one hopes, the gene of in-

TABLE 9.4 Some disease genes identified by positional cloning

Year	Disease	Chromosome
1986	Chronic granulomatous disease	X
	Duchenne muscular dystrophy	X
	Retinoblastoma	13
1989	Cystic fibrosis	7
1990	Wilms tumor	11
	Neurofibromatosis type 1	17
1991	Fragile X syndrome	X
	Familial polyposis coli	5
1992	Myotonic dystrophy	19
1993	Huntington disease	4
	Neurofibromatosis type 2	22
	Wilson disease	13
	Adrenoleukodystrophy	X
1994	Hereditary breast cancer (*BRCA1*)	17
	Polycystic kidney disease	16
	Achondroplasia	4
1995	Hereditary breast cancer (*BRCA2*)	13
	Ocular albinism	X
	Ataxia telangiectasia	11
	Bloom syndrome	15
1996	Friedreich ataxia	9

BOX 9D   *Is It a Gene?*

Question: Could the following sequence from a normal individual represent part of an exon of a gene? (From Table 6.4, we note that stop triplets are ATT, ATC, and ACT.)

GCCAATCTCTGAGATCATCGAACACTCCGGTTAACGATTTCGGG
1,2,3...

Answer: No. Examine the sequence in the three possible reading frames:

Starting with 1:
GCC AAT CTC TGA GAT CAT CGA ACA CTC CGG TTA ACG ATT TCG

Starting with 2:
CCA ATC TCT GAG ATC ATC GAA CAC TCC GGT TAA CGA TTT CGG

Starting with 3:
CAA TCT CTG AGA TCA TCG AAC ACT CCG GTT AAC GAT TTC GGG

In each reading frame, the sequence includes one or more stop triplets (blue). This would not be possible if the whole sequence were in an exon.

---

terest. The big question is whether or not researchers can recognize in lengthy and boring flows of A's, T's, G's, and C's the gene of interest—or, for that matter, any gene at all. Several different methods are available to attack this problem. For example, if a segment of cloned DNA has been sequenced, one can simply ask if it contains a long stretch of bases without a stop triplet. The rationale for this methodology is plain: A typical exon of a typical gene is about 200 bases long, and it must *not* have a stop triplet, at least not in a normal individual, or it could not code for a complete polypeptide. On the other hand, if bases are randomly ordered (as might be expected in an intron or in intergenic DNA), about every twentieth triplet should be a stop (3 stops out of 64 possible triplets). A stretch of DNA without a stop triplet is called an **open reading frame (ORF)**. Of course, researchers must check each of the three possible reading frames to see if any one of them is free of stop codons (Box 9D).

There are many other things a researcher can do to see if cloned DNA contains hints of a gene. Some of these methods are technically tedious, but the idea behind them is usually not difficult to understand. We briefly sketch three such methods.

Zoo Blotting. The corresponding genes in related organisms are often similar in sequence and structure. This is because evolution is conservative: What works for a species in a given era of geologic time is often preserved as that species evolves and splits over millions of subsequent years. The similarity of the globin genes in humans, monkeys, cattle, mice, birds, and fish is an often-cited case, but there are hundreds of other molecular examples.

On the other hand, noncoding DNA is less well preserved over evolutionary time, because there is no selection pressure; this DNA does not serve the organisms in any obviously useful way, so mutational changes are not weeded out. They accumulate as eon succeeds eon until once-similar segments of DNA bear little resemblance to each other. Thus, if a piece of cloned human DNA can be shown to be similar to that of other species, it has a good probability of being a gene—some gene. The experimental technique is to simply use a cloned human DNA segment as a probe to see if it hybridizes in a Southern blotting process with DNA from other species—a so-called **zoo blot.** Figure 9.12 shows some results for the Duchenne muscular dystrophy gene. As another example, the Myriad Genetics team showed that the presumptive *BRCA1* gene hybridized in a zoo blot with the DNA isolated from several other mammals (mouse, rat, rabbit, sheep, and pig).

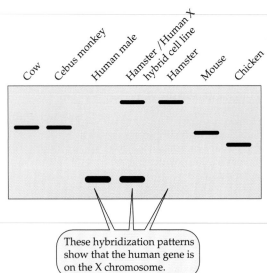

The fact that the human DNA probe hybridizes to all the DNA samples suggests that it may be part of a gene conserved in evolution.

These hybridization patterns show that the human gene is on the X chromosome.

Figure 9.12 A zoo blot used in the investigation of the gene for Duchenne muscular dystrophy. DNA samples from different species were treated with a restriction enzyme. The fragments were electrophoresed, Southern blotted onto a nylon membrane, probed with a segment of human DNA from the presumptive (at the time) Duchenne muscular dystrophy gene, and autoradiographed as shown here. Each lane represents a different source of DNA. (After Monaco et al. 1986.)

CpG Islands. For reasons that are not well understood, the promoters of about half of human genes are in regions where the DNA is much richer in the dinucleotide CG (GC on the other strand) than other regions. The "p" in CpG, which stands for phosphate, is used to distinguish the CG pair along the length of a *single* DNA strand from a CG complementary pair on *opposite* DNA strands.

The genes that are near such **CpG islands** include housekeeping genes, such as those for energy production, that are needed in every cell. A few restriction enzymes—for example, *Sac*II, which recognizes CCGCGG—tend to cut in these regions but very rarely does *Sac*II cut outside CpG islands. Incidentally, an estimate of the number of CpG islands leads to a calculation of 67,000 for the number of human genes (Fields et al. 1994; also see Box 9C).

Exon Trapping. This method of gene identification is ingenious. We noted in Chapter 6 that an intron is spliced out of messenger RNA molecules by recognizing specific base sequences at the two ends of the intron. Letting × and o represent these left and right splice sites within the intron, here is what happens when the intron is removed:

Let's say we insert a piece of foreign DNA into the intron above at about the place where the "t" of the word "intron" falls. If this piece of DNA contains an exon and adjoining nucleotides on its left and right it will look like this:

Exon 1   Intron   Foreign exon   Intron   Exon 2
═════×───o═══════════════×───o═════

Note that splice sites accompany the foreign exon, thus creating two smaller introns, each complete with a pair of splice sites. After removing these intron regions, the mature mRNA will be longer than before because it includes the new exon:

Exon 1   Foreign exon   Exon 2
═════════╪═══════════╪═════════

It is said that the foreign exon has been trapped; thus the name **exon trapping**.

On the other hand, if no exon and consequently no pair of adjoining splice sites are present in the insert, the foreign DNA will be removed along with the original intron between the single left (x) and right (o) splice sites. The resulting mature RNA will look no different than before:

Exon 1   Foreign DNA   Exon 2   Exon 1   Exon 2
═════×──────────o═════ → ═════╪═════

Thus, whether an exon is present or not determines whether the resulting messenger RNA is longer or shorter (measurable by electrophoresis).

The final proof that a disease gene is in hand is this: It is mutated in persons with the disease and is not mutated in normal persons. Although direct sequencing of the gene in affected and normal persons is tedious, that step may be necessary in order to find a single-base substitution or a frameshift mutation involving just a few bases. The cystic fibrosis gene was nailed down when sequencing showed that many affected persons had a deletion of three adjacent bases in the cloned gene that should have coded for amino acid 508, a phenylalanine. If the mutational event is larger (e.g., a gene with an extended trinucleotide repeat), then simple electrophoresis of DNA samples may work: DNA from affected persons may show a slower-moving band than in normal persons due to a larger restriction fragment.

In the identification of *BRCA1*, researchers led by Mark Skolnick at Myriad Genetics generated and searched YAC, BAC, and cosmid clones. These clones contained DNA segments that spanned about 600,000 bases demarcated by genetic linkage markers that flanked the gene. In these clones, they were able to find three nearby base sequences that looked genelike and to merge them into one long open reading frame that seemed to code for a polypeptide 1,863 amino acids long. In each of five different pedigrees, they identified a distinctive mutation that was present in all affected members of the pedigree but not in their normal relatives (Table 9.5). These results formed the strongest evidence that what the researchers had isolated was really a gene responsible for breast cancer in the family members (Nowak 1994). Subsequently more than 100 different mutations have been identified in the *BRCA1* gene. Using similar methods, in 1995 researchers identified a second breast cancer gene, *BRCA2*, on chromosome 13. *BRCA2* accounts for about the same number of cases as *BRCA1*.

To summarize briefly, a disease gene in vivo is often cloned in vitro using (1) genetic linkage studies based on family pedigrees, and (2) physical mapping techniques involving ever smaller segments of DNA. In the last step, researchers recognize the gene by its mutated alleles. As of early 1997, 84 human genes had been positionally cloned by these techniques, but the number will increase rapidly as the Human Genome Project gains momentum (Bassett et al. 1997).

More on the Human Genome Project

This chapter has been largely about the **Human Genome Project** (**HGP**), a challenging, decades-long, multibillion-dollar, international plan to sequence every last base of human DNA. Additional ingredients of the HGP are the mapping and sequencing of model organisms (Box 9E).

As described in this chapter, genetic and physical mapping studies have established throughout the human genome thousands of relatively closely spaced sites, which could be considered the index to an immensely long book. The systematic sequencing of bases in and around each site has begun, thus filling out the detailed information for the index entries (van Heyningen 1996). Much of the sequencing is being done in about 20 major institutes,* each specializing in one or a few chromosomes, or in biologically interesting regions of those chromosomes. The two largest centers are at Washington University in St. Louis and the Sanger Centre near Cambridge, England. The French government is building a $200 million sequencing center near Paris. Most groups have agreed to publish their sequence data promptly on the Internet and to refrain from seeking early patents on potentially useful segments of DNA (Marshall and Pennisi 1996).

Despite early doubts about the goals of the colossal project and worry about draining funds from other desirable research, most (but not all) investigators worldwide now agree that the known and possible benefits are worth the costs. In any event, the enterprise is moving forward pretty much on schedule since its inception in 1989 (Collins and Galas 1993; Roberts 1993). The target year for completion is 2005. The American Society of Human Genetics (representing about 4,500 physicians, scientists, and genetic counselors) strongly supports the project:

> The mapping and sequencing of the human genome will provide extensive new knowledge about the

*Some prominent sequencing centers in the United States are listed in Table 9.6.

genes involved in inherited diseases, birth defects, later-onset diseases with a genetic component—including heart disease, neurological and behavioral disorders, and many forms of cancer—and even susceptibility to infection. This knowledge will have profound impact on the understanding, detection, prevention, and treatment of these disorders. (Caskey et al. 1991)

The cloning of the genes for Huntington disease, amyotrophic lateral sclerosis (Lou Gehrig's disease), neurofibromatosis types 1 and 2, myotonic dystrophy, and fragile X—all of these before 1994—and other disease genes more recently were helped by techniques and data developed through the HGP. Economic spin-offs will be increasingly possible when the methods are applied to farm crops and animals. Knowledge of evolutionary processes will be deepened when the DNA sequences of different species are compared. And as is usual in research, unexpected results and insights are anticipated.

Interest in the Human Genome Project is centered in Europe, Japan, and the United States under the umbrella of the Human Genome Mapping Organization (HUGO). In this country, organizational and financial support comes from the National Institutes of Health and the Department of Energy. The head of the National Center for Human Genome Research is Francis Collins, a superb gene hunter, who helped in the cloning of the genes for cystic fibrosis, neurofibromatosis, and Huntington disease. In addition to coordinating research efforts, encouraging the development of new technology, and overseeing the massive job of data analyses, these organizations and others are also tackling the troublesome issue of gene patenting and the possible misuse of personal genetic information.

From its beginning, the HGP has been concerned with these nontechnical issues. James Watson, the first director of human genome research in the National Institutes of Health, established a group to investigate the ethical, legal, and social implications of the HGP. The first head of this so-called **ELSI** working group was Nancy Wexler, a neurophysiologist who was instrumental in the search for the Huntington disease gene. The ELSI group consists of medical geneticists, ethicists, theologians, lawyers, social scientists, and others with special expertise. It receives 3–5% of the annual bud-

TABLE 9.5	The first mutations identified in *BRCA1*	
Pedigree number	Change in or starting at codon[a]	Description of mutation
1901	24	Frameshift mutation (deletion of 11 bases) produces a stop at codon 36
2082	1,313	Nonsense mutation (C to T) changes glutamine to stop
1910	1,756	Frameshift mutation (insertion of 1 base) produces a stop at codon 1,829
2099	1,775	Missense mutation (T to G) changes methionine to arginine
2035	Not applicable	Unknown regulatory mutation blocks transcription of the gene

Source: Miki et al. 1994.
[a]The normal allele has 1,863 codons.

BOX 9E *The Genomes of Others*

Researchers are determining the complete base sequence of many organisms to compare their genes and chromosomes. Starting with some smaller genomes, researchers have tested out their sequencing operations on ever-larger scales. The first organism to be completely known was the tiny bacteriophage ΦX174. Its sequence of 5,386 bases was determined by one of the inventors of DNA sequencing, Frederick Sanger, in 1977. Over 140 viral genomes have been sequenced, including the AIDS virus (about 9,000 ribonucleotides), the smallpox virus (186,000 bases), and the Epstein-Barr virus (160,000 bases), which is involved in mononucleosis.

The first bacterium to be completely sequenced was *Haemophilus influenzae*, which is responsible for meningitis and ear infections in children. Its sequence of 1.8 million bases (Mb) was reported in 1995 by researchers at The Institute for Genomic Research (TIGR), a private laboratory in Maryland (Marshall 1994). Researchers at TIGR used novel methods that relied on sophisticated computer programs that arranged in proper order about 25,000 short, randomly cut, overlapping, sequenced DNA segments. This same research group has also completely sequenced other bacteria, including *Helicobacter pylori* (1.7 Mb), which causes stomach ulcers (Wade 1997), and the unusual organism *Methanococcus jannaschii* (1.7 Mb). The latter is an archaeon found in near-boiling hydrothermal vents deep in the Pacific Ocean (Kreeger 1995). (The Archaea are a taxonomic group of unicellular organisms that are distinct from both prokaryotic and eukaryotic organisms.) TIGR has recently announced that it plans to use these powerful techniques on the human genome, which is thousands of times bigger (Venter et al. 1998).

A milestone was reached in January 1997 with the announcement of the completion of the 4.6-Mb genome of *Escherichia coli* by researchers at the University of Wisconsin, and independently by a Japanese group at the Nara Institute of Science and Technology (Moxon and Higgins 1997; Pennisi 1997). Far more is known about this workhorse of genetic research than about any other life form. Research on *E. coli* has led to many of the key concepts of molecular biology, including DNA replication, transcription, translation, the genetic code, gene regulation, and DNA repair.

Sequencing the microbes that cause tuberculosis, cholera, syphilis, gonorrhea, Lyme disease, malaria, ulcers, and other diseases is either complete or is underway and likely will have been completed by the time you read this.

The most remarkable sequencing accomplishment has been the genome of baker's yeast (*Saccharomyces cerevisiae*). The complete 12-Mb sequence parceled into 16 chromosomes and including about 6,000 genes has been determined by a consortium of several U.S., Canadian, Japanese, and over 90 European laboratories (Goffeau et al. 1996). Yeast, the first eukaryotic organism to have its DNA completely known, has a compact genome with little repetitive DNA and few introns. Although the use of yeast in the brewery may require genes absent in people, many yeast genes do have human counterparts. For example, the recently identified human genes causing ataxia telangiectasia and Bloom syndrome (see Table 7.2) code for DNA repair enzymes. These functions were suggested by yeast genes with corresponding sequences. Similarly, the function of the neurofibromatosis 1 gene in humans is the same as a yeast gene that regulates cell division.

Another sequencing project is that of the 1-millimeter nematode worm *Caenorhabditis elegans*, with a genome of 100 Mb. Of that, 30 Mb have been determined, a greater quantity than for any other organism. For the fruit fly *Drosophila melanogaster*, over 2.5 Mb of the 120-Mb genome have been determined. For *Mus musculus*—the laboratory mouse—whose genome is about the same size as that of humans, fine-scale genetic mapping is complete, physical mapping is progressing, and sequencing of interesting regions has begun.

The sequencing of plant DNAs is also progressing, especially that of rice and other crops of economic importance, and also one of no economic importance at all, wall cress (*Arabidopsis thaliana*). This 5-inch-tall weed, which can be easily grown and propagated in the laboratory, and has become the plant of choice for molecular genetic investigations. With about 70 Mb of DNA, it has one of the smallest plant genomes.

Of the species mentioned here, five (*E. coli*, *S. cerevisiae*, *C. elegans*, *D. melanogaster*, and *M. musculus*) are "model" organisms that have been included within the purview of the Human Genome Project. The organizers of the HGP recognized that comparative information from organisms with shared evolutionary histories would be valuable in understanding basic biological mechanisms.

---

get of the HGP to support public conferences, workshops, special investigative commissions, and research proposals (E. Marshall 1996). Areas of study include the use of genetic testing in clinical practice and the access to genetic information by parties other than patients and their doctors. A third area is the understanding of the power and limitations of genetics by the public and by health professionals, whose training may not have included much modern genetics. Maintaining confidentiality of medical data and ensuring freedom in personal decisions are already familiar problems in health care, insurance, and employment. But they will become more difficult as the increased precision and enormous volume of personal health data become a reality. Although we delay a full discussion of the prospects for and problems of genetic testing until a later chapter, we consider the following questions prompted by the Human Genome Project:

- Can my genetic information be separated from my general medical file to help insure confidentiality?

- Will my DNA sample be stored for future use? Will it be destroyed when I die? Can it be used for research without my consent?

TABLE 9.6 U.S. genome sequencing centers receiving awards in 1996 from the National Center for Human Genome Research

Institution	Funding (million $)	Research
Washington University St. Louis, MO	6.7	Large-scale sequencing on chromosomes 22 and X
Whitehead Institute/MIT Genome Center, Cambridge, MA	4.1	Chromosomes 9 and 17; robotics; maps for sequencing
The Institute for Genomic Research Rockville, MD	3.2	Chromosome 16 sequencing; software development
Stanford University Palo Alto, CA	2.5	Chromosomes 4 and 21; directed sequencing; DNA chips
Baylor College of Medicine Houston, TX	1.3	Chromosome X sequencing dyes; DNA purification methods
University of Washington Seattle, WA	1.0	Chromosome 7; streamlined high-accuracy sequencing

Source: Marshall and Pennisi 1996.

- Will information about me gained through DNA testing be revealed to my close family members?

- Will I be given a written report about my genes?

- How well does an analysis of my genes predict my future?

- Does my doctor know any more about genetics than I do?

Part of the testing problem is that the cloning of a gene and the identification of mutations that lead to disease allow fairly rapid commercial development of diagnostic tests. But significant prevention and treatments of a disease will take much, much longer to devise if truly effective measures are ever realized. What is a person to do about a positive diagnosis? For example, Myriad Genetics and other companies have introduced commercial testing for mutations in the *BRCA1* gene that lead to breast cancer in about 85% of cases and to ovarian cancer in about 45% of cases (A. Marshall 1996). Critics of marketing the *BRCA1* test emphasize the need for expert counseling to help patients and their families decide on testing, interpret the results, and cope with uncertainty. A multitude of questions intrude upon a woman at risk (Kahn 1996): In what way am I involved? Should I be tested at all? Is the test 100% accurate? Does the presence of a mutant allele mean certain breast or ovarian cancer? At what age? Is a prophylactic double mastectomy an option? Should I also have my ovaries removed? Will those operations reduce the cancer risk to zero? What about my sister? What about my 10-year-old daughter? Will insurers or employers demand the results of the test? Can I conceal the results by being tested under a false name or code number? Will some form of treatment—short of disfiguring or sterilizing surgery—be available in my lifetime?

If the HGP holds to its scheduled completion date of 2005, and indications are that it will be close (Wade 1998), *all* genes will be within grasp, but most of them will not be well understood. (And a lot of technicians will be out of sequencing jobs.) What comes next?

Summary

1. Granting patents on the commercialization of DNA sequences is controversial. Is a gene cloned in laboratory glassware a natural product or is it a new, nonobvious, and economically useful commodity—the three criteria for a patent?

2. Traditional genetic linkage maps are based on the frequency of crossing over between genes, measured by recombinant gametes from doubly heterozygous parents. For genes relatively near each other, the observed percent recombination is proportional to the map units between them.

3. Before the advent of DNA markers, few human genes were mapped; most were blood group or other polymorphic genes. The problem was finding double heterozygotes of known phase and the low expectation that two random genes were within mappable distance of each other.

4. DNA markers (RFLPs, VNTRs, and microsatellites) are so extensive, so polymorphic, and so easily scored that they revitalized human genetic mapping studies. Now thousands of DNA markers have been mapped with respect to each other and to many disease-causing genes.

5. The human genome of about 3 billion bases is spaced over about 3,700 genetic map units.

6. Physical maps are based on the substance of chromosomes: stainable bands, rearrangements or deletions of material, long or short segments of cloned DNA that are organized in proper order (contigs), and the actual sequence of A's, T's, G's, and C's.

7. Physical mapping began with assigning genes to specific chromosomes using mouse-human hybrid cells containing just one or a few human chromosomes. Fluorescent in situ

hybridization (FISH) can pinpoint the position of a gene by bathing a metaphase spread with a labeled probe visible under ultraviolet light.

8. Contigs can cover a large span using yeast artificial chromosomes (YACs) with cloned human DNA inserts of 500 kb, or they can cover shorter spans using cosmids with inserts of about 40 kb. The clones are ordered using sequence tagged sites (STSs). Researchers have assembled detailed physical maps of all human chromosomes based on STS-labeled clones.

9. Information on the order of bases is now being gathered in quantity. Sequence data known to be that of genes (rather than that of introns or intergenic DNA) can be obtained from cDNA, which is obtained by reverse transcription from mRNA in various tissues. These genes are not specifically those that cause disease when mutated.

10. Disease-causing genes are often located in the "raw" genome by positional cloning, using many of the genetic and physical mapping methods outlined in this chapter.

11. Hints that a particular sequence is a protein-encoding gene come from zoo blotting, exon trapping, CpG islands, and open reading frames (ORFs). The final proof that researchers have isolated a disease gene is that it is mutated in persons with the disease.

12. Well on its way to completion in the year 2005, the Human Genome Project involves mapping and sequencing human DNA and the DNA of other organisms. The ELSI group is concerned with ethical, legal, and social dimensions of the huge enterprise.

Key Terms

bacterial artificial chromosome (BAC)	linked genes
candidate gene	map unit
contig	open reading frame (ORF)
cosmid	parental gamete
coupling phase	phage artificial chromosome (PAC)
CpG island	physical map
DNA marker	positional cloning
ELSI	recombinant gamete
exon trapping	repulsion phase
expressed sequence tag (EST)	restriction fragment length polymorphism (RFLP)
flanking marker	sequence tagged site (STS)
fluorescence in situ hybridization (FISH)	somatic cell fusion
functional cloning	yeast artificial chromosome (YAC)
gene mapping	zoo blot
genetic linkage map	
Human Genome Project (HGP)	

Questions

1. In Figure 9.2, how could you fairly easily determine the frequency of the four egg types at the bottom of the figure? Assume a random population of eggs from many mothers of the given genotype. Note that in most mothers no crossover will occur between the two loci. Microscopes and molecular technology are not needed.

2. Why is a double heterozygote needed in order to map two genes on the same chromosome by observations on recombination?

3. What kinds of gametes, in what proportions, can be made by a person who develops from a egg carrying genes *A* and *B* and a sperm carrying *a* and *b*, if the two loci are
   (a) on different (nonhomologous) chromosomes?
   (b) adjacent to each other on chromosome 1?
   (c) 8 map units apart on chromosome 1?
   (d) at opposite ends of chromosome 1?

4. The total genetic length of all human chromosomes is about 3,700 map units, and the total number of base pairs is about 3×10^9. On average, about how many base pairs correspond to 1 map unit?

5. Modern gene mapping and the HGP owe much to a paper written in 1980 by David Botstein of MIT and his colleagues (including Mark Skolnick, who cloned the *BRCA1* gene). They wrote, "Although it is possible to detect linkage among simple Mendelian traits in humans, no method of systematically mapping human genes has been devised, largely because of the paucity of highly polymorphic marker loci." They went on to show how RFLPs solve the problem. Explain why DNA markers (such as RFLPs) let researchers outline the human genome with thousands of signposts.

6. The restriction enzyme *Eco*RI cuts DNA at a specific six-base sequence (GAATTC). What is the average length of DNA fragments generated when *Eco*RI acts on DNA (assuming the base sequence is random and all GAATTC sites are cut)?

7. Variation in the lengths of DNA fragments after electrophoresis and Southern blotting underlies the VNTRs and STRs used for DNA fingerprinting (see Chapter 8) as well as the first RFLPs used in mapping. The variation, however, comes about for different reasons. Discuss.

8. Using the scale shown in Figure 9.3, approximately how long in map units is the longest human chromosome? The shortest?

9. Crossing over is quite a bit more frequent in females than in males (for reasons that are not well understood). The genetic lengths shown in Figure 9.3 are the averages for the two sexes. While the average length of chromosome 1 is about 290 map units, the difference in the genetic lengths of chromosome 1 in females and males is actually 140 map units. What is the genetic length of chromosome 1 in females? In males?

10. In Figure 9.3, why is there an AC/TG repeat site exactly at each end of every chromosome.

11. Owerbach et al. (1980) examined 15 human–mouse hybrid cell lines by Southern blotting for the presence or absence of DNA sequences complementary to the human insulin gene. The data for six of the lines are given below. Which chromosome carries the gene?

Hybrid cell line	Insulin sequences detected (+) or not (−)	Human chromosomes present
W-8	+	6 7 10 11 14 17 18 20 21 X
T-5	+	4 5 10 11 12 17 18 21
D-5	+	3 5 11 14 15 17 18 21
W-2	−	8 10 12 15 17 21 X
T-2	−	2 5 6 10 12 18 20 21 X
T-8	−	17 18 20

12. Would it be possible to assign human genes by means of human–mouse cell hybrids if mouse chromosomes were randomly lost and human chromosomes retained?

13. Would you expect that genes found to be on the same chromosome by cell fusion methods will always be found to be linked by traditional pedigree linkage studies? Explain.

14. Describe the abnormal polypeptide produced by the *BRCA1* mutation in family 1910 in Table 9.5. Note that the mutant allele has a one-base insertion in codon 1,756, which leads to a stop at codon 1,829. (The normal polypeptide has 1,863 amino acids.)

15. A family has several individuals with breast cancer, suggesting a hereditary form of the disease, which might be due to any one of hundreds of different mutations in either *BRCA1* or *BRCA2*. What method of testing would insure that a mutation in one of these genes would be found, if it existed?

Further Reading

Many popular books on the human genome project have been written; a sensible, well-documented one is by Cook-Deegen (1994). Many of the technical issues are presented in plain language by McConkey (1993) and by Glick and Pasternak (1994). For details of positional cloning see Brook (1994) and for the details of FISH see Griffin (1994). Topics of *Scientific American* articles include the analysis of expressed genes by Haseltine (1997), genetic testing by Beardsley (1996), and somatic cell hybridization by Ruddle and Kucherlapati (1974) and by Ephrussi and Weiss (1969). The magazine *Science* for October 25, 1996, the annual "Genome Issue," has many articles of interest, including one by Lander on what comes after the Human Genome Project.

Internet Sites

The Internet pours forth information at ever faster rates. Although some Internet sites are more self-serving than helpful, here are ten that we found useful in writing this chapter:

BioTech (including a variety of educational resources)
http://biotech.chem.indiana.edu/pages/contents.html

cDNA Map of the Human Genome (from *Science*, 25 October 1996)
http://www.ncbi.nlm.nih.gov/SCIENCE96/

Généthon Human Genome Research Centre (English version)
http://www.genethon.fr/genethon_en.html

Genetics Societies (descriptions, policies)
http://www.faseb.org/genetics/mainmenu.html

HealthGate (including access to MEDLINE, a huge biomedical database)
http://www.healthgate.com

Information for Genetic Professionals, University of Kansas Medical Center
http://www.kumc.edu/GEC/prof/geneprof.html

National Center for Human Genome Research (the single most useful site)
http://www.nchgr.nih.gov

Nature magazine
http://www.america.nature.com

Online Mendelian Inheritance in Man (extremely detailed information)
http://www3.ncbi.nlm.nih.gov/omim/

Science magazine
http://www.sciencemag.org

CHAPTER 10

Complicating Factors

In the remote rain forests of New Guinea live a group of Melanesians called the Fore people. Until recently, they practiced ritual cannibalism of dead relatives as a way of honoring them. But what attracted the attention of geneticists in the 1950s was the high frequency and unusual transmission pattern of a strange disease that plagued the Fore people. They called it **kuru,** or laughing death.

Signs of brain damage—uncoordinated movements, trembling, loss of balance, inability to talk or swallow, emotional instability, and uncontrollable laughter—would suddenly appear. No treatment existed, and death came within a year after the onset of symptoms. About 75% of kuru victims were adult women, and the rest were mostly children of both sexes above the age of 4. Kuru was rare among adult men and very young children and totally absent in outsiders who lived among the Fore. But a few Fore people developed kuru while living far from their villages.

Because it did not behave like a standard infectious disease, and because it seemed to run in families, some scientists suggested that kuru might be inherited. To account for its unusual age and sex distribution, they proposed an autosomal disease-causing allele that was dominant in females and recessive in males, with the onset of symptoms occurring early in homozygotes and late in female heterozygotes. The main problem with this scheme lay in explaining how such a lethal allele could become so common.

Is the Trait Really Genetic?

A trait or disorder that runs in families can be attributed to genetic factors (simple or complex), to shared environmental factors (diet, living habits, chemical pollutants, etc.), or to some combination of genetic and environmental factors. Gene interactions can be complex, and the methods for investigating genetic traits are not always straightforward. Thus, the first question to ask about any phenotype—even if it obviously runs in families—is whether it is genetically determined at all. At least two important criteria, taken together, are needed to establish a genetic cause:

1. *The trait must occur more often among genetic relatives of probands than among the general population.* If the trait has a one-gene mode of inheritance without complicating factors, its genetic cause can be fairly easily established. But if two or more genes are involved or if complications occur, then it becomes increasingly difficult to compute expected ratios, and the genetic causation becomes less and less clear.

2. *The trait must not spread to unrelated persons exposed to similar environmental situations.* This criterion helps exclude infections, environmental poisons, nutritional deficiencies, and other conditions that may run in families or be prevalent in local populations. But some infectious agents may mimic biological inheritance. For example, syphilis, AIDS, and some other diseases can be transmitted from mother to fetus across the placental barrier.

These two criteria are necessary but not always sufficient. Additional, and usually less forceful, observations supporting a genetic basis may apply in some situations:

3. *The trait appears at a characteristic age without any obvious precipitating event.* Many genetic disorders are present at birth, but so are developmental abnormalities not related to the genotype of the newborn. Other genetic disorders, such as Huntington disease, appear at a later age but with some variation in the age of onset.

4. *The trait varies in frequency among separate populations.* Human populations differ in the frequencies of many alleles. Racial or ethnic characteristics may indicate genetic differences, but it is often difficult to disentangle genetic from environmental causations.

5. *Identical twins (who have identical genotypes) share the trait more often than do nonidentical twins (who, like ordinary siblings, have about 50% of their genes in common).* Also, if the trait is genetic, adopted twins will resemble their biological parents more than their adopting parents. Substantial difficulties arise in the interpretation of twin studies, however, as presented in Chapter 11.

6. *A human trait resembles one known to be inherited in experimental or farm animals.* Scientists can gain knowledge of a human genetic disease by studying animals that appear to have the same inherited defect.

For several years after kuru was discovered by scientists, its pattern of transmission remained ambiguous, and in the end the genetic hypothesis turned out *not* to stand the test of time. The first clue was the similarity of kuru symptoms to those of *scrapie*, a fatal infectious disease of sheep and goats. The progress of the infection in scrapie, mad cow disease (a fatal neurological disease of cows), kuru, and another human illness called *Creutzfeldt-Jakob disease* is similar. An extremely long period separates the time of exposure and the onset of symptoms. Brain damage leads to loss of coordination and mental deterioration and usually ends in death. The immune system of infected animals does not respond effectively, and unusual fibrils can be found in infected brain tissue. The infectious agent of scrapie—called a **prion**—is very strange, consisting largely or entirely of protein with little or no nucleic acid. The infectious nature of kuru was finally confirmed by inoculating chimpanzees with the brain tissue of kuru victims. After year-long incubation periods, the animals developed a fatal disease closely resembling kuru.

An explanation for the unique age and sex distribution of kuru among the Fore tribes lay in the ceremonial cannibalism of close kin (Gajdusek 1977; Zigas 1990). Preparing for the rites, women and young children of both sexes handled the dead body, including the brain, which was cooked and eaten. The infectious agents probably entered the women and young children through breaks in their skin or through the mucous membranes of the eyes or nose. Men and older boys seldom took part in the ritual. Cannibalism has since been abandoned by the Fore people, and the incidence of kuru has been much reduced.

In the remainder of this chapter, we will describe many of the difficulties encountered in trying to establish a genetic basis for various traits, as well as the problems of interpreting human genetic data. We will also present some recent molecular findings that shed new light on a few of these old problems.

Variations in Phenotypic Expression

From observations of gene transmission in plants and animals, it is known that possession of a gene does not guarantee that it will be expressed the same way in every individual. When a gene is "turned on," its RNA and polypeptide products interact with elements of the internal environment (including other gene products and metabolites) and external environment (including nutrition, medicines, climate, pollutants, exercise, and mental stress). Some genes, such as those determining the blood groups, are always expressed in essentially the same way, regardless of the "background" genotypes or outside conditions. Other genes, such as that for brachydactyly (short fingers and toes), may be expressed in different ways or not at all, depending on the presence or absence of factors that are often not specifically identifiable.

Sex Differences

The sex hormones can exert a powerful influence on the internal environment. The product of any gene exists in one or the other of the two different chemical milieus that characterize the two sexes. Although this does not seem to affect how most genes are expressed, some inherited traits do differ between males and females. In general, we are considering here the effects of *autosomal* loci, whose pattern of inheritance would be distinct from sex-linked genes.

An example of such **sex-influenced inheritance** is *pattern baldness*, in which premature hair loss occurs on the front and top of the head but not on the sides (Figure 10.1). Its inheritance is difficult to study because hair loss may be caused by many factors, both genetic and environmental. In addition, hair loss begins at different ages and varies in extent. Pattern baldness affects mainly males; females who carry the same alleles may show only thinning of hair rather than complete loss. Male hormones, the androgens, are implicated in the expression of the trait. For example, a full head of hair, sometimes luxuriant, is seen in men who do not mature

sexually because of testicular injury prior to puberty. But after androgen treatment, these men sometimes lose their head hair. Likewise, females with tumors of the adrenal gland may secrete large amounts of androgens and become bald; removal of the tumor promotes growth of head hair. Thus, expression of this type of baldness requires both the possession of a specific allele (which appears to be common) and the presence of male hormones.

Father-to-son transmission (as in the Adams family) and other data tend to rule out X-linked inheritance in favor of autosomal inheritance, but the exact mode or modes of transmission are still not clear. It has been suggested that the autosomal gene responsible for pattern baldness, B_1, acts dominantly in males and recessively in females. If so, the genotype/phenotype correspondences are as follows:

Figure 10.1 Pattern baldness in the Adams family. Each man fathered the next one in this list. (A) John Adams (1735–1826) at about age 65, second U.S. president. (B) John Quincy Adams (1767–1848) at age 52, sixth U.S. president. (C) Charles Francis Adams (1807–1886) at age 41, diplomat. (D) Henry Adams (1838–1918) at about age 45, historian. (A, lithograph by Nathaniel Currier, courtesy of the New York Historical Society. B, engraving by Francis Kearny, courtesy of the New York Historical Society. C, lithograph by Nathaniel Currier, courtesy of the Library of Congress. D, photograph by his wife, Marian Hooper Adams, courtesy of the Massachusetts Historical Society.)

Genotype	Female phenotype	Male phenotype
B_1/B_1	Pattern bald	Pattern bald
B_1/B_2	Nonbald	Pattern bald
B_2/B_2	Nonbald	Nonbald

The particular pedigree in Figure 10.2 is consistent with this view, but it is also consistent with other genetic hypotheses (see question 3).

Traits affecting structures or processes that exist in only one sex are called **sex-limited traits**. The manner of gene transmission is irrelevant here; it is the interaction of gene products with the internal environment of the body that results in the sex difference. For example, *hydrometrocolpos*, a condition characterized by the accumulation of fluid in the uterus and vagina, occurs in women homozygous for a certain autosomal recessive gene. Obviously, males can never exhibit this symptom, even if homozygous. Still another example can be seen in dairy cattle, where both the quantity and quality of milk are known to be influenced by autosomal genes inherited equally through both parents. Because bulls cannot express these economically important genes, breeders must evaluate their genetic worth by assessing the milk production of their female relatives. (A prize bull can father thousands of offspring after his death through frozen sperm and artificial insemination.)

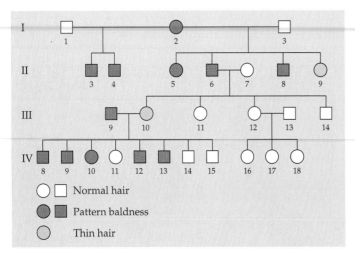

Figure 10.2 Part of a pedigree of pattern baldness, presumed to be dominant in males ($B_1/–$) but recessive in females (B_1/B_1). Female I-2 was reported to be "very bald on the top and front of the head" from an early age. Her sons, II-6 and II-8, and her great-grandsons, IV-8, IV-9, IV-12, and IV-13, became bald by age 25 in the same pattern. (The age of onset was not known for II-3 and II-4, her sons by a previous marriage.) If pattern baldness is recessive in females, then II-5 should not have been affected. Since she was reported to be chronically ill and did not become bald until rather late in life, other factors may have been responsible. (From Osborn 1916, with original numbering of the pedigree.)

Variable Expressivity

The expression of sexual traits is just one factor that interacts with gene products. More generally, the totality of both genotype and environment can modify the way a particular gene product is utilized. **Variable expressivity** refers to the differences in the observed effects of a given genotype in different individuals—for example, the degree of severity of a certain inherited syndrome. Thus, expression—although it always occurs in some form—may be represented as a *continuum* rather than as an all-or-none phenomenon. A sex-influenced trait can be considered a specialized case of variable expressivity.

In experimental organisms, a number of genes that have been studied affect the expression of other genes in clearly defined ways. **Suppressor mutations** reduce the effect of another mutant gene to a normal or near-normal phenotype. **Modifiers** are genes that change the degree of expression of other genes. In mice, mink, and other mammals, for example, many genes are responsible for coloring the fur various shades of black, brown, or yellow, but a modifier gene called *dilute* reduces their intensities, leading to faded grays, tans, or buffs.

Siamese cats and Himalayan rabbits present a familiar example of *environmentally caused variation* in a gene-controlled trait (Figure 10.3). These animals are homozygous for a mutant allele of a coat color gene that produces light or white fur all over the body except on the extremities. The paws, ears, nose area, and tail have dark fur, the particular color variation being determined by other coat color genes. The development of the dark fur in these animals is dependent on temperature; the lower the temperature, the darker the fur and the more extensive the dark areas. Apparently, the pigment-producing enzyme made by this mutant allele works best at the somewhat lower temperatures that occur farthest from the core of the body. If a patch of white fur is shaved from the side of a Himalayan rabbit and an ice pack is applied (or if the rabbit is simply put in a cold environment), the new hair that grows back will be dark.

Variable expressivity is the rule in humans, but it is not usually possible to pinpoint specific modifying genes or specific environmental agents. Nearly all mutant phenotypes, especially those due to dominant genes, can be shown to vary in some respect from one affected person to another. A good example in humans is the extremely variable autosomal dominant disease *neurofibromatosis type 1* (*NF1*), or *von Recklinghausen disease* (discussed in Chapter 7).

Variable Age of Onset. Another aspect of variable expressivity is the age at which the phenotypic consequences first appear. Although many genetic condi-

Figure 10.3 Some animals, such as Siamese cats, possess a pigment-producing enzyme that works best at the somewhat lower temperatures of their extremities. (Copyright J. F. Glisson; courtesy of Hatcher R. Granville, Rich-Hat Cattery, Atlanta, and the Siamese Cat Society of America.)

tions, such as albinism or polydactyly, are present at birth, others only appear at a later age. For example, even though the metabolic error in phenylketonuria can be diagnosed at birth by a simple urine test, there are no overt clinical abnormalities for a few months. The change of an apparently healthy baby into a severely affected one (if untreated) occurs gradually over the first year of life, with variation from patient to patient in the age of onset of various symptoms (**variable age of onset**).

Exceedingly variable in this regard, and also quite common, is a group of disorders known collectively as *diabetes mellitus*. The basic defect involves the hormone *insulin*, which is needed for proper use of the sugar glucose. In diabetic patients, glucose builds up in the blood and is excreted in the urine. Excluding pregnancy-related diabetes and diabetes that occurs as part of several dozen other medical syndromes, there are two major forms of the disease (Atkinson and Maclaren 1990). In the more severe juvenile onset form, insulin is absent because the body's immune system has destroyed the hormone-secreting cells in the pancreas. These patients, who are usually thin, require regular insulin shots for survival. In the less severe adult onset form, insulin is present but does not work normally. These patients are often obese, but their symptoms can often be managed through diet, exercise, and weight control, without requiring insulin shots.

The juvenile onset form is also known as *type 1* or *insulin-dependent diabetes mellitus* (*IDDM*). The adult- or maturity-onset form (usually over age 40) is *type 2* or *noninsulin-dependent diabetes mellitus* (*NIDDM*). Only about 10% of diabetics are type 1. In people under 25 years of age, the frequency of diabetes in many populations is about 1 in 1,000, rising to about 1 in 10 or even 1 in 5 over the age of 60. Neither form is

simply inherited. Instead, each general type is due in part to a different genetic system involving several genes.

In both types, the expression of symptoms requires the interaction of certain mutant alleles (at different loci) and particular dietary and health-related factors. These include viral infections, obesity, diet, and exercise levels. Variation in the age of onset exists within each form, with clinical features beginning anytime from childhood to old age. For example, a type 2 form called *maturity-onset diabetes of the young* (*MODY*) has an early rather than a late onset. Without treatment, diabetes leads to progressive debilities: blindness, kidney failure, nerve damage, gangrene in the feet, heart attacks, and strokes. Treatment can lessen the severity of the symptoms and in some cases allow a near-normal life-style.

In recent years, scientists have identified several genes involved in susceptibility to diabetes. Fairly clear findings deal with the relatively rare MODY phenotype, which is inherited as an autosomal dominant. Mutations of three different genes—*GCK* on chromosome 7p (which codes for glucokinase), *ADA* on 20q (which codes for adenosine deaminase), and an unknown locus on 12q—account for about 75% of MODY pedigrees. And some families with rare forms of type 2 diabetes have mutations in genes that code for insulin or insulin receptor, or have mutations in mitochondrial genes (Raffel et al. 1997).

Geneticists have also analyzed type 1 diabetes and found two genes showing the most clear-cut linkages. The immune system's *HLA* locus (on chromosome 6p) accounts for the largest share (about 40–70%) of genetic susceptibility to type 1 diabetes—though not all by itself. And a locus near the chromosome 11p gene that codes for insulin predisposes to a much smaller amount of genetic susceptibility. Other gene regions with minor effects are located on chromosomes 15q, 11q, 6q, and 2q, with additional loci under study.

Huntington disease (*HD*), a severe autosomal dominant neurological disorder (see Box 5C), is also characterized by variable age of onset. Only occasionally do the first signs of disease appear in children; about 60% of cases are diagnosed between the ages of 35 and 50, with the remainder occurring equally before and after this age range. Recall from Chapter 7 that scientists have mapped, cloned, and sequenced the *HD* gene, whose mutant alleles all contain a variable number of extra repeats of the CAG trinucleotide. Although the function of the protein product is unknown, researchers discovered that the age of onset of HD is highly correlated with the number of CAG repeats: the more repeats, the earlier the age of onset. But how these repeats come about and why the early-onset cases are usually transmitted through affected males (rather than affected females) remain a mystery.

Incomplete Penetrance

An extreme form of variable expressivity is the total absence of expression in persons known to carry a particular autosomal dominant allele. Such nonexpression of a gene that usually has a detectable phenotype is called **incomplete penetrance**. It may occur for many reasons, both genetic and environmental. As applied to individuals, complete versus incomplete penetrance is an *all-or-none* phenomenon: Either a particular genotype is expressed in some form or another or it is not expressed. As diagnostic methods improve and the threshold for detecting abnormal phenotypes is lowered, however, the criteria by which we label a particular genotype as expressed or not expressed may also change.

In the case of a dominantly inherited trait, incomplete penetrance is manifested as "skipped generations." This situation in a pedigree of *polydactyly**** (extra digits) is seen in Figure 10.4. Since the condition is fairly rare, it is highly likely that all six affected individuals possessed copies of precisely the same dominant allele. Individual II-6 must have received the allele from her father and transmitted it to her daughter, although she herself was not affected. A two-generation "skip" is seen on the right side of the pedigree. (Although dominant inheritance is fairly clear in this pedigree, a lower degree of penetrance might obscure the mode of inheritance.) Polydactyly also illustrates another aspect of variable expressivity. Figure 10.5 il-

*Reference to a polydactylous giant is found in I Chronicles 20:6: "And yet again there was war at Gath, where was a man of great stature, whose fingers and toes were four and twenty, six on each hand, and six on each foot: and he also was the son of a giant."

lustrates a case in which six well-formed digits appear on both hands and both feet, although each appendage reveals slightly different bone development. The precise cause of incomplete penetrance is usually unknown. But in some special cases, it is found that a specific allele at one locus masks the characteristic effects of an allele at another locus. This phenomenon is called **epistasis**. In the mouse, for example, the genotype *b/b* is phenotypically brown and *B/–* is phenotypically black, but neither brown nor black pigment appears in the fur if the mouse is also homozygous *c/c* at another locus. The genotype *c/c* leads to an albino mouse, which develops *no* fur pigment, so color alleles at other loci are irrelevant. We say that the genotype *c/c* is epistatic to alleles at the *B* locus and other coat color loci. Epistasis should not be confused with dominance, which is the masking of one allele by another allele at the same locus.

One Gene with Several Effects

Given the complexity of embryonic development, it is not surprising that a single abnormal gene can have multiple, sometimes seemingly unrelated consequences. One gene product can be involved in many biochemical processes affecting different pathways of growth and maturation. The resulting developmental cascade of effects is called **pleiotropy**—literally, "many turnings." As with variable expressivity, pleiotropy is the rule rather than the exception, especially when a given phenotype is carefully examined at different levels. Consider the multiple effects of insulin deficiency in *juvenile-onset diabetes*, as shown in Figure 10.6. Note that the excess utilization of fats for energy results in the production of acid substances (acidosis) in such high concentrations that the body is unable to neutralize them completely, a situation that can even cause coma or death.

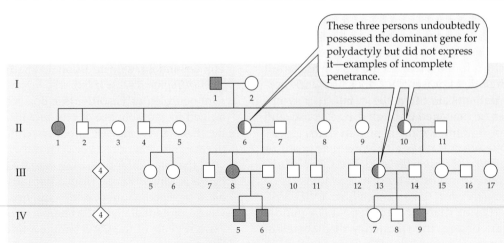

Figure 10.4 A pedigree of polydactyly. (From Neel and Schull 1954.)

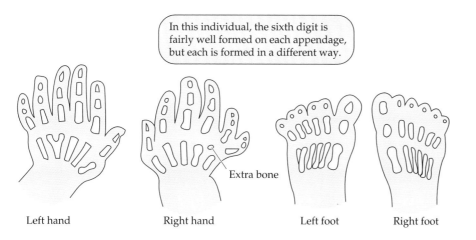

In this individual, the sixth digit is fairly well formed on each appendage, but each is formed in a different way.

Extra bone

Left hand   Right hand   Left foot   Right foot

Figure 10.5 X-rays of a 7-day-old Norwegian boy (not from the pedigree in Figure 10.4) with polydactyly of both hands and both feet. A fork in one of the bones of the palm of the left hand and in the left foot is clearly seen. Curiously, the extra small bone in the palm of the right hand also occurred in the right hand of the polydactylous mother. (Redrawn from Sverdrup 1922.)

One Phenotype from Different Causes

We have just described how one gene can have multiple effects. Conversely, any one of several different genes can result in similar effects. Polydactyly (Figures 10.4 and 10.5) is a case in point. Extra digits can be due to mutant alleles of any of about a half dozen different genes, usually showing dominant inheritance with incomplete penetrance. (We consider here only the phenotypes in which polydactyly is the sole observable malformation; other cases are known in which it is associated with more serious syndromes.) In one form, a reasonably functional digit appears on the little-finger side of the hand (or little-toe side of the foot). In a second form, an extra digit in the same position is much reduced in size. Other types of polydactyly involve an extra thumb, or a thumb with an extra joint, or a duplicated index finger. The frequencies of the various types of polydactyly among newborns differ considerably, the extra little finger being the most common. Polydactyly is much more common in blacks than in whites.

Another feature of polydactyly is that not all cases are genetic. Errors of development brought about by slight changes in the prenatal environment can also produce bone malformations. Such environmental mimics of a characteristically genetic condition are called **phenocopies**. A person with polydactyly who has no affected relatives—a so-called sporadic case—*may* be a phenocopy, but other causes of sporadic cases are possible (such as a newly occurring dominant mutation).

The term used to described multiple genetic causes of the same, or nearly the same, phenotype is **genetic heterogeneity**. From the point of view of an affected person, determining a precise causation can be important. For example, *hemophilia A* results from a defect in so-called clotting factor VIII, whereas *hemophilia B*, with a similar phenotype, results from a defect in clotting factor IX. The symptoms of hemophilia A can be alleviated by using a concentrate of factor VIII; but this treatment would be of no use to a person suffering from hemophilia B.

Genetic heterogeneity is always expected for complex anatomical, physiological, or behavioral traits. The ability to hear, for example, depends on the proper functioning of hundreds or perhaps thousands of genes. *Deafness*, of a greater or lesser degree, can result when one of these genes mutates to an allele that impairs the development of the intricate structures of the ear or the nerve connections to the brain. Hearing loss is quite common, affecting about 1 in 250 children under the age of 19 years. (Adult-onset hearing loss is also common and probably hereditary to a considerable extent. But it is far less understood.) About 35–50% of all childhood deafness is thought to arise from genetic factors, with the remainder resulting from nongenetic factors, including certain viral infections of the mother during pregnancy (measles and rubella, especially), premature birth, or other types of environmentally imposed harm to the fetus or newborn (Fischel-Ghodsian and Falk 1997).

Among the inherited cases of childhood deafness, 60–70% involve deafness alone. About two-thirds of these *nonsyndromic* examples appear to be simple autosomal recessives,* perhaps involving as many as 35 different loci. About one-third are autosomal dominants, and a few are X-linked or mitochondrial. An autosomal dominant gene (*DFNA1*) on chromosome 5q, whose mutation caused progressive deafness in an eight-generation Costa Rican kindred, is homologous to the *diaphanous* gene in fruit flies and mice and the *Bnilp* gene in yeast. The product of this highly conserved locus encodes a protein that helps to organize actin, a protein that (in humans) stiffens the hair cells of the inner ear (Lynch et al. 1997).

---

*Groce (1985) provides a fascinating description of hereditary deafness that once existed at very high frequency in Chilmark, Massachusetts, on the island of Martha's Vineyard. Affected individuals were not regarded as handicapped and were fully integrated into daily life. Indeed, everyone in the village (with or without hearing loss) grew up using sign language!

Figure 10.6 Pleiotropic consequences of insulin deficiency occurring in juvenile-onset diabetes mellitus. (Redrawn from Neel and Schull 1954.)

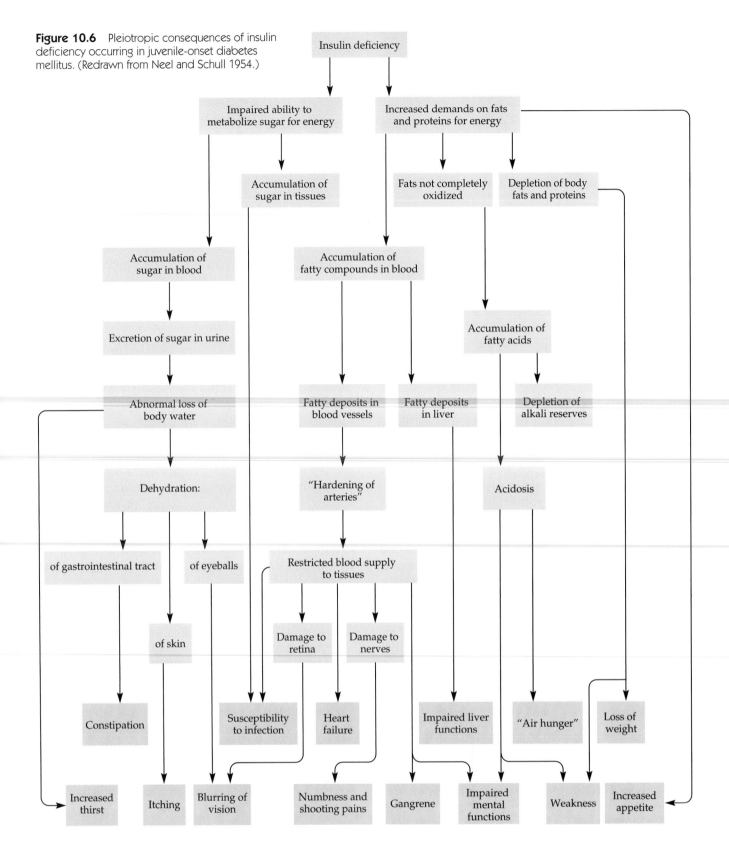

Since 1992, over 30 nonsyndromic deafness loci have been mapped (Pennisi 1997). Roughly 30% of all cases of hereditary childhood deafness are associated with other abnormalities, however. Researchers have tallied up about 200 different forms and cloned over 25 loci associated with such *syndromic* hearing loss. With refinements of testing procedures, these syndromes may be found to have multiple causations.

Also, studies of hereditary hearing loss in mice are enhancing our understanding of deafness in humans (Steel and Brown 1994).

The most common example of inherited deafness, accounting for about 3% of all deaf children, is *Waardenburg syndrome* (*WS*). This autosomal dominant disorder is seen frequently in Kenya, South Australia, and South America. Its phenotype is highly variable. Indeed, only 20–25% of affected individuals show a hearing loss. The main features are wide-set eyes, a wide bridge of the nose, eyebrows that meet in the midline, and a white forelock (Figure 10.7). Other defects may include patches of unpigmented skin, eyes that differ in color, prematurely gray hair, and white eyelashes. The normal allele, which is located on chromosome 2q, may somehow be involved with the migration of neural crest cells that become nerve and pigment cells. WS is homologous to the *splotch phenotype* in mice. Recent studies reveal that in many cases, both phenotypes are caused by the same developmental gene—called *PAX-3* in humans, *Pax-3* in mice, and *paired box* in fruit flies. The gene product is a DNA-binding protein that is thought to regulate other genes during early development (Chapter 15).

Diversity from Within and Between Loci

So far, in discussing the genotypes and phenotypes associated with simply inherited human traits, we have usually considered one locus with two alleles (one normal and one mutant), three genotypes (homozygous normal, heterozygous, and homozygous mutant), and either two phenotypes (if one allele is completely dominant) or three phenotypes (if the two alleles are codominant). True, we discussed *multiple alleles* in Chapter 4 and have occasionally mentioned that multiple alleles exist for certain disease-causing loci. But the implication has generally been that the different alleles are found in different populations and that a given mutant phenotype results from homozygosity for the *same* mutant allele.

But this scheme greatly oversimplifies the true picture. In recent years, with rapid advances in molecular techniques, geneticists have been able to isolate and chemically dissect the mutant alleles associated with a number of human disorders. What they discovered is that some mutant phenotypes thought to result from homozygosity of *one* mutant allele at a given locus actually arise from the combination of *two* different mutant alleles at that locus. Thus, some mutants are genotypically heterozygous rather than homozygous. Such an individual is called a **compound heterozygote**. This situation has been found, for example, with phenylketonuria, familial hypercholesterolemia, cystic fibrosis, Duchenne muscular dystrophy (Chapter 5)—indeed, with most of the disorders we have discussed so far. Rather than trying to understand just a few genotypes and many phenotypes at a given locus, we are now presented with many genotypes, each giving rise to one or a few variant phenotypes.

(A)   (B)

Figure 10.7 Three people (including a mother and daughter) with Waardenburg syndrome. Note the wide-set eyes, the wide bridge of the nose, the white lock at the hairline in the middle of the forehead, and (in B) growth of the eyebrows toward the midline. (A from Partington 1959. B reproduced by kind permission of Professor A. E. H. Emery.)

Far more extreme than phenotypic variants of the *same* disease associated with compound heterozygotes is the occurrence of apparently *unrelated* disorders associated with mutations in the same allelic series. For example, Romeo and McKusick (1994) point out that different mutations of a single gene (called *RET*, on chromosome 10q) have been found in patients with Hirschsprung disease (colon obstruction and enlargement due to absence of nerves) and different forms of endocrine (adrenal and thyroid) cancer. They add that there exist "well over 100 other loci with two or more diverse clinical disorders." *Modifier genes* at different loci probably also contribute to the striking phenotypic diversity seen in such cases. Although well known in many experimental organisms, however, they are hard to identify in humans. Certainly, all of these findings greatly complicate the study of genetic disorders. But knowledge of multiple alleles and their contribution to phenotypic diversity helps to explain much of the variable expressivity and incomplete penetrance noted years ago with many disease-causing genes.

This chapter is mainly devoted to the effects of single genes; traits associated with multiple loci are discussed in the next chapter. However, we will mention here an unusual inheritance pattern for *retinitis pigmentosa* (*RP*), a type of hereditary blindness caused by degeneration of photoreceptors (rods and cones) in the retina. RP can be transmitted as an autosomal dominant, an autosomal recessive, or an X-linked trait, with only one locus involved in any particular pedigree. But recently, a few families have been found in which the inheritance of RP involved *two* photoreceptor-specific genes rather than one. In these pedigrees, individuals heterozygous for a certain mutant allele at either the one locus or the other were unaffected, but the double heterozygotes had RP. (No homozygotes for either locus were observed.) Nobody knows whether this newly discovered mode of transmission, called **digenic inheritance**, is widespread in humans or other organisms. If so, it could add yet another level of complication to the analysis of Mendelian inheritance.

Collecting and Interpreting Genetic Data

Even when the alleles of a single, fully penetrant gene lead to uniform phenotypes (and no phenocopies exist), geneticists still may not observe a clear-cut Mendelian ratio confirming a specific genetic hypothesis.

Biases in Collecting Data

Departures from standard ratios of a simple trait may occur even in large samples, because raw data can be biased by the methods of collection. For example, suppose that 100 engineers are asked to tabulate the sex of persons in their own sibships, including themselves.

Assuming an average sibship size of three, typical data might be as follows:

Males	Females	Total	% Males
189	111	300	63.0%

The percentage of males seems unusually high. But the puzzle is solved by realizing that the engineers themselves—the *probands*—are included in the ratio, and they are predominantly male. Assume, for example, that 90 of the engineers are men and 10 are women. Subtracting these probands from the raw data, leaving only their brothers and sisters, provides an unbiased sample for calculating the sex ratio:

	Males	Females	Total	% Males
Raw data	189	111	300	63.0%
Probands	−90	−10	−100	—
Unbiased sample	99	101	200	49.5%

Looking at the raw data, we might propose some scheme to account for the seemingly unusual sex distribution. On more careful analysis, however, we see that the raw data are simply skewed by the method of collection: An excess of males was "forced" into the observations. This example illustrates **ascertainment bias**, the term *ascertainment* referring to the method used to discover and record data. The bias does not mean that the collected information is useless or wrong, only that it must be handled with care to avoid drawing unwarranted conclusions.

Ascertainment bias can be important in the case of autosomal recessive inheritance, for which we might reasonably expect to find 25% affected children from heterozygous parents. In fact, the reasonable expectation may *not* be 25% affected or even close to 25%. In trying to establish ratios that prove autosomal recessive inheritance, there are two main problems: dominance and small families. Incomplete data can also cause difficulties.

Dominance.   If the normal allele is truly dominant, then a heterozygote cannot be distinguished from a normal homozygote by his or her phenotype alone. Without further analyses, a heterozygote can only be identified as the child or the parent of an affected person (Figure 10.8). If the recessive phenotype is very obvious, then the cases that come to medical attention are for the most part affected children from unaffected heterozygous parents (Figure 10.8B). *But those heterozygous parents who happened not to produce any affected children are never found.* Yet the expected 25% affected is valid only if *all* children from *all* heterozygous parents are accounted for. In other words, geneticists often locate heterozygote × heterozygote parents indirectly (through their affected child) rather than directly (by looking at the parents themselves). This method leads to a bias, because at least one child must be affected before the $A/a \times A/a$ matings can be ascertained. Thus, some families—the $A/a \times A/a$ matings with no affected offspring—are cut off from observations.

Small Families. The problem of overlooking $A/a \times A/a$ families who happen not to produce any a/a offspring is minimal in large sibships. For example, what is the probability that no a/a child will show up in a sibship of size 15? Simple calculations show that in the case of recessive inheritance among families with 15 offspring each, only about 1% of parents—that is, $(3/4)^{15}$— would be missing by virtue of having all $A/-$ children. But if $A/a \times A/a$ parents produce only *one* child, then 75% (rather than merely 1%) would be undetectable when the one and only child is unaffected. Geneticists use such probabilities in making statistical corrections for ascertainment bias.

Incomplete Data. For recessive inheritance in humans, where experimental matings are not possible and where the geneticist must make use of whatever data are available, the simple Mendelian ratios may not be the expected values. Furthermore, genetic surveys may be incomplete. Geneticists may use medical or dental records, death certificates, or other data that were collected for other purposes and with various degrees of thoroughness and care over an extended period. Questionnaires directed to practicing physicians may yield different degrees of completeness, depending on the phenotype in question: A serious condition that is easy to treat will bring affected persons to doctors more often than a mild condition that is difficult to treat. Reliance on cases reported in the scientific literature may also be biased: Families with one or a few affected persons may be missing because they are less interesting than families with many affected.

Thus, when we analyze patterns of inheritance, some fairly sophisticated statistics are often needed. Even the most refined mathematics, however, cannot correct for not knowing how affected persons first came to the attention of the data collector. Poor data yield unreliable conclusions regardless of how they are manipulated.

Interpreting Data and Testing Hypotheses

We have already noted that observed data are usually not *exactly* the same as ideal expectations based on some hypothesis; there are **deviations** between the two to a greater or lesser amount. For example, Mendel counted 787 tall and 277 short plants out of 1,064 offspring from heterozygous parents (Chapter 4). But these numbers are not exactly 798 (3/4 of 1,064) and 266 (1/4 of 1,064), as predicted by the law of segregation:

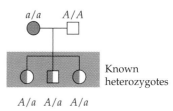

(A) Persons with a normal phenotype have an affected parent

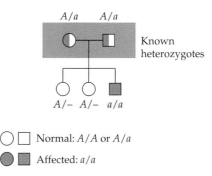

(B) Persons with a normal phenotype have an affected child

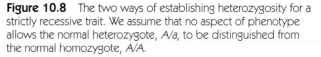

Figure 10.8 The two ways of establishing heterozygosity for a strictly recessive trait. We assume that no aspect of phenotype allows the normal heterozygote, *A/a*, to be distinguished from the normal homozygote, *A/A*.

Class	Observed number	Expected number	Deviation
Tall	787	798	−11
Short	277	266	+11
Total	1064	1064	0

The deviations of −11 in one phenotypic class and +11 in the other could be due to chance fluctuations, or they could be due to a wrong hypothesis. Because the deviations are "small," they are likely to be due to chance and therefore inconsequential. So we ignore them and state with some assurance that the observations are consistent with the Mendelian hypothesis. But had the deviations been "large," we would propose that the *expected numbers* are not appropriate and seek some other hypothesis to give a better fit.

Coin tossing provides another example. Suppose that you flip a penny 20 times, getting 15 heads and 5 tails. Here is the corresponding table:

Class	Observed number	Expected number	Deviation
Heads	15	10	−5
Tails	5	10	+5
Total	20	20	0

Here the expectation of 10 heads and 10 tails is calculated from the hypothesis of a true coin, fairly flipped.

Are the deviations (–5 and +5) sufficiently small that we consider them mere chance? Or are the deviations sufficiently large that we are suspicious of the coin or of the way it was tossed?

Some rules regarding probabilities allow us to decide between these possibilities. Essentially, we want to use the size of the deviations to evaluate whatever hypothesis is set forth to explain the observations. As the deviations become larger, at what point do we begin to doubt the hypothesis? An answer of sorts is provided by a statistical tool (not described here) whose end result is a **probability** P.* The value of P decreases as the deviations increase. Technically, P is the probability of getting deviations *this large or larger* by chance alone, if the hypothesis is correct. Conceptually, P reflects the credibility of the hypothesis: The smaller the value of P, the less plausible the hypothesis. We then make a decision based on the value of P.

If P is low, then we *reject* the hypothesis. We say that the observed results are not consistent with the hypothesis, that the hypothesis is not believable, or that the large deviations between observed and expected results are significant—that they are not likely to be due to mere chance. In many biological applications, "low" is defined as 0.05 or smaller. This is called the 5% level of significance.

If P is high, then we *accept* the hypothesis. We say that the observed results are consistent with the hypothesis, that the hypothesis seems like a good one, or that the deviations are not significant. "High" is defined as more than 0.05.

In summary, we want to know whether some idea we have is worth pursuing, and we gather some evidence to help us decide. The reasoning is similar to that in a court trial: The hypothesis is that the accused is innocent. If the evidence supporting the hypothesis is incredible—has low P—then the hypothesis of innocence is *rejected*. The jurors, of course, have no mathematical means for reaching a decision but must grapple with concepts like "reasonable doubt."

In the coin-tossing experiment, the P value turns out to be low. But this statistical procedure does not tell us *what* is wrong with the coin experiment or even that there *is* something wrong, only that we should be suspicious. A low probability value does not mean that the hypothesis is false, only that it is *likely* to be false. Mistakes can be made. We might reject a hypothesis that is really true or accept a hypothesis that is really false. Good statistical procedures are designed to minimize these sorts of errors. They help us make judgments about evidence, just as the adversary system does for jurors in a court of law. And just as a "hung jury" may cause a new trial, so a borderline decision based on our statistical method may cause an investigator to gather new evidence.

Statistical analyses provide practical, standardized tests applicable to any discipline. For better or worse, statistics now pervades our world, from medical prognoses to weather forecasts to election-night predictions. In a sense, statistical thinking flowers when knowledge is scant and wilts as knowledge increases. Yet probabilities do help to guide the direction of many human activities.

For about 60 years following the rediscovery of Mendel's laws, human geneticists struggled with the problems of collecting and analyzing data on simply inherited traits and on traits determined by the interaction of multiple loci. Just dealing with such "standard" genes in humans kept them fully occupied.

Meanwhile, other geneticists developed increasingly sophisticated statistical, chemical, and physical means of analyzing the behavior of genes and chromosomes in a wide variety of organisms. The more closely they looked, the more they began to find intriguing exceptions to standard genetic behavior. Some of these unexpected behaviors of genes and chromosomes were quickly recognized in a wide variety of organisms and provided explanations for unsolved genetic mysteries. But others seemed so bizarre, and were often so difficult to analyze, that they were thought to be oddities confined to one or a few experimental organisms. More and more often, however, some of these apparently quirky types of genes and chromosomes are turning up in a broad array of organisms, including humans.

Sex Chromatin and the Lyon Effect

Biologists have puzzled over the behavior, structure, and evolution of sex chromosomes ever since their discovery a century ago. Sex chromosomes are unique in several ways. For one thing, although small imbalances among *autosomal* genes and chromosomes are very poorly tolerated, causing severe defects or death, the sex chromosomes are much more flexible in this regard. Even in species where the X is very large, many organisms (including humans) can survive with too many or too few X chromosomes. Equally puzzling is the *normal* chromosomal makeup of females (who have two X chromosomes), compared with that of males (who have one X and a small, nearly geneless Y chromosome). Because females have twice as many X-linked alleles as males do, should not one or the other sex suffer from a relative excess or deficiency of X-linked gene products? By what mechanism do the two sexes compensate for this twofold difference in dosage of X-linked genes?

---

*Details about the *chi-square method* can be found in most general genetics textbooks and upper-level human genetics texts.

In mammalian somatic cells, it turns out that only one X chromosome is normally functional. Thus, there must be a way of silencing one X chromosome in females. The discovery of the silencing mechanism is a good example of serendipity in science—accidentally finding something new and different while looking for something else. During the 1940s, two Canadian scientists noted a darkly staining mass in the nuclei of some interphase cells in cat brains (Figure 10.9A). Investigating further, they discovered that the interphase cells from females exhibited dark-staining spots in their nuclei, but cells from males did not. Examination of human nerve cells showed the same sex difference. They correctly speculated that each spot consists of one tightly condensed (heterochromatic) X chromosome.

Because of its correlation with sex, the spot was named the **sex chromatin**, although the term **Barr body** is also used. It can be seen in many types of somatic cells, usually lying against the inner surface of the nuclear membrane. Cells with one or more Barr bodies are said to be *chromatin-positive*, while those lacking any are *chromatin-negative*.

Detection of Sex Chromatin in Humans

The simplest and most commonly used method of checking for sex chromatin is the *buccal smear*: Cells are gently scraped from inside the cheek, spread on a glass slide, stained, and examined with a light microscope. Usually, only 20–70% of the buccal cells from normal females contain a Barr body, although all cells are presumed to carry it.

In the late 1950s, a new twist to the sex chromatin story developed when researchers discovered that some humans have too many or too few sex chromosomes. People with *Klinefelter syndrome* have the karyotype 47,XXY and are male; conversely, people with *Turner syndrome* have the karyotype 45,X and are female. (Their phenotypes will be described in Chapter 13.) Quite surprisingly, these syndromes showed for the first time that the human Y chromosome plays a major role in sex determination. Until then, it had been thought that the critical factor was (as in the fruit fly) the number of X chromosomes, with two X chromosomes leading to femaleness and one to maleness, and with the Y chromosome needed only for fertility in males.

What about the sex chromatin in people with these sex chromosome abnormalities? Cells from males with one extra X chromosome (47,XXY) are chromatin-positive, whereas cells from Turner (45,X) females are chromatin-negative. They are exceptions to the original "rule" that the sex chromatin is found only in females. Further investigations showed that more than one sex chromatin occurs in persons with additional X chromosomes, but not in persons with additional Y chro-

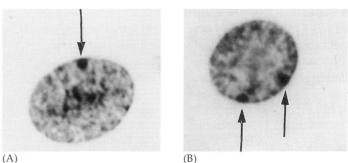

(A) (B)

Figure 10.9 Sex chromatin (Barr) bodies, indicated by arrows, in nuclei of female cells. (A) Nucleus in a buccal smear from inside the cheek of a 46,XX female. One Barr body lies against the nuclear envelope. (B) Nucleus in a buccal smear from a 47,XXX female. Two sex chromatin bodies are visible. (Courtesy of Murray L. Barr, University of Western Ontario.)

mosomes. The more precise rule states that *the number of Barr bodies is one less than the number of X chromosomes*:

Number of Barr bodies	Karyotype(s)	
	Female	**Male**
0	45,X	46,XY; 47,XYY
1	46,XX	47,XXY; 48,XXYY
2	47,XXX	48,XXXY; 49,XXXYY
3	48,XXXX	49,XXXXY
4	49,XXXXX	

The Inactive X Hypothesis

What happens to a chromosome that forms a Barr body? Are its genes still functional? Is the same X chromosome (maternal versus paternal) condensed in every cell? What determines whether or not a given chromosome will be condensed? Based primarily on studies with mice, the following **inactive X hypothesis** was proposed in 1961 by English geneticist Mary Lyon:

1. The condensed X chromosome found in somatic cells of normal female mammals is genetically inactive.

2. Inactivation first occurs early in development (around the 32- to 64-cell stage), at which time either the maternal or the paternal X chromosome is equally likely to be inactivated.

3. This inactivation event occurs independently and randomly in each embryonic cell. But once the decision is made for a given cell, it is irreversible: The same X chromosome will be inactivated in all its descendant cells, producing a clone from each of the original cells.

The net effect of this whole process—called **lyonization**—is to equalize the phenotypes in males and females. Such a phenomenon is known as **dosage compensation**. Although normal females possess twice as many X chromosomes as males, they generally have the same number of active X-linked genes and the same amount of X-linked gene products.

Evidence for Lyonization

Evidence for lyonization was sought in females *heterozygous* for X-linked genes; these females should have clones of cells in which one allele or the other is active, but never cells with two active alleles. If entire clones remain in place during development, good-sized phenotypic "patches" will result, each patch expressing only one X-linked allele of a gene. If cells from different clones intermingle during development, however, then an intermediate or finely variegated "salt-and-pepper" phenotype will be observed. One line of evidence given by Lyon involved the mottling or dappling on the fur of female mice heterozygous for a number of X-linked coat color genes. This effect is also seen in the familiar calico or tortoise shell cats (Figure 10.10), whose patches of black or orange hair indicate which X-linked gene is active in each heterozygous cell. With rare exceptions, calico cats are female.

Because humans lack fur, we cannot observe calico people. Perhaps the next best trait, however, is phenotypic expression of the recessive X-linked trait *anhidrotic ectodermal dysplasia*. Some affected males were described by Charles Darwin as the "toothless men of Scinde"; they lack sweat glands, teeth, and most of their body hair. Studies of females heterozygous for this gene reveal considerable variability in phenotypic expression, taking the form of irregular skin patches with few or no sweat glands (Figure 10.11) and some regions of the jaw with missing or defective teeth.

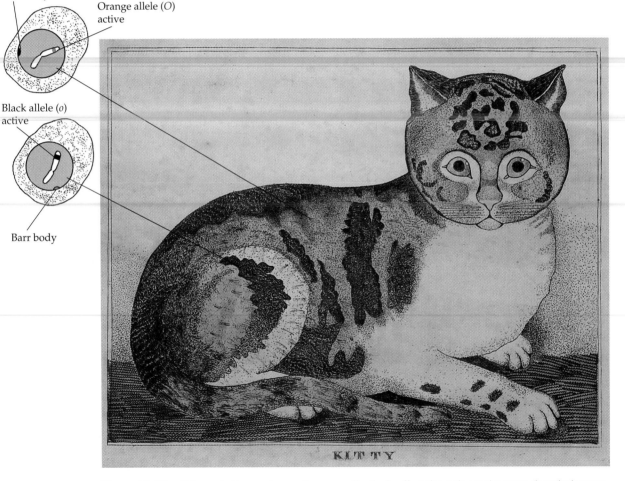

Barr body

Orange allele (*O*) active

Black allele (*o*) active

Barr body

Figure 10.10 Calico cat. Many autosomal genes collectively affect the color and pattern (banded versus solid) of the individual hairs on a cat. Here we consider only one X-linked locus and its role in forming the light and dark patches of fur on a calico cat. This female is heterozygous for a dominant allele (*O*) that leads to orange (or yellow) hairs and a wildtype recessive allele (*o*) that produces black (or dark-speckled) hairs. In any one cell, either one allele or the other is active; the inactive allele is on the X chromosome forming the Barr body. The size of each patch of fur depends on how many cells descended from each cell present at the time of the random inactivation decision. (Painting by George White. The Metropolitan Museum of Art, Gift of Mrs. E. C. Chadbourne, 1952. (52.585.29) Photograph © 1981 The Metropolitan Museum of Art.)

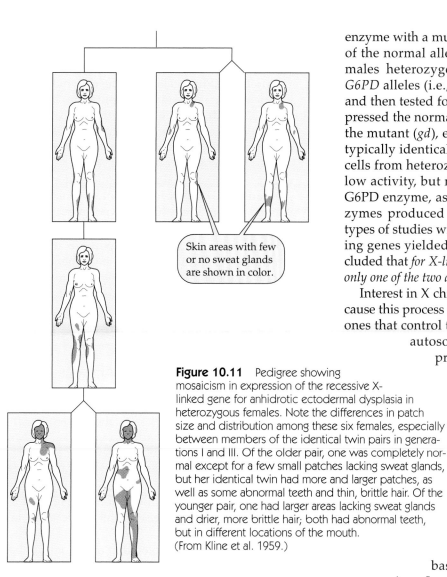

Figure 10.11 Pedigree showing mosaicism in expression of the recessive X-linked gene for anhidrotic ectodermal dysplasia in heterozygous females. Note the differences in patch size and distribution among these six females, especially between members of the identical twin pairs in generations I and III. Of the older pair, one was completely normal except for a few small patches lacking sweat glands, but her identical twin had more and larger patches, as well as some abnormal teeth and thin, brittle hair. Of the younger pair, one had larger areas lacking sweat glands and drier, more brittle hair; both had abnormal teeth, but in different locations of the mouth. (From Kline et al. 1959.)

enzyme with a much lower level of activity than that of the normal allele (*Gd*). When skin cells from females heterozygous for the normal and mutant *G6PD* alleles (i.e., *Gd/gd*) were individually cloned and then tested for enzyme activity, some clones expressed the normal allele (*Gd*) and others expressed the mutant (*gd*), even though all clones were genotypically identical (*Gd/gd*). (2) Individual red blood cells from heterozygous females also show high or low activity, but never intermediate activity of the G6PD enzyme, as is found in heterozygotes for enzymes produced by autosomal genes. (3) Similar types of studies with other X-linked enzyme-producing genes yielded similar results. Researchers concluded that *for X-linked genes in an XX mammalian cell, only one of the two alleles remains active.*

Interest in X chromosome inactivation is keen because this process might be the same or similar to the ones that control the selective turning on and off of autosomal genes and also parental imprinting (discussed later). In recent years, scientists have found several differences between active and inactive X chromosomes. For one thing, the heterochromatic, inactive X replicates later in the cell cycle than do the other chromosomes. Also, gene transcription on the inactive X is repressed by **methylation**, the addition of methyl (CH_3) side groups to cytosine bases of DNA in gene control regions. In addition, the histone proteins complexed with DNA on the inactive X contain many fewer attached *acetyl* (CH_3CO) side groups than do the histones associated with active X chromosomes. But exactly how these differences are involved in initiating and maintaining X inactivation is not clear.

We do know that X inactivation starts and spreads out from an *X inactivation center* in the long arm (band Xq13) near the centromere. A gene called *XIST* (for *X inactive-specific transcript*), recently found in this region, is essential to the inactivation process. Quite surprisingly, *XIST* is expressed from the *inactive* (rather than the active) X chromosome only, and it codes for an RNA but not a protein product. This RNA remains in the nucleus, tightly coating the inactive X's DNA.

Researchers postulate that the inactivation process involves the following steps: initiation, which requires at least two X chromosomes; counting of the total number of X chromosomes in a nucleus; choice of one X to remain active, with "blocking" of that chromosome's X inactivation center; spreading of inactivation on the other X; and maintenance of inactivation

Note also in Figure 10.11 that the two pairs of identical twins in generations I and III show differing patterns of skin patches, a phenomenon called **mosaicism**. Many other human X-linked genes are also known to show mosaic expression in heterozygous females.

One of the best systems for studying lyonization makes use of an X-linked gene controlling production of the enzyme *glucose-6-phosphate dehydrogenase (G6PD)*. This enzyme, which occurs in nearly all plants and animals, acts in glucose metabolism. It turns out that human 46,XX females with *two* doses of the normal X-linked *G6PD* allele produce no more of the enzyme in their cells than do 46,XY males with *one* dose of the normal allele, although one might expect the former to produce twice as much. Furthermore, 47,XXY Klinefelter males and 47,XXX females produce no more G6PD than do normal males and females, suggesting that regardless of how many X chromosomes are present in a given cell, only one X is active.

Even more to the point were these observations: (1) A recessive allele (*gd*) of this gene produces a defective through cell division (Willard 1995, 1996). It appears that the X inactivation center is required for all except the final step.

Other Considerations

Studies of X chromosome inactivation have revealed some additional twists to the story. Here we present several of them.

1. *Part of the X chromosome is not inactivated.* If all but one X chromosome is completely inactivated in every individual, then why do Turner females (born with one X chromosome only) differ from regular females, and why do Klinefelter males (with two X chromosomes and one Y) differ from regular males? To explain these abnormal phenotypes, it was suggested that some X-linked genes must always escape inactivation. These genes would be expressed doubly in regular females and Klinefelter males (both with two X chromosomes) and singly in regular males and Turner females (both with one X chromosome). We now know that this is indeed the case: About 20 X-linked genes do *not* exhibit mosaicism in heterozygous females (Disteche 1995, 1997). Nearly all of these loci are in the short arm of the human X chromosome, over half being clustered near its tip. How they escape X inactivation is a mystery. Their products, however, show no obvious connection to the phenotypic features of Turner or Klinefelter syndrome. So maybe there are additional X-linked genes (influencing stature and fertility, for example) that escape inactivation.

2. *Germ cells behave differently.* In female fetuses, future germ cells undergo lyonization along with somatic cells. But as the germ cells develop into oocytes and enter meiosis, their inactivated X chromosomes become *reactivated*. Thus, each egg cell ends up with an active X. In males, for unknown reasons, the one X chromosome is briefly inactivated in all primary spermatocytes just as they enter meiosis.

3. *Post-inactivation cell selection may occur in females heterozygous for some very severe X-linked disorders.* Lesch-Nyhan syndrome is one example. In the tissues affected by such a disorder, clones in which the mutant allele is active are unlikely to survive. Thus, the remaining cells are much more likely to have an active normal allele, and it will appear as though there was nonrandom X chromosome inactivation from the very beginning.

4. *Some genes get reactivated as mammals age.* Studies of female mice carrying a certain chromosomal abnormality have shown that some inactivated genes may not stay that way. This work involved a piece of autosome that was stuck into the middle of an X chromosome. Normally, autosomal genes do not ly-

onize. But when moved to the X chromosome, they get inactivated right along with the X-linked genes. In this experiment, the displaced piece of autosome carried a wildtype allele for the enzyme tyrosinase, which leads to normal pigmentation. Its recessive allele (which, when homozygous, causes albinism) was present on a normal autosome. Thus, the fur was mottled with colored and albino patches, similar to that of calico cats. But as these animals aged, they became more darkly pigmented—apparently because the displaced tyrosinase gene was reactivated in some cells. Other experiments with mice have given similar results.

X chromosome inactivation is perhaps the first and best-known example of unexpected phenomena that explain ambiguous patterns of gene expression. But it is definitely not the only one.

Dynamic Mutations and the Fragile X Syndrome

Recall from Chapter 7 that in 1991, geneticists discovered a new kind of mutation called a *trinucleotide repeat expansion*. It was first detected in the gene for the fragile X syndrome, but since then about ten other disorders have been traced to the expansion of certain triplet base sequences within their genes (see Table 7.1). These **dynamic mutations** show great variability in phenotype and age of onset. They also provide examples of **anticipation**, in which the severity of a disorder increases and its age of onset decreases from one generation to the next.

Decades ago, it was noted that males outnumber females by at least 25% among the mentally retarded at institutions. In 1969, a secondary constriction was found on the long arm of the X chromosome of four retarded males in one family. Since this cytological defect was associated with chromosome breakage, it was called a *fragile site* (see Figure 2.4), and the related X-linked mental retardation was dubbed **fragile X syndrome (FXS)**. Not until the late 1970s was FXS recognized as the most common form of inherited mental retardation.* Recent research, using DNA tests, suggests that the prevalence of FXS is about 1 in 4,000 to 5,000 males (Sherman 1996; Turner et al. 1996).

Most FXS males are moderately to severely retarded; yet some are only mildly retarded, and a few

---

*Down syndrome, although several times more common than FXS, occurs randomly and usually does not run in families (Chapter 13). Actually, "the map of the X chromosome is now littered, from one telomere to the other, with genes for mental handicap, alone or in combination with other features," and *nonspecific* X-linked mental retardation (i.e., reduced learning capacity alone) is much more common than FXS (Turner 1996). But such phenotypes are very difficult to diagnose and analyze.

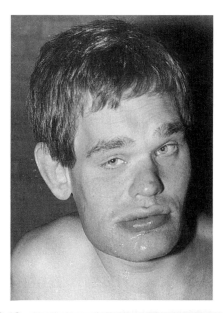

Figure 10.12 Fragile X syndrome in a 26-year-old male. Note the coarse features and the big head with relatively large forehead, jaw, and ears. Not all affected males show such obvious characteristics, however. (From De Boulle et al. 1993.)

males with this fragile site may show normal intelligence. Most affected adult males also have some of the following traits (Figure 10.12): a long face with a big jaw, prominent ears, a high-arched palate, large testicles, and flat feet. Lax joints and a heart valve defect may occur too. Affected males tend to be tall as children but short as adults. Their speech is usually delayed in development and often high-pitched and repetitive. Young males may be hyperactive and irritable. "Autistic-like features, such as poor eye contact, hand flapping, and hand biting, are often seen by 2 to 5 years of age" (Hagerman 1996). FXS males are fertile, and a few have produced offspring (see question 14).

Recent studies based on DNA tests suggest a prevalence of FXS in females of about 1 in 8,300. These affected females are heterozygous, carrying the fragile site on one of their X chromosomes, and represent 50–70% of all full-mutation (Chapter 7) carrier females. They exhibit a wide range of phenotypes, from mild learning disabilities with no obvious physical features to severe retardation plus most of the physical traits seen in affected males.

Genetics of the Fragile X Syndrome

Long before the unusual molecular nature of the *FXS* gene was discovered, geneticists noted that something weird happens to the expression of fragile X syndrome across generations. In trying to make sense of it, they assumed the following: There is a *familial mental retardation locus,* now called **FMR1**, located in or very close to the fragile X site (called **FRAX-A**) at band Xq27.3. Its abnormal allele is expressed cytologically as the

fragile site and phenotypically as the fragile X syndrome. Because carrier females produce 50% offspring who possess this allele and 50% who do not, the segregation of *FMR1* is normal. But unlike all other X-linked genes, it shows a high degree of nonexpression (nonpenetrance) in males, who should express all their hemizygous genes. Even more surprising, affected offspring tend to cluster in certain generations.

As shown in Figure 10.13, the inheritance of fragile X syndrome in the general population mainly involves normal carrier females and **transmitter males**. The latter have the abnormal *FMR1* allele but do not express it: They are phenotypically unaffected and do not show the fragile X site in any of their cells. Likewise, their *carrier daughters* show no retardation and few or no fragile sites. The key fact of this pedigree is that in generation II, 41% of the males are transmitters and only 9% are mentally retarded, whereas in generation IV, 13% of the males are transmitters and 37% are mentally retarded. So among males who supposedly inherit the same allele, four times as many are affected in generation IV as in generation II. (Although not shown here, a similar difference is seen among heterozygous females: About 3.5 times as many are mentally subnormal in generation IV as in generation II.) Indeed, depending on the generational position of

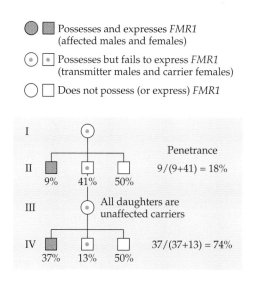

Figure 10.13 Inheritance and penetrance of fragile X syndrome, with emphasis on male members of a pedigree. Compare the frequencies of retarded males and transmitter males in generation II with those in generation IV. Note that the generation I and III carrier females should possess exactly the same fragile X allele, yet they give rise to greatly differing percentages of affected male progeny. (Note: If females had been included in this pedigree, their frequencies would be: in generation II, 5% affected, 45% unaffected carriers, and 50% unaffected noncarriers; in generation IV, 17% affected, 33% unaffected carriers, and 50% unaffected noncarriers.)

BOX 10A   *A Pedigree of Fragile X Expansion*

Let us consider inheritance of the *FMR1* allele—from premutation through full mutation—in a family with fragile X syndrome, whose pedigree is shown in the accompanying figure.

In generation I, an unaffected transmitting male with 80 repeats in his *FMR1* allele is married to a female whose two X chromosomes have normal *FMR1* alleles. In generation II, their two daughters have each received their father's premutation allele, but one is reduced to 75 repeats and the other is increased to 83 repeats.

In generation III are these daughters' offspring. The older daughter's four children include one homozygous normal female (with 30 repeats on each X), two normal carrier daughters (each with an increased number of premutation repeats), and an affected son (with a greatly expanded allele containing at least 580 repeats). The younger daughter (II-3) has an unaffected transmitter son whose *FMR1* allele is slightly reduced and an unaffected carrier daughter whose *FMR1* allele is expanded to 250 repeats.

In generation IV, one carrier female's children include a homozygous normal female and a mildly affected daughter who carries an 800-repeat full mutation. The three children of the other premutation carrier include a homozygous normal female, a retarded daughter with a 500-repeat full mutation, and a retarded son with 300 to 1,000 repeats.

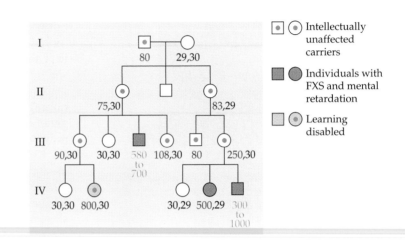

Intellectually unaffected carriers

Individuals with FXS and mental retardation

Learning disabled

The numbers beneath each symbol (two for females, one for males) indicate the number of repeats in that person's X chromosome(s). The numbers in red designate the alleles descended from the original premutation allele. Note that there are many different outcomes as it passes through several generations. (Adapted from Cronister 1996.)

FMR1 in a pedigree, its *penetrance* ranges from 18% to 74% in sons of carrier females and from 10% to 34% in daughters of carrier females.

Molecular Biology of Fragile X

A partial explanation of the unusual inheritance pattern of fragile X came when the fragile X site was isolated and mapped in 1991 (Chapter 7). Scientists were amazed to find that the locus varied in length, not only among fragile X males, but also among other members of an affected family. Even within an affected person, different cells can vary in allele size. This instability was traced to a *sequence of three nucleotide bases, CGG, that are repeated a variable number of times.* People with a normal allele have about 6 to 55 (most commonly 30) copies of the triplet. Unaffected carriers of the intermediate-sized *FMR1* allele (called a *premutation*) usually have about 55 to 200 copies. And individuals with the fragile X syndrome (and the *full mutation*) usually have over 200 copies. Recall also from Chapter 7 that (1) the interspersion of a few AGG triplets in the normal and premutation alleles acts as a stabilizer, preventing great increases in size, and (2) the methylation of cytosines in control regions of the full mutation prevents its transcription.

Although premutation carriers are unaffected, their children and grandchildren are at high risk of being affected. When transmitted through males, the gene's copy number stays constant or decreases. Only with *transmission through females* do the increases in triplet copy number (first from normal to premutation, then to full mutation) occur. The exact mechanism for these changes is not known. But in succeeding generations, this gene somehow grows longer, and its abnormal phenotype becomes more severe (Box 10A).

How the full mutation leads to retardation is not understood. The normal gene product, called *FMR protein*, is expressed in many tissues but predominantly in the brain and testes. It acts as an mRNA-binding protein in ribosomes, but its exact role is unknown. What is clear is that the absence of normal FMR protein causes the fragile X syndrome, for which no cure or highly effective treatment exists.

The *FMR1* gene is highly conserved, showing 97% homology between humans and mice and even occurring in certain worms and yeast. Such findings suggest that the normal allele must have an important cellular function. An animal model for FXS has recently been created by experimentally knocking out the *FMR1* gene in mice and observing the phenotypes that result

(Chapter 18). Affected mice have oversized testes, are hyperactive, and exhibit learning deficits, all of which makes them promising subjects for further studies of FXS.

Before the discovery of the CGG amplification, counseling of fragile X families based on cytological findings was very difficult. Even now, with direct diagnosis by DNA analysis, the forecasts are not always clear-cut. Male fetuses with over 230 repeats will be retarded, and males and carrier females with fewer than 167 repeats will probably be mentally normal. But males with about 200 repeats are borderline cases, and their phenotypes cannot be predicted with absolute certainty. Similarly, carrier females with over 200 repeats have at least a 50% risk of being mentally retarded (Nussbaum and Ledbetter 1995).

Also complicating the picture in both sexes is the occurrence of *mosaicism*. In perhaps 20% of FXS males, some cells have methylated full mutations but other cells have unmethylated, functional premutations. Phenotypes may vary according to the proportions of each cell type, some males even attaining the low end of the normal IQ range. A less common kind of mosaicism is a mixture of cells that all have the full mutation, but some are unmethylated. Such individuals may be "high functioning" FXS males who produce some FMR protein and are only mildly retarded (Sutherland and Mulley 1997). In carrier females, of course, lyonization will randomly give rise to clones of cells in which either the normal or the FXS allele is inactivated. Here, too, the relative proportions of each type of cell will affect the phenotype.

Also on the X chromosome, the recently identified *FMR2* locus lies next to the *FRAX-E* fragile site and is associated with the rare and relatively mild FRAX-E retardation syndrome.* Other examples of dynamic mutations—including the genes for Huntington disease and myotonic dystrophy—are discussed in Chapter 7.

Genomic Imprinting

Gregor Mendel reported that the outcomes of *reciprocal crosses* were identical. That is, the phenotypes and phenotypic ratios seen among the progeny were the same whether a particular trait was inherited through the male or the female parent. For almost a century, experiments with a wide variety of plants and animals have verified this concept—with the exception of reciprocal crosses involving sex-linked genes. In fact, some invertebrates (such as male bees and wasps) regularly develop by *parthenogenesis*—that is, from an unfertilized egg—and thus have maternal chromosomes only. And while there has never been a scientifically authenticated "virgin birth" occurring naturally among mammals, biologists have always assumed that—as with garden peas and parthenogenetically developed animals—the mammalian genome normally inherited from a female parent is functionally equivalent to that inherited from a male.

Mammals Need Both a Mom and a Dad

In the 1980s, however, researchers were surprised to discover that the two parental genomes are *not* always functionally equivalent in mammals. In mouse zygotes, the two pronuclei (one male, one female) look somewhat different from each other and can be individually sucked out of the cell. By removing a pronucleus from one mouse zygote and putting it into another one (from which a pronucleus has also been removed), researchers can produce a zygote containing two female pronuclei or two male pronuclei. But early embryonic development from such zygotes is abnormal. The two pronuclei come together and form the usual diploid number of chromosomes, just as normal fertilization would. Yet viable offspring are never produced. Purely female-derived embryos have poorly developed extraembryonic membranes, and purely male-derived embryos show highly retarded development of the embryo proper (Sapienza 1990). Thus, mammalian offspring need both a mom and a dad for normal growth and development.

In human zygotes, the same situation apparently prevails. An occasional pregnancy produces, instead of a normally developing fetus, a uterine growth called a *hydatiform mole*. Such placental masses contain extraembryonic membranes but no fetal tissue, and their chromosomal composition is entirely paternal—presumably the result of the fertilization of an egg without a pronucleus and the subsequent doubling of the sperm chromosomes. Conversely, *ovarian teratomas*, which are embryonic tumors that lack placental tissue, are derived from nonovulated oocytes and have two maternal sets of chromosomes. Human *triploids* possess three complete chromosome sets (Chapter 14) and rarely survive past early pregnancy. Triploid early abortuses with two sets of paternal chromosomes and one set of maternal chromosomes usually have a large placenta and some molelike features, whereas triploid abortuses with two sets of maternal chromosomes and one set of paternal chromosomes have an underdeveloped placenta.

Do mammalian zygotes need to have an *entire* genome from each parent, or will certain critical chromosomes, or chromosome parts, or groups of genes, or even single genes do the trick? Experiments with some mouse embryos that possess rearranged chromosome parts (Chapter 14) show that certain genes or groups of

---

*Two other fragile sites exist near the tip of Xg, and about two dozen rare heritable fragile sites are on other chromosomes, but none of these seem to be associated with a particular phenotype.

genes are active when contributed by one parent but not active when contributed by the other parent.

This phenomenon, whereby progeny phenotypes differ according to whether a particular gene (or group of genes) came from the mother or the father, is called **genomic imprinting**. It can be considered a form of *silencing* or temporary gene inactivation that makes a locus functionally haploid. The somewhat confusing terminology is as follows: An *imprinted allele* is one whose expression is changed or silenced when it passes through a particular sex. Somehow, the DNA in this allele gets marked or tagged, altering its genetic information. An allele is said to be *paternally imprinted* if it is not expressed when inherited from the father; it is *maternally imprinted* if it is not expressed when inherited from the mother (Peterson and Sapienza 1993).

What percentage of mammalian genes might be subject to imprinting? Estimates vary widely—from a fraction of a percent to as much as 10–20% of the genome—but so far, fewer than 20 imprinted genes have been identified in mice and humans (Barlow 1995). Interestingly, they seem to be clustered in certain regions of a few chromosomes (11 and 15 in humans) rather than being randomly distributed. Unlike nonimprinted genes, they are rich in GC nucleotide pairs and short repeated sequences of 25 to 75 base pairs; they also have few and small introns.

One important mouse gene encodes insulin-like growth factor II (IGF-II), and is transmitted as an autosomal recessive. The gene is expressed when inherited from a male mouse but "silent" when inherited from a female. IGF-II is necessary for normal embryonic growth. When males with a mutated form of this gene are mated to normal females, the heterozygous offspring are viable, fertile, and normally proportioned—but only 60% of normal size. The reciprocal cross, however, does not give rise to any of these proportionate dwarfs; instead, all progeny are phenotypically normal. In humans, the counterpart gene called *IGF2* is inherited on chromosome 11 and also maternally imprinted (see later).

How and Why Does Imprinting Occur?

Nobody knows for sure. But a strong possibility is that during gamete formation in mammals, some genes are differentially altered or "tagged" by the attachment of methyl groups to certain cytosine groups in DNA. This kind of process is also involved in X chromosome inactivation, which occurs in the cells of all female mammals, and also in the fragile X syndrome. But it may not be the only answer to the riddle of imprinting. Some changes in the degree of chromosome compaction, for example, have also been noted in association with gene activity and imprinting. Timing of replication is a third possible factor: For all known imprinted genes, the paternal alleles usually replicate

earlier in S phase than do the maternal ones. (With unimprinted genes, the two alleles replicate at the same time.) Finally, there may be modifier genes that affect the establishment and maintenance of imprinting processes in various tissues at various times.

Whatever its nature, the imprinting procedure would have to be *erased* some time during the embryo's development, so that each individual could *reimprint* its genes according to its own sex during gametogenesis (Figure 10.14). Recent studies show that zygotic DNA is methylated, but demethylation in embryonic cells occurs during early cleavage divisions. Shortly after implantation in the uterine wall, the embryo's somatic cells are methylated anew; germ cells get methylated later, as they develop in gonads. Recent evidence suggests that an *imprinting center* on chromosome 15q has a key role in resetting the imprint during gamete formation.

Complicating matters even more is the fact that imprinting may not be completely stable: In some individuals or circumstances, and in certain somatic tissues or at certain times, an imprinted allele may get reactivated, resulting in the expression of *both* alleles rather than just one allele at that locus. Studies of imprinting in certain rare cancers raise further complications: (1) Different genomes may be imprinted in different ways; (2) within an individual, some target cells may not be imprinted; and (3) imprinting may be controlled by *imprinting genes*, and polymorphisms among the alleles of these genes may cause differences in the imprinting of target genes (Sapienza and Hall 1995).

Imprinting occurs in mammals but in no other vertebrates. How did it come about, and why has it survived? Theories abound, many revolving around developmental strategies (Jaenisch 1997). One interesting theory is that some imprinted genes are just *innocent bystanders* in a process that was originally targeted to the inactivation of transposon DNA (Chapter 6), which comprises over 35% of human DNA and can cause abnormalities in the structure and expression of genes. Transposons, which are very rich in CG dinucleotides, can be thought of as genomic parasites: "We suggest that cytosine methylation of the type found in mammals is part of a nuclear host-defense system that evolved primarily to counter the threats posed by parasitic sequence elements" (Yoder et al. 1997). Although evidence can be marshaled for various hypotheses about imprinting, no hypothesis has actually been proved.

Imprinting and Human Disorders

In a review of imprinting, Hall (1990) lists about a dozen human diseases suspected on a clinical basis to be imprinted and whose genes reside in chromosomal segments homologous to known imprinted regions of mouse chromosomes. She also lists about 100 addi-

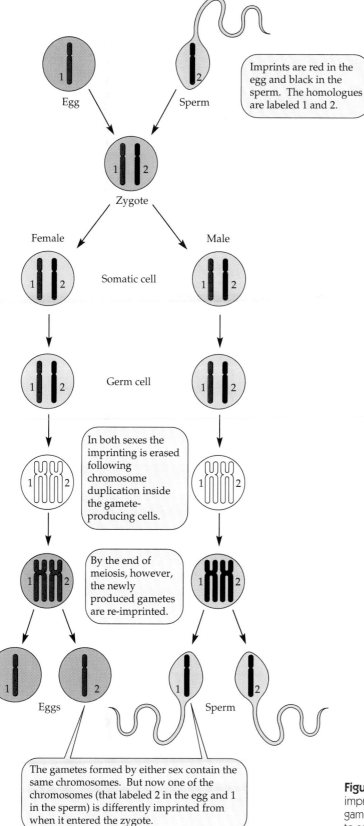

Imprints are red in the egg and black in the sperm. The homologues are labeled 1 and 2.

Egg

Sperm

Zygote

Female

Male

Somatic cell

Germ cell

In both sexes the imprinting is erased following chromosome duplication inside the gamete-producing cells.

By the end of meiosis, however, the newly produced gametes are re-imprinted.

Eggs

Sperm

The gametes formed by either sex contain the same chromosomes. But now one of the chromosomes (that labeled 2 in the egg and 1 in the sperm) is differently imprinted from when it entered the zygote.

tional diseases that should be considered candidates for imprinting because their genes likewise reside in these chromosomal regions homologous to imprinted segments in mice.

There is some evidence for imprinting in a few dozen human disorders. The age of onset or severity or segregation ratios of these conditions seem to differ according to which parent contributed the gene (Reik 1989). *Huntington disease (HD)* (Chapter 7) provides the best example. Significantly more persons with juvenile- or adolescent-onset HD inherit the gene from an affected father than from an affected mother. And more cases of late-onset HD (after age 50) can be traced to an affected mother than to an affected father. Conversely, early-onset *myotonic dystrophy* (Chapter 7) and *fragile X syndrome* are more serious and more frequent in the offspring of affected mothers.

The most astonishing examples of imprinting in humans involve two rare disorders whose physical and genetic features in some ways present a mirror image. Both involve growth and behavioral defects. Newborns with *Prader-Willi syndrome (PWS)* have poor muscle tone and feeding problems. As children, however, they become compulsive overeaters and obese. They are mildly to moderately retarded and short in stature. They also have small hands and feet, tiny external genitals and underactive gonads, and some characteristic facial features (Figure 10.15A). The *Angelman syndrome (AS)* is quite different (Figure 10.15B). Affected children tend to be hyperactive, with movements that are jerky, repetitive, and puppetlike. Despite severe mental and motor retardation and absence of speech, they seem happy and laugh excessively. They also show abnormal brain waves and suffer from seizures. They have an abnormally small head with a big jaw and a large mouth with a prominent tongue.

Figure 10.14 The processes of imprinting, erasure, and re-imprinting presumed to occur between two generations of gametes. As genes pass from one sex to another (i.e., mother to son or father to daughter), the imprints will change even though the alleles remain the same. Consider what happens to two homologous autosomes if a particular zygote is female (left column); now compare that to what happens if the zygote is male (right column). (Adapted from Hoffman 1991.)

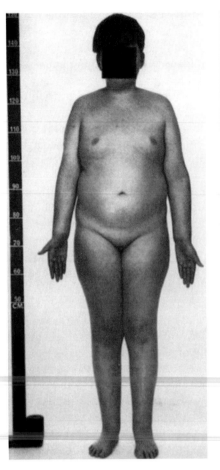

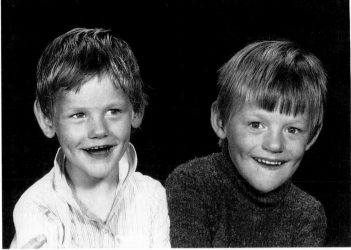

Figure 10.15 Two syndromes that result from differential imprinting of the same region on chromosome 15q. (A) A male with Prader-Willi syndrome has a deletion of this region in the *paternal* chromosome 15. Note the obesity, tiny penis, and small hands and feet. (Photo from Connor and Ferguson-Smith 1993; courtesy of Blackmore Scientific Publications.) (B) Two brothers with Angelman syndrome have a deletion of this same region in the *maternal* chromosome 15. Facial features include a wide mouth, thin upper lip, pointed chin, and prominent jaw. (Photo courtesy of Devlin and Robb et al.)

Both syndromes often involve microdeletions in the q11–13 region of chromosome 15. The two loci are independent but very closely linked, and many of these deletions overlap. What makes them so fascinating, however, is their pattern of transmission. When the deletion is inherited from the father, it results in the Prader-Willi syndrome; but when inherited from the mother, it results in the Angelman syndrome. In other words, the differing phenotypes caused by loss of this one chromosomal region suggest that the normal *PWS* gene is expressed from the paternal chromosome only and the normal *AS* gene is expressed from the maternal chromosome only (Horsthemke et al.1995).

Several candidate genes have been identified for both the Prader-Willi and Angelman syndromes. Up to five possibilities exist for PWS, but their possible roles in PWS are unclear. The situation is more promising for Angelman syndrome: In 1997, several cases of AS were reported to be associated with mutations of the gene *UBE3A*. Its normal enzyme product is thought to involve the degradation of certain proteins during brain development. Complicating this gene's discovery was the fact that cultured cells taken from these Angelman patients showed expression by *both* alleles of this gene rather than just the maternal allele.

But *UBE3A* is imprinted in the brain. "There are a large and growing number of examples of genes for which imprinted expression is either tissue-specific, specific to developmental stage, species-specific, promoter-specific, or partial" (Kishino et al. 1997).

Another condition that shows imprinting is the *Beckwith-Weidemann syndrome* (*BWS*). This rare disorder, of which 85% of cases are sporadic and 15% familial, is characterized by overgrowth and in some cases a predisposition to childhood cancers. Affected individuals are greatly overweight at birth and have a large, protruding tongue. Abdominal wall defects and earlobe creases are common, and some body parts may be larger on one side than on the other. The most frequently occurring cancers affect the kidney (Wilms tumor*) and the liver.

A cluster of genes on chromosome 11p15.5 is associated with BWS (Reik and Maher 1997). The two main candidate genes for BWS are closely linked and oppositely imprinted: *IGF2* (a growth enhancer) is maternally imprinted and paternally expressed, while *H19* (a growth suppressor whose RNA transcript is not

---

*Wilms tumor occurs with several different genetic disorders and is associated with several different loci on chromosome 11.

translated) is paternally imprinted and maternally expressed. Various chromosomal abnormalities and mutations involving either or both loci have been observed in sporadic BWS cases. But how these (and perhaps other) loci interact in both normal and abnormal situations is not yet clear, and several scenarios have been proposed.

We do know that loss of the maternal allele at the *H19* locus plays a role in the development of Wilms tumor. Certain other types of childhood cancers (in BWS and other disorders) also show intriguing patterns of selective loss of either maternal or paternal chromosome segments during the course of their malignant growth. And some of these tumors exhibit patterns of inheritance that suggest imprinting. Recent studies of the relation of the parental origin of chromosomes to cancer development and growth have provided new twists to an extremely complex area of research.

One emerging concept is called **loss of imprinting (LOI)**, a process by which an imprinting pattern gets reversed (Feinberg 1993, 1995). For example, maternal chromosome 11 normally has an inactive *IGF2* locus, an unmethylated *H19* promoter, and an active *H19* locus. Conversely, paternal chromosome 11 normally has an active *IGF2* locus, a methylated *H19* promoter,

and an inactive *H19* locus. With just one active allele at each locus, a balance exists between expression of growth promoter and expression of growth suppressor. But in the cells of many Wilms tumors associated with BWS, the maternal chromosome has reversed its imprinting and methylation pattern. As a result, *both* alleles of the *IGF2* locus produce growth factor, but *neither* allele at the *H19* locus produces growth-suppressing RNA! The resulting imbalance apparently predisposes these cells to tumor development (Issa and Baylin 1996).

Inheritance Patterns of Imprinted Alleles

What do pedigrees of imprinted genes look like? They resemble standard inheritance patterns, but with phenotypic expression among children varying according to the sex of the transmitting parent. Consider the hypothetical case of a rare dominant allele, called *D*, that is imprintable. Figure 10.16 shows how the expression of this allele would vary from generation to generation, depending on whether it is (A) paternally imprinted or (B) maternally imprinted.

In the two pedigrees, the structures of generations II to IV are identical with regard to the number, sex, and birth order of offspring. They are also arranged so that the sister and brother II-1 and II-2 are both af-

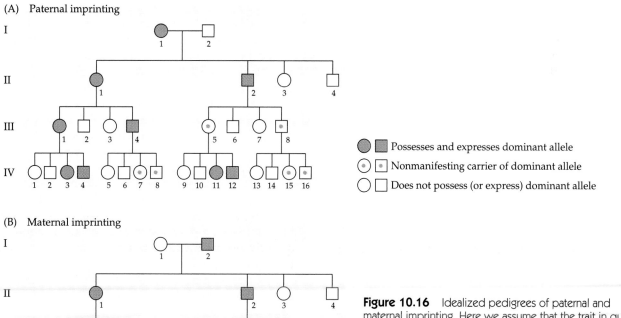

Figure 10.16 Idealized pedigrees of paternal and maternal imprinting. Here we assume that the trait in question is inherited as an autosomal dominant, but with the following conditions. (A) When the dominant allele is paternally imprinted, children are not affected if they receive it from the father. (B) Conversely, when the dominant allele (at a different locus) is maternally imprinted, it is not expressed if inherited from the mother. (Adapted from Hall 1990.)

fected in both pedigrees. But in generation I, the affected, transmitting parent is female in (A) and male in (B). And in generations III and IV, there is an interchange of affected versus nonmanifesting individuals between (A) and (B). For example, III-1 and III-4 are affected in (A) but nonmanifesting carriers in (B). Likewise, III-5 and III-8 are nonmanifesting carriers in (A) but affected in (B). We make the following additional observations:

1. In each generation, we expect roughly equal numbers of affected males and females and roughly equal numbers of nonmanifesting males and females, regardless of which parental sex transmits the imprinted gene.

2. This type of pedigree is easily distinguishable from pedigrees of mitochondrial inheritance. In the latter, only the progeny of females—and never any descendants of males—express the particular trait.

3. Either pedigree (A or B) could be interpreted as a simple autosomal gene with incomplete penetrance. Thus, geneticists need more extended pedigrees to identify with certainty the phenomenon of imprinting.

Other Unexpected Variations in Gene Structure and Function

As the preceding discussions indicate, scientists are beginning to correlate some variations in phenotypic expression of specific alleles with irregularities in DNA structure. Following are a few additional examples.

Genes Within a Gene

In Chapter 7, we described type 1 neurofibromatosis (NF1), one of the most common autosomal dominant diseases. Although it is virtually 100% penetrant, NF1 exhibits both a high mutation rate and extremely variable expressivity (Gutmann and Collins 1995). Doctors are unable to predict how severe and disfiguring it will be in any particular individual, nor can they treat it very effectively. By acting as a *tumor suppressor* (Chapters 7, 17), the *NF1* gene plays an important role in controlling cell division. Interestingly, it also appears that most new *NF1* mutations are of paternal origin. Whether this effect is due to an increased mutation rate in older men, or to imprinting, or to some other phenomenon is not known.

In 1990, researchers isolated the *NF1* gene, which resides on chromosome 17. They determined that the *NF1* gene is huge (about 300,000 base pairs) and actually includes three smaller genes within intron 27. Although such **embedded genes** were known to occur in lower organisms, this was only the second such example found in humans.* Whether these genes play any role in the development or variability of neurofibromatosis is not known. But with regard to the genomes of various species, genes embedded within the introns of other genes are now known to be commonplace (Fields et al. 1994).

Jumping Genes

As we pointed out in Chapters 6 and 7, genes that don't stay put were first discovered and analyzed by corn geneticists in the 1940s. Although most mainstream geneticists accepted the presence of jumping genes in corn, they did not think that such exotic gene behavior occurred in other organisms. That viewpoint changed in the 1960s and 1970s with the discovery of similar genes—called **transposable elements** or *transposons*—in bacteria and fruit flies. We now know that jumping genes are quite widespread in plants and animals.

The first known example of jumping DNA in humans that gives rise to a distinct phenotype involved hemophilia A (Chapter 7). In 1988, researchers analyzed the DNA sequences of two different mutations of the X-linked *clotting factor VIII gene*. They found that in both cases, a piece of a special kind of repeated DNA—called *LINE-1* for long *interspersed elements*—was inserted from a different chromosome into the mutant gene. Both of the new insertions produced an enzyme called *reverse transcriptase*. This enzyme, quite surprisingly, allows RNA to be copied into DNA—the reverse of the usual situation (Chapter 6).

Another example of a jumping gene in humans, reported in 1991, involved the *neurofibromatosis* gene (Chapter 7). As if the presence of embedded genes within this locus were not amazing enough, Francis Collins's research group also discovered that one *NF1* mutation occurred when a small piece of DNA jumped from its usual location into the middle of the *NF1* gene. This DNA insertion was a type of repetitive DNA called *Alu* (Chapter 6) and does not itself produce a protein product. But its insertion into the *NF1* gene led to an abnormal NF1 protein product.

Transposon DNA makes up a surprisingly large fraction of human DNA and often causes abnormal mutant phenotypes when moving around. It is not known whether the persistence of jumping genes in many organisms means that they play some important role in evolution.

---

*Another human example is the X-linked *factor VIII gene*, which causes hemophilia A and which contains a single embedded gene of unknown function.

In Summary

In the beginning of this chapter, we discussed some problems in interpreting observations about traits that run in families but do not exhibit clear-cut phenotypes or patterns of inheritance. We then described some unusual or newly discovered types of gene structure or behavior—including X chromosome inactivation, dynamic mutations, and genomic imprinting—that might help explain some of the old problems. These **epigenetic** mechanisms of gene regulation involve "modifications in gene expression that are brought about by heritable, but potentially reversible, changes in chromatin structure and/or DNA methylation. ... Epigenetic effects usually involve gene silencing" (Henikoff and Matzke 1997). And they are not rare: In mammalian cells, for example, over 50% of the DNA is silenced by various epigenetic processes. Whether future molecular studies will unravel all such mysteries we cannot say. But it is clear that such exceptions to Mendel's standard rules, especially as they apply to human illnesses, will continue to provide a challenge to geneticists and physicians.

Summary

1. A trait that is genetic will run in families and will not spread to unrelated persons in the same circumstances, but these conditions alone may not be enough to establish a genetic hypothesis.

2. Sex-influenced genes are expressed differentially in the two sexes.

3. Variable expressivity of a gene, including variable age of onset, results from variation in environments and in genotypes.

4. Incomplete penetrance of a gene refers to the nonexpression of a characteristic phenotype. This outcome may be caused in some cases by epistasis, the masking of a phenotype by the action of a nonallelic gene, or it may result from epigenetic changes.

5. A pleiotropic gene affects various and apparently unrelated aspects of an individual's phenotype due to a development cascade of reactions.

6. Genetic heterogeneity refers to multiple genetic causes of nearly the same phenotype. A phenocopy is a nongenetic condition that mimics a genetic condition.

7. In traits involving multiple alleles, some phenotypes previously thought to be homozygous are found to be compound heterozygotes for two different mutant alleles. One locus may also give rise to two or more different clinical disorders.

8. Digenic inheritance occurs when affected individuals are double heterozygotes for recessive mutations in two different genes.

9. Indirect methods of data collection produce biases that require special methods of statistical analysis. The deviations of observed data from those expected on the basis of some hypothesis can be used to test the credibility of that hypothesis.

10. X-linked traits are usually expressed equally in both sexes, even though females have twice as many X-linked alleles as males. In mammals, the dosage of X-linked genes is equalized by the inactivation (lyonization) of one X chromosome in every female body cell. Thus, females heterozygous for X-linked genes are phenotypic mosaics.

11. A fragile site on the X chromosome is associated with familial mental retardation (fragile X syndrome). The disorder often increases in severity and degree of penetrance in successive generations of a family, owing to an unstable mutant allele that grows in size.

12. A mammalian zygote does not develop properly unless it contains both a maternal and a paternal genome. In both sexes, some genes become differentially imprinted (silenced) during gamete formation and are expressed differently during development. Thus, in pedigrees of imprinted genes, the offspring phenotype varies according to which parent contributed a particular allele.

13. Epigenetic modifications (usually silencing) of gene expression involve heritable but potentially reversible changes in chromatin structure and/or DNA methylation. Examples include X chromosome inactivation and imprinting.

14. Small genes embedded within the introns of other genes have been discovered in recent years.

15. Jumping genes (transposable elements) are found in many plants and animals. They have no known function, but by inserting their repeated DNA sequences into new locations, they can cause mutations and other damage to the genome.

Key Terms

anticipation	lyonization
ascertainment bias	methylation
Barr body	modifier
compound heterozygote	mosaicism
digenic inheritance	phenocopy
dosage compensation	pleiotropy
dynamic mutations	prion
embedded gene	sex chromatin
epigenetic	sex-influenced inheritance
epistasis	sex-limited trait
fragile X syndrome (FXS)	suppressor mutation
genomic imprinting	transmitter male
inactive X hypothesis	transposable element
incomplete penetrance	variable age of onset
loss of imprinting	variable expressivity

Questions

1. Assume that either the recessive genotype d/d (at the D locus) or the recessive genotype e/e (at the E locus) can produce deafness. If these loci are on different chromosomes, what is the probability of a deaf child from the mating of two unaffected individuals who are double heterozygotes, that is $D/d\ E/e \times D/d\ E/e$?

2. Assume that pattern baldness is an autosomal trait that is dominant in males (bald = B_1/B_1 or B_1/B_2; nonbald = B_2/B_2) but recessive in females (bald = B_1/B_1; nonbald = B_1/B_2 or B_2/B_2). What are the expected genotypes and phenotypes of the children of a bald woman and a nonbald man?

3. Is the pattern of transmission of baldness in Figure 10.2 consistent with the hypothesis of an X-linked recessive? Here, assume that $b/(Y)$ = bald man; $B/(Y)$ = nonbald man; b/b = bald woman; $B/-$ = nonbald woman.

4. In mice, $B/-$ is black and b/b is brown. At another locus, $C/-$ is colored and c/c is colorless (albino). The albino genotype is epistatic to all genotypes at the B locus. What offspring phenotypes, and in what proportions, are expected from two doubly heterozygous black mice—that is, $B/b\ C/c \times B/b\ C/c$?

5. In mice, the recessive allele d (*dilute*) interacts with B genotypes in this way:

   $B/-\ D/-$ = black
   $b/b\ D/-$ = brown
   $B/-\ d/d$ = gray
   $b/b\ d/d$ = light chocolate

   What offspring, and in what proportions, are expected when a triple heterozygote is crossed with a complete homozygote—that is, $B/b\ C/c\ D/d \times b/b\ c/c\ d/d$?

6. In rabbits, the three alleles C, cb, and c give these phenotypes:

   C/C or C/c^b or C/c = full color
   c^b/c^b or c^b/c = Himalayan
   c/c = albino

   What are the genotypes of the parents if a litter of rabbits contains all three phenotypes?

7. In humans, a few families have been reported in which both parents are albino but all their children have normal pigmentation. Some families are also known in which both parents are deaf but all their children have normal hearing. What explanation is reasonable for these cases? Suggest genotypes for the albino parents and their offspring.

8. Consider a case of digenic inheritance in which the double heterozygotes $A/a\ B/b$ are affected with a certain disorder, but all other genotypes involving these two loci show normal phenotypes.

   (a) What fraction of affected offspring is expected from the mating $A/a\ B/b \times A/A\ B/B$?
   (b) What fraction of affected offspring is expected from $A/A\ b/b \times a/a\ B/B$?
   (c) List all eight genotypes that show a normal phenotype.

9. True or false: Every female should have exactly 50% of her cells expressing her maternal X and 50% expressing her paternal X. Explain.

10. Why are phenotypic differences between members of identical female twin pairs likely to be greater than those between identical male twin pairs? (*Note*: Identical twins, being derived from a single zygote, have identical genotypes.)

11. Linder and Gartler (1965) studied certain benign uterine tumors up to several centimeters in diameter that were removed from women who were heterozygous A/B for these two alleles at the $G6PD$ locus. In a given individual, some tumors consisted of all A cells and others of all B cells, but none were mixtures of A and B cells. On the other hand, pieces of adjacent uterine tissue as small as 1 mm in diameter contained both A cells and B cells. What can be concluded about the origin of the uterine tumors?

12. In Figure 10.14, what offspring phenotypes would be expected if a transmitter male in generation IV married a normal, noncarrier woman?

13. Cronister (1996) reports that affected fragile X syndrome males produce sperm containing the premutation, rather than the full mutation present in their somatic tissues. What offspring are expected from a mating between an FXS male and a normal female?

14. A case exists of identical twin females who are heterozygous for the fragile X full mutation; one twin is mentally retarded, while the other exhibits normal intelligence. Offer an explanation.

15. In Figure 10.16, what offspring phenotypes would be expected if IV-4 married IV-9 (his second cousin)? Consider separately the situations of (a) paternal imprinting and (b) maternal imprinting.

Further Reading

Good summaries of genetic heterogeneity in humans appear in Mueller and Cook (1997) and in Beaudet et al. (1989). For more details on data collection and analysis, see Mange and Mange (1990) or any upper-level human genetics text. The entire August 1997 issue of *Trends in Genetics* (Vol. 13, no. 8) is devoted to epigenetics and contains some short articles on gene silencing. Heard et al. (1997), Willard (1995, 1996), Gillis (1994), and Lyon (1992, 1995) discuss X chromosome inactivation. The fragile X syndrome is reviewed by Sutherland and Mulley (1997), Richards and Sutherland (1992), Chapter 14 of Gardner and Sutherland (1996), Hagerman and Cronister (1996), Nussbaum and Ledbetter (1995), and a short book by Dykens et al. (1994). Genomic imprinting is reviewed by Bartolomei and Tilghman (1997), Sapienza (1990), Lalande (1996), Tycko (1994), Hall (1990, 1992), Ohlsson et al. (1995), and Sapienza and Hall (1995).

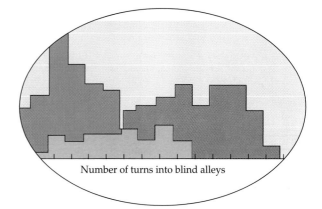
Number of turns into blind alleys

CHAPTER 11

Quantitative and Behavioral Traits

Soon after birth, identical twins Jim Lewis and Jim Springer were adopted into different Ohio families. Thirty-nine years later, they were reunited for the first time as part of a large, carefully designed twin study conducted by Thomas Bouchard and colleagues at the University of Minnesota (Franklin 1989). The twins discovered they had led astonishingly similar lives (Figure 11.1). Like their given names, some of the coincidences were probably the result of blind chance: Each was brought up with an adopted brother named Larry and a family dog named Toy; each married and divorced a woman named Linda and remarried one named Betty; their first sons were named James Allan and James Alan.

Other coincidences could have been influenced by inborn tendencies: In school, both liked math and disliked spelling; both enjoyed carpentry and mechanical drawing; both were trained in law enforcement and were deputy sheriffs in different Ohio

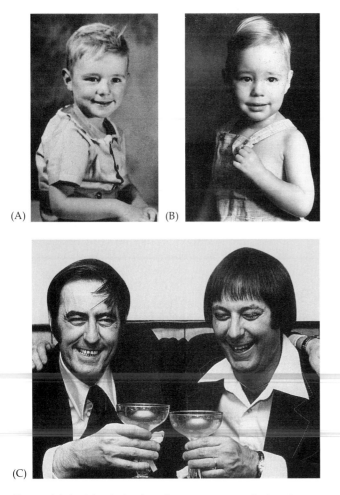

(A) (B)

(C)

Figure 11.1 Identical twins who were separated when they were about 1 month old and brought up in different families. (A) Jim Lewis at age 2. (B) Jim Springer at age 3. (C) The 39-year-old twins at their reunion in 1979. (A and B courtesy of Thomas J. Bouchard, Jr.; C courtesy of Ira Berger/New York Times Pictures.)

towns; both drove similar blue Chevrolets and vacationed at the same stretch of Florida beach; they shared nearly identical patterns of drinking and chain-smoking. The twins thought that these events were weird, even downright spooky (but see Question 1).

Studying similarities and differences in identical twins raised apart focuses on questions related to heredity and environment. Our major concerns here are physical and behavioral characteristics (e.g., stature, personality) that are gene-influenced to some degree. The genetic basis is usually subtle, not following simple Mendelian rules of inheritance. Rather than finding one gene with a major effect, researchers often find many genes with individually small effects.

Some students may find it odd to bring together the concepts of *genetics* and *behavior* at all. Certainly, behavior genetics is a relatively new field of study. This is because early behavioral scientists were uninterested in genetics, and early geneticists could not easily study ill-defined behaviors. But in the 1930s, Robert Tryon, at the University of California, Berkeley, began systematic genetic studies of the behavior of rats. Starting from a single group of animals, Tryon was able to breed one line that made few errors in running a maze and another line that performed poorly (Figure 11.2). The ability to separate out maze-bright or maze-dull rats by selective mating over several generations demonstrated genetic variation for a behavioral trait.

In ensuing decades, behavioral geneticists experimented with many other animals, including fruit flies, bees, birds, mice, cats, dogs, and monkeys. Investigations of human conditions as diverse as normal personality traits and devastating disorders such as Alzheimer disease leave no doubt that some of the behavioral differences among us depend in part on genetic differences. The relevant genes affect hormones, neurotransmitters, and cell receptors, as well as the form and function of sense organs, nerves, and muscles—systems that are also important in many physical characteristics. Of course, environmental factors also influence the same behavioral traits, and their modifying effects may be greater than the effects of genetic factors. We especially want to emphasize that evidence for genetic influence, even strong genetic influence, never rules out the possibility of environmental influence. Any disease (such as phenylketonuria) that is clearly genetic in origin but is nevertheless treatable to a large degree shows the relevance of both genes and environment.

Genetic and Environmental Variation

Complex physical and behavioral traits are rarely "either/or" situations like many of the Mendelian characteristics that we have considered. Rather, they vary *continuously* and are measured by length, weight, time, patterns, color gradations, activity levels, test scores, or some other suitable scale. And when it comes to intelligence or personality, the definitions of the traits themselves may be imprecise and the resulting measurements correspondingly uncertain. In particular, studies of human intelligence have sparked considerable academic, social, and political controversy. Yet these difficulties have not stopped geneticists, anthropologists, psychologists, educators, and others from studying such **quantitative variation**. They have tried to measure how much of the variation depends on people being conceived with different genotypes and how much depends on people living in different environments.

This is the old nature/nurture question, which seeks to partition the underlying causes of observed differences. The question is asked because it deals with interesting or important human traits. Robert Plomin (1990) of Pennsylvania State University, writes:

> Some of society's most pressing problems, such as drug abuse, mental illness, and mental retardation, are behavioral problems. Behavior is also key in health as well as illness, in abilities as well as disabilities, and in the personal pluses of life, such as sense of well-being and the ability to love and work.

A person's phenotype unfolds during development and maturation, when genes and gene products interact with each other and with life's circumstances. The latter consist of all environmental factors both before and after birth and every influence, whether obvious or subtle, that may help make one person different from another. These happenings are sometimes summarized in a cause-and-effect diagram:

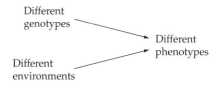

It is not proper to ask whether a trait is "due to genes" or "due to the environment," since a person cannot exist without both. Rather, the nature/nurture question is one of degree. We can ask, for example, how much of the observable phenotypic variation in hair color is due to people having different hair color alleles and how much is due to hair being exposed to different environments. In a sense, we are asking how boldly, relative to each other, the two arrows in the preceding diagram should be drawn.

Because the genes provide the initial guidelines for the development of a new person, this diagram is sometimes described in the following way: The genotype of an individual determines a *range* of possible phenotypes, and within that predetermined range, a specific phenotype is molded by environmental influences. The range determined by genes may be narrow or broad, depending on the trait under consideration. For example, one's ABO blood type is gene determined within a very narrow range. In fact, we usually do not consider environmental influence at all. But researchers have noted that the strength of the blood-typing reactions may change with age, and very rarely, blood type B may be acquired later in life.

At the other extreme, infectious diseases are primarily caused by contact with external agents and are seemingly independent of a person's genotype. Still, we all know people who, through all adversity, never catch a cold. Studies of twins and adoptees reinforce the suggestion that disease resistance has a genetic component, expressed through the manifold cells

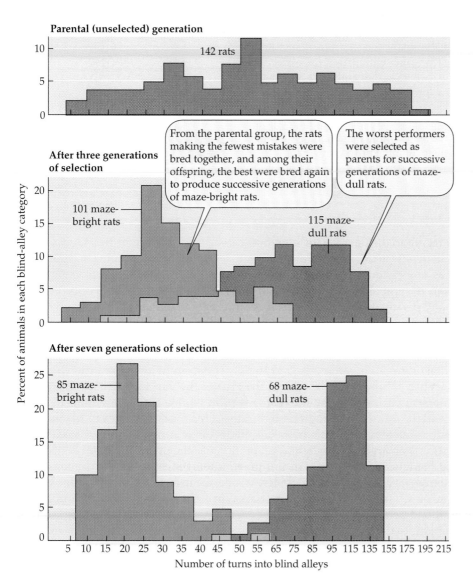

Figure 11.2 The effects of selective breeding on maze learning, represented as histograms (adapted from Tryon 1940). The horizontal scale along the bottom is total number of errors—turns into blind passageways—made by a rat running a maze. For example, an animal that made 8 errors would be included in the first step of the histogram. The gray areas represent overlap in the performance of the two groups.

BOX 11A *Sir Francis Galton*

A precocious child, Francis Galton wrote to his older sister, "I am four years old and I can read any English book … and cast up any sum in addition and can multiply by 2, 3, 4, 5, 6, 7, 8, [9], 10, [11]" (Galton 1909). His letter shows erasures of the 9 and 11, perhaps indicating a young twinge of guilt. He later studied medicine and mathematics, explored southwest Africa for the Royal Geographical Society, and established a widespread network of self-recording weather stations. He was the originator of many important statistical tools, including the ideas of correlation and regression, used today to study biological variability, (Crow 1993). The first to utilize twins to study stature and intelligence, his classic 1875 paper on the subject was entitled "The History of Twins as a Criterion of the Relative Powers of Nature and Nurture." He also began the use of fingerprints by developing a system to catalog and compare them. Curious about everything, but passionate about counting and measuring, Galton wrote in his autobiography (1909):

Many mental processes admit of being roughly measured. For instance, the degree to which people are bored, by counting the number of their fidgets. I not infrequently tried this method at the meetings of the Royal Geographical Society, for even there dull memoirs are occasionally read. … I have often amused myself with noticing the increase in that number as the audience becomes tired. The use of a watch attracts attention, so I reckon time by the number of my breathings, of which there are fifteen in a minute. They are not counted mentally, but are punctuated by pressing with fifteen fingers successively. The counting is reserved for the fidgets.

Galton shared and appreciated Mendel's zeal for counting things, and he felt a sentimental bond because of their common birth year (1822). Galton, however, never recognized the fundamental importance of Mendel's

work (although he was 78 when Mendel's paper was rediscovered). More controversially, he was the founder of *eugenics*, the study of ways to improve the genetic endowment of humankind, to which we return in Chapter 20.

Sir Francis Galton (1822–1911)

and reactions of the immune system. In trying to discover one characteristic that is largely or only affected by environmental variables, Plomin has suggested "niceness," the degree to which a person is sympathetic, trusting, and cooperative. Still, many personality traits are increasingly found to be influenced to lesser or greater degree by genetics. Thus, for no characteristic of interest are differences between people entirely of genetic or entirely of environmental origin.

The analysis of quantitative traits makes use of **biometry**, which is statistics applied to biological variation. Although the computations are beyond the scope of this book, we will describe why the nature/nurture problem is a statistical concept. The question was first seriously addressed by the inventive and many-sided Sir Francis Galton (Box 11A), a contemporary of Gregor Mendel and a half first cousin of Charles Darwin. Galton measured many physical and behavioral characteristics in humans and inquired into their inheritance.

Heritability

A useful way to describe the origins of variability is by the algebraic equation

$$V_P = V_G + V_E$$

This equation can be understood on two levels. Most simply, it says that the total phenotypic variation (V_P) that is observed for a given trait in a given population has two components: variation in the genotypes (V_G) and variation in the environments (V_E). The V's can also be taken to represent a more precise measure called *variance*. This estimate of variability is the one that statisticians usually compute to express how diverse a group is. The more widespread the values in the group, the greater the variance. This is illustrated by the histograms in Figure 11.2. The variance that is represented in these histograms relates to the number of mistakes made by rats in running a maze. The parental generation has the greatest variance, since the rats in this group span the entire range of maze-running ability. Within the selected group of maze-bright rats, the variation in error rate becomes less and less in successive generations. Similarly, the maze-dull rats show progressively less variation. Histograms like these can be a starting point for estimating the separate genetic and environmental components of variation.

The formula $V_P = V_G + V_E$ assumes that genetic and environmental sources of variation are independent of each other. This assumption is not always true, however, because *interactions* may occur. For example, environmental factors reinforce genetic factors whenever artistic or musical youngsters are given special training in art or music. A second example of genetic-environmental interaction involves the effect of cigarettes on death from diseases of the heart and blood vessels. The cardiovascular death rate in smokers who also have a family history of heart attack is much greater than the summed risks of smoking, on the one hand, and family history, on the other. In cases such as these, corrective terms can be included in the variability equation.

A quantity called **heritability in the broad sense (H_b)** expresses the proportion of phenotypic variation that is due to the genotypic differences among members of a population. Mathematically, it is the genetic variance divided by the total phenotypic variance:

$$H_b = V_G / V_P$$

Suppose that all members of a population have the same genotype, as is approximately true in an inbred line of mice. Then $V_G = 0$, and any observed variability in a trait must be attributed to environmental variation; thus, the heritability in this restricted case would be zero. Conversely, suppose that all members of a population grow up under precisely controlled environmental conditions (like plants in a greenhouse), in which case $V_E = 0$ (or near 0). Here, any variability is attributed to genotypic differences, and the heritability would be 100%.

In human populations, it is not usually possible to control either the genotypes or the environments, and values for V_G and V_E must be estimated indirectly. These calculations can be done approximately by measuring the correlations between the phenotypes of relatives—identical twins, fraternal twins, sibling pairs, parents, and offspring. Heritability values for, say, height and fingerprint patterns are in the range of 80–90%, while heritabilities for weight, blood pressure, and IQ scores are in the range of 40–60%. Although the heritability of a particular trait is a rough measure of the genetic differences between people compared with environmental differences, we qualify this statement later in the chapter.

Another quantity, **heritability in the narrow sense (H_n)** is the proportion of variation in a population that is due to just a *part* of the genetic variation—the so-called additive part, symbolized V_A. Because V_A is less than (or possibly equal to) V_G, H_n is less than (or equal to) H_b. The narrow-sense heritability provides animal breeders with the best prediction of offspring phenotypes from knowledge of parental phenotypes. For example, breeders can more easily alter through selective mating the butterfat percentage of cow's milk (H_n = 60% for this trait) than they can increase the total yield of milk (H_n = 30%). To the extent that a desired phenotype is due to a favorable environment or to the

extent that the effect of a desirable or undesirable allele is masked by genetic dominance or epistasis, artificial selection can make no predictable headway.

The Genetic Component of Variation

The analysis of quantitative traits began during the first decade of the twentieth century, when the Danish botanist Wilhelm Johannsen calculated the separate contributions of genes and environment to the weight of seeds. (He also coined the terms *gene, genotype,* and *phenotype*.) Later, investigators in Sweden and in the United States showed that Mendelian rules were adequate to explain continuous variation in seed color and in the dimensions of flowers. The explanations supposed that several genes affected the trait in question, each by a small amount. This type of heredity was called **multiple gene inheritance** (or *multifactorial* or *polygenic* inheritance). Modern nomenclature favors **quantitative trait loci (QTL)**. Note that the idea of multiple *genes* is quite different from that of multiple *alleles*. The former refers to many genes affecting one phenotypic characteristic, the latter to many alleles of one gene. We will now present a polygenic model for the inheritance of any quantitative trait, ignoring, for the moment, environmental influence.

An Additive Model

Consider the variability in a quantitative trait that can be traced to genetic differences, and assume the following:

   1. The trait is affected by, say, three genes (*G, H, I*), each with two alleles: *G, g; H, h; I, i.*
   2. Each allele indicated by an uppercase letter acts similarly, the combined effects simply adding together.

Suppose, for example, that the genotype *g/g h/h i/i* has the phenotype 40 units (in combination with a standard environment and a constant genetic background at other loci). Let each uppercase allele add 5 units above this. Thus, each of the seven phenotypes is specified by noting the number of uppercase alleles (Table 11.1). For example, a genotype with two uppercase alleles has the phenotype 50 units, regardless of whether the uppercase alleles are both at the same locus (e.g., *G/G h/h i/i*) or at different loci (e.g., *G/g H/h i/i*). Altogether there are 27 possible genotypes generated by combining any genotype at *G* (3) with any at *H* (3) with any at *I* (3).

Using either the gamete-by-gamete (checkerboard) method or the gene-by-gene method for solving genetic problems, we can use the information in Table 11.1 to predict offspring phenotypes from any set of parental genotypes. For example, we can ask what

TABLE 11.1 An additive model of polygenic inheritance involving three genes

Number of uppercase alleles	Phenotype (arbitrary units)	Representative genotype	Number of different genotypes
0	40	g/g h/h i/i	1
1	45	G/g h/h i/i	3
2	50	G/G h/h i/i	6
3	55	G/g H/h I/i	7
4	60	G/G H/H i/i	6
5	65	G/G H/H I/i	3
6	70	G/G H/H I/I	1
		Total	27

phenotypes, and in what proportions, are expected among the progeny of two triple heterozygotes, *G/g H/h I/i*. Since the same offspring phenotype may result from several different genotypes, some simplifications are possible. For example, a triple heterozygote can make eight types of gametes, but we need not resort to an 8 × 8 checkerboard. The gametes have either zero, one, two, or three uppercase alleles, so a 4 × 4 checkerboard will do. The fractions of the gametes are worked out in Figure 11.3, and the relevant checkerboard is set up in Figure 11.4.

You may also wish to work out offspring distributions for some of the other matings involving three genes with additive alleles. You will discover that the average of the offspring phenotypes is always the same as the average of the two parents. This is characteristic of genetic situations based on additive alleles

and provides important information to breeders. To the extent that V_P is due to V_A, they can be assured of producing, on the average, offspring as superior as their hand-picked parents.

The additive model of gene action underlying quantitative traits may involve any number of genes. A more general formula would involve more than two alleles for quantitative trait loci, with perhaps one allele dominant to the others. There are further refinements that could be added to the model: gene interactions, such as epistasis, or gene linkage, or major and minor genes making different contributions to the phenotype.

If we assume that a trait is also influenced by environmental factors, then the histogram in Figure 11.4B would become smoothed out, and intervening phenotypic values (such as 46 to 49 units) would be filled up. The distribution would be broadened at the left and right by individuals with all lowercase or all uppercase alleles living in correspondingly extreme environments. Thus, as few as three additive genes, accompanied by environmental variations, can account for quite a range of phenotypes among the offspring from a particular mating. Even more variation would be expected among the members of a large population.

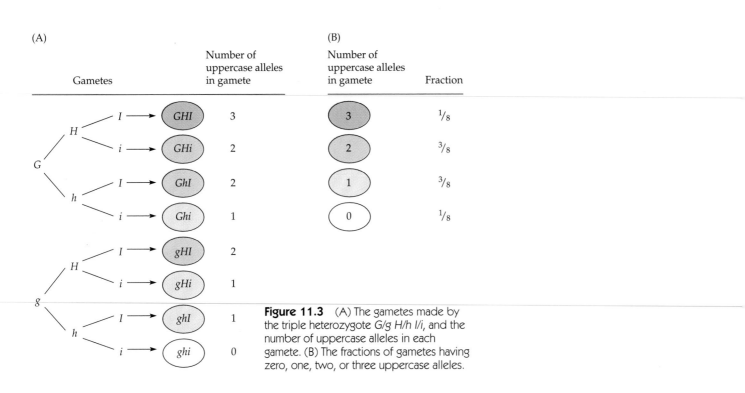

Figure 11.3 (A) The gametes made by the triple heterozygote *G/g H/h I/i*, and the number of uppercase alleles in each gamete. (B) The fractions of gametes having zero, one, two, or three uppercase alleles.

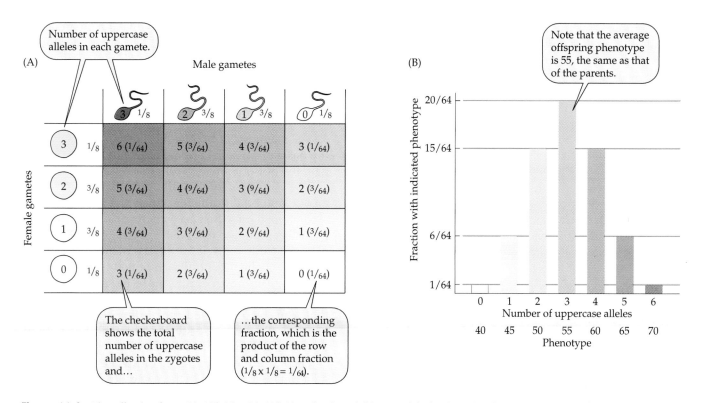

Figure 11.4 The offspring from *G/g H/h I/i × G/g H/h I/i* under the additive model of polygenic inheritance. (A) Checkerboard. The total number of uppercase alleles in the zygotes and the corresponding fractions (product of row and column fractions) are given within the body of the checkerboard. (B) Summary of expected phenotypes. Zygotes with the same number of uppercase alleles are combined; the corresponding phenotypic value is obtained from Table 11.1.

Skin Color

The degree of skin darkness in people seems to have evolved in response to varying environmental conditions, thereby producing populations that are fairly homozygous for different genes controlling the rate of production of skin pigments called melanins (Chapter 15). The mechanism of action of the control genes is not well understood, nor is the nature of the environmental agents responsible for selection. One suggestion, however, involves the ability of ultraviolet light to convert precursor substances into vitamin D under the surface layers of the skin. Lighter surface layers are conducive to greater vitamin D synthesis, compensating peoples living in far northern or far southern latitudes for weaker sunlight. Since vitamin D is needed for bone formation, and its deficiency leads to rickets, strong selective pressures could exist if dietary vitamin D were restricted.

The American geneticist Charles B. Davenport studied skin color inheritance in the families of black-white matings in Louisiana, Bermuda, and Jamaica. He suggested that *two* genes with additive alleles were involved (Table 11.2). His measurements of skin darkness were obtained by matching a skin patch to a continuous color scale, but his cut-off points between phenotypic classes were necessarily arbitrary. (You will also recognize that "white" skin is not the white of this page, nor is "black" skin the color of this ink.) Nevertheless, Davenport was able to show generally good agreement between his observations and the predictions of a simple two-gene model.

More recent studies have used a spectrophotometer to measure the percentage of light reflected from skin:

TABLE 11.2 Davenport's hypothesis for skin color variation in black-white crosses

Number of uppercase alleles	Phenotype		Possible genotypes
	Percentage black	Shorthand description	
0	0–11	White	*a/a b/b*
1	12–25	Light	*A/a b/b, a/a B/b*
2	26–40	Medium	*A/a B/b, A/A b/b, a/a B/B*
3	41–55	Dark	*A/A B/b, A/a B/B*
4	56–78	Black	*A/A B/B*

The darker the skin, the less light reflected. In one study, the analyses of averages and variances in a British population suggested that more than two genes were involved, probably three or four. The investigators were able to roughly calculate the components of variance: V_E was about 35% of the total variance in skin color, V_G (also V_A in this case) was about 65%. Thus, in this population, in either the broad or narrow sense, the heritability of skin color was about 65%.

In summary, in the simplest cases, the genetic contribution to variation in continuous traits may be explained on the basis of several QTLs, each following Mendelian rules and each having a small, additive, phenotypic effect.

Twins in Genetic Research

Twins occur in a little over 1% of pregnancies that do not involve fertility drugs. They are always a source of family interest and general curiosity. More often than with single births, they also cause pregnancy complications for their mothers and are subject to health-related problems as newborns (often premature) and infants. Because identical twins share the same genotype, they give human geneticists just a little of what a highly inbred line gives to animal or plant geneticists. Human twin data have been especially important in the study of the nature/nurture question.

The Biology of Twinning

There are two types of twins: identical, or **monozygotic (MZ)**, and fraternal, or **dizygotic (DZ)**. Monozygotic twins derive from a single zygote (one egg fertilized by one sperm) that divides into two separate cell masses within the first two weeks of development. MZ twins are the same sex, their genotypes being identical except for possible somatic mutation (as is thought to happen with some regularity in the development of the cells of the immune system; Chapter 18). *Conjoined* twins may arise in those very rare instances when the cell masses remain partially joined.*

Dizygotic twins result from two zygotes (two eggs separately fertilized). DZ twins are like-sexed about half the time, and they have only half their alleles identical, on the average—the same as for sibs born at different times. Although DZ twins or sib pairs could share *all* their genes or *none* of their genes, depending on the vagaries of chromosomal segregation and crossing over, these extreme events are virtually impossible. We all know sib pairs, nonetheless, who are very similar or very different in some aspects of their phenotype.

The diagnosis of MZ versus like-sexed DZ twins on the basis of physical appearance and mannerisms is usually, but not always, reliable. Visual appearance can be supplemented with information on blood groups and other traits for which one or both parents are heterozygous. Just *one* genetic difference between twins is enough to establish that they are DZ. When no genetic difference is found, the twins may be either MZ or DZ. The more traits examined without finding a difference, however, the more likely it is that the twins are MZ. Since there are so many polymorphisms that can be examined, including DNA markers (Chapter 9), monozygosity can be established with little doubt, if deemed necessary.

The pattern of occurrence of MZ versus DZ twinning can be contrasted in three ways (Bulmer 1970): by race, by maternal age, and by inheritance patterns.

Racial Variation. The frequency of MZ twinning is remarkably constant throughout the world, with about 4 MZ twin pairs per 1,000 pregnancies. On the other hand, the frequency of DZ twinning varies widely (Table 11.3), being greatest among African groups and least among Asians. This suggests that MZ twinning results from a random accident of early development equally likely to occur in any embryo. DZ twinning, however, is influenced by whatever environ-

*Conjoined twins are often called *Siamese twins* after a famous exhibition pair, Chang and Eng, who were joined in the lower chest region by a tough, flexible ligament 3–4 inches thick and extending 5–6 inches between them. Born in Siam (Thailand) in 1811, the twins were bright, resourceful, and wry. At one New York performance, they refunded half the admission fee to a one-eyed man because they said he could see only half of what other viewers could. They eventually settled in North Carolina as farmers, married sisters, fathered a total of 21 children, and died within hours of each other at age 62 (Wallace and Wallace 1978).

TABLE 11.3 Approximate twinning rates in different populations, ranked by DZ rate

Population	Twin pairs per 1,000 pregnancies[a]	
	MZ	DZ
Blacks		
Nigerians	5	40
South African blacks	5	22
U.S. blacks	4	12
Caucasians		
Italians	4	9
Swedes	3	9
U.S. whites	4	7
Asians		
Koreans	5	6
Chinese	5	3
Japanese	5	3

[a]Data are average values primarily from tables in Bulmer, 1970.

mental and genetic differences exist between different racial groupings.

Maternal Age Variation. Nearly constant MZ rates but quite variable DZ rates are also observed as a function of maternal age. In the Italian population depicted in Figure 11.5, mothers between 30 and 40 years have a higher rate of DZ twinning than those who are older or younger. It is evident that to produce DZ twins, mothers must ovulate two eggs in the same menstrual cycle. Ovulation is under the control of gonadotropic hormones secreted by the pituitary gland. Gonadotropin levels increase from adolescence through the entire reproductive period, paralleling the increase in DZ twin rates up to about age 37. The sharp drop after this age seems to be due to a general failing of ovarian function as menopause approaches.

A further indication of the pivotal role of gonadotropins on reproduction comes from women with certain infertility problems. Treatment with gonadotropins (or some other drugs) increases the number of eggs that are ovulated and that could be fertilized during any one menstrual cycle. Especially striking is the high frequency (10–40% in different studies) of twins, triplets, quadruplets, quintuplets, and even higher multiplicities that result from the use of various ovulation stimulants.

Inheritance Patterns. MZ twinning shows no tendency to recur in the same mother or to run in families. A woman with one set of DZ twins, however, is somewhat more likely to have another set of DZ twins than a "nontwinning" mother. Furthermore, the twinning rate among the close female relatives of a DZ-producing mother is slightly increased, suggesting a small genetic component of DZ twinning, but not of MZ twinning (Figure 11.6). The nature of the genes involved in DZ twinning is unknown, but they may act by increasing the level of gonadotropins. The genes are expressed only in females but can be transmitted by either sex.

Analysis of Twin Data

If geneticists assembled inbred lines of mice, each line consisting of just two animals, they would not have particularly good material for quantitative genetic analysis. Yet this is the type of situation that human geneticists must deal with in twin studies. Some information of interest can be obtained, but the relatively small numbers of twins limit the reliability of the data. Further difficulties arise because human geneticists have little control over the environmental variables they might wish to evaluate. Indeed, it is not always clear what environmental factors are relevant to variation in complex traits such as intelligence.

The analysis of twin data for a quantitative trait, such as adult height, proceeds in the following way: For each twin pair, the difference in height is obtained. Then the data from all MZ twins (or from all DZ twins, like-sexed for a proper comparison) can be combined into a single figure that is the average of these differences (Table 11.4). Alternatively, we can compute the variance statistics V_{MZ} (or V_{DZ}) to express how much MZ twins (or DZ twins) differ from each other, on average.

To apply twin data to the nature/nurture question, we need to know how V_{MZ} and V_{DZ} are related to the components of variability V_E and V_G in a typical human population. Because identical twins have the same genotype, whatever variability exists between them must be due to environmental differences. Therefore, the quantity V_{MZ} is related to V_E. These variances are not equal, however, since environmental influences on identical twins are different from those affecting random, unrelated individuals, the basis of the V_E statistic. The same family environment and similar experiences based on being the same age and looking exactly alike (usually) suggest that V_{MZ} is less, to some unknowable degree, than V_E.

These difficulties are minimized but not eliminated when we examine identical twins who have been raised apart. Four studies have focused on such twin pairs: an ongoing study of personality by Thomas J. Bouchard, Jr., and colleagues at the University of Minnesota, with about 50 pairs, and older studies in the United States (19 pairs), England (44 pairs), and Denmark (12 pairs) (Rose 1982). We ignore the data in

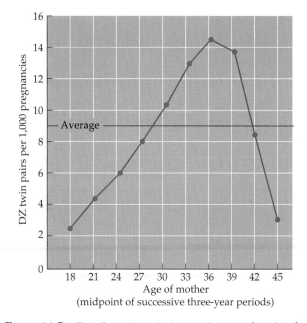

Figure 11.5 The dizygotic twinning rate by age of mother for Italian births, 1949 to 1965. The monozygotic twinning rate was almost constant, at about 4 per 1,000, for all maternal age-groups. (Adapted from Bulmer 1970.)

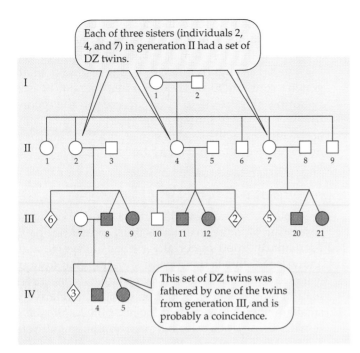

Each of three sisters (individuals 2, 4, and 7) in generation II had a set of DZ twins.

This set of DZ twins was fathered by one of the twins from generation III, and is probably a coincidence.

Figure 11.6 An unusual pedigree of unlike-sexed DZ twins (shown as solid symbols). Hard to understand is the set of DZ twins *fathered* by III-8. He may have received heritable factors that foster DZ twinning from his mother and passed them on to his children. But he could not have influenced the apparent multiple ovulation by his wife that led to his DZ twin children. Thus, the twin set in generation IV is probably coincidental. (After Gedda 1961.)

The difference in variances, $V_{DZ} - V_{MZ}$, that one can calculate using twins is somewhat related to V_G. The relationship is based on the fact that DZ twins differ in both genotype and environment, whereas MZ twins differ in environment only. You might think that the subtraction of one from the other should cancel out the environmental variation, leaving just the genetic component. But there are good reasons why this difference is not equal to V_G, the genetic variance between *random* persons. One is that the genetic difference between fraternal twins or sibs is not as great as that between unrelated persons. In addition, the environmental influences on the two types of twins may not be equal, so the subtraction would not cancel that source of variability. For example, family and friends sometimes treat MZ twins more uniformly than they treat like-sexed DZ twins, exposing MZ twins to more similar experiences.

Twin data applied to many quantitative traits confirm that most of those traits are influenced by variation in both genotype and environment. For the traits in Table 11.4, it appears that the genetic component of height is greater than that of weight, which is greater than that of IQ; and this conclusion is borne out by many other sets of data as well. Twin data can provide a quantity similar to heritability in the broad sense. For example, the broad-sense heritability for height in the studied twin populations is about 84%.

Recognizing all the difficulties involved in twin studies, we can at least state the following: To whatever extent MZ twins differ, environmental variation is a factor in the trait under study; to whatever extent DZ twins differ more than MZ twins, genetic variation is a factor.

Threshold Traits. Some phenotypes have the following attributes.

1. They are discontinuous, being either present or absent.

2. They appear, from twin studies, to have a genetic component.

3. They do not show simple Mendelian inheritance.

a different British study by Sir Cyril Burt (Box 11B). But even for identical twins raised apart, the differences between them would not be expected to be a very good estimate of V_E. For example, separated MZ twins are sometimes raised in families related to each other. In addition, adopting families generally tend to be at least middle class in socioeconomic level. Thus, although the individual case histories of twins raised apart make fascinating reading (Rosen 1987), the collected data are not extensive, providing interesting but not precise conclusions (see later).

Examples of such traits include a number of birth defects, such as cleft lip/cleft palate (separately or together); behavioral abnormalities, such as schizophrenia; and metabolic disorders, such as diabetes, asthma, rheumatoid arthritis, and obesity (Table 11.5). As in the case of other complex traits, it is likely that multiple QTLs as well as many environmental influ-

TABLE 11.4   Average differences between pairs of twins and pairs of siblings for three quantitative traits

	MZ twins		DZ twins (50 pairs)	Siblings (50 pairs)
	Raised together (50 pairs)	Raised apart (9 pairs)		
Height	1.7 cm	1.8 cm	4.4 cm	4.5 cm
Weight	4.1 lb	9.9 lb	10.0 lb	10.4 lb
Stanford-Binet IQ	5.9	8.2	9.9	9.8

Source: Newman et al. 1937.

BOX 11B Did Burt Cheat?

The reputation of English psychologist Sir Cyril Burt (1883–1971) has had its ups and downs. Burt had a lifelong interest in childhood education, theories of intelligence, and the role of genes in determining mental processes. His studies supported the idea that intelligence is a highly heritable trait. During his lifetime, he was much admired for his keen intellect, extraordinary knowledge, mathematical brilliance, and clinical skill, especially with children.

That bright view of Burt was darkened soon after his death. First, Princeton psychologist Leon Kamin discovered in 1974 that some figures in a series of Burt's papers were very suspicious. For example, correlation values for twins showing about 80% heritability of IQ remained the same despite increasing sample sizes. Then, a 1976 front-page story in the *London Sunday Times* reported that two of Burt's associates may not have existed. Finally, there came the biography of Burt by English psychologist L. S. Hearnshaw (1979). The book recounted how Burt could be not only charming and generous, but also cantankerous and jealous. He was fond of creating pseudonyms to write articles

supporting his own work or attacking that of others, which he then published in his own journals. Hearnshaw documented how Burt had spoiled his own distinguished career—perhaps as the result of mental instability and personal setbacks later in life—by perpetrating several scientific frauds, including fabricated research on MZ twins raised apart.

More recent books (Fletcher 1991; Joynson 1989) have turned the matter upside-down again. They present critiques of Hearnshaw's biography, providing alternative explanations that do not involve Burt in fraudulent activity. For example, at least some of the constant correlations appear to be related to the partial reanalysis of old data that had been misplaced during the bombing of England in the early 1940s. Some twin data were only gradually rediscovered and reported in the 1950s and 1960s. As for the missing associates, they may have been prewar, unpaid social workers; Burt credited them with the work in postwar years, even though they were no longer around.

The arguments by Joynson and Fletcher may or may not be true, but their defense of Burt seems to be as

plausible as the idea that a distinguished researcher would crudely make up data to support a firmly held belief. If he did invent some twins, would he not also change correlation values to render them more believable? Chances are the truth of the matter will never be known (Mackintosh 1995).

In any event, it is certainly true that the methods that Burt used were poor by today's standards for evidence or even by the standards of his own time. Details of his intelligence-testing methods never appear in any of his dozens of papers on the subject. Clearly, his results on IQ heritability are inadequately supported; but vagueness is not the same as fraud. Furthermore, although it is important to clarify the actions of a scientist who may have been unjustly maligned, the matter little affects the science of behavioral genetics. Removing Burt's data from the literature on IQ does not alter the conclusion of numerous other studies showing that genetic factors play a significant role in the development of mental functions.

---

ences are involved. The discontinuity in expression—the trait being either present or absent rather than being continuously variable—is explained by the concept of a **threshold**. This explanation implies that an essentially normal phenotype results from many different gene-environment combinations; beyond a critical accumulation of known or unknown genetic and/or environmental stresses, however, the phenotype is abnormal.

The genetic component in threshold traits is indicated, in part, by twin studies. Twins are said to be **concordant** if they both have the trait in question and **discordant** if one has the trait and the other does not. For the traits in Table 11.5, the concordance among MZ twins is 35–95%, whereas the concordance among DZ twins is 5–25%. Although some investigators have disputed the role of genes in the occurrence of schizophrenia, data on adopted children support a genetic component in this disease also. For example, adopted children are more likely to develop mental disorders if their biological mother was schizophrenic than if their biological mother was not. For the two types of diabetes noted in Table 11.5, different MZ twin concordances have confirmed that the diseases' underlying

genetic systems are quite different (Hall and Lopez-Rangel 1997; Hrubec and Robinette 1984).

Identifying the specific genes and environmental factors that modify threshold and other complex traits is difficult—much more difficult than finding the single gene that underlies disorders such as sickle-cell disease or cystic fibrosis through simple inheritance. A major advance is the construction of fine-resolution maps (as in Figure 9.3), which make it possible to identify chromosomal segments where QTLs might reside. Also needed are many families with substantial numbers of both affected and nonaffected individuals on which DNA analysis can be performed (Figure 11.7). The laboratory work is tedious—determining, for example, the genotype at 300 or more microsatellite repeat loci or other DNA markers and correlating the results with the presence or absence of the disease (Kahn 1996). Another difficulty that is inherent to this research is genetic heterogeneity; genes that influence the disease phenotype in one family may be different than those in another family. Investigators in this area must be well versed in statistical analysis, always on the lookout for false positive results and other numerical flukes.

TABLE 11.5   Three threshold traits

| Trait | Percent concordance | | Phenotype |
	DZ twins	MZ twins	
Cleft lip/cleft palate	5	35	A variable abnormality affecting the formation of the upper lip and roof of the mouth during gestation. There is a vertical fissure in the midline of the lip (25% of cases), palate (25%), or both (50%). Modern surgery can usually correct this defect both functionally and aesthetically. Although it often accompanies other syndromes, the frequency of this condition by itself is about 1 per 1,000 births.
Schizophrenia	15	45	A variable psychiatric disorder sometimes accompanied by delusions and hallucinations. Patients may lose all interest in their surroundings, or may make illogical and inappropriate responses to people or happenings around them. The criteria for diagnosis vary; thus frequency estimates range from about 1 in 80 to about 1 in 200 Americans. The biochemical and structural differences between the brains of schizophrenics and normal persons is being actively investigated. The likelihood that specific genes are involved is becoming stronger.
Diabetes mellitus			Both types result in high levels of glucose in the blood, and may be characterized by constant thirst, frequent urination, vision difficulties, slowly healing sores, and coma.
Type 1	5	40	In type 1, which affects young people, an insulin deficiency prevents glucose from entering cells. This deficiency stems from the autoimmune destruction of the beta cells of the pancreas—the source of insulin. The frequency is about 1 in 400. Several genes are known to contribute to the type 1 phenotype.
Type 2	25	95	Type 2 affects older people. Insulin is present, but other proteins needed to transport glucose into cells do not seem to function properly. Obesity is an important predisposing factor in type 2. The frequency is about 1 in 50.

Behavioral Traits

Trying to sort out genetic and environmental influences on human behavioral traits such as intelligence, sexual orientation, or personality is especially difficult. One problem, noted previously, is inadequate knowledge of, and control over, important sources of environmental variation. In addition, some behavioral traits lack objective standards of measurement that all investigators agree on.

A further difficulty is the ideological frame in which the nature/nurture question is often put. Studies of IQ, for example, are construed to support either a *hereditarian* or an *environmentalist* outlook on society and then perhaps used to influence political or social policy. Hereditarians emphasize the role of genes in fixing the limits of behavioral characteristics, whereas environmentalists see the newborn as much more pliable, to be shaped in large degree by societal forces. The debate is sometimes bitter, both inside and outside the academic community (Box 11C). Thus, it is important to understand that hereditarians are not necessarily racists, nor are environmentalists necessarily utopian dreamers. Common sense suggests that neither dogma can be 100% true. Both hereditarians and environmentalists recognize that human achievement can be enhanced by a supportive family, good schooling, and a decent job, in addition to and regardless of any genetic predispositions. With regard to be-

havioral problems, Bruce Pennington, a psychologist at the University of Denver, has written (1997):

> Many nongeneticists are resistant to genetic explanations because they fear that a genetic etiology means that the disorder is immutable. Instead, exactly the opposite is often the case; understanding the genetics of a disorder can lead to fundamental advances in its treatment.

Human geneticists have repeatedly emphasized this point. In fact, the authors of one of the first textbooks on human genetics (quoted in Chapter 1) wrote nearly the same thoughts as those above half a century ago.

A Genetic Component to Human Behavior

Behavior is what people do and how they do it. The mental or muscular responses by which such actions are expressed depend on environmental inputs to our sense organs. The various stimuli are coordinated and controlled by immensely complicated networks of nerve impulses and by hormonal messengers. Our activities are therefore based on aspects of anatomy and physiology that we know can be affected by gene-encoded proteins. It should therefore be anticipated that genetic variation can lead to behavioral variation.

Several simply inherited traits have a clear-cut behavioral component as part of the phenotype. One of the most distressing is the *Lesch-Nyhan syndrome*, caused by an X-linked recessive gene (Chapter 16).

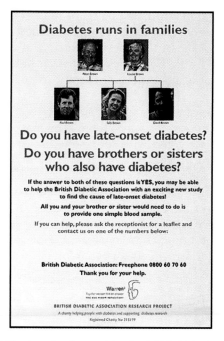

Figure 11.7 An advertisement by the British Diabetic Association for pairs of sibs with late-onset diabetes. Researchers determine whether affected sibs share any DNA marker allele more often than the expected 50%. If so, a gene predisposing to diabetes may be in the nearby chromosomal segment.

The normal allele controls the production of an enzyme needed in the metabolism of purines. Boys hemizygous for a mutant allele compulsively bite and mutilate their lips and fingers. How the enzymatic abnormality brings about the bizarre behavioral component is still a mystery.

Another example of a single-gene behavioral phenotype involves *porphyria*, which may have affected King George III of England (1738–1820). The unusual abnormalities of this rare autosomal disorder stem from a defect in the synthesis of heme, the iron-containing part of the hemoglobin molecule. The primary physical symptoms are abdominal pain, constipation, vomiting, and wine-red urine. (The word *porphyria* derives from the Greek, meaning "purple.") In the case of George III, neurological symptoms—visual disturbances and restlessness—were first noted when he was age 50. Then after three weeks, he became delirious, with convulsions and a prolonged stupor. The king recovered from this bout but suffered from several subsequent attacks of what was then diagnosed as insanity (Macalpine and Hunter 1969). The illness of George III provided the early impetus for research in psychiatry.

Mental impairment is not an uncommon characteristic of serious single-gene diseases. One example is

BOX 11C *The Fight over Violence*

When genetic researchers depart from matters of bodily function and turn toward human behavior, politics often enters the picture. Voices become especially strident when investigators suggest that genetic factors influence crime and aggression. As one example, the National Institutes of Health had planned to support a 1992 conference at the University of Maryland entitled "Genetic Factors in Crime: Findings, Uses and Implications." The conference draft had been peer-reviewed and approved by the ethical, legal, and social implications working group (ELSI) of NIH's National Center for Human Genome Research.

However, the conference was denounced as racist—as a veiled attack on African Americans, who make up 46% of the U.S. prison population. Critics complained that connecting genetics to crime was bogus because no data supported such a linkage (Anderson 1992). Organizers of the conference conceded that an

advertising brochure probably oversimplified some issues. The brochure stated that genetic research "holds out the prospect of identifying individuals who may be predisposed to certain kinds of criminal conduct, of isolating environmental features which trigger those predispositions, and of treating some predispositions with drugs and unintrusive therapies."

Bernadine Healy, then director of NIH, froze the funds for the conference and asked the meeting organizers to form a multiracial advisory panel to examine the objections. The conference was eventually held, but not until three years later. It was retitled "The Meaning and Significance of Research on Genetics and Criminal Behavior" and broadened to cover race relations, the social responsibilities of scientists, and the role of poverty and unemployment. For every speaker investigating predispositions to antisocial behavior, other speakers were invited to cast doubt on the findings or warn of possible misuse of the genetic information.

The conference was mostly civil, but with spirited give and take. Invited speakers objected again and again to the conference being held at all. At one point, outside protesters stormed the auditorium and seized the podium to emphasize their view that genetic behavioral research was racist pseudoscience (Roush 1995). It took 2 hours to return the meeting to order. With regard to the purpose of the meeting, few data were presented to show any influence of genes on criminal activity. Some speakers noted, however, that such an influence should be expected, because genes are known to influence human physiology, including hormonal and nervous responses that are known to affect behavior. In the end, the conference participants found a limited area of agreement. Moreover, protesters saw no means to stop people from trying to understand possible genetic bases of antisocial activities, and researchers were sensitized to possible serious misinterpretation and dangerous abuse of their findings.

phenylketonuria (Chapter 1). Another is *fragile X syndrome*, which stems from a peculiar mutated region, an expanded trinucleotide repeat, near the tip of the long arm of the X chromosome (Chapter 7). The more copies of the repeat, the more severe the retardation. The situation is similar with Huntington disease and with other neurological diseases associated with trinucleotide repeats (see Table 7.1). It is not known, however, how the repetitions bring about the behavioral phenotypes.

The mental deterioration (progressive loss of memory, confusion, and anxiety) of *Alzheimer disease* also has clear genetic components. As we will see in Chapter 13, a diagnostic feature is deposits of a short protein called β-amyloid in the brain. This protein is derived from a longer protein coded by a gene on chromosome 21. Three other genes (in addition to environmental factors) are now implicated in Alzheimer disease.

Geneticists are also studying the reading disabilities known collectively as *dyslexia*. Although the defining criteria vary, dyslexia affects perhaps 5–10% of schoolchildren. They have unexpected difficulties in learning to read, despite normal intelligence, customary instruction, and typical patterns of family and social support. Dyslexia sometimes runs in families, with children of a dyslexic parent having a 30–40% risk. In one study (Blakeslee 1994), researchers showed that dyslexia was linked to a DNA marker in the region of chromosome 6 encompassing the so-called *HLA* genes that affect immune functions (Chapter 18). This finding, which has recently been replicated by other researchers,[*] is interesting because dyslexic children sometimes have asthma, hay fever, or other immune system problems. Most forms of dyslexia, however, probably emerge from multiple biological and environmental factors that are not individually identifiable.

Williams syndrome is an odd and rare condition that occurs in about 1 in 20,000 to 50,000 births (Blakeslee 1996; Finn 1991). Both gifted and inept at the same time, persons with Williams syndrome have a variety of distinctive physical characteristics. Although mentally retarded as judged by their performance on IQ tests (scores between 50 and 70), they are surprisingly good in conversation, having large vocabularies and good recall of names and faces. They are also extremely friendly to both family members and total strangers, constantly seeking out human contact. But persons with Williams syndrome cannot easily visualize and manipulate objects or tell left from right, spatial abilities that are controlled by a specific area in the back of the brain. They are terrible at arithmetic. All these traits are associated with a heterozygous deletion of a short piece of chromosome 7 (Chapter 14).

If single-gene mutations or small deletions can affect behavior in obvious ways, there is no reason to doubt that multiple genes with subtle effects can affect various mental processes, even if specific loci and their products cannot be identified. This argument applies both to normal characteristics (e.g., familiar aspects of personality) and to pathological processes (e.g., schizophrenia). In addition to gene mutations, gene imbalance (as occurs in chromosomal aberrations) can also have important effects on behavior. For example, patients with Down syndrome have distinguishing personality features, such as friendliness, joviality, love of music, and ability as mimics. Patients with Turner syndrome tend to do very poorly on tests that require them to visualize forms in space, although the distribution of their IQ scores is not unusual. In school, they often do poorly in arithmetic and mathematics despite otherwise normal or superior performance.

In summary, there are real genetic contributions to variations in complex behavioral traits, expressed on a continuing basis throughout life (Mann 1994). The specific relationships between genotypes, environments, and the resulting phenotypes are often obscure, but possible interactive neurological and hormonal pathways for gene action are increasingly active areas of investigation.

Intelligence

All of us have a general idea of the meaning of intelligence (a dictionary synonym is "mental acuteness") by which we sometimes judge ourselves and others. Many psychologists emphasize abstract reasoning ability, which includes thinking rationally, solving problems, understanding the basics of a complex situation, and responding effectively to new environments. These meanings are sufficiently broad and applicable to such a wide variety of human endeavors that no obvious yardstick for measuring intelligence immediately presents itself. Some psychologists would even broaden the definition further to include such attributes as musical and artistic talent, physical ability, common sense, insight into oneself, and skill in interpersonal relationships (Winn 1990). Apart from trying to identify and measure the various aspects of intelligence and determining the relative importance of nature and nurture, there is much interest in the relevant neural and biochemical networks, as well as in the evolution of intelligence in human and other primate species.

Galton, with his passion for measuring, appears to have made the first attempts to test mental abilities. At

---

[*]Be wary of research that has not been independently replicated. For example, in the first edition of this book we described a gene influencing dyslexia linked to DNA markers on chromosome 15. That linkage could not be confirmed by the same researchers when they expanded their family base, although other investigators still suggest a chromosome 15 gene (Pennington 1997).

the 1884 International Exposition in London, he set up a booth where visitors paid threepence to be scored in several ways. Some of Galton's tests dealt with sensory perception (e.g., accuracy in discriminating different weights) or with quickness of reaction, which he felt correlated with a person's intellect.

In Paris, beginning in 1904, Alfred Binet and his colleagues at the Sorbonne were given the task of identifying schoolchildren who needed special education to improve their academic situation. Binet's easily administered tests proved fairly successful in predicting how well a child performed in subsequent schooling. The test items were verbal, numerical, or pictorial problems of increasing difficulty for older children. Although Binet hoped to measure "general intelligence" rather than knowledge accumulated through schooling and other experiences, he recognized that test construction depended on trial-and-error standardization of many different test items rather than on a theory of intelligence. Psychologists at Stanford University modified the Binet procedure to evaluate white middle-class American children. Test questions were selected so that, overall, boys and girls performed equally well and were scaled to average 100 for each age-group. Gould (1996) gives a detailed history of IQ testing, including accounts of its misuse by the American eugenics movement.

Do various IQ tests actually measure intelligence? In addition to the generalized Stanford-Binet, Wechsler, and Scholastic Aptitude Tests (SATs),* other quite different types of tests purport to measure one or another aspect of mental ability. Researchers point to the considerable degree of correlation between many of these tests and suggest that some common capability is being examined. The shared factor is what psychologists call general intelligence, designated g. This idea of intelligence is thus derived from the statistical analysis of test scores. It would be more satisfying, of course, if the flow of ideas were reversed, so that test construction was derived from an analysis of intelligence. In fact, the reality and importance of g underlies numerous arguments over mental testing. Although many, perhaps most, educational psychologists believe that intelligence can be measured accurately and reliably (Arvey et al. 1994), others think that intelligence is too diverse and abstract a concept to be condensed into a single, labeled factor (Gould 1996).

*The various Wechsler scales, in addition to verbal tasks, include performance components (such as assembling an object) to judge aspects of mechanical ability. The Wechsler Adult Intelligence Scale (WAIS) is the most widely used IQ test for adults in the United States, having been standardized against age-groups from 16 to 75. Many of you will be familiar with the SATs, which are taken by more than a million college-bound students every year (Jensen 1981).

We emphasize again that intelligence is not a genotype; it is a very complex phenotype that develops under the influence of genes and the experiences of a lifetime. An IQ test provides a simplified assessment of an aspect of that phenotype. Although IQ measurements are related to what it means to be intelligent, the score is neither comprehensive nor invariant. Furthermore, a high IQ score does not ensure high achievement in life's endeavors, nor does a low score guarantee failure. All of us are acquainted with the exceptions in both directions—persons with great or little perseverance, honesty, courage, charisma, or other significant traits that influence success or failure. Still, there is little doubt that in our culture, greater achievement usually favors those with more of the qualities represented by intelligence.

Heritability of IQ. In this section we ask, To what degree do IQ variations *within* middle-class whites (the group on whom the tests were standardized) arise from genetic differences? It is, of course, the measured IQ scores themselves that are at issue, not intelligence in its broader context. A quite different question is raised in the next section: To what degree does the difference in average IQ *between* black and white Americans arise from genetic differences? The answers to the "within" and "between" questions need not be the same. In any event, numerical results such as heritability values are only starting points in assessing how social and educational changes can help improve intellectual development.

Many populations have yielded values for the heritability of IQ, although some individual studies are based on small numbers. A 1981 summary of 111 studies that met certain criteria is given in Figure 11.8. The data are presented as the correspondences (correlation coefficients) in scores between pairs of individuals on a variety of tests that purport to measure an aspect of intelligence. The pairs of individuals were raised either together or separately; they shared none of their genes (top line), or an increasing proportion of their genes (middle lines), or all of their genes (bottom two lines).

The statistical significance of these data is hard to assess, partly because the results in any one line are so heterogeneous. Overall, it appears that neither genetic nor environmental influences can be neglected. For example, if environmental influences were absent, then zero correlation should be found between unrelated persons reared together (first line); but the 23 studies summarized show that there *is* an association. The last two lines in Figure 11.8 lead to a similar conclusion: The MZ twins raised together or apart do differ, and that difference must necessarily arise from *environmental* factors. On the other hand, the influence of *genotypic* factors on IQ is strongly suggested by the

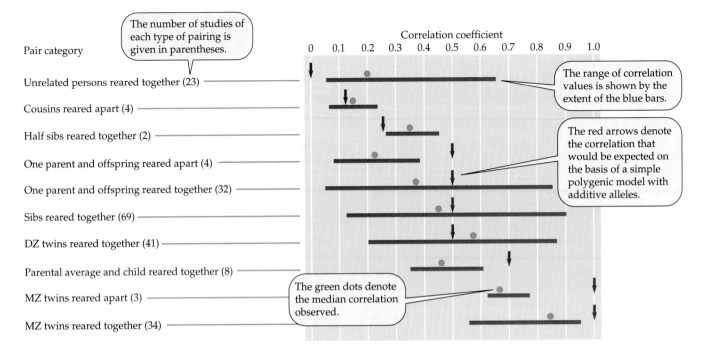

Figure 11.8 Correlation coefficients between pairs of individuals given a variety of IQ tests. In rough terms, the similar trend of dots and arrows suggest the existence of genetic factors. That the dots and arrows are not superimposed suggests the existence of environmental factors. (Redrawn and abridged from Bouchard and McGue 1981.)

higher correlation between MZ twins reared together (about 0.84) than between either DZ twins reared together (0.57) or sibs reared together (0.45).

Estimating heritability values from correlations is complicated and involves assumptions about the study population that may be only approximately true. Typical calculations for children and adolescents show that heritability of IQ in the broad sense is about 50% (Plomin and DeFries 1998; Scarr and Carter-Saltzman 1983). For adults, the heritability values are higher (Bouchard et al. 1990). A Swedish study of 110 identical twin pairs and 130 fraternal twin pairs—all 80 or more years old—yielded a heritability of 62% (McClearn et al. 1997). Researchers ascribe some of the remaining IQ variability to purely environmental variations and some to associations between genotypes and environments. This latter source of variation arises, for example, when parents with "favorable" genotypes not only transmit favorable alleles to their children, but also (because of the favorable genotypes) provide their children with an enriched environment conducive to achieving high IQ scores.

We wish to reiterate the limits in the meaning of heritability. These cautions have been set forth by many persons, including University of California psychologist Arthur Jensen in the 1969 paper that sparked heated arguments over the race-IQ issue.

• *Population versus individual*. Heritability is not defined for an individual. It would make no sense to say, for example, that 60 points of a person's IQ score are due to his genes and 40 points to his environment. Rather, heritability is a measure of the genetic variability of persons within a population. In groups with relatively homogeneous environments, the heritability of a trait will be larger than for groups in heterogeneous environments. Perhaps this tendency is one reason why heritability estimates among British populations are often higher than those estimated for the more diverse cultural and educational settings of the United States.

• *Known versus unknown genes*. Heritability values do not depend on knowing the metabolic actions of whatever genes and alleles are involved. Although it would be nice to understand the relevant biochemistry, the heritability values do not depend on this knowledge. Undoubtedly, many genes—perhaps hundreds—with individually small effects influence the type of mental processes involved in IQ measurements.

• *Constant versus changeable IQ*. Many times we have noted that an environmental change can modify the expression of a gene-influenced phenotype. The significant heritability of IQ does not rule out strong environmental modifications of IQ. Binet himself emphasized the value of remedial programs for special-needs pupils who were identified by his tests. Binet further noted that the intelligence of such students was thereby increased (quoted in Gould 1996). Although there is a high correlation between IQ scores for the same person taken at different ages, the correlation is not perfect. Furthermore, the correlation applies to existing circumstances of schooling and middle-class values, and novel circumstances might affect the IQ phenotype more. No one is predestined to have and maintain a particular IQ value.

The possible social consequences of variation in intelligence are major themes of *The Bell Curve: Intelligence and Class Structure in American Life* (Herrnstein and Murray 1994), a long and controversial but clearly written book. The authors (and others) support the idea that low IQ is correlated with poverty, school dropout, unemployment, unstable family life, poor parenting ability, welfare dependency, and crime. They accept IQ heritability in the neighborhood of 60%. Yet they sometimes write as if the acknowledged 40% of variation in intelligence that is environmentally influenced did not exist and as if IQ was unchangeable. They claim that little that has been done or that could now be done will improve low intelligence and the resulting social problems (others disagree). Pessimistic about the future for people at the bottom end of the bell curve, they advocate public policies that often appear mean-spirited. It is not news that the world is unfair, but should not public policy aim to make it fairer?

Racial Differences in IQ.   In this area of investigation, the facts do not always speak for themselves. The same set of data may be interpreted by different scientists to mean different things, and these viewpoints may be oversimplified or exaggerated by journalists seeking eye-catching copy on scientific matters that impinge on public policy. The major undisputed fact about the IQ scores among white and black children is that the average racial difference is about 15 points. Nevertheless, there is considerable overlap, so the IQ score of a particular individual is no guide to his or her race. And of course, in a democratic society, individual qualities are important, not group averages, a point that all responsible hereditarians and environmentalists agree on. The reasons for the average IQ difference between blacks and whites are difficult to assess, and most investigators, including Herrnstein and Murray (1994), think that the underlying causes are unknowable on the basis of currently available data. The many interrelated environmental variables that could affect test scores include the following:

• *Standardization.* Because blacks were not included in the development or standardization of the Stanford-Binet tests, cultural differences, especially language differences, could lead to lower scores for blacks. Many educational psychologists, however, believe that intelligence tests are not culturally biased against American blacks (Arvey et al. 1994).

• *Discrimination.* Social and educational opportunities that are available to whites of any socioeconomic level have often been denied to blacks. Predominantly black schools, for example, tend to be inferior to predominantly white schools. Difficult to measure are the effects of hundreds of years of racial prejudice rooted in slavery—prejudice that is sometimes blatant and sometimes subtle. As newspapers often remind us, acts of discrimination arise in the streets, restaurants, stores, workplaces, police departments, corporate boardrooms, and elsewhere. The resulting stresses are a burden on the physical and mental health of black people, including those who are highly successful by anyone's standards (Staples 1996).

• Socioeconomic status. One measure of socioeconomic status (SES) combines, into a single index, data on father's and mother's education, father's occupation, number of siblings, and reading materials and appliances in the home. Studies show that blacks in general have lower SES ratings than whites, and people with low SES ratings typically get lower IQ scores.

• *Motivation.* What factors allow a test taker to do as well as possible? How is performance affected by a subject's self-esteem and by the individual's perception of how the results will be used? How important are the attitudes and expectations of teachers, friends, and parents? These elements have been shown to play significant roles in taking IQ tests (Goleman 1988).

These variables suggest that some or all of the average IQ difference between blacks and whites could be due to *environmental* differences between the two groups.

The argument that at least some of the average IQ difference could be due to *genetic* differences flows from the concept of races: subdivisions of humankind that have come to differ, to a greater or lesser degree, in the frequencies of some of the alleles they possess (Chapter 12). Blacks and whites do differ in traits controlled by single genes—for example, hemoglobin variants and blood types. These differences have occurred through the processes of evolution—such as mutation and selection. There is no reason to suspect that genes that affect behavioral processes could not also come to have different frequencies. But the idea that genetic factors for IQ performance *could* differ between races certainly does not mean that they *do* differ. Jensen suggests that most of the black-white difference in IQ is due to genetic factors. His argument is that even when black and white children are matched for environmental factors, most of the IQ difference remains. Others believe that the matching can never be fairly done (Bodmer and Cavalli-Sforza 1970). In short, the arguments for a genetic contribution to the average black-white difference in IQ are weak.

Personality

It has been noted with amusement that parents of single children tend to be environmentalists, while parents of more than one—seeing the striking differences that may develop even in a steady environment—tend to be hereditarians. For example, many parents take the credit—perhaps rightly so—for providing the proper environment that produces a delightful first child: a warm home life, enriching toys, good schools, and so on. Yet a second or third child that they bring up in substantially the same way often turns out to have a completely different personality. One child may be outgoing and talkative, another shy and quiet, a third uniquely a worrywart or a risk taker or a slob.

The systematic study of genetic influences on personality traits has expanded over the last decade or two. A large, ongoing investigation is being done by Thomas Bouchard and colleagues at the University of Minnesota (Holden 1987). They have assembled and compared four groups of twins: MZ twins raised together, MZ twins raised apart, DZ twins raised together, and DZ twins raised apart. (One remarkable set of MZ twins raised apart, among the group of about 50 such sets, was described at the beginning of this chapter.) All twins were given batteries of tests having thousands of items, including one self-appraisal called the Multidimensional Personality Questionnaire. This questionnaire allowed investigators to evaluate such personality traits as social potency (a tendency toward leadership or charisma), social closeness (the need for intimacy, aid, and comfort), and traditionalism (respect for authority and rules).

From this study and from several others with similar intent, researchers have concluded that many different personality traits have sizable genetic components (Table 11.6). On the average, about 50% of observed personality diversity can be ascribed to genetic diversity. Furthermore, and surprisingly, very little of the remaining 50%—the environmental influence—seems to be due to children sharing a common family environment. Rather, many of the environmental influences that affect the development of personality traits tend to come randomly from outside the home, that is, the nonshared environment. Even allegedly shared family experiences may be perceived differently by each child and produce nonshared responses. For example, strict discipline by parents might affect quite differently a naturally shy child with many personal interests and an outgoing sibling with many social interests. Behavioral geneticists have indeed noted that MZ twins brought up in the same family are not much more alike than MZ twins brought up in different families. In short, it appears that parents can take some credit or blame, but not a lot either way, for providing suitable environments that influence the development of personality traits in their children.

A gene that may influence personality (specifically, the trait called *novelty seeking*) has been identified by two separate groups of researchers in Jerusalem and in Bethesda, Maryland. Their work involves **dopamine**, one of dozens of **neurotransmitters** in the brain. A neurotransmitter is a small chemical that is released from the end of one nerve cell and then binds to receptor molecules at the beginning of the next nerve cell in a neural network (Figure 11.9). The receptor molecules are large gene-encoded protein molecules. By this mechanism, the neurotransmitter sets in motion chemical changes that advance (or sometimes retard) electrical impulses along the nerve pathways. Dopamine is known to influence the way people and animals respond to many drugs: cocaine, amphetamines, heroin, alcohol, and nicotine.

In humans, there are five distinct (but somewhat similar) receptor molecules, D1 through D5, that bind dopamine. These **dopamine receptors** differ in their spatial distribution in the brain and in how they respond to stimulation. Whereas the D4 receptors seem to be implicated in the novelty seeking trait, the D2 receptors may be involved in alcoholism. The five dopamine receptors are encoded by independent (but somewhat similar) genes, several of which have been isolated, cloned, and sequenced. The *D2* gene is on the long arm and the *D4* gene on the short arm of chromosome 11.

Since dopamine is known to mediate exploratory behavior in experimental animals, it made sense for researchers to try to correlate novelty seeking and a gene involved in dopamine activity. In some but not all studies, people with a particular allele of the *D4* **dopamine receptor gene** (designated *DRD4*) tend to be high on the scale for novelty seeking (Bower 1996).

TABLE 11.6 Heritabilities of some personality traits

Personality trait	Heritability
Constraint	0.58
Harm avoidance	0.55
Negative emotionality	0.55
Social potency	0.54
Stress reaction	0.53
Absorption	0.50
Well-being	0.48
Alienation	0.45
Traditionalism	0.45
Aggression	0.44
Control	0.44
Positive emotionality	0.40
Social closeness	0.40
Achievement	0.39

Source: Tellegen et al. 1988.

(A)

(B)

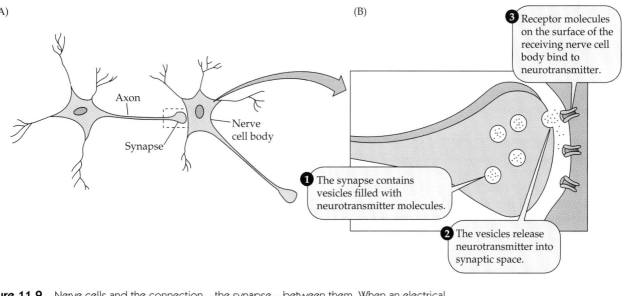

Figure 11.9 Nerve cells and the connection—the synapse—between them. When an electrical impulse gets to the synapse at the end of the axon, neurotransmitter molecules from vesicles in the knob are released into the space separating the two cells. The binding of the transmitter molecules to the receptor molecules causes additional chemical reactions that may initiate further electrical conduction through the cell body and axon of the next cell.

Such persons are impulsive, exploratory, fickle, excitable, quick-tempered, and extravagant. On the other hand, their opposites without the allele tend to be reflective, rigid, loyal, stoic, slow-tempered, and frugal. But alleles of the *DRD4* gene account for only about 10% of the variation in this trait. Thus, other genes and/or environmental factors must also influence novelty seeking.

Genes that interact with a different brain neurotransmitter, serotonin, may also affect personality, in this case anxiety-related traits (Goldman 1996).

Alcoholism

People view alcoholism variously as a biological problem, a psychosocial abnormality, or a moral weakness. Occurring in about 10% of American men and 4% of American women at some time during their lives, alcohol addiction can have disastrous consequences (Desmond 1987). Alcohol is responsible for a large proportion of fatalities caused by accidents on the highways, in the home, and on the job. The annual cost in the United States is estimated at more than $100 billion.

Alcohol intoxication depresses the central nervous system, producing uninhibited behavior, poor judgment, mood changes, incoordination, slurred speech, blackouts, and coma. In advanced stages of addiction, alcoholics are obsessed with alcohol to the exclusion of everything else—family, friends, work, and food—all the while denying they have a problem. Long-term alcohol abuse is associated with many degenerative dis-

eases, especially of the liver, but also of the stomach, esophagus, pancreas, heart, and brain. In affected women who drink heavily during pregnancy, alcohol diffusing across the placenta can affect the fetus, producing *fetal alcohol syndrome*, with symptoms ranging from mild to profound physical and mental abnormalities.

The development of an alcoholic is influenced by many personal, social, and cultural factors interacting with each other. In addition, some genetic variation for alcoholism is supported by two lines of evidence: (1) higher concordance in MZ twins than in DZ twins in some (but not all) studies and (2) increased risk for children of alcoholics even when adopted into nonalcoholic families (Goodwin 1991).

Three separate adoption studies were undertaken in the late 1970s in Denmark, Sweden, and the United States. The results were generally consistent with each other. The Swedish study (Table 11.7), for example, investigated persons who were born to single mothers and who were adopted at a very early age by nonrelatives. The table shows that when adopted males had an alcoholic biological parent, they were 2.3 times more likely to develop alcoholism than when they had nonalcoholic biological parents. A lower rate of alcoholism among females (included in other investigations) is not well understood but may be due in part to societal roles, to a lower rate of identification, or to biological factors related to the metabolism of alcohol.

Researchers have tried to find physiological or neurological characteristics that would identify those at increased risk for alcohol abuse. Enzymes that have been examined include two in the liver, alcohol dehydrogenase and aldehyde dehydrogenase, which break down alcohol via the following pathway:

$$\text{Alcohol} \xrightarrow{\text{Alcohol dehydrogenase}} \text{Acetaldehyde} \xrightarrow{\text{Aldehyde dehydrogenase}} \text{Acetate}$$

No consistent differences have been demonstrated between alcoholics and nonalcoholic controls with regard to these enzymes, but an interesting racial trait has come to light. More than 80% of Asians possess a variant, slow-acting form of the second enzyme, aldehyde dehydrogenase. It is thought that this difference (leading to a buildup of acetaldehyde) accounts for the marked facial flushing, accelerated heart rate, and other symptoms of distress that many Asians experience when drinking even small amounts of alcoholic beverages.

In a 1990 study, researchers found a significant genetic difference in the D2 dopamine receptor between a group of 35 alcoholics and a control group of 35 nonalcoholics. The researchers identified two alleles of the *DRD2* gene, *A1* and *A2*. The *A1* allele was present in 69% of the alcoholics but in only 20% of the nonalcoholics. Because susceptibility to alcoholism is affected by so many biological, psychological, and social factors, it was surprising that alleles of this one gene seemingly discriminated so large a proportion of affected versus nonaffected persons. Furthermore, what the *A1* allele might be doing differently than *A2* was not explained and is still not known.

Indeed, this striking difference has not generally been supported by additional studies, which are more negative than positive in associating alcoholism with the *A1* allele of the dopamine receptor gene. Problems in comparing different studies include the different criteria used to define alcoholism (e.g., age of onset and severity of illness) and the way control groups are set up. One particular study casts serious doubt on the association of alcoholism and the *DRD2* gene. Using cloned *DRD2* alleles, researchers could find no mutational differences between alcoholics and nonalcoholics (Gejman et al. 1994). Interestingly, this group of researchers included several who reported the original association of the *A1* allele of *DRD2* with alcoholism. In addition, a substantial number of the DNA samples used in this later study were from the original *A1/A2* study, thus maximizing the likelihood of actually finding a genetic difference if one existed. In defense of a possible genetic difference, the molecular investigators only checked the coding regions of *DRD2*. Although unlikely, a mutational difference could exist in the promoter or introns of the gene.

Genetic investigations of alcoholism are continuing and might cast light, not just on this serious disease, but also on broader aspects of addictive behavior, neurological disorders, and brain physiology. Twin and adoption studies show that genes are involved in alcoholism, but we know little about what those genes are and how they act.

Homosexuality

Opinions on homosexuality span a wide range, and the very mention of the word evokes strong emotional or religious responses from some people. We agree with many others that homosexuality is not a perversion, illness, or sin, but a natural human variation growing out of complex developmental processes that can lead to loving and durable relationships. It was not until 1980, however, that the American Psychiatric Association decided that homosexuality is not a disease and removed it from the *Diagnostic and Statistical Manual of Mental Disorders*. Interestingly, persons who see homosexuality as a "matter of choice" (i.e., having a largely environmental basis) often have a more negative attitude toward gays and lesbians than those who believe that homosexuals are "born that way," (i.e., have a biological basis). For this reason, the gay community favors a biological interpretation involving hormone chemistry, brain structure, neural circuitry, or genetic factors. Their experiences suggest that when others view homosexuality as "a fact of life," they are more tolerant and accepting. The issue of causation is unresolved, however, leading to vigorous debates on scientific inquiry, legal matters, and public policy (Friedman and Downey 1994). In any event, the law protects to some degree people with any sexual orientation, even if the private beliefs of others are discriminatory and hurtful.

TABLE 11.7 Alcoholism among male adoptees

Biological parent	Number in sample	Percentage of adopted sons[a] who were alcoholic	
Alcoholic father	89	39.4	avg = 34.0%
Alcoholic mother	42	28.6	
Nonalcoholic father	723	13.6	avg = 14.6%
Nonalcoholic mother	1,029	15.5	

Source: Bohman 1978.

[a]Swedish males born to unwed mothers between 1930 and 1949, adopted as infants by nonrelatives and followed through 1972. Alcoholism was determined by official records of alcohol-related fines, arrests, clinic visits, hospitalizations, etc.

Several studies of twins have suggested that homosexuality is partly genetic. In pooled data, about 57% of MZ twin brothers of gay men were themselves gay, whereas only about 24% of DZ twin brothers of gay men were gay. Similarly, about 50% of MZ twin sisters of lesbian women were themselves lesbian, whereas only about 16% of DZ twin sisters of lesbians were lesbian. Other studies confirm that male homosexuality is clustered in family lines with the brothers, uncles, and male cousins of gay men more frequently gay than in the general population. Virtually all of these male relatives are related to the gay index cases through maternal lines rather than through paternal lines, suggesting that possible genes influencing sexual orientation are on the X chromosome (Figure 11.10).

X chromosome inheritance of homosexuality has been confirmed by examining DNA marker sites of gay men and their close male relatives (LeVay and Hamer 1994). These sites are polymorphic dinucleotide repeats, VNTRs, or RFLPs spaced along the X chromosome within band Xq28, near the tip of the chromosome. A total of 33 out of 40 pairs of gay brothers had exactly the same alleles of five nearby DNA markers, suggesting that this region also contains one or more genes (currently unidentified) leading to homosexuality. This region of the X has about 4 million base pairs, enough for hundreds of genes (among which are the loci for hemophilia A and the two X-linked color vision abnormalities). Of course, the fact that 7 pairs of gay brothers were *not* concordant in this region means that other genes or environmental influences are at work. More recent studies by the same team of investigators (Angier 1995) have extended these observations, showing, for example, that the *heterosexual* brothers of gay men tend *not* to possess the same Xq28 sequences. Confidence in the idea that a gene influencing sexual orientation resides near the tip of the long arm of the X chromosome should, however, wait until other groups of researchers confirm the result.

A genetic explanation leaves unresolved another issue. Since homosexuality clearly reduces the number of offspring left by gay and lesbian people, why are these characteristics so frequent (2% more or less) in the general population? One would expect natural selection to reduce the frequency of homosexuality more and more every generation. Explanations that involve increased advantage to the near heterosexual relatives of homosexual individuals are convoluted at best. Perhaps new mutations contribute to homosexuality every generation, but nobody really knows.

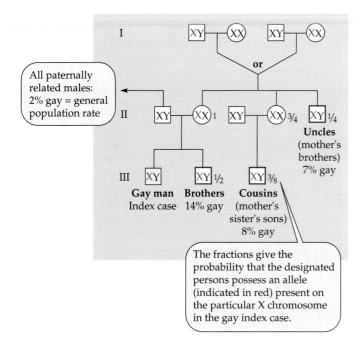

Figure 11.10 The frequency of homosexuality in some pooled maternal male relatives of typical index cases. Beneath the bold symbols we indicate the relationship and the frequency of homosexuality. The fractions to the side of the bold symbols give the probability that the designated persons possess an allele present on *the particular X chromosome* in the gay index case. Note that the frequency of homosexuality drops as the likelihood of sharing an X chromosome allele drops. See also questions 11 and 12. (Homosexual frequencies are from from LeVay and Hamer 1994.)

Summary

1. Behavioral and physical differences among people are influenced by both genetic and environmental factors, a situation sometimes represented as the nature/nurture question. The traits may be quantitative, varying continuously over a broad range.

2. The heritability of a trait in the broad sense is the proportion of phenotypic variation that is due to all genetic differences among members of a particular population at a particular time. A high heritability estimate does not rule out the possibility of altering phenotypes by environmental changes.

3. The genetic component of quantitative variation, called multiple gene inheritance or quantitative trait loci (QTL), involves the cumulative effects of several or many genes with varying effects that are each inherited according to Mendelian rules.

4. Monozygotic (MZ) twins are genetically identical, whereas dizygotic (DZ) twins are equivalent to sibs born at the same time. MZ twinning seems to result from random events of early development. On the other hand, DZ twinning tends to run in families and vary between racial groups.

5. Twin data provide only rough estimates of genetic and environmental contributions to quantitative and threshold

traits. Environmental influences are suggested whenever MZ twins differ. Genetic influences are suggested whenever like-sexed DZ twins differ among themselves more than do MZ twins.

6. Identifying quantitative trait loci (QTL) is difficult, but is becoming possible with the help of fine-resolution chromosome maps.

7. Variations in many human behavioral traits are influenced by genetic differences, although specific genes cannot always be identified and the pathways between the gene products and the behavioral phenotypes are usually obscure. Williams syndrome is an example of a specific gene and a known protein product being associated with an unusual behavioral phenotype.

8. IQ tests were originally developed in France to identify schoolchildren who needed help, and IQ testing continues to be useful for that purpose. IQ tests are standardized by trial-and-error methods rather than by theories of intelligence. Mental abilities are varied and complex, although many psychologists use the statistical construct, g, as a good measure of intelligence.

9. Within white, middle-class populations (on whom most tests have been standardized), heritability of IQ in the broad sense is about 50%. These calculations are based on correlations in IQ between pairs of individuals of various degrees of relatedness and involve assumptions that may be only partly true.

10. Some or all of the difference in standardized IQ scores between blacks and whites could be due to environmental and cultural differences between races. Some could reflect genetic differences, but the evidence for this is weak.

11. Personality differences have both genetic and environmental components, the latter due primarily to influences outside the family environment. Some genes that affect the function of brain neurotransmitters (dopamine and serotonin) influence personality traits.

12. Alcoholism develops under the varied effects of both environmental and genetic factors. The role of specific genes in affecting predisposition to alcoholism is unclear.

13. Homosexuality in males may be influenced by unidentified genes in a segment of the long arm of the X chromosome.

Key Terms

biometry
concordant twins
discordant twins
dizygotic (DZ) twins
dopamine
dopamine receptor
dopamine receptor gene
heritability in the broad sense (H_b)

heritability in the narrow sense (H_n)
monozygotic (MZ) twins
multiple gene inheritance
neurotransmitter
quantitative trait loci (QTL)
quantitative variation
threshold

Questions

1. The list of coincidences that bonded the Jim twins (see beginning of chapter) becomes less weird when we recognize that the list is *highly selective*. What does this mean?

2. The two selected groups of rats, maze-bright and maze-dull, depicted in Figure 11.2, did not diverge much more from each other after seven generations of selection. Why do you suppose this was true?

3. Which population would tend to show the higher heritability for a particular trait?

(a) A population that is relatively homozygous for genes affecting the trait or one that is more heterozygous?
(b) A population that is in a relatively uniform environment for factors affecting the trait or one in a more heterogeneous environment?

4. In Table 11.1, what six genotypes have phenotype 60 (four uppercase alleles)?

5. Although the inheritance of human eye color has not been worked out, it certainly involves more than one gene. Galton distinguished eight eye colors; by combining his two darkest categories, we can fit his scheme to the model of polygenic inheritance in Table 11.1:

Galton's classification	Assumed number of uppercase alleles
Light blue	0
Blue	1
Blue-green	2
Hazel	3
Light brown	4
Brown	5
Dark brown and black	6

(a) What are the genotypes of the lightest-eyed parents who could produce a child with brown eyes (five uppercase alleles)?
(b) Assume that anyone with zero, one, or two uppercase alleles is said to be blue-eyed and anyone with three or more uppercase alleles is said to be dark-eyed. (Eye color thus becomes a threshold trait.) What is the darkest-eyed child possible from two "blue-eyed" parents?

6. Under Davenport's model of skin color inheritance (Table 11.2), are these statements true or false?

(a) When one parent is white (zero uppercase alleles), the progeny can be no darker than the other parent.
(b) When one parent is light (one uppercase allele), the progeny can be no darker than the other parent.

7. It has been postulated that an egg and a polar body may very rarely be separately fertilized by two sperm to yield an unusual type of twin pair. Comment on the degree of similarity of these twins compared with regular DZ or MZ pairs. Assume that one sperm fertilized the mature egg and the second sperm fertilized the second polar body (sister to the egg). Also assume, for simplicity, no crossing over during meiosis.

8. Below are the Stanford-Binet IQ scores of the 19 pairs of MZ twins reared apart (Newman et al. 1937). The average of all the IQ scores is 95.7, and the average difference within pairs is 8.2 points. What conclusions are warranted?

106	95	91	85	101	90	93	127	102	116
105	94	90	84	99	88	89	122	96	109

102	88	115	97	78	92	106	96	116
94	79	105	85	66	77	89	77	92

9. *Very* low IQ can be a consequence of homozygosity for a single recessive allele (as in untreated PKU) or the result of a developmental birth defect. On the other hand, *moderately* low IQ reflects, to some extent, multiple gene inheritance. Generally, the sibs of the severely retarded have higher IQ scores than the sibs of the moderately retarded. Explain.

10. Assume that in country A, all schools are *equally* bad, learning is *never* encouraged, and all other environmental factors relevant to taking IQ tests are *constantly* unfavorable. Thus, $V_E = 0$ with regard to IQ. Assume that in country B, all schools are *equally* good, children are *impartially* encouraged in mental tasks, and all other environmental factors relevant to IQ are *constantly* good. Here, too, $V_E = 0$. What is the heritability of IQ within A and B? Comment on the cause of the likely difference in IQ scores between the two countries.

11. See if you can calculate the probabilities of getting the red X shown at the sides of the boldfaced males in Figure 11.10. (For convenience, we speak of the "red X" when we really mean a specific allele on the red X.) *Hints:* Start from the gay index case in generation III. His X chromosome (red) must have come from his mother (probability 1) in generation II. This same X chromosome came from either the mother's father (probability $\frac{1}{2}$) or from the mother's mother (probability $\frac{1}{2}$) in generation I. Then calculate the probabilities going forward to the remaining male relatives.

12. The penetrance of a gene is a measure of how often the gene is expressed when it is present. From Figure 11.10, we can make rough penetrance calculations for the putative gene influencing homosexuality. For example, for the brothers of the gay index case, the gene is expressed in 14% of brothers, and the likelihood of the gene being present is 50%. Therefore, the penetrance is roughly $\frac{14}{50}$, or 28%. Make penetrance calculations on the basis of the male cousins and uncles.

13. Do you agree or disagree with the following statement by Koshland (1987)?

The debate on nature and nurture in regard to behavior is basically over. Both are involved, and we are going to have to live with that complexity to make our society more humane for the individual and more civilized for the body politic.

14. Comment on the following statement by Nelkin and Lindee (1995), who examine the grandiose claims pervasive in American culture for the power of DNA:

In supermarket tabloids and soap operas, in television sitcoms and talk shows, in women's magazines and parenting advice books, genes appear to explain obesity, criminality, shyness, directional ability, intelligence, political leanings, and preferred styles of dressing. There are selfish genes, pleasure-seeking genes, violence genes, celebrity genes, gay genes, couch-potato genes, depression genes, genes for genius, genes for saving, and even genes for sinning. These popular images convey a striking picture of the gene as powerful, deterministic, and central to an understanding of both everyday behavior and the "secret of life."

Further Reading

A good book for lay persons is by Steen (1996), and a more detailed text is by Plomin et al. (1997). Sherman et al. (1997) provide a short authoritative summary of behavior genetics, and Plomin and McClearn (1993) edit a more technical collection. The journal *Behavior Genetics* publishes current research relating to both animals and humans. The issue of *Science* for June 17, 1994, includes a special report entitled "Genes and Behavior." In addition, the entire issue of *Trends in Genetics* for December 1995 is devoted to quantitative inheritance. Background information on quantitative inheritance is included in Bodmer and Cavalli-Sforza (1976), and background information on psychological testing is included in Anastasi (1988). An article by Wright (1995) is an interesting summary of twin research.

Material on intelligence testing and its interpretation is voluminous, varied, and sometimes acrimonious. Sample both Jensen (1969), Herrnstein and Murray (1994), their detractors (Gould 1996; Lewontin et al. 1984), and rebuttal (Arvey et al. 1994; Murray 1994). Scarr (1987) includes a personal account of her involvement in a controversial field. We recommend an article by McInerney (1996) that reviews 75 reviews of *The Bell Curve*.

Relevant *Scientific American* articles include Plomin and DeFries (1998) on IQ, Brown and Pollitt (1996) on malnutrition and intelligence, Greenspan (1995) on behavior in the fruit fly, LeVay and Hamer (1994) on homosexuality, Bodmer and Cavalli-Sforza (1970) on intelligence and race, and Macalpine and Hunter (1969) on the madness of King George III.

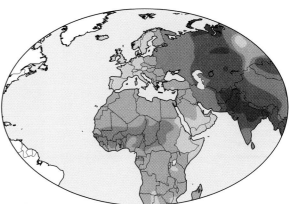

CHAPTER 12

Population Genetics
and Evolution

A wealth of evidence indicates that we and the creatures around us have evolved gradually from other creatures over the course of millions of generations. The idea that this evolution came about in large part by the mechanism of natural selection was the most important scientific insight of the 1800s.

A Brief History

The notion of natural selection occurred independently to two well-read Victorian naturalists, one much better known than the other, although each of them set forth convincing documentation to back up his claims. The better-known man, Charles Darwin (1809–1882), was the son of a wealthy country doctor (Figure 12.1A); he did not have to work to earn a living. As a young man, he took a five-year voyage of discovery on a surveying ship, the *Beagle*. Based on his findings, Darwin then investigated and later wrote at length about the mysteries of the origin of species by natural selection, or, as he first called his concept,

(A) (B)

Figure 12.1 (A) Charles Darwin at age 60. (B) Alfred Russel Wallace at age 55.

"descent with modification." The lesser-known and younger man, Alfred Russel Wallace (1823–1913), was the son of a financially insecure businessman (Figure 12.1B). Wallace, too, traveled widely as a young man, exploring the Amazon region for four years and the islands of Indonesia for eight. During the latter trip, he collected and preserved over 125,000 plants and animals. Indeed, he earned his living by selling beautiful and exotic specimens to museums and by lecturing and writing about his many tropical adventures.

While in Indonesia, Wallace (1855) published a paper in which he wrote, "Every species has come into existence coincident both in space and time with a pre-existing closely allied species." After reading this article, Darwin wrote Wallace that he agreed with almost every word and that he was preparing a major work on the subject. Although he gave no details to Wallace, Darwin had already arrived at a reasonable theory to explain evolution. In fact, he had been struggling with his manuscript, *On the Origin of Species by Means of Natural Selection*, for over a dozen years, but he was reluctant to publish anything. Indeed, he dreaded the controversy he knew his ideas would cause.

Through the intercession of friends, the similar views of Wallace and Darwin were eventually read at a meeting of the Linnean Society of London, a group of eminent natural historians. Wallace meanwhile continued to work in the Indonesian islands, gathering evidence for his own important additional works on biogeography of island populations. In 1859, Darwin finally published *On the Origin of Species*, whose first printing of 1,250 copies sold out in one day.

The two men were honorable, generous, and cordial, but Wallace was usually content to bask in the light of England's better-established scientists. Dubbed a moon to Darwin's sun, Wallace was often deferential; for example, his popular 1889 book summarizing the principles of natural selection was entitled *Darwinism*.

The notion that species evolved over time actually predated the work of Darwin and Wallace by many years. In the early 1800s, for example, the Frenchman Jean-Baptiste de Lamarck argued strongly for evolution and thought quite reasonably that it occurred by the inheritance of so-called acquired characteristics. But convincing scientific support for this particular mechanism for evolution was not brought forward by Lamarck, nor by anyone else. On the other hand, Darwin and Wallace marshaled considerable evidence that evolution occurred by natural selection. This mechanism for evolution, coupled with the workings of heredity, can account for the step-by-step adaptive changes characteristic of the evolution of populations over space and time.

In the early decades of the 1900s, the ideas of Darwin and Wallace were integrated with the discoveries of Mendel into a mathematics of evolutionary processes. The first genetic theory of successive generations of a breeding group was set forth succinctly in 1908 by G. H. Hardy, an English mathematician, and quite independently by Wilhelm Weinberg, a German physician (discussed later).

Later population genetics theory was derived in large part by two Englishmen, J. B. S. Haldane and Sir Ronald Fisher, and an American, Sewall Wright. Along with most other scientists (since about 1800), they agreed that populations of organisms have changed and diverged over eons to produce the untold number of current species. Each man, however, put his special stamp on the theories behind the facts; that is, they differed among themselves, sometimes vigorously, on the mechanisms that might explain how evolution occurred—and is still occurring. The evolutionary mechanisms that they argued about—in addition to natural selection—were *mutation, migration,* and *drift,* factors that by themselves lack the "management skills" of selection. That is, these factors act in a more random way to bring about change. The mathematical models that they constructed were abstruse and became better understood through the writings of others. An especially important work was the 1937 book *Genetics and the Origin of Species,* by Theodosius Dobzhansky of Columbia University.

Recent work in evolutionary theory has emphasized variation at the molecular level, in the nucleotide sequences of DNA. Some base substitution mutations, for example, are likely to have no phenotypic manifestation because they are in an intron or in the third position of a codon that does not change the resultant amino acid. Such mutations might therefore be expected to be selectively neutral and perhaps be useful to researchers in tracing evolutionary lineages.

We will not broach here the question of biological evolution versus special creation.* Rather, we start with the knowledge that gradual, cumulative, and nonmiraculous changes over many millions of generations can account for the wonderful complexity of adaptations and the astonishing diversity of species, including our own. Evolution—the fundamental thread tying together all fields of biology—has been repeatedly investigated and thoroughly confirmed. The documentation, presented in many books on evolutionary biology (e.g., Dawkins 1986; Futuyma 1995; Strickberger 1990) includes the following kinds of data:

---

*Regarding this issue, the United States Supreme Court has ruled that it is unconstitutional to require public school teachers to discuss special creation whenever they discuss evolution (Edwards v. Aguillard 1987). The justices argued that special creation reflects a religious doctrine (the Book of Genesis); teaching it would therefore violate the separate status of church and state mandated by the First Amendment. A prior court case emphasized that so-called "creation science" is not science in any sense of the word (McLean v. Arkansas Board of Education 1982).

1. The fossil record of ages past, showing temporal sequences of more or less continuous forms, albeit with gaps

2. The geographical distributions of past and present species over wide or narrow ranges (intensely studied by Wallace)

3. The use of artificial selection by plant and animal breeders to generate a prodigious variety of pets, flowers, livestock, crops, and so forth (intensely studied by Darwin)

4. The retention of developmental stages of remote ancestors during the embryology of current species (Figure 12.2)

5. The similar underlying structures of some functionally dissimilar body parts as revealed by studies of comparative anatomy

6. The greater similarity of cellular and molecular structures (chromosomes, nucleotide sequences, amino acid sequences, etc.) among more closely related species

7. The nearly universal genetic code and other similarities and subtle differences at the biochemical level among all living creatures

Population Genetics

Evolutionary ideas deal with populations and require a different kind of thinking than that used to analyze pedigrees. A **population** is simply a collection of individuals. We present first some basic concepts in the field of *population genetics,* which is based partly on logical extensions of Mendel's ideas. To describe the proportion of a certain genotype, phenotype, or allele within a population, we use the term *frequency,* which is similar in meaning to the term *probability.* We want to calculate the frequency of a particular type within a population.

Suppose that all members of a population have their genotypes tested, giving these results:

Genotype	Number	Frequency
B/B	114	57%, or 0.57
B/b	56	28%, or 0.28
b/b	30	15%, or 0.15
Total	200	100%, or 1.00

Each **genotype frequency** is the proportion that that particular genotype is of the whole, expressed as a percentage or, more usually, as a decimal fraction. For example, freq(*B/B*), the frequency of the *B/B* genotype, is $^{114}/_{200}$ = 0.57. Note that the sum of all genotype frequencies must necessarily be 1, a rule that can provide a check on arithmetic.

If each genotype in a population corresponds to a different phenotype, then genotype frequencies and **phenotype frequencies** are the same. If *B* is dominant

Stage of embryology

Early

Gill pouch

Middle

Late

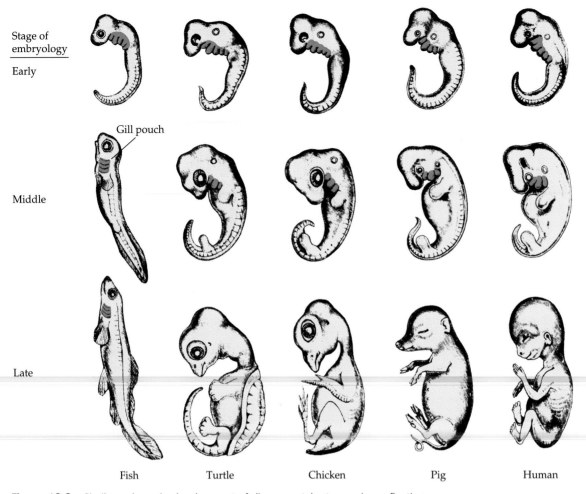

| Fish | Turtle | Chicken | Pig | Human |

Figure 12.2 Similar embryonic development of diverse vertebrate species, reflecting a common evolutionary origin. All possess gill pouches in early embryology, although only those in fishes go on to develop into adult organs for extracting oxygen from water. The fishlike embryonic period of all vertebrates is consistent with their possession of common genes, modified over time to adapt the adults to other life-styles.

to b, however, then the genotypes B/B and B/b are phenotypically indistinguishable; thus, we have

Phenotype	Number	Frequency
Dominant	170	0.85
Recessive	30	0.15
Total	200	1.00

A little more arithmetic is needed to obtain what are called the **allele frequencies**, the proportions of the total number of alleles represented by B or b. We note that each individual possesses two alleles per locus, so among the 200 parents, there are 400 alleles to be tallied as either B or b. Assuming that B/B's and B/b's are distinguishable from each other (no dominance), the 114 B/B homozygotes together possess 228 B alleles. The 56 heterozygotes add to the tally of both allelic forms, 56 of B and 56 of b. The total number of B alleles is therefore 228 + 56, and the total number of b alleles is 60 + 56. Allele frequencies can then be calculated:

For the B allele: freq(B) = (228 + 56)/400 = 284/400 = 0.71
For the b allele: freq(b) = (60 + 56)/400 = 116/400 = 0.29
Sum of allele freq = 1

To make this calculation, homozygous and heterozygous genotypes must be distinguishable from each other. We will see later in this chapter what can be done to estimate allele frequencies when this condition is not met. The allele frequencies are often symbolized more briefly as p = **freq(B)** and q = **freq(b)**. Using this notation, $p + q = 1$.

An easily remembered relationship exists between allele frequencies and genotype frequencies:

$$p = \text{freq}(B) = \text{freq}(B/B) + (\tfrac{1}{2})\text{freq}(B/b)$$
$$q = \text{freq}(b) = \text{freq}(b/b) + (\tfrac{1}{2})\text{freq}(B/b)$$

In words, the top equation says that the frequency of the B allele equals the frequency of the homozygous genotype plus half the frequency of the heterozygous genotype. This is a commonsense expression, a weighted average. That is, the alleles in the homozygote are *all B*, while the alleles in the heterozygote are *one-half B*. (If you like algebra, you might want to formally derive the formulas.)

Assuming that the genotype frequencies given earlier (0.57, 0.28, 0.15) apply to a *parent* generation, can we now predict the proportions of the various genotypes among their *children*? The answer is yes, but we must first know what matings occur among the parents. In particular, we seek the proportions of the six possible matings:

1. $B/B \times B/B$	4. $B/B \times B/b$
2. $B/b \times B/b$	5. $b/b \times B/b$
3. $b/b \times b/b$	6. $B/B \times B/b$

The frequency of mating 1, for example, is the proportion that this cross is of all crosses in the population. But to find its value, we need to know the rules of the mating game. For example, if "like attracts like," it may be that the only matings that occur are 1, 2, 3, and 4, in which both mates have the same phenotype. But many other mating schemes can be envisioned. In short, we need to know the **mating frequencies**.

One mathematically convenient and sometimes realistic assumption, called **random mating**, is that mating occurs without regard to a person's genotype or phenotype. This assumption is true (or very nearly true) for many human traits—blood groups or serum cholesterol levels, for example. It means that an individual is equally likely to mate with any member of the opposite sex. Random mating frequencies depend only on the frequencies of the two pertinent genotypes (which are multiplied together). For example, for mating 1 above, we calculate

$$freq(B/B \times B/B) = freq(B/B\ \male) \times freq(B/B; \female)$$
$$= (0.57)(0.57) = (0.57)^2$$

The expression freq($B/B\ \male$) means the frequency of B/B among males, and freq($B/B\ \female$) means the frequency of B/B among females. We are assuming that these are equal, that is, that genotypes are distributed similarly in the sexes.

Matings 4, 5, and 6 are a bit different from 1, 2, and 3 in that the two mates are genotypically different from each other. As a consequence, one must add together the frequencies of *reciprocal* crosses. For example, the mating $B/B \times b/b$ (mating 6) means either $B/B\ \female \times b/b\ \male$ or the reverse, $B/B\ \male \times b/b\ \female$. Thus,

$$freq(B/B \times b/b) = freq(B/B\ \female) \times freq(b/b\ \male) +$$
$$freq(B/B\ \male) \times freq(b/b\ \female)$$
$$= (0.57)(0.15) + (0.57)(0.15)$$
$$= 2(0.57)(0.15)$$

The factor 2 is present in the random mating frequencies of matings 4, 5, and 6. It is not present in matings 1, 2, and 3 because the spouses have the same genotype; reversing the sex is not a different situation.

For the specific numbers assumed, the random mating frequencies for the three genotypes are summarized in Table 12.1. Note that just as the two possible *alleles* in a population have frequencies that add to 1, and the three possible *genotypes* have frequencies that add to 1, so, too, do the six *matings* have frequencies that add to 1.

Using the mating frequencies in Table 12.1, we can find the frequency of any particular genotype among the offspring. Note, for example, that the genotype b/b is expected among the children of matings 2, 3, and 5. From these parents, the expected fractions of b/b children, based on Mendel's first law, are

Mating	Probability of b/b offspring
2. $B/b \times B/b$	$\frac{1}{4}$
3. $b/b \times b/b$	1
5. $b/b \times B/b$	$\frac{1}{2}$

Based on this and similar tables for other offspring genotypes, we could then predict the overall frequencies of the three genotypes among the offspring. These calculations are not conceptually difficult, but they are a bit tedious. There is an easier way to get at the answer.

The Hardy-Weinberg Law

G. H. Hardy was an eminent English mathematician (Box 12A) and Wilhelm Weinberg was a German clinical physician working in the early part of this century. In 1908, Hardy and Weinberg applied the recently rediscovered rules of Mendel to randomly mating populations in a manner somewhat similar to that used in our earlier discussion. They expressed their results in what has become known as the **Hardy-Weinberg law**. For two alleles of an autosomal gene, B and b, the Hardy-Weinberg law can be stated in two parts:

TABLE 12.1 Random mating frequencies of three genotypes

Mating	Frequency under random mating	
1. freq($B/B \times B/B$)	$(0.57)^2$	= 0.3249
2. freq($B/b \times B/b$)	$(0.28)^2$	= 0.0784
3. freq($b/b \times b/b$)	$(0.15)^2$	= 0.0225
4. freq($B/B \times B/b$)	$2(0.57)(0.28)$	= 0.3192
5. freq($b/b \times B/b$)	$2(0.15)(0.28)$	= 0.0840
6. freq($B/B \times b/b$)	$2(0.57)(0.15)$	= 0.1710
	Total	= 1.0000

Note: The assumption is that genotypes B/B, B/b, and b/b, have frequencies 0.57, 0.28, and 0.15, respectively, and that individuals with these genotypes mate randomly with respect to each other.

BOX 12A How Hardy, a Mathematician, Got into Biology

G. H. Hardy's 1908 article that helped establish the Hardy-Weinberg law took the form of a letter to the editor of the American journal *Science*. It has the tone of a kindly grandfather, beginning:

> I am reluctant to intrude in a discussion concerning matters of which I have no expert knowledge, and I should have expected the very simple point which I wish to make to have been familiar to biologists.

Hardy had been invited to discuss the topic by his Cambridge University colleague Reginald Punnett (of Punnett squares, or checkerboards), with whom he played cricket. Punnett was critical of other biologists who believed that a dominant allele once introduced into a population would spread until the frequency of the dominant phenotype reached ¾. Punnett the biologist, however, could not prove that the others were wrong, so he presented the problem to his distinguished mathematical friend.

Hardy loved pure mathematics, which he considered useless but beautiful; he disliked practical applications, which he considered dull and ugly. This may explain why he published this problem-solving paper not in the obvious place, the English journal *Nature*, but across the ocean, far away in the American journal *Science*. It has been suggested that perhaps Hardy didn't want it to be seen by his colleagues of pure mathematics. "It must have embarrassed him that his mathematically most trivial paper became not only far and away his most widely known, but has been of such distastefully practical value" (Crow 1988).

---

1. Under so-called Hardy-Weinberg conditions, the genotype frequencies in a population become, after *one* generation,

$$\text{freq}(B/B) = p^2$$
$$\text{freq}(B/b) = 2pq$$
$$\text{freq}(b/b) = q^2$$

where p and q are allele frequencies, that is, $p = \text{freq}(B)$ and $q = \text{freq}(b)$. These equations mean that the genotype frequencies can be predicted from the allele frequencies in any generation. Here are the conditions under which the Hardy-Weinberg law is applicable:

(A) Approximately random mating

(B) A fairly large population (several hundred or more)

(C) A negligible amount of mutation between the B and b alleles

(D) A negligible amount of migration into and out of the population

(E) A negligible amount of selection, with all genotypes about equally viable and equally fertile

2. As long as the Hardy-Weinberg conditions prevail, the allele and genotype frequencies do not change. That is, generation after generation, the frequencies of the three genotypes remain constant at p^2, $2pq$, and q^2. You will recognize these three terms as the square of the binomial, $(p + q)^2$. That is, the square of allele frequencies gives the genotype frequencies. The constancy over time is called an *equilibrium*, meaning no net change in frequencies. Of course, the individuals in succeeding generations are not the same; it is the expected frequencies of their genotypes that stay the same.

Strictly speaking, the Hardy-Weinberg conditions are more stringent than we have indicated. They include an *infinitely* large population, *no* mutation, *no* migration, and *no* selection. But for practical purposes, the law can be applied where these conditions are only approximated. There is no living population that exactly meets the mathematical model. But there are many genes in many populations (probably most genes in most populations) that meet the conditions sufficiently well that the Hardy-Weinberg law becomes usable and useful.

A proof of the Hardy-Weinberg law is based on the equivalence of random mating of genotypes and random combination of gametes at fertilization. If it is merely chance which genotypes combine, then it is merely chance which allele (B or b) contained in an egg and which allele (B or b) contained in a sperm combine. We set up a checkerboard for the entire population at once, rather than for a particular mating (as we have done before). As in any other checkerboard, we label the rows and columns with the parental gametes—in this case, two sorts of eggs and two sorts of sperm with frequencies p and q:

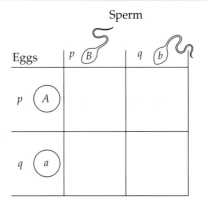

The offspring genotypes are written in the four squares of the checkerboard by combining the alleles and multiplying the frequencies heading the rows and columns. Thus, among the children, freq(B/B) = p^2 [from top left square]

freq(b/b) = q^2 [from bottom right square]

freq(B/b) = $pq + pq = 2pq$ [from other squares]

The second part of the Hardy-Weinberg law, that which asserts constant frequencies over successive generations, is shown by constructing a checkerboard for the *next* generation. To label its rows and columns, we need to first calculate the allele frequencies (equal to the gamete frequencies) among the children from the previous checkerboard. Recall that an allele frequency is equal to all the homozygotes and half the heterozygotes. Thus,

freq(B) = $p^2 + (\frac{1}{2})(2pq) = p^2 + pq = p(p + q) = p(1) = p$

freq(b) = $q^2 + (\frac{1}{2})(2pq) = q^2 + pq = q(p + q) = q(1) = q$

We see that the allele frequencies are unchanged; thus, the headings and body of the next checkerboard are unchanged from the last. Simply stated, the same little checkerboard above is repeated over and over, showing the constancy of genotype frequencies at p^2, $2pq$, and q^2.

The Meaning of the Hardy-Weinberg Law. The Hardy-Weinberg equilibrium is a statement of no change—of no evolution. Over long periods of time, it is unlikely to be realistic, because we know that populations do evolve. But over short periods, the Hardy-Weinberg stability may apply. The maintenance of alleles at constant frequencies also means that genetic variability within a population is not blended away over successive generations. This Mendelian notion of discrete hereditary units was an important contribution to evolutionary thought.

A population evolves when the conditions of the Hardy-Weinberg law are not met: by departures from random mating, by unpredictable changes that occur because a population is small, by mutation, by migration, and especially by natural selection. These mechanisms lead to allele frequency changes, with corresponding changes in the average phenotype of a population. The accumulation of modifications over time brings about the evolutionary patterns that have been documented by paleontologists and others. Although the grand sweep of changes over millions of bygone years cannot be correlated with specific genes, more recent evolutionary events can sometimes be studied by analyzing allelic changes. The emergence of antibiotic-resistant bacteria is a well-known example. Such analyses may use the Hardy-Weinberg law, which can be modified as needed.

It is important to note that the frequency of an allele remains constant under Hardy-Weinberg conditions, whether it is dominant or recessive. Dominance is *not* an evolutionary force; it is merely a label for how alleles express themselves in a heterozygote. An allele does not *automatically* come to have a frequency of $\frac{1}{2}$ or any other number. Rather, the frequency of an allele depends primarily on how it affects reproductive success or failure. Some dominant alleles impair the ability of persons to have offspring, as do some recessive alleles. The frequency of such an allele then goes down because it hampers reproduction, not because it is dominant or recessive. The actual frequency of a hereditary disease in human populations, far from being 75% (when due to a dominant allele) or 25% (when due to a recessive), may be 1 in thousands of people to 1 in millions. The low frequency persists whether it is inherited in a dominant or a recessive pattern. As Hardy says in his 1908 paper, "There is not the slightest foundation for the idea that a dominant character should show a tendency to spread over a whole population, or that a recessive should tend to die out."

The frequency of an allele can theoretically be any number from 0 to 1 inclusive. Its actual frequency will be the result of a long evolutionary history. The genotype frequencies predicted from the Hardy-Weinberg law will vary accordingly. A graphical display of some representative values is presented in Figure 12.3. As freq(A) varies from 1 to 0 (reading from top to bottom), the frequency of the A/A homozygote does likewise; but since the genotypic proportion is a squared function, the decrease in freq(A/A) is rapid. For example, as freq(A) = p decreases from 1 to 0.5, freq(A/A) = p^2 decreases from 1 to 0.25. If the frequency of an allele

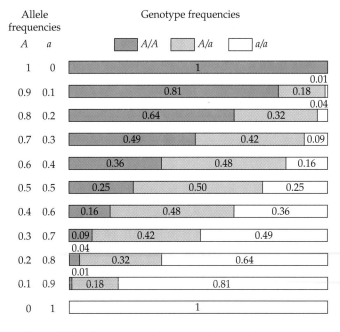

Figure 12.3 The expected genotype frequencies in a Hardy-Weinberg population for different values of the allele frequencies. A fuller geometrical interpretation for three of the lines is given in Figure 12.4.

is quite low (say, 1 in 100), then the corresponding homozygote is rare indeed (1 in 10,000). The frequency of heterozygotes is largest when the two alleles are equal in frequency and decreases to 0 at both extremes. In the middle third of the range of allele frequencies, heterozygotes are the largest genotypic class. The features of the Hardy-Weinberg equilibrium are also nicely shown by a geometrical scheme based on checkerboards (Figure 12.4).

Some Applications

Hardy-Weinberg arithmetic and algebra can give some interesting results. A few examples follow.

The MN Blood Groups. In addition to the well-known ABO and Rh blood differences, there are dozens of other genetic systems that result in certain chemicals (antigens) on the surface of red blood cells. For example, the MN blood groups, which have no medical significance, are based on a simple two-allele system, the alleles represented by M and N. (More formally, M and N should be superscripts on a basic gene designation, but the informal system is almost always used.) The alleles lead to corresponding antigens on red blood cells. The heterozygote has both antigens, so the alleles are said to be codominant. The phenotypes simply indicate which antigens are present:

Genotype	Phenotype
M/M	M
M/N	MN
N/N	N

Table 12.2 (second column) shows typical data on the MN blood group distribution in the United States. Heterozygotes are the most common type, and more persons are M than N. The 500 typed persons have a total of 1,000 M or N alleles. The frequencies of the alleles are

$$\text{freq}(M) = p = (2 \times 160 + 240)/1,000 = 0.56$$
$$\text{freq}(N) = q = (2 \times 100 + 240)/1,000 = 0.44$$

If the tested population met the conditions of the Hardy-Weinberg law, then the expected genotype (or phenotype) frequencies would be (third column)

$$\text{expected freq}(M/M) = p^2 = (0.56)^2 = 0.3136$$
$$\text{expected freq}(M/N) = 2pq = 2(0.56)(0.44) = 0.4928$$
$$\text{expected freq}(N/N) = q^2 = (0.44)^2 = 0.1936$$

Multiplying these expected frequencies by 500 gives the expected number of the three genotypes (fourth column). As you can see, the expected and observed numbers are in good agreement, suggesting that this population is mating at random with respect to their

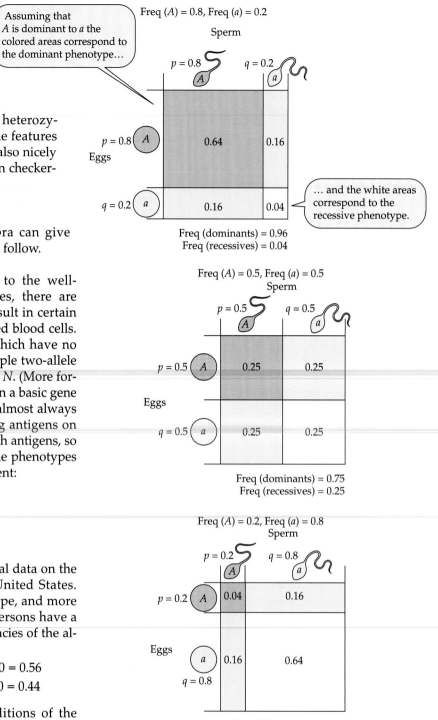

Figure 12.4 Checkerboards used to represent three Hardy-Weinberg equilibriums. The frequencies of the parental gamete types, A and a, correspond to allele frequencies, since each gamete carries one allele. The linear dimensions of the row and column headings are proportional to the allele frequencies in the parents, and the areas within the checkerboards are proportional to the genotype frequencies in the offspring.

TABLE 12.2 Agreement of observed MN blood group numbers with Hardy-Weinberg expectations

Genotype	Observed number	Expected frequency	Expected number
M/M	160	0.314	157
M/N	240	0.493	246
N/N	100	0.194	97
Total	500	1	500

MN types and meets the other conditions of the Hardy-Weinberg law. Since people are generally unaware of this characteristic and since it has no medical significance, there was really no reason to suspect otherwise.

The Problem of Dominance. Among 250,000 randomly chosen infants, perhaps just one will be affected with the fairly mild genetic disorder alkaptonuria, whose genotype is k/k (see Table 5.3). The diagnosis is simple—the darkening of urine in the diapers—but there is currently no test to tell which of the remaining 249,999 infants are normal carriers, K/k, and which are normal homozygotes, K/K. These observations are summarized as follows:

Genotype	Number	Frequency
K/–	249,999	$0.999\,996 \approx 1$
k/k	1	$0.000\,004 = 4 \times 10^{-6}$
Total	250,000	1

When we try to calculate the frequency of the alkaptonuria allele, $q = \text{freq}(k)$, a problem arises: We do not know the number of heterozygotes. Therefore, the previous method, which requires this information, cannot be used. Perhaps some assumptions about the population might aid in calculating q. Because the disease is very rare and relatively benign, a reasonable supposition is that the distribution of genotypes approximates the Hardy-Weinberg formulation. Making this assumption, we would then expect

$$q^2 = \text{freq}(k/k) = 4 \times 10^{-6}$$

Taking the square root of both sides, we get an estimate of the frequency of the recessive allele for alkaptonuria:

$$q = \sqrt{4 \times 10^{-6}} = 2 \times 10^{-3} = 0.002$$

Thus, among all the alleles of this gene, 1 in 500 gives rise to the defective enzyme if the population meets the Hardy-Weinberg conditions. What we are doing is equating an *observed* frequency (4×10^{-6}) with an *expected* frequency (q^2), given some hypothesis about the population. If the hypothesis does not apply, we err in proceeding with the calculation.

Although the Hardy-Weinberg law seems to be acceptable, the frequency estimate of 0.002 is not very good for another reason. If, by chance, the sample of 250,000 includes just a few more affected individuals, our estimate might be twice as big. On the other hand, if the sample does not happen to include even one alkaptonuric, then we have an allele frequency of 0. This range of possible estimates arises because we are looking at very small numbers of affected individuals. If we could identify and tally all of the normal *heterozygotes* (of which there are many more), our potential error would be much less. But we have no choice in this case, and we accept $q = 0.002$ as the best possible estimate of the allele frequency under the circumstances.

Having estimated the allele frequency, however roughly, we can now estimate the proportion of heterozygotes in the population:

$$\text{freq}(K/k) = 2pq = 2(0.998)(0.002) \approx 0.004$$

Thus, about 1 person in 250 is a carrier, a somewhat surprising result when we started with the information that only 1 in 250,000 has the metabolic disorder. For every person with alkaptonuria, there are about 1,000 carriers.

X-Linked Genes. Genes on the X chromosome, but not on the Y, will be present twice in females but only once in males. For this reason, the equilibrium frequencies under Hardy-Weinberg assumptions must be stated separately for the two sexes:

	Genotype	Hardy-Weinberg equilibrium frequency
Among females:	H/H	p^2
	H/h	$2pq$
	h/h	q^2
Among males:	H/(Y)	p
	h/(Y)	q

H and h are used here to represent the two alleles of any X-linked gene (e. g., hemophilia); p and q represent the frequencies of H and h, respectively, in either sex. Among females, the Hardy-Weinberg frequencies are just as they would be for an autosomal gene. A male's *one* allele, however, either H or h, determines his X-linked phenotype. Therefore, the frequencies of male genotypes or phenotypes are the same as the allele frequencies. For example, approximately 1 in 10,000 males is afflicted with classical hemophilia and therefore has the genotype $h/(Y)$. This proportion tells us that $q = \text{freq}(h) = 1/10,000 = 10^{-4}$. The Hardy-Weinberg formulation suggests why the trait is exceedingly rare (but not absent) among females. An affected female must have the genotype h/h and the expected frequency $q^2 = (10^{-4})^2 = 10^{-8} = 1$ in 100 million. If the total population of the United States (more than 250 million) were one big randomly mating Hardy-Weinberg population, we would expect to find just a few female hemophiliacs. Indeed, a few cases of homozygous female bleeders, with bleeder fathers and carrier mothers, have been reported.

An Exception: Inbreeding

Populations do not always mate randomly as assumed in our previous discussions. For some traits, people mate *assortatively*—that is, like phenotype with like. Human groups tend to mate assortatively with regard to stature (e.g., tall with tall), skin color, intelligence, and certain other quantitative traits.

Not only do matings occur between couples with similar *phenotypes*; matings also occur between couples with similar *genotypes*—that is, between those who are related to each other. Such couples may possess copies of precisely the same gene inherited from a common ancestor. A mating between relatives is said to be **consanguineous** ("with blood"),* and the offspring are said to be **inbred**. As we explain shortly, inbred individuals are more likely than others to be homozygous for any gene. This effect is seen for all forms of inbreeding, but the likelihood of homozygosity depends on the strength of the inbreeding.

In humans and other species with separate sexes, the closest form of inbreeding is parent-child or brother-sister mating. Such unions, called **incestuous**, are usually prohibited by law and by religion. The incest taboo probably evolved early in prehistory to minimize adverse effects on offspring and to help maintain the stability of family units. Because incest involves only family members and because those involved feel guilt and shame, the matter is not often resolved in open court, and its frequency is unknown.† In some cultures at some periods of history, however, brother-sister marriages were encouraged among royalty (but not among the general public). For example, Cleopatra may have been a child of a brother-sister mating (Figure 12.5). Plant and animal geneticists use programs of continued brother-sister matings to produce so-called *inbred strains* of corn, flies, mice, and so on. After a moderate number of generations of sib mating, the organisms become homozygous for most or all of their genes.

Uncle-niece and aunt-nephew marriages are very rare in human groups except where social custom dictates this form of consanguinity. In southern India, for example, Hindu men favor mar-

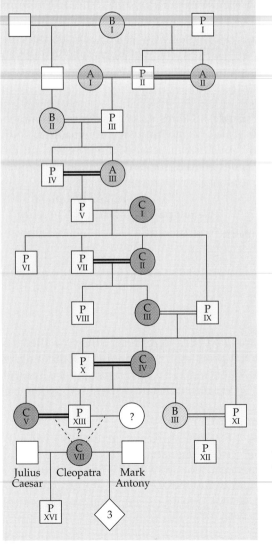

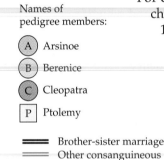

Names of pedigree members:

- (A) Arsinoe
- (B) Berenice
- (C) Cleopatra
- [P] Ptolemy

══ Brother-sister marriage
═ Other consanguineous marriage

*Although consanguinity derives, not from shared *blood*, but from shared *alleles*, the use of the former term persists in some everyday expressions: "bloodline," "full-blooded," "blue-blood," and so on.

†Here we define incest narrowly as sexual intercourse between sibs, half sibs, parent-child, grandparent-grandchild, or uncle-niece (aunt-nephew). Social scientists usually define incest more broadly to include abuse of young children within the home by older persons, especially adults, including step-relatives.

Figure 12.5   A pedigree of the Ptolemies, kings of Egypt from 323 to 30 B.C. The dynasty was often turbulent, marked by intrigue, intrafamily violence, and war. The famous Cleopatra is C-VII. It is unclear whether she was a child of a brother-sister marriage or whether she was not inbred at all. She was to marry her younger brother (half brother?), but he met an untimely death in war. In the manner of her forebears, Cleopatra dispatched another brother (or half brother) by poison. She and her son by Caesar, P-XVI, were the last of the Ptolemaic line. (Pedigree based on Weigall 1924.)

riage with a sister's daughter, but may not marry a brother's daughter. In one study of caste groups representing different socioeconomic levels (from well-to-do to so-called untouchables), about 19% of all marriages were between uncle and niece and another 22% were between first cousins (Reddy 1987).

Marriage between first cousins is the most frequent type of human consanguineous mating, in part because cousins may be close in age and their families are often known to each other. The fraction of first-cousin marriages in western Europe and North America is approximately 0.5%, but it is higher, sometimes much higher, elsewhere, as already noted for southern India. In the largely Muslim countries of northern Africa and western Asia, marriages contracted between persons who are related as second cousins or closer account for between 20% and 55% of the total, and among these, first-cousin marriages predominate (Bittles et al. 1991).

Note that two people who are related to each other share one or more ancestors. Sibs have the same mother and father, and first cousins share a set of grandparents (Box 12B). Two people in the same close ethnic group may have many identifiable common ancestors within a half dozen or so generations back. Extending even further back in time, we are all related one way or another. Consider that the maximum

number of ancestors n generations back for one person is $2n$. Thus, you have 2 parents, 4 grandparents, 8 great-grandparents, 16 great-great-grandparents, and so on. This number rapidly exceeds the entire population of the world at some time in the not-too-distant past and shows that all of us must have many ancestors in common.

Measuring Inbreeding. However consanguinity occurs, it increases the likelihood of homozygosity among the inbred progeny. A person can be homozygous without being inbred, of course, as is clear from the Hardy-Weinberg formulation: The probability of homozygosity in a randomly mating population is equal to p^2 (for A/A) plus q^2 (for a/a). Inbreeding pushes the proportion of homozygotes above this value. The *excess homozygosity* (above $p^2 + q^2$) comes about because of shared ancestry; a specific allele from a joint forebear can be successively replicated and transmitted through *both* lines of descent to mates who are related, and thence to their offspring. Two alleles that have a common history are said to be **identical by descent**. They not only code for the same polypeptide, but also trace back to exactly the same segment of DNA in a common ancestor.

The usual measure of inbreeding, called the **coefficient of inbreeding**, was invented by the much hon-

BOX 12B *Cousins, Close and Far-Flung*

Two people are related to each other if they have one or more shared ancestors. Sibs, of course, have the same mother and father, but most other relatives have common ancestry more remote than the parental generation. In Western civilizations, the system of naming related persons, beyond close relatives, is based on the term *cousin* and a few modifiers. In the pedigree here, first cousins (G and H) have a common set of grandparents (A and B), and second cousins (I and J) have a common set of great-grandparents (A and B again).
For first cousins, one each of their parents are sibs; for second cousins, one each of their parents are first cousins. The term *removed* pertains to a difference in generations: thus, G and J are first cousins once removed, and G and L are first cousins twice removed.

If D and E had the same father but different mothers (or vice versa), they

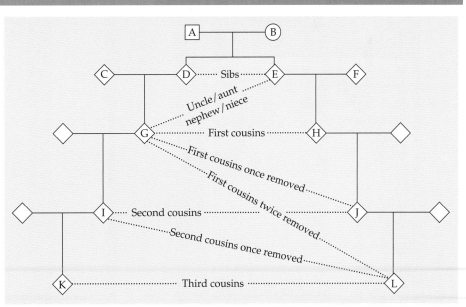

would be *half sibs*, and their children would then be *half* first cousins. If G and H had two common sets of grandparents, they would then be *double* first cousins; this relationship would

arise, for example, if their fathers were brothers and their mothers were sisters. Note that the diamond symbol is used when the sex of the individual is not relevant.

ored and long-lived American geneticist Sewall Wright (Figure 12.6). The coefficient measures the probability that alleles are identical by descent. Its value for the children of parents related in various ways is as follows:

Relationship between parents	Coefficient of inbreeding of their children = f
Siblings	¼
Uncle-niece or aunt-nephew	⅛
First cousins	1/16
First cousins, once removed	1/32
Second cousins	1/64

Note that the inbreeding coefficient, usually designated by f, decreases by a factor of ½ for each looser degree of relationship. The value $f = 1/16 = 0.0625$ for a person whose parents were first cousins means that about 6% of the time a pair of alleles—for any gene being considered—is expected to be identical by descent.

The Effects of Inbreeding.

Whether inbreeding is good or bad depends on whether increased homozygosity is good or bad. For the human species, in which matings are usually between unrelated or only remotely related persons, inbreeding often has adverse effects. This is likely due to increased homozygosity for recessive alleles with a detrimental effect on well-being. We can measure these effects either in terms of particular recessive genotypes or in terms of general health.

For a population mating at random and conforming to the other conditions of the Hardy-Weinberg law, we

have shown that the frequency of a recessive phenotype is q^2, where q is the frequency of the recessive allele. For example, the alkaptonuria allele has freq(k) = $q = 1/500$. Therefore, the *random* frequency of alkaptonuria is

$$q^2 = \text{freq}(k/k) = (1/500)^2 = 4.0 \times 10^{-6}$$

When a person is *inbred*, we need to know the person's coefficient of inbreeding, f, as well as the frequency of the allele q in order to calculate the probability of homozygosity. The general formula to calculate homozygosity for a recessive allele k is

$$\text{freq}(k/k) = qf + q^2(1 - f)$$

Notice that the expression for freq(k/k) reduces to q^2 when $f = 0$ (as it must because $f = 0$ corresponds to random mating). For children of first cousins having an inbreeding coefficient of 1/16, we would expect the frequency of alkaptonuria to be $(1/500)(1/16) + (1/500)(1/500)(15/16) = 1.3 \times 10^{-4}$. This frequency is about 32 times greater than the random frequency of alkaptonuria.

Using the general formula, we can calculate the likelihood of homozygosis among the offspring of any consanguineous mating and for any allele frequency. We can show, for example, that the rarer the allele, the greater is the likelihood of homozygosity from consanguineous mating relative to random mating (see question 10). This is because it is extremely unlikely that very rare alleles would ever come together in an individual *independently* of each other. But related persons can easily "co-inherit" the identical allele from a common ancestor. Thus, inbreeding is usually bad for people because it tends to reveal detrimental recessive alleles (for hemochromatosis, phenylketonuria, albinism, cystic fibrosis, Tay-Sachs, etc.) that are usually hidden in heterozygotes. Recessive mutations that do not result in specific, recognizable diseases may also cause harm by their small cumulative effects when made homozygous.

Although information on incest is difficult to collect, such data are often quite dramatic. One investigation from Canada included 21 mothers who, at the time of contact, were pregnant from an incestuous relationship (Baird and McGillivray 1982). Of the ensuing 21 children (7 from father-daughter matings and 14 from brother-sister matings), 43% had severe malformations or retardation without known cause or (in one case) a specific autosomal recessive disorder. Extensive data from lesser degrees of relationships are also available. This type of information can be used to estimate how many harmful recessive alleles an average person carries in a *heterozygous* state (thereby causing the carrier no damage). This estimation is done by extrapolating the amount of mortality or disease among children with known f values to what it would

Figure 12.6 Sewall Wright (1889–1988) was a leading population geneticist. He was still publishing papers at the age of 92, when this picture was taken. (Courtesy of William B. Provine, Cornell University.)

be if $f = 1$ (completely homozygous). Although this calculation is not very precise, the results suggest that the average person carries several recessive genes in heterozygous state that would lead to a premature death *if homozygous* (Bittles and Neel 1994).

The Prognosis for a First-Cousin Marriage. What should be said to first cousins who contemplate marriage and are anxious about possible harm to their offspring? Popular opinion holds that such children are likely to suffer malformations or be less intelligent than their peers. But first cousins carry only a small additional risk of having children with genetic defects of greater or lesser severity. All couples, whether related or not, face the likelihood (perhaps 2–3%) of having a child with a serious defect, depending on the criteria used to define "serious." Another few percent of children may fall outside a commonly accepted definition of "normal" but not have a serious defect.

A genetic counselor would certainly obtain a detailed family history from both of the cousins. If there is a suggestion of a deleterious recessive allele in one of them or in an ancestor, a specific probability calculation can be made. The couple can then consider both the risk, severity, and burden of the disorder. More often than not, however, there will be nothing in the family history on which to base a specific calculation, because rare recessive detrimental alleles will usually remain hidden in heterozygous condition generation after generation (see cystic fibrosis pedigree in Chapter 5). Only vague statements of the small but real increased risk (a few percent) from first-cousin consanguinity can then be made. The couple will have to evaluate this not very satisfactory information and come to a decision primarily on the basis of their own feelings.

Mechanisms of Evolution

When you encountered the work of Darwin and Wallace for the first time, you may have thought, "I could have done that," because the basic ideas seem so obvious and simple. Evolution by **natural selection** (often shortened to **selection**) is based on inherited differences among the members of a species that are competing for food, shelter, and mates. Some individuals are more successful than others, and they are able to transmit more of their "success" genes to the next generation. The average phenotype of a species slowly shifts toward the successful forms—those that are better suited to the environment. Over eons, this action of selection is a major cause of the exquisite adaptations that we see in living things. We discuss here (in an oversimplified way) how this happens, and we also consider other forces—*mutation, migration,* and *drift*—that contribute to evolutionary changes.

Mutation

Occurring in the recent or remote past, **mutations** are the ultimate source of the variation on which evolution acts. In Chapter 7, we noted that mutations can occur in any cell at any time, but many of them happen at the time of DNA replication. We noted that mutations may be caused by random thermal motions, chemicals, or radiation, but the source of a particular spontaneous mutation is usually unknowable. Mutations can affect any aspect of phenotype: physical, biochemical, or behavioral. Most mutations result in *detrimental* alleles, because the change alters a system already attuned to its internal and external environments. (By analogy, how likely is it that a *random* change in the innards of an adequate TV set will improve the picture?) Occasionally, a mutation might be beneficial initially (especially if it produces only a small change in the phenotype), or it might become beneficial in a later generation as a result of an environmental change. For example, mutations resulting in penicillin resistance are of no use to bacteria unless they encounter penicillin in their environment.

The time required for a mutation in an egg or sperm to show up in an individual's phenotype varies. For example, an autosomal mutation to a *dominant* allele will be immediately evident if transmitted to an offspring. On the other hand, the number of generations that elapse between an autosomal *recessive* mutation and its eventual expression in a homozygote depends on the rate at which the particular mutation recurs, on the pattern of matings, and on chance events. A high mutation rate and inbreeding, for example, will hasten the formation of the homozygotes in which the recessive allele will be expressed.

Mutation Rates. Determining a spontaneous mutation rate is inherently difficult because the events to be counted are very rare and, in the case of recessive mutations, generations removed from their phenotypic expression. These problems can be overcome by using experimental organisms, especially *haploid* microorganisms, in which the problem of dominance and recessiveness vanishes. In mice, researchers have developed tester stocks that are homozygous recessive for a half dozen different genes affecting fur color, eye color, or other easily scored phenotypes. With extensive mouse-rearing facilities and patience, a mutation to a dominant allele at any of these loci can be identified by the sudden appearance of the dominant phenotype among progeny generations.

For humans, experimental methods are not possible; instead, investigators must ferret out informative families from among worldwide populations. Even though mutations are rare, there are a lot of us to use as a base population, and large collections of hospital records may be available. It is possible to tally at least small numbers of mutant phenotypes and thereby obtain approximate values for some mutation rates. A mutation rate for a given gene is often expressed as *the number of new mutations per million gametes per generation.*

For example, we can identify new dominant mutations that arise from recessive alleles ($b \rightarrow B$) by noting persons with the dominant (usually abnormal) trait who have normal parents. In practice, an investigator examines all the births in a given region during a given time period. Then the mutation rate is the number of affected newborns with normal parents divided by twice the total number of newborns. The factor 2 occurs because an affected newborn arises from two gametes, only one of which carries the new mutant allele.

In one study, investigators saw 2 achondroplastic dwarfs with normal parents among 69,277 infants born at a Boston hospital over a 10-year period (Nelson and Holmes 1989). This dominant condition produces abnormal bone development, resulting in unusually short arms and legs, a normal-sized but swayback trunk, and an overlarge head (Figure 12.7).* It is the

*Achondroplastic dwarfs are quite different from normally proportioned pituitary dwarfs, who are deficient in growth hormone (see Figure 4.5). An organization called The Little People of America was formed in 1960 to help meet some of the social difficulties faced by all types of dwarfs.

Figure 12.7   An apparent achondroplastic dwarf, Giacomo Favorchi, painted by the Dutch artist Karel van Mander around 1600. (Courtesy of Statens Museum für Kunst, Copenhagen.)

most common form of dwarfism. Severe cases are stillborn or die in infancy, but persons who survive the first year of life often become healthy adults of normal intelligence (Angier 1994; Mawer 1998). The mutation rate, based on the Boston data, is $2/(2 \times 69,277) = 14$ mutations per million gametes per generation, or 14×10^{-6}. Clearly, the rate cannot be known very precisely because of the small number of mutations on which the calculation is based. A gene (dubbed *FGFR3*) responsible for achondroplasia has been identified on chromosome 4 and cloned. It encodes one of many receptors for growth factors that are needed to achieve normal embryonic development. It is intriguing that virtually all cases of achondroplasia involve mutations in exactly the same nucleotide, a "hot spot" within the gene (Bellus et al. 1995). This situation is very different from many other disorders (e.g., cystic fibrosis, or familial breast cancer), for which hundreds of different mutations have been documented.

Mutation rates for human genes range from a low of about 1 mutation to a high of about 100 mutations per million gametes per generation. Published values are likely to be overestimates of the true mutability of human genes, simply because the traits must be documented in several to many families for a calculation to be made. In contrast, some hereditary phenotypes are known in just one or a few families. These latter, truly rare diseases undoubtedly have much lower mutation rates. Thus, an average rate for a whole gene of 1 mutation per million gametes per generation (1×10^{-6}) may be nearer the truth than the higher values.

The published rates are also atypical in that the mutations lead to easily identifiable, usually severe phenotypes. An "ordinary" mutation may be more benign, leading to an amino acid substitution that has little, if any, effect on the functioning of its protein. Some researchers, especially the Japanese mathematical geneticist Motoo Kimura (1979), believe that most base substitution mutations occurring over evolutionary time may be neutral or nearly neutral in their effect on phenotype.

All in all, a gene is a remarkably stable chemical entity, because a probability of 10^{-6} can be interpreted to mean that a single gene followed from gamete to gamete will change only once in a million generations—about every 30 million years. Nevertheless, the total number of mutations among all of the 80,000 or so genes per individual is substantial and quite adequate to provide the raw material for evolution.

Selection

With regard to selection, the important variable to consider is **Darwinian fitness**, or simply **fitness**, a measure of the average number of offspring left by an individual or class of individuals. Fitness depends on *survival* to reproductive age and *fertility* once reproductive age is reached. For example, Tay-Sachs disease (Chapter 16) represents zero fitness due to infant death, whereas Turner syndrome (Chapter 13) represents zero fitness due to infertility. Zero Darwinian fitness would also apply to members of religious orders who practice celibacy. But consider persons with Huntington disease, a crippling condition characterized by progressive brain destruction; they may have near-average Darwinian fitness because the devastating dominant allele often does not manifest itself until after they have produced children.

The greater the overall fitness of individuals, the more their genes tend to be represented in future generations. This idea is the salient point of evolution by natural selection. Fitness is not usually an all-or-none condition, however; a difference of a percentage point in health or fertility may be significant over long stretches of time. The catchword *survival of the fittest* usually means just somewhat greater procreation by a particular group of individuals. In addition, fitness—a phenotypic trait—relates to prevailing environments. If surroundings change, the fitness of individuals may change. This relation is particularly striking when severe genetic diseases are treatable: The phenylketonuria genotype is nearly lethal in one dietary situation but nearly normal in another.

That natural selection occurs is well documented. When species are exposed to new environmental conditions, the differential reproduction of the more fit types leads to new genetic strains, such as poison-resistant rats, DDT-resistant mosquitoes, and penicillin-resistant bacteria (Box 12C). And in industrial areas with soot-covered tree trunks, pale-colored moths of many species have evolved to melanic (dark) strains that are apparently better able to camouflage themselves and avoid being eaten by birds. This is the standard textbook example of evolution in action, enhanced by the observation that moths have evolved back to pale forms in a few now smoke-free areas with effective pollution control. However, the lines of evidence supporting bird predation as the means of selection have been seriously questioned (Sargent et al. 1997). In addition, there are no good explanations for

the parallel increase and decrease of melanic moths in rural areas of North America, where industrialization has never been a factor.

New strains of animals and plants have also been bred by **artificial selection**. Through this process, breeders have enhanced certain economical or aesthetic traits for the benefit of people. The target species itself may or may not benefit, but if overall fitness is reduced, as often happens, the breeding stock can be coddled. Chickens laying more eggs? Mink with unusual colors? Leaner pork? Higher protein corn? Fancier goldfish? Some progress toward any goal is almost always possible—in either direction: Cow's milk can be selected for either higher or lower butterfat; dogs, for long legs or short, for short fur or long. The implication of these examples is that mutation has supplied plenty of phenotypically expressed genetic variation to be exploited by artificial (or natural) selection.

Now let us examine quantitatively the simplest aspects of selection: changes in allele frequencies of a *single* gene in a *constant* environment. In reality, the reproductive fitnesses of various organisms depend on a complex interaction of their total genotypes with varied and changing environments. Three specific types of simple selection are outlined in Table 12.3. In case I, the dominant allele A is weakly deleterious. If A were lethal (fitness = 0), it would be reduced to near-zero frequency in one generation. Note that "lethal" need not mean that the allele kills its carrier, only that it prevents reproduction. Each generation, all new cases of the dominant trait must arise from recurrent mutation ($a \rightarrow A$), since only a/a's have progeny.

Case II considers selection against a recessive allele. In the extreme situation illustrated in Table 12.3 of a **recessive lethal**, A/A and A/a are equally fit, but a/a has zero fitness. Notice that when the potentially harmful a alleles are present in heterozygotes, they are temporarily protected from the action of negative selection.

How much is the frequency of a recessive lethal allele reduced in one generation? We assume that the population meets all the Hardy-Weinberg conditions

BOX 12C   *Why Should I Take All My Pills?**

When your doctor prescribes an antibiotic to treat an infection, you are told to take the entire supply—down to the last pill in the bottle. Even if you feel better after taking just some of the pills, the doctor insists that you continue to the end. This directive is based on good evolutionary biology. After a short time, the drug may have killed many of the bacteria, but the few survivors are those that are more resistant to the drug. If the pills are stopped before all the bacteria are killed, the resistant survivors will contribute their drug-fighting genes to succeeding generations. The result could be unfortunate for your well-being: the same adverse symptoms as before, but an illness that requires a more potent dose of the antibiotic or a different antibiotic. Either course of action may have undesirable side effects.

*Adapted from Berra 1990.

TABLE 12.3 Simple models of different types of natural selection

Case	Genotype A/A	A/a	a/a	Type of selection
I	0.9	0.9	1	Weak selection against a dominant phenotype
II	1	1	0	Strong selection against a recessive (lethal) phenotype
III	0.8	1	0.3	Selection against both homozygotes; the heterozygote has highest fitness

Note: The numbers between 0 and 1 are fitness values. They express the relative numbers of offspring left by each genotype. For example, in case I, *A/–* individuals leave 0.9 offspring for every 1 left by *a/a*.

except for the indicated selection. Under these circumstances, the frequency of a recessive lethal, freq(*a*) or simply *q*, is reduced in one generation from *q* to $q' = q/(1 + q)$. (The quantity q' is read *q* prime; the algebra is worked out in Mange and Mange 1990.) The change is a small one when *q* is small to begin with as would be the case for any lethal allele. For example, if *q* = 0.0200, then the frequency of a recessive lethal will be reduced in one generation to $q' = 0.0196$. Using this formula, we can show that *the rarer the recessive lethal becomes, the less effective selection is in reducing its frequency further.*

The reason why selection against a recessive lethal is progressively less effective becomes clear when we examine the ratio of heterozygotes to homozygous recessives. In a population of 1 million, these ratios are as follows (using Hardy-Weinberg arithmetic):

q = freq(a)	Number of A/a	Number of A/a	Ratio A/a:a/a
0.2	320,000	40,000	8:1
0.02	39,200	400	98:1
0.002	3,992	4	998:1

Thus, as the allele gets rarer (reading from top to bottom), *A/a* heterozygotes are not reduced in numbers nearly as quickly as *a/a* homozygotes are. In the last line, there are nearly 1,000 times more phenotypically normal heterozygotes, which *mask* the recessive lethal from selection, than homozygous recessives, which *expose* the lethal to selection.

Eugenic proposals in the early part of this century suggested that our genetic endowment could be much improved if we got rid of rare heritable diseases by sterilization of affected persons. As far as recessive traits are concerned, we can see that such a plan could not succeed. For example, the nonreproduction of a recessive genotype would yield allele frequency changes over three generations of 0.0100 to 0.0099 to 0.0098 to 0.0097. This tiny reduction would represent the history of about 100 years of sterilization of *a/a* individuals!

As a corollary, treating rare recessive lethal genotypes such as phenylketonuria, so that these individuals can lead productive and fertile lives, will increase the frequency of the causative alleles only very slowly. Thousands of years may be involved for appreciable frequency changes. Moreover, the snail's pace of evolutionary change characteristic of recessive *lethals* is slower yet for the more common recessive *detrimental* alleles, which have fitness values greater than 0 but less than 1. Substantial reductions in the frequency of severe recessive diseases might be achieved by identifying heterozygous persons (e.g., by biochemical tests) and counseling them as to risks. Voluntary modification of reproduction by heterozygotes could achieve quickly what sterilization of affected persons could not.

For case III in Table 12.3, there is a well worked out human example—that of sickle-cell disease in regions of the world where malaria is prevalent. Here the fitness of sickle-cell heterozygotes is higher than that of either homozygote. The reasons for this odd circumstance, a **superior heterozygote**, are as follows: The malarial parasite, which spends part of its life cycle inside red blood cells, flourishes less well when these cells possess some sickle-cell hemoglobin. Therefore, heterozygotes, *A/S*, are less susceptible to malaria than are normal homozygotes, *A/A*. Although persons with sickle-cell disease, *S/S*, are also resistant to malaria (as you would expect), they often die young of sickle-cell complications, so their malarial resistance is of little benefit. As a result of losing a certain proportion of *both* homozygotes, a balance is struck, leading to stable intermediate allele frequencies. This so-called **genetic equilibrium** occurs in a population when the frequency of the sickle-cell allele is about 15%. Because of the equilibrium, sickle-cell disease, despite being a serious disease, is not eliminated in regions where malaria occurs, but recurs in about 1 in 45 newborns (Figure 12.8).

In regions of the world where malaria is not a problem, the fitness of *A/A* homozygotes is as high as that of *A/S* heterozygotes, because the absence of malaria erases the selective advantage of being heterozygous. Thus, only the *S* alleles in persons with sickle-cell disease (*S/S*) are lost through selection. In the United States, for example, sickle-cell disease acts like a simple recessive detrimental, and freq(*S*) continues to decrease at a slow rate in successive generations. It still remains a serious health problem among U.S. black populations, however.

Malaria   Sickle–cell disease

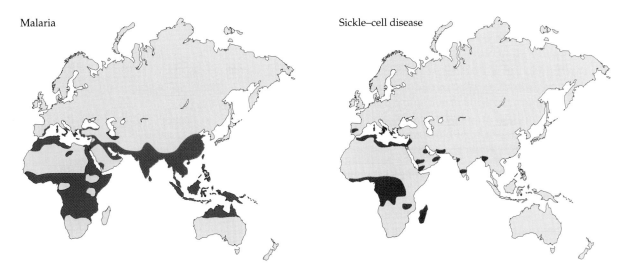

Figure 12.8 The geographical distribution of malaria before about 1930, compared with the distribution of sickle-cell disease. (After Motulsky 1960.)

In summary, selection is the organizing principle of evolution, producing allele frequency changes that gradually allow populations to be better adapted to their environments. It is primarily selection that has produced the multitude of creatures embellished with their multitude of marvelous structures and behaviors. We are awestruck at the resulting variety and complexity of living things: orchid petals, cactus spines, firefly flashes, cobra fangs, hummingbird wings, human brains. Darwin, too, was dazzled. He ended the first (1859) edition of *The Origin of Species* with these words:

> There is grandeur in this view of life, with its several powers, having been originally breathed into a few forms or into one; and that, whilst this planet has gone cycling on according to the fixed law of gravity, from so simple a beginning endless forms most beautiful and most wonderful have been, and are being, evolved.

Migration and Gene Flow

We have described how mutation provides the raw material for evolution and how selection can gradually modify allele frequencies over successive generations. Now we briefly consider allele frequency changes due to migration between populations and, in the next section, changes due to haphazard events that occur just because population numbers are small.

Migration is the movement of people (or other organisms) from one place to another. Prehistoric migrations (e.g., Asians into the New World) often involved movement into previously uninhabited regions, where new environments may have imposed new selective pressures. More recent migrations have been into al-

ready occupied territory (e.g., Europeans and Africans into the New World). Although the immigrants may remain genetically isolated from their neighbors, usually some interbreeding occurs, leading in subsequent generations to mixed populations. When the original groups differ in allele frequencies, it may be possible to document the amount of **gene flow**, that is, the introduction of new alleles into a population. To the extent that a population receives new alleles, the genetic variability within it is increased. But gene flow between populations causes the populations to become more similar to each other than they would otherwise be.

Analysis of gene flow has been used to complement historical knowledge of migration for various groups:

American blacks: a two-way mix of Africans (of various cultures) and Europeans (of various cultures, but primarily English)

Brazilian subgroups: a three-way mix of native American Indians, Sudanese and Bantu Africans, and Europeans from Portugal, Italy, and Spain

Hawaiian subgroups: multiple mixtures of native Polynesians, whites, Japanese, Chinese, and Filipinos

For example, about 400,000 slaves were brought to the United States, mostly during the eighteenth century. They came from western regions of Africa, extending from Senegal in the north, through Benin areas, to Bantu-speaking regions in the south. As noted in Figure 12.9, the percentages from each area are approximately known from molecular genetic data based on variations in the β-globin gene. In the United States, the offspring of slaves and Europeans and their subsequent descendants were, and are, almost always considered black. Thus, by this artificial definition, the mixing has been virtually all in one direction. But by examining the frequencies of a particular allele in African, European, and American black groups, it is possible to calculate the proportions that the parent populations have made to the mixed group. In the southern United States, about 4–10% of black ancestry is white, and in northern populations that figure rises to 20–30% (Mange and Mange 1990).

Genetic Drift

In general, **genetic drift**, or simply **drift**, refers to unpredictable changes in allele frequencies that are increasingly more likely the smaller the population. Whether a particular allele is "good" or "bad" from the point of view of selection, its frequency may change abruptly—up or down—from one generation to the next. Drift can occur because of chance departures from Mendelian expectations in a few families or because a newly formed isolated group is not representative of the original population. The latter form of drift is called *founder effect*, and it can result in an unusually high frequency of a genetic disease in an ethnic population. Another special case is the *bottleneck effect*, in which a community's sample of alleles is changed when the population is reduced in numbers by natural catastrophe, infectious disease, or other factors.

In Table 12.4, we illustrate drift for a very small population of jelly beans. We pick a founder group of four jelly beans (representing, say, the alleles in one male and one female) from a large base population with "allele frequencies" of 0.7 (red) and 0.3 (black). Such a sample of four jelly beans is more likely than not to be unrepresentative of the base population. It is not even possible, of course, to get exactly 0.7 and 0.3 with four jelly beans. One can get close, that is, 0.75 (3 red) and 0.25 (1 black), but that result is expected to happen only about 41% of the time (as calculated in the table). Note that it would not be unusual with such a small sample to completely eliminate one allele or the other: About 24% of the time the four jelly beans are expected to be all red, and about 1% of the time they are expected to be all black. The major point of this exercise is to illustrate that in small groups, chance plays a significant role. Furthermore, a small population brings about haphazard fluctuations in *all* genes. Some may go up in frequency, while others may go down.

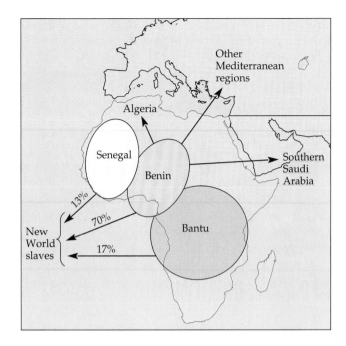

Figure 12.9 Areas and migration of three African groups—Senegal, Benin, and the Bantu-speaking region—carrying different chromosome segments surrounding the β hemoglobin locus. All three regions were raided for slaves for the New World. Blacks in Mediterranean regions and in Saudi Arabia, however, are exclusively of the Benin type, having followed established caravan trade routes across the Sahara. (Map after Labie et al. 1986; New World data from Nagel 1984.)

The importance of drift in evolutionary history is disputed. Fisher, who downplayed its role, thought that most of evolution occurred in large populations by the kinds of selective systems we considered earlier. Such evolutionary changes are very slow, to be

TABLE 12.4 Genetic drift illustrated by picking four jelly beans at random from a large bowl with 70% red and 30% black

Jelly beans picked				Probability factors	Probability calculations	Overall distribution
1st	2nd	3rd	4th			
●	●	●	●	.7 .7 .7 .7	$(.7)^4 = .2401$	4 red, 0 black
●	●	●	●	.7 .7 .7 .3		
●	●	●	●	.7 .7 .3 .7	$4(.7)^3(.3)^1 = .4116$	3 red, 1 black
●	●	●	●	.7 .3 .7 .7		
●	●	●	●	.3 .7 .7 .7		
●	●	●	●	.7 .7 .3 .3		
●	●	●	●	.7 .3 .7 .3		
●	●	●	●	.3 .7 .7 .3	$6(.7)^2(.3)^2 = .2646$	2 red, 2 black
●	●	●	●	.7 .3 .3 .7		
●	●	●	●	.3 .7 .3 .7		
●	●	●	●	.3 .3 .7 .7		
●	●	●	●	.7 .3 .3 .3		
●	●	●	●	.3 .7 .3 .3	$4(.7)^1(.3)^3 = .0756$	1 red, 3 black
●	●	●	●	.3 .3 .7 .3		
●	●	●	●	.3 .3 .3 .7		
●	●	●	●	.3 .3 .3 .3	$(.3)^4 = .0081$	0 red, 4 black

sure, but long periods of time have been available. Wright, an advocate of drift, emphasized that evolution could occur more rapidly when a species is broken up into groups of moderate size, between which there is a limited amount of migration. When populations are too small, however, drift leads inevitably to the loss of alleles, to the reduction of variability, and to the consequent inability of populations to respond to environmental change. Wright (1930) wrote that at intermediate sizes, "there will be continuous kaleidoscopic shifting of the prevailing gene combinations, not adaptive itself, but providing an opportunity for the occasional appearance of new adaptive combinations of types which would never be reached by a direct selection process." With regard to human evolution, drift was more important in prehistory when our species was broken up into small hunting camps, tribes, and clans.

Human Evolution

A **species** is a group whose members are able to breed with each other but unable to produce viable and fertile offspring with members of a different species. The average phenotype of a species can change over time in response to the evolutionary forces already noted. Another main theme of evolution is the divergence of two species from a common ancestral stock. During this branching process, two geographically isolated groups of the same species come to have genetic differences because they must respond to different sets of environmental conditions. Eventually, each group evolves unique gene combinations that lead to the intersterility that distinguishes them as separate species.

Human beings all around the planet now constitute a single species. Tracing our evolution from preexisting species is unfinished work, and hardly a year goes by without several announcements of important new fossil discoveries (e.g., Culotta 1995). Different interpretations of bony fragments enliven the paleontological and general literature. But the broad outline of current knowledge points to the separation of a humanlike (hominid) lineage from apelike ancestors about 4 to 6 million years ago in Africa (Figure 12.10). A number of species that were intermediate between the apelike ancestors and the more recognizably human types have been grouped in a genus called *Australopithecus*; these creatures were small-brained, but unlike fossil or modern apes, they walked upright on two legs. The well-known Lucy skeleton from

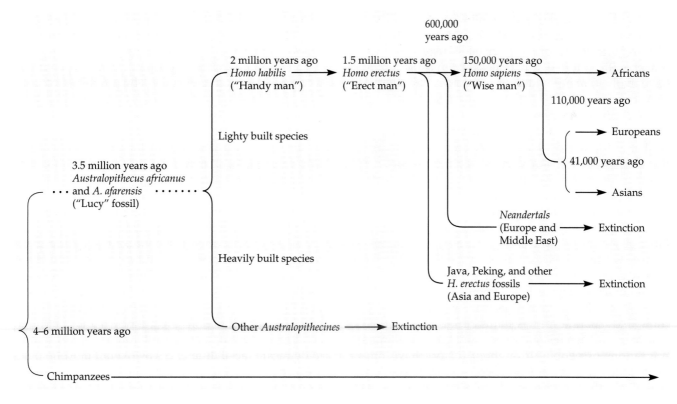

Figure 12.10 Some generally, but not universally, accepted landmarks in the evolution of human beings and closely related species. The discovery of new fossils and the compilation of new molecular data may well lead to modifications of this evolutionary tree. Migrations between human groups, especially in the last 50,000 years, complicate this simple presentation.

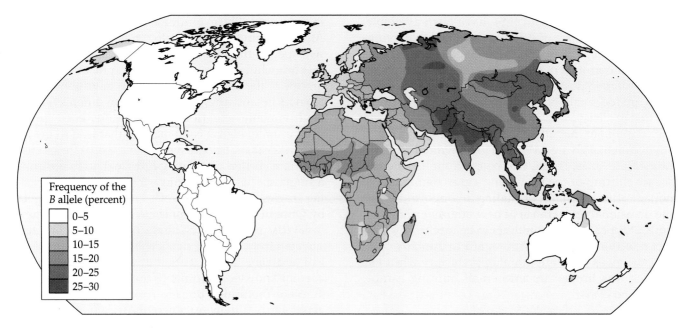

Figure 12.11 The distribution of the *B* allele of the *ABO* gene among indigenous populations of the world. (From Mourant et al. 1976.)

Ethiopia represents an early australopithecine (about 3.5 million years old), but earlier fossils of two-legged creatures (about 4 million years old) have recently been found in Kenya (Tattersall 1997; Wilford 1995). (Hominid fossils in the 2- to 4-million-year range have been found only in Africa.) Although some australopithecine branches seem to have been dead ends, one branch with a more lightly built skeleton is thought to have evolved about 2 million years ago into populations of the genus *Homo* that are our direct ancestors. A succession of species—*H. habilis, H. erectus, H. sapiens*—represent increasing brain size and greater use of the hands for making tools, brandishing weapons, and carrying food and babies. Other hominid species not in our direct ancestral line also inhabited the earth throughout these periods and perhaps as recently as 30,000 years ago (Wilford 1996).

Just as separate species can differentiate from each other given isolation and enough time, so, too, can subgroups within a species. **Races** can be defined as the subdivisions of a species that have come to differ, to a greater or lesser degree, in the frequencies of the alleles they possess. But human races have not achieved anything like separate species status because (1) the time available has been relatively short and (2) the continual migrations of people throughout history have prevented the degree of isolation necessary for speciation. It is likely that few, if any, human groups have remained completely isolated genetically for more than a few generations. Races depend on finer distinctions than species, but the differences often call attention to themselves and invite some sort of classification.

Although there is theoretical agreement on what constitutes separate species, there is no single handy criterion by which to define races. Consequently, there are many opinions as to how the job should be done, or even if it should be done. Among anthropologists, the "lumpers" recognize relatively few different races—commonly just Africans, Caucasians, and Asians—whereas the "splitters" come up with as many as 60 or more divisions. The problems of classification relate to the number and placement of lines between more or less merging categories, and no consensus is likely. Perhaps no practical purpose is served by drawing lines at all. But racial groups are of scientific interest, primarily for investigating how they evolved. That is, the measurement of variability within and between divisions of humankind provides insight into the processes of biological and cultural evolution (Cavalli-Sforza et al. 1994).

The traditional measurements of racial differences relate to hair and skin color and to the size and shape of various parts of the body. These traits are substantially influenced by environmental variables and depend on genes (not particularly noteworthy ones) whose precise functions are unknown and that interact during development in unknown ways. Attempts to delineate racial differences by overall morphologies are also made difficult because the traits are measured along a continuous scale; the uninterrupted sequence of data makes division arbitrary.

To avoid some, but not all, of these problems, anthropologists have increasingly turned to simply inherited traits, which are not influenced too much by the environment. As early as 1918, antibodies anti-A and anti-B were used by U.S. army physicians to survey different groups of soldiers. Since then, blood samples have been channeled into researchers' test tubes from literally millions of people living in virtually every habitat around the globe. Using the Hardy-Weinberg law to calculate allele frequencies, masses of blood group and plasma protein data have pointed up simple differences between human races. More recently, other kinds of simply inherited traits have been added to the repertoire of tools. These include various enzymes, hemoglobin types, histocompatibility types, and especially DNA markers of both nuclear and mitochondrial DNA. Using these characteristics, we know now that the genetic variability among the individual members of any race, however defined, is greater than the average differences between races.

Extensive tables and maps of population differences are organized by Cavalli-Sforza et al. (1994) and by Mourant et al. (1976). Since racial prejudice is widespread, we emphatically state that these data provide no basis for the "superiority" of one group over another. For example, current knowledge of the worldwide distribution of the *B* allele of the *ABO* gene is presented in Figure 12.11. The maximum frequencies of about 30% appear in central Asia. Frequencies decrease westward into Europe, diminishing to about 5% and even less in the Basque populations of France and Spain. The gradual changes in the distribution of the *B* allele across Europe seem to be due to the Mongolian invasions of the fifth through fifteenth centuries. The allele is absent among Indian populations of the New World and among the aboriginal populations of Australia.

Unfortunately, the collection of data on human genome variation has always been done in a haphazard manner by a multitude of worldwide researchers according to their own particular, often narrow interests. For many years, L. L. Cavalli-Sforza of Stanford University, has been spearheading an effort to coordinate the accumulation of consistent data via the Human Genome Diversity Project. Although some people have worried about possible discrimination against ethnic groups who contribute cell samples, the project has recently gained support from the U.S. National Research Council, and initial funding has been forthcoming (Cavalli-Sforza 1998; Steinberg 1998).

The Origin of Major Human Groups

Fossils show that about 1.5 to 2 million years ago, *Homo erectus* migrated from Africa and spread throughout the Old World. But researchers disagree strongly about the subsequent evolution of the major human groups (Figure 12.12). The **multiregional theory** (also called *regional continuity*) holds that *H. sapiens* evolved from *H. erectus* independently in several geographically wide-

spread areas—Africa, Europe, Asia, Australia—over the last million years (Thorne and Wolpoff 1992). This theory assumes that intercontinental migrations provided enough gene flow between groups to maintain a single species at any time. But some experts simply do not consider it likely that parallel patterns of evolution could continue over such long periods of time. In contrast, the **out-of-Africa theory** states that *H. sapiens* arose only in Africa, spread over the world from there relatively recently (beginning 100,000 or so years ago), and replaced any preexisting *H. erectus* groups in various geographical locations (Wilson and Cann 1992). The weight of evidence seems to favor the out-of-Africa viewpoint, but neither theory is likely to be 100% true (Gibbons 1997; Nei 1995).

Interesting observations on this controversy rely on *mitochondrial DNA* (*mtDNA*). Recall (Chapters 2 and 3) that mitochondria are small organelles that generate most of a cell's chemical energy. Each cell has several hundred mitochondria, and each mitochondrion has about ten copies of a DNA circle with about 16,600 nucleotides. The base sequences encode transfer RNAs, ribosomal RNAs, and a few dozen enzymes that help

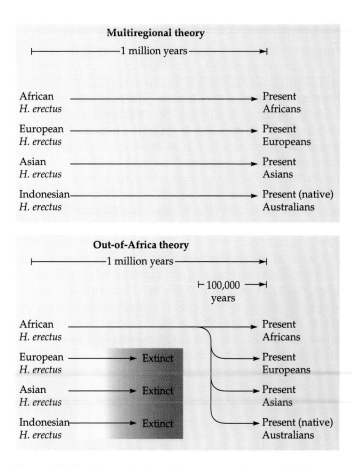

Figure 12.12 The multiregional theory versus the out-of-Africa theory of the origin of modern human groups. (After Cavalli-Sforza et al. 1994.)

synthesize the energy-carrying molecules of ATP. Mitochondrial DNA is relatively easy to purify (partly because there are so many copies per cell), and its simple structure (no introns or repetitive sequences) makes it ideal for some types of analysis. Although an individual has about 10^{15} molecules of mtDNA, all of the molecules seem to be identical. Recall further that all mitochondria in a fertilized egg come from the egg only; the mitochondria that are present in the tail section of a sperm are excluded during the formation of the zygote. Thus, the genes in mtDNA follow an exclusively maternal inheritance pattern generation after generation, without any reassortment or crossing over.

The mutation rate of mtDNA is high, measuring about one base substitution per seven bases per million years (or higher in some studies). Since these rates are much higher than that for nuclear DNA, mtDNA is a sensitive indicator of evolutionary modification. Researchers at the University of California, Berkeley, and elsewhere have examined variation in mtDNA sequences from hundreds of different human populations from around the world. Using computer programs to construct evolutionary trees, they have been able to trace all present-day human mtDNA sequences to a small number of women, perhaps one woman, who lived 150,000 to 200,000 years ago, the presumed oldest time that our species could have begun to spread over the earth. Many other females and males lived contemporaneously with the ancient mitochondrial source, and many other females and males of that time have had modern descendants, but this woman was the source of all mitochondria in the present human population. Data from mitochrondrial DNA suggest an African origin of all present human groups because African branches of the trees reach back further in time and African populations are more variable in DNA sequences than others. Similar conclusions arise from work with nuclear genes, including DNA sequences on the Y chromosome (Cann 1997; Fischman 1996; Wade 1997).*

Critics of these interpretations claim, however, that the mitochondrial dating techniques underestimate evolutionary time spans. They are also critical of the statistical programs that generate the branching trees, claiming that many other equally likely trees could come out of the computer analyses—trees that suggest an origin of modern humans older than 200,000 years ago and a place other than Africa. These critics emphasize aspects of the fossil record and suggest that human races evolved over perhaps 1 million years, according to the multiregional model. Recent versions of the multiregional theory also allow some gene flow from later migrations out of Africa. As the amount of this postulated later gene flow increases, the multiregional and out-of-Africa theories become more complex and more like each other (Tishkoff et al. 1996).

Most archaeologists are in agreement with a fascinating aspect of human evolution based on recent mitochondrial DNA sequencing. In a carefully controlled study, Svante Pääbo and colleagues at the University of Munich, as well as a second group of researchers at Pennsylvania State University, examined DNA extracted from a Neandertal arm bone. This fossil came from the original type specimen found in 1856 in the Neander valley (German *Tal*, "valley") near Düsseldorf, Germany. The Neandertals were a distinct group of archaic humans with very muscular and thickset bodies. They inhabited Europe and western Asia from about 300,000 years ago up to, but not beyond, 30,000 years ago, coexisting with modern humans during the latter part of that time.

A long-standing question is whether Neandertals interbred with the lineage leading to modern humans. Using PCR, the researchers amplified 379 bases of mitochondrial DNA. They found an average of 27 differences between it and modern humans, whereas pairs of modern humans differ from each other on the average in only 8 bases. Using a reasonable mutation rate for mitochondrial DNA, Pääbo and his colleagues suggest that Neandertals split from the lineage leading to modern humans about 600,000 years ago (Kahn and Gibbons 1997; Ward and Stringer 1997). Furthermore, they conclude that Neandertals became extinct without contributing mitochondrial genes to modern humans. It is likely that the two groups evolved completely independently. They did not interbreed—or if they did, the unions were infertile or produced sterile offspring.

Finally we note that DNA disintegrates over time. All that remains after some thousands of years are highly damaged, short fragments. Retrieval of DNA sequences older than 100,000 years may well be impossible. In addition, since PCR is capable of amplifying part of a single molecule of DNA, the procedure is very sensitive to contamination by, perhaps, the DNA from a few skin cells sloughed off from the examiner. Furthermore, the PCR replication process is not 100% accurate, especially when there are just a few templates to start with. Reports of authentic amplification of ancient DNA millions of years old, (e.g., DNA from amber-embedded insects) must be viewed with skepticism. Although we discussed some too-good-to-be-true cases in the first edition of this book, such claims have now been largely discredited (Lindahl 1997).

---

*The man who possessed the ancestral Y chromosome sequence is sometimes referred to as Adam, and the woman who possessed the ancestral mitochondrion as Eve. These Biblical terms are purely metaphorical. Note that any nuclear gene (or DNA marker) in modern humans can in theory be traced back to a single ancestral form at some time in the past. The ancient man or woman who is the common ancestor in this sense will differ from gene to gene. This is because chromosomal reassortment and crossing over shuffles nuclear (but not mitochondrial) genes every generation.

Summary

1. Independently of each other in the mid-nineteenth century, Darwin and Wallace developed the idea that gradual evolution by natural selection can account for the vast array of complex living forms. Many diverse lines of evidence support the concept of evolution.

2. Population genetics extends the Mendelian rules of inheritance from families to larger groups, that is, to mixed populations that can be described by the frequencies of genotypes, phenotypes, alleles, and matings.

3. The Hardy-Weinberg law states that population statistics remain constant when evolutionary forces are negligible. For p = freq(A) and q = freq(a), the equilibrium genotype frequencies of A/A, A/a, and a/a become p^2, $2pq$, and q^2, respectively, in one generation.

4. Some aspects of evolutionary theory begin with modifications of the Hardy-Weinberg law. The frequency of a particular allele at a given time depends primarily on evolutionary forces in the past and not on whether the allele acts in a dominant or recessive fashion.

5. Hardy-Weinberg arithmetic can be applied to many human traits to obtain statistical information on allele and genotype frequencies. The Hardy-Weinberg law can be extended to situations involving X-linked genes.

6. A mating between relatives is said to be consanguineous, and the resulting children are inbred. Inbred persons may be homozygous for alleles identical by descent from a common ancestor. The rarer a recessive allele, the greater is the risk of its appearing homozygously in inbred persons as compared to noninbred persons.

7. Whether inbreeding is beneficial or detrimental depends on the consequences of increased homozygosity. On the average, inbreeding in humans has small, adverse effects on the offspring of first cousins, the most common consanguineous mating.

8. Mutations provide the raw material for evolution. Measuring the rates of mutations in humans is imprecise because of the rarity of mutational events for a given gene. A typical human gene mutates at the rate of about one mutation per million gametes per generation.

9. Selection—natural or artificial—occurs when various genotypes in a population differ in fitness, which is defined as the ability to leave offspring. Superior fitness requires a combination of higher than average survival to reproductive age and/or greater fertility.

10. Selection against a rare recessive phenotype decreases the frequency of the recessive allele only very slowly, because most rare recessive alleles are hidden in heterozygotes. The rarer the allele, the more effectively are the alleles concealed. Selection in favor of the heterozygote stabilizes the allele frequencies at intermediate levels.

11. Migration may lead to gene flow, the introduction of new alleles into a population. Drift refers to haphazard changes in allele frequencies that become more likely the smaller the population. Drift may have been important in human prehistory, when groups were smaller.

12. Hominid species have evolved over the last 5 million years. Our species, *Homo sapiens*, is divided into racial groups, sometimes distinctive but often ill defined. Allele frequencies in various races differ to varying degrees and provide information on past migrations in space and time.

13. The multiregional theory and the out-of-Africa theory compete as explanations for the origin of major human groups. The analysis of mitochondrial and nuclear DNA has generally favored the out-of-Africa theory.

Key Terms

allele frequency	migration
artificial selection	multiregional theory
coefficient of inbreeding	mutation
consanguineous mating	natural selection
drift (genetic)	out-of-Africa theory
fitness (Darwinian)	p and q
gene flow	phenotype frequency
genetic equilibrium	population
genotype frequency	race
Hardy-Weinberg law	random mating
identical by descent	recessive lethal allele
inbred offspring	species
incestuous mating	superior heterozygote
mating frequency	

Questions

1. In a certain population, there were 100 people of genotype A/A, 400 of genotype A/a, and 500 of genotype a/a. Calculate the frequencies of the alleles A and a.

2. For the allele frequencies that you calculated in question 1, what would be the genotype frequencies in the next generation if the population now meets the conditions of the Hardy-Weinberg law? (Note that these values turn out to be different from the genotype frequencies in question 1. Evidently, the parental population in question 1 had not previously met one or more of the Hardy-Weinberg conditions.)

3. For each of the populations below, calculate the expected frequencies of the phenotypes for the MN blood group, assuming that the conditions of the Hardy-Weinberg law are met.

	Allele frequency	
Population	**M**	**N**
Bushmen of Botswana	0.60	0.40
Amish of Indiana	0.50	0.50
Polynesians of Easter Island	0.40	0.60

4. A group of 100 people splits away from a larger population and establishes a separate society. With respect to the MN blood types, the emigrants number: type M = 41, type MN = 38, type N = 21.

   (a) What are the allele frequencies? Do you need to assume the Hardy-Weinberg law to do these calculations?
   (b) If this group and their descendants now meet the conditions of the Hardy-Weinberg law, what are the expected frequencies of the MN phenotypes in subsequent generations? (Although the group is fairly small, assume that genetic drift is negligible.)

5. Consider the sickle-cell gene, S, and its normal allele, A. In an adult African population, an investigator finds the following numbers of the three genotypes:

Genotype	Number
A/A	605
A/S	390
S/S	5
Total	1,000

What numbers (out of 1,000) should the investigator have expected if this population had met the conditions of the Hardy-Weinberg law?

6. In a certain population, 1 person in 10,000 is albino (*c/c*). What fraction of this population is expected to be heterozygous? Do you need to assume the Hardy-Weinberg law?

7. Among human males, about 10% are color-blind because they carry an allele of an X-linked gene. What percentage of women are expected to be color-blind? Assume the Hardy-Weinberg law.

8. Among French Canadians in Quebec, the frequencies of the *ABO* alleles are roughly freq(*A*) = 0.3, freq(*B*) = 0.1, and freq(*O*) = 0.6. What are the expected Hardy-Weinberg frequencies of the ABO blood types? Note that this question extends the Hardy-Weinberg law to a case with *three* alleles of a gene. As with two alleles, the expected frequency of any homozygote is the square of its allele frequency; the expected frequency of any heterozygote is twice the product of the corresponding allele frequencies.

9. Draw the pedigree of John and Mary, who are double first cousins.

10. The frequency of an allele for a recessive disease is freq(*e*) = $1/1,000$.

   (a) What is the expected frequency of the disease among the offspring of unrelated parents, that is, under random mating?
   (b) What is the expected frequency of the syndrome among the offspring of first cousins in this population?
   (c) How many times more frequent is the *e/e* genotype in case (b) than in case (a)?

11. In families with both normal-height and achondroplastic children, the fathers tend to be younger when they beget the normal-height babies than when they beget those with dwarfism. Give a possible explanation.

12. Assume that a rare dominant allele has incomplete penetrance. Thus, some affected children might have normal-appearing parents even when a parent possesses the dominant allele. Would the calculated mutation rate then be too high or too low?

13. Assume that all human genes stopped mutating completely. How would you know that this unusual phenomenon occurred?

14. Explain why selection against a recessive lethal becomes less and less effective as the allele becomes rarer and rarer.

15. The effectiveness of selection against a recessive lethal could be speeded up if we could identify heterozygotes and could counsel them about the personal and social risks of procreation. Why would this procedure be potentially much more effective in reducing the recessive trait than just nonreproduction by affected homozygotes?

16. Give two reasons why the frequency of the sickle-cell allele among American blacks is less than that among their African ancestors.

17. Recall from Chapter 8 that insulin, with amino acid chains labeled A and B, is formed from a longer polypeptide called proinsulin when a segment (called C) is removed and discarded. Kimura (1979) compared the amino acid differences between species for insulin and for the C segment. He found far fewer differences for the A and B chains than for the C segment. Does this make evolutionary sense? Why?

18. Why is mitochondrial DNA useful in studying evolution?

Further Reading

Cavalli-Sforza et al. (1994) provide an extensive treatment of all topics in this chapter. Keats (1997) and Hartl (1988) give short introductions to population genetics, and Cavalli-Sforza (1998) gives a short account of the DNA methods used to investigate recent human evolution. Crow (1987) presents historical insights, and Neel (1994) includes an interesting chapter on why and how to do a study of human inbreeding. Jameson (1977) reprints important papers in evolutionary genetics, including ones by Hardy and Weinberg. Mawer (1998) explores in a novel what it is like to be an achondroplastic dwarf.

The autobiographies by Darwin (1969 reprint) and by Wallace (1908) provide a unique introduction to evolutionary thought. General accounts of evolution include the informative books by Dawkins (1986) and Strickberger (1990). Gould (1995) provides a collection of entertaining, stimulating, and authoritative essays on evolutionary topics. Critiques of the creationist standpoint are explored by the Committee on Science and Creationism (1984) and in more detail by Berra (1990) and Futuyma (1995). Of many *Scientific American* articles on evolutionary topics, those by Tattersall (1997), Grimaldi (1996), Pääbo (1993), Thorne and Wolpoff (1992), Wilson and Cann (1992), Ross (1992), Cavalli-Sforza (1991), and Stringer (1990) are of current interest.

PART 4

Cytogenetics and Development

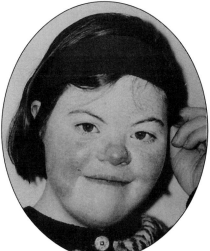

CHAPTER 13

Changes in Chromosome Number

One year after Mendel published his rules of inheritance for garden peas, the English physician John Langdon Down (1866) reported some interesting observations. As medical superintendent of an asylum for the severely retarded, he had noticed that about 10% of the residents resembled each other and could be easily distinguished from the rest of the patients: "So marked is this, that when placed side to side, it is difficult to believe that the specimens compared are not children of the same parents."

This condition, originally called mongolism, is now known as Down syndrome (DS). Individuals with Down syndrome have weak reflexes, loose joints, and poor muscle tone. All parts of the body are shortened owing to poor skeletal development. The face is broad and flat with a small nose, irregular teeth, and abnormally shaped ears. The eyes may be close-set with narrow, slanting eyelids; a large, furrowed tongue may protrude from a mouth framed by rather thick lips (Figure 13.1A). Hipbones are abnormally shaped and aligned, and

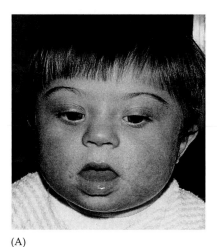 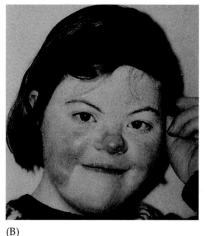

(A)   (B)

Figure 13.1 Children with Down syndrome. (A) One-year-old boy. (B) Sixteen-year-old girl. (Photographs courtesy of Paul Polani, University of London.)

the feet often display a sizable gap between the first and second toes; the little finger is often short and curved inward. Some highly unusual dermatoglyphic (Greek *derma*, "skin"; *glyph*, "carving") patterns of hand creases, fingerprints, and footprints are also associated with Down syndrome. An affected individual will probably not exhibit all of these traits, however, and the "characteristic" facial features may become less obvious with age.

Defects of the heart, digestive tract, kidneys, thyroid gland, and adrenal glands are also common. Obesity is common in adults. Males have poorly developed genitals and are almost always sterile. In females, ovarian defects and irregular menstruation are the rule; but fertility is possible, and over two dozen live births have been recorded. (About 40% of these babies had Down syndrome.)

About 15% of babies with Down syndrome die in their first year, often from heart abnormalities. Susceptibility to infection (especially pneumonia) is also common, probably owing to defects in the immune system. Leukemia is 15 to 20 times more frequent in people with Down syndrome than in the general population, and other types of cancer are more common too. But antibiotics and other medical improvements have dramatically extended the mean life expectancy, and over 50% of affected individuals will survive into their 50s.

Motor development is delayed, and toilet training may require several years. Most children learn to talk, but speech is usually thick and harsh sounding, perhaps in part because of hearing defects. IQ scores range widely, from 20 to 85 (i.e., from profound retardation to the low normal range). In general, however, the learning abilities of affected children may be com-

pared to those of unaffected 6- to 8-year-old children. Although clumsy with their hands, they respond well to early intensive stimulation and training for simple tasks. Some individuals with Down syndrome learn to read and write, attend regular schools, and find employment. With regard to their personality traits:

> The popular stereotype is that of a placid, affectionate, good-natured, music-loving, moderately intellectually disabled individual. ... In fact, Down syndrome children exhibit a great variety of individual personalities and, like their chromosomally normal peers, are not immune to deviant behaviors, conduct disorders, or psychosis. (Epstein 1995)

What causes the basic defects? Dr. Down could not possibly have guessed that they resulted from the presence of an extra chromosome, because his paper preceded the discovery of chromosomes by about 15 years. Indeed, 90 years passed before geneticists determined the correct human chromosome number, and not until 1959 did they realize that people with Down syndrome have a triple dose of the smallest chromosome. By then, decades of work—mainly with fruit flies—had shown how departures from the normal diploid number might come about.

This chapter and the next deal with such *changes in quantity* of otherwise normal genetic material, changes involving hundreds or even thousands of genes. This type of variation contrasts with most of the genetic variability we have discussed up to now, which arises from *changes in quality* (mutations) of individual genes in organisms with a normal chromosome number. Cells with normal chromosome sets have **euploid** karyotypes (Greek *eu*, "good"; *ploid*, "set"). For humans, haploid gametes and diploid zygotes are the euploid conditions. In this chapter, we discuss **aneuploid** organisms—those with unbalanced sets of chromosomes owing to an excess or deficiency of individual chromosomes. It is this *imbalance* among genes (the quantitatively abnormal ratio of genes present), rather than the nature of individual genes, that causes the abnormal phenotype or even death of an aneuploid organism.

Aneuploidy can arise by a variety of mechanisms. The first such mechanism to be observed was **nondisjunction**, in which two homologous chromosomes or sister chromatids fail to separate (disjoin) into daughter cells. If this happens during *meiosis*, a gamete may end up with one too many or one too few chromosomes. Union of this aneuploid gamete with a euploid one leads to aneuploidy in the zygote and its descendant cells. During a *mitotic* division, nondisjunction occurs when the two sister chromatids of a chromosome fail to separate, and this can lead to a mixture of euploid and aneuploid cells within one individual. Experiments with a vast range of species have shown that nondisjunction is a major cause of aneuploidy. And aneuploidy, with a total frequency of about 1 in 250 live births, is the source of much grief to humans.

Mistakes in Cell Division

Nondisjunction was first discovered around 1913 by Calvin Bridges, an undergraduate researcher at Columbia University. He had set up a standard kind of cross involving an X-linked eye color trait in fruit flies, but got a few unexpected male and female progeny in the F_1 generation. After ruling out other possibilities, Bridges proposed that the unexpected females developed when an egg with two X chromosomes was fertilized by a Y-bearing sperm, and microscopic analyses verified that these females were indeed XXY. The exceptional males, on the other hand, had one X chromosome and no Y, so they must have come from the fertilization of a no-X egg by an X-bearing sperm. (Unlike humans, fruit flies with two X's and one Y are female, while those with one X and no Y are male.)

How were the XX and no-X eggs formed? Bridges suggested that they could have arisen during the first meiotic division when the two X chromosomes failed to disjoin (i.e., separate) into two daughter cells. He coined the term *nondisjunction* to describe this phenomenon. Here, nondisjunction apparently occurred when the centromeres of homologous X chromosomes moved independently to the same pole at anaphase. The discovery of nondisjunction was a great genetic breakthrough, an exception that proved the rule of Mendelian inheritance and provided direct physical evidence that genes exist on chromosomes. Later research showed that nondisjunction can occur in any organism and in any dividing cell and can involve any chromosome. Consequences vary widely, depending on where and when the misdivision takes place.

Meiotic Nondisjunction

Consider the most common situation, in which only one pair of chromosomes misbehaves. For simplicity's sake, we will focus on spermatogenesis, in which all four products of meiosis are functional. Standard terminology is as follows: For a gamete, **nullisomic** and **disomic** mean, respectively, the presence of *no* homologous chromosomes and of *two* homologous chromosomes rather than one.* For a zygote, **monosomic** and **trisomic** refer, respectively, to the presence of *one* ho-

*Recall that n = the haploid number of chromosomes (23 in humans).

mologous chromosome and of *three* homologous chromosomes, rather than the usual pair.

Nondisjunction may happen during either the first or second meiotic division; the timing makes a difference, as Figure 13.2 shows. If nondisjunction occurs during meiosis I, all the sperm derived from that primary spermatocyte will be abnormal. Half of them will contain *neither* member of the given chromosome pair, and half will carry *both*. In humans, for example, these sperm will end up with a total of 22 ($n-1$) or 24 ($n+1$) chromosomes rather than the usual $n = 23$. It follows that after the fertilization of normal eggs, the resulting zygotes will have either $2n - 1 = 45$ or $2n + 1 = 47$ chromosomes, rather than $2n = 46$ chromosomes. If nondisjunction occurs in a secondary spermatocyte undergoing meiosis II, only two of the four sperm will be abnormal. As before, these two will bear either one too few or one too many chromosomes (Figure 13.2B).

Normal development of the zygote in all diploid species depends on the presence of exactly two of each chromosome type; any deviation from this pattern causes abnormalities or death. Nondisjunction of sex chromosomes is better tolerated than that of autosomes, however, partly because of the Lyon effect. Recall from Chapter 10 that only one X chromosome is fully active in any cell. Because all additional X chromosomes are almost entirely inactivated, they have less effect on phenotype than does the presence of too many (or too few) autosomes.

In addition, an *excess* of chromosomes is tolerated better than a *deficiency*. Thus, among autosomal aneuploidies, trisomies are more viable than monosomies. But only those trisomies involving the smallest or most heterochromatic (i.e., least gene-rich) chromosomes are able to survive at all. Among these, the greater the imbalance among autosomal genes, the more abnormal the phenotype. In humans, the only viable autosomal trisomy is Down syndrome, which involves the smallest chromosome. And even with trisomy 21, only about 25% of conceptions survive to birth (Tolmie 1997).

What causes the phenotypic abnormalities seen in aneuploid organisms? Is it the small effects of many genes added together, or is it just a few scattered genes that are highly deleterious or lethal when present in anything but two doses? In fruit flies (and probably other organisms too), it is the former situation. Also, the effects of having a gene (or group of genes) in triplicate are less severe than having just one copy.

Mitotic Nondisjunction

In the cases discussed so far, an abnormal chromosome number is established in the zygote stage, the result of fertilization involving an aneuploid gamete. Consequently, *all* the cells of the body are expected to

(A) First division nondisjunction

(B) Second division nondisjunction

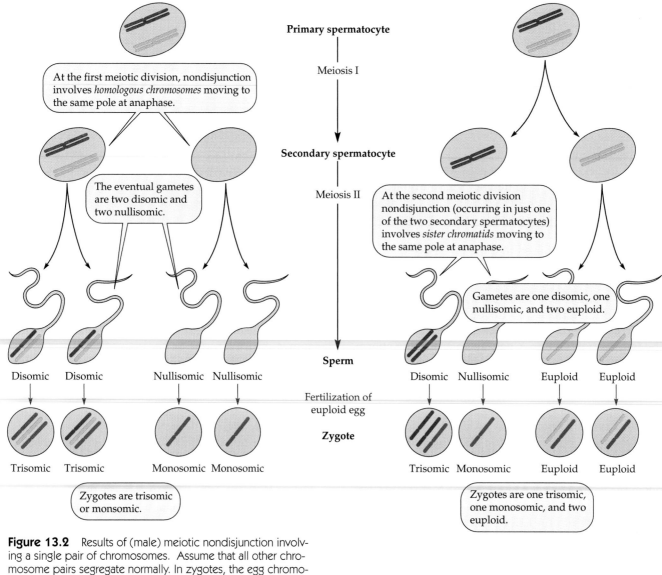

Figure 13.2 Results of (male) meiotic nondisjunction involving a single pair of chromosomes. Assume that all other chromosome pairs segregate normally. In zygotes, the egg chromosome is shown in lavender.

be aneuploid. In mitotic nondisjunction, however, the zygote may be normal, with aneuploidy occurring sometime during its later development. If the resultant cells survive and continue to divide, they give rise to descendant line(s) of aneuploid cells. The individual then has a mixture of cells with different chromosome numbers and is called a *mosaic*. This anomaly can occur in somatic cells or in cells destined to give rise to gametes. The latter case, which is called **germinal mosaicism**, can lead to aneuploidy among offspring as well.

The earlier that nondisjunction takes place, the larger the proportion of aneuploid cells that might be found in the mosaic. If nondisjunction occurs during the *first* cleavage division of the zygote, as shown in

Figure in 13.3A, then *all* the daughter cells will be aneuploid; half will be monosomic and half trisomic. If it occurs during the *second* cleavage division (Figure 13.3B), only *half* of the resultant cells will be aneuploid. Nondisjunction in the *third* cleavage division will initially make one-fourth of the cells aneuploid. And so on. If nondisjunction occurs late in development, only a tiny group of aneuploid cells will be formed, and they may escape detection.

Actually, the proportion of aneuploid cells in an individual after birth does not necessarily reflect what happened in the embryo before birth, because aneuploid cells have a lower probability of survival than do normal cells. Indeed, embryos with high proportions of abnormal cells may not survive at all, as shown by

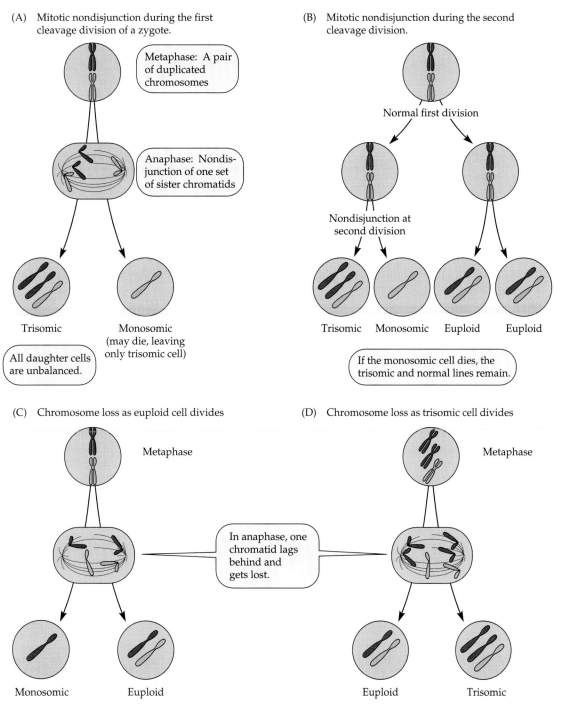

(A) Mitotic nondisjunction during the first cleavage division of a zygote.

Metaphase: A pair of duplicated chromosomes

Anaphase: Nondisjunction of one set of sister chromatids

Trisomic

Monosomic (may die, leaving only trisomic cell)

All daughter cells are unbalanced.

(B) Mitotic nondisjunction during the second cleavage division.

Normal first division

Nondisjunction at second division

Trisomic   Monosomic   Euploid   Euploid

If the monosomic cell dies, the trisomic and normal lines remain.

(C) Chromosome loss as euploid cell divides

Metaphase

In anaphase, one chromatid lags behind and gets lost.

Monosomic   Euploid

(D) Chromosome loss as trisomic cell divides

Metaphase

Euploid   Trisomic

Figure 13.3 How mosaicism arises from mitotic nondisjunction (A and B) or from chromosome loss (C and D).

the high frequency of aneuploidy in spontaneously aborted fetuses (Chapter 14).

Mosaics can also arise through other processes. The most frequent of these is **chromosome loss**, whereby a chromosome lags so far behind during anaphase that it is not included in either daughter nucleus. Consequently, one of the daughter cells will contain one less chromosome than the other. If the original cell was euploid, one of the daughter cells will be monosomic

(Figure 13.3C). If the original cell was trisomic, then following the loss of one of the three chromosomes, one daughter cell will be normal rather than trisomic (Figure 13.3D). Clearly, this event could lead to mosaicism in an individual who started out as a trisomic; indeed, it has been documented in several Down syndrome mosaics. Like nondisjunction, chromosome loss may occur during any cell division, mitotic or meiotic, and can involve autosomes or sex chromosomes.

In any given cell division, more than one chromosome may undergo nondisjunction or loss, but the result is usually lethal. Those few cases that do survive beyond birth generally involve either mosaicism or autosomal aneuploidy in conjunction with a sex chromosome aneuploidy. We will see in Chapter 17, however, that specific single or multiple aneuploidies characterize certain types of cancer.

Nondisjunction of Autosomes

Long before the cause of Down syndrome was known, studies of this condition in twins (Chapter 11) suggested a genetic role. Such studies are usually done by finding one affected member of a twin pair, then determining (1) the type of twinning and (2) whether or not the other member is also affected. Among identical twins, it was found that if one member had Down syndrome, then almost always the other one did too.* In contrast, if one member of a pair of fraternal twins had Down syndrome, the other was almost always unaffected.

Because the pattern of transmission did not fit the usual models for single-gene inheritance, however, it was suggested in the 1930s that Down syndrome might result from a chromosomal abnormality. Cytological methods then available were too primitive to test this hypothesis, but three years after the normal human chromosome number was established came the discovery that somatic cells of people with Down syndrome contain 47 chromosomes. The extra chromosome was one of the G group (Figure 13.4), at first thought to be the next-to-smallest—that is, chromosome 21; so **trisomy 21** became another name for Down syndrome. Improved techniques later showed that the extra chromosome was actually the *smallest* member of the G group, but rather than trying to change the name, geneticists live with the inconsistency. That is, chromosome 21 rather than 22 is actually the smallest.

Saying that people with Down syndrome have an extra chromosome does not really explain how their abnormalities come about. Does chromosome 21—which contains only about 1.7% of the total DNA in humans—carry genes involved with the development of the brain, skeleton, and other organ systems that are affected? More specifically, does chromosome 21 carry genes for those enzymes, hormones, blood groups, and so on, that might be expressed in excess in people with trisomy 21? Unfortunately, not enough is known about networks or cascades of biochemical events in normal human development to say how an extra chromosome

interferes with these processes. Certainly, there must be complex interactions among the gene products of many chromosomes.

Not all of chromosome 21 has to be present in triplicate to produce Down syndrome. In a few rare individuals, the only extra chromosomal material is the distal (farthest from the centromere) half of the long arm. As geneticists use cytogenetic and molecular techniques to dissect and characterize chromosomes, some additional clues are beginning to emerge (Box 13A). Of the 1,000 or so genes thought to be on chromosome 21, at least 30 genes of known function have been identified. Among these, many lie near the tip of the long arm in bands 21q22.13 to 21q22.2—the critical region that, when triplicated, is most often associated with Down syndrome (Korenberg 1993).

Molecular studies reveal that the long arm of chromosome 21 has dozens of special DNA marker sequences called RFLP sites (Chapter 9), as well as several cytological markers. These studies allow researchers to start answering some other key questions about Down syndrome: In which parent did the nondisjunction occur? Does the phenotype of the affected child vary with parental origin of the extra chromosome? What is the correlation, if any, between crossing over and nondisjunction on chromosome 21? Does the maternal age effect (discussed shortly) result from increased nondisjunction in older women, or is it

Figure 13.4   G-banded metaphase spread from an individual with Down syndrome, showing three chromosomes 21 (circled). (From Epstein 1995; photo courtesy of Steven A. Schonberg.)

---

*It can happen that after the splitting of an embryo to form identical twins with trisomy 21, one or the other twin loses a chromosome 21. Alternatively, nondisjunction occurring in a mitotic division after separation of karyotypically normal twins could give rise to Down syndrome in one twin.

BOX 13A *Phenotypic Map of Chromosome 21*

By studying those rare individuals with Down syndrome who are trisomic for only part of chromosome 21, scientists have begun to match up specific regions of the chromosome with specific features of Down syndrome. After analyzing the detailed cytogenetic and clinical descriptions for each patient with partial trisomy, they combine the data. Of the 25 phenotypic traits they studied, we include thirteen.

In the diagram, the bright red lines designate regions that (when trisomic) produce the trait and thus are likely to include loci affecting the trait. Regions shown in lighter red may also contribute to that phenotype, but the involvement of the open regions is not clear. Note that the latter fall mostly in the midsection of 21q, which is the region that is least understood. The numbers refer to chromosome bands. (Adapted from Korenberg et al. 1994, Korenberg 1995, and Gardner and Sutherland 1996.)

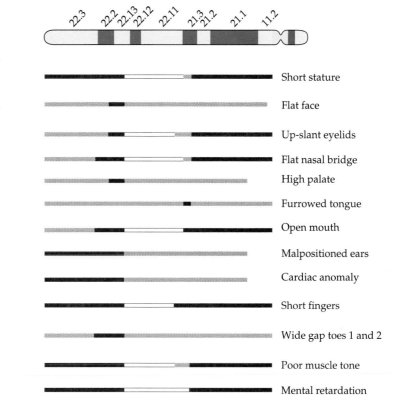

Short stature
Flat face
Up-slant eyelids
Flat nasal bridge
High palate
Furrowed tongue
Open mouth
Malpositioned ears
Cardiac anomaly
Short fingers
Wide gap toes 1 and 2
Poor muscle tone
Mental retardation

caused by decreased destruction of their aneuploid embryos? Is the Down syndrome phenotype associated with just a few key genes on chromosome 21, or (as with fruit flies) is it a consequence of a generalized imbalance of genes? Are some parents at high risk for nondisjunction, and if so, can they be identified in advance?

Frequency and Origin of Down Syndrome

Down syndrome occurs in about 1 out of every 800 to 1,000 live births in all ethnic groups. It afflicts families of wealth or intellectual achievement as well as the poor or uneducated. Nearly all cases (95–98%) occur just once within a given family and thus are called *sporadic Down syndrome*. The 2–5% of cases that run in families (so-called translocation Down syndrome) are discussed in Chapter 14.

Over a century ago, it was noted that babies with Down syndrome are often the last-born members of a sibship. This observation suggested that birth order, maternal age, paternal age, or a combination of these might be implicated. In the 1930s, a **maternal age effect** was shown to be the critical factor in most cases: Women over 35 years of age produced over half of all babies with Down syndrome, although they accounted

for only about 15% of all births. Of course, older women have older husbands and more children, but statistical analyses can disentangle such closely correlated factors and measure their individual effects. These studies indicate that there is no paternal age effect.

The well-documented maternal age effect in the incidence of Down syndrome has focused the greatest attention on the mother, particularly the older mother. This is because oocytes may spend several decades in prophase of meiosis I, whereas sperm are continually being formed anew. Although it is still the prevailing hypothesis, nobody knows exactly what causes nondisjunction in women of any age. Many possible environmental factors (thyroid disorders, viral infections, radiation, caffeine, drugs, hormones, contraceptives, reduced frequency of coitus, etc.) have been suggested and then usually rejected for lack of substantial and convincing evidence.

Complicating the issue even more is the fact that different chromosomes exhibit different trisomy rates in spontaneously aborted fetuses. For unknown reasons, trisomy of chromosome 16 is vastly more frequent than that of any other autosome. It comprises 20–35% of all trisomies found in recognized spontaneous abortions and is estimated to occur in about 1.1% of all human conceptions (Hassold et al. 1996). Generally speaking, trisomy for a chromosome in group G, E, or F is much more likely to be found than is trisomy for a group A or B chromosome. This suggests that there are many chromosome-specific influences on this phenomenon (Eichenlaub-Ritter 1996; Warburton and Kinney 1996). For example, chromosome size is probably correlated with the number of genes and thus with the severity of imbalance in the trisomic embryo. It may also directly affect nondisjunction frequencies.

Molecular methods often reveal variations that distinguish the maternally derived chromosomes from the paternally derived chromosomes. Such studies of chromosome 21 trisomy have recently shown that only about 4–5% of sporadic cases of Down syndrome are paternal in origin. Among these, most of the nondisjunction events occur during meiosis II. Among the 91–92% of sporadic cases caused by maternal nondisjunction, however, the situation is reversed: Most are meiosis I errors. The remaining 2–5% of sporadic cases can be traced to mitotic nondisjunction in somatic cells of the early embryo (Antonarakis 1993).

Table 13.1 shows how the frequency of Down syndrome births varies among different maternal age-groups. Note that beyond age 35, the frequency rises rather dramatically, so that a 45-year-old woman is almost 70 times more likely to produce a child with Down syndrome than is a 20-year-old woman. Studies of spontaneous abortions have shown that among conceptions, the incidence of Down syndrome is around four times higher than among live births. Thus, trisomy 21 is highly lethal in utero. More generally speaking, a wide variety of studies have shown that human meiosis is highly prone to error. Indeed, some studies suggest that over 50% of all ovulated oocytes may be aneuploid in women aged 40 to 45 years (Hassold et al. 1996).

Yet despite decades of research, the causes of nondisjunction in humans are poorly understood. Only in the past decade has it become possible to study female meiosis fairly directly by analyzing the chromosomes in "reject" (i.e., unfertilized) metaphase II oocytes obtained from *in vitro fertilization* (*IVF*) procedures (Chapter 20). A great deal of recent molecular and cytological evidence indicates that aberrant meiotic recombination (i.e., abnormal rates or locations of exchanges) is a major factor in all human trisomies. In some cases, it apparently leads to the premature separation of oocyte bivalents during meiosis I. But other data point instead to the precocious separation during meiosis I of the centromeres holding sister chromatids together (Angell 1997). Unfortunately, it is not known whether IVF-associated oocytes retrieved from artificially stimulated ovaries (in women who have trouble conceiving) are actually comparable to "regular" oocytes.

With the advent of prenatal testing in the late 1960s, the routine screening of pregnant women over the age of 35 became possible. As a result, fewer babies with Down syndrome have been born to older women, and the frequency of Down syndrome among live births has dropped considerably from the 1 in 700 frequency that occurred before prenatal testing was possible. Whereas "over-35" mothers used to produce a large proportion of babies with Down syndrome, they now account for many fewer of them. For example, a recent study estimated that in the Atlanta area, at least 56% of women over 35 years of age have prenatal testing done. If the fetus has trisomy 21, about 90% of those women terminate the pregnancy. As a result, the rate of trisomy 21 babies born to women over 35 years old in that area has been cut in half (Bishop et al. 1997; Yoon et al. 1996).

Indeed, most babies with Down syndrome are now being born to women under 35 years of age, who, although their risk factor is very much lower, vastly outnumber older mothers. Because it is not feasible to do prenatal tests on *all* women under the age of 35, researchers have sought some

TABLE 13.1 Estimated risks of live births of babies with Down syndrome for maternal ages 15 to 50.

Maternal age in years	Prevalence[a]	Maternal age in years	Prevalence[a]
15–19	1,560	35	355
20	1,540	36	280
21	1,520	37	220
22	1,490	38	170
23	1,450	39	130
24	1,410	40	97
25	1,350	41	73
26	1,280	42	55
27	1,200	43	41
28	1,110	44	30
29	1,010	45	23
30	890	46	17
31	775	47	13
32	660	48	9
33	545	49	7
34	445	50	5

Source: Data from Gardner and Sutherland 1996, and Hecht and Hook 1994.
[a]Prevalence = 1 in *x* births, where *x* is the number in the columns.

additional clues to help them identify those younger mothers who are at increased risk of producing babies with Down syndrome. In 1984, it was noted that early in pregnancy, the concentration of a substance called **α-fetoprotein (AFP)** is often lower than average in the blood serum of mothers who later give birth to babies with Down syndrome. When considered along with the levels of two or three other substances in maternal serum (Chapter 19), a test for abnormal AFP levels can lead to the detection of about 60% of Down syndrome fetuses in younger mothers (Wald and Kennard 1997). Unfortunately, prenatal ultrasound scanning is not a reliable detector of Down syndrome.

Down Syndrome and Alzheimer Disease

Ironically, delayed development of people with Down syndrome is usually followed by some aspects of premature aging—often involving brain abnormalities of the type seen in people with **Alzheimer disease (AD)** (Figure 13.5). These include *amyloid plaques* (extracellular deposits of protein from the degeneration of nerve cell endings) and **neurofibrillary tangles** (clusters of protein filaments inside nerve cells). The relations among these defects are not clear, however. All chromosomally normal people with Alzheimer disease and all adults with Down syndrome over age 40 have plaques and tangles in their brains. Yet autopsies show that in the latter group, only 8% have AD by age 49, and 75% are affected after 60 years of age. Similarly, not all chromosomally normal individuals with plaques and tangles develop AD. Perhaps, rather than a simple cause-and-effect connection, a certain *threshold* number of plaques and tangles must accumulate before dementia sets in.

Several proteins occur in the plaques and tangles present in AD patients and in older people with Down syndrome, but the **β-amyloid** protein of plaques has commanded the most attention. It is the short degradation product of a larger protein known as **amyloid precursor protein (APP)**. APP occurs in the cell membranes of most tissues, but its normal function is unknown. In 1991, a gene coding for APP was found on chromosome 21q. It appears to be a relatively minor player, however: Worldwide, only about 20 families have been found with both early-onset hereditary Alzheimer disease and mutations in the *APP* gene!

Researchers are also studying the **tau protein** found in neurofibrillary tangles. Normally, it interacts with the tubulin protein in microtubules of neurons, stabilizing their structure. Under certain abnormal circumstances, however, it cannot bind to microtubules and instead aggregates to form neurofibrillary tangles that kill the nerve cell. Whether the amyloid plaques or the neurofibrillary tangles are the primary cause of Alzheimer disease has been hotly debated for over a decade. Currently, it appears that plaques are necessary but, by themselves, not sufficient to cause AD.

Animal Models for Down Syndrome and Chromosome 21 Disorders

Any disorder found only in humans is often difficult to study. If a condition occurs naturally or can be experimentally induced in other animals (preferably in small laboratory mammals whose physiology and genetics are well understood), then scientists can much more readily study its development and do preliminary tests of experimental treatments.

Charles Epstein and colleagues at the University of California in San Francisco have produced mice that are trisomic for mouse chromosome 16, which carries several genes that are homologous to human chromosome 21 genes. (Several other genes at the very tip of human 21q have their homologues on mouse chromosomes 3, 10, or 17, rather than on mouse 16.) Although the mouse fetuses do not survive quite until birth, their multiple defects can be studied in great detail. Many of these trisomic mice exhibit the same type of heart defect that is found in some people with Down syndrome; they also show defects in the nervous system and immunological abnormalities.

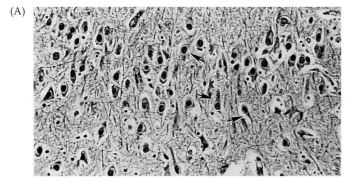
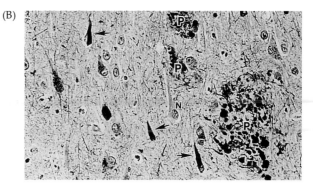

Figure 13.5 Plaques and tangles. (A) Normal brain tissue. Arrows point to neurons. (B) Brain tissue from an adult with Down syndrome. P, plaques; N, neurons. Arrows point to tangles. (Micrographs courtesy of National Down Syndrome Society.)

By combining trisomic mouse cells with normal diploid mouse cells, Epstein and his co-workers have also produced *mosaic mice* that survive beyond birth, and these, too, are being intensively studied. But researchers stress that because mouse chromosome 16 is a much larger chromosome than human chromosome 21, the two trisomies cannot be identical. Thus, a *partial trisomy* of mouse chromosome 16 that carries genes similar to those found on the distal region of human chromosome 21q would be much better for comparative studies.

There is good evidence that differences in gene dosage lead to proportional differences in the amount of enzyme produced. Indeed, at least eight loci on chromosome 21 show about 50% increased activity when trisomic. But the effects of aneuploidy must also involve dosage differences for cell receptors, regulatory genes, homeotic and other developmental genes, and genes that produce subunits for important structural proteins such as collagen.

Researchers have also been trying to produce animal models for Alzheimer disease. One method is to inject the human *APP* gene into mouse zygotes and then recover a few *transgenic* mice (Chapter 8) that have incorporated the human locus into their genome. Initial reports in 1991 suggested that the brains of several *APP* transgenic mice contained some deposits resembling the β-amyloid plaques and tangles seen in people with AD. But some of the results could not be verified, so the excitement over these experiments was short-lived. Recent reports seem more promising, however. Researchers have developed a few different transgenic mouse models that exhibit some learning and memory problems as well as amyloid plaques and nerve cell degeneration in their brains (Finn 1997; Mattson and Furukawa 1997). Yet none of these transgenic mice have produced neurofibrillary tangles.

Researchers have also constructed transgenic mice that carry a different human chromosome 21 locus called *SOD1* (super*o*xide *d*ismutase). This gene is associated with *familial amyotrophic lateral sclerosis* (*FALS*), or *Lou Gehrig's disease*, in several families.* Dominant mutations of the *SOD1* gene somehow damage the nerves carrying impulses from the brain to skeletal muscles, thus leading to paralysis and death. Its normal product, the enzyme copper/zinc superoxide dismutase (CuZnSOD), converts toxic superoxide ions to oxygen and hydrogen peroxide. Some transgenic cells that overexpressed the normal *SOD1* allele had abnormally low levels of serotonin (a neurotransmitter) and prostaglandins (a membrane

lipid with hormonelike effects), as do people with Down syndrome. The neuromuscular junctions of certain transgenic muscles were also abnormal. These and other results suggest that "a relatively small change in CuZnSOD activity appears to have a major effect on the ability of cells or tissues to handle severe oxidative challenges" (Epstein 1995). Transgenic mice carrying normal human *SOD1* alleles do not show any symptoms of FALS, but those carrying certain mutant *SOD1* alleles do exhibit significant neurodegenerative effects.

Mosaicism and Down Syndrome

Roughly 2% of all live-born people with Down syndrome are mosaics, possessing both normal and trisomic cells. These individuals show great phenotypic variability in all respects. The degree of abnormality varies, depending on what proportion of tissues end up being trisomic and which specific tissues these are. If cells of the nervous system turn out to be euploid, for example, the individual may have normal intelligence. Those individuals with only a tiny proportion of aneuploid cells may have a completely normal phenotype and escape detection entirely.

In some families, multiple cases of trisomy 21 can be traced to aneuploidy in a mosaic parent's gonadal tissue. These instances of *germinal mosaicism* are important to detect early, because unlike the usual sporadic pattern, the occurrence of nondisjunction in the mosaic parent's trisomic germ cells can lead to multiple Down syndrome births in a family. Indeed, about half of the gametes resulting from meiosis in trisomic oocytes or spermatocytes may contain two rather than one chromosome 21.

A small fraction of cases of Down syndrome may be due to mitotic nondisjunction in the embryo rather than to meiotic nondisjunction in a parent. Recall that nondisjunction during the first cleavage division (Figure 13.3A) would result in one trisomic daughter cell and one monosomic daughter cell. The latter usually dies, leaving the trisomy 21 and its descendant cells to make up the entire embryo. But nondisjunction at the second or a later cleavage division (Figure 13.3B) will produce an embryo with normal and trisomic cells and, initially, one monosomic cell. This embryo will be a mosaic and will most likely develop into an individual with some characteristics of Down syndrome.

Other Autosomal Aneuploids

The only other autosomal trisomies that occur with any significant frequency in newborns, **trisomy 18** and **trisomy 13**, were first reported in 1960. Rough estimates for live-birth frequencies are 1 in 6,000 for trisomy 18 and 1 in 12,000 for trisomy 13. There is a maternal age effect, but it is not so striking as that found for Down syndrome. Atypically, trisomy 18 results

---

*Familial cases account for only 10% of all cases of amyotrophic lateral sclerosis (ALS), however. The other 90% are sporadic, and may or may not have similar causes.

mostly from maternal errors in meiosis II rather than in meiosis I (Fisher et al. 1995). Both syndromes are associated with gross abnormalities of nearly all systems, leading to early death. In fact, they share many characteristics, with considerable overlap: Profound mental and developmental retardation, heart defects, kidney defects, low-set and malformed ears, small eyes, and a small, receding jaw are almost always observed. Prominent heel bones and deformities of various joints are often present. Other traits also are associated with these two syndromes, both of which may be detectable by prenatal ultrasound scanning.

The most distinctive additional features of *trisomy 18 syndrome* are the occurrence of simple arch patterns on three or more fingertips (rare in normal individuals) and a long skull that bulges in the back. Because of abnormally increased muscle tone, the fists are tightly clenched, with the index finger bent sideways across the third finger (Figure 13.6B), and the limbs and the hip joints are so stiff that they can hardly be moved. Many babies with trisomy 18 have "rocker bottom" feet with big toes that are short and bent upward. Approximately half have extra folds of skin on the neck. Distinctive facial features—including a round face, small and wide-spaced eyes, and a small mouth (Figure 13.6A)—also help with diagnosis. For unknown reasons, about 80% of all trisomy 18 births are female. Birth weight is low, and the mean survival time is about 2.5 months; nearly all cases die during the first year.

For *trisomy 13 syndrome*, the most distinctive additional traits, occurring in about 75% of all cases, are a small head, cleft palate or lip (or both), wine-colored birth marks, and extra fingers and toes (Figure 13.7). Unusual dermatoglyphics are the rule too. Almost all babies with trisomy 13 appear to be deaf, and some are blind. Extreme "jitteriness" and seizures are not uncommon. Males may have undescended testes, and females may have a uterus divided into two parts. The fingers may be abnormally bent. The nose is often large and triangular, and the eyes are frequently wide-spaced and defective. The mean survival time is about 3 months, most affected infants dying within a year.

In conclusion, we emphasize that these disorders result from imbalances of normal chromosomal material, not from gene mutation. That so many systems of the body are deranged when one small chromosome is present in excess suggests that normal metabolism and development are finely attuned not just to the nature of the gene products but to their concentration as well. This finding is also true of other species that

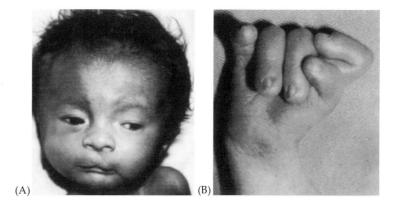

Figure 13.6 Trisomy 18. (A) Child with trisomy 18. (B) The hand of an infant with trisomy 18 demonstrates typical flexion deformities. (From Summitt 1973.)

have been studied. Furthermore, the relative rarity of autosomal aneuploidies among live-born children indicates that most imbalances of even small chromosomes interfere so seriously with metabolism that embryonic development is aborted in early pregnancy:

> Very minor imbalances may cause defects that are not readily detectable in early infancy, and some chromosomal "defects" may be without effect. However, as a first principle, anything but 100% of the normal amount of (at least autosomal) genetic material produces a less than 100% normal phenotype. … As a

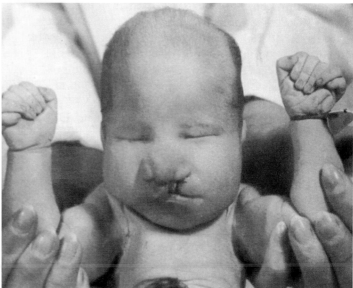

Figure 13.7 Trisomy 13. This infant was not photographed as a living patient but was found in the Department of Anatomy at the University of Western Ontario, where it had been preserved for 30 years. The features typical of trisomy 13 leave no doubt as to the diagnosis. Note the cleft lip, extra fingers, large triangular nose, and wide-spaced eyes. (Photograph by Murray J. Barr; from Valentine 1986.)

very general rule, if the imbalance consists of less than 1% of HAL [the total *haploid autosomal length*], the conceptus is often viable in utero, and live birth frequently results. If the excess is greater than 2%, [spontaneous] abortion is likely. Imbalance involving *deficiency* ... is generally much less survivable. (Gardner and Sutherland 1996)

Sex chromosome aneuploidy, however, presents a somewhat different story.

Nondisjunction of Sex Chromosomes

In the late 1950s, scientists discovered that the human Y chromosome plays a major role in sex determination. Until then, it had been thought that the critical factor was (as in the fruit fly) the number of X chromosomes—two X's leading to femaleness and one to maleness, with the Y chromosome needed only for fertility in males.

But soon after the human chromosome number was established, the role of the human Y chromosome in sex determination was clarified by two previously known syndromes. People with **Klinefelter syndrome** have one too many sex chromosomes: Their karyotype is 47,XXY, and their phenotype is male. Conversely, people with **Turner syndrome** have one too few sex chromosomes: Their karyotype is 45,X, and their phenotype is female. Thus, two X's do not always make a female, and one X does not always make a male. Findings on other karyotypes rounded out the picture: In humans, the presence or absence of a Y chromosome is much more important to male development than is the number of X chromosomes.

Later, some 47,XYY males were found among the inmates of penal institutions, raising the question of whether an extra Y chromosome may somehow predispose its host to criminal behavior. In the rest of this chapter, we describe the major sex chromosome anomalies and some aspects of their influence on behavior.

Klinefelter Syndrome: 47,XXY Males

In the 1940s, Harry Klinefelter and his colleagues at Massachusetts General Hospital described a syndrome occurring in males and usually not detected until after puberty. Later, cytogeneticists Patricia Jacobs and J. A. Strong (1959) showed that males with Klinefelter syndrome usually, but not always, have the **47,XXY karyotype**.

The signs almost always observed in these males (Figure 13.8) are very small testes (about one-third normal size), absence of sperm, and androgen (male sex hormone) deficiency. The penis and scrotum are usually of normal size, however. Additional but much more variable symptoms include poorly developed male secondary sexual characteristics (such as scanty facial hair) and breast development in some. There is

an increased risk of breast cancer in adults. Sexual behavior is normal among young males with Klinefelter syndrome, who experience spontaneous erections and (usually spermless) ejaculations. A few have fathered children, however. (In these rare cases, so far all offspring have been chromosomally normal.) Many marry and maintain sexual relations, but impotence is common among older men.

Males with Klinefelter syndrome often have unusually long limbs and average about 2–4.5 inches taller than unaffected males; their hands and feet may be large too. But many males with Klinefelter syndrome show normal height and intelligence and function well in society. Indeed, many 47,XXY males exhibit no symptoms of Klinefelter syndrome except infertility. These individuals may live their entire lives without the slightest inkling that they are in any way unusual. But others exhibit mild retardation or learning disabilities, which—together with an increased tendency to emotional and social problems—may account for their clear overrepresentation in mental and penal institutions.

For the more serious cases, surgical removal of breast tissue can relieve some psychological stress.

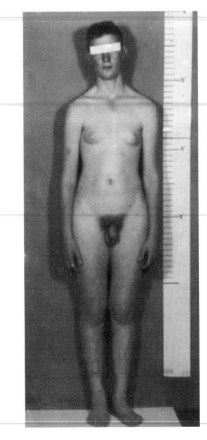

Figure 13.8 A 47,XXY male with Klinefelter syndrome. Note the breast development and female pattern of pubic hair growth. (Photograph by Earl Plunkett; from Valentine 1986.)

Although the sterility is unalterable, treatment with testosterone does promote development of the sex organs, body hair, musculature, deeper voice, and so on; it may also improve social adjustment, behavior, and learning ability of some males with Klinefelter syndrome. A warm and stimulating family and school environment early in life also seems to help mental and emotional development.

Klinefelter syndrome is not rare in the general population. Its overall frequency is 1 in 600 to 1,000 live-born males and about 1 in 300 among spontaneous abortions. In subpopulations of tall men (over 183 cm, or 6 feet), the frequency of males with Klinefelter syndrome may be as high as 1 in 260. Among residents of mental institutions, it is even greater—about 1 in 100; and roughly the same percentage is found in penal institutions. Perhaps 1 in 20 male patients seen in fertility clinics has Klinefelter syndrome. Because of underdevelopment of secondary sexual characteristics, Klinefelter males also appear relatively frequently among hospital patients. The differences in incidence among various groups of males point up the importance of carefully defining each test group and of not trying to project the results from any one group onto the general population.

Within families, Klinefelter syndrome occurs randomly. The extra chromosome is thought to arise from nondisjunction, with slightly more than half (56%) of cases occurring during oogenesis in the affected man's mother. Experts disagree on whether or not there is a significant maternal or paternal age effect (Robinson and de la Chapelle 1997; Willard 1995).

Klinefelter Variants. About 80–85% of males with Klinefelter syndrome exhibit the usual 47,XXY karyotype. But some are now known to be karyotypically normal, and about 10% have abnormal karyotypes that are different from 47,XXY, including 48,XXXY, 48,XXYY, 49,XXXXY, and 49,XXXYY. Klinefelter syndrome involving single cell lines of 48 or 49 chromosomes seems to be associated with more extreme problems of all sorts, including severe mental retardation. Indeed, most cases are detected through chromosome surveys in mental and penal institutions.

How do all these unusual karyotypes arise? To account for the karyotypes with 48 or 49 chromosomes, two nondisjunctional events are required. For example, a 48,XXXY zygote can arise in several ways, as shown in Figure 13.9: a Y-bearing sperm fertilizing an XXX egg; an XY sperm fertilizing an XX egg; or an XXY sperm fertilizing an X-bearing egg. Often the pattern of inheritance of X-linked genes or special DNA sequences called RFLPs (Chapter 9) can distinguish among the various possibilities.

Several types of mosaicism collectively account for about 10% of males with Klinefelter syndrome. As with any mosaic, the phenotypes of the individuals differ, depending on what proportions of the cells are aberrant and also on how the aberrant cells are distributed among the various tissues. Mitotic nondisjunction or chromosome loss, or both, must be invoked to explain the origin of these different mosaics. Perhaps some began as aneuploid zygotes after meiotic nondisjunction in the parents.

The 47,XYY Karyotype

In 1961, some researchers at the Roswell Park Memorial Institute in Buffalo, New York, discovered the first 47,XYY male by chance—as the father of a child with Down syndrome. He was tall and of average intelligence, and he had no serious physical problems. His chromosomal constitution resulted from paternal nondisjunction during meiosis II, producing a YY sperm. Several other **47,XYY karyotypes** were re-

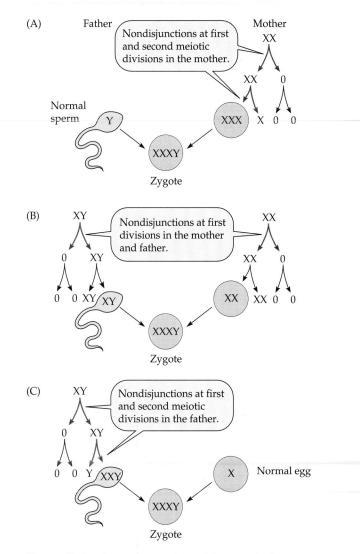

Figure 13.9 Three ways, each requiring a total of two nondisjunctional events, by which a 48,XXXY zygote could be formed.

ported later, but they did not begin to attract widespread interest until 1965. Studying the chromosomes of 197 mentally subnormal males with dangerous, violent, or criminal tendencies in an institution in Scotland, Jacobs and colleagues found seven males with the 47,XYY karyotype—a surprisingly high frequency of 1 in 28 (3.5%) that was not observed among other groups of males they tested. Their preliminary report also noted that the average height of the seven 47,XYY inmates was 186 cm (6 feet, 1 inch), compared with a mean height of 170 cm (5 feet, 7 inches) among the 46,XY males in the same institutional sample. Jacobs, Brunton, and Melville (1965) commented,

> At present it is not clear whether the increased frequency of XYY males found in this institution is related to their aggressive behavior or to their mental deficiency or to a combination of these factors.

Other geneticists began testing tall institutionalized males and finding additional 47,XYY males. Unfortunately, not all investigators were careful to define their samples or to study and describe control males from the same populations. (Of course, one cannot help but find criminals when testing is restricted to prisons.) Based sometimes on individual cases found in preselected environments and without much knowledge of how 47,XYY males in the general population behaved, some highly extravagant claims were made about the phenotypic effects of an extra Y chromosome. Soon articles linking Y chromosomes to violent or criminal behavior began appearing in the popular press as well as in journals read by psychologists, sociologists, and lawyers. The so-called XYY syndrome was also sensationalized in a few murder trials, where it was suggested that the defendants' 47,XYY karyotypes rendered them less responsible for their actions.

Keeping in mind that different researchers may not define even a simple trait like tallness in the same way and that results may vary widely from study to study, we present some estimates of the 47,XYY frequencies. Among males at birth in the general population, the incidence of 47,XYY is on the order of 1 in 1,000 males; among tall males in the general population, it is perhaps 1 in 325. As samples narrow down to tall mental patients or tall penal inmates and then to tall mental-penal inmates, frequency estimates increase to about 1 in 30.

Why are XYY males overrepresented in institutionalized populations? Many of the early claims about XYY males were based on small numbers of cases found in highly selected populations without adequate control studies of 46,XY males in the same selected populations. Or controls, when used, were not always well matched. Most early studies had another serious defect: The researchers who worked up psychological profiles on these men often knew what their karyotypes were, perhaps leading to an unconscious bias in the evaluations. The only way to eliminate such bias is to conduct **double-blind studies**, in which those who do the karyotyping know nothing of the subjects' histories, and those who compile the histories know nothing about their subjects' karyotypes. Only after both sets of data are independently and fully completed should they be combined to see if there is a correlation between behavioral profiles and karyotypes.

One large-scale *retrospective study* (see Box 1A) that avoided many of the usual pitfalls made use of extensive records kept by the Danish government on every 26-year-old male reporting to his draft board. These data included the results of a physical exam and intelligence test as well as educational history, social class, and criminal convictions. To save money, investigators restricted their search to 4,139 tall males. Among this group, they found 12 XYY karyotypes, a frequency of 1 in 345.* The remaining tall, karyotypically normal 46,XY males made up the control group.

How did the 12 tall 47,XYY males compare with their tall 46,XY counterparts? The mean height of the XYY males was 4 cm (1.5 inches) greater than that of the controls. Five of the XYY males (42%) had criminal records, compared with only 9% of the controls, but their crimes were not violent, and their sentences were light. The mean educational level of the XYY males and their test scores on the army intelligence test were significantly lower than those of tall control males. But even among the control males, those with criminal records had a significantly lower mean educational level and lower test scores than those without criminal records. This finding suggests that tall males with lower intellectual functioning are more likely to be convicted of crimes, regardless of their karyotypes:

> If the syndrome is as common as recent surveys suggest, there are probably a large number of undiagnosed XYY males who are neither in prisons nor in frequent barroom brawls. A particularly acceptable social adjustment … would be as a pro-football linebacker, but it is also possible that the majority of individuals with XYY chromosomes are not sufficiently large or aggressive for this vocation. (Nora and Fraser 1989)

Turner Syndrome: 45,X Females

An American endocrinologist, Henry Turner (1938), first described several grown females who lacked breast development and other secondary sexual characteristics. They also failed to menstruate and were sterile. Although the body form was reasonably well proportioned, these women were very short (adult height under 150 cm, or 4 feet, 11 inches). Their necks were short and webbed, and their forearms showed

---

*Also identified were 16 47,XXY males and a few 46,XY males with other chromosomal abnormalities.

greater than normal angling away from the body when the palms faced forward (Figure 13.10). Most females with Turner syndrome have (instead of ovaries) primitive streak gonads that lack both germ cells and hormone-producing tissue. Their oviducts, uterus, and vagina remain small and immature, and the external genitals are also infantile. Up to 10% menstruate and ovulate, however, and in rare cases have given birth.

Other defects might include widely spaced nipples on a broad chest, a narrowed aorta (the large vessel that carries blood away from the heart), kidney abnormalities, and some minor skeletal deformities. The hairline at the back of the neck may be low. In contrast to the childlike body form, the face often looks old. Indeed, premature aging sometimes occurs, and life expectancy may be reduced owing to defects of the heart and other organs. Females with Turner syndrome are often identified at birth by characteristic skin folds on the back of the neck, by swelling of the hands and feet, by abnormally large fingerprint patterns, and by low birth weight. But many women with Turner syndrome show no phenotypic abnormalities even as adults, except for infertility and below-average height—the latter despite normal levels of growth hormone (Robinson and de la Chapelle 1997; Saenger 1996).

Although some women with Turner syndrome may show a slight to moderate decrease in IQ, most are completely normal. Some seem to have trouble with a certain type of space perception, however. Their general behavior is normal during childhood. But failure to undergo puberty, coupled with small size, makes it difficult for them to keep up with their peers socially as teenagers. Hormone treatments may enhance skeletal growth and promote the development of breasts and other secondary sex characteristics. This, plus corrective surgery if necessary, may alleviate some of their social problems and also make normal sexual relations possible.

In 1959, it was reported that females with Turner syndrome have a 45,X karyotype. In about 80% of cases, it is the maternal X chromosome that is retained, and there is no evidence for a maternal or a paternal age effect. Other observed karyotypes include structural aberrations of the X and Y chromosomes (including isochromosomes, deletions, or translocations, which are discussed in Chapter 14) as well as ring X chromosomes (Robinson and de la Chapelle 1997).

About 1 in 2,500 live-born females have Turner syndrome. But it is highly lethal in embryos, being the most common karyotype among miscarriages, accounting for about 20% of all chromosomally abnormal miscarried embryos. Indeed, it appears that over 99% of all nonmosaic 45,X zygotes are lost during the first three months of pregnancy. Yet the few that survive are not so severely affected, especially with regard to mental capacities, as are the other live-born chromosomal aneuploidies.

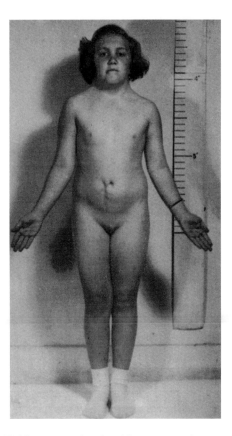

Figure 13.10 A 45,X female with Turner syndrome. This 14-year-old girl is 142 cm (4 feet 8 inches) tall. Note the lack of sexual development, the webbed neck, the broad chest with wide-spaced nipples, the old-looking face, and the bent forearms. (Photograph by Earl Plunkett; from Valentine 1986.)

Skuse et al. (1997) studied the behavior of 80 females with Turner syndrome and concluded that those who retained a paternal X showed significantly better social adjustment than those who inherited a maternal X. The authors propose the existence of an X-linked locus for social cognition that is maternally imprinted (Chapter 10). They suggest further that such a system might also explain why 46,XY males are more prone than 46,XX females to behavioral problems. Not surprisingly, this hypothesis has generated a great deal of interest and controversy.

Mosaicism and Turner Syndrome. Recent molecular analyses using FISH and PCR techniques indicate that mosaicism accounts for a large proportion of live-born females with Turner syndrome but a lower percentage of miscarriages. The most common types of mosaicism involve 45,X cells in combination with 46,XX or 47,XXX or 46,XY cells or the triple mosaic karyotype of 45,X/46,XX/47,XXX. Phenotypes of mosaic individuals range from fully affected to normal, perhaps depending partly on the proportion and distribution of the different cell lines.

We can calculate that at least 60%–80% of 45,X live-borns have a second cell line, compared with only 15%–20% of 45,X spontaneous abortions. Thus, sex chromosome mosaicism is probably at least 3–5 times as common among liveborn as among spontaneously aborted 45,X conceptuses. This implies that the presence of a second cell line confers a selective advantage to otherwise 45,X fetuses and increases their likelihood of surviving to term. (Hassold et al. 1992)

45,X Females Are Different from 46,XX Females. Given that one X chromosome is normally inactivated, we might expect these two karyotypes to be phenotypically equivalent. But there are reasons for the disparity. First, recall from Chapter 10 that the inactivation of one X chromosome in the somatic cells of 46,XX females is not complete. At least 20 (and perhaps many more) X-linked genes escape inactivation, thus being expressed in double rather than single dose in normal 46,XX females. But Turner females, lacking a second X chromosome, can only express these genes in single dose. Moreover, in 46,XX fetuses, as oocytes enter meiosis, their inactivated X chromosomes are *reactivated*; apparently, two active X's are needed for normal development of egg cells. Turner females, lacking the second X, are virtually always sterile.

A Gene for Short Stature. Among 45,X females who survive, loss of a whole X chromosome is always associated with short stature—as is loss of just the tip of Xp or Yp. The two short arms share a small region of homology called the **pseudoautosomal region (PAR1)**, and all X-linked genes within this region escape inactivation. Thus, geneticists reasoned, PAR1 might contain at least one important locus affecting height.

The frequency of short stature in human populations is about 3%. Many different genes must be involved, but most of them remain unknown. In 1997, an international team of researchers reported the discovery of a major PAR1 locus involved in linear growth (Rao et al. 1997; Zinn 1997). They had studied the DNA of 36 people (including some Turner females) with short stature and various sex chromosome abnormalities causing the loss of a tiny piece of PAR1. In particular, all these individuals lacked a 170-kb segment in Xp22 or Yp11.3. This segment was present in their relatives with normal height and in 30 unrelated individuals with normal height and different sex chromosome abnormalities in Xp22 or Yp11.3. Within this segment, the researchers identified a new locus they called *SHOX* (for *s*hort stature *ho*meobo*x*-containing gene).* This gene, which is highly conserved across

species, codes for two alternatively spliced mRNAs and proteins that are expressed in different tissues. Their exact functions are unknown. Researchers estimate that about 1% of all patients with unexplained short stature may carry a *SHOX* mutation.

Poly-X Females

Females with the **47,XXX karyotype**, first reported in 1959, present no distinctive phenotype aside from a tendency to be tall and thin, and many of them seem to be completely normal. Their frequency among newborns in the general population is roughly 1 in 1,000 to 2,000. The origin of the extra X chromosome in 47,XXX females is usually nondisjunction in meiosis I of the mother. A maternal age effect has also been noted. For example, the risk is 1 in 2,000 live births at age 35, but 1 in 310 at age 45. As with other chromosomal anomalies, mosaics are also observed.

Mental retardation is uncommon, but 47,XXX females often are delayed in motor and language development and learning skills. Some triple-X females are socially awkward and show a tendency to exhibit psychological problems. Although some 47,XXX females have menstrual difficulties, many menstruate regularly and are fertile. Most of their children are phenotypically and karyotypically normal. This outcome is surprising, because we would expect that half the eggs from a 47,XXX female would be XX, resulting in progeny that are 47,XXX or 47,XXY. Perhaps, prior to meiosis in XXX females, one of the X chromosomes is lost by lagging during mitotic divisions in the germ line. (A similar deficiency of aneuploid offspring has also been noted from 47,XYY fathers, who often appear to have only 46,XY primary spermatocytes.)

Females with more than three X chromosomes are extremely rare; only about 40 cases of the 48,XXXX karyotype are known. These females are usually tall and severely retarded, and they also exhibit a wide range of physical abnormalities. Although some show incomplete sexual development, about half of the known cases underwent normal puberty and beginning of menstruation. About two dozen 49,XXXXX females have been found in populations of mentally retarded individuals, and none in the general population. They are usually short and exhibit a variety of physical abnormalities, including incomplete sexual development.

Summary

1. Nondisjunction is either the failure of homologous chromosomes to separate (disjoin) during the first meiotic division or the failure of sister chromatids to separate during mitosis or the second meiotic division. In either case, the two daughter cells will have unbalanced chromosome sets; that is, they will be aneuploid.

---

*A *homeobox* (Chapter 15) is a special sequence of about 180 bp found within many developmental genes. It encodes a DNA-binding protein that acts as a gene regulator.

2. Meiotic nondisjunction in a parent leads to aneuploidy of gametes, which can lead to aneuploidy in all cells of an off-spring.

3. Mitotic nondisjunction leads to mosaicism within an individual. The degree of abnormality expressed depends on when nondisjunction occurs, which tissues are affected, and the viability of the aneuploid cells relative to the normal cells.

4. Aneuploidy often causes gross abnormalities or death of cells or organisms as a result of generalized gene imbalance. Sex chromosome aneuploidy is better tolerated than autosomal aneuploidy, and trisomies are better tolerated than monosomies. Aneuploidy for large autosomes is usually lethal.

5. In humans, only three relatively common autosomal aneuploidies are observed among newborns: trisomy 21, trisomy 18, and trisomy 13. All affected individuals are mentally retarded and exhibit malformations of many organ systems.

6. Trisomy 21, or Down syndrome, now accounts for about 5% of the retarded population. Sporadic Down syndrome has a frequency of about 1 in 800 live births. Advanced maternal age is a strong determining factor, and over 90% of all trisomy 21 cases are due to nondisjunction in the mother, usually in meiosis I. Rare multiple affected siblings in a family may be due to nondisjunction in a phenotypically normal parent with germinal mosaicism for trisomy 21.

7. Many older people with Down syndrome develop Alzheimer disease, which is characterized by dementia due to degenerative changes in the brain. β-amyloid protein, which accumulates around abnormal nerve cells, is derived from a larger molecule called amyloid precursor protein. The latter is produced by a gene on chromosome 21. Nearby on the same chromosome is a gene associated with a few (but not all) cases of familial early-onset Alzheimer disease.

8. Animal (mouse) models for the study of chromosome 21 genes have been generated by means of partial trisomies and transgenes.

9. In humans, the presence of a Y chromosome determines maleness, irrespective of how many X chromosomes are present. The absence of a Y chromosome determines femaleness.

10. Males with a 47,XXY karyotype usually have Klinefelter syndrome, characterized by small testes, infertility, and underdeveloped secondary sexual characteristics. They also tend to be tall and may (or may not) exhibit mild mental deficiency and/or some breast development.

11. Males with the XYY karyotype tend to be tall, and some may show below-average intelligence; but they have no other distinctive traits. Misconceptions about 47,XYY males arose partly because data from small and selective studies were prematurely generalized. Very little is known about 47,XYY males in the general population.

12. Females with the 45,X karyotype have Turner syndrome. They are short, have primitive streak gonads, and lack secondary sexual characteristics, but usually have normal intelligence. Most 45,X conceptions spontaneously abort; the 45,X karyotype is highly lethal in utero. Most live-born females with Turner syndrome are mosaics.

13. Maternal age effects are most strongly associated with autosomal trisomies and an elevated frequency of meiosis I

errors. Paternal age effects are much smaller. They often result from errors in meiosis II for autosomes, or from the nondisjunction of sex chromosomes during either meiotic division. Nearly all autosomal nondisjunction is maternal in origin.

Key Terms

45,X karyotype	germinal mosaicism
47,XXX karyotype	maternal age effect
47,XXY karyotype	monosomic
47,XYY karyotype	neurofibrillary tangle
α-fetoprotein (AFP)	nondisjunction
amyloid precursor protein	nullisomic
(APP)	pseudoautosomal region
aneuploid	(PAR1)
β-amyloid	tau protein
chromosome loss	trisomic
disomic	trisomy 13
double-blind study	trisomy 18
euploid	trisomy 21

Questions

1. Diagram the loss of one member of a chromosome pair during (a) meiosis I and (b) meiosis II of spermatogenesis. Compare the results with those in Figure 13.2.

2. Describe the zygote formed by the fertilization of a disomy 21 egg by a nullisomy 21 sperm.

3. Over 35% of all karyotyped spontaneous abortions exhibit autosomal trisomies. But autosomal monosomies, the reciprocal product of trisomies caused by nondisjunction, are rarely found in karyotyped spontaneous abortions. Suggest a possible reason.

4. With reference to chromosome 21, what kinds of eggs could be produced by a female with Down syndrome? As part of your answer, include diagrams that show how these eggs are formed.

5. At least one case has been reported of a pair of otherwise identical twins (same blood types, etc.) of which one member has Down syndrome and the other is unaffected. How might this situation have come about?

6. Halliday et al. (1995) report slightly higher rates of full-term Down syndrome births from older women who received good prenatal obstetric care. Suggest a possible reason for this finding.

7. In one published study, a few females with Turner syndrome showed (X-linked recessive) red-green color vision defects, whereas none of the much larger sample of tested males with Klinefelter syndrome did. Color blindness is usually expressed much more frequently in males than in females. What is the genetic explanation for these observations?

8. In other studies, green-shifted color perception has been noted in both females with Turner syndrome and males with Klinefelter syndrome having both mothers and fathers with normal color vision. Does this give any information on the nondisjunctional events that occurred in the parents? (Assume no new gene mutation.) Explain.

9. A pair of twins had identical blood groups and could successfully exchange skin grafts (another criterion for genotyp-

ic identity). One twin was a normal male, but the other was a female with Turner syndrome. How can this be explained?

10. Diagram all types of nondisjunction of the sex chromosomes that can occur in a 46,XX female and a 46,XY male during meiosis I alone, meiosis II alone, and both meiosis I and II. Assume that no more than one pair of centromeres fails to disjoin in any one division.

11. Referring to the diagrams you created for question 10, list three ways by which a 47,XXY zygote produced by a single nondisjunctional event could be the progeny of karyotypically normal parents.

12. Why are 47,XXY males and 46,XY males (who both have one active X chromosome) phenotypically different?

13. The most common type of mosaicism found in males with Klinefelter syndrome is XXY/XY. What two types of postzygotic events could account for this?

14. The phenotypes of males with Klinefelter syndrome can vary widely. Data from one study showed that those who were identified at fertility clinics were generally more masculine than those found at an endocrinology (hormone study) clinic:

Sample	Poor growth of facial hair	Breast development	Feminine pattern of pubic hair
Subfertile males	32%	26%	32%
Endocrinology patients	88%	63%	54%

How can the relationship between phenotype of these males with Klinefelter syndrome and their method of ascertainment be explained?

15. A few prospective studies (see Box 1A) have followed the development of 47,XYY males from birth through adolescence, but these investigations have been very controversial. List all the advantages and disadvantages you can think of for such 47,XYY studies.

Further Reading

A number of books and articles cover the material in Chapters 13 and 14—that is, aneuploidy in general. Books include Gardner and Sutherland (1996), Obe and Natarajan (1994), Vig (1993), and Epstein (1986). Articles on aneuploidy include Tolmie (1997), Hassold et al. (1996), Jacobs and Hassold (1995), Eichenlaub-Ritter (1996), Bishop et al. (1996), and Epstein (1988). On sex chromosome anomalies, see Robinson and de la Chapelle (1997) and Willard (1995).

Many articles and books are available on Down syndrome alone. Some written for the lay public include Selikowitz (1997), Cunningham (1996), Pueschel (1990), and Patterson (1987). More technically oriented are Epstein (1991, 1995), Epstein et al. (1995), and Stewart et al. (1988). Alzheimer disease is discussed by Roses and Pericak-Vance (1997), Berg et al. (1994), Nadel and Epstein (1992), Pollen (1993), and Selkoe (1991).

CHAPTER 14

Other Chromosomal Abnormalities

In 1961, Dr. J. C. P. Williams, a physician in New Zealand, described some young patients with heart defects who also share an odd array of physical and behavioral traits. In addition to their cardiac problems, which often include a narrowing of the aorta, these children exhibit pixie-like features: short stature, turned-up nose, wide mouth, small chin, large ears, and puffy eyes (Figure 14.1). They may also have a hoarse voice and prematurely aging skin. Although they do very poorly on IQ tests (mean = 58) and in drawing, reading, writing, and arithmetic, many possess remarkable verbal and musical abilities as well as a talent for remembering names. They also tend to be very sensitive and sociable and have a strong need for order in their lives (Lenhoff et al. 1997).

In 1993, Ewart et al. reported that these and other *Williams syndrome* traits are associated with the loss of a tiny piece of one chromosome 7. The larger the deletion, the more extensive the abnormalities. Some of the physical problems arise from the absence of one allele of a gene that codes for

Figure 14.1   Four young children with Williams syndrome. (Courtesy of the Williams Syndrome Association.)

Also, through detailed analyses of rare aberrations, geneticists have been able to find the locations of some important genes that had eluded them for years.

However they arise and whatever they may contribute to genetic analyses, such defects inflict great pain and suffering on human carriers and their families. Nearly all of the chromosomal syndromes involve mental retardation and growth defects. (This "classic" combination of mental deficit and developmental abnormalities is not surprising, given that about one-third of our genes are involved with brain development and many others are involved with organ development.) Seizures are very common, as are unusually low birth weight and failure to thrive. Heart and other vascular defects, abnormalities in the sex organs, and unusual arrays of fingerprint patterns are frequently observed. Different syndromes are recognized by different combinations of abnormalities.

the protein elastin. Elastin molecules form elastic fibers that provide stretchiness to blood vessels, skin, lungs, and other tissues. Next to this gene sits one that might explain some of the behavioral characteristics. It codes for a kinase protein, an enzyme involved in intracellular communication and strongly expressed in the brain (Frangiskakis et al. 1996). The detrimental effects occur because only one allele is present and expressed at each of several affected loci. This unbalanced situation describes many such abnormalities in chromosome structure.

In the preceding chapter, we discussed aneuploidy, abnormalities involving too many or too few *whole* chromosomes in a diploid set. Here we concentrate on what happens when *parts* of chromosomes are lost, gained, or moved to new positions in the genome. Some of these structural abnormalities occur when chromosomes break and the resulting parts fail to reattach properly. Instead of rejoining to each other, the two ends from one break may attach to the ends from other breaks. Depending on the number and positions of breaks, many kinds of rearrangements can be formed.

Such changes have been extensively analyzed in fruit flies, corn, mice, and other experimental organisms, where chromosome breaks can be induced by radiation, by a wide variety of chemicals, and by certain viruses. But detailed analyses of structural aberrations in humans have become possible only recently as technical advances have allowed investigators to analyze the fine banding and molecular structure of human chromosomes. These developments have led to the discovery of many new chromosomal syndromes.

Deletions and Duplications

Pieces of chromosomes can be lost or gained, resulting in *partial monosomy* or *partial trisomy*. The loss of a chromosome piece—that is, a **deletion**—can take place in several ways. As shown in Figure 14.2A, a single break can lead to the loss of a chromosome tip, which is what usually happens in humans. Less frequent in humans is the occurrence of two breaks with loss of the intervening segment and rejoining of the two flanking portions, one of which contains the centromere (Figure 14.2B). But if two breaks occur near the ends of a chromosome and the centromere is on the middle piece, the **acentric** fragments (i.e., those lacking a centromere) will get lost and a centric ring may be formed by union of the two broken ends of the middle piece (Figure 14.2C). Deletions may also occur as a by-product of rearrangements between different chromosomes. Generally speaking, most deletions are newly arising; that is, chromosomally normal parents produce a child with a deletion. Fortunately, the chance that they will have another similarly affected child is slight, probably less than 0.5%.

Except in the X chromosome, deletions of more than a band or two (which may contain hundreds of genes) tend to be lethal even when heterozygous (i.e., when paired with a normal, nondeleted chromosome). Deletion-carrying offspring who do survive to birth often have severe phenotypic defects. Deletions in the short arm of chromosomes 4, 5, 11, and 18 and in the long arm of 7, 11, 13, 15, 18, 21, and 22 are the ones most often seen in infants. But deletions for every chromosome tip and many interstitial (internal) seg-

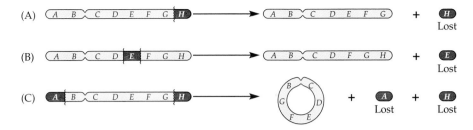

Figure 14.2 Origin of deletions. (A) One break near a chromosome tip. A small piece is lost. (B) Two breaks followed by the loss of a small internal piece. (C) Two breaks followed by the loss of both tips and formation of a ring chromosome. Chromosome fragments that are acentric (i.e., lacking a centromere) will be lost from the nucleus in the subsequent cell division.

ments are known too. Some chromosomes (4, 5, 17, 22) seem to be particularly susceptible to breakage. Or perhaps breaks in these chromosomes are simply less deleterious than those occurring in some other chromosomes and thus lead to a type of sampling bias.

Deletions encompass a wide range of sizes: (1) large ones detectable by classic cytogenetic methods; (2) tiny ones called *microdeletions* (see later), detectable by high-resolution banding methods; (3) extremely tiny ones detectable only by FISH or other molecular methods; and (4) submicroscopic ones involving the loss of a single locus and detectable only by molecular methods. We will begin with the largest type of deletion.

Relatively common deletions that cause the *Wolf-Hirschhorn syndrome* involve the short arm of chromosome 4 (i.e., 4p–).* Phenotypically, this deletion is often associated with improper fusion of the midline of the body. Also, affected infants usually have wide-set and bulging eyes, a broad nose, low-set and malformed ears, a very small lower jaw, and a cleft palate (Figure 14.3). Heart, lung, and skeletal abnormalities are common, and in males the penis may be improperly fused. Birth weight is very low. These babies generally fail to thrive, and most of them die young.

Another cytologically visible deletion (about 1 in 50,000 live births) is the loss of much of the short arm of chromosome 5. Because newborns with 5p– usually have a high-pitched mewing cry like that of a kitten, this aberration is called the *cat-cry syndrome* or *cri du chat syndrome* (Figure 14.4). Other features are variable, but they usually include a small head with a round face, wide-set eyes with epicanthal folds on the upper lids (covering the inner corners of the eyes), low-set ears, and slow growth. Most affected individuals survive beyond childhood, but with severe mental retardation. Deletion sizes differ from case to case, but the crucial region absent in all children with the cat-cry

syndrome is a tiny segment near the middle of 5p15. By correlating the particular traits exhibited by patients with specific deletions in chromosomes 4 and 5, researchers have begun to assemble *phenotypic maps* of these two chromosomes (Estabrooks et al. 1995).

More frequently seen is *Williams syndrome*, described at the beginning of this chapter. This syndrome affects about 1 in 10,000 to 20,000 newborns. It results from deletions in 7q,* which almost always include the elastin gene and usually several nearby genes as well.

Over 65% of patients with *Prader-Willi syndrome* (*PWS*) show cytogenetically visible deletions in 15q near the centromere.† Recall from Chapter 10 that PWS is a disorder characterized in infants by poor muscle tone, poor reflexes, poor feeding, and underdevelopment of the gonads and external genitals.

*Specifically, 7q11.23. DNA sequence analyses identified an especially large number of *Alu* repeats in this region. Some of the deleted segments show *Alu* repeats at both ends, suggesting that the deletions may result from unequal crossing over between mismatched *Alu* elements (Chapter 7).

†Another 15% have tiny microdeletions in this region, and 20% show a phenomenon called uniparental disomy; both of these situations will be discussed later in the chapter. Finally, a few rare individuals with PWS have totally normal chromosomes and mutations in the *PWS* allele.

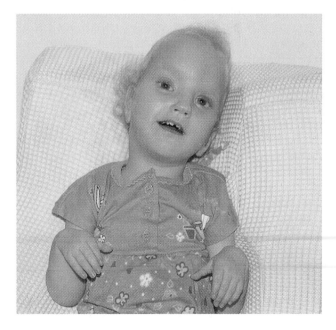

Figure 14.3 A patient with Wolf-Hirschhorn syndrome, associated with 4p monosomy. Note the prominent forehead, wide-set eyes, and large, low-set ears. (Courtesy of the Longo Family.)

*Recall that p refers to the short arm of a chromosome and q to the long arm. A minus sign following the p or q means that the arm is shorter than normal; a plus sign means that it is longer than normal.

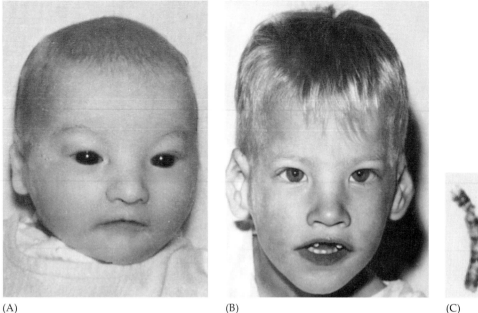

(A)   (B)   (C)

Figure 14.4   A patient with the cat-cry syndrome, associated with a 5p– deletion. (A) Affected new-born, showing moon face and wide-set eyes. (B) Same child at 4 years; the cat-cry and moon face have disappeared, but the epicanthal skin fold has persisted. The ears are somewhat misshapen and low set. (Photos from Valentine 1986.) (C) Chromosomes 4 and 5 from an affected individual, showing a deletion (arrow) of part of the short arm of one chromosome 5. (Chromosome photographs courtesy of Irene Uchida.)

Children with PWS are mentally retarded and extremely obese; they are also very short, with small hands and feet. Large deletions of the same chromosome band are also associated with 65–75% of *Angelman syndrome* (*AS*) cases, but up to 30% of AS cases are karyotypically normal and probably due to mutations or uniparental disomy. Children with AS show an array of symptoms that include profound mental retardation, jerky movements, unusual facial features, poor muscle tone, and epilepsy. They also laugh excessively but do not speak. Recall that the *PWS* locus is paternally expressed and the *AS* locus is maternally expressed.

Smith-Magenis syndrome, which is associated with deletion of part of 17p11, involves both physical and behavioral abnormalities. Facial defects (including cleft palate), short stature, and mental handicap are standard, and nearsightedness and hearing loss are common. Sleep disturbances, reduced pain sensitivity, self-injurious behavior (such as pulling out fingernails and toenails), and the unusual habit of self-hugging are frequently observed.

Large deletions of chromosome 18 are also known. Loss of the short arm (18p–) results in growth retardation and mental retardation as well as distinctive facial features. When part of the long arm is lost (18q–), these anomalies occur with somewhat different defects of the ears, eyes, mouth, and facial structure (Figure 14.5). Heart and minor skeletal defects are also frequent. Kline et al. (1993) have constructed a preliminary phenotypic map for the distal tip of 18q.

Geneticists have described many 21q deletions, which are called *partial monosomy 21*. Based on molecular studies of six cases, Chettouh et al. (1995) present a map of 21 features associated with partial monosomy 21. Their results indicate that the region between the genes *APP* and *SOD1* (in bands 21.3 to 22.11, shown in Box 13A) is critical for the expression of mental retardation, facial anomalies, poor muscle tone, and certain other traits.

Deletions of parts* of Xp or Xq have been reported in women who show some but not all characteristics of Turner syndrome. Their phenotypes range from aspects of Turner syndrome to having only minor menstrual abnormalities. Deletions of Yp that include loss of the key testis-determining gene, *SRY* (Chapter 15), leads to an absence of masculinization; that is, affected individuals become phenotypic females. Loss of Yq does not prevent maleness per se, but affected individuals may show some abnormalities of testicular development and be sterile; they may also exhibit short stature.

*Deletion of *all* of Xq is lethal because, with the loss of the *XIST* locus in the Xq inactivation center (Chapter 10), both X's will be expressed, and functional disomy for Xp is lethal.

Pieces of chromosomes can be gained as well as lost. These gains may result when, following three chromosome breaks, a segment of one chromosome is inserted elsewhere in the homologous chromosome or into a different chromosome (Figure 14.6A). Such gains are often called **duplications**, but keep in mind that the process yields three copies of the segment (i.e., partial trisomy) in a diploid cell. Duplications can also arise in meiosis through errors in chromosome replication, through errors in crossing over (Figure 14.6B), or by normal crossing over between chromosomes that are heterozygous for certain other structural abnormalities. When the resultant **unbalanced gamete** (here, one having an extra piece of a chromosome) combines with a normal gamete, the zygote will possess an extra batch of genes, ranging from just a few genes to an entire chromosome arm.

The phenotypic effects of duplications vary according to their size and position. In general, however, small duplications are less harmful than are deletions of comparable size. Thus, there seem to be many fewer well-defined human syndromes associated with duplications alone. The wide variety of different breakpoints and duplication sizes also makes this true.

Geneticists believe that small duplications—often occurring through unequal crossing over—have played an important role in the evolution of all species, providing additional pieces of DNA that are

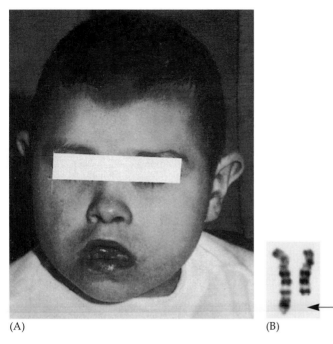

(A)   (B)

Figure 14.5 (A) A child with a deletion (arrow) of one-fourth to one-half of the long arm of chromosome 18. The downturned mouth, absence of the midline indentation that runs between the nose and the mouth, and the peculiar shape of the ears are characteristic of these patients. The carp mouth trait also occurs in certain other chromosomal syndromes. (Photograph by F. Sergovich, from Valentine 1986.) (B) Chromosomes from an affected individual. (Chromosome photographs courtesy of Irene Uchida.)

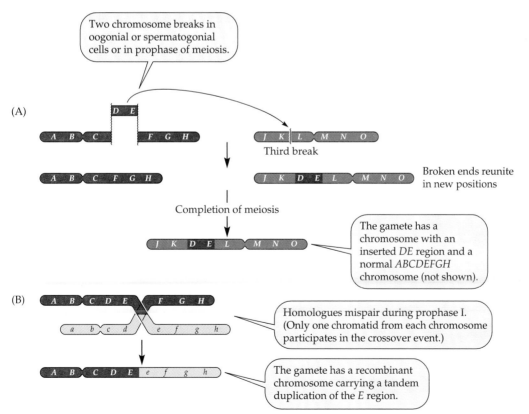

(A)

Two chromosome breaks in oogonial or spermatogonial cells or in prophase of meiosis.

D E

A B C   F G H   J K L M N O

Third break

A B C F G H   J K D E L M N O   Broken ends reunite in new positions

Completion of meiosis

J K D E L   M N O

The gamete has a chromosome with an inserted *DE* region and a normal *ABCDEFGH* chromosome (not shown).

(B)   A B C D E   F G H

         a b c d   e f g h

Homologues mispair during prophase I. (Only one chromatid from each chromosome participates in the crossover event.)

A B C D E e f g h

The gamete has a recombinant chromosome carrying a tandem duplication of the *E* region.

Figure 14.6 Origin of duplications. (A) Two breaks in one chromosome, with insertion of the middle segment into a third break in a different chromosome. *Note:* The normal homologues of these two chromosomes are not shown here. (B) Rare incorrect (out-of-register) synapsis followed by unequal crossing over. Such small duplicated segments have been observed in fruit flies, in which very precise banding analysis is possible. In humans, this phenomenon accounts for the formation of some regions with many duplicated segments (see Figure 7.11). Gametes with duplicated regions, when combined with normal gametes, produce zygotes with extra genetic material. Gametes with deletions can also be produced in these processes, but these have not been indicated in the diagrams.

then free to mutate to become new genes. For example, it appears that the independent but structurally similar genes controlling the production of certain protein chains in human hemoglobin all arose from duplications of a single ancestral gene (Chapter 6).

Supernumerary Chromosomes and Isochromosomes

Roughly one person in 1,500 carries an extra structurally abnormal chromosome that may or may not be associated with phenotypic effects. Such chromosomes, whose origins are often difficult to identify except by FISH and molecular methods, are called **supernumerary** (or *marker* or *accessory*) **chromosomes** (Figure 14.7). They are highly variable in size and composition, and some give rise to mental or physical abnormalities. A few syndromes are fairly well known.

Up to 50% of all supernumerary chromosomes are *inverted duplication 15* (Webb 1994). These always contain two centromeres (one of which may be inactive) and two copies of the acrocentric short arm, but otherwise vary greatly in size. The tiniest ones contain very little chromatin between the centromeres and usually cause no harm. But the medium-size (G group size)

and large (bigger than G group) ones, which contain some 15q euchromatin between the centromeres, are not so benign. They can give rise to the *inverted duplication 15 syndrome*, which includes mental retardation, developmental delay, seizures, behavioral problems, and minor facial abnormalities. A key factor associated with degree of abnormality is the presence of genetic material from 15q13, the same band involved in the Prader-Willi and Angelman syndromes.

The *cat eye syndrome* is associated with a supernumerary chromosome that has a good part of chromosome 22 in duplicate. Phenotypically, patients often have a break in the iris that enlarges the pupil, but the most consistent features are skin tags (small flaps of tissue) and pits in front of the ears. Cardiac and urinary tract defects, a closed anus, and some mental handicap also occur in a significant fraction of cases.

A special type of duplication, whose mechanisms of formation are not well understood, is a metacentric chromosome consisting of two identical arms. Such a mirror-image chromosome is called an **isochromosome** (Greek *iso*, "alike," "equal"). The classic hypothesis for mode of formation is misdivision of the centromere. Instead of the usual lengthwise separation between duplicate chromatids, the centromere seems to divide crosswise, thereby joining two identical chromatid arms and forming one short-arm isochromosome and one long-arm isochromosome. After the cell divides, the two daughter cells will contain, in addition to the normal homologue, an isochromosome for either the long arm or the short arm. But misdivision of the centromere is now considered less likely than other proposals (Figure 14.8). Breakage directly through the centromere or misreplication during S phase of cell division could also produce an isochromosome.

Isochromosomes for the long arm of (acrocentric) group D or G chromosomes, when present along with a normal homologue, lead to effective trisomy for that chromosome, even though the total number of chromosomes is only 46. Indeed, a few cases of Down syndrome are known to have an *isochromosome of 21q* plus a normal chromosome 21. People with a supernumerary *isochromosome 18p* exhibit a fairly well defined syndrome featuring low birth weight, an abnormally small head with distinctive facial features, poor muscle tone, bent fingers, and moderate to severe mental retardation. Isochromosomes have also been reported for several other chromosomes. Indeed, isochromosome Xq is the most common structural abnormality of the X, and the phenotype of affected females resembles that of Turner syndrome (Robinson and de la Chapelle 1997).

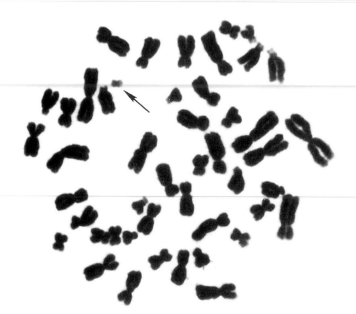

Figure 14.7   The arrow indicates a harmless supernumerary chromosome found in a student of normal phenotype and intellect who karyotyped herself as part of a class project. Further study of her family revealed that she inherited the extra chromosome from her normal mother, and that her normal sister was also a carrier. The student (who became a successful physician) later passed it on to a normal son. In this case the tiny extra chromosome is harmless because these acrocentric arms encode only ribosomal RNA, whose excess has no apparent phenotypic effect. But not all cases of supernumerary chromosomes are so benign. (Courtesy of Stanton F. Hoegerman.)

Microdeletions and Microduplications

In the past few years, scientists have uncovered a new class of chromosomal changes that cannot be identi-

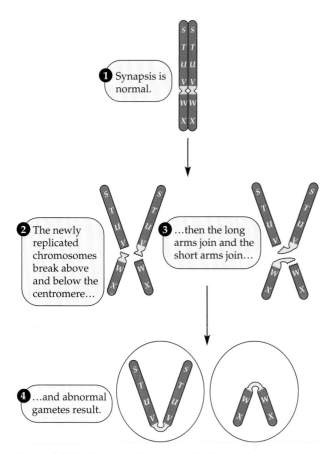

Figure 14.8 One possible mechanism for isochromosome formation in a postreplication chromosome. Breaks occur in both chromatids, but on opposite sides of the centromere. Following the exchange of the short and long arms, each isochromosome has an intact centromere.

fied by standard light microscopy. The associated disorders were originally thought to be transmitted as simple Mendelian traits, often showing autosomal dominant inheritance. But high-resolution banding analyses have revealed that in a few cases they are associated with extremely tiny deletions or duplications. Sometimes such **microdeletions** or **microduplications** are detectable only by FISH and other molecular techniques. These syndromes are often associated with additional, unrelated features resulting from the loss or addition of genes located next to the locus in question. Such disorders are also called **contiguous gene syndromes** because they involve a group of closely linked genes (Ledbetter and Ballabio 1995). The 20 or more known microdeletion syndromes, for example, usually involve mental retardation. The great phenotypic variability seen in these syndromes probably reflects the great variability (10 to 100 genes) in the size of the microdeletions and microduplications.

Deletions (some cytogenetically visible and some microdeletions) in 15q near the centromere are associated with the Prader-Willi and Angelman syndromes.

Tiny internal deletions in 13q are associated with *retinoblastoma* (tumor of the retina of the eye, discussed in Chapter 17); their incidence is about 1 in 100,000 live births. A deletion in 11p causes the *WAGR syndrome*, a complex that includes *W*ilms tumor (a cancer of the kidney), *a*niridia (absence of the iris of the eye), ambiguous *g*enitals and *g*onadoblastoma (tumor of the gonad), and mental *r*etardation. Such cancer-related deletions are discussed in Chapter 17.

The most common autosomal deletion site in humans, 22q11, is associated with a wide variety of heart abnormalities. Two conditions in particular, *DiGeorge syndrome (DGS)* and *Shprintzen velocardiofacial syndrome (VCFS)*, exhibit overlapping deletions and variable but similar phenotypes. Indeed, they are probably variants of the same disorder. DiGeorge syndrome also features absence or incomplete development of the thymus (Chapter 18) and parathyroid glands, as well as palate abnormalities and certain facial characteristics.* Sometimes psychiatric symptoms occur too. The causes of DGS vary, however. About 90% of cases involve 22q11 deletions, but some cases may instead be associated with environmental or maternal factors.

Patients with VCFS usually have heart defects, cleft palate, learning problems, speech defects, and characteristic facial features. Abnormalities of the parathyroid and thymus glands may also occur. At least 80% of patients with VCFS harbor either cytologically visible deletions or microdeletions in 22q11.

The first known example of a microduplication syndrome in humans was reported in 1991. *Charcot-Marie-Tooth (CMT) disease*, named after the physicians who first described it in 1886, is a fairly common neuromuscular condition. Type 1 CMT (CMT1) is usually inherited as an autosomal dominant, fully penetrant disorder with variable phenotype and variable age of onset. (Autosomal recessive and X-linked forms occur too, and there are at least three different loci associated with CMT1.) Molecular analyses of a chromosome 17 locus reveal that the defective allele is not a point mutation or a deletion. Rather, it is a submicroscopic duplication near the centromere in 17p, originating from unequal crossing over (Chapter 7).

Translocations

Breaks in two or more nonhomologous chromosomes followed by reattachments in new combinations can lead to the formation of **translocations** (Figure 14.9). Compared to other chromosomal abnormalities, translocations are relatively common among newborns and

*Wilson et al. (1993) suggest the acronym CATCH-22 to describe the phenotypes associated with 22q11 deletions: *C*ardiac abnormality, *A*bnormal facial features, *T* cell deficit due to thymic underdevelopment, *C*left palate or palatal dysfunction, and *H*ypocalcemia.

also among people who are mentally retarded. In some cases, the rearranged chromosomes may be transmitted through many successive generations.

If the rearrangement of chromosome parts involves the exchange of two acentric pieces that reattach to two centric pieces, and no chromosomal material becomes lost, the translocation is **reciprocal**. Because all of the genetic material is still present but in a different arrangement, a heterozygote for such a translocation is **balanced**. The phenotype is usually normal, but fertility may be reduced (see later). If a breakpoint happens to occur within a gene, however, it may disrupt normal function and behave like a mutation. In the general population, about 1 person in 625 carries a reciprocal translocation.

Translocations between and among acrocentric chromosomes (i.e., numbers 13, 14, 15, 21, and 22) represent a special case known as a **Robertsonian translocation**, in honor of the investigator who first found such translocations in grasshoppers. Robertsonian translocations are a common structural abnormality in humans, present in about 1 in 1,000 newborns. About 75% of these join together chromosomes 13 and 14, constituting "the most common single rearrangement in the human race" (Gardner and Sutherland 1996). Figure 14.10A shows the long arms of two acrocentric chromosomes connected at the centromere, as are the two largely heterochromatic tips. The tiny chromosome so formed is usually lost. The loss has no phenotypic consequences, since, as long as other acrocentrics' short arms are present, no essential material is lost. But the longer translocated chromosome contains the full complement of essential genes from the two chromosomes. A gamete with this bal-

anced chromosome, when combined with a normal gamete, gives rise to a balanced heterozygote. The resulting individual has a total of only 45 chromosomes, yet is phenotypically normal.

Insertional translocations require three breaks, with at least two in one chromosome. Then a deleted chromosomal fragment moves from its normal location to a new site on the same or a different (nonhomologous) chromosome. Insertional translocations occur less frequently than do reciprocal or Robertsonian translocations. The carrier has a normal phenotype, but may produce some unbalanced gametes. With reference to all of these types of rearrangements, **cryptic translocations** are by definition so small that they cannot be detected by standard cytogenetic techniques. Rather, they are identified only by the use of FISH or other molecular methods.

Collectively, translocations involve all chromosome arms more or less at random—with the exception of a few high-affinity combinations such as 13q with 14q, 11q with 22q, 9 with 22, and 9 with 15. Autosomal translocations are sometimes involved in specific types of cancer; balanced reciprocal translocations, for example, occur in leukemias and lymphomas. They will be discussed further in Chapter 17.

Behavior of Translocations in Meiosis

The meiotic behavior of reciprocal translocations, when heterozygous with structurally normal chromosomes, is rather complicated.* Because prophase I synapsis involves gene-by-gene pairing, the translocated chromo-

---

*Homologous reciprocal translocations, however, present no pairing complications.

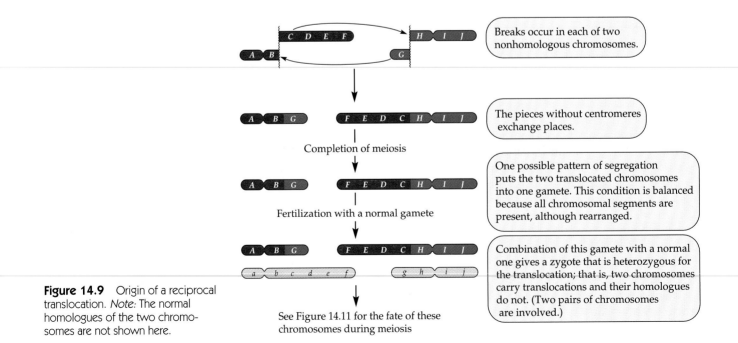

Figure 14.9 Origin of a reciprocal translocation. *Note:* The normal homologues of the two chromosomes are not shown here.

Breaks occur in each of two nonhomologous chromosomes.

The pieces without centromeres exchange places.

Completion of meiosis

One possible pattern of segregation puts the two translocated chromosomes into one gamete. This condition is balanced because all chromosomal segments are present, although rearranged.

Fertilization with a normal gamete

Combination of this gamete with a normal one gives a zygote that is heterozygous for the translocation; that is, two chromosomes carry translocations and their homologues do not. (Two pairs of chromosomes are involved.)

See Figure 14.11 for the fate of these chromosomes during meiosis

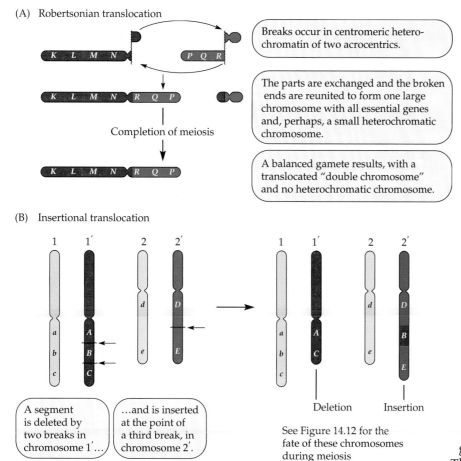

(A) Robertsonian translocation

Breaks occur in centromeric hetero-chromatin of two acrocentrics.

The parts are exchanged and the broken ends are reunited to form one large chromosome with all essential genes and, perhaps, a small heterochromatic chromosome.

Completion of meiosis

A balanced gamete results, with a translocated "double chromosome" and no heterochromatic chromosome.

(B) Insertional translocation

A segment is deleted by two breaks in chromosome 1′…

…and is inserted at the point of a third break, in chromosome 2′.

Deletion   Insertion

See Figure 14.12 for the fate of these chromosomes during meiosis

Figure 14.10 Special types of translocations. (A) Robertson-ian translocation. Two breaks occur very near the centromeres of acrocentric chromosomes, whose short heterochromatic arms carry no essential genes. Thus, a gamete that carries only the chromosome with the joined long arms is essentially balanced. (B) Insertional translocation. Three breaks occur and the deleted fragment gets inserted into the third break; the insertion het-erozygote is balanced. (Here we show an *interchromosomal* insertion, with two breaks in one chromosome and one in a non-homologous chromosome, but all three breaks could occur in one chromosome.)

somes must form some unusual patterns to match up with their partners. Reciprocal translocations will arrange themselves into a distinctive crosslike shape (Figure 14.11A), from which pairs of chromosomes seg-regate in three possible ways during anaphase I. If seg-regation is alternate (i.e., if chromosomes that are diag-onally opposite move to the same pole), then the resultant gametes and zygotes are balanced. Note that among these balanced gametes, half will carry the two translocated chromosomes and half will carry the two normal chromosomes. If separation is not alternate (i.e., if it occurs in the vertical plane or in the horizontal plane), then the resultant gametes will be abnormal. Any zygote formed by its combination with a normal gamete will be unbalanced and abnormal.

The three kinds of segrega-tion are not equally frequent, but enough unbalanced gametes are produced to reduce the fer-tility of translocation heterozy-gotes. That is, the unbalanced offspring do not always survive to birth. In experimental plants and animals, such reduced fer-tility is passed on to about half the offspring (i.e., to transloca-tion carriers). Such a dramatic decline in reproductive fitness may not be so obvious in hu-mans because they have few off-spring and are likely to "re-place" miscarriages, stillbirths, or early infant deaths with addi-tional pregnancies. Yet some translocation heterozygotes will show a clear reduction in fertility, and most are likely to exhibit a higher-than-normal rate of loss of progeny.

Insertional translocations may give rise to loop formations (Figure 14.12) during meiosis, when normal gene-by-gene pairing is disrupted. The gametes formed will be of four types: a normal haploid karyotype, a balanced insertion, a deletion, and a duplica-tion. Whether any of the resultant unbalanced zy-gotes will survive depends on the size of the deletion or duplication. Although it may seem counterintuitive, carriers of short insertions are much more likely to produce abnormal children (risk of 10–50%) than are carriers of long insertions. This is because the larger insertions usually lead to spontaneous abortion in early pregnancy, thereby reducing the risk for an ab-normal child to 0–5% (Gardner and Sutherland 1996).

Translocation Down Syndrome

Recall that nearly all cases of Down syndrome occur sporadically in families whose other members are un-affected. The affected individuals have the karyotype 47,XX,+21 or 47,XY,+21, the extra chromosome coming from nondisjunction in one of the karyotypically nor-mal parents, usually the mother. In contrast to this sit-uation, about 4% of people with Down syndrome have 46 rather than 47 chromosomes, one of which is abnor-mally long. This long chromosome is a Robertsonian translocation, usually involving chromosomes 14 and 21. Because nondisjunction is not the cause, no mater-nal age effect is seen here. Although such cases are rare, they are important to identify because this form of Down syndrome may recur in the same family.

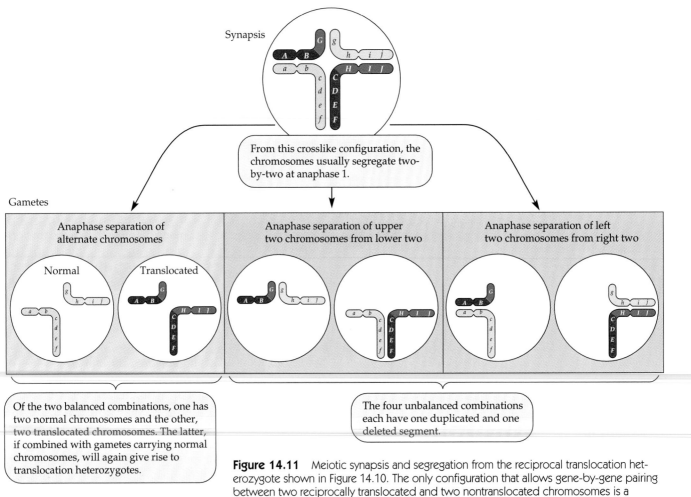

Synapsis

From this crosslike configuration, the chromosomes usually segregate two-by-two at anaphase 1.

Gametes

| Anaphase separation of alternate chromosomes | Anaphase separation of upper two chromosomes from lower two | Anaphase separation of left two chromosomes from right two |

Normal   Translocated

Of the two balanced combinations, one has two normal chromosomes and the other, two translocated chromosomes. The latter, if combined with gametes carrying normal chromosomes, will again give rise to translocation heterozygotes.

The four unbalanced combinations each have one duplicated and one deleted segment.

Figure 14.11   Meiotic synapsis and segregation from the reciprocal translocation heterozygote shown in Figure 14.10. The only configuration that allows gene-by-gene pairing between two reciprocally translocated and two nontranslocated chromosomes is a crosslike configuration. (For simplicity, we have not shown the chromosomes in their actual doubled state.)

A carrier parent of an individual with translocation Down syndrome has only 45 chromosomes, one of which is the combined 14q21q. As shown in Figure 14.13A, three chromosomes in these translocation heterozygotes contain nearly all of the material present in the original four chromosomes. The tiny tips of chromosomes 14 and 21 are lost, but the fact that the translocation carrier is unaffected means that no essential genes are missing.

Now consider the kinds of gametes that a carrier might produce (Figure 14.13B). During meiosis, two chromosomes will usually proceed to one pole and one to the other; thus, six types of gametes are possible. Of these, two types are balanced: one bearing the normal chromosomes 14 and 21 and another carrying the single large 14q21q translocation. When combined with structurally normal chromosomes during fertilization (Figure 14.13C), the former will yield a normal zygote, and the latter will give rise to a translocation heterozygote with a normal phenotype. Thus, the translocation may pass undetected through a number of generations.

Among the types of unbalanced gametes produced by a carrier, only one will lead to a viable zygote after combining with a gamete with normal chromosomes. This zygote will have an extra dose of chromosome 21 and will develop Down syndrome. A carrier parent may therefore produce up to three kinds of offspring: phenotypically and karyotypically normal, phenotypically unaffected translocation heterozygote, and translocation Down syndrome. The theoretical risk of having a child with Down syndrome is thus 1/3, but the actual observed frequencies are much lower: about 5–10% when the carrier is the mother and about 1–2% when the carrier is the father. (Some people are known to be carriers of Robertsonian chromosomes derived, in part, from a chromosome 13. These individuals are also at elevated risk of having offspring with trisomy 13.) Although the multiple occurrence of Down syndrome in a family may suggest the presence of a 14q21q translocation, it is not a certainty. In fact, most cases of multiple Down syndrome are due to multiple instances of sporadic nondisjunction in karyotypically normal parents.

1. If both translocated segments (i.e., the *XIST*-bearing segment and the "*XIST*-less" segment) remain active, then the intact (untranslocated) X is inactivated and the genome is functionally balanced. That is, the cell has one intact inactive X and, although it has been separated into two portions, the equivalent of one whole active X.

2. If the *XIST*-bearing translocated segment gets inactivated, the intact X must then remain active. But the translocated "*XIST*-less" segment of the X is also active, so now the cell has *two* active alleles (rather than just one) for every locus on this segment of the X.

An embryo with an X-autosome translocation will randomly contain both types of cells (types 1 and 2). Those embryos with a substantial fraction of type 2 cells cannot survive unless the doubly active segment of the X is very small. But if all or nearly all the X type 2 cells die, the embryo can survive, although it may show phenotypic abnormalities. Nearly all of the male and about half of the female translocation carriers will probably be infertile, and for fertile females, the risk of structural or functional aneuploidy in a live-born child could be 20–40% (Gardner and Sutherland 1996).

When *Y-autosome translocations* involve the (mostly) genetically inert long arm of the Y (occurring in about 1 in 200 newborns), the translocation carrier may be phenotypically unaffected. In such cases, the autosome is usually an acrocentric chromosome 15 or 22 with a break on its short, heterochromatic arm. In rare cases where the Y long arm is translocated to a nonacrocentric chromosome, the observed phenotypic abnormalities may be associated with breaks in the autosome's euchromatin. Translocations of the short arm of the Y almost always lead to infertility and other abnormalities, including mental retardation. Because the short arm contains the testis-determining region—more specifically, the *SRY* gene (Chapter 15)—the carriers are male. A few such cases involving breaks in chromosome 15q12–13 have also given rise to Prader-Willi syndrome.

X-Y translocations are rare and can give rise to 46,XX males if the translocated Y segment contains the *SRY* gene. These and other types of X-Y translocations will be discussed in Chapter 15.

Other Abnormalities

Four additional types of chromosomal aberrations are seen in humans. Three are compatible with survival beyond the fetal stage, and one is not.

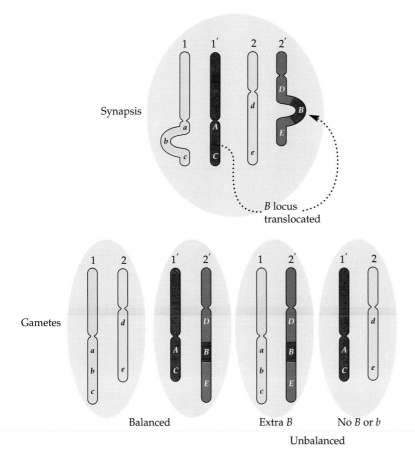

Figure 14.12 Meiotic synapsis and segregation from the insertional translocation heterozygote shown in Figure 14.10B. To allow for gene-by-gene pairing of the four chromosomes, two loops might form. One is on normal chromosome 1 when paired with its now-deficient homologue 1', and the other is the insertion on chromosome 2' when paired with the other normal chromosome (2). Four types of gametes can be formed: one (1 + 2) will be completely normal; one (1' + 2') will be a balanced insertional heterozygote; one (1 + 2') will have a duplicated segment; and one (1' + 2) will have a deleted segment.

Translocations Involving Sex Chromosomes

Sex chromosome aneuploidy is much more benign than autosomal aneuploidy. The reasons lie partly with the paucity of genes on the Y, but mostly with dosage compensation (lyonization) of the X chromosome (Chapter 10). Recall that inactivation begins at an X inactivation center containing the *XIST* gene and spreads in both directions through the X chromosome.

X-autosome translocations present an interesting situation. There are two X chromosome segments, one carrying the *XIST* locus and the other lacking it. The latter cannot be inactivated. (Also, in X-autosome translocations with *XIST*-bearing segments, the inactivation can spread to the adjacent autosomal region.) Consider the following two inactivation alternatives for cells containing balanced X-autosome translocations:

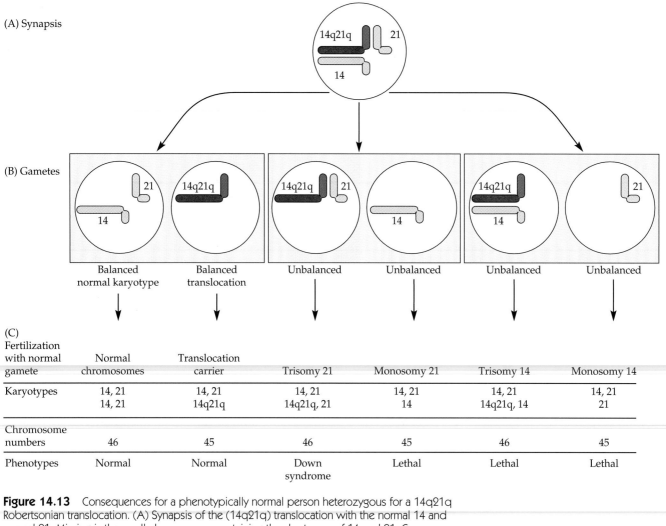

	Normal chromosomes	Translocation carrier	Trisomy 21	Monosomy 21	Trisomy 14	Monosomy 14
Karyotypes	14, 21 14, 21	14, 21 14q21q	14, 21 14q21q, 21	14, 21 14	14, 21 14q21q, 14	14, 21 21
Chromosome numbers	46	45	46	45	46	45
Phenotypes	Normal	Normal	Down syndrome	Lethal	Lethal	Lethal

Figure 14.13 Consequences for a phenotypically normal person heterozygous for a 14q21q Robertsonian translocation. (A) Synapsis of the (14q21q) translocation with the normal 14 and normal 21. Missing is the small chromosome containing the short arms of 14 and 21. Compare with Figures 14.10A and 14.11. (B) Segregation (diagonal, up-down, left-right) gives rise to six types of gametes, of which two are balanced and four are unbalanced. (C) Fertilization with normal gametes results in two zygotes producing normal phenotypes, one zygote leading to Down syndrome, and three lethal zygotes. Of the two normal phenotypes, one is karyotypically normal (46 chromosomes) and one is again heterozygous for the translocation, like the parent (45 chromosomes).

Inversions

When two breaks occur in one chromosome and the intervening segment gets inverted before the broken ends rejoin, the resultant aberration is known as an inversion (Figure 14.14A). The reversed segment may or may not include the centromere. Heterozygosity for an inversion does not ordinarily produce a distinctive phenotype. Instead, its presence is detected by unusual meiotic events or unusual segregation patterns in organisms heterozygous for the inversion. Although estimates differ widely, the total incidence of inversions in humans might be as high as 1–2%. The particular chromosomes involved and the locations of breakpoints appear to be highly nonrandom. Inver-

sions within centromeric heterochromatin of chromosomes 1, 9 (the most common), 16, and Y are fairly frequent, causing the position of the centromere in these chromosomes to vary considerably. Because these inversions pose no known genetic risks, some cytogeneticists consider them to be **variant chromosomes** rather than abnormal chromosomes. Inversion variants not involving centromeric heterochromatin have also been described for chromosomes 2 (the most common), 3, 5, and 10. Other inversions are much less frequent (Gardner and Sutherland 1996).

Complete gene-by-gene pairing during synapsis of an inversion heterozygote requires an inverted chromosome and its normal homologue to form a loop during prophase I (Figure 14.14B). What happens then depends on (1) whether or not crossing over occurs within the inversion loop and (2) whether or not the centromere is included within the inversion. When no crossovers occur within the inversion loop, no unbalanced chromosomes are formed and there is no effect on fertility.

But when crossing over occurs within the inversion loop, whether or not it includes the centromere, the two crossover chromatids end up with both duplications and deletions. In addition, crossover chromatids (from inversions that do not include the centromere) may be either **dicentric** (having two centromeres) or acentric (having no centromere). Dicentrics break and acentrics get lost, resulting in highly unbalanced gametes and zygote death. (For details, see Mange and Mange 1990 or any general genetics textbook). The overall effect on fertility varies with the frequency of crossing over within the inversion. Thus, larger inversions (with more crossing over) may give rise to more defective gametes and greater infertility. But the viability of a zygote containing a recombinant gamete actually depends on the total genetic content of the *noninverted* segments, because it is the latter that get duplicated or deleted following recombination.

Recent cytogenetic studies indicate that partial (rather than complete) pairing can also occur in inversion heterozygotes. For example, when the inverted region is short, the outer (distal) noninverted segments may pair normally, while the inner inverted region and its counterpart on the normal homologue loop out (or lie adjacent but unsynapsed) and thus do not recombine.

Ring Chromosomes

Recall that when two breaks occur in one chromosome, ring chromosomes can be formed (see Figure 14.2C). *Acentric rings*, lacking a centromere, usually get lost in subsequent cell divisions; but *centric rings*, because they have a centromere, can be passed on to daughter cells. Loss of the distal segments of both arms leads to partial monosomy, so that individuals heterozygous for ring chromosomes often exhibit major phenotypic abnormalities, including mental retardation. But little or no chromatin is lost with telomere-to-telomere fusion, resulting in rare normal or near-normal phenotypes. Although they are relatively uncommon in general, rings have been reported for every human chromosome.

Ring chromosomes can be unstable in both meiosis and mitosis. At meiosis, crossing over between the ring and its normal homologue can disrupt the ring. In somatic cells, a single sister chromatid exchange in a ring chromosome will produce a double-sized *dicentric ring*. During anaphase, the dicentric ring will break, and both daughter cells may receive centric rings of different sizes. This situation can also occur with sister chromatid exchange in the mitotic cells of embryos:

> Thus, at mitosis daughter cells arise that are partially or totally aneuploid for the chromosome in question. … These cells may die; some, however, survive in the mosaic state and presumably make an unfavorable contribution to the phenotype. This continuous generation and loss of cells seriously undermines the growth *rate*, although it may not greatly influence the quality of growth. The result is the general ring syndrome—whichever autosome is concerned—of marked growth retardation, borderline to moderate mental deficiency, minor dysmorphogenesis, and, perhaps, intact fertility. (Gardner and Sutherland 1996)

It appears that males heterozygous for a ring chromosome are less fertile. In most cases of known ring chromosome inheritance, it was transmitted by the mother.

Polyploidy

Entire chromosome sets may also be present in excess, a situation called **polyploidy** (Greek *poly*, "many"; English *ploid*, "set"). An organism may be **triploid**, with three representatives of each chromosome, or **tetraploid**, with four of each chromosome, and so on. Triploids may arise when two sperm fertilize one egg. Tetraploids arise, following fertilization, in two possible ways: (1) when a cell undergoes two successive S phases during one interphase or (2) when a metaphase cell skips anaphase and telophase and returns directly to interphase.

Polyploidy is quite common in plants and often has great commercial value for farm crops such as wheat. Because of irregularities in synapsis during meiosis, many

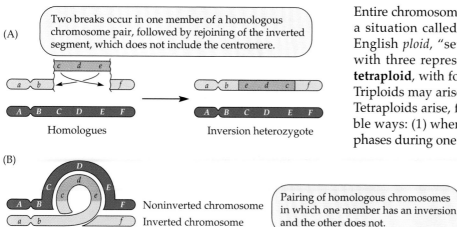

Figure 14.14 (A) Origin of an inversion. (B) Pairing configuration of an inversion heterozygote. (For simplicity, we have not shown the chromosomes in their actual doubled state.)

polyploids are infertile. (This inability to set seeds can give rise to seedless varieties of fruits; the common banana, for example, is a triploid.) Although during synapsis no more than two chromosomes can be paired for any given segment of a chromosome, different segments of the same chromosome may be paired with different partners. The resultant gametes will contain one representative of certain chromosomes and multiple representatives of others, leading to zygote lethality.

Polyploidy is rare among animals and lethal in humans (Figure 14.15). It seems that polyploid mammals are unbalanced in terms of the ratio of autosomes to metabolically active X chromosomes. In humans, triploidy is very common among embryos that are spontaneously aborted during the first three months of pregnancy, and more than 99% are lost by the end of the second trimester. Many are incompletely formed, but others show few or no gross abnormalities, and there is no distinctive phenotype. The only defect always seen is an enlarged placenta, sometimes with grape-shaped cysts. Growth of the embryos is stunted, resulting in low birth weight of the few live-born triploids. Intersexuality is also the rule. Webbing of the fingers and toes is common, and various abnormalities of the head, skeleton, heart, kidneys, and brain have also been reported. The few babies with triploidy who have survived beyond birth are mosaics with some diploid tissues. They almost always show severe mental and motor retardation.

Triploid embryos generally have one parent's genome in double dose and the other parent's genome in single dose. Does it matter which parent contributes the double dose? Yes, it does. Recall from our discussion of imprinting in Chapter 10 that human triploids with two sets of paternal chromosomes have a large placenta, whereas those with two sets of maternal chromosomes have an underdeveloped placenta. Indeed, the former develop the phenotype we have just described, and the latter develop so poorly that they usually abort in very early pregnancy.

Uniparental Disomy

The basic rules of inheritance assume that an individual gets one chromosome of each homologous pair from the mother and the other one from the father. So when unaffected parents have a child affected with a recessive disorder, we know that (barring nonpaternity or a new mutation) both parents must be heterozygous for the mutant allele. Recently, however, some cases have been reported in which a child with a recessive disorder has only *one* carrier parent, the other parent being homozygous normal at that locus. The child's karyotype looks normal, but molecular analyses show that the child has received *both* mutant alleles from the carrier parent and *no* member of that particular allele or chromosome from the other parent.

This odd phenomenon is called **uniparental disomy**, or **UPD** (*uni* meaning "one," and *disomy* referring to the presence of exactly two chromosomes of a

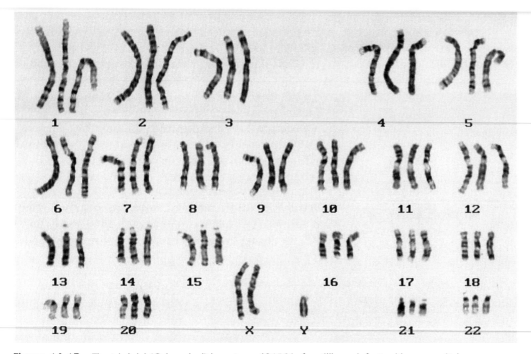

Figure 14.15 The triploid (Q-banded) karyotype 69,XXY of a stillborn infant with congenital malformations. Quinacrine staining allows each set of three homologous chromosomes to be distinguished from others. (Courtesy of Irene Uchida.)

homologous pair).* UPD can also lead to abnormalities when the chromosome or chromosome segment in question contains *imprinted* loci and an individual has two copies of the imprinted (silenced) gene or genes. (For example, some cases of Prader-Willi and Angelman syndromes have been traced to UPD of chromosome 15.) As shown in Figure 14.16, there are several ways by which UPD can arise. In all cases, it requires at least two events of abnormal cell division. Nondisjunction in both parents, followed by the complementary union of a disomic and a nullisomic gamete, is one possible cause (Figure 14.16A). Additional scenarios, involving chromosome loss or duplication in zygotes or in embryonic cells, can also be invoked (Figure 14.16B–D).

Uniparental disomy has been reported for at least 14 human chromosomes, almost half of all cases involving chromosome 15. Examples include a few cases of cystic fibrosis (chromosome 7), Beckwith-Wiedemann syndrome (chromosome 11), a female with Duchenne muscular dystrophy (X chromosome), and homozygotes for various chromosomal aberrations. The actual frequency of this phenomenon in the human population is unknown, and it was originally thought to be uncommon. But some geneticists suggest that it may be more common than previously realized and should be considered in the following cases:

- Rare homozygous recessives having only one heterozygous parent

- Rare recessive disorders accompanied by unexpected features such as mental retardation or growth retardation

- Females (with normal fathers) affected with rare X-linked recessive disorders

- Father-to-son transmission of X-linked recessive disorders

- Transmission of a very rare recessive disorder from parent to child

- Transmission of an apparently balanced chromosomal aberration from a healthy parent to a child with developmental defects

Frequencies of Chromosomal Abnormalities

Compared to all other species whose chromosomes have been studied, humans show an extraordinarily high rate of chromosomal abnormalities. Karyotype data are available for several categories of cells and phenotypes in humans, ranging from gametes to newborns to adults (Table 14.1). Some of the frequency estimates are fairly crude due to the different methods of sampling a given group, the different karyotyping techniques used over a period of years, and the small sample size for some groups.

Gametes and Fetuses

The chromosomes in human gametes are very difficult to study, but researchers are improving their techniques. Quite surprisingly, even normal fertile males and females may produce about 10% karyotypically abnormal gametes (Hook 1992). There is considerable variation in the estimates from different surveys, however, and it may be that some of the observed structural defects are technical artefacts (Jacobs 1992).

Chorionic villus sampling (*CVS*), to be described in Chapter 19, provides a way to detect certain chromosomal and metabolic disorders in fetal tissue during early pregnancy. Studies indicate that about 3% of 10-week-old fetuses have chromosomal abnormalities. This frequency is a little higher than the roughly 2% of chromosomal abnormalities found with *amniocentesis*, a prenatal test that is done at approximately the fifteenth or sixteenth week of pregnancy. Preferential spontaneous abortion of chromosomally abnormal fetuses probably accounts for the difference in frequency between weeks 10 and 15+ of pregnancy.

Abortions, Stillbirths, and Neonatal Deaths

Getting reliable data on spontaneous abortions (commonly called miscarriages) is very difficult. For one thing, it is not always clear whether an abortion (defined as termination of pregnancy before 20 to 22 weeks of development or when embryonic weight is less than 400–500 g) is spontaneous or induced. Yet the distinction is critical, because the former show a much higher frequency of abnormalities of all types than do the latter.

At least 15% of all recognized pregnancies end up as *detectable* spontaneous abortions, and the actual frequency of early embryonic loss may be as high as 50–60% (Jacobs 1992).* Researchers can now detect tiny amounts of a hormone (human chorionic gonadotropin) secreted by a week-old implanted embryo, which allows them to measure very early pregnancy loss. What is still not known is how much loss occurs *before* implantation, that is, during the first week of development. Estimates range widely, from about 15% to 55%. Researchers basically agree, however, that about 90% of spontaneous abortions occur during the first trimester of pregnancy and that the embryo is usually less than 8 weeks old.

---

*Note that *bi*parental disomy is the normal state of affairs.

*Some investigators think that at least 70% of all human conceptions are spontaneously aborted, most of them before the first missed menstrual period.

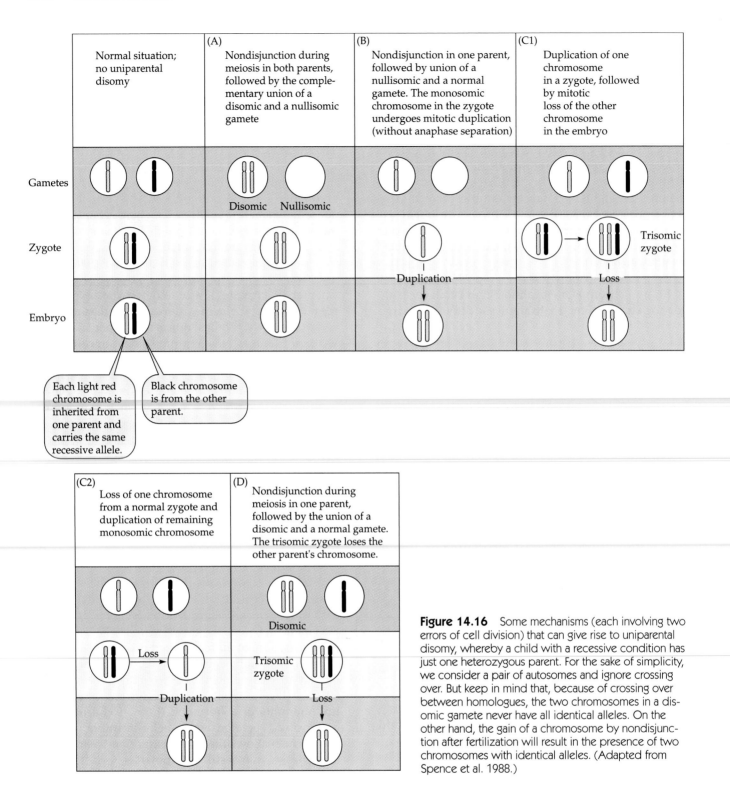

Figure 14.16 Some mechanisms (each involving two errors of cell division) that can give rise to uniparental disomy, whereby a child with a recessive condition has just one heterozygous parent. For the sake of simplicity, we consider a pair of autosomes and ignore crossing over. But keep in mind that, because of crossing over between homologues, the two chromosomes in a disomic gamete never have all identical alleles. On the other hand, the gain of a chromosome by nondisjunction after fertilization will result in the presence of two chromosomes with identical alleles. (Adapted from Spence et al. 1988.)

Fetal loss occurs for a variety of reasons. Overall, scientists estimate that at least one-half of all known spontaneous losses involve chromosomal abnormalities. In the youngest abortuses (up to 4 weeks old), the frequency of abnormalities is at least 90%; among those aged 5 to 8 weeks, it is about 60%; and in those 9 to 12 weeks of age, it ranges from 12% to 32%. The single most common (9%) chromosomal aberration is 45,X; indeed, it is thought that fewer than 1% of all 45,X zygotes ever reach full term. As a group, trisomics are the most frequent (27%) class of abnormality, with trisomy 16 by far the most common among

TABLE 14.1 Approximate frequencies of chromosomal abnormalities (excluding FISH-detected deletions) in different populations

Population	Frequency of abnormalities (%)
Sperm of normal fertile males	≥10?
Eggs of normal fertile females	≥10?
All zygotes	10–30
Preimplantation embryos from in vitro fertilization	20–25
All clinically recognized pregnancies (35–28 weeks)	≥8
All miscarriages (spontaneous abortions):	≥48
From unrecognized pregnancies (≤5 weeks)	Unknown (33–67?)
From all recognized pregnancies (5–28 weeks)	≥30–35
Stillbirths (328 weeks gestation)	5–6
Live births	0.5–1
Infant and childhood deaths	5–7
Patients with congenital malformations	4–8
Patients with congenital heart defects	13
Patients with 33 birth defects and mental retardation	6
Mental retardation (IQ ≤69), excluding fragile X	3–35
Imprisoned adults	0.4–3
True hermaphrodites (both ovarian and testicular tissue)	25
Males with defects in sexual differentiation	≤25
Females with defects in pubertal development	≤27
Females with primary ovarian deficiency (no eggs)	65
Infertile males	2–15
Adults with multiple (33) miscarriages	2–5

Source: Data from Chandley 1997; Eichenlaub-Ritter 1996; Garber et al., 1997; Gardner and Sutherland 1996; Hassold et al. 1996; Hook 1992; and Jacobs 1992.

these. Polyploids are the next most frequent class of abnormality (10%), while structural abnormalities occur in only 2% of spontaneous fetal deaths (Jacobs 1992). Autosomal monosomies are lost very early in development, probably before implantation (i.e., during the first week or two of development).

In contrast to spontaneous abortions, only about 5% of induced abortions exhibit chromosomal abnormalities. But here, too, the earlier the gestational age, the more frequent the abnormalities. About half of the aberrations are autosomal trisomies. Triploidies and 45,X are also fairly frequent.

Roughly 1% of all recognized pregnancies end in *stillbirths*, whose observed rate of chromosomal abnormalities is about one-tenth of that found in spontaneous abortuses. But as shown in Table 14.2, the observed types of defect vary from one developmental stage to another. Among stillbirths, for example, trisomy 18 is by far the most common aberration. A striking maternal age effect has been noted in these studies: About 35% of stillbirths and neonatal deaths from mothers over age 40 had karyotypic aberrations, whereas only 6% of those born to mothers under 40 showed chromosomal abnormalities.

Newborns and Adults

There is a further tenfold decrease in the frequency of chromosomal abnormalities found in live births (compared to stillbirths). Here the rate of abnormal karyotypes is 1 per 120 births (Gardner and Sutherland 1996). Among the various abnormal karyotypes, about 35–50% are considered of clinical importance; these include the autosomal trisomies, Klinefelter and Turner syndromes, and the unbalanced translocations that are accompanied by congenital malformations.

No comprehensive karyotypic studies of randomly chosen adults have been reported. But small surveys have detected a variety of chromosomal abnormalities, as well as an increase in mosaicism for sex chromosome aneuploidy (e.g., the number of 45,X cells), in both aging males and females. Among couples who have had three or more spontaneous abortions, the rate of chromosomal abnormality is greater than in randomly selected adults.

Studies of people who are mentally retarded reveal many types of abnormalities. Down syndrome (trisomy 21) accounts for about 10% of all retarded children, and the fragile X syndrome comes in second. Also, among institutionalized males who are mentally subnormal, about 3% are either 47,XXY or 47,XYY. Balanced translocations and inversions also appear more frequently among people who are mentally subnormal than among newborns.

Conclusions on Fetal Loss

It is clear that chromosomal abnormalities constitute a significant medical problem. Yet the grief and the burden to families and to society would be vastly greater if a large proportion of karyotypically abnormal conceptions were not spontaneously aborted early in pregnancy. In fact, live-born humans still show a much higher rate of chromosomal abnormalities than that found in any other species—even after spontaneous abortion eliminates over 95% of defective conceptions. The reasons for this situation are not clear. Perhaps it is the price we pay for human evolution. Or perhaps a high rate of mutation, chromosomal abnormalities, and spontaneous abortions leads to the spacing out of births in a species that requires an extremely long period for child raising. In any case, as Dorothy Warburton (1987) has pointed out:

> Low fertility and embryonic or fetal loss often cause disappointment and grief when they occur. Few couples realize how common such problems are, and

TABLE 14.2 Frequencies and distributions of chromosome abnormalities in different populations, assuming 15% spontaneous abortions and 1% stillbirths

| | Percent abnormal | | | | | | |
	45,X	Trisomic	Triploid	Tetraploid	Structural	Other	Total
Spontaneous abortions	8.6	26.8	7.3	2.5	2.0	0.7	47.9
Stillbirths	0.3	3.8	0.6	—	0.4	0.6	5.7
Live births	<0.01	0.3	—	—	0.6	0.02	0.9
All clinically recognized pregnancies	1.3	4.3	1.1	0.4	0.8	0.2	8.0
Likelihood of survival to birth (%)	0.3	5.8	0	0	74	11.5	11.6[a]

Source: Adapted from Jacobs 1992.
[a]This total is not expected to be the sum of the figures to its left.

many think that they, or their physicians, are somehow inadequate. ... Except for maternal age, almost no genetic or environmental factors are known to alter the rates of human chromosomal abnormalities. Thus, physicians and the public should be taught that some degree of reproductive loss is normal and that the best insurance against being childless may be not to allot too short a period for childbearing.

Chromosomal Abnormalities and Gene Mapping

Chromosomal abnormalities provide useful tools for narrowing down the locations of specific genes. **Deletion mapping** hinges on the principle that the absence of a chromosome segment should be correlated with reduced expression, nonexpression, or abnormal expression of genes on that segment. Males with X chromosome deletions, for example, usually have some kind of defect or disorder. Duplications have also been used to assign genes to specific regions of chromosomes. From analyses of enzyme levels in several individuals who carry different duplications or deletions in the same chromosome, it is sometimes possible to determine the location of the gene responsible for that enzyme. For example, two loci on chromosome 21—the locus encoding *superoxide dismutase (SOD) enzyme* and the locus encoding antiviral protein—show dosage effects: Trisomics produce excess product, and monosomics have a reduced amount.

Studies of individuals with translocations or small deletions have also been important in gene mapping. For example, translocation and deletion mapping, combined with molecular techniques, finally allowed geneticists to localize the X-linked recessive Duchenne muscular dystrophy (*DMD*) gene in 1986 (Chapter 5).

The first clue to its general position came in the 1970s, with reports of a few rare DMD females who had X-autosome translocations. Although the autosomal breakpoints varied from case to case, every DMD female had a breakpoint at Xp21—the large band in the middle of the short arm.*

Several years later, researchers began to analyze the DNA of a few rare DMD males who suffered from additional X-linked recessive disorders. One of these patients, Bruce Bryer (Figure 14.17), suffered from four rare and severe inherited conditions: Duchenne muscular dystrophy (a muscle-wasting disease), retinitis pigmentosa (an eye disorder), McLeod red cell phenotype (an abnormal blood type), and chronic granulomatous disease (an immune disorder). From the time of his birth in 1966, in Spokane, Washington, he gamely battled the effects of these illnesses, attending school whenever possible and even becoming an accomplished organist. He died at age 17 after a car accident.

When Bruce Bryer's case was discovered by human geneticists, only muscular dystrophy was known to be X-linked (Chapter 5), and its gene location was still a mystery. Was it possible that Bruce was missing a section from his one X chromosome that contained the gene loci responsible for all four conditions? Because the total loss of several important genes is usually lethal, no such males had ever been reported. But when Uta Francke (then at Yale University) first looked at Bruce's X chromosome from a culture of his blood cells, it seemed as though a tiny piece might be deleted from the short arm (see Figure 14.17). It took her and many other researchers almost three years of molecular tests to prove that this was indeed the case and that all four genes occupy neighboring loci in normal X chromosomes. This discovery led to the development of molecular tests for predicting whether or not a male fetus carries a defective allele of the muscular dystrophy gene.

Another DMD male in this study also had glycerol kinase (GK) deficiency and adrenal hypoplasia (AH), and he had a similar-sized, partially overlapping deletion in the same region. By fitting together all the vari-

*Recall from our earlier discussion of X-autosome translocations that the structurally normal X chromosome, containing an intact *XIST* locus, is usually inactivated in situations like this.

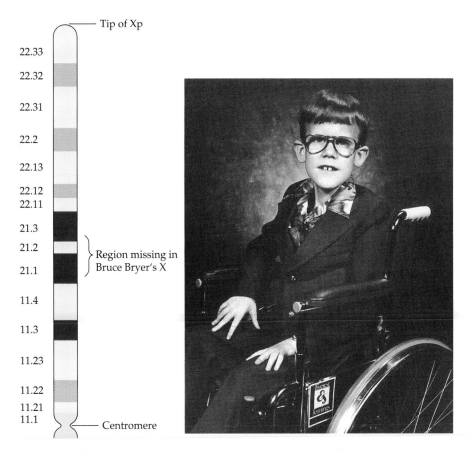

Figure 14.17 Bruce Bryer and a diagram of his X chromosome. Shown here is the standard Q-banding pattern of the X chromosome short arm, as presented by Francke et al. (1985). The bracket indicates the missing section in Bruce Bryer's X chromosome. This deletion included the loci for Duchenne muscular dystrophy, chronic granulomatous disease, retinitis pigmentosa, and McLeod's red cell phenotype. Figure 14.18 shows the fine structure of the region and the order of some of these genes. (Photograph © Eastern Washington State Historical Society; courtesy Muscular Dystrophy Association.)

ous puzzle pieces of cytogenetic, phenotypic, and molecular data, a team of Dutch, American, and Canadian researchers was able to come up with a concise description of the chromosomal region involved in these diseases (Figure 14.18). They determined the relative positions of the various loci and also constructed a map of the molecular landmarks in this region, covering about 4 million DNA base pairs.

Perhaps because of its huge size, mutations of the *DMD* gene vary from family to family. Only about 25% of cases are caused by a base substitution (mutation of a single base) within the DNA sequence that constitutes the gene. In about 70% of cases, a *DMD* mutation results from a deletion (and in about 5% of cases a duplication) of a long stretch of bases within that region of the DNA. Over 90% of Becker muscular dystrophy patients also show deletions in their DNA. Most of the *DMD* and *BMD* deletions occur in two "hot spots" within this huge gene.

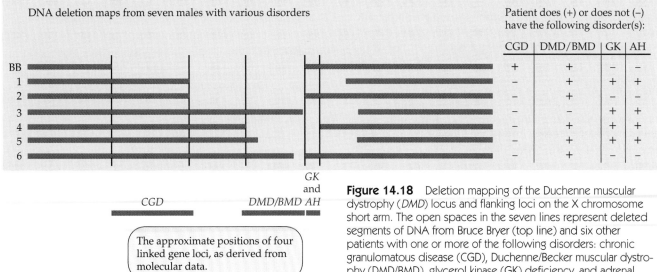

Figure 14.18 Deletion mapping of the Duchenne muscular dystrophy (*DMD*) locus and flanking loci on the X chromosome short arm. The open spaces in the seven lines represent deleted segments of DNA from Bruce Bryer (top line) and six other patients with one or more of the following disorders: chronic granulomatous disease (CGD), Duchenne/Becker muscular dystrophy (DMD/BMD), glycerol kinase (GK) deficiency, and adrenal hypoplasia (AH). Not included in this study were the genes for Bruce Bryer's retinitis pigmentosa (which probably lies just to the left of *CGD*) and McLeod red cell phenotype (which lies between *CGD* and *DMD*). (Adapted from van Ommen et al. 1986.)

Quite surprisingly, the size of a *DMD* deletion is not correlated with the severity of the disease. DMD patients with large deletions may have phenotypes similar to those of DMD boys with no detectable deletions, and the same is true for BMD patients. But the intragenic location of a deletion does matter; those causing defects in one end of the protein product are more often associated with DMD, whereas those causing defects in the central part of the protein product are more often associated with BMD. And the most important factor is whether the deletion causes a "reading frameshift" in the DNA, resulting in a much shortened and nonfunctional protein product (Chapter 7). Over 90% of DMD patients, but very few BMD patients, have frameshift mutations.

Comparable studies of other important disease loci and any closely linked chromosomal abnormalities will continue to add greatly to our understanding. Meanwhile, the discovery of so many deletions associated with monogenic disorders such as Duchenne muscular dystrophy has muddied the distinction between simple point mutations (such as a base substitution, deletion, or addition) and tiny chromosomal aberrations. As Epstein (1988) points out, aneuploidy may be a continuum affecting single genes as well as chromosome segments and whole chromosomes.

Summary

1. Chromosomal rearrangements arise when segments of broken chromosomes reattach in different ways. Detached acentric pieces of chromosomes are lost.

2. Deletions of more than a few genes are usually lethal, even when heterozygous with a normal chromosome.

3. Deletions and duplications are often the by-products of meiotic segregation from other rearrangements. Extra chromosomal pieces (duplications, trisomies) are phenotypically less harmful than the loss of chromosomal material (deletions, monosomies). Microdeletions and microduplications are also associated with several severe disorders.

4. Supernumerary chromosomes or marker chromosomes are highly variable in composition and phenotypic effects. Isochromosomes consist of a duplicated chromosome arm and may be associated with abnormalities due to resulting imbalance. Isochromosomes involving the long arms of acrocentric chromosomes can give rise to trisomy.

5. Meiotic segregation from heterozygous deletions, duplications, and translocations leads to some unbalanced gametes. Translocation Down syndrome is associated with specific types of chromosome 21 rearrangements. X-autosome translocations can lead to abnormalities of X chromosome inactivation and of sexual development.

6. Crossing over within the pairing loop of inversion heterozygotes usually leads to imbalance in some gametes. In some cases, dicentrics and acentrics are also formed.

7. Ring chromosomes are unstable during cell division, giving rise to several different cell lines in one individual.

8. Polyploidy is rare in live-born animals, but triploidy accounts for about 7% of early human spontaneous abortions.

9. Uniparental disomy occurs when an individual has two chromosomes of a homologous pair from one parent and none from the other. The two chromosomes may include both parental homologues or two copies of exactly the same chromosome. Thus, a child with only one carrier parent may be affected with a recessive disorder.

10. For unknown reasons, humans show an extremely high rate of chromosomal abnormalities. Among early spontaneous abortuses, the frequency of errors (mostly trisomies, polyploids, and 45, X) is at least 50%. The observed spontaneous abortion rate of 15–20% would be much higher if very early (preimplantation) losses could be detected.

11. About 6% of stillbirths and 1% of live births exhibit chromosomal abnormalities. Of the latter, about one-half consist of structural aberrations (mostly translocations), and about one-third are trisomies.

12. Many persons who are mentally subnormal are also karyotypically abnormal, having either Down syndrome or the fragile X syndrome.

13. Traditional pedigree studies involving chromosome aberrations have proved very useful, when combined with molecular methods, in localizing genes to specific regions of the chromosomes and also in constructing phenotypic maps.

Key Terms

acentric	microduplication
contiguous gene syndrome	polyploidy
cryptic translocation	reciprocal translocation
deletion	ring chromosome
deletion mapping	Robertsonian translocation
dicentric	supernumerary chromosome
duplication	tetraploid
insertional translocation	triploid
inversion	unbalanced gamete
isochromosome	uniparental disomy
microdeletion	variant chromosome

Questions

1. Diagram the synaptic configurations (with gene-by-gene pairing) of a pair of chromosomes *heterozygous* for:

   (a) an interstitial deletion.

   (b) a tandem duplication.

2. Diagram, with letters representing loci, the synaptic configurations that would be seen in chromosome pairs *homozygous* for:

   (a) an inversion.

   (b) a reciprocal translocation.

3. Among the parents of people with Down syndrome, a few have 45 chromosomes, including a Robertsonian translocation between the two chromosomes 21.

   (a) What types of gametes would they form?

   (b) Could they produce unaffected progeny?

4. A published case history describes a woman with 45 chromosomes, one of which was a Robertsonian translocation between the two chromosomes 15. She was childless, having had 12 spontaneous abortions. What was the reason for this?

5. How can a metacentric chromosome be changed to an acrocentric one?

6. List two ways by which a triploid could arise.

7. Is there any way by which fertilization between two aneuploid gametes produced by individuals with the same type of translocation heterozygote could produce a euploid zygote? (*Hint:* See Figure 14.11.)

8. Women with one normal X and a second X that is an isochromosome of the long arm are often mosaics, possessing a 45,X cell line as well. How could this come about? (*Hint:* Consider a scheme similar to that in Figure 14.8, but with both breaks immediately above the centromere.)

9. Vidaud et al. (1989) report a case of father-to-son transmission of hemophilia A, with nonpaternity ruled out. How could this happen? Be specific.

10. In Figure 14.16A, describing a mechanism for uniparental disomy, we assume no crossing over between homologues in either parent. How would the outcome be changed with the occurrence of crossing over?

11. How can uniparental disomy be useful in chromosome mapping?

12. Lupski et al. (1991) report a mating between two people affected with Charcot-Marie-Tooth syndrome. Both of their children were affected, the second one very severely and with onset before the age of 1 year. What is the most likely explanation for the second child's phenotype?

13. The cat eye syndrome is sometimes called tetrasomy 22p. Why?

14. This question is based on the following diagram of the meiotic behavior of an inversion heterozygote.

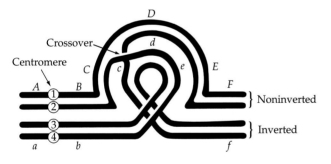

(a) Diagram and describe the four gametes resulting from crossing over between chromatids 2 and 4 within the inversion.
(b) Suppose that the centromere in this chromosome pair is between loci *D/d* and *E/e* rather than between *A/a* and *B/b*, but that the same crossover occurs. Diagram and describe the four gametes that would result. *Hint:* Rearrange the centromeres in the "anaphase separation" diagram above.
(c) Suppose that the tip of the short arm of this chromosome pair contains more loci (say, *W/w*, *X/x*, *Y/y*, and *Z/z* lying to the left of *A/a*) than does that in (a) and (b). Assuming the same crossover and the centromere within the inversion loop as in (b), diagram and describe the four resultant gametes. Comment on the relative viability of recombinant zygotes that would result here, compared to those from (b).

15. Suggest an explanation for the formation of the benign supernumerary chromosome in Figure 14.7.

Further Reading

For details about chromosomal abnormalities, consult books by Gardner and Sutherland (1996), Obe and Natarajan (1994), Vig (1993), Therman and Susman (1995), and Epstein (1986). Articles by Spinner and Emanuel (1997), Robinson and de la Chapelle (1997), Ledbetter and Ballabio (1995), and Chandley (1997) are also very informative. For reviews of frequencies of chromosomal abnormalities in germ cells, spontaneous abortions, live births, and other categories, see Hook (1992), Garber et al. (1997), Jacobs (1992), and Hassold et al. (1996).

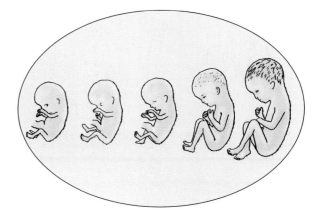

CHAPTER 15

Development

How does a single cell, the zygote, get transformed into an exquisitely complex organism made up of so many distinctive cells, tissues, and organs? And how does each species, including ours (Figure 15.1), develop a unique body pattern and shape? Questions about the nature of development have fascinated and baffled scientists for ages. During the past century in particular, embryologists and geneticists struggled to explain how such striking differences can arise among the cells of the body, given that their nuclei remain genetically identical. Important new insights came in the 1970s and thereafter, when the development of new high-powered molecular techniques allowed the isolation and analysis of mutations affecting early development. Much of this recent knowledge in developmental genetics comes from intensive studies of a few representative animals, including fruit flies, free-living roundworms, zebra fish, and mice. From these and other subjects, scientists have learned that, to an amazing extent, many similar genes operate in a vast array of evolutionarily dissimilar organisms. In

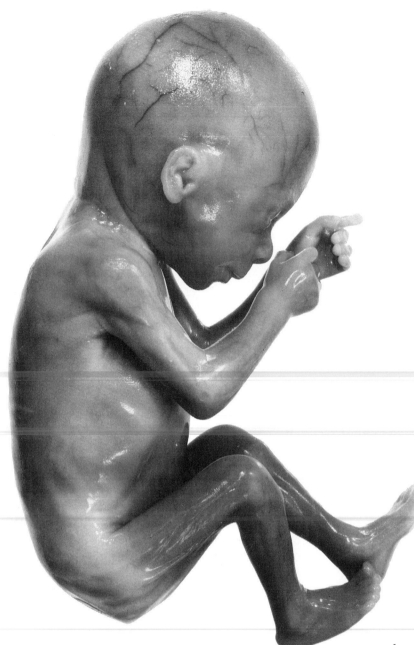

same developmental process inside the uterus. Shortly after natural fertilization in the upper part of an oviduct, the zygote begins to divide as it slowly migrates down the oviduct to the uterus, yielding two cells, then four, then eight, and so on, without any intervening periods of cell growth. During these **cleavage divisions**, the total amount of embryonic cytoplasm remains constant as it is partitioned among smaller and smaller cells. Metabolic needs are met by yolk and other materials originally laid down in the egg.

After about three days, a solid ball of approximately 16 cells enters the uterus. Further division and the absorption of fluid convert it by the fourth day into a hollow **blastocyst** of 100 to 200 cells. Among these, a tiny clump (the *inner cell mass*) destined to form the embryo proper can already be distinguished from the remaining cells, which will contribute to the embryo's surrounding membranes (Figure 15.2). After floating free for about two days, the blastocyst attaches to and begins to penetrate the spongy lining of the uterus, whose rich blood supply will nourish the developing embryo. By the end of the second week, *implantation* is complete. Inside the rapidly growing blastocyst, the embryo begins to absorb nutrients diffusing from maternal tissue.

By the third week, the inner cell mass has become a flat, disk-shaped embryo (Figure 15.3). In a process called *gastrulation*, massive, highly ordered cell migrations and interactions have led to the formation of three embryonic layers: *ectoderm*, *mesoderm*, and *endoderm*. From these layers, all the organs of the body will form. First, a portion of mesoderm forms a central rod called the *notochord*. Above it, primitive neural tissue arises from ectoderm, becomes grooved, and then folds together to form a *neural tube*, forerunner of the brain and spinal cord. The endoderm will later give rise to cells lining the respiratory and digestive tracts and also to glandular cells in associated organs.

Another important cell type (unique to vertebrates) is the *neural crest*, which arises from ectoderm at the topmost region of the neural tube. Neural crest cells

Figure 15.1   A human 17-week embryo, actual size. Note the absence of subcutaneous fat and the delicate fetal skin, through which the blood vessels of the scalp can be seen. (From Moore and Persaud 1998.)

humans, mutations of these genes are often associated with developmental defects or with cancer. Before describing some of these discoveries, however, we will briefly summarize the major events in human development.

Human Embryonic Development

Whether conceived naturally or in a glass dish (Chapter 20), normal human embryos go through the

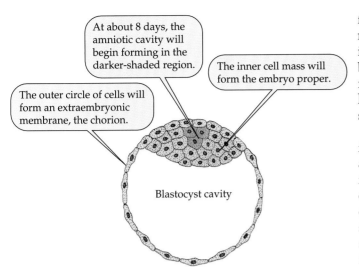

The outer circle of cells will form an extraembryonic membrane, the chorion.

At about 8 days, the amniotic cavity will begin forming in the darker-shaded region.

The inner cell mass will form the embryo proper.

Blastocyst cavity

Figure 15.2 A human blastocyst (diagrammatic cross section) at about 4.5 days. The blastocyst consists of about 200 cells and is drawn here about 400 times actual size.

later migrate to many parts of the body to form pigment cells, parts of the nervous system, some endocrine glands, some connective tissues, and some facial structures. Small mesodermal blocks called *somites* develop on either side of the notochord and developing neural tube; the somites will ultimately differentiate into muscle, skeleton, blood cells, and excretory and reproductive organs. Blood vessels and primitive blood cells soon make up a simple *cardiovascular system* that distributes oxygen and nutrients throughout the embryo while removing waste materials. The cardiovascular system is the first organ system to become

functional. It links the embryo with specialized maternal-fetal tissue called the *placenta*, which is embedded in the uterine wall. Here the embryonic and maternal bloods flow very close to each other but remain separated by cell membranes. The placenta also produces hormones that maintain pregnancy and prevent menstruation.

No signs of sexual development appear until the fourth week, when tiny bulges (the genital tubercle and labioscrotal swellings) arise that will much later form the external sex organs. Also, *primordial germ cells* can be distinguished in the wall of a primitive cavity called the yolk sac (Figure 15.3B). These cells, the descendants of which are destined to become eggs or sperm, migrate during the fifth week into the newly forming internal gonadal ridges.

The most crucial and sensitive developmental period spans weeks 3 through 8 (Figure 15.4)—a time when every major organ system begins to take form and the presence of alcohol, drugs, viruses, or radiation may lead to malformations. Thus, a missed menstrual period should alert a woman to the possibility of pregnancy and the need to limit her exposure to these *teratogens* (*teratos*, "monster;" *gen*, "producing"). During week 4, the flat, disk-shaped embryo bends to form a C-shaped cylinder with a relatively large head (Figure 15.5). By the seventh week, rudimentary eyes, ears, nose, jaws, heart, liver, and intestines are distinctly visible, and the development of other structures (lungs, kidneys, sex organs, and skeleton) is also well under way. During this period, a critical process called *induction* will cause changes in the developmental fate of one group of cells in response to chemical signals from a nearby group of cells. We will discuss the nature of these signals later in the chapter.

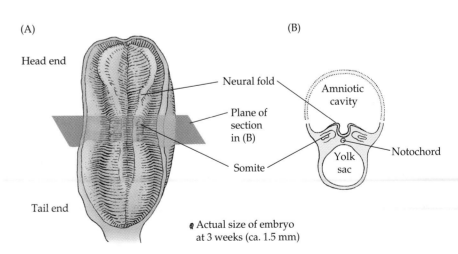

(A)

Head end

Tail end

(B)

Neural fold

Plane of section in (B)

Somite

Amniotic cavity

Yolk sac

Notochord

Actual size of embryo at 3 weeks (ca. 1.5 mm)

Figure 15.3 A human 3-week embryo. (A) Top view, looking down from the amniotic cavity on what had been the inner cell mass. (B) Cross section of an embryo at the level indicated by the plane in (A). The yolk sac will get much smaller as the embryo grows and expands into the amniotic cavity. (Adapted from Moore and Persaud 1998.)

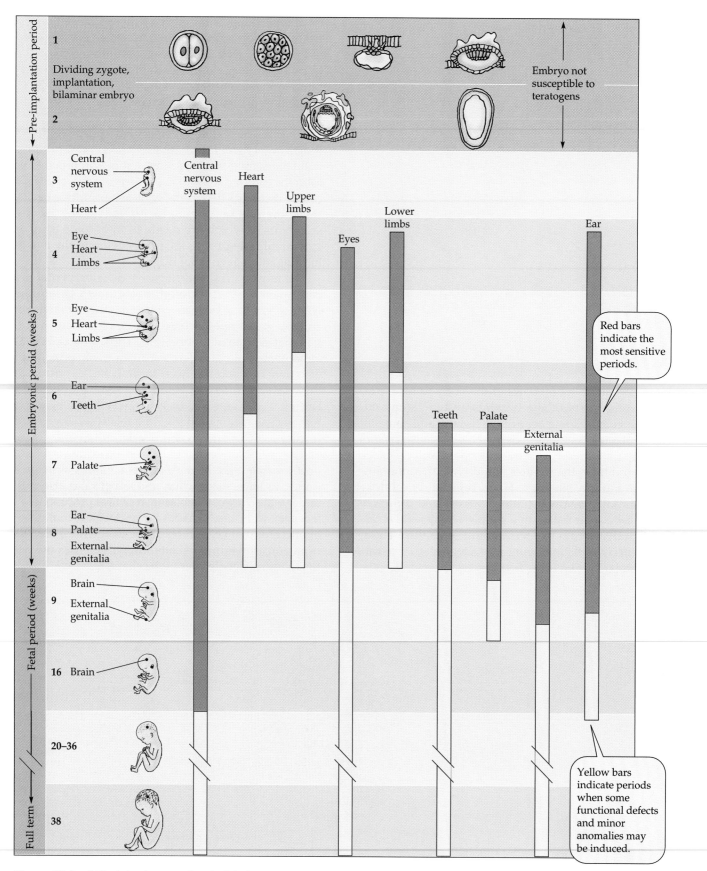

Figure 15.4   Critical developmental periods in humans. Red bars indicate periods when teratogenic substances may cause major abnormalities. Yellow bars indicate less sensitive periods. Black dots indicate the most common sites of action of teratogens in the embryo or fetus. (Adapted from Moore and Persaud 1998.)

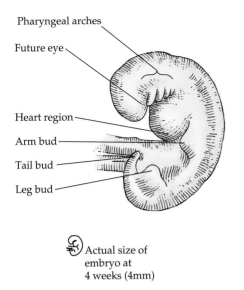

Actual size of
embryo at
4 weeks (4mm)

Figure 15.5 A human 4-week embryo, without its surrounding membranes. (Adapted from Moore and Persaud 1998.)

By the end of the eighth week, the embryo is about 30 mm (1.2 inches) long and weighs roughly 5 g (1/6 ounce). Its tail has disappeared. Limbs with fingers and toes are clearly formed, and the head constitutes almost half of its total length (Figure 15.6). The embryo is fully encased in two transparent membranes: an inner *amnion* pressed close to an outer *chorion*. Anchored to the placenta by an umbilical cord, the embryo moves quite freely in the fluid-filled amniotic cavity. The amniotic fluid acts as a protective shock absorber; it also prevents the embryo from forming adhesions with the surrounding membranes and promotes normal symmetry in development. The fetus will both drink and breathe this fluid.

From the end of the eighth week of development until birth at 38 weeks, the fetus will increase about 650-fold in weight (to 3,180 g, or 7 pounds) and about 12-fold in length. Meanwhile, development of all the

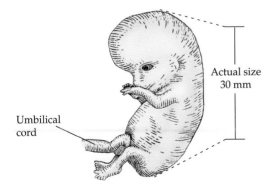

Actual size
30 mm

Umbilical
cord

Figure 15.6 A human 8-week embryo without its surrounding membranes. All major organ systems are developing. The toes and fingers have separated, the webbing that once was between them having now disappeared. The tail is gone too. (Adapted from Moore and Persaud 1998.)

organ systems continues. Because the respiratory system is the last to develop, lung function sets the limits of viability; brain development is crucial too. A 26-week-old fetus may survive if born prematurely, even though it has attained only 30% of the normal birth weight. And with increased knowledge, improved technology, and extraordinary effort and expense, it is becoming possible to keep alive some babies born even earlier than this—the earliest being a 500-g, 22-week-old fetus. Such heroic measures present an unresolved ethical dilemma, however, because most of the very premature survivors will suffer from serious problems for years to come.

Genetics of Embryonic Development

Over a century ago, scientists began to study early development in frogs and sea urchins by experimentally deleting specific regions of early embryos, or by transplanting specific regions from one place to another, and observing the results. Different organisms sometimes yielded quite different outcomes, however, so it was not always possible to generalize about the "rules" of development in all organisms. Further experiments in the 1920s revealed that a tiny critical region in frog embryos (called the "organizer"), when grafted onto another embryo, could by cell-to-cell contact induce the formation of a second embryo in the new host. Clearly, cell-to-cell interactions are extremely important in early development, but the exact nature of the inducing substance(s) eluded researchers for decades. At the time, few geneticists were involved in such studies. In recent years, however, researchers have identified at least 15 organizer-specific genes in vertebrates (Lemaire and Kodjabachian 1996).

Starting in the early 1970s, geneticists began an intensive study of many mutations affecting embryonic and larval development in *Drosophila*, the fruit fly. Using the new tools of molecular biology they, too, did deletion and transplantation experiments—but by manipulating genes rather than groups of cells. They found that three sets of genes control early development in fruit flies: *maternal effect* genes, *segmentation* genes, and *pattern formation* genes. Although many of these genes' phenotypic effects had been studied for decades, their precise modes of action in the egg and early embryo were not worked out until the late 1980s. Human homologues of these genes often play very similar roles in human development. For their combined efforts, American geneticists Edward B. Lewis and Eric Wieschaus and German researcher Christiane Nüsslein-Volhard received a Nobel prize in 1995.

Maternal Effect Genes

Long before the advent of molecular genetics, it was known that the egg contains stores of nutritive material great enough to support the early embryo through at least its first few cleavage divisions. (Many types of eggs contain large amounts of yolk, but human eggs have very little.) In addition, cytoplasmic elements in the egg were known to exert some genetic influence over the very early embryo. In snails, for example, the direction of shell coiling (left versus right) is controlled by maternal gene products that function during egg development. These egg proteins (rather than the snail's own genotype) determine the spindle position and thus the planes of cleavage in the zygote and embryo.

Later studies in a wide variety of organisms showed that maternal effects involve the stockpiling of various materials in the egg cell. These include everything needed for cleavage divisions: nucleotides and enzymes for DNA replication, as well as tubulin for spindle formation. Another category of reserves involves constituents needed for protein synthesis, including ribosomes and messenger RNA molecules that encode genetic information transcribed from maternal DNA. No wonder the egg is so much larger than other animal cells!

In insects, the first developmental step occurs when **maternal effect genes** in the mother set up important gradients (or other partitions) of substances in the egg cytoplasm. These maternal gene products will govern the expression of the zygotic genes required for normal head-to-tail and dorsal-to-ventral axis formation in the developing embryo. How is this known? Scientists have analyzed many mutants of maternal effect genes in fruit flies and seen how such mutants can alter the fates of future embryonic cells. In every case, the effects are best explained by the existence of one or more gradient systems in the egg (Nüsslein-Volhard 1996).

Segmentation and Pattern Formation Genes

In all organisms, at some point the embryo's own genes become activated and take over from the maternal genes. In humans, this shift first occurs between the four- and eight-cell stages; in fruit flies, it occurs after 13 division cycles. Establishment of the basic body plan is accomplished by two kinds of events. First the **segmentation genes** interact with one another in a cascade of steps that gradually subdivide the embryo into a linear series of progressively smaller but roughly similar segments. Then the **pattern formation genes** assign a different developmental potential to each subunit (Figure 15.7).

Such pattern formation occurs through the action of **homeotic genes**, which act by "naming," or specifying, the structures unique to each subunit. They were originally identified when fruit flies carrying mutant alleles displayed bizarre phenotypes, such as legs sprouting from the fly's head in place of the antennae,

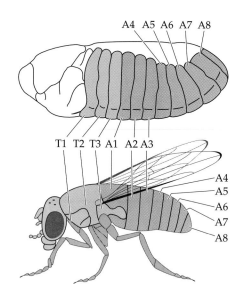

Figure 15.7 Segmentation in the fruit fly (*Drosophila*) embryo (top) and adult (bottom). Solid lines indicate the boundaries between individual thoracic (T) and abdominal (A) segments. In the adult, each thoracic segment has a unique set of appendages: T1 has legs only; T2 has legs and wings; T3 has legs and halteres, which are small, stalklike balancing organs.

or wings emerging from the head region where eyes are usually found (Figure 15.8). Thus, when the normal function of these genes is disrupted by mutation, the fate assigned to various bands is inappropriate to their location.

In the mid-1980s, geneticists isolated and sequenced the genes and gene products of some segmentation and homeotic loci. All were found to be complex and interrelated. In particular, they shared a highly similar sequence of about 180 DNA base pairs that was dubbed the **homeobox**. Further analyses of numerous other maternal, segmental, and homeotic genes revealed that a homeobox is present in many of them. The 60-amino-acid sequences (*homeodomains*) encoded by these homeoboxes recognize and bind to the DNA of control regions of genes or interact with other DNA-binding proteins. Specifically, these homeodomain-containing proteins are **transcription factors** (see also Chapter 17) that regulate the turning on or turning off of other genes by binding to their enhancer or promoter regions.

Even more exciting was the finding that these master sequences exist in a wide range of organisms, from worms to mammals and from fungi to higher plants. Not all animals have segmentation patterns as visible as those of insects, but many are segmented in one way or another—such as the structure of their central nervous systems, muscles, and (where they exist) backbones. Homeobox sequences have remained highly conserved throughout evolution. The stunning

(A)

(B)

Figure 15.8 Homeotic mutants of *Drosophila*. (A) This four-winged fruit fly has several mutations that collectively change the third thoracic segment (T3) into another second thoracic segment (T2). As a result, it has wings where there should be halteres. (Photo courtesy of E. B. Lewis.) (B) This fruit fly has a mutation that causes legs to sprout from the area of the head where the antennae should be. (Photograph courtesy of J. Haynie.)

similarity between the genes of highly disparate organisms* demonstrates that at least some developmental genetic mechanisms are almost universal.

> The homeobox family now numbers more than 500, continues to grow, and includes genes from organisms broadly representative of both the plant and animal kingdoms. In plants, the first homeobox genes were identified because they regulate leaf development. In yeast and other fungi, homeobox genes regulate mating type. ... Homeodomains per se must be very ancient because yeast, plants, and animals probably have a billion years of evolution separating them. (Kornberg and Scott 1997.)

Fruit flies have eight homeobox (*HOM*) genes in a complex located on chromosome 3. In mice and humans, 38 related homeobox genes (called *Hox* in mice, *HOX* in humans) have been identified. They are distributed among four different gene clusters (called *Hox*/*HOX-A*, *-B*, *-C*, and *-D*) on separate chromosomes. These four clusters represent duplications of an ancestral complex (*HOM-C*) that is also present in fruit flies, as well as some divergence within and between clusters. Each cluster contains 9 to 11 fairly small genes that fall within 13 closely related groups called **paralogues** (Figure 15.9). No cluster contains representatives of every paralogue, however. Note that the vertebrate *Hox-B* cluster most closely resembles the fruit fly *HOM-C* complex, both of which lack paralogues 10–13.

Another fascinating discovery was that—comparing, say, fruit flies and mice—the homologous homeobox-containing genes remain in the same relative order within a complex. Furthermore, *these genes are expressed both temporally and spatially along the embryo's anterior-to-posterior (head-to-tail) axis in the order of their chromosomal position.* The 3'-most fruit fly gene (*lab*) and the mouse *Hox* genes *a1* and *b1*—that is, those lying the closest to the 3' end in the cluster—are the first to be expressed during development. Their expression starts in the most anterior body region controlled by this gene complex and gradually extends backward to the tail region as development proceeds. Then the second gene (*pb*) in the fruit fly (and genes *a2* and *b2* in the mouse) kick in, with their domain of expression starting a little more posteriorly but also spreading tailward. And so on through the entire complex, with the final fruit fly gene (i.e., that in the most 5' position) being expressed last of all, and only in the tail region. Likewise, the mouse genes *a13*, *c13*, and *d13* are expressed last and in the most posterior region. Thus, each body region is defined by a different array of homeobox genes expressed in its cells—the

*For example, some human and mouse homeodomains (identical to each other) differ by only one amino acid (out of 60) from the homeodomain encoded by a corresponding gene (*Antennapedia*) in fruit flies. Indeed, some mammalian *HOX*/*Hox* genes are so similar to their homologues in fruit flies that they can substitute for one another (Gilbert 1997). Considering the entire genome, such homologies allowed Banfi et al. (1996) to identify and map 66 novel human cDNAs that show significant homology to known mutations of fruit fly genes. Almost 50 of these human cDNAs were homologous to fruit fly loci involved in development and reproduction.

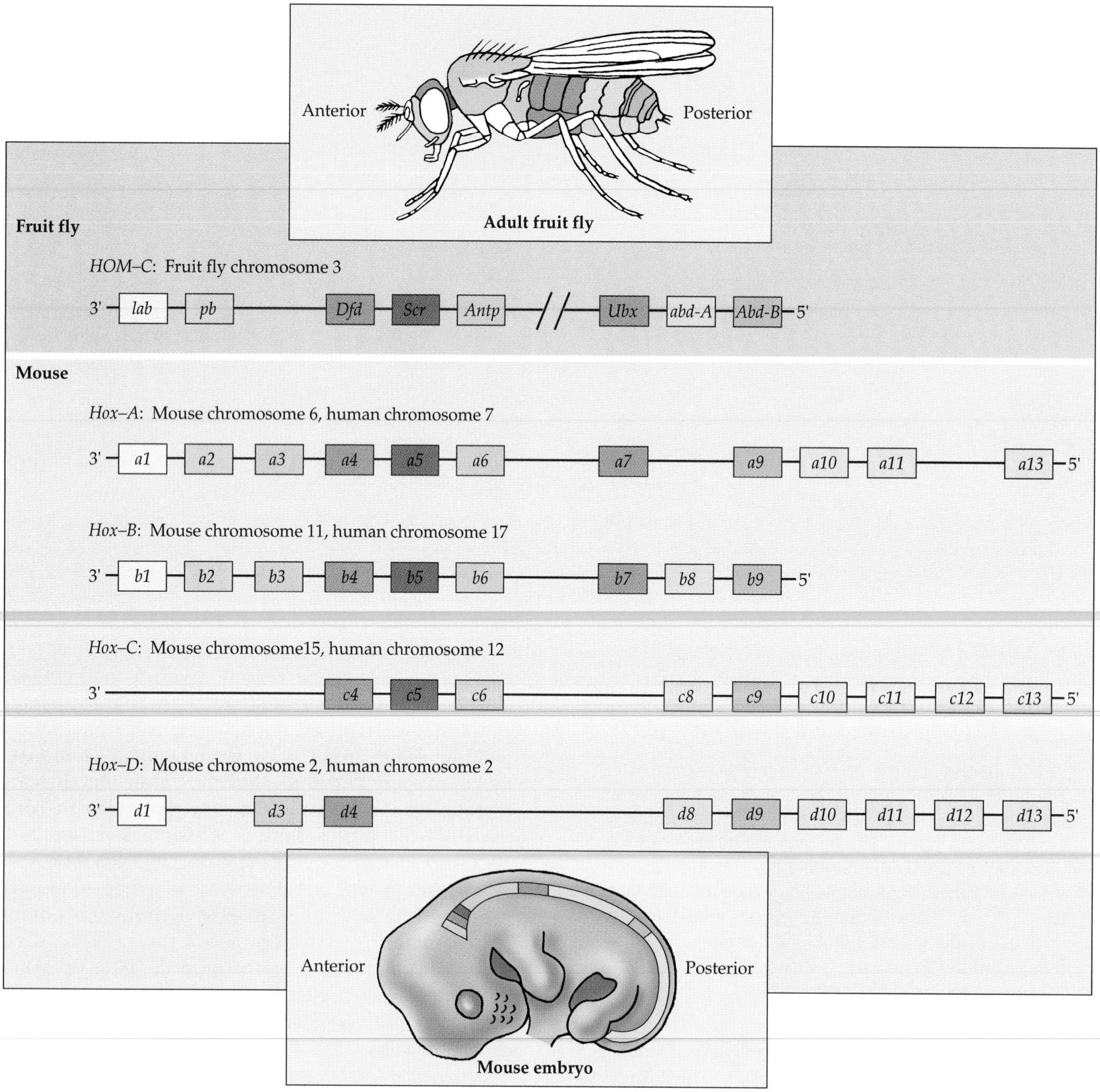

Figure 15.9 The homeobox (HOM/Hox) gene family in fruit flies and vertebrates. In the fruit fly, eight linked genes comprise the HOM-C complex: labial (lab), proboscipedia (pb), Deformed (Dfd), Sex combs reduced (Scr), Antennapedia (Antp), Ultrabithorax (Ubx), abdominal-A (abd-A), and abdominal-B (abd-B). Each is assigned a number (1–9), with the space designated as number 3 indicating a separation between the two subgroups in this complex. Color-coding shows each gene's domain of activity in forming the adult fly. Their head-to-tail order matches the 3'-to-5' order of the genes on the chromosome. The four mammalian Hox/HOX clusters (A, B, C, and D), along with the HOM-C complex, represent related copies of an ancestral gene cluster that is thought to have existed before insects and mammals diverged. The mammalian clusters have extra genes (numbered 10–13), all of whose homeoboxes most closely resemble that of Abd-B and probably resulted from independent evolutionary duplications. Each gene within a cluster is positioned in a vertical line with its closest homologues in the other clusters (for example Abd-B, Hox-a9, Hox-b9, Hox-c9, and Hox-d9). Such groups are called paralogues. Their homeoboxes resemble each other more closely than they resemble those of the other genes in their own HOM/Hox/HOX complex; thus the order of paralogues is conserved during evolution. A diagram of a mouse embryo (bottom) roughly depicts its central nervous system (forebrain, midbrain, hindbrain, and spinal cord). Here, too, color-coding shows that the anterior-to-posterior domains of Hox gene activity (starting in the hindbrain) match the 3'-to-5' order of paralogues. (Adapted from Wolpert et al. 1998 and Coletta et al. 1994.)

fewest genes acting in the most anterior region and most genes of these complexes acting in the most posterior region. And in each region the particular array of transcription factors encoded by these genes initiates a unique cascade of gene activity. Unfortunately, we know relatively little about which genes are targets of homeobox gene action, but intensive research is underway to identify them.

Homeotic Genes in Vertebrates

Here we present a small sample of what scientists have learned about the involvement of homeobox genes in the embryonic development of vertebrates.

Backbone. The development of the backbone provides an excellent example of how specific combinations of *Hox/HOX* genes determine specific kinds of vertebrae. By selectively inactivating or deleting both copies of individual *Hox* genes in mice, researchers have recently determined the role of each *Hox* gene in the development of the backbone. The absence or inactivation of a particular *Hox* gene can convert one kind of vertebra into a *more anterior* kind. For example, when *Hox-b4* (Figure 15.9) is inactivated in mouse embryos, what should have been their second cervical (neck) vertebra instead becomes more like the first cervical vertebra. Similarly, the deletion of *Hox-c8* causes the first lumbar (lower back) vertebra to behave like a thoracic (chest) vertebra by forming a rib. The same kind of phenomenon occurs in fruit flies.

Hox expression in mice can also be experimentally altered by treating embryos with **retinoic acid**, a derivative of vitamin A that is normally produced only in certain cells of the body. This small molecule diffuses across cell membranes. When combined with intracellular retinoic acid receptors (RARs), it acts as a transcription factor, with particular specificity for development of the anterior-posterior axis and the limbs. Excess retinoic acid, when applied to embryos, interferes with the normal patterns of expression of *Hox* genes in at least two ways. Depending on the time of treatment, it can cause anterior *Hox* genes to be expressed in more posterior regions and posterior *Hox* genes to be expressed more anteriorly. In addition, excess retinoic acid inhibits the normal migration of neural crest cells from the anterior region of the neural tube. It also has several other detrimental effects on embryos.

Retinoic acid and vitamin A are known to cause abnormalities in the embryos of many animals, including humans. Retinoic acid (brand name Accutane®), given orally, has been widely used as a treatment for severe acne—which raises serious concerns, because many of the users are women of childbearing age. Indeed, some birth defects associated with Accutane use have already been recorded. "The malformed infants had a characteristic pattern of anomalies, including absent or defective ears, absent or small jaws, cleft palate, aortic arch abnormalities, thymic deficiencies, and abnormalities of the central nervous system" (Gilbert 1997).

Limbs. The development of a limb in three dimensions—anterior to posterior, proximal (close to the main body axis) to distal (away from the body), and dorsal (back side) to ventral (underside)—is an immensely complex process. Here we will sketch only a few highlights. Homeobox genes are important regulators of vertebrate limb development, and dozens of mutations affecting limb structure have been identified in mice, chickens, frogs, and fish.

In these and other vertebrates, two pairs of *limb buds*, each made up of a few hundred mesodermal and ectodermal cells, develop at specific sites along the anterior-posterior axis. The pectoral fin in fish and the forelimb in amphibians, birds, and mammals all form at the level of the first thoracic vertebra. This is the most anterior boundary of expression of the *Hox-b8* and *Hox-c6* genes. Pelvic fin and hindlimb buds form at the *Hox-c9* anterior expression boundary. Retinoic acid (produced normally by a small group of embryonic cells) seems to be involved in the activation of these *Hox* genes. Other *Hox* regulators include **signaling molecules** called *fibroblast growth factor*, *bone morphogenetic proteins*, *Wnt protein*, and *sonic hedgehog protein*. They are encoded by non-homeobox-containing genes called *FGF*, *BMP*, *WNT*, and *SHH*.* A screen for homeobox-containing genes active in chicken limb bud mesoderm turned up 23 different *Hox* genes. Some are expressed in wings only, some in both wings and legs, and some in legs only.

Along their proximal-distal axis, all vertebrate limbs consist of several zones defined by the bones they contain. In the forelimbs, these include the proximal region (humerus), the middle region (radius and ulna), and a distal region of wrist bones (carpals) plus back-of-the-hand bones (metacarpals)—if they exist—and up to five different digits (phalanges). Figure 15.10 shows these bones, whose formation begins in the following manner. As a limb bud pushes outward, the *progress zone* of mesodermal cells at its tip (lying just under a thickened ridge of ectodermal cells) produces and leaves in its wake successive groups of daughter cells with increasingly more distal fates. The first group of cells released (i.e., those lying next to the body wall) will give rise to the humerus. The second layer of cells laid down will form the radius and ulna. This pattern of fate determination continues until the

---

*Many vertebrate genes and proteins are named after their counterparts in fruit flies, which often have whimsical names. For example, the vertebrate *sonic hedgehog* gene is homologous to the *hedgehog* (*hh*) gene in fruit flies, and the vertebrate *radical fringe* gene is homologous to the *fringe* gene in fruit flies.

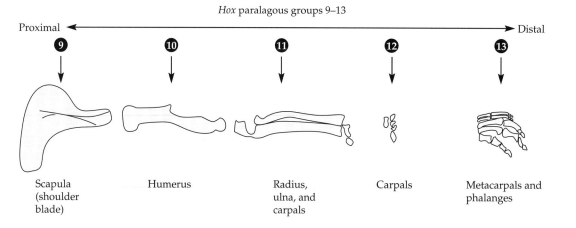

Hox paralagous groups 9–13

Proximal ← → Distal

❾ ❿ ⓫ ⓬ ⓭

Scapula (shoulder blade)   Humerus   Radius, ulna, and carpals   Carpals   Metacarpals and phalanges

Figure 15.10 Diagram of vertebrate (mouse) limb segments and the *Hox* paralogue groups involved in the development of each segment. The higher-numbered paralogues are expressed in the more distal limb regions. (Adapted from Davis et al. 1995.)

most distally released cells that are laid down will become phalanges. During this period, changing arrays of intercellular signaling proteins are produced that collectively induce and maintain these developmental stages. After the progress zone produces the presumptive phalangeal cells, it stops dividing. Later, each of the these limb compartments will expand by cell division and differentiate into a particular part of the limb.

Numerous limb mutations are known in chickens and mice. Some arose spontaneously, while others were produced by various experimental techniques. Their expression patterns have been determined by the systematic deletion or inactivation of *Hox* loci—singly and in small combinations. Many defects of limb development are known in humans too, but their genetic bases are not so well understood. Recently, however, Muragaki et al. (1996) reported on a nearly identical *HOX-D13* mutation in three families with hand and foot abnormalities. *Synpolydactyly* (*syn*, "union"; *poly*, "many"; *daktylos*, "finger"), which involves both fusion and duplication of fingers, is inherited as an autosomal dominant with incomplete penetrance. Figure 15.11 shows the hands of two affected individuals, a heterozygous male and his homozygous mother. In the son, branching of the middle meta-

carpal gave rise to an extra middle finger. His mother, the offspring of a first-cousin marriage, shows much more serious defects. Her hands and feet are very small, with very short digits. On the hand, digits III, IV, and V are joined together and share a single knuckle formed by the transformation of metacarpals to short carpel-like bones. She also has two extra carpal bones and some shortened phalanges. Her feet have only three digits, caused by the replacement of three metatarsals by a single tarsal-like bone.

The cause of these abnormalities is a mutation of *HOX-D13*, located on chromosome 2q. It is not a simple base substitution, nor does it affect the homeodomain in this gene. Rather, it is an amplification (specifically, the insertion of 21 extra base pairs) in a triplet repeat polyalanine-coding tract. (Many homeobox genes have regions of triplet repeats encoding tracts of polyalanine or polyglycine in their homeo-

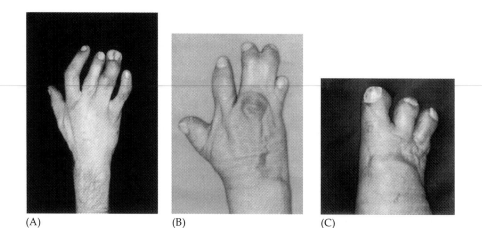

(A)   (B)   (C)

Figure 15.11 Synpolydactyly in hands and feet, caused by a *HOX-D13* mutation. (A) The heterozygous son has a branched metacarpal giving rise to an extra finger (IIIa) that fused with his fourth finger (IV). Surgery has partly separated digit III from the fused IIIa/IV. (B) The hand of his homozygous mother shows a more extreme phenotype. Fused digits III, IV, and V share a single knuckle due to various abnormalities, including the transformation of metacarpals I, II, III, and V to short, carpal-like bones. (C) The foot of the homozygous mother is also highly abnormal, due partly to the replacement of metatarsal III, IV, and V with one tarsal-like bone. (From Muragaki et al. 1996.)

domain proteins.) The result is seven extra alanines near the N terminus of the HOX-D13 protein, which somehow alters its function. The other two families studied had similar insertions (one of 24 and the other of 30 base pairs) in the same polyalanine coding tract.

In contrast to experimental knockouts, a naturally occurring vertebrate *Hox* mutation was also found in mice. Mortlock et al. (1996) describe a deletion in *Hox-a13* that causes *hypodactyly* (hypo, "under"), the presence of fewer than the normal number of digits. It is inherited as an incompletely dominant autosomal trait. Heterozygotes have a shortened first digit and some fusions of carpals on all four paws, but are otherwise normal and fertile. Homozygotes, however, show more serious malformations and usually die before birth; those that survive are infertile. The mutation, a 50-base-pair deletion in the triplet repeat region of *Hox-a13*, creates a frameshift that produces a defective protein. Perhaps the deletion was caused by mispairing and unequal crossing over.

Studies of limb development in experimental animals have yielded many insights of a general nature that apply to humans as well:

> One of the most important findings in modern developmental biology concerns the homologies of processes as well as those of structures... This can be seen with the development of the limb... Fly limbs and vertebrate limbs have little in common except their function. However, both fly and vertebrate limbs appear to be formed through the same developmental pathway... Thus, it seems like nature figured out how to make a limb only once, and both arthropods (*Drosophila*) and vertebrates (chicks and mice) use that process to this day. (Gilbert 1997)

Central Nervous System, Neural Crest, and Pharyngeal Arches. *Hox/HOX* genes are involved in the differentiation of the central nervous system, especially the hindbrain and spinal cord. (Forebrain and midbrain development do not involve *Hox/HOX* genes.) Early in development, the *hindbrain* of the mouse gets partitioned into eight balloon-like segments called *rhombomeres*, from which certain cranial nerves will arise. Ultimately, the hindbrain will give rise to the parts of the brain called the *cerebellum* (which coordinates complex muscular movements), the *pons* (fiber tracts that in mammals connect the cerebrum and cerebellum), and the *medulla* (a reflex center controlling involuntary activities). *Hox-b1* is expressed in rhombomere 4 only, *Hox-b2* from rhombomere 3 posteriorly through the remaining rhombomeres and spinal cord, *Hox-b3* from rhombomere 5 posteriorly, and *Hox-b4* from rhombomere 7 posteriorly. The more 5′ members of the complex are expressed only in the posterior regions of the spinal cord.

Recall that the neural crest cells derive from ectoderm at the very top of the neural tube. They migrate widely throughout the length of the body, giving rise to an amazing array of cell and tissue types. In the head region, for example, neural crest contributes to many facial structures containing nerve, cartilage, bone, and other connective tissues. Also in the head region, neural crest enters other segmented embryonic structures called *pharyngeal arches* (shown in a human embryo in Figure 15.5) that give rise to various structures in the lower face and neck. In other regions of the body, neural crest produces pigment cells, nerves, glandular tissue and a variety of connective tissues.

Several *Hox-a* loci are involved in development of rhombomere and neural crest derivatives. With regard to rhombomere derivatives, for example, experimental inactivation of *Hox-a1* produces mice lacking certain inner ear structures and parts of the hindbrain, and their neural tube often fails to close. Using "knockout" technology (see Box 17B), researchers produced some homozygous *Hox-a3⁻/Hox-a3⁻* mice, all of which died at birth or shortly thereafter. These mutant homozygotes exhibited a set of developmental abnormalities of neural crest derivatives. For example, they had short, thick necks, and the thymus and other neck glands (including the thyroid) were missing or reduced in size. Serious heart and blood vessel defects were probably the cause of death.

Particularly intriguing to researchers is the fact that the abnormalities appearing in *Hox-a3⁻/Hox-a3⁻* mice are strikingly similar to those seen in a human birth defect called *DiGeorge syndrome*. Babies with this disorder exhibit similar head and facial abnormalities, lack thymus and parathyroid glands, have too little thyroid tissue, and suffer from cardiovascular and immune system defects. So far, there is no clear-cut genetic correspondence between the two phenotypes, however. DiGeorge syndrome is usually linked to interstitial deletions in human chromosome 22q, but may sometimes be associated with defects in 10p. Yet the human counterpart of the mouse *Hox-a3* locus is known to be on chromosome 7, and chromosome 7 has not been linked to DiGeorge syndrome. Perhaps the DiGeorge phenotype is associated with two or more (non-*HOX-a3*) loci involved with neural crest migration and differentiation, or with genes controlled by *HOX-a3*.

Other Homeobox-Containing Genes. It turns out that most genes with homeoboxes are not the homeotic (*HOM/Hox/HOX*) genes described in the previous section (Wolpert et al. 1998). Several other families of homeobox-containing genes, each with two or more critical functional domains, are important in embryonic development. Here we briefly describe one of them.

Pax/PAX Genes. A group of transcription factor proteins with two or three DNA-binding domains has been identified in fruit flies, fish, mice, and humans. The key domain is called *paired* because it is also encoded by the fruit fly segmentation gene *paired*. Its protein product includes a *paired domain* of about 130 amino acids. In noninsect animals, this gene family was dubbed *Pax*, for *pai*red bo*x*. A second conserved motif (though not present in all *Pax/PAX* loci) is the homeobox. A third conserved motif may also be present (Gruss and Walther 1992). Nine *Pax/PAX* genes are known in mice and in humans. Unlike the *Hox/HOX* genes, they are dispersed throughout the genome rather than occurring in clusters (Stapleton et al. 1993). *Pax/PAX* loci are expressed mainly in the nervous system, but also play a role in the development of the ear, eye, urogenital system, limb bud, muscle, some glands, and B cells of the immune system. Exactly how they bring about their effects is not yet understood, since their target genes have not yet been identified.

In mice and humans, mutations of *Pax/PAX* genes have been linked to some serious disorders, possibly including cancer (Read 1995; Stuart et al. 1994). In mice, mutations of the *Pax3* locus give rise to the very severe "splotch" phenotype: deformities of neural crest derivatives plus defects in structures of the central nervous system (including spina bifida) and of the heart. In humans, mutations of the *PAX3* locus on human chromosome 2q are associated with the less severe *Waardenburg syndrome type 1* (see Figure 10.7). It is a dominantly inherited condition that involves defects of neural crest migration, including deafness, pigmentation anomalies, and wide-set eyes (Zoghbi and Ballabio 1995).

The highly conserved *Pax6* locus has been identified in organisms ranging from sea urchins to mammals. In mammals, *Pax6/PAX6* is expressed in the developing eye and nose as well as in the developing brain. In mice and rats, deletions or mutations in *Pax6* give rise to the dominantly inherited "Small eye" phenotype. Heterozygotes have smaller than normal eyes, but homozygotes lack both eyes and a nose and die shortly after birth. Quite amazingly—considering the fact that insect and vertebrate eyes are totally different types of structures—fruit flies have a homologous gene, called *eyeless* (*ey*), that is involved in eye development.

Fruit flies homozygous for mutations of *eyeless* show partial or total absence of their compound eyes. Furthermore, both the fruit fly *ey* gene and the (transgenic) mouse *Pax6* gene, when artificially activated in tissues (imaginal discs) that usually form other structures, induce the development of normal, light-sensitive eyes on fruit fly wings, legs, and antennae! Thus, *ey* appears to be a "master control gene that initiates the eye morphogenetic pathway and is shared be-

tween vertebrates and invertebrates" (Quiring et al. 1994). In humans, deletions or mutations at the homologous *PAX6* locus on chromosome 11p cause dominantly inherited *type 2 aniridia* (*an*, "lacking"; *irid*, "iris"). The phenotype of affected individuals is variable, ranging from partial (Figure 15.12) to total absence of the iris of the eye and other eye abnormalities that may collectively lead to total blindness (Hanson and Van Heyningen 1995).

Numerous other homeobox-containing genes have been identified. They are collectively expressed in various regions of the embryo, including anterior head regions, nervous system and sensory organs, skeleton, gut, heart, lungs, liver, and pancreas. Here, too, astonishing conservation of domain structure and function has been found over a wide range of organisms. For more information about these other classes of homeodomains, see Manak and Scott (1994).

Signaling Molecules

Cells in multicellular organisms need to communicate with each other to coordinate their myriad activities, beginning with the earliest stages of embryonic development. They do so by using

> hundreds of kinds of signaling molecules, including proteins, small peptides, amino acids, nucleotides, steroids, retinoids, fatty acid derivatives, and even dissolved gases such as nitric oxide and carbon monoxide. Most of the signaling molecules are secreted from the *signaling cell* by exocytosis… Others are released by diffusion through the plasma membrane, while some remain tightly bound to the cell surface and influence only cells that contact the signaling cell.

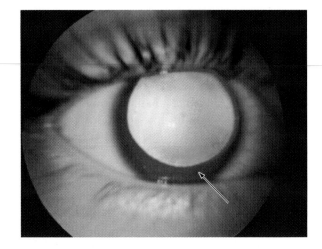

Figure 15.12 Phenotype of an individual heterozygous for a dominant *PAX6* mutation that causes aniridia. The defective iris consists of just a small rim of tissue (arrow). The abnormally large "pupil" in the middle of the eye looks bright because light has been reflected from the retina at the back of the eye. (Photograph from Day 1990.)

Regardless of the nature of the signal, the *target cell* responds by means of a specific protein called a *receptor*. It specifically binds the signaling molecule and then initiates a response in the target cell. (Alberts et al. 1994)

Signals can be short-range, affecting only nearby cells or in some cases even the signal-producing cells themselves. Or they can be long-range, affecting target cells in other parts of the body—such as signals transmitted via the axons of nerve cells or the blood-borne hormones produced by endocrine cells.

Earlier in this chapter, we mentioned some effects of retinoic acid on embryonic development and alluded to the role of several other kinds of signaling molecules. Here we will describe a few representatives of the major families of intercellular signaling proteins that are important during early development. *Growth factors* are signaling molecules that, among other things, regulate the proliferation of specific kinds of cells. They play an important role in *induction*, the change in the developmental fate of one group of cells brought about by signals from a nearby group of cells.

Fibroblast Growth Factor Family. Ten different *fibroblast growth factor (FGF)* loci encode proteins that stimulate growth in many different cell types.* They also function during embryonic formation of the mesoderm, blood vessels, nerve cell extension, spinal cord, midbrain, facial structures, and limb buds. In later development, they regulate the growth of long bones and closure of the skull bones. *FGF receptors* (encoded by *FGFR* genes) have an extracellular domain for binding the FGF protein, a transmembrane segment, and an intracellular domain that transmits the signal to cytoplasmic and nuclear molecules (Chapter 17).

Many mutations affecting both FGF and FGF receptors are known in mice and humans. They often cause abnormalities of the face and/or limbs, including most kinds of human dwarfism. About 95% of people with a dominantly inherited disorder called *achondroplasia* (see Figure 12.7) have the same mutation in the chromosome 4p *FGFR3* gene, which encodes fibroblast growth factor receptor 3. About 50% of cases of a milder but genetically heterogeneous condition, *hypochondroplasia*, are associated with a different *FGFR3* mutation (Bellus et al. 1995). And two other mutations in this gene, affecting adjacent amino acids, can cause a lethal disorder called *thanatophoric dysplasia* (*thanatos*, "death"; *dysplasia*, "abnormal growth"). Symptoms include extremely short limbs and ribs, with death resulting from breathing difficulties.

Over 100 heterogeneous human syndromes involving *FGFR* loci are known, many of which are familial. Their total frequency is about 1 in 2,900 births. Here we describe several severe autosomal dominant malformation syndromes. *Craniosynostosis* (*cranio*, "skull"; *syn*, "union"; *ostosis*, "bone formation") is the abnormal development and premature fusion of the flat bones of the skull, leading to skull, facial, and other anomalies (Angier 1994). Muenke and Schell (1995) point out that four craniosynostosis syndromes have been linked to *FGFR* mutations: Apert (AS), Crouzon (CS), Pfeiffer (PS), and Jackson-Weiss (JWS) syndromes. Their main features are widely spaced protruding eyes, a tower-shaped skull, a beaked nose and an underdeveloped midface. What best distinguishes the four syndromes, however, are limb abnormalities.

The Apert and Crouzon syndromes are the most important clinically, each accounting for about 4.5% of all craniosynostosis cases (Cohen et al. 1997). *Apert syndrome* (Figure 15.13) involves severe malformations of the skull, hands, and feet, as well as some abnormalities of the skin, brain, skeleton, and other internal organs. Most notably, an oversized brain is housed in a misshapen skull, and several digits are fused in the hands and feet. Intelligence ranges greatly, from normal through varying degrees of mental deficiency. Because affected individuals rarely reproduce, over 98% of cases are sporadic, arising through new mutation. The *Crouzon* and *Pfeiffer syndromes* are shown in Figure 15.14.

All four syndromes have recently been shown to involve mutations in three of the four mammalian *FGFR* loci: *FGFR1* on chromosome 8p, *FGFR2* on chromosome 10q, and *FGFR3* on chromosome 4p. Indeed, scientists were astonished to find that these three different genes can carry identical mutations! They encode the same (proline to arginine) amino acid substitution at the same position (position 4) in a conserved sequence of 16 amino acids that occurs in all mammalian FGFR proteins (Bellus et al. 1996). The three identical mutations give rise to three different conditions: Pfeiffer syndrome (*FGFR1*), Apert syndrome (*FGFR2*), and a condition called nonsyndromic craniosynostosis (*FGFR3*).

Additional mutations—and additional surprises—have also been found at these loci. Table 15.1 summarizes some of the phenotypic heterogeneity associated with two of these loci: *FGFR2*, in which mutations can give rise to four different syndromes, and *FGFR3*, in which mutations can also give rise to four different disorders. Note, too, that Pfeiffer syndrome is associated with mutations in both *FGFR1* and *FGFR2*. These findings suggest that, for unknown reasons, specific nucleotides in the *FGFR* loci have an unusually high mutation rate.

---

*Although first identified in fibroblasts, these growth factors are now known to act on many cell types. The same is true of most growth factors, whose original names imply tissue specificity but which were later found to have wider effects.

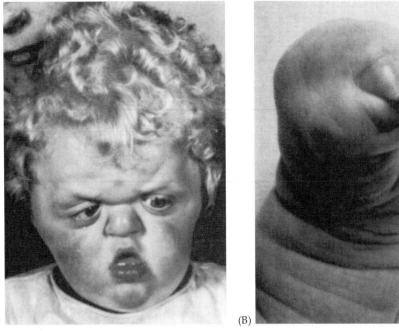
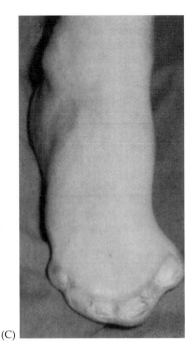

(A) (B) (C)

Figure 15.13 Some features of Apert syndrome. (A) The forehead is prominent and high, and the eyebrows are interrupted by a bulge resulting from improper fusion of the underlying skull bones. The eyes protrude somewhat, and the eyelids slope downward on the outside. The nose is short and broad. (B) Severe but typical syndactyly of the hand, in which the middle three digits are completely fused and share a single fingernail. (C) Severe syndactyly of the foot, in which all the digits are extremely short and fused together. (From Wilkie et al. 1995.)

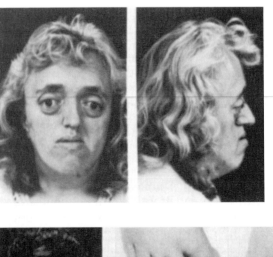

(A)

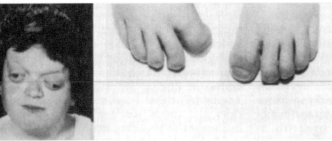

(B)

Bone Morphogenetic Protein Family. Effects of the family of secreted signaling molecules called *bone morphogenetic proteins* (*BMPs*) are much broader than the name implies. Collectively, more than 20 examples are found in a wide range of animals, from sea urchins to mammals, and they play a role in the regulation of a wide variety of processes, including cell division, cell migration, cell differentiation, and cell death. Indeed, vertebrate BMPs influence the development of nearly all tissues and organs, especially the kidneys, heart, lungs, gut, skin, limbs, teeth, hair, and germ cells (Hogan 1996).

In humans, eight *BMP* loci are distributed among six different chromosomes. The human *BMP5* locus, which resides on chromosome 6, is homologous to the

Figure 15.14 Crouzon and Pfeiffer syndromes. (A) This woman, who has Crouzon syndrome, belongs to a five-generation kindred with 11 affected members. Note the wide-set and protruding eyes, beak-like nose, short upper lip, underdeveloped upper jaw, and malocclusion (poor bite) between upper and lower jaws. Individuals with Crouzon syndrome have normal hands and feet. (From Preston et al. 1994.) (B) This person with Pfeiffer syndrome has severe craniofacial abnormalities, including wide-set, protruding eyes that appear rather asymmetrically positioned. Also, her big toes are both quite broad. (From Rutland et al. 1995.)

TABLE 15.1 Mutations in three fibroblast growth factor receptor loci and the phenotypes associated with them[a]

Disorder	Number of known mutations in		
	FGFR1 (chromosome 8p)	*FGFR2* (chromosome 10q)	*FGFR3* (chromosome 4p)
CRANIOSYNOSTOSES			
Pfeiffer syndrome	1[b]	13	—
Apert syndrome	—	2[b]	—
Crouzon syndrome	—	20	—
Jackson-Weiss syndrome	—	3	—
Nonsyndromic craniosynostoses	—	—	1[b]
CHONDRODYSPLASIAS			
Achondroplasia	—	—	2
Thanatophoric dysplasia	—	—	8
Hypchondroplasia	—	—	1

[a]Note that Pfeiffer syndrome is associated with mutations in both *FGFR1* and *FGFR2*, and that mutations in *FGFR2* or *FGFR3* can give rise to several different phenotypes. For details see Cohen et al. (1997), Bellus et al. (1996), and Muenke and Schell (1995).

[b]This number includes the occurrence of an identical mutation in a conserved sequence present in all three loci.

short ear locus in mice. Mutations in the latter are associated with many different skeletal and soft-tissue defects (Kingsley 1994). The human *BMP4* gene, located on chromosome 14q, is so similar to a homologous fruit fly gene called *decapentaplegic* (*dpp*) that it can substitute for *dpp* in fruit fly embryonic development! The BMP4 protein functions in neural tube polarity and in the formation of vertebrae, muscle, and dermis. Also, it is probably involved in the normal cell death of certain neural crest cells and the loss of webbing between digits in hands and feet. At least one human disorder has been linked to *BMP* mutations. Some cases of inherited *fibrodysplasia ossificans progressiva* (involving bone and other defects, sometimes including mental retardation) are associated with overexpression of *BMP4*.

Much remains to be learned about BMPs and their receptors, as well as their interactions with various homeobox loci and with other genes that regulate embryonic development.

Hedgehog Family. Among the many fruit fly mutations affecting early embryonic development, one type produces larvae of about half the normal length whose undersides have a solid patch rather than multiple bands of tiny, hooklike projections. Dubbed *hedgehog* (*hh*), this segmentation locus is extremely important in both short-range and long-range patterning processes. In embryos, the *hh* locus is itself activated by a preceding complex cascade of gene activity. Then the hedgehog protein interacts with another secreted signaling protein called *wingless*, and they maintain each other's

continued production. Together they become involved in the specification of various structures in the larval segments, followed by key roles in the development of epidermis, eyes, wings, and legs of the adult fruit fly.

At least five highly conserved vertebrate homologues of *hh* have recently been cloned. In mammals they include loci called *sonic hedgehog* (*Shh*/*SHH*), *Indian hedgehog* (*Ihh*/*IHH*), and *desert hedgehog* (*Dhh*/*DHH*). We will concentrate on the *SHH* locus, which plays critical roles in development of the notochord, neural tube, motor neurons, somites, limbs, gut, and left-right asymmetry (of the heart, for example). It is also expressed in fetal liver, lung, and kidney tissues. *SHH* is activated by certain growth factors and is in turn a key activator of various other growth factor genes as well as *HOX* genes.

Deletions or mutations of the human *SHH* gene, located on human chromosome 7q, are associated with a dominantly inherited birth defect called *holoprosencephaly type 3*. Holoprosencephaly (*holo*, "whole"; *prosencephalon*, "forebrain") affects 1 in 250 spontaneously aborted fetuses and 1 in 16,000 live births. Incomplete development and division or closure of anterior midline structures results in abnormalities of the brain and face (Figure 15.15). Even within families, the phenotype is highly variable, ranging from mild to severe or lethal. The latter may include missing eyes or cyclopia (a single eye opening and a nose that is absent, or present as a tubular structure above the eye) and/or midline clefting. Milder cases may include a small head, abnormally spaced eyes, cleft lip and/or palate, and tooth abnormalities.

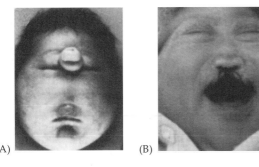

Figure 15.15   Three examples of the great phenotypic variability associated with holoprosencephaly (but not including the most malformed types of cases). (A) This infant has, instead of a nose, a knoblike proboscis between closely separated eyes. (B) This infant has an abnormally small head, very closely spaced eyes, missing nose bones, and a cleft lip. (C) This very mildly affected woman, mother of the baby in (B), has normal facial features except for one missing tooth (an incisor) in the midline of her upper jaw. (From Roessler et al. 1996.)

In all organisms studied, the receptor for the SHH protein is a membrane-bound protein encoded by a gene called *Patched* (*Ptc*/*PTC*). In humans, this locus resides on chromosome 9q, and mutations of *PTC* have been linked to certain malignant cancers. *Basal cell nevus syndrome*, a rare autosomal dominant disease, involves developmental defects of the skeleton (polydactyly, spina bifida, jaw and rib abnormalities) as well as a predisposition to basal cell carcinoma (a skin cancer) and medulloblastoma (a brain tumor). Much more common are the noninherited *basal cell carcinomas* caused by *PTC* mutations in skin cells.

Adhesion Molecules and Genes

In the formation of tissues, cells do not just stick together in any old way. Rather, they are intricately organized into very distinctive and diverse patterns (Gumbiner 1996). The processes by which embryonic organs take shape involve many complex interactions. Cells bind to each other, move, migrate, divide, differentiate, and die in precisely patterned ways. And in adults, cell interactions and movements are key factors in wound healing, inflammation (response to cellular injury), and cancer metastasis (spread of cancer cells from one tissue to another). How are such processes coordinated and regulated? Research begun in the 1970s has uncovered key proteins known as *adhesion molecules*. Those occurring on the cell surface are called **cell adhesion molecules (CAMs)**.

CAMs are widespread among animals. Broadly speaking, two major classes (the cadherins and the immunoglobulin superfamily) bind cells to other cells, while two different classes (the integrins and selectins) bind cells to substrates. The classes are distinguished by their particular molecular structures and functions, and they exhibit different binding specificities. All can bind to elements on other cells, but cadherins and the immunoglobulin family of CAMs bind most tightly to similar CAMs on neighboring cells. Integrins and selectins, however, bind to elements unlike themselves. Integrins also serve a major function in binding cells to elements of the extracellular matrix, allowing cells to adhere and migrate along substrates. Adhesion molecules in general associate with proteins of the cytoskeleton on the inside of the plasma membrane, thus making a direct route for transmission of signals from outside to inside the cell. Indeed, CAM-mediated cell-cell and cell-substrate adhesions set up intricate patterns of chemical signals that cause cells to behave in specific ways.

Leukocyte cell adhesion molecules, each one a complex of a light (β) chain and a heavy (α) chain, are a type of integrin. They exist on the surface of some white blood cells and under certain circumstances cause their aggregation. *Leukocyte adhesion deficiency*, an autosomal recessive condition, usually involves abnormalities in the β subunit of leukocyte adhesion molecules. Affected individuals suffer recurrent bacterial infections, defects in pus formation and wound healing, and other problems associated with adhesion-dependent functions of immune cells. The locus encoding the β chain is on human chromosome 21q.

Neural cell adhesion molecules, members of the immunoglobulin superfamily, are membrane-associated glycoproteins whose structure resembles antibody molecules (Chapter 18). They form cell-cell bonds that are extremely important in both early embryonic development (especially between nerve cells and between nerve and muscle cells) and in adults. A subgroup called L1 is required for the outgrowth of certain axons. NCAM is another important subgroup. Researchers mapped one *NCAM* locus to human chromosome 11q and also determined that it is highly conserved in vertebrates.

Cell adhesion defects have been implicated in *Kallmann syndrome*, a genetically heterogeneous disorder that can be inherited as an X-linked recessive, an autosomal recessive, or an autosomal dominant. Its incidence is about 1 in 10,000 among males and 1 in 50,000 among females. Genes associated with the autosomal types have not yet been mapped, but the one associated with the X-linked (and most frequent) type has been cloned and mapped to a site near the tip of Xp. Affected males show an odd combination of two symptoms: small testes and no sense of smell. These defects result from a deficiency of gonadotropin-releasing hormone in the hypothalamus and from missing olfactory bulbs and tracts in the brain. It turns out, however, that there is an embryological connection between these traits. The cells that normally produce the gonadotropin-releasing hormone actually arise in the developing nose of the fetus and then normally migrate to the hypothalamus. The formation of olfactory bulbs and tracts is normally induced by the olfactory nerve, whose cells also arise in the developing nose and migrate to the brain (Ballabio and Zoghbi 1997).

In people with Kallmann syndrome, these two types of embryonic cells (secretory and nerve) never make it to their usual destinations in the brain and thus never function properly. Scientists have analyzed the DNA of several patients, and from the nucleotide sequence of the gene, they have deduced the structure of the protein product. Certain parts of its 680-amino-acid sequence are very similar to known cell adhesion molecules. Thus, it appears that the defects of embryonic cell migration that characterize Kallmann syndrome result from defects in neural CAM-like proteins and genes.

Programmed Cell Death and Suicide Genes

Odd as it may seem, embryos regularly produce many more cells than exist in the newborn organism. In fact, **apoptosis** (*apopt*, "falling off"; *osis*, "process"), also called **programmed cell death**, is an important feature of normal embryonic development. It is an orderly process whereby certain types of cells are eliminated at specific stages of development:*

> [Programmed] cell death is seldom chaotic. Rather, it is a subtly orchestrated disassembly of critical molecu-

lar structures, permitting the doomed cells to vanish with minimum disruption to the surrounding tissue. (Wyllie 1998.)

Familiar examples from nature include the loss of a tadpole's tail as it becomes a frog and the loss of most of a caterpillar's tissues during its metamorphosis into a butterfly or moth. In addition to removing normal but unneeded cells from embryos or adults, however, apoptosis provides a defensive mechanism for ridding the body of cells seriously damaged by mutation, viral infection, malignancy, radiation, or other causes. In some cases, the dead cells, rather than quickly disappearing, are incorporated into certain tissues or structures. An example is formation of the protective outer layer of our skin, the epidermis.

Apoptosis also controls the shape and functioning of many organs in both the developing embryo and the adult. For example, early in development our hands are paddle-shaped, but after the loss of parallel rows of cells, the remaining structures form our fingers. Incomplete loss of such cells may leave webbing between the digits. And in certain parts of the vertebrate brain and spinal cord, 40–85% of the nerve cells normally die before birth of the organism! The carefully regulated interplay between cell proliferation and cell death likewise plays an important role in the regulation of the cell cycle.

Recent studies in several different organisms indicate that the "condemned" embryonic cells do not just die at random. Instead, it appears that by expressing specific **cell death genes**, they actively commit suicide. Mutations of these genes may block cell death, in which case the affected cells survive and function normally. In the well-studied roundworm *Caenorhabditis elegans*, exactly 131 specific cells among its 1,090 embryonic somatic cells are normally destroyed before hatching. Over a dozen different *ced* (cell *d*eath) genes are known to direct several stages of this process, from actual death to engulfment by neighboring cells (Figure 15.16). The most important of these are *ced-3* and *ced-4*. In addition, a different *ced* gene, called *ced-9*, actually protects cells against destruction. In humans, a cancer-causing *oncogene* (see Chapter 17) called *bcl-2* encodes a protein that prevents apoptosis.

Indeed, it appears that all cells contain some genes that promote apoptosis and other genes that prevent it. "The two [types] produce a balanced cocktail of proteins to keep certain cells alive and force others to self-destruct" (Hart 1994). But each cell type may possess its own special array of activators and inhibitors. It also turns out that many different signals, originating inside or outside a cell, can activate these genes. Exactly what do the products of these genes do? Some are protein-cutting enzymes called *proteases*. Others are *endonucleases* that chop up chromosomal DNA into

---

*Cell death by apoptosis is quite different from cell death following injury. The latter process involves the swelling and rupture of the injured cells, which are then devoured by white blood cells that have swarmed to the area. With apoptosis there are no signs of inflammation. The dying cells shrink rather than swell, and inside their nuclei the chromatin disintegrates into fragments that condense to form distinct clumps near the nuclear membrane. Finally, neighboring cells (and perhaps some scavenger white blood cells) ingest the dead cells.

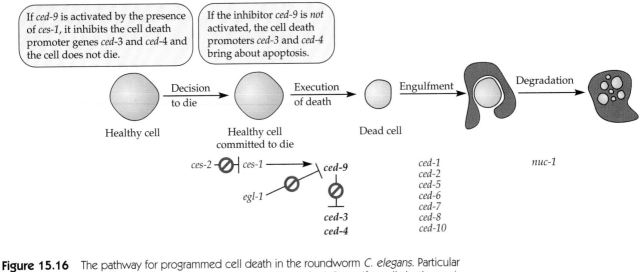

If *ced-9* is activated by the presence of *ces-1*, it inhibits the cell death promoter genes *ced*-3 and *ced*-4 and the cell does not die.

If the inhibitor *ced-9* is *not* activated, the cell death promoters *ced*-3 and *ced*-4 bring about apoptosis.

Figure 15.16 The pathway for programmed cell death in the roundworm *C. elegans*. Particular stages of cell death are associated with mutations in 14 different loci: 2 *ces* (for cell death specification), 1 *egl* (for egg laying), 10 *ced* (for cell death), and one *nuc* (for nuclease). The most important genes are *ced-9* (an inhibitor) and the cell death promoters *ced-3* and *ced-4*. Red "stop" symbols indicate inhibition of cell death (by genes *ces-2*, *egl-1*, and *ced-9*). The arrow indicates promotion of cell death (by all the other genes). (Adapted from Ellis et al. 1991.)

nucleosome-length (i.e., 180-bp) pieces. Yet others act as membrane-bound death receptors or as intracellular transcription factors.

Researchers have discovered more than 25 genes that act as cell death promoters or inhibitors in mammals. Although no *ced-4* homologues have yet been identified in fruit flies or mammals, a number of *ced-3* homologues exist. The first one found in mammals, called *ICE* (for *i*nterleukin-*c*onverting *e*nzyme), can trigger either cell death or inflammation. It belongs to a family of over 10 *ced-3*-homologous loci collectively known as *caspases* (for *Cys-Asp-ases*), which code for protease enzymes that chop up proteins. One homologue of the apoptosis inhibitor *ced-9* prevents cell death, and is important in lymphocyte and neuronal development. Over half a dozen members of this family are known. They include a locus that inhibits the just-described inhibitor, thereby triggering cell death! As Gilbert (1997) points out:

> Here we see a principle that underlies much of development: activation is often accomplished by inhibiting a repressor. This might be explained by the fact that in each nucleus, most of the genes are repressed. To activate a particular gene, an inhibitor of this repression is needed. Similarly, inhibition is often accomplished by the suppression of that inhibitor of the repressor. (Developmental biologists get used to talking in double and triple negatives.)

Researchers hope that the study of apoptotic genes will help us understand developmental processes that not only span the course from early development to aging, but also include pathological conditions such as cancer and neurodegenerative diseases (e.g., Parkinson, Alzheimer, and Huntington diseases). Some of the secondary tissue damage associated with heart attacks and strokes is caused by excessive cell death. Apoptosis has also been implicated in osteoporosis, in AIDS, in certain immune system and intestinal disorders, and in some types of heart and kidney disease (Barr and Tomei 1994; Duke et al. 1996).

Sexual Development

Most of us have strong opinions about what constitutes maleness and femaleness, and our daily lives are permeated by personal and societal expectations for sexual development and behavior. In both plants and animals, there is great variability in modes of sex determination, systems of mating, and mating behavior. In most cases, an individual's sex is determined genetically. However, there are some fascinating examples among fish, lizards, turtles, crocodiles, and birds in which environmental factors (such as temperature or local sex ratio) play a major role, even causing sex reversal under some circumstances.

Some sexually reproducing plants and animals maintain both sexes in the same body and practice either self-fertilization or cross-fertilization. But for those plants and animals (including humans) in which the two sexes are normally housed in separate bodies, individuals whose body structure or behavior cannot be classified as completely male or completely female have always been regarded with special interest.

Indeed, the concept of male and female joined harmoniously in one body has been idealized in mythology and art (Box 15A). But in real life, the occurrence of both male and female physical traits in one body is not so ideal. We now know that some of these human sexual abnormalities result from variations in sex chromosome constitution (Chapter 13) or in prenatal hormone levels (Chapter 16) or other factors.

In the 1950s, scientists began the first clinical studies of human sexual behavior. These sources of information, augmented by recent molecular genetic studies, have led to a deeper understanding of how maleness and femaleness normally develop in humans. Unlike many other developmental genes, however, those involved in sexual development are *not* broadly conserved in evolution. Thus, because of the tremendous differences in sexual systems, scientists are unable to use information gleaned from simpler organisms to identify homologous genes in mammals (Marx 1995). Instead, over the past few decades they have had to struggle mightily to uncover the roles of about a half-dozen genes that operate in early sexual development. Along the way they encountered several apparent breakthroughs that later turned out to be frustrating detours. The picture that is finally emerging involves a series of steps, some occurring before birth and some after.

Prenatal Development

The first step in sex determination occurs at the time of fertilization, when either an X-bearing or a Y-bearing sperm enters the egg. This establishment of *chromosomal sex* sets the stage for future changes, although the two sexes are physically indistinguishable at first. By the sixth week of development in humans, both 46,XX and 46,XY embryos have the same sex organs. These include a pair of "unisex" gonads called *undifferentiated gonads*, which have developed from *genital ridges* that lie on one side of the interim kidneys.* During this transformation, a nuclear receptor encoded by *SF1* (steroidogenic *factor 1*) on chromosome 9q activates certain genes associated with hormone production. It is also involved in the development of mature gonads and adrenal glands. *WT1* (the gene for Wilms *tumor suppressor protein 1* on chromosome 11p), is active in the development of both gonads and kidneys.

There are also two sets of primitive duct systems: the male-type *Wolffian ducts* and the female-type *Müllerian ducts* (Figure 15.17A). These duct systems will eventually form the rest of the internal sex organs. Externally, as seen in the figure, both 46,XX and 46,XY embryos develop a knoblike *genital tubercle* and a *urogenital groove* surrounded by specialized folds and swellings of tissue. Thus, in the beginning, the internal and external sex organs of any embryo are equipped

---

*During development, the urinary and genital systems are very closely associated, especially in males. Along the way, two pairs of kidneys form. The interim kidneys degenerate by the end of the third month, but their ducts become the Wolffian ducts of the reproductive system. The permanent kidneys start developing in the fifth week and are functional by about the ninth week of development.

BOX 15A *Hermaphrodites in Myth and Art*

The term *hermaphrodite* comes from Hermaphroditus, the son of the Greek god Hermes and the Greek goddess Aphrodite. Myth has it that while bathing, Hermaphroditus grew together with the nymph Salamacis, thus combining in one body both male and female characters. Hermaphroditic gods abound in the myths and art of India and Persia too; Ardhanarisvara, for example, represents the joining of the male god Siva with the female Sakti. And the creation stories of many other cultures (including one Hebrew version of Adam's creation) feature a hermaphrodite who becomes bisected into male and female sexes (Mittwoch 1986).

From the second century B.C., a Greek statue of a hermaphrodite, with breasts and male genitals. (Courtesy of the Musée de Louvre, Paris.)

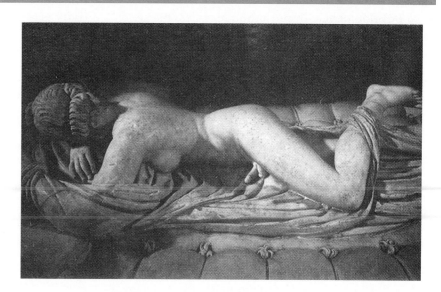

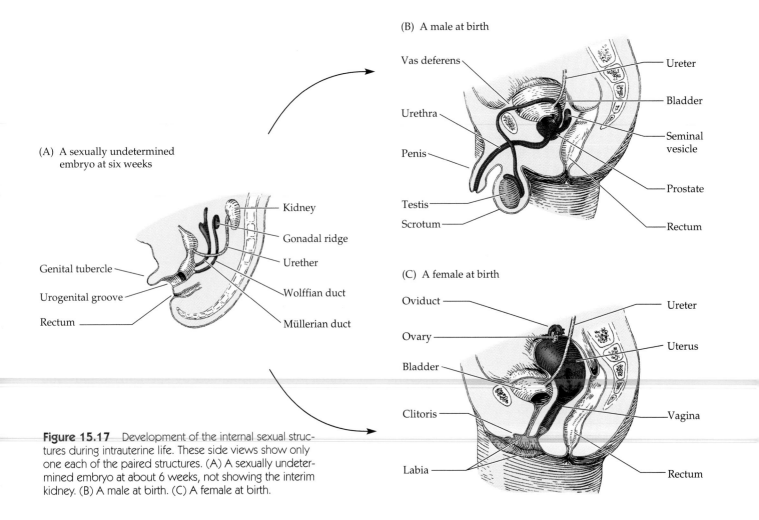

(A) A sexually undetermined
    embryo at six weeks

Genital tubercle

Urogenital groove

Rectum

Kidney

Gonadal ridge

Urether

Wolffian duct

Müllerian duct

(B) A male at birth

Vas deferens

Urethra

Penis

Testis

Scrotum

Ureter

Bladder

Seminal
vesicle

Prostate

Rectum

(C) A female at birth

Oviduct

Ovary

Bladder

Clitoris

Labia

Ureter

Uterus

Vagina

Rectum

Figure 15.17 Development of the internal sexual structures during intrauterine life. These side views show only one each of the paired structures. (A) A sexually undetermined embryo at about 6 weeks, not showing the interim kidney. (B) A male at birth. (C) A female at birth.

to form either a male or a female. But unless male hormones intervene at later stages of fetal life, it will develop along female lines.

Male Development.   In presumptive males, the second step in sex determination occurs by the seventh week of development, when the **SRY gene** on the short arm of the Y chromosome produces a protein called **testis-determining factor** (**TDF**). (The discovery of this gene is described in a later section.) TDF triggers the inner parts of the gonadal ridges to begin developing into male gonads, the *testes*. The *SRY* gene is necessary but not sufficient for differentiation of the mammalian testis, however. Autosomal and X-linked genes are also known to be involved. One key player is *SOX9* (*SRY-box 9*), a gene on chromosome 17q. It shares with *SRY* a special DNA-binding sequence domain (or box) of 79 amino acids. Their protein products cause 70–80° bending of the DNA near the sites they bind, probably bringing into contact chromosomal regions that would otherwise be separated.

The third step in male development occurs when somatic cells in the fetal testes begin to produce two different hormones. One cell type produces **anti-**

Müllerian hormone (**AMH**), which causes the Müllerian ducts to regress by apoptosis. The human *AMH* gene, which resides on chromosome 19q, functions only in the presence of a Y chromosome. Its activity is regulated by the *SF1* gene product. Also required for this process of regression is a normal **AMH receptor**, encoded by a gene (*AMHR*) on chromosome 12q.* Meanwhile, the male sex hormone **testosterone**, produced by a different type of cell in the immature testes, acts to maintain the Wolffian ducts.

If the appropriate tissues are responsive to these hormones, the following changes occur by the seventh week. Internally, the Müllerian ducts begin to degenerate. Under the local influence of testosterone, the Wolffian ducts begin to form the *prostate gland, seminal vesicles*, and *vasa deferentia* (singular, vas deferens), tubes connecting the whole system (Figure 15.17B). In addition, neural pathways in the *hypothalamus* (part of the brain) that control menstrual cycling in females are

*Rare mutations in either the *AMH* or the *AMHR* gene produce males with *persistent Müllerian duct syndrome*. Affected individuals are otherwise phenotypically normal males (sometimes with undescended testes) who also have a uterus and oviducts.

altered in male embryos. Finally, a very potent derivative of testosterone (called *dihydrotestosterone*) organizes the shaping of external genitals. (Conversion of testosterone to dihydrotestosterone is catalyzed by two variants of the enzyme 5-α-reductase, to be discussed later.) The genital tubercle and surrounding folds lengthen and fuse to form a *penis*, while nearby swellings develop and fuse to form the *scrotum*.

Female Development. In presumptive females, the products of various genes (including those controlling the production of female hormones) cause the outer parts of the genital ridges to begin to develop as *ovaries* around the twelfth week. A recently cloned gene on Xp, called *DAX1* (for *d*osage-sensitive sex reversal *a*drenal hypoplasia congenita on the *X* chromosome, gene 1), plays a key role in this process. It encodes a nuclear hormone receptor that is related to *SF1*, and also functions as an "anti-testis" gene by opposing the effects of *SRY*. Other X-linked loci and autosomal loci, yet to be discovered, probably affect sexual development too (Eicher and Washburn 1986). We know, for example, that two X chromosomes, with genetic determinants on both arms, are required for ovarian maintenance. In addition, several autosomal loci are known to affect normal ovarian development.

Quite independently, the Wolffian tubes (in the absence of testosterone secretion) regress, and the Müllerian tubes (in the absence of anti-Müllerian hormone secretion at an appropriate time) form the female ducts. The forward parts become the *oviducts* (Fallopian tubes), and the end parts fuse to form the *uterus* and most of the *vagina* (Figure 15.17C). The external genitals also develop: a *clitoris* from the genital tubercle, small and large *labia* ("lips," or folds) from the surrounding folds and swellings, and the lower part of the vagina from the urogenital groove. Thus, the sex of a fetus is clearly evident by the end of the fourth month. The main steps, and the major genes involved in determining the sexual phenotypes, are summarized in Figure 15.18.

Postnatal Development

On the basis of the appearance of its external genitals, a newborn's sex is assigned and recorded on the birth certificate. Sex-specific names, clothing, toys, and games usually follow, often along with sex-specific treatment. Boys may be encouraged to be energetic and independent, whereas little girls are often rewarded for cleanliness, passivity, and dependence. And despite recent sociocultural changes, different roles and aspirations may be projected for the two

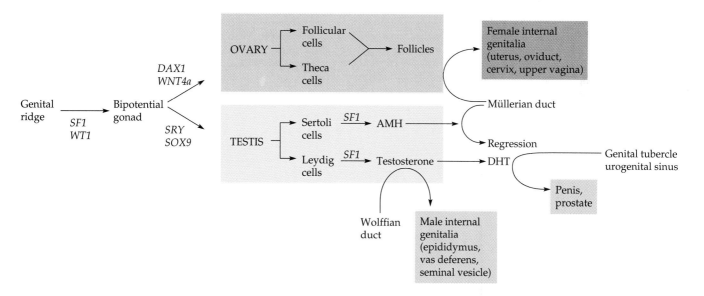

Figure 15.18 Postulated cascades giving rise to the development of male and female phenotypes in mammals. A few of the key genes are noted here. Development of the undifferentiated (bipotential) gonads from the genital ridge requires the *SF1* and *WT1* genes. Conversion of the undifferentiated gonads into ovaries requires the genes *DAX1* and *WNT4a*, and conversion into testes requires *SRY* and *SOX9*. Production of the hormone estrogen by the ovaries stimulates development of the Müllerian ducts into female internal genital organs; the clitoris develops from the genital tubercle. In males, the testes (influenced in part by the *SF1* gene) produce two hormones: anti-Müllerian hormone causes regression of the Müllerian ducts, and testosterone stimulates development of the Wolffian ducts into male internal genital organs. Also, the conversion of testosterone to dihydrotestosterone (DHT) stimulates receptive tissues to form the male prostate gland and penis. (From Gilbert 1997, adapted from Marx 1995.)

sexes by certain political and religious philosophies, by the various media, and especially by advertisers. Males are usually portrayed as doers and rescuers— dependable, rational, and stoic; females are often portrayed as helpers, mediators, observers, or victims— unreliable, intuitive, emotional.

How well do such assumed differences in behavior hold up under close scrutiny? Many researchers, after surveying and analyzing research on sex differences, have concluded that girls are not necessarily more social or suggestible, nor are they better at rote learning or worse at analytical thinking. Whether girls are more compliant and nurturing or less active and competitive is not clear. But there is considerable evidence that by the age of 12 or 13, girls excel in verbal ability, while boys excel in visual-spatial ability and mathematics and are more aggressive (Kimura 1992). Are the observed differences in these traits related to differences in brain structure and/or function? This idea is still a matter of hot debate (Box 15B). But even if it were true, the expression of these traits varies so widely within both sexes and shows so much overlap between the sexes that *one could never predict, on the basis of sex alone, how any given male or female will behave.*

BOX 15B *Gender Identity and Sex Roles*

Do boys and girls develop different patterns of behavior because they are treated differently, or are they treated differently because they show different behavioral patterns from the time of birth? Both physiological and environmental factors definitely play a role, but are behavioral differences due more to genetic differences between the sexes or more to differences in upbringing?

In birds and many mammals, including primates, the sexually stereotyped behaviors expressed in courting, mating, and caring for young are definitely controlled by hormones and neural pathways. Even so, normal females may occasionally exhibit mounting and thrusting behavior, and normal males may sometimes crouch to be mounted or display other aspects of female mating behavior. In humans, hormones also control the most basic differences between the sexes: gamete production, pregnancy, and lactation.

But much less is known about the relationship of mating behavior to hormones or brain circuitry in humans. How can these processes be studied? One approach is to look at the behavior patterns of people born with anatomical and hormonal abnormalities. These include 46,XX females who were masculinized during fetal life by hormone imbalance or by hormone treatment, but then surgically corrected, provided with appropriate hormone treatment, and raised as normal girls. Compared with control groups of females, the fetally masculinized group contained more "tomboys" and more late-marrying individuals. Observed heterosexual, homosexual, or bisexual tendencies vary from study to study, some showing no differences from controls and others showing some difference.

Studies of hermaphrodites and other individuals with ambiguous genitals are also inconclusive. In many cases, the chosen sex of rearing predominated well beyond puberty, regardless of the chromosomal sex or the development of inappropriate secondary sexual characteristics during puberty (e.g., breasts in "males" or virilization in "females"). These studies suggest that environmental factors are more important than genetic or hormonal factors. But other cases are known (mainly with 5-α-reductase deficiency) in which similar types of individuals who were raised as females abruptly changed their sexual identity and sex roles when virilization occurred at puberty.

One dramatic case, the attempted change in sex identity of an infant, was undertaken following a botched circumcision. In 1963, the penis of an 8-month-old baby boy (one of a pair of identical twins) was accidentally destroyed, and after much agonizing and consulting with specialists at Johns Hopkins University, the parents decided to raise him as a female. At 17 months, the child was given a new name, new hairdo, new clothes, new toys and games, and thereafter was treated quite differently from "her" twin brother. Four months later, a sex change operation was performed. Early reports suggested that these changes were having their hoped-for effect and that—aside from some tomboyish tendencies—she seemed to be developing into a normal female.

But a recent follow-up report (Diamond and Sigmundson 1997) tells a much different story. Even as a small child she preferred more "male-like" activities and clothes, and even tried to urinate standing up. By age 9, she suspected she was a boy. As a young teenager, she often refused estrogen treatments, rebelled against psychotherapy, and rejected the idea of having surgery to construct a vagina. She was teased mercilessly about her unusual behavior and considered suicide. At age 14, after finally learning about her medical history, she chose to change her identity back to maleness. Breasts were removed, a penis was surgically constructed, testosterone treatments were administered, and the transition to male behavior (including dating) was made relatively easily. At age 25, he married and adopted the children of his wife.

These case histories argue strongly in favor of nature over nurture. But the wide variety of outcomes from studies of individuals with various kinds of sexual abnormalities suggests that gender identity may not develop automatically and unswervingly on the basis of chromosomal sex or even genital sex. Recent studies implicate the importance of hormonal influences on the prenatal brain, suggesting that

"plasticity may be limited to a finite, and perhaps relatively brief, intrauterine androgen exposure. … In other words, the organ that appears to be critical to psychosexual development and adaptation is not the external genitalia, but the brain." (Reiner 1997.)

For additional information on sexual development, see Kandel et al. (1995), Kimura (1992), Bleir (1984), and Fausto-Sterling (1995).

Anthropologists also find that sexual norms of behavior vary tremendously from one culture to another. For example, traits such as emotional dependency, the ability to do hard physical labor, and the tendency to gossip have each been assigned to different sexes in different societies. Such cultural attributes are superimposed, layer by layer, on a basic physiological distinction between the sexes: males inseminate, whereas females menstruate, gestate, and lactate. The developmental events of puberty that underlie these irreducible sexual functions are described in the following paragraphs.

Between the ages of 10 and 15, the production of certain hormones sets off a chain of events, called *puberty*, that culminates in sexual maturity and the ability to reproduce. Triggered by genetic and environmental factors that are poorly understood, puberty starts when substances produced by the hypothalamus cause the tiny *pituitary gland* just beneath it to begin secreting *gonadotropic hormones*. These, in turn, stimulate certain cells in the gonads to make sex hormones, which bring about striking changes (secondary sex characteristics) in many sensitive tissues throughout the body. Both sexes produce the same hormones, but in much different proportions. (Actually, the different sex hormones are very similar to each other. They belong to a group of chemicals called *steroids*, which are derived from cholesterol.) Under the influence of nerve pathways set up in the hypothalamus during embryonic development, female hormone levels vary considerably during the menstrual cycle. In adult males, the cycling of gonadotropic hormone production is greatly reduced but not eliminated.

The two major gonadotropic hormones are named for their function in females. *Follicle-stimulating hormone (FSH)* activates a few egg follicles in the ovary to undergo maturation each month. With the increased production of *luteinizing hormone (LH)* in midcycle, one follicle is stimulated to release its egg and then to become a *corpus luteum* (yellow body). Before ovulation, the maturing follicle secretes **estrogen**,* the female sex hormone, which leads to development of the uterine lining. Other effects of estrogen at puberty include increased growth of the sex organs and breasts, widening of the pelvis, and deposition of subcutaneous fat. Following ovulation, the corpus luteum secretes the hormones estrogen and *progesterone*. The latter stimulates further thickening of and increased blood supply to the uterine lining in preparation for implantation of a fertilized egg. Both sex hormones also act on the hypothalamus and pituitary gland to decrease production of FSH and LH as part of the menstrual cycle. A

third gonadotropic hormone, *prolactin*, is secreted mainly after pregnancy and stimulates milk production; actually, small amounts of prolactin are secreted at all times by both sexes.

In males and females at puberty, the *adrenal glands* atop the kidneys make an **androgen**, or male hormone, that is partly transformed into testosterone. In females, this hormone is responsible for the growth of pubic and underarm hair and the development of oil and sweat glands in the skin. An excess of adrenal androgens in females (due to adrenal tumors, for example) can cause the abnormal development of male secondary sex characteristics such as facial hair, large muscles, and deepening of the voice. It can also enlarge the clitoris.

In pubertal males, FSH and LH stimulate the maturation of germ cells into sperm in the *seminiferous tubules* of the testes. A little bit of estrogen is produced by the testes and the adrenal glands, but in the presence of normal amounts of androgens, it has no obvious effects. However, normal teenage males often show slight breast development at puberty, which usually regresses. When such tissues persist, excess estrogen may be the cause. Very high concentrations of testosterone are necessary within the testes for complete sperm maturation. Some (non-*SRY*) Y-linked genes are needed for normal sperm production and fertility (Mittwoch 1992; Lahn and Page 1997). At puberty, testosterone also brings about most of the masculine changes and acts on the pituitary and hypothalamus to inhibit the production of LH and FSH.

Some Errors in Sexual Development

McKusick's catalog of inherited phenotypes (1997) lists several dozen aberrations in sexual development. For some of these, the modes of inheritance and the basic biochemical defects are well established, but for others the evidence remains scanty or ambiguous. It has been suggested that of about 50 known "sex genes," about 20 are X-linked, 2 are Y-linked, and the remainder are autosomal.

Usually, ovaries develop only in 46,XX embryos, and testes develop only in XY embryos. But when neither or both occur, or when gonadal sex does not correspond to chromosomal sex, the result is a **hermaphrodite**, of which there are numerous types.

True hermaphrodites, with both ovarian and testicular tissue, are rare (Figure 15.19). Although their external genitals are often ambiguous, roughly two-thirds of them are raised as males, irrespective of their chromosomal sex. They often produce eggs, but rarely sperm, and they are almost always sterile. Most true hermaphrodites have a uterus, and some menstruate. "Perhaps surprisingly, 21 pregnancies have been reported among 10 true hermaphrodites... All but one

---

*At least six different estrogens are known, but we will not distinguish among them. Likewise, several different androgens (male hormones) exist, among which testosterone is the most important.

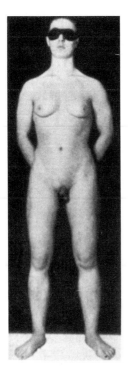

Figure 15.19   A person with true hermaphroditism, having both breasts and a penis. Many true hermaphrodites, however, have ambiguous or predominantly female-type external genitals. (From Overzier 1963.)

Defects of Androgen Target Cells

Androgens produced by fetal testes normally permit and enhance the development of the Wolffian ducts into male-type structures, but problems arise when the tissues on which they act fail to respond because of receptor or enzyme defects. One such type of *male pseudohermaphroditism* is described in a case report about a 44-year-old housewife, typically female in appearance:

> The patient's history revealed that she had developed normally but had never menstruated. ... She had been married for 20 years with normal coitus, libido, and orgasm, but had no pregnancies. ... The hands were rather large; breasts were well developed. There was no vestige of hair on the face, and no axillary [underarm] or pubic hair other than a trace of down on the vulva . ... The vagina ended in a blind pouch 8 to 20 cm in depth. (Morris 1953.)

A pelvic tumor contained "tubules of a testicular type." Morris called the clinical syndrome *testicular feminization,* and suggested that this patient probably never had any female internal organs. Yet affected individuals are definitely female from the psychosexual aspect—with sex urges like those of other women and strong desires for childbearing.

Testicular feminization is an X-linked recessive trait occurring in individuals who are chromosomally male (46,XY). The gene is located on the long arm of the X chromosome, near the centromere. Affected individuals produce plenty of androgen, which circulates in their bloodstream. But the **androgen receptors** that normally allow the hormone to enter the appropriate cells are absent or defective. Thus, unable to respond to the hormone's masculinizing effects, affected individuals develop as phenotypic females (Figure 15.20) who are sterile. The disorder, which occurs with a frequency of about 1 in 65,000 46,XY individuals, has also been called *complete androgen insensitivity syndrome.* Similar conditions have also been described in mice, rats, and cows; indeed, the mouse and human loci appear to be the same.

Although the testes are intact and functional inside the abdomen, no Wolffian structures are formed in the fetus. Nor are there oviducts, uterus, and upper vagina—which means that the response to Müllerian-inhibiting hormone produced by the fetal testes is appropriate. Because the external genitals are typically female and the short, blind lower vagina looks normal from the outside, the newborn is declared a girl and raised as such. At the time of puberty, estrogens (converted from testosterone) stimulate breast growth and other female secondary sexual characteristics. Normal male levels of testosterone also circulate in the bloodstream of these women, but to no avail: The tissues

... was 46,XX, and usually testicular tissue had been removed prior to pregnancy" (Simpson 1997). There are a few reports of apparent autosomal recessive mutations for true hermaphroditism.

Most true hermaphrodites are 46,XX, but 46,XY karyotypes and *mosaics* containing both 46,XX and 46,XY tissue are known too. How testicular tissue manages to develop in a 46,XX individual is not clear, but one possibility is that during embryonic development, some XY cells were present but later disappeared or became very rare. In other cases, part of the father's Y chromosome might have been attached to another chromosome. In some cases, however, the cause remains unknown.

In contrast to true hermaphrodites, whose gonadal sex is mixed, *pseudohermaphrodites* have either testes or ovaries, but the external genitals are the opposite of the gonadal sex, or ambiguous, or abnormal in some other way. There are two main pseudohermaphroditic classes:

Type	Gonads	External genitals
Male pseudo-hermaphrodites	Testes	Female-like or abnormal
Female pseudo-hermaphrodites	Ovaries	Male-like or abnormal

In the remainder of this section, we describe several disorders that can produce pseudohermaphroditism.

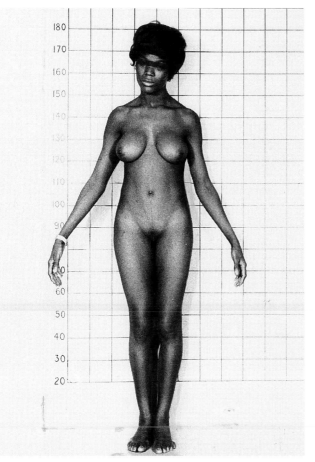

Figure 15.20 The testicular feminization syndrome in a 17-year-old 46,XY individual. Note the well-developed breasts, female-type body build, and absence of pubic hair. This woman is about 5 feet, 11 inches tall. (From Zourlas and Jones 1965.)

that would usually be stimulated to maleness do not respond. Insensitivity to androgens also prevents the growth of sexual hair in the pubic and underarm areas, although hair on the head is often luxuriant. If affected individuals have not developed hernias during childhood (from the attempted descent of the testes into nonexistent scrotal sacs), what finally brings them to the doctor's office at the time of puberty is the failure to menstruate.

Deficiency of 5-α-Reductase

In normal male fetuses, two variants of the enzyme *5-α-reductase* catalyze the conversion of testosterone to *dihydrotestosterone* in tissues destined to form the male external genitals. Type 1 is encoded by a gene on chromosome 5, and type 2 is encoded by a gene on chromosome 2p. A recessive mutation of the latter gene has been described in various ethnic groups, including a large, inbred kindred from the Dominican Republic (Imperato-McGinley and Gautier 1986). Affected members exhibit *male pseudohermaphroditism* due to deficiency of type 2 5-α-reductase. (It has no

obvious effect in 46,XX females, who show normal sexual phenotype and fertility.) Males homozygous for a mutation of the responsible gene have normal internal male organs, but their external genitals are ambiguous, and some of these 46,XY children are raised as females.

At puberty, there is no breast development or menstruation. Nor do normal male patterns of facial and body hair growth occur, and the prostate gland also fails to enlarge. But in the presence of testosterone (and of type 1 5-α-reductase, which is not deficient in this disorder), the penis enlarges, the voice deepens, and some semen is produced. Affected individuals also attain normal male height and muscular development. Most subjects who were raised as girls begin to feel that they are males at the time of puberty. Indeed, some marry and hope to sire children. This sudden change suggests that gender identity may not be fixed in early childhood, as has been assumed. Instead, under the influence of sex hormones on the brain, it may remain flexible until the time of puberty.*

In Chapter 16, we will describe some additional examples of male and female pseudohermaphroditism that are caused by enzyme deficiencies.

Sex Reversal and the *SRY* Gene

Much of what we know about the genetics of sexual development has been learned from rare cases of **sex reversal**. Most notably, 46,XX males and 46,XY females provided the key to geneticists' 32-year search for the elusive *SRY* gene (Weissenbach 1995).

About 1 in 20,000 male births is a phenotypic male with a 46,XX karyotype. Such *sex-reversed males* look entirely normal as boys, and often as men too. But all are sterile, and some adults show some of the clinical features associated with 47,XXY *Klinefelter males* (Chapter 13). These include small testes with abnormal tissue structure, decreased male hormones, and reduced male secondary sexual characteristics. Intelligence is normal, and mean height is shorter than 46,XY males but taller than 46,XX females.

Phenotypic females with a 46,XY karyotype are born with roughly the same frequency. These *sex-reversed females* also look normal as children, show normal intelligence, and are often taller than 46,XX females. They have Fallopian tubes, an underdeveloped uterus, and poorly differentiated (usually streak) gonads that render them sterile. Tumors may develop in the gonads. The absence of menstruation and secondary sexual characteristics can be treated fairly effectively with hormones. Both 46,XX males and 46,XY females may lead normal lives, including marriage.

---

*On the other hand, given the low status of females in some Latin societies, affected individuals may prefer maleness for practical rather than biological reasons.

How do sex-reversed males and females come about? Mutations or rearrangements of the various genes associated with sex determination are one cause. But many years ago it was proposed that sex reversal can occur when a tiny but critical bit of Y chromosome is lost (from 46,XY females) or gained (by 46,XX males). The missing or extra piece was presumed to carry the gene for testis-determining factor (TDF).

Recall that the human Y chromosome is roughly one-third the size of the X chromosome. It carries about 30 genes and gene families (Lahn and Page 1997), compared to the many known genes on the X. Yet despite their great dissimilarity, the X and Y chromosomes pair during meiosis. This is because the tips of their short arms share a tiny region of homology (Figure 15.21). During male meiosis, a synaptonemal complex and one chiasma always form in this region, resulting in a regular exchange of material between the tips of Xp and Yp. In 46,XY males the homologous region is present in *duplicate* (rather than singly, as with regular X-linked loci), and any genes within it are inherited as though they are autosomal—hence the term **pseudoautosomal region**, or **PAR1**.* PAR1 contains nine known genes.

Normal pairing between the X and Y, but *without* regular crossing over, also extends into the neighboring nonpseudoautosomal (and genetically nonhomologous) region, often just beyond the centromere of the Y chromosome. Very rarely, an "illegitimate" crossover

occurs in this latter region, resulting in the transfer of the TDF-encoding locus to the X chromosome and loss of that locus from the Y chromosome in that particular spermatocyte. If the resulting sperm fertilizes a normal egg, a 46,XX male or a 46,XY female will be conceived (Figure 15.22).

Researchers focused their search for the TDF-encoding gene by analyzing and comparing the DNAs from several dozen 46,XX males (with only a tiny piece of a Y chromosome attached to one of their two Xs) and a few 46,XY females (with a tiny deletion of their Y chromosome). Both types of abnormalities were located in the same minute region of the Y chromosome where the gene for the testis-determining factor must reside.

In 1987, an international team of scientists reported that they had finally found the gene for testis-determining factor, but the excitement was short-lived. Further research indicated that this new gene, called *ZFY* (for zinc *f*inger on the *Y*), did not meet all the necessary criteria; for one thing, it turned out to have a homologue on the X chromosome. Three years later a group of British investigators (Sinclair et al. 1990, Goodfellow and Lovell-Badge 1993) identified a nearby gene, located very close to the pseudoautosomal boundary, which they called **SRY** (sex-determining *r*egion of the *Y* chromosome). *SRY* codes for a

---

*A second and much smaller pseudoautosomal region (called PAR2) has recently been identified at the tips of the *long* arms of X and Y. But no genes have yet been found in this region, nor is crossing over in PAR2 obligatory.

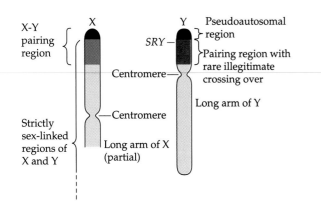

Figure 15.21 Fine structure of the human X chromosome (short arm) and the Y chromosome. The tiny *pseudoautosomal region* (black) carries nine homologous genes. During male meiosis, one obligatory crossover occurs here. The *X-Y pairing region* (black plus dark blue and dark red) includes a quarter of Xp and most of Yp. The dark blue and dark red regions carry different (nonhomologous) genes, but rare "illegitimate" crossovers can occur here too. The *strictly sex-linked region* (light blue and light red areas) makes up the most of the X and Y.

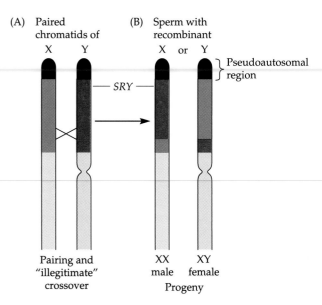

Figure 15.22 Results of "illegitimate" crossing over between the X and Y chromosome short arms during spermatogenesis. Only one chromatid for each chromosome is shown. (A) An illegitimate crossover (indicated by chiasma) occurs in the nonpseudoautosomal X-Y pairing region. (B) The recombinant chromosome gives rise, after fertilization, to an XX male (with *SRY*) or an XY female (lacking *SRY*). *Note:* A recombinant chromosome is named according to the *centromere* it carries. Thus, in (B), the recombinant chromosome on the left has the X centromere (not shown) and is called an X. The one on the right has the Y centromere and is called a Y.

transcription factor closely resembling some DNA-binding proteins known to turn other genes on or off.

Additional evidence also points to *SRY* as the testis-determining factor: (1) Essentially the same Y chromosome DNA sequence is detected in all males, but not in the females, of a wide variety of mammals. (2) In XY embryos of mice, the *Sry* gene* is expressed only in cells of the gonadal ridges, and at a time just before testes should begin to form. (3) In mice, the *Sry* DNA sequence is present in the male-determining region of normal Y chromosomes, but missing in a mutant Y chromosome (with a tiny deletion) that has lost its male-determining ability. (4) If mouse *Sry* gene sequences are transplanted into XX zygotes, some (but not all) of these prospective females develop into males. The sex-reversed mice possess male rather than female sex organs and exhibit normal male copulatory behavior. (5) In humans, at least two sex-reversed 46,XY women were found to have mutations (rather than cytologically observable deletions) within a key part of their *SRY* genes.

Although they seem to have found the long-sought "master switch" that activates a cascade of steps leading to maleness, scientists sound a strong note of caution: Not all sex-reversed cases can be explained by these findings. Indeed, most 46,XY females possess the *SRY* gene, and the causes of their sex reversal are unknown. Likewise, some 46,XX males lack any detectable Y chromosome DNA. Certainly, non-Y-linked mutations must affect later steps in the cascade of events leading to the development or nondevelopment of gonads—but they have not yet been clearly identified. Thus, to fully understand the complex process of sex determination in both males and females, researchers will have to find and analyze the other genes—known to exist on both sex chromosomes and autosomes—that function in sexual development.

Other types of sex reversal have also elucidated the functions of additional sex-determining genes. For example, mutations of *SOX9* cause *campomelic dysplasia* (*kamp*, "bending"; *melos*, "limb"), a usually lethal form of dwarfism. Infants born with this condition have severe abnormalities of the long bones (including bowing), skull, face, spine, ribs, pelvis, trachea and bronchi. In addition to these major defects, they also exhibit what is called *autosomal sex reversal*: 75% of the 46,XY individuals develop as females or hermaphrodites.

Defects in the *DAX1* gene have provided crucial information about its normal developmental role. Males born with mutations or deletions of *DAX1* have a condition called *adrenal hypoplasia congenita*. Under-

development of their adrenal glands causes a severe and (if untreated) potentially lethal deficiency of certain adrenal hormones. Males who reach the age of 14 also exhibit *hypogonadotropic hypogonadism* (Muscatelli et al. 1994; Zhang et al. 1998). This underdevelopment of their gonads at the time of puberty is due to the later effects of *DAX1* mutations on the hypothalamus and pituitary glands, reducing their output of gonadotropic hormones. A different condition, *dosage-sensitive sex reversal*, results from a rare duplication of the locus. Males who possess a normal Y chromosome and an X chromosome with two copies (rather than the usual single copy) of *DAX1* develop as phenotypic females. These cases, as well as experiments with sex-reversed transgenic mice carrying two copies of *DAX1*, reveal the anti-testis function of this locus (Swain et al. 1998).

The various phenomena described in this chapter indicate that scientists have made some exciting beginnings in describing the genetics of normal development as well as developmental disorders. It is now clear that development is regulated by a complex and ever-changing pattern of gene expression. Although scientists are still a long way from answering all the questions about how genes control development—including the question of what regulates the regulators—they have found some important clues and approaches to this major unsolved biological problem. It appears that at the molecular level, development in widely differing groups of animals operates by mixing and matching the same few kinds of genes and proteins. The challenge is to explain how a few types of genes can give rise to such different organisms.

Summary

1. The human zygote undergoes repeated divisions to form an embryo and surrounding membranes. The most critical and sensitive developmental period spans weeks 3 through 8, when all the major organ systems are forming.

2. Several hierarchies of gene action control early development. These include the maternal effect genes of the mother, as well as the segmentation and pattern formation genes present in the zygote. Pattern formation (homeotic) genes contain a highly conserved region called the homeobox. Its protein product, the homeodomain, regulates groups of genes by binding to their DNA.

3. The *HOM-C/HOX* homeobox-containing gene clusters, most intensively studied in fruit flies and mammals, encode transcription factors that play a key role in early embryonic development. They control axis formation and regulate the development of many structures, including the central nervous system, the backbone, and the limbs. The genes within a cluster are expressed in a head-to-tail direction within the embryo in the same order as their relative 3'-to-5' positions on the chromosome.

4. Other groups of homeobox-containing genes are known, some of which also contain different functional domains.

---

*Recall that the names of mouse genes that are equivalent to human genes often are not fully capitalized.

Groups of developmental genes without homeoboxes also exist. These, too, are often highly conserved in a wide range of animals.

5. The types of proteins encoded by developmental genes include growth factors, transcription factors, signaling factors, adhesion molecules, and receptor molecules.

6. Some conserved genes control the development of functionally similar (but structurally different) organs in groups as diverse as insects and mammals. For example, the *Pax6/PAX6* gene in mammals and the homologous *ey* gene in fruit flies both control eye development.

7. Numerous developmental defects, especially involving the skeleton, are associated with mutations of genes encoding fibroblast growth factor receptors in mammals.

8. Cell movements and tissue differentiation that give rise to animal body form are regulated by various types of adhesion molecules. Another normal feature of development, programmed cell death, is orchestrated by a series of cell death genes.

9. Many sex-linked and autosomal genes control sex differentiation. But unlike most other developmental genes, they are not generally conserved across broad evolutionary groups.

10. In very early mammalian development, the sex organs have the potential to form either a male or a female. Presence of a Y chromosome (with its *SRY* gene) triggers the development of testes. The hormone testosterone, produced by the embryonic testis, stimulates the development of male sex organs. The presence of two X chromosomes, along with the absence of a Y chromosome and male hormones, leads to the development of female sex organs.

11. Physical changes at puberty are regulated by two main types of hormones. Gonadotropins are secreted by the pituitary and act on the gonads. Sex hormones are secreted by the gonads and adrenal glands, and act on many parts of the body. Both sexes secrete all these hormones, but in different proportions.

12. True hermaphrodites, which are usually 46,XX, have both ovarian and testicular tissue. Pseudohermaphrodites, either 46,XX with ovaries (female) or 46,XY with testes (male), have external genitals that are ambiguous or characteristic of the opposite sex. Either type of individual may be raised as a male or female.

13. Testicular feminization occurs when 46,XY individuals, with testes that produce normal amounts of testosterone, develop into apparently normal (but sterile) females. The absence of androgen receptors renders their tissues unresponsive to male hormone.

14. The tips of Xp and Yp are homologous and make up a tiny, pseudoautosomal region that (during male meiosis) regularly pairs and includes a crossover. Next to the pseudoautosomal region of Yp is a region that harbors the *SRY* gene encoding the testis-determining factor. Illegitimate crossing over in this region may give rise to sex-reversed 46,XY females or 46,XX males.

Key Terms

androgen
androgen receptor
anti-Müllerian hormone (AMH)
apoptosis
blastocyst
cell adhesion molecule (CAM)
cleavage division
estrogen
hermaphrodite
homeobox
homeotic gene
maternal effect gene
paralogue

pattern-formation gene
programmed cell death
pseudoautosomal region
pseudohermaphrodite
retinoic acid
segmentation gene
sex reversal
signaling molecule
SRY gene
testis-determining factor (TDF)
testosterone
transcription factor

Questions

1. Distribution of the various products of maternal genes within an egg cell is not uniform. Some types of eggs show *gradients* of concentration of these developmentally significant materials, whereas others exhibit a *mosaic* partitioning of important substances. What effect does this unequal distribution of substances have on the developing blastocyst and embryo as it undergoes repeated cleavage divisions?

2. Researchers studying limb development in mice by means of gene knockout experiments often found that the loss or inactivation of single *Hox* loci resulted in fairly mild symptoms. Severe phenotypic effects usually required the simultaneous knockout of two or more loci. How could these results be explained?

3. List several ways in which apoptosis can be advantageous to organisms over the long course of evolution.

4. Since 1968, female athletes at the Olympic Games have been required to take simple tests that indicate the number of X chromosomes and (later) the presence of Y chromosome DNA. This policy was intended to detect any male members of female teams. Can these tests unequivocally accomplish their purpose? Explain.

5. A few cases have been described of males who possess oviducts and a uterus as well as the usual male internal organs. The condition is probably inherited as an autosomal recessive. What type of error in fetal development could account for this phenotype?

6. Why are males generally more vulnerable than females to errors in sexual development?

7. What mode(s) of inheritance could explain the following pedigree (Bowen et al. 1965) for partial androgen insensitivity?

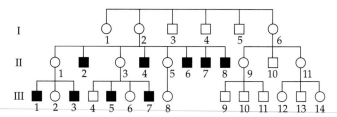

Further Reading

Wolpert (1992) is an interesting and readable book on embryonic development, written for the lay public. Gilbert (1997) is a comprehensive textbook of animal development, including the genetics of embryonic development. For further details on developmental genetics, see books by Wolpert et al. (1998), Akam et al. (1994), and Gerhart and Kirschner (1997). Raff (1996) provides a fascinating overview of the evolution of animal form. Good articles include McGinnis and Kuziora (1994), De Robertis et al. (1990), Scott (1994), McGinnis and Krumlauf (1992), Manak and Scott (1994), and Nüsslein-Volhard (1996). Gumbiner (1996) describes cell adhesion. On programmed cell death, see Duke et al. (1996), Ellis et al. (1991), and Jacobson et al. (1997).

Prenatal human development is clearly presented by Moore and Persaud (1998). Clayton-Smith and Donnai (1997) and Cohen et al. (1997) present overviews of human malformations, and Muenke and Schell (1995) summarize much of what is known about skeletal disorders associated with mutations of fibroblast growth factor receptors.

Sex differentiation in animals, including humans, is summarized in books by Hunter (1995) and Wachtel (1994) and in articles by Jiménez et al. (1996), Goodfellow and Lovell-Badge (1993), Werner et al. (1996), and Mittwoch (1986). Disorders of sexual differentiation are described by Shafer (1995) and Simpson (1997).

PART 5

Genetic Disease

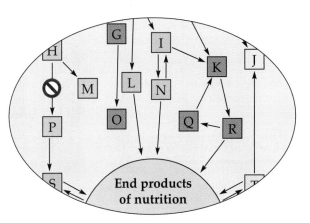

End products
of nutrition

CHAPTER 16

Metabolic Disorders

I n 1887, an American neurologist, Bernard Sachs, described the case history of an infant who had a strange condition and died at the age of 2 years:

The little girl … was born at full term, and appeared to be a healthy child in every respect; [her] body and head were well proportioned, [her] features beautifully regular. Nothing abnormal was noticed until the age of two to three months, when the parents observed that the child was much more listless than children of that age are apt to be … and that [her] eyes rolled about curiously. … The child would ordinarily lie upon [her] back, and was never able to change [her] position; muscles of head, neck, and back were so weak that [she] was not able either to hold [her] head straight or to sit upright. … [She] could not be made to play with any toy, did not recognize people's voices, and showed no preference for persons around [her]. During the first year of [her] life, the child was attracted by the light, and would move [her] eyes, following objects drawn across [her] field of vision; but later on absolute blindness set in. Hearing seemed to be very acute … the slightest touch and every sound were apt to startle the child. . . . The child never learned to utter a single sound. … [T]he child grew steadily weaker, [she] ceased to take [her] food properly, [her] bronchial troubles increased, and finally, pneumonia setting in, [she] died.

The Concept of Inherited Metabolic Disease

Later in the chapter, we discuss the genetics and biochemistry of the devastating and untreatable illness described by Bernard Sachs and now known as Tay-Sachs disease. Its cause, like that of phenylketonuria (PKU), described in Chapter 1, was ultimately traced to a missing or defective enzyme. Thus, both conditions are examples of inherited metabolic disease, or inborn errors of metabolism, which were first described around the turn of the twentieth century.

Garrod's Inborn Errors of Metabolism

At the time that Mendel's work was being discovered, English physician and biochemist Archibald Garrod (Figure 16.1) was studying a rare condition (incidence of 1 in 250,000) known as *alkaptonuria*.* Otherwise

*Recall that *uria* means "in the urine."

Figure 16.1 Archibald Garrod (1857–1936) served from 1915 to 1919 as a colonel in the British Army Medical Services in Malta. The black armband is a sign of mourning for the loss of all three of his sons during and just after World War I. Although Garrod's medical career was extremely successful and distinguished, his pioneering insights into the nature of metabolic disorders and the principle of biochemical individuality were not appreciated during his lifetime. (Photo from Bearn 1993.)

healthy newborns with this disorder usually had one striking feature: Their urine, upon standing in contact with air, turned black. (In their later years, affected individuals show darkening of portions of the ears and sclerae, or whites of the eyes; cartilage and tendons are similarly affected. The deposits in cartilage lead to arthritis. Males may develop black stones in the prostate.) The dark substance in the urine was the oxidation product of alkapton, now called *homogentisic acid*, which contains a six-carbon ring structure. Other researchers had shown that the excretion of homogentisic acid in urine was increased when people with alkaptonuria were fed excess protein or the amino acids tyrosine and phenylalanine, both of which contain benzene rings. Yet *unaffected* individuals who ingest excessive amounts of homogentisic acid never excrete any of it in their urine.

Garrod suggested that homogentisic acid is an ordinary product of metabolism that is broken down in normal people but is not degraded in people with alkaptonuria. This **metabolic block** results in homogentisic acid being excreted intact in the urine. More specifically, Garrod thought that people with alkaptonuria do not properly metabolize ring-containing fractions of proteins because they lack a special enzyme that is present in unaffected individuals. He proposed that during the normal breakdown of proteins, phenylalanine is converted to tyrosine, which in turn is changed to homogentisic acid and then to simpler products (Figure 16.2).

Garrod's analysis did not stop with the concept of metabolic blocks, however. The brilliance of his proposal lay in the genetic connection that he and population geneticist William Bateson were able to make. Garrod noted that alkaptonuria tended to occur in several sibs of a family whose parents were unaffected and also that many affected children were the offspring of first-cousin marriages. This pattern of inheritance fit the requirements for a rare recessive trait, the first human application of Mendel's newly discovered laws.

Garrod classified alkaptonuria and three other conditions that he had studied as **inborn errors of metabolism**. He suggested that these four genetic conditions must represent only a tiny fraction of the errors of metabolism that exist in human populations, pointing out that many of them may produce no obvious phenotypic effects. Citing known variations in hemoglobins and in muscle proteins as examples of chemical diversity within and between species, Garrod correctly prophesied in 1909 that

it is among the highly complex proteins that such specific differences are to be looked for. … [We] should expect the differences between individuals to be … subtle and difficult of detection. … Even those idiosyncrasies with regard to drugs and articles of food

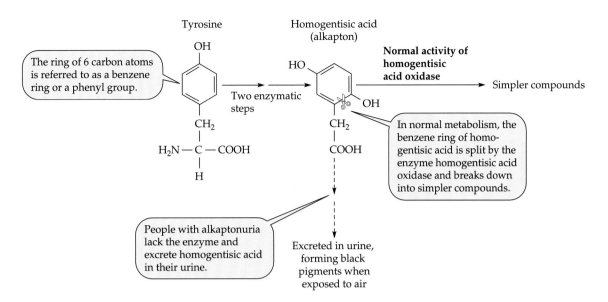

Figure 16.2 The metabolic block in persons with alkaptonuria. The splitting of the benzene ring by homogentisic acid oxidase during normal metabolism was proposed in 1902 by Garrod. His idea that people with alkaptonuria lacked this enzyme because of a genetic defect was far ahead of its time.

which are summed up in the proverbial saying that what is one man's meat is another man's poison presumably have a chemical basis.

But Garrod's ideas, like Mendel's, were largely ignored for over 30 years (Bearn 1993). During that period, studies of pigment formation in flowers, in the fur of animals, and in the eyes of fruit flies laid the groundwork for more extensive research on the nutritional requirements of the pink bread mold *Neurospora*. In a series of experiments done at Stanford University and reported in 1941, George Beadle and Edward Tatum irradiated wildtype *Neurospora* to produce a class of mutants that could survive only if specific chemicals were added to the nutrient medium. By collecting and analyzing dozens of these strains, they were able to show that each one had a single metabolic block—presumably due to an enzymatic defect—that could be traced to the mutation of a single gene (Figure 16.3). Although Beadle and Tatum were not the first to conceive of the one gene–one enzyme hypothesis, their relatively simple testing system greatly accelerated research in this developing field of biochemical genetics and earned them a Nobel prize as well. (Later work refined their proposal to a **one gene–one polypeptide** concept, which we now know to be an oversimplification too, given the recent discoveries of some overlapping, nested, split, and shuffled genes.)

By the mid-1940s, it was clear that (1) the biochemical processes that take place in an organism are genetically controlled; (2) every biochemical pathway can be resolved into a series of individual steps, each mediated by a different enzyme; and (3) each enzyme is usually encoded by one or a few genes. In fact, by analyzing groups of mutant strains that have different blocks in the same metabolic pathway, it is possible to determine the exact order in which these metabolic steps normally occur (see question 8). Indeed, we can generalize to say that if a condition is simply inherited, it is usually caused by an abnormality in a single protein molecule.*

Since 1950, the discovery of metabolic diseases has increased greatly, in large measure owing to the development of new methods for identifying metabolites of various kinds. At least 250 different human metabolic disorders (most quite rare) are known, distributed among many categories (Table 16.1). In recent years, our understanding of these diseases and their genetic bases has also been vastly enriched by molecular techniques. Indeed, only recently has the story of alkaptonuria been updated (Scriver 1996). In 1958, researchers identified the deficient enzyme responsible for alkaptonuria, *homogentisic acid oxidase* (also called *homogentisate dioxygenase* or *HGO*). Finally, in 1995, a Spanish team cloned the gene encoding this enzyme and mapped it to chromosome 3q (Fernández-Cañón et al. 1996; Scazzocchio 1997). They also found that two different missense mutations accounted for all homozygotes or compound heterozygotes with alkaptonuria in two unrelated Spanish pedigrees.

Detection of Metabolic Disease in Newborns

Certain symptoms tend to be common to a wide variety of inherited metabolic diseases. But because some

---

*Not all single-gene disorders involve enzyme defects, however. Many involve receptors, hormones, transport proteins, immunoglobulins, and structural proteins such as collagens.

TABLE 16.1 Some additional examples of metabolic disorders

Disorder	Chromosome[a]	Defective/deficient protein	Clinical features
AMINO ACID METABOLISM			
Maple sugar urine disease	19q	Branched-chain keto-acid dehydrogenase	Sugary-smelling urine; breathing/feeding problems, mental retardation, death
Ornithine transcarbamylase deficiency	Xp	Ornithine transcarbamylase	Irritability, severe mental and physical retardation, coma. Fatal if untreated
CARBOHYDRATE METABOLISM			
Hereditary fructose intolerance	9q	Fructose-1-phosphate aldolase	Hypoglycemia and vomiting after fructose intake; liver and kidney damage; fatal if untreated
Galactosemia	9p	Galactose-1-phosphate uridyl transferase	Inability to digest milk and milk products; vomiting, enlarged liver, jaundice, cataracts; mental retardation and death if untreated
NUCLEIC ACID METABOLISM			
ADA deficiency	20q	Adenosine deaminase	Toxicity to lymphocytes, causing severe combined immunodeficiency disease; fatal if untreated
Gout (one of many types, with various causes)	Xq	Phosphoribosyl pyrophosphate synthetase	Superactive enzyme leading to excess of purines and uric acid; possibly deafness
ORGANIC ACID METABOLISM			
Glutaric acidemia type I	19p	Glutaryl-CoA dehydrogenase	Brain damage, causing seizures and involuntary movements of head, trunk and arms; fatal
MCAD deficiency	1p	Medium chain acyl-CoA dehydrogenase	Variable; may resemble Reye syndrome; hypoglycemia, vomiting, lethargy, liver damage; sudden death may occur
LIPOPROTEIN AND LIPID METABOLISM			
Familial LCAT deficiency	16q	Lecithin:cholesterol acyltransferase	Corneal opacities, anemia, kidney problems
METAL METABOLISM			
Hemochromatosis	6p	HLA-H protein	Excess iron damages liver, heart, pancreas, glands, skin, joints; fatal if untreated
Menkes (steely hair) disease	Xq	P-type ATPase membrane cation transporter	Copper deficiency; grayish, broken hair; abnormal facial features, cerebral degeneration, bone damage, arterial rupture; fatal
Wilson disease	13q	P-type ATPase cation transport	Excess copper; liver disease, ring around iris, tremor, emotional and behavioral effects
LYSOSOMAL ENZYMES			
Hurler syndrome (Mucopolysaccharidosis I)	4p	α-L-iduronidase	Enlarged liver, spleen; skeletal deformities, coarse facial features, large tongue, hearing loss, corneal clouding; heart disease, mental retardation, death
Hunter syndrome (Mucopolysaccharidosis II)	Xq	Iduronate sulphate sulfatase	Severe form: coarse facial features, short stature, skeletal deformities, joint stiffness, mental retardation, death
Gaucher disease, type I (adult, or chronic, form)	1q	Glucocerebrosidase	Variable; enlarged spleen, bone defects and fractures, arthritis
PEROXISOMAL ENZYMES			
Adrenoleukodystrophy[b]	Xq	Possibly ALD membrane transport protein	Variable; excess of very long-chain fatty acids; adrenal insufficiency; childhood form includes dementia, seizures, paralysis, loss of speech, deafness, and blindness; fatal
Zellweger syndrome	7q	Peroxin-1? (No peroxisomes)	Abnormal skull and facial features; weakness; defects of eyes, brain, liver, kidneys, heart; fatal

TABLE 16.1 (Continued)

Disorder	Chromosome[a]	Defective/deficient protein	Clinical features
HORMONES			
X-linked ichthyosis	Xp	Steroid sulfatase	Dry, scaly "fish-skin"; mild corneal opacity
Goiterous cretinism (one of many types of hypothyroidism)	?	Iodotyrosine dehalogenase	Dwarfism, mental retardation, goiter, coarse skin and facial features
VITAMINS			
Transcobalamin II deficiency	22q	Transcobalamin II	Failure to thrive, weakness, anemia, immune deficiency, neurological disease
Biotinidase deficiency	3p	Biotinidase	Seizures, poor muscle tone, hair loss, skin rash, developmental delay, uneven gait, visual problems, conjunctivitis
BLOOD PROTEINS			
von Willebrand disease, type I	12p (AD)	von Willebrand factor	Varies greatly; bruising; bleeding from gums, cuts, and gastrointestinal tract; skin hematomas; heavy menstrual bleeding
Hemophilia A	Xq	Factor VIII, subunit a	Spontaneous bleeding, especially into large joints and muscles; hematomas, chronic arthritis; can be fatal if untreated

[a]All of these disorders are recessively inherited except for those noted as AD (autosomal dominant) or XD (X-linked dominant).
[b]Featured in the 1993 movie *Lorenzo's Oil.*

syndromes can be treated if recognized early enough, it is important that physicians and parents be alert to these diagnostic signals.

There are several ways by which inherited metabolic disorders may come to attention. In a few disorders, the metabolic defect acts during early intrauterine development, leading to abnormalities of facial structures, enlargement of organs (such as the liver), sexually ambiguous genitals, biochemical abnormalities, and other findings that are clearly expressed at birth. In many disorders, however, the metabolism of the mother (who may be a carrier but is not herself affected) makes up for the deficiency in the unborn child. These newborns may appear to be normal and are discharged from the hospital at the usual time. But intake of foods containing substances that cannot be properly metabolized or fuller expression of the metabolic deficiency leads sooner or later to abnormalities in the newborn, infant, or child.

Some disorders are characterized by an acute and potentially fatal episode 1 to 3 weeks after birth. The child may stop feeding, become difficult to arouse, and develop vomiting, seizures, a peculiar odor, and any of a variety of other findings. Still other disorders exhibit few if any outward abnormalities; damage to the nervous system, liver, or other organs progresses silently. And some metabolic disorders are predispositions that remain unexpressed for potentially long periods until the individual is exposed to an anesthetic,

sedative, antimalarial drug, or other medication or chemical that can elicit an idiosyncratic response. Another important sign is a family history characterized by the death of sibs in early life from unknown causes. But once recognized in a family, a metabolic disorder can be anticipated in subsequent pregnancies or births, carefully monitored, and perhaps treated.

The mode of inheritance is a good clue to the basic defect in genetic disorders. Recessive traits usually involve enzymes (especially enzymes that break down substances to simpler compounds) or peptide hormones. They tend to show a fairly uniform phenotype and an early age of onset. Dominant traits, on the other hand, usually involve nonenzymatic or structural proteins or enzymes that are involved in complex control systems. They tend to be much more variable in phenotype and often have a later (i.e., adult) age of onset. Thus, biochemical analyses of dominant disorders are more difficult, and relatively fewer are known.

Here we present a few examples of inborn errors of metabolism, starting with the case described at the beginning of this chapter.

Tay-Sachs Disease

Dr. Sachs referred his sick little patient to an ophthalmologist, who discovered a peculiar *cherry-red spot* on the retina of each eye (Figure 16.4A). In the next few years, Sachs encountered several more children with

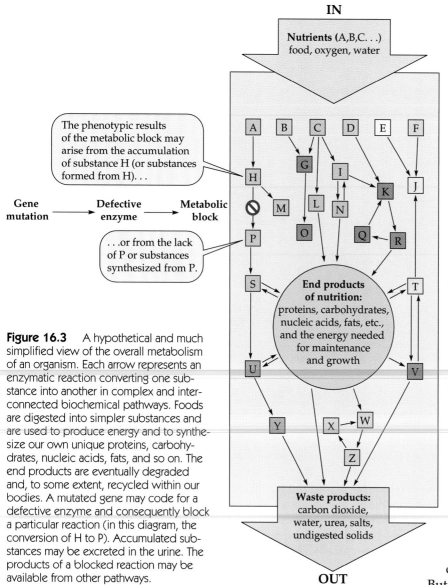

The phenotypic results of the metabolic block may arise from the accumulation of substance H (or substances formed from H)...

Gene mutation → Defective enzyme → Metabolic block

...or from the lack of P or substances synthesized from P.

Figure 16.3 A hypothetical and much simplified view of the overall metabolism of an organism. Each arrow represents an enzymatic reaction converting one substance into another in complex and interconnected biochemical pathways. Foods are digested into simpler substances and are used to produce energy and to synthesize our own unique proteins, carbohydrates, nucleic acids, fats, and so on. The end products are eventually degraded and, to some extent, recycled within our bodies. A mutated gene may code for a defective enzyme and consequently block a particular reaction (in this diagram, the conversion of H to P). Accumulated substances may be excreted in the urine. The products of a blocked reaction may be available from other pathways.

completely immobile and may need to be institutionalized. Most patients die by the age of 4.

TSD occurs in all ethnic groups, but predominantly among Ashkenazi Jews (the descendants of Jews who settled in eastern and central Europe). Because of unusual founder effects, high rates of TSD are also found among Moroccan Sephardic Jews, French Canadians in Quebec, some non-Amish Pennsylvania Dutch, and southwest Louisiana Cajuns (descendants of French Acadians expelled from Canada in the 1700s). Carrier screening in high-risk groups and the use of prenatal diagnosis have led to a drastic reduction in the frequency of affected homozygous newborns in recent years, however (Chapter 19).

In North American Jewish populations, the incidence (before screening programs) was roughly 1 in 4,000 births—much higher than the 1 in 320,000 births found in non-Jewish populations. Roughly 1 in 30 Jews is a heterozygote, whereas only 1 in about 250 non-Jews is heterozygous (Gravel et al. 1995). Why this extremely harmful allele has been maintained at such a high frequency is not known (Diamond 1991). But since two different mutant alleles remain at high frequency in the Jewish population, it is probably not a chance effect. Perhaps, as is the case with sickle-cell disease (Chapter 12), the carriers have some selective advantage.

Structural and Biochemical Abnormalities

The cerebrum of affected individuals becomes greatly enlarged by the swelling of individual nerve cells. This problem is caused by the accumulation of a specialized type of lipid known as a *ganglioside*. Within individual nerve cells, this material collects in the form of *membrane-bound cytoplasmic bodies* (Figure 16.4B), which fill up and expand the cell (Rutledge and Percy 1997).

Gangliosides, found mostly in the brain, are important components of normal membranes. These large molecules are synthesized by the addition of several simple sugars to a long-chain lipid and degraded in the lysosomes by the removal of these sugars—one by

the same malady, including a sib of the first case and four sibs in another family. A search of the medical literature for descriptions of this condition uncovered two reports published in the 1880s by an English ophthalmologist named Warren Tay. He described a family in which the cherry spots occurred in several members, all of whom died by the age of 3 years (Tay 1881). Within a decade, about two dozen cases were recorded, almost exclusively among Jewish families.

This still untreatable illness is called infantile *Tay-Sachs disease* (TSD). Usually, the first symptom noticed is the "startle reaction" to sharp noises. After 1 year of age, the child's condition degenerates rapidly; generalized paralysis, blindness, gradual loss of hearing, severe feeding difficulties, and some enlargement of the head are the rule by 18 months. By 2 years, the child is

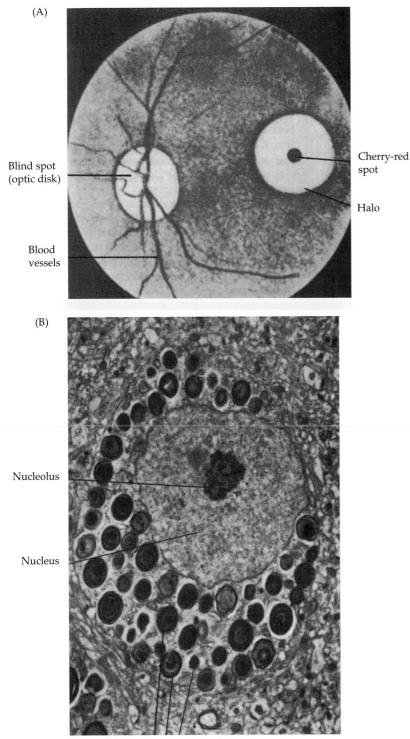

(A)

Blind spot (optic disk)

Blood vessels

Cherry-red spot

Halo

(B)

Nucleolus

Nucleus

Membranous cytoplasmic bodies

Figure 16.4 Manifestations of Tay-Sachs disease in the retina of the eye and in nerve cells of the brain. (A) The cherry-red spot on the retina can be seen through the pupil of the eye. The spot is caused by the absence of cells that usually overlay the fovea centralis, the region of sharpest vision. Neither the red spot nor the pronounced whitish halo is seen in normal retinas. (B) Darkly staining membrane-bound cytoplasmic bodies in a brain cell of a deceased patient contain excess lipid. The structures are absent in normal nerve cells. (A from Sloan and Fredrickson 1972; B from Terry and Weiss 1963.)

one—from the lipid part of the molecule. This latter process is defective in Tay-Sachs disease, resulting in the massive accumulation of a ganglioside called G_{M2}. The basic defect is a missing *hexosaminidase* enzyme that normally splits off the terminal sugar group (hexosamine) from the rest of the G_{M2} molecule. There are two forms of the enzyme in the tissues of unaffected children, but only one form in the tissues of Tay-Sachs patients. The form missing in Tay-Sachs children is *hexosaminidase A*, or *Hex-A* (Figure 16.5). The other form, found in both unaffected children and children with Tay-Sachs disease, is *hexosaminidase B* (*Hex-B*).

Molecular Biology and Genetic Variants

The Hex-A enzyme actually consists of two subunits, α and β chains (Figure 16.6), which are encoded by genes on chromosomes 15q and 5q, respectively. TSD results from a deficiency of α chains caused by mutations in the *HEX-A* locus. Over 54 *HEX-A* mutations have been found, and at least 40 are associated with infantile TSD (Gravel et al. 1995).

Affected Ashkenazi Jews usually make no complete messenger RNA and no α chains for this locus. Two mutant alleles account for 95% of infantile TSD cases in this population. The most common allele, occurring in about 80% of carriers, has a four-base-pair *insertion* that causes a frameshift and a premature termination of the polypeptide. (This mutation is also the predominant *TSD* allele in the Louisiana Cajun population.) The other mutant, a *base substitution*, probably results in defective splicing of the messenger RNA. It is found in about 16% of Jewish carriers. These two mutations occur less frequently in other populations.

Most affected French Canadians also make no α chains, but they have a large *deletion* at the 5′ end of the gene. Some affected persons who belong to neither of these populations make altered α chains that form defective Hex-A with low activity or low stability. Other variants have also been described in populations around the world. In all groups, some affected individuals are **compound het-**

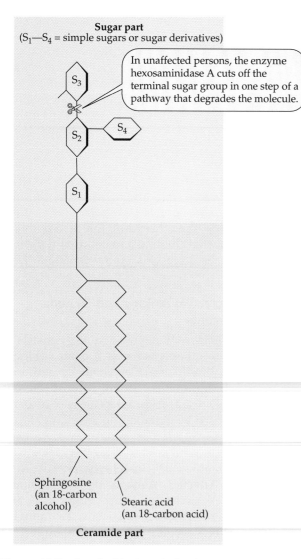

Sugar part
(S_1—S_4 = simple sugars or sugar derivatives)

In unaffected persons, the enzyme hexosaminidase A cuts off the terminal sugar group in one step of a pathway that degrades the molecule.

S_3

S_2 S_4

S_1

Sphingosine (an 18-carbon alcohol)

Stearic acid (an 18-carbon acid)

Ceramide part

Figure 16.5 Ganglioside G_{M2}, the lipid that accumulates in the cells of persons with Tay-Sachs disease. This large, complex molecule is a component of the membranes of brain cells. Infants with Tay-Sachs disease are homozygous for an autosomal recessive allele and lack the enzyme necessary to perform this step.

erozygotes (i.e., heterozygous for two different mutant alleles) rather than being homozygous for one particular allele.

Other types of TSD (juvenile, adult, chronic) are known. In these conditions, the symptoms have a later onset or are less severe, and the responsible enzymes show weak activity. Affected persons may make α chain precursors that do not associate well with β chains and are not converted to the mature form. Or they may carry a classic *TSD* allele plus a milder mutant allele, yielding a protein with some enzymatic activity.

Heterozygous carriers of the Tay-Sachs allele have Hex-A levels that are intermediate between those of the two homozygotes. Thus, they can be detected by measuring the level of this enzyme in a small blood sample. In Chapter 19, we describe the screening programs set up to identify carriers in Jewish populations. The hexosaminidase assay also allows for the prenatal diagnosis of *TSD* homozygotes. In recent years, TSD has been almost eliminated from U.S. Jewish populations—by detecting dual-carrier couples *before* they produce an affected offspring and by diagnosing Tay-Sachs disease during early pregnancy. Current techniques also make use of DNA analyses to detect the mutations directly.

The Hex-B enzyme, consisting of β subunits only, is produced by the chromosome 5 β chain locus. Mutation at this locus causes absence of both Hex-A and Hex-B enzyme activity and is associated with another fatal but very rare lipid storage disease that is virtually identical to Tay-Sachs disease. *Sandhoff disease* is characterized by the massive accumulation of a different form of G_{M2} ganglioside in the cerebrum and also in other organs. At least a dozen different mutations are known.

Phenylketonuria and Some Related Conditions

Phenylketonuria (phenylketones in the urine), or PKU, is one of the most common defects of amino acid metabolism and is also among the most common inherited metabolic diseases affecting brain development. It once caused hopeless mental and physical degeneration; until the early 1960s, it accounted for about 1% of the severely retarded people in institutions. If detected within a few weeks of birth, however, it is treatable by dietary therapy.

Recall from Chapter 1 that this condition is inherited as an autosomal recessive. The metabolic defect is an inability to convert the amino acid *phenylalanine* to *tyrosine* (Figure 16.7A), owing to the virtual absence of a liver enzyme called *phenylalanine hydroxylase* (*PAH*). Some of the phenylalanine that accumulates in the body fluids (cerebrospinal fluid, blood plasma, and sweat) is converted to phenylpyruvic acid, which in turn is metabolized to several other derivatives. But it is the phenylalanine itself that appears to cause the brain damage. Because tyrosine is deficient in these individuals, so, too, are its derivatives, including the pigment melanin. Thus, untreated affected persons tend to have lighter hair and skin compared to their relatives, and they often have blue eyes.

Phenylketonuria affects about 1 in 10,000 newborns in the United States and Europe and about 1 in 16,500 newborns in China. It is more frequent in certain European populations and their descendants (1 in 2,600 among the Turks, 1 in 4,000 among the Irish, and 1 in 5,300 among the Scotch and Scandinavians), but rare among Ashkenazi Jews, Finns, African Amer-

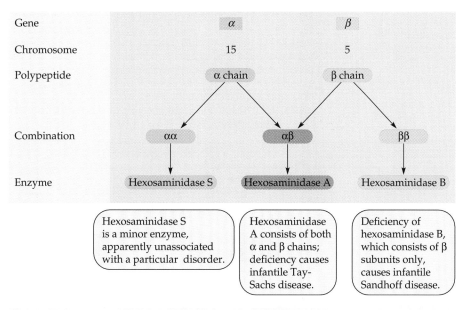

Gene		α	β	
Chromosome		15	5	
Polypeptide		α chain	β chain	
Combination	$\alpha\alpha$	$\alpha\beta$	$\beta\beta$	
Enzyme	Hexosaminidase S	Hexosaminidase A	Hexosaminidase B	

Hexosaminidase S is a minor enzyme, apparently unassociated with a particular disorder.

Hexosaminidase A consists of both α and β chains; deficiency causes infantile Tay-Sachs disease.

Deficiency of hexosaminidase B, which consists of β subunits only, causes infantile Sandhoff disease.

Figure 16.6 The α- and β-hexosaminidase system. Independent genes produce the α chains and β chains needed for the breakdown of ganglioside G_{M2}. Unaffected individuals produce all three enzymes. (Adapted from Sandhoff et al. 1989.)

icans, and Japanese. The frequency of carriers likewise varies from group to group, but among Caucasians it averages about 1 in 50. Why such detrimental alleles are so common is unknown.

The only clear-cut trait shown by newborn babies with PKU is the accumulation of excess phenylalanine in the blood plasma—a condition called *hyperphenylalaninemia*— starting several days after birth.* No consistent clinical signs occur during the first month or two of life. After two or three months, however, rapid and progressive deterioration of central nervous system development sets in. The brains of untreated persons with PKU tend to be smaller than normal, although they show no gross structural defects. At the cellular level, however, there is inadequate development of *myelin*, a fatty substance that normally forms an insulating sheath around certain nerve fibers and thereby speeds up the transmission of nerve impulses.

Babies and children with untreated PKU (now a rarity) are hyperactive and uncoordinated; some never sit, walk, or develop bowel and bladder control. The majority never learn to talk and do not progress beyond a mental age of 2 years. They are extremely agitated, with awkward and jerky movements caused by abnormally increased muscle tone. Behavioral problems range from depression, anxiety, and agoraphobia (the morbid fear of public places and open spaces) to violent and destructive temper tantrums. In addition to nervous system disorders, anomalies of the skin (eczema), teeth (defective enamel), and bones (small skull, growth retardation) may occur. About 75% of untreated people with PKU die before the age of 30.

Unfortunately, the precise cellular actions of this biochemical defect are still unknown. But in 1990, scientists reported the first useful animal model for PKU. By treating mice with a mutagen, they produced three different germ line mutations in the mouse *Pah* locus. Two true-breeding descendant lines exhibit severe PKU-like symptoms, and the other one shows a mild form of hyperphenylalaninemia. All should provide excellent opportunities for a detailed study of the developmental and physiological effects of the *Pah* mutations (McDonald and Charlton 1997).

Dietary Treatment

In the 1950s, physicians began to treat PKU by restricting the dietary intake of phenylalanine.* Synthetic mixes of amino acids are supplemented with vitamins, minerals, fats, carbohydrates, fruits, vegetables, and certain other foods. The diet is nutritious but distasteful to those who must endure it. Patients need to be continually monitored, and diets are adjusted to maintain the optimal concentration of serum phenylalanine: Either too much or too little can lead to brain damage.

Therapy begun after the age of 3 to 6 months shows little or no benefit, apparently because the brain has already been irreparably damaged. But when started within the first month after birth, and properly controlled, the low-phenylalanine diet is very effective. Hyperactive and difficult babies become alert and responsive and usually develop normal or near-normal intelligence. The practice now is to encourage continuation of the strict diet for at least 12 to 15 years and preferably throughout life, because there is some decline in IQ following cessation of treatment. Compliance is especially important for females of childbearing age who have PKU (discussed shortly).

Neonatal Screening and PKU Variants

To get affected infants on the diet before brain damage occurs, they must be identified within a few weeks

---

*Not all babies with abnormal test results will have PKU. In most cases, the higher values are temporary or related to other conditions.

*The sugar substitute Equal® (aspartane), for example, is on the long list of forbidden foods or food products.

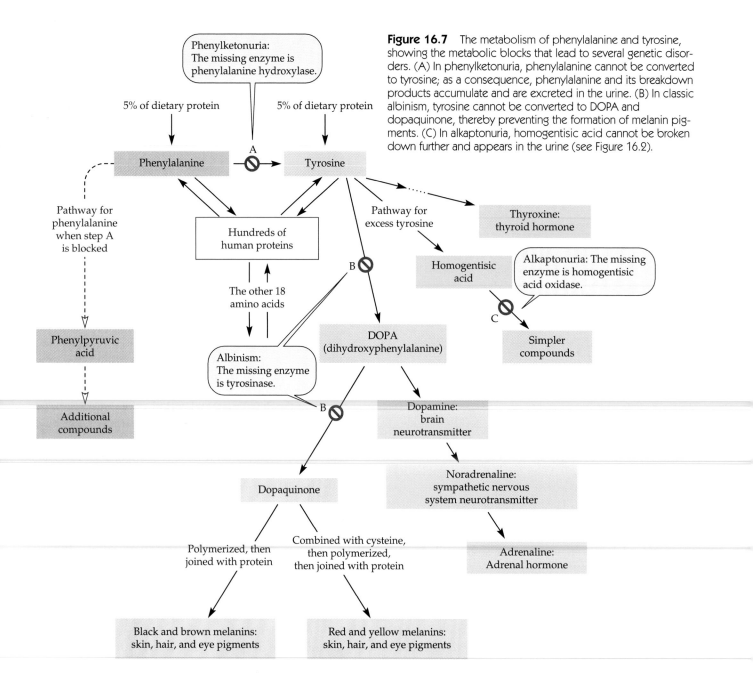

Figure 16.7 The metabolism of phenylalanine and tyrosine, showing the metabolic blocks that lead to several genetic disorders. (A) In phenylketonuria, phenylalanine cannot be converted to tyrosine; as a consequence, phenylalanine and its breakdown products accumulate and are excreted in the urine. (B) In classic albinism, tyrosine cannot be converted to DOPA and dopaquinone, thereby preventing the formation of melanin pigments. (C) In alkaptonuria, homogentisic acid cannot be broken down further and appears in the urine (see Figure 16.2).

after birth. This goal became possible in 1963 with the development of a simple, reliable, and inexpensive test for excess serum phenylalanine (Chapter 19). Most infants are now tested for this disorder before they leave the hospital in which they were born.

Even before early screening for PKU became possible, it was known that some individuals homozygous for mutant PKU alleles developed normal or near-normal intelligence. Until the early 1960s, virtually all cases of PKU were ascertained because of their mental retardation, and all had high levels of phenylalanine in their blood—that is, hyperphenylalaninemia (Figure 16.8). But did all individuals with hyperphenylalaninemia become mentally retarded? Physicians sus-

pected that some people with hyperphenylalaninemia actually developed normally. It soon became clear that this was the case, and screening programs began turning up new classes of individuals with *intermediate* levels of serum phenylalanine. Clinicians were then faced with the vexing problems of deciding which individuals should undergo dietary restriction and what concentrations of phenylalanine are safe at what ages.

Further studies revealed that the metabolism of phenylalanine involves a pathway of several enzymes, with each complete or partial deficiency leading to a different type of hyperphenylalaninemia. Even classic PKU itself is not a single genetic entity. The five main types of hyperphenylalaninemia (Figure 16.9) are all

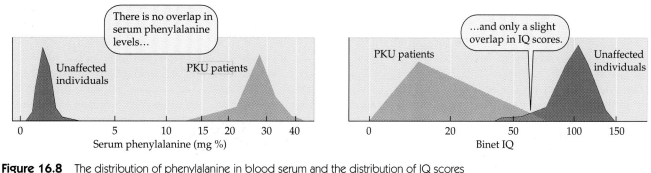

Figure 16.8 The distribution of phenylalanine in blood serum and the distribution of IQ scores in persons with PKU and in unaffected individuals. (The nonlinear measurement scales reflect a certain kind of statistical analysis.) This study was done before screening programs had begun, so that most affected individuals were in institutions. Following the adoption of PKU screening programs in the 1960s, some newborns with intermediate levels of serum phenylalanine were discovered, leading to problems of classification and treatment. (Redrawn from Penrose 1951.)

inherited as autosomal recessives. *Classic phenylketonuria* is due to a virtually complete absence (less than 1%) of activity of the enzyme *phenylalanine hydroxylase (PAH)* (number 1 in Figure 16.9). It accounts for roughly 60% of all hyperphenylalaninemia cases. Partial PAH deficiencies lead to *mild hyperphenylalaninemia* of two types (numbers 2 and 3). They account for about 35% of all cases of hyperphenylalaninemia and require either no treatment or only early dietary restriction. In up to 5% of hyperphenylalaninemia cases, the PAH enzyme is normal, but one of two other enzymes (numbers 4 and 5) in that metabolic pathway is nonfunctional. These two rare conditions are resistant to PKU dietary treatment: Even after phenylalanine levels are under control, brain function continues to deteriorate.

the greatest number of PKU mutations. Different alleles tend to occur in different populations, but usually a small number of alleles predominates in any one region. For example, just five mutant alleles may account for most PKU cases in Europe (Eisensmith et al. 1992).

Individuals with PKU can be homozygotes for a particular mutant allele, but about 75% are *compound heterozygotes* for different *PAH* mutant alleles (Scriver et al. 1994; Scriver, Kaufman, et al. 1995). Specific combinations of alleles have been correlated with certain clinical PAH-deficient phenotypes. Broadly speaking, the lower the tested PAH enzyme activity levels, the higher the observed blood phenylalanine levels and the lower the IQ. But there are some cases in which af-

Molecular Biology and Genotype/Phenotype Correlations

The *PAH* locus resides near the tip of chromosome 12q. It is a large gene (90 kb long, with 13 exons) encoding a polypeptide of 452 amino acids. The locus is highly conserved, with human and rat *PAH* genes showing over 90% homology. More than 325 different disease-associated *PAH* mutations have been identified (Kayaalp et al. 1997), giving rise to tremendous diversity of genotypes and great variability of phenotypes. Scientists can take advantage of this diversity to study the origins and migration patterns of various populations (Zschocke et al. 1997).

Most *PAH* variants are single-base substitutions (i.e., missense, nonsense, or splicing mutations). Most cause PKU, some result in non-PKU hyperphenylalaninemia, and a few are functionally benign "silent polymorphisms." Exon 7, which probably encodes the enzyme's active site, harbors by far

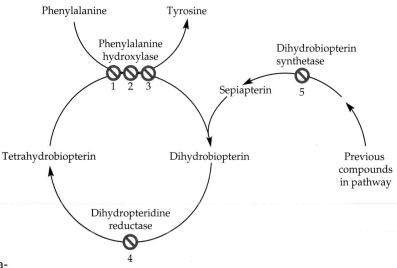

Figure 16.9 Simplified scheme of phenylalanine metabolism, showing the key enzymes and substrates. Locations of the five metabolic blocks known to cause the various types of hyperphenylalaninemia are indicated by numbers 1-5 and described in the text. (Adapted from Bickel 1987.)

fected individuals with identical genotypes exhibit different clinical phenotypes or where individuals with the same blood phenylalanine levels have quite different IQs. Such inconsistencies may be due to genotypic variation at other loci.

Despite its effect on the brain, PAH is expressed only in the liver. Thus, its activity or inactivity cannot be detected prenatally in fetal amniotic cells. But molecular techniques make it possible to detect the mutant DNA itself. Because many RFLP sites are linked to the *PAH* gene, most families already known to be at risk for PKU can be effectively screened and diagnosed. Yet for those families with uninformative segregation patterns and for carriers without a family history of PKU, this kind of analysis alone will not work. Direct detection of the mutant alleles in the DNA of some carriers or homozygotes is also necessary. Such analyses—using oligonucleotide probes specific for each mutation—are relatively simple, quick, and cheap. They can also be automated, an advantage that opens the door to carrier screening in the general population.

Maternal Phenylketonuria

Before the advent of treatment for PKU, affected females were usually profoundly retarded, were often institutionalized, and rarely had babies. But as effective diets rescued more and more of them from retardation and allowed them to lead normal lives that included motherhood, they began to produce non-PKU offspring who were retarded. When mothers with PKU go untreated during pregnancy, over 90% of their babies are mentally retarded and about 75% have a small head (microcephaly). In addition, these babies often have low birth weight and heart defects. Despite the fact that they are heterozygotes for *recessively* inherited PKU, many also exhibit hyperphenylalaninemia.

The cause of the newly observed *maternal PKU* soon became clear. Most of these mothers had abandoned the diet therapy by their early teens, after which their serum phenylalanine levels increased greatly. This dietary change seemed to have limited ill effects on them, but it caused irreparable harm to their developing offspring by transferring high phenylalanine levels across the placenta (Rouse et al. 1997). Indeed, if these women reproduce at average rates, the incidence of PKU-related mental retardation could return to its original level within just one generation!

The problem can be prevented if women with PKU remain on diet therapy through adulthood or return to it before they become pregnant. But many females of childbearing age are unaware of their PKU until after they have produced an abnormal baby. Oddly enough, many have no recollection of ever being on a special diet, and their parents often see no reason to tell them. As mentioned in Chapter 1, the treatment of maternal PKU raises many medical, legal, and ethical questions.

Albinism

Although not too common in humans, the absence of coloration is such a striking trait that most of us have seen it at some time. *Albinism* has been noted by writers as far back as the first century A.D. Indeed, from the description of his birth in ancient Old Testament writings, it appears that Noah might have been an albino. European explorers were fascinated by the "white Negroes" they saw in Africa (Figure 16.10) and by the "moon-eyed people" found in some indigenous tribes of Central and North America.

Albinism occurs in a wide variety of animals, from insects to mammals. The basic defect is usually the absence of melanin pigments, inherited as an autosomal recessive. Garrod (1909) regarded albinism as an inborn error of metabolism caused by the lack of an end

Figure 16.10   An albino child in a Liberian village. (Copyright Richard Dranitzke/Science Source/Photo Researchers.)

product rather than by the excess of some biochemical intermediate. He suggested that an intracellular enzyme was missing. Again he was correct. Complete, or classic, albinism is now known to be caused by the absence of the functional enzyme *tyrosinase* (Figure 16.7B). But the biochemistry of melanin formation is very complex, and other types of albinism are also known, including a few X-linked or dominant types. In mice, over 65 different genes (with more than 150 known mutations) affect the skin and fur colors. Although genetic influences on human pigmentation may be equally complex, relatively few loci associated with albinism have been identified so far in humans.

Tyrosinase is active only in specialized cells known as **melanocytes**. These arise in the neural crest (Chapter 15) during very early embryonic development and later migrate to their final destinations. Melanocytes occur in the lower layer of the skin epidermis and the hair bulbs as well as in certain parts of the eye, ear, nervous system, and mucous membranes. Inside the melanocytes, tyrosinase is restricted to small specialized structures known as *melanosomes* (see Figure 7.2). During normal pigment formation, melanin is produced in these melanosomes, which are transferred to the adjacent epidermal cells and come to lie over the nucleus of each cell. By capping the epidermal nuclei and absorbing much of the ultraviolet radiation that enters the cell, melanin shields the DNA from the mutagenic effects of sunlight and other sources of ultraviolet radiation.

People with albinism have normal melanocytes with structurally normal melanosomes that lack only pigment. Among unaffected persons, skin color is determined not by the number of melanocytes—which appears to be about the same in all races—but rather by the number and distribution of melanosomes produced by these cells, as well as by the type of melanin. In all races, the rate of production of melanosomes varies with amount of sunlight, hormonal changes, and genetic constitution.

Albino Phenotypes

In humans, several kinds of albinism are known, including even a few with normally pigmented skin. Although the severity of the clinical features varies, all types of albinism involve reduced or absent eye pigmentation and structural abnormalities of the eye. Severe squinting, poor vision, and *nystagmus* (rapid, jerky, involuntary movements of the eyeballs) also occur. All affected individuals lack stereoscopic depth perception as a result of misrouting of optic nerve fiber pathways in the brain. This impairment is due to a lack of pigment in certain nerve cells during early development.

Albinism exists in two major but partly overlapping phenotypic groups (Table 16.2). *Oculocutaneous albinism*

(*oculo*, "eye"; *cutaneous*, "skin"), or *OCA*, involves the absence or partial deficiency of melanin in the eyes, skin, and hair, as well as the visual defects previously described. It is a heterogeneous category involving mutations at several different loci and phenotypes with skin pigmentation ranging from absent to nearly normal. But with the much less frequent *ocular albinism* (*OA*)—which can be either X-linked or autosomal—defects are limited to the eyes; the skin and hair are normally pigmented or only slightly underpigmented. About ten kinds of OCA and four kinds of OA have been described (Barsh 1995; King et al., 1995, 1997).

First we will consider three subgroups of OCA. Among these, the most serious clinical form, the classic one studied by Garrod, is *tyrosinase-negative OCA* (*OCA1*). Absence of the enzyme tyrosinase leads to the total lack of melanin in the eyes, skin, and hair. Extreme sensitivity to sunlight causes excessive roughness, wrinkling, and folding of the skin as well as a high frequency of skin cancer. The irises of the eye are gray to blue, containing no visible pigment. Affected individuals are usually cross-eyed and very nearsighted, often to the point of legal blindness. Worldwide, the prevalence of OCA1 is about 1 in 30,000 to 40,000. But its frequency varies among different ethnic groups. In the United States, for example, the frequency is about 1 in 28,000 among blacks and about 1 in 39,000 among whites.

In a variant of OCA1, tyrosinase exhibits partial (5–10%) activity, leading to the formation of some pigment in hair, skin, and eyes. And in another (rare) OCA1 variant, tyrosinase is temperature-sensitive, active only at temperatures below 35°C (95°F). The skin is unpigmented, as is scalp and underarm hair, but pigment does form in hair on the arms and legs. This phenotype is comparable to Siamese cats, in which dark fur is produced only at the cooler extremities.

Another major type of oculocutaneous albinism, *OCA2*, is also known as *tyrosinase-positive OCA*, because the *TYR* gene and tyrosinase are normal. The nature of the metabolic error is poorly understood, but probably involves a defect in the transport of tyrosine into the melanosomes. This type of albinism is the most common worldwide, although less frequent in Caucasians (about 1 in 36,000 in the United States) than in blacks (1 in 10,000 in the United States). Among Nigerians, the frequency is about 1 in 5,000 to 15,000, and among South African blacks, it is 1 in 3,900. It is also quite prevalent among certain Native American tribes (about 1 in 140 to 240 for the Tele Cuna, Hopi, Jemez, and Zuni), perhaps owing to the special place of albinos in these cultures.

Because both types of OCA are autosomal recessives, matings between two tyrosinase-negative or two tyrosinase-positive albinos are expected to produce all albino progeny. But because the two autosomal loci are inde-

TABLE 16.2   Classification of some kinds of albinism

Mutant phenotype[a]	Gene	Gene product	Gene location	Enzyme activity
OCULOCUTANEOUS ALBINISM, DEFECTS IN EYES AND MELANOCYTES				
OCA1A, classic tyrosinase-negative OCA: no pigment in hair, skin, or eyes	*TYR*	Tyrosinase	11q	None
OCA1B, yellow OCA: reduced pigment in hair, skin, and eyes	*TYR*	Tyrosinase	11q	Partial
OCA1TS: no skin pigment; pigmented hair in cooler parts of body (arms and legs) only	*TYR*	Tyrosinase	11q	Temperature-sensitive
OCA2, tyrosinase-positive OCA: reduced pigment in hair, skin, and eyes; skin pigment in form of freckles and moles; linked to PWS and AS	*P*	P polypeptide	15q	Partial
OCA3, brown OCA: reduced pigment in hair, skin, and eyes	*TRP1*[b]	DHICA oxidase (glycoprotein 75)	9p	None
OCULOCUTANEOUS ALBINISM, DEFECTS IN EYES, MELANOCYTES, AND OTHER CELL TYPES				
Hermansky-Pudlak syndrome (HPS): reduced pigment in hair, skin, and eyes; platelet defects, bruising and bleeding, colitis, lung fibrosis	*HPS*	Organelle membrane protein	10q	Partial
Chediak-Higashi syndrome (CHS): reduced pigment in hair, skin, and eyes; susceptibility to bacterial infections	*CHS*	Organelle membrane protein	1q	Partial
OCULAR ALBINISM (DEFECTS IN EYES ONLY)				
OA1, X-linked (Nettleship-Falls) OA: pigment normal in skin and hair but not in eyes	*OA1*	Unknown	Xp	Unknown
Autosomal recessive OA; part of OCA1 spectrum: slight loss of pigment in skin and hair	*TYR*	Tyrosinase	11q	Partial
Autosomal recessive OA; part of OCA2 spectrum: slight loss of pigment in skin and hair	*P*	P product	15q	Partial

Source: Adapted from King et al. 1997 and Oetting et al. 1996.

[a]All phenotypes involve visual defects. Recent molecular genetic findings have led to changes in nomenclature and groupings. Note, for example, that mutations in the tyrosinase or P product genes can give rise to oculocutaneous or ocular albinism.

[b]Tyrosine-related protein 1.

pendent, matings between tyrosinase-positive and tyrosinase-negative albinos produce all unaffected offspring.

Two autosomal recessive OCA disorders in which pigmentation defects occur in several cell types are Hermansky-Pudlak syndrome (HPS) and Chediak-Higashi syndrome (CHS). In addition to the previously described ocular defects and considerable variation in skin and hair color, both disorders involve abnormalities of blood platelets that cause internal bleeding. There are some differences, however, between the two conditions. In patients with HPS, the accumulation of a nonmelanin pigment in various body tissues leads to fibrosis (especially of the lungs) and inflammation of the colon. Those affected with CHS have abnormal lysosomal granules in certain cells and exhibit severe immune deficiencies. They are highly susceptible to bacterial infections and often die in childhood. Although rare in most populations, HPS is relatively common (1 in 1,800 births) in Puerto Rico and in one village in the Swiss Alps. CHS is also rare.

Biochemistry and Molecular Biology

As shown in Figure 16.7B, tyrosine is normally converted to *dihydroxyphenylalanine* (*DOPA*), a key biological compound. In melanocytes, DOPA is converted to *dopaquinone*, which finally is polymerized with proteins in the melanosomes to form *black and brown melanins*. Alternatively, dopaquinone may combine with cysteine before being conjugated with melanosome protein to form *red and yellow melanins*. (Humans usually have a mixture of these two types.) The first two steps in this scheme (tyrosine → DOPA → dopaquinone), plus a third step further along the pathway to black and brown pigments, require the enzyme tyrosinase. The *TYR* gene, homologous to the albino (*c*) locus in mice, resides on chromosome 11q and encodes a polypeptide of 529 amino acids. Over 75 different OCA mutations have been found at this locus. Most of them are associated with the classic 1A phenotype, and most affected individuals are compound heterozygotes.

So far, only a few other human genes connected with melanin synthesis have been mapped or cloned. Some of them are homologous with pigment genes in mice. In 1992, scientists mapped the *P* gene, associated with the OCA2 (tyrosinase-positive) phenotype, to chromosome 15q. It lies in the same region as the loci for Prader-Willi syndrome and Angelman syndrome. (Indeed, about 50% of individuals with PWS and AS show hypopigmentation of the hair, skin, and eyes.) Over 20 mutations have been found at this locus.

In 1995, researchers mapped the *HPS* gene to chromosome 10q and the *CHS* gene to chromosome 1q. Functions of their gene products are unknown, but both HPS and CHS proteins are associated with the membranes of organelles (including melanosomes) and may be involved with protein sorting (Ramsay 1996). Also in 1995, the gene associated with OA1 was cloned from near the tip of Xp. Its protein product, which is expressed almost exclusively in the retina, is unrelated to any other known protein. In 1996, the OCA3 (brown) phenotype was associated with a mutation of the *TRP1* (*t*yrosine-*r*elated *p*rotein 1) gene on chromosome 9p. It encodes an enzyme (DHICA oxidase) that normally helps to form black melanin. Researchers have mapped at least ten other human genes that are homologous to known mouse pigmentation loci, but some of them are not associated with any known clinical conditions. Conversely, the loci of several albino phenotypes in humans remain unidentified.

Lesch-Nyhan Syndrome

In 1962, a seriously ill boy was referred to Johns Hopkins School of Medicine. There he was examined by a physician, William Nyhan, and by a medical student, Michael Lesch. The child suffered from various problems: extremely high concentrations of uric acid in the urine and blood, mental retardation, spasticity (excess muscle tension and awkward movements), choreoathetosis (constant and involuntary writhing movements of the legs and arms), and self-mutilation. After ruling out other conditions known to be associated with excess production of uric acid, Lesch and Nyhan concluded that they were seeing a new metabolic disease.

Boys* with *Lesch-Nyhan syndrome* (*LNS*) look healthy at birth and seem to develop normally for

several months. Nevertheless, early signs of anomaly are sometimes noted during the first weeks of life, such as the presence of orange "sand" (uric acid crystals) in the diapers or the occurrence of colic or uncontrolled vomiting. The latter may cause dehydration, which is followed by further concentration of uric acid, kidney stones, obstruction of the urinary tract, and other problems.

Within a few months, delays in motor development set in. Within a year, the muscles develop excess tone and the boy begins to exhibit uncontrollable writhing movements of the hands and feet. Involuntary spasms of the arms, legs, ankles, neck, and trunk prevent the boy from walking unassisted and often lead to dislocated hips or club feet. Feeding problems contribute to the patient's small size and frail condition and perhaps to the neurological symptoms as well. Poor muscle control also prohibits the child from speaking clearly or from being toilet trained. Seizures occur in about half of these children.

After 2 or 3 years of age, some boys begin to exhibit the most unusual trait associated with this disorder: compulsive biting of the fingers, lips, tongue, and inside of the mouth. Also typical is aggressive behavior toward others (hitting, pinching, swearing, and spitting). Patients with Lesch-Nyhan disease were sometimes classed as severely retarded, but some researchers suspect that their poor scores are partly due to difficulties in communication. These boys often seem alert, cheerful, responsive, and genuinely eager to control their destructive tendencies. The degree of self-mutilation varies considerably among patients and also in the same patient under different circumstances (Seegmiller 1997).

Treatment for LNS includes high fluid intake, adequate nutrition, and use of the drug allopurinol. The drug quite effectively reduces excess uric acid along with the kidney damage it causes. Unfortunately, the behavioral problems remain mostly untreatable, except by the use of physical restraints to prevent self-mutilation. Although survival into the 20s or 30s is known, most affected individuals die before then.

Biochemistry

Lesch and Nyhan determined that the basic error was a block in purine metabolism. As shown in Figure 16.11, the purine nucleotides needed for DNA and RNA production can be either synthesized from scratch or rescued from degraded nucleic acids and recycled by means of a salvage pathway. In unaffected humans this latter pathway works very efficiently, recycling 90% of free purines; but in affected persons, it fails to function, and excess purines are converted to uric acid.

Affected individuals show 0–1% of normal activity of an enzyme called *hypoxanthine-guanine phosphoribo-*

---

*Affected females are extraordinarily rare.

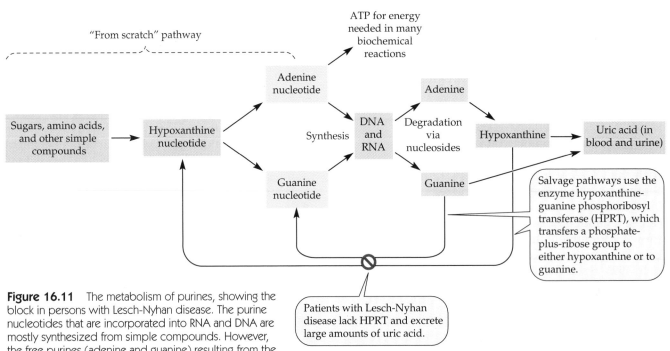

Figure 16.11 The metabolism of purines, showing the block in persons with Lesch-Nyhan disease. The purine nucleotides that are incorporated into RNA and DNA are mostly synthesized from simple compounds. However, the free purines (adenine and guanine) resulting from the degradation of old nucleic acids can be recycled. In this metabolic scheme, most arrows represent several chemical reactions.

syl transferase (*HPRT*). In unaffected individuals, this enzyme functions in all cells. It is normally most active in the brain, especially in *basal ganglia*—paired masses of gray matter that are embedded in the middle of the cerebrum and that control certain movements of skeletal muscles. (It is also highly active in testes, which usually atrophy in affected individuals.) No anatomical defects have been found in the brains of LNS patients, however. How the absence of HPRT can lead to such highly stereotyped behavior remains a mystery. Recent studies implicate an associated neurotransmitter deficiency—a greatly reduced density of dopamine-containing neurons or terminals in certain brain regions of affected individuals (Ernst et al. 1996; Nyhan and Wong 1996).

Partial deficiency of HPRT (1–20% activity) leads to a very severe form of gout, with high uric acid production. Affected persons (about 1 in 200 males with gout) have a normal life span and are not sterile. Although about 20% show some neurological symptoms (e.g., retardation, spasticity, or seizures), none of them self-mutilate.

Genetics and Molecular Biology

Total HPRT deficiency is inherited as an X-linked recessive. The trait is rare (1 in 100,000 to 380,000 births) and not associated with any particular ethnic groups. Due to lyonization, carrier females are mosaics. Although clinically normal, they may show some sub-

tle defects in purine metabolism, including increased levels of uric acid in their urine. Only three females with full-blown LNS are known. One resulted from a total deletion of the *HPRT* locus on one X chromosome and the nonrandom inactivation of the other X in all cells (Sculley et al. 1992).

In humans, the *HPRT* gene lies near the tip of Xq, closely linked to the fragile X and deutan loci. Although the gene is large, its actual coding region (nine exons) is of modest size and gives rise to a 218-amino-acid protein product. Sequence comparisons with mouse, rat, and Chinese hamster proteins reveal about a 95% homology among the four species.

Armed with the complete sequences for both the *HPRT* gene and the protein, researchers can analyze specific mutations in persons with Lesch-Nyhan disease or gout. Over 70 different mutations have been identified, and the changes are highly diverse. Indeed, many are unique, occurring only in a single family. Roughly 88% are point mutations or very tiny frameshift mutations; only about 12% are major deletions or rearrangements (Rossiter and Caskey 1995).

Carrier Detection and Prenatal Diagnosis

Because the HPRT enzyme is detectable in all cells of the body, including amnion and chorion cells, prenatal diagnosis by amniocentesis or chorionic villus sampling is an option for women known to carry a defective allele. The occurrence of mild forms of HPRT deficiency complicates the process of prenatal diagnosis, however. Great care must be taken to distinguish fetuses affected with Lesch-Nyhan disease from those with *partial* HPRT deficiency, because termination of pregnancy for the latter type is unwarranted.

Recent work with HPRT-deficient (*HPRT⁻*) mice has raised the possibility of diagnosing *HPRT⁻* embryos within a few days of conception (i.e., before implantation). Researchers were able to detect the absence of HPRT activity in single mouse cells removed from the eight-cell preimplantation embryos derived from carrier female mice. Experimental procedures like these have already been extended to humans for the preimplantation diagnosis of several inherited disorders, as well as for sex determination (Chapter 19).

Animal Models and Possible Gene Therapy

No naturally occurring animals with phenotypes corresponding to the Lesch-Nyhan syndrome are known, but researchers can now study genetically engineered *HPRT*-deficient mice. The production of *HPRT⁻* alleles can be accomplished by chemical or retroviral means or by the selection of preexisting spontaneous mutants. Scientists have also used some clever tricks to substitute *HPRT⁻* alleles for normal *HPRT* alleles, and to inactivate normal alleles. Such *transgenic* mice and *knockout* mice (Chapter 17) allow researchers to incorporate almost any biochemical mutant into laboratory mice, making it possible to study many other inborn errors of metabolism known to afflict humans. The results are not always predictable, however. For example, although experimentally produced *HPRT⁻* mice show no enzyme activity whatsoever, they express neither gout nor the serious behavioral abnormalities associated with LNS. The most likely reason is that mice, unlike humans, probably have alternative metabolic pathways that can offset their HPRT deficiency.

The devastating nature of LNS in humans, the lack of effective treatment for its neurological symptoms, and the special characteristics of the *HPRT* gene (i.e., being expressed in all cells, being a reasonable size, and being readily selectable) make this disorder a good candidate for *somatic cell gene therapy*. The technical and ethical problems associated with this issue will be discussed in Chapter 20. Here we will only mention that a few preliminary steps have been taken. Some cultured cells from LNS patients have been "corrected" in vitro by introduction of *HPRT⁺* DNA. Some cultured and genetically corrected bone marrow cells were transplanted back into one LNS patient; they apparently "took," but no improvement in behavioral symptoms was seen (Rossiter and Caskey 1995).

Congenital Adrenal Hyperplasia

Some cases of male and female pseudohermaphroditism (Chapter 15) have been traced to defects in the synthesis of testosterone in the testes or of additional androgens (male hormones) in the adrenal glands, or both. In these patients, the adrenal glands grow excessively large in an attempt to compensate for the deficiency; hence the name *congenital adrenal hyperplasia* (*CAH*). (*Hyperplasia* refers to excess production and growth of normal cells in a tissue or organ.) CAH is a group of diseases.

Biochemistry and Phenotypes

Recessive autosomal mutations affecting about half a dozen different enzymes are known, each blocking a specific step in the multibranched biochemical network. These metabolic pathways are summarized very simply as follows:

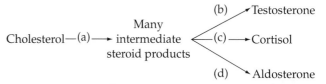

Testosterone is the principal androgen controlling the development of male sex organs, secondary sexual characteristics, sperm, and body growth. The hormones *cortisol* and *aldosterone* control, respectively, sugar and salt balance in the body. Cortisol also acts as an inhibitor, regulating the amount of a substance that stimulates the adrenal cortex to produce precursor substances. Thus, a deficiency of cortisol leads to an overproduction of hormone precursors and an accumulation of certain hormone products.

If step (a) is blocked in a 46,XY fetus, no testosterone is produced and a gonadal male with female genitals—a *male pseudohermaphrodite*—will result. These infants also suffer severe loss of salt and water because of the absence of cortisol and aldosterone. They may die within a few weeks.

If the mutant block occurs somewhere in the intermediate steps of the pathway, there may be a biochemical detour around it that results in the production of some testosterone. Hence, the genitals of these male pseudohermaphrodites may be *partly* masculinized. In such cases, high blood pressure rather than extreme salt and water loss is usually observed. Early diagnosis is important so that the deficient hormones can be prescribed; following a decision on whether to raise the infant as male or female, the genitals can be surgically corrected.

When the enzyme block occurs at (c) or (d) in a fetus, some of the backed-up intermediates are shunted into the testosterone branch, and an excess of male hormone is produced. The result is excess virilization in affected males and females. In fact, 90–95% of all cases of CAH are caused by a deficiency of the enzyme *21-hydroxylase* operating in one or both of these pathways (Donohoue et al. 1995).

The worldwide frequency of *21-hydroxylase deficiency* is about 1 in 10,000 to 18,000 births. It is the

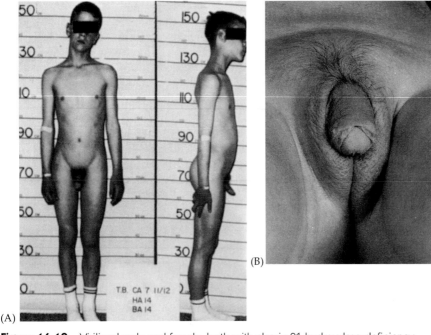

Figure 16.12 Virilized male and female, both with classic 21-hydroxylase deficiency. (A) This 7-year-old 46,XY male has the type of congenital adrenal hyperplasia that leads to an excess of testosterone and precocious puberty (including pubic hair and enlarged penis), but no enlargement of the testes. His height is already about 155 cm (5 feet, 1 inch). (From Bongiovanni et al. 1967.) (B) The ambiguous genitals (enlarged clitoris/phallus), pubic hair, and pigmented labioscrotal folds in a 26-month-old 46,XX female pseudohermaphrodite. (From New and Levine 1973.)

most common cause of ambiguous genitals in newborn females (New et al. 1997). In the two rare *classic* forms (salt-wasting and simple virilizing), there is an early onset of virilization (Figure 16.12). The *nonclassic* form (attenuated, late onset) is much more frequent—found in up to 1% of all Caucasians and 3% of Jews of European origin. Worldwide, nonclassic 21-hydroxylase deficiency may be the most common autosomal recessive disorder known in humans. These cases are highly variable in phenotype and age of onset, the most striking effects occurring in females. Individuals are born without symptoms, but young children of either sex may show premature development of pubic hair. Later-onset cases may show no deficiency until late childhood or puberty. Symptoms may include advanced bone age (leading to short stature), severe acne, and excess hairiness. Men may have low sperm counts, and women may exhibit infrequent menstruation, male pattern baldness, and ovarian cysts. Only the external genitals and secondary sexual characteristics are affected, however. Internal sex organs (i.e., gonads and Wolffian or Müllerian duct systems) are entirely normal.

Genetics and Molecular Biology

The three known forms of 21-hydroxylase deficiency are inherited as autosomal recessives. The enzyme 21-

hydroxylase is actually one of a diverse group of *heme* proteins called *cytochrome P450*. Its encoding gene, *CYP21*, resides on chromosome 6p within the major histocompatibility complex *HLA* (Chapter 18) in a segment containing several other genes.* Most *CYP21* mutations result from the duplicated nature of this region. Sequence similarities among these loci can lead to mispairing and unequal crossing over, generating duplications and deletions in the meiotic products. About 20–25% of the *CYP21* mutations associated with classic 21-hydroxylase deficiency are deletions, and homozygosity for these deletions always results in the salt-wasting disorder. Another (poorly understood) process called *gene conversion* somehow transfers deleterious point mutations from the *CYP21P* pseudogene to the *CYP21* locus. Regular point mutations may also occur. Altogether, over 15 deleterious mutations of this gene have been identified.

For families harboring CAH, it is possible to detect heterozygote carriers as well as affected fetuses. In pregnancies at risk for CAH, prenatal treatment is started between weeks 5 and 9 of fetal development—that is, before a diagnosis can be made by amniocentesis or chorionic villus sampling (CVS). The pregnant female is given a glucocorticoid hormone that crosses the placenta and substitutes for the deficient cortisol. It also inhibits the production of excess adrenal androgens by the fetus, thereby reducing virilization. If the fetus is later found (by CVS or amniocentesis) to be unaffected, treatment is stopped.

Population screening of newborn babies has also been done, using the same drops of dried blood that are collected to test for PKU (Chapter 19). Affected babies are treated with a glucocorticoid, and in salt-losing cases with a mineralocorticoid. Affected 46,XX females who are successfully treated with the appropriate hormones and with surgery (if needed to correct ambiguous genitals) may lead normal lives that include childbearing.

---

*They include (see Figure 18.16) two related *C4* loci that produce substances called serum complement, as well as a nonfunctional copy (a *pseudogene*) of *CYP21*. Geneticists think that there was originally just one each of the *C4* and *CYP21* loci, but that a duplication occurred long ago, giving rise to two copies of each locus.

Pharmacogenetics and Ecogenetics

Box 16A presents one dramatic example of how undiagnosed metabolic disease can be mistaken for other disorders—or even for foul play. But not every inborn error of metabolism causes death, mental retardation, or even serious illness. In fact, an individual can be born with a metabolic defect and live for years, perhaps a lifetime, without any health problems. If the affected biochemical pathway is a relatively unimportant one or needed only under special circumstances, no ill effects may ever be noticed.

As technology introduces new substances into our environment, however, it is inevitable that some people will be unable to metabolize some of them because of the presence of previously unrecognized enzyme variants. In recent years, the discovery of more and more cases of this type has spawned new fields of study called **pharmacogenetics** and **ecogenetics**. These new areas of research deal with the genetic variation leading to differences in the ways that individuals metabolize drugs, foods, and other substances in their environment. The results of such studies will play an increasingly important role in medicine.

Glucose-6-Phosphate Dehydrogenase Deficiency

One example, known since ancient times (reviewed by Beutler 1993), is a common X-linked recessive inborn error of sugar metabolism involving abnormalities of the enzyme *glucose-6-phosphate dehydrogenase (G6PD)*. A few kinds of G6PD deficiency are of great medical interest—not only because of their striking effects, such as *acute hemolytic anemia* caused by the breakdown of red blood cells (*hemo-*, "blood," *-lytic*, "destruction"), but also because they are surprisingly common, affecting about 400 million people (mostly males) worldwide (Luzzatto and Mehta 1995). But virtually all the affected people live in (or come from) tropical or subtropical countries where malaria is endemic. Here the mutant alleles have some selective advantage because heterozygous females are more resistant to malaria than are $G6PD^+$ (unaffected) homozygotes.

The G6PD enzyme occurs in all cells of the body. In red blood cells, it functions only under special circumstances in a minor biochemical pathway, where it acts on glucose-6-phosphate. Coupled to this reaction is a second step (catalyzed by a different enzyme) that generates *reduced glutathione*. The two steps are summarized as follows:

$$\text{Glucose-6-phosphate} \xrightarrow{\text{G6PD}} \text{Intermediate} \xrightarrow{\substack{\text{Other} \\ \text{enzymes}}} \text{Reduced glutathione}$$

Reduced glutathione is necessary for maintaining the integrity of the cell membrane. But because this pathway is scarcely utilized, under usual conditions even a low activity of G6PD is sufficient to keep red blood cell membranes intact. Only in "emergency" situations, when reduced glutathione is rapidly depleted, does the need arise for a normal full-strength, quick-acting G6PD enzyme to reverse this reaction.

Rapid depletion of reduced glutathione can occur after infections or when certain drugs (such as the antimalarial drug *primaquine*) or foods (such as *fava beans*) are ingested. In persons with certain mutant

---

BOX 16A *Metabolic Illness and Criminal Justice*

In 1990, Patricia Stallings was convicted of first-degree murder for poisoning her baby and sentenced to life imprisonment. Her ordeal began in 1989, when she brought her critically ill 3-month-old son to a hospital in St. Louis. The attending physician suspected that the baby's symptoms (severe vomiting, gastric distress, and labored breathing) were caused by poisoning with ethylene glycol, a substance found in antifreeze. This diagnosis was confirmed by laboratory analysis of the infant's blood. The baby recovered from the crisis and was placed in a foster home. But after one parental visit, during which Stallings had fed her child a bottle of milk, he became acutely ill again and died. The hospital laboratory and a commercial laboratory then analyzed his blood and the empty milk bottle. After both labs reported the presence of ethylene glycol, Stallings was charged with murder and jailed to await trial.

Meanwhile, she had become pregnant again. While in custody she gave birth to another son, who was immediately placed in foster care. But about two weeks later, he began to exhibit the same severe symptoms that killed his brother. This time, however, the diagnosis was *methylmalonic acidemia (MMA)*—a rare, autosomal recessive disorder of amino acid metabolism. Thus, the possibility existed that her other child had died from MMA rather than from poisoning. Yet the crucial evidence was withheld during her trial, and she was convicted and sentenced to life imprisonment.

After hearing about the case, two St. Louis University experts in metabolic disorders decided to do their own analyses of the dead baby's blood and urine. (The incriminating bottle had disappeared.) All their tests for ethylene glycol were negative, but the samples gave clear evidence (i.e., large amounts of accumulated methylmalonic acid and other substances) that the child did have MMA. When these results were presented to the prosecuting attorney, he withdrew the charges and Patricia Stallings was released. But what about the original lab reports? When reviewed, all were found to be in error for one reason or another.

G6PD alleles, the low activity of the G6PD enzyme cannot meet the challenge: The concentration of reduced glutathione falls below required levels, and the red blood cells rupture. Jaundice in affected newborns is another consequence of G6PD deficiency. Depending on genetic and environmental circumstances, these attacks of hemolytic anemia vary from mild to acute. They can sometimes be fatal if blood transfusions are not given.

The locus encoding G6PD lies near the tip of Xq. Researchers have molecularly analyzed over 60 different mutations of this gene; virtually all are point mutations leading to a single amino acid change. Only two small deletions, and no major rearrangements, have been identified, "suggesting that complete absence of *G6PD* may be lethal" (Glader and Naumovski 1997). The active enzyme consists of either two or four identical polypeptides, each with 515 amino acids.

There are two perfectly *normal* alleles of the G6PD gene: *B*, which is common throughout the world, and *A*, which is confined to black Africans and their descendants. The amino acid sequences encoded by these two variant alleles are identical except at one position. *Favism* (acute hemolytic anemia caused by eating fava beans) is often but not exclusively associated with a mutation of the *B* allele known as the Mediterranean or *B⁻* allele. Despite its severe enzyme deficiency and clinical effects,* this mutant is extraordinarily common in Greece, Sardinia, northwest India, and southern Italy and among the Sephardic Jews of Israel. *Primaquine sensitivity* among blacks of African descent is associated with a mutation of the *A* allele known as *A⁻*. Partly because the enzyme's activity is somewhat higher than in the case of the *B⁻* mutant, it can maintain reduced glutathione in young blood cells but not in old ones. The frequency of the mutant allele is over 20% among males in many parts of Africa and about 10–15% among black American males. (Such high allele frequencies mean that some homozygous affected females also occur.)

The Porphyrias

Because their symptoms mimic so many other known conditions, the group of disorders known as the porphyrias may be misdiagnosed (T. Evans 1997; Marion 1995). One clue, in some cases, is a change in the color of the urine—to a reddish or brownish hue—when it is exposed to light and air. This is due to the excessive excretion of metabolic precursors of *heme*, the red-colored part of a hemoglobin molecule. (These precursors are toxic to the skin, liver, and nervous system.) Heme belongs to the *porphyrin* class of molecules, which forms the base of all respiratory pigments in animals and plants. The heme synthetic pathway involves eight enzymes, each encoded by a different gene. Enzymatic defects are known for each step except the first, giving rise to seven different forms of porphyria (McGovern et al. 1997). A few of these disorders, of which we discuss two, provide an exception to the general finding that enzyme deficiencies behave as recessives.

One of the most common porphyrias is *acute intermittent porphyria (AIP)*. People who carry a dominant allele for AIP may live for decades without any inkling of the potentially fatal condition they harbor. When given barbiturates (e.g., sleeping pills) and certain other sedatives or anesthetics, however, they may suffer attacks characterized by severe pains in the abdomen and limbs, muscular weakness, and mental instability. Sometimes abdominal surgery is performed, but when nothing is found amiss, the physician may suggest that the problem is entirely psychosomatic, especially in view of the patient's seemingly abnormal behavior. Ironically, barbiturate-type anesthetics may be used during the operation and more sedatives prescribed thereafter; life-threatening symptoms that include delirium, excruciating pain, and paralysis may follow. Other factors known to trigger these episodes include infections, increased levels of steroid hormones (especially estrogens), and severe dieting or starvation. Yet, for unknown reasons, about 90% of people who carry AIP mutations never suffer any ill effects.

For AIP, the basic metabolic error is reduced activity of the enzyme *porphobilinogen (PBG) deaminase*.* The gene resides on chromosome 11q and contains 14 exons. Its two separate promoters generate two primary transcripts, from which alternative splicing produces two distinct mRNAs. One is active in all tissues, while the other is active only in red blood cells. Over 100 different mutations are known (Puy et al. 1997), most occurring in only one or a few families. Although their defective enzymes exhibit only half the enzymatic activity shown by normal controls (because they have one fully functional allele and one nonfunctional allele), only 10% of mutant carriers ever express any symptoms of AIP. The increased use of barbiturates in recent years has led to the detection of AIP in many populations—notably in Scandinavia, Lapland, and the United Kingdom. AIP is usually expressed after puberty and more often in women than in men (Kappas et al. 1995). Treatment includes avoidance of triggering substances, a high-carbohydrate diet, and prompt treatment of infections; medical alert bracelets are recommended.

---

*In people with favism, reactions may also be triggered by certain infections as well as by sulfa drugs, chloramphenical, vitamin K, and aspirin. Other oxidizing substances, such as mothballs, can cause severe reactions too.

*Also known as *hydroxymethylbilane* (HMB) *synthase.*

Another autosomal dominant disorder, *variegate porphyria* (*VP*), is especially common (1 in 330) among whites in South Africa (Jenkins 1996). Affected adults, in addition to suffering acute drug-induced attacks like those already described, have a variety of skin problems. These include blistering and fragility of those areas exposed to sunlight, with frequent infection and scarring of the lesions, as well as excess hairiness and overpigmentation. Here the deficient enzyme is *protoporphyrinogen oxidase* (*PPO*). In 1995, the *PPO* gene was mapped to chromosome 1q; it contains 13 exons.

Several cases of genetic sleuthing on variegate porphyria are particularly interesting. Dean (1957) traced its inheritance in a huge South African kindred back to one very prolific seventeenth-century Dutch settler. Researchers have recently identified a specific mutation of *PPO* present in 26 out of 27 South African families with VP that they studied, including at least one family descended from the founder (Meissner et al. 1996). They suggest that this particular allele, whose enzyme product exhibits only 5% of normal activity, may predominate in South Africa.

Macalpine and Hunter (1969) reviewed the medical history of King George III, the "mad" British monarch who reigned in the late 1700s, and hypothesized that he had variegate porphyria. In addition to the mental and physical symptoms, they note references to the king's dark, discolored urine. Other descendants of Mary, Queen of Scots, also suffered intermittent "colics," progressive paralysis, and madness. Among the sufferers was her son, King James VI, who claimed to have urine the color of wine. These posthumous diagnoses have been challenged by some experts, but now that the *PPO* gene is known, it might be possible to test living relatives for the possible presence of an inherited mutation.

Other Drug Reactions

Dozens of examples are now known in which a drug or other factor that benefits most people can cause severe reactions or even death in some unlucky patients. Here we describe a few such drug reactions.

Succinylcholine Sensitivity.

The muscle relaxant *succinylcholine* is used before the administration of general anesthesia to allow the easy insertion of a tube into the windpipe. It is also used in patients undergoing electric shock treatment. Normally, the relaxing effect lasts only 2 to 3 minutes, because the enzyme *plasma cholinesterase* quickly inactivates the drug by splitting it apart. But about 1 in 2,000 whites of European extraction and about 1 in 50 Alaskan Eskimos is homozygous for autosomal recessive mutations of the cholinesterase gene. Because the variant enzymes with low activity are much less effective at inactivating the

drug, paralysis lasts for hours and may be fatal if the patient's breathing is not maintained by a respirator during that time. The *CHE1* gene, which maps to chromosome 3q, has four exons and three introns. The enzyme, whose normal function is unknown, consists of four subunits, each with 574 amino acids. It occurs in most tissues but is most active during early development (especially in brain and germ cells) and in nervous system tumors. About ten different allelic variants have been described.

Malignant Hyperthermia.

A rare but very dramatic and potentially fatal drug reaction can occur during general anesthesia. Susceptible patients develop muscle rigidity, *hyperthermia* (high fever), a high pulse rate, abnormal heartbeat rhythms, and other serious symptoms. Triggering agents include all halogenated ethers (i.e., those containing bromine, chlorine, fluorine or iodine) as well as the muscle relaxant succinylcholine, but a crisis may not occur every time a susceptible person is anesthetized. The death rate, once as high as 80%, is now about 10%, thanks to improved diagnosis plus treatment with a drug called dantrolene. *Malignant hyperthermia* (*MH*) is also known to occur in pigs, dogs, cats, and horses. In humans, where it was first recognized in 1960, estimates of its incidence range from 1 in 15,000 to 1 in 150,000. It is more common in the young than in older people, in males than in females, and in whites than in blacks or Asians (D. Evans 1997).

MH is not a single disorder. Although most cases are inherited as a dominant trait, some may be transmitted as a recessive trait, and about 20% of cases seem to be sporadic (i.e., nongenetic). The basic defect is abnormal regulation of calcium in skeletal muscle cells, but several different causes could lead to the same outcome. Indeed, two separate loci have already been identified. One of them, first studied in pigs, controls the *ryanodine receptor* (*RYR*) of skeletal muscle cells. This receptor acts as a channel for transporting calcium through membranes and is defective in people (and pigs) with MH (Kalow and Grant 1995). In 1990, the human *RYR1* gene was mapped to chromosome 19q. More than 10 mutations are known, but they account for only a minority of MH cases. In 1997, researchers identified a locus on chromosome 1q, called *CACNL1A3*, associated with MH in a French family. It encodes a different calcium channel in skeletal muscle (Hogan 1997). Other linkage studies indicate that a third MH locus might exist on chromosomes 17q, and the heterogeneity of this disorder suggests that additional loci might also be found.

Poor Metabolizers.

One of the most common enzyme deficiencies of humans occurs in so-called *poor metabolizers* (*PMs*). These people show adverse side effects to ordinary doses of over 30 commonly prescribed medications—including some β-blocker blood pressure drugs (such as *debrisoquine*), drugs for angina and abnormal heart rhythms, asthma medicines, cough medicines, and antidepressants. (Unaffected individuals handle debrisoquine 10 to 200 times more effectively.) PM frequencies vary considerably among different groups—for example, 2% of American blacks, 5–10% of Caucasians, and 19% of San Bushmen of Africa.

The recessively inherited deficiency is in a *cytochrome enzyme*, P4502D6. The P450 cytochromes are important liver enzymes that act on a broad array of substrates and play a major role in the detoxification of drugs and environmental pollutants. The *P450* gene superfamily includes about 60 different loci, of which just 6 may account for the metabolism of most drugs. The key gene associated with PM, called *CYP2D6*, resides on chromosome 22q. More than 16 allelic variants are known, many of which cause splicing errors. This great variability and high total frequency of mutant alleles can create major problems for drug companies and regulatory agencies as well as for patients and physicians. On the other hand, some poor metabolizers may show a reduced risk to certain cancers.

Isoniazid Sensitivity. In the 1950s, a common enzyme variant was discovered among tuberculosis patients being treated with the drug *isoniazid*. Many of them inactivated this drug so slowly that they required only a fraction of the normal dose. In fact, the standard dose produced toxic side effects. Here the difference in response is caused by a variant of the *N-acetyl-transferase* enzyme, which is inherited as an autosomal recessive trait. Its gene, called *NAT2*, is on chromosome 8p. About 25 alleles have been identified, three of which are extremely common. Homozygous recessive *slow acetylators* are widespread (40–70%) among Caucasians, but less common (10–20%) among Asian populations. In cases like this, where two or three gene variants are well established, the mutant alleles are presumed to have some selective value, such as the ability to better metabolize certain types of food or other substances in the environment.

Lung Disease and Ecogenetics

Emphysema means, literally, "inflated tissues." In pulmonary emphysema, the lungs become permanently inflated owing to loss of elasticity in the walls of their tiny air sacs, the alveoli. Affected people cannot exhale properly, and the exchange of oxygen and carbon dioxide across alveolar walls is reduced. As the disease progresses, damaged alveoli combine to form larger (and even less efficient) air sacs and also become infiltrated with fibrous tissue. Because of damage to alveolar capillaries, the heart must work much harder to pump blood through the lungs. Infections also occur more frequently. The patient becomes more and more incapacitated and dependent on extra oxygen.

What causes the alveoli to lose their elasticity? Environmental factors (especially smoking) have been implicated, and the inherited deficiency of an enzyme called α_1-*antitrypsin* (*AAT*) may also be a key element. AAT is a small glycoprotein (394 amino acids) produced in liver cells and secreted into the blood. From there it diffuses into body tissues, mainly the lungs, and functions as a *protease inhibitor*. It counterbalances the action of certain lysosomal *proteases* (intracellular enzymes that digest proteins), which in this case are released by infection-fighting white blood cells in the lungs. Proteases are useful in digesting bacteria and other dead cells. But they can also attack elastin fibers in the walls of the alveoli if they are not bound and inactivated by α_1-antitrypsin.

The AAT gene, called *PI* (for *protease inhibitor*), maps to chromosome 14q. This locus is highly polymorphic, with over 70 known alleles whose frequencies vary among different populations. Most of these variants (about 95% total) are normal ones, a group called *PI*M*. (The asterisk is part of the designated genotype.) The three most common *PI*M* allelic subtypes collectively occur in 98–100% of nonwhites and 82–96% of whites. The most clinically significant alleles are *PI*Z*, *PI*S*, and *PI*Q0*. The first two are point mutations. They produce defective polypeptides that tend to get hung up and form cytoplasmic inclusions in the liver cells, rather than being secreted into the bloodstream. The third type, a rare "null" mutant, makes no polypeptide at all.

PI*M/PI*Z heterozygotes produce about 57% of normal levels of AAT, usually enough to protect the lungs. But homozygotes for both the *PI*Z* and *PI*Q0* alleles suffer from severe lung disease. *PI*Z/PI*Z* homozygotes, who account for most people with classic AAT deficiency, produce only 15% of the normal amount of AAT. Their frequency is about 1 in 3,100 in Scandinavians and 1 in 6,700 in North American whites, but they are rare or absent in black and Asian populations (Cox 1995). Affected smokers show a much earlier onset of emphysema than do affected nonsmokers. Those who smoke as teenagers may be disabled in their 30s, and their life span may be 23 years less than that of nonsmokers (Pierce 1997). Some *PI*Z/PI*Z* homozygotes also develop liver disease, which may occasionally be serious enough to require a liver transplant.

All AAT disease phenotypes develop from an interplay between genotype and environment. Thus, AAT deficiency is an **ecogenetic** disorder, with cigarette smoking and other chemical irritants being major hazards for people having this inherited tendency to lung disease. As studies of disease susceptibilities become more sophisticated, scientists expect to find many more examples of ecogenetic traits.

Why Do Drug-Metabolizing Enzymes Exist?

Not all drug reactions are so striking as the ones just discussed, but there is no longer any doubt that metabolic variability must be taken into account in the dispensation of drugs. Garrod's prophesy of a chemical basis for "idiosyncrasies with regard to drugs and articles of food" is amply borne out. Indeed, even some differences in body odor may have a genetic basis (Box 16B).

But how did cytochromes and other drug-metabolizing enzymes (DMEs) arise? Since drug detoxification may actually account for less than 1% of all DME functions, we need to consider the other DME roles. Because DMEs existed very early in evolution, even before the emergence of plants and animals, they probably carried out some very important cellular functions. Scientists now think that with the evolution of plants and animals, DMEs also began to play a critical role in animal-plant "warfare" and/or cooperation (Nebert 1997):

> There is ample evidence that the evolution of DME genes and DME receptor genes in animals has occurred because of the interaction of animals with plants. ... For example, there was an "explosion" of new animal cytochrome P450 genes in the *CYP2* family ... when animals first came onto land and began ravaging stationary terrestrial plant forms. Presumably, plants had to defend themselves by evolving new genes (and thus new metabolites) to make them less palatable or more toxic, and animals responded by evolving new DME genes in order to adapt to the constantly changing plants. ... The synthesis and degradation of plant metabolites by animal DMEs undoubtedly have evolved to the present-day metabolic activation and detoxification of innumerable environmental pollutants, carcinogens, and drugs.

Summary

1. Archibald Garrod first proposed the occurrence of gene-determined inborn errors of metabolism and described several examples in humans.

2. All metabolic events are mediated by enzymes under genetic control. Thus, a mutation in an enzyme-producing gene is likely to induce a metabolic block in one step of a biochemical pathway. The consequences of an excess of precursors or lack of end products depend on the pathway involved.

3. Enzymatic disorders are usually inherited as recessive traits, although many so-called homozygotes are actually compound heterozygotes, carrying two different mutant alleles of the same gene.

4. Tay-Sachs disease is an autosomal recessive error of lipid metabolism caused by deficiency of the Hex-A enzyme. Resulting accumulation of a certain lipid in nerve cells leads to progressive mental deterioration, blindness, paralysis, and death in early childhood. Tay-Sachs disease occurs in all ethnic groups, but mostly among Ashkenazi Jews and French Canadians.

BOX 16B *The Fish-Odor Syndrome*

"What have we here? a man or a fish? dead or alive? A fish: he smells like a fish; a very ancient and fish-like smell; a kind of, not of the newest, Poor-John.* A strange fish!" (William Shakespeare, 1611)

So cried the jester Trinculo when he discovered the savage and deformed slave Caliban, in Shakespeare's *The Tempest* (II.ii.24–27). This quote could also apply to some unfortunate individuals who actually smell like rotting fish or, to some people, like stale urine (Mitchell 1996). They suffer from an unusual condition called *trimethylaminuria*, more simply known as the *fish-odor syndrome*. Although they show no other clinical symptoms, their extremely offensive body odor usually makes them social outcasts from an

*A Poor-John is a salted, dried hake (a fish related to Atlantic cod).

early age, experiencing ridicule, rejection, isolation, and employment difficulties: "Affected persons may be deeply disturbed, depressed, and even suicidal, with psychosocial problems in school" (McKusick et al. 1997).

The syndrome results from a defect in the liver enzyme FMO3 (*f*lavin-containing *m*ono-*o*xygenase), which normally converts the smelly protein trimethylamine to nonsmelly trimethylamine *N*-oxide. The trimethylamine is produced by bacteria in the gut, acting on breakdown products from the digestion of certain substances, including fish and other seafoods, eggs, liver, soybeans, milk, and choline- or lecithin-containing health foods or drugs. In affected individuals, the unprocessed trimethylamine is excreted in their breath, sweat, and urine. Although the resulting stench cannot be completely elimi-

nated, it may be reduced by avoidance of the trimethylamine-producing substances, by using acidic soaps and lotions, and perhaps by intermittent suppression of gut bacteria through limited antibiotic treatment. In addition, psychiatric counseling may help affected people to deal with their serious social and psychological problems.

The trait is inherited as an autosomal recessive, with the *FMO3* locus residing on chromosome 1q. Homozygotes for the mutant allele are affected, and heterozygotes exhibit reduced levels of the FMO3 enzyme activity. In 1997, Dolphin et al. sequenced the coding regions of the *FMO3* DNA from an affected individual, identifying a single missense mutation possessed by all affected members of that pedigree.

5. Classic phenylketonuria (PKU), a defect of amino acid metabolism, leads to profound mental retardation unless it is detected near the time of birth and treated with a low-phenylalanine diet. Properly treated, these children have intelligence in the normal range. Many alleles are known; they differ among ethnic groups and in clinical effects.

6. Women with PKU who become pregnant while not on the restrictive diet give birth to babies who suffer brain damage from the high levels of phenylalanine in the mother's blood. Unless women with PKU return to dietary therapy before pregnancy, the success in preventing mental retardation by treating homozygotes will be totally offset by maternal PKU occurring in their children.

7. Albinism is the inherited absence of melanin pigments. All albinos have visual problems as a result of melanin deficiencies in the eyes; most have pigment deficiencies in the skin and inner ear too. The classic tyrosinase-negative oculocutaneous albinism (OCA1) is caused by the deficiency of an enzyme that normally converts tyrosine to DOPA. Tyrosinase-positive oculocutaneous albinism (OCA2) may be associated with faulty transport of tyrosine. About a dozen other types of albinism exist.

8. Boys with Lesch-Nyhan syndrome (an X-linked recessive disorder of purine metabolism) are retarded, lack motor control, and exhibit stereotyped biting behavior. Partial HPRT deficiency leads to a milder condition characterized by severe gout but no retardation or behavioral problems. Mutations of the *HPRT* gene are very heterogeneous.

9. There are numerous types of male and female pseudohermaphroditism. The most common inherited forms are collectively known as congenital adrenal hyperplasia. Most cases of CAH are caused by a deficiency of the enzyme 21-hydroxylase. Different mutant alleles, in various combinations, lead to different forms of CAH.

10. Pharmacogenetics and ecogenetics deal with inherited enzyme variants that lead to individual differences in the metabolism of drugs, diet, or other environmental substances. In particular, inefficient inactivation of a drug can lead to serious illness.

11. G6PD deficiency is an X-linked recessive defect of sugar metabolism that causes hemolytic anemia when affected individuals eat fava beans or take certain drugs, such as primaquine. *G6PD* mutant alleles are most common in populations where malaria occurs because heterozygotes are more resistant to malaria than are unaffected homozygotes.

12. The porphyrias, errors of porphyrin metabolism, involve sensitivity to barbiturates and certain other sedatives. Porphyrias are inherited as autosomal dominants—an exception to the rule that the mutations causing enzyme disorders are usually recessive.

13. Some other pharmacogenetic examples include succinylcholine sensitivity, malignant hyperthermia, poor metabolizers, and isoniazid sensitivity.

14. Some kinds of lung disease are ecogenetic disorders, involving an interplay between environmental factors and inherited deficiencies of the enzyme α_1-antitrypsin.

Key Terms

compound heterozygote
ecogenetics
inborn error of metabolism
melanocyte
metabolic block
one gene–one polypeptide
pharmacogenetics

Questions

1. Sometimes the intermediate substance just preceding a metabolic block is *not* excreted in excess. Suggest an explanation.

2. In a progress report on an early newborn screening program in Massachusetts, researchers state that 27 cases of phenylketonuria were detected out of 217,752 babies tested. This frequency of roughly 1 in 8,000 is considerably higher than that estimated for the country as a whole. Suggest an explanation.

3. A yellow variant of albinism phenotypically resembles tyrosinase-positive albinism, but affected individuals have a distinctly yellowish cast to the hair and the ability to tan slightly when exposed to sunlight. In addition, the plucked hair follicles of these individuals give equivocal results with the so-called "hair bulb test." After incubation with tyrosine, some hair bulbs show a small amount of pigmentation, but others do not. After incubation with cysteine (along with tyrosine and DOPA), some yellow pigment is formed, but no black pigment develops. Refer to Figure 16.7 and suggest where the biochemical block may occur in this mutant.

4. Females who are heterozygous at the *G6PD* locus nevertheless show levels of enzyme activity that range from near-deficiency to within the normal range. Why?

5. Most children up to the age of 2 years are able to digest the milk sugar lactose because they have large amounts of the enzyme *lactase*, which splits lactose into glucose and galactose. But after the age of 2, most of the world's people no longer produce much of this enzyme and thus become lactose intolerant. Lactose-intolerant children who continue to drink milk may become malnourished and perhaps even die. (Eating cheese and yogurt presents no problem, however.) The vast majority of northern Europeans, white Americans of European descent, and a few dairying tribes in Africa possess a mutation that causes the continued production of lactase through adulthood. But 70% of U.S. blacks and nearly all Africans and Asians are lactose intolerant. Comment on the usefulness of aid programs that involve the shipping of large amounts of dried milk to Africa and Asia.

6. Hunter syndrome and Hurler syndrome are recessive lysosomal storage disorders of mucopolysaccharide metabolism. When researchers put cultured Hunter syndrome cells in medium from Hurler syndrome cell cultures, or Hurler syndrome cells in medium from the Hunter syndrome cell cultures, both metabolic defects are corrected. What does this suggest about possible allelism of the two defective mutants?

7. Females heterozygous for HPRT deficiency are mosaics for the expression of this enzyme's activity. Some of their fibroblast cells and hair follicles test positive for this enzyme

and some test negative. However, when red or white blood cells are sampled from the same females, all show normal HPRT activity. Suggest a reason.

8. The occurrence of several biochemical mutants involving the same metabolic pathway allows researchers to determine the relative positions of individual steps in a pathway. The general principles are: (1) The substance just prior to a block accumulates, and (2) for organisms like bacteria and fungi, the mutant organisms cannot grow on any substances prior to the block but will grow if provided with substances just after the block. For example, the following table summarizes the growth responses of several different mutant strains of the bacterium *Salmonella typhimurium* in several different media. Wildtype *Salmonella* can grow on minimal medium, but none of the four mutant strains can. In this table, + means growth and 0 means no growth. Diagram the correct order of all the steps in this biosynthetic pathway, indicating the location of the metabolic block imposed by each mutation. *Hint:* The fewer substances a mutant can grow on, the further along is its metabolic block.

Further Reading

The classic book by Harris (1980) very lucidly presents the general principles of human biochemical genetics. Rimoin et al. (1997) and Scriver et al. (1995) are gold mines of information, containing dozens of detailed chapters written by experts on all types of metabolic disorders. Clarke (1996) and Fernandes et al. (1995) provide shorter guides to inherited metabolic diseases. Nebert (1997) explores the interesting question of why drug-metabolizing enzymes exist. Motulsky (1960) discusses metabolic variants and the role of infectious diseases in human evolution. On pharmacogenetics and ecogenetics, see Weber (1997) and Grandjean (1991). T. Evans (1997) provides an account of one family's terrifying experiences with porphyria.

			Growth response			
				Minimal medium plus		
Mutant strain	Accumulated product	Minimal medium	Indole	Anthranilic acid (Ant)	Tryptophan (Trp)	Indole glycerol phosphatase (IGP)
2	Ant	0	+	0	+	+
6	Indole	0	0	0	+	0
3	IGP	0	+	0	+	0
8	None of the 4 compounds	0	+	+	+	+

9. Sachse et al. (1997) studied the cytochrome P450 *CYP2D6* alleles associated with the metabolism of debrisoquine (and perhaps 25% of all drugs) in a Caucasian population. They found that just five different mutations accounted for all the poor metabolizers (7% PM) among the 589 unrelated people in their sample. They suggest that it would be useful to test for poor metabolizers by genotyping patients for the most common PM alleles, so that "individual differences in drug disposition could be compensated for in part by dose adjustment." What do you think are the advantages and disadvantages of this plan?

10. In addition to those *CYP2D6* alleles that give rise to normally active metabolizers and poor metabolizers, a few alleles are associated with increased metabolic activity. How should drug dosages be adjusted for these "ultrarapid" metabolizers?

11. Some people who are poor metabolizers or slow acetylators may show increased rates of some types of cancer. Offer an explanation for this effect.

Because of scientific, social, political
concerns, the race to clone a gene
became an exciting and

CHAPTER 17

The Genetic Basis of Cancer

In the 1970s and 1980s, First Ladies Betty Ford and Nancy Reagan, former child star Shirley Temple Black, and feminist Gloria Steinem, among others, courageously broke the taboo on a hush-hush subject. They stepped forward to help demystify breast cancer by talking about the ordeal of their own mastectomies (Wallis 1991). In the 1990s, other well-known people have continued our awareness of cancer by publicizing their affliction. Washington DC Mayor Marion Barry, Gulf War General H. Norman Schwarzkopf, Senator Robert Dole, and golfer Arnold Palmer have discussed their bouts with prostate cancer (Trafford 1995). Skater and Olympic gold medalist Scott Hamilton has spoken of his testicular cancer. Willing to give up privacy, these people have helped soften the stigma of cancer for others. They have encouraged many people to have regular checkups, to explore treatment options, to compare treatment centers, and to get second opinions. They have fostered openness and a more optimistic view of cancer progression.

Is this more hopeful outlook warranted? An official United States War on Cancer began in 1971 (Box 17A). But many of the returns on this multibillion dollar investment—especially in prevention—still lie in the future, for the toll from all types of cancer combined has remained high and substantially unchanged from 30 years ago (Bailar and Gornik 1997; Stolberg 1998). True, the general trend of cancer mortality has been downward in the 1990s (about –0.5% per year), but the trend had been upward in the 1970s and 1980s (about +0.4% per year). Mortality rates from some specific cancers have been decreasing steadily owing to earlier detection (uterine and cervical cancer) or causes still being discussed (cancers of the stomach and colon/rectum). Notably, the death rate due to cancer among children has decreased over 50% since 1970 due to better treatments. But since cancer is primarily a disease of aging, the improved outlook for children (fewer than 2,000 deaths per year) does not affect national mortality statistics very much. The increases in the incidence of skin cancer and prostate cancer (among other types) and the huge increase in lung cancer (first noticed among men in the 1930s and among women in the 1960s) have overshadowed the decreases. The tally of new cases of all forms of cancer among Americans is now about 1.4 million annually (Table 17.1). Deaths from cancer number about 560,000 per year and account for about 23% of all deaths—more than any other cause of death except heart disease.

Among women, the biggest killers are (in order) cancer of the lung, breast,* and colon/rectum (Figure 17.1). Lung cancer overtook breast cancer as a cause of death in women in 1987 and increases its sorry grip yearly. Among men, the biggest killers are (in order) cancer of the lung, prostate, and colon/rectum. Because the progression of prostate cancer is slow, the mortality rates have been much less dramatic than its incidence.

*Despite most women's great fear of breast cancer, they are eight to nine times more likely to die of heart disease (44,000 deaths per year from breast cancer versus 376,000 deaths from heart disease).

Of the more than 200 different kinds of human cancer, about 85% are **carcinomas**, solid tumors derived from epithelial tissues—skin, linings of the respiratory, digestive, urinary, and genital systems, plus various glands, the breasts, and nervous tissue. Carcinomas of the prostate, breast, lung, and colon/rectum account for more than half of all human cancers in the United States (top four rows in Table 17.1). Also classified as solid tumors, **sarcomas** (2% of all cancers) are composed of closely packed cells derived from bone, cartilage, muscle, and fat. Two types of malignancies affect white blood cells. **Leukemias** (4% of all malignancies) are cancers of the white blood cells manufactured in the bone marrow, and **lymphomas** (4% of all cancers) affect the white blood cells present in the spleen and lymph nodes. These and some other cancer types are listed in Table 17.2.

Cancerous Cells

Although most forms of cancer show little tendency to run in families, the diseases are nonetheless primarily genetic. This apparent contradiction stems from the

TABLE 17.1 Estimated number of new cancer cases and deaths for major cancer sites in the United States in 1997

Cancer site	Number of cases	Number of deaths	Some major causes
All sites	1,382,400	560,000	
Prostate	334,500	41,800	Testosterone (estrogen?)
Breast	181,600	44,200	Ovarian hormones
Lung	178,100	160,400	Tobacco
Colon/rectum	131,200	54,900	Animal fat, low fiber, (alcohol?)
Lymphoma	61,100	25,300	Unknown
Bladder	54,500	11,700	Tobacco
Melanoma (skin)	40,300	7,300	Ultraviolet light (sunburning)
Uterus	34,900	6,000	Estrogen
Mouth and throat	30,800	8,400	Tobacco, alcohol
Kidney	28,800	11,300	Tobacco (pain relievers? diuretics?)
Leukemia	28,300	21,300	X-rays
Pancreas	27,600	28,100	Tobacco
Ovary	26,800	14,200	Ovulation
Stomach	22,400	14,000	Salt, tobacco
Brain/nerves	17,600	13,200	Trauma, X-rays
Thyroid	16,100	1,200	(Iodine excess?)
Liver	13,600	12,400	Hepatitis virus, alcohol (tobacco?)
Esophagus	12,500	11,500	Alcohol, tobacco

Source: Parker et al. 1997.

BOX 17A *The Beginnings of the War on Cancer*

The National Cancer Act of 1971 was ceremoniously signed into law during the Christmas season by President Richard M. Nixon before 100 guests in the State Dining Room. Popularly called the War on Cancer, the legislation built on people's fears of the dreadful and far-reaching disease. It promised enhanced resources for research and patient care and a more direct link between the National Cancer Institute and the White House. Several months before, columnist Ann Landers had urged her readers to write their Washington representatives to support an attack on cancer, and they had responded with an unprecedented avalanche of mail on Capitol Hill. With the 1972 presidential elec-

tion approaching, it did no harm for Nixon to preempt the cancer issue, however humanitarian his motives may have been.

Expectations for the conquest of cancer within the decade ran high (Greenberg 1986, 1991). After all, well-funded national projects had allowed Americans to build an atomic bomb in the 1940s and to walk on the moon in the 1960s. Congress thus resolved that a similar undertaking should eliminate cancer by 1976, America's 200th anniversary. But cancer proved much too powerful an opponent, and the timetable was completely unrealistic. The major obstacle to victory was a lack of understanding of the biology of the disease and a lack of emphasis

on the requisite basic research. Instead, early priorities were on treatment by the standard techniques of surgery, radiation therapy, and chemotherapy. Although there have been some successful campaigns in dealing with childhood cancers, these afflictions account for but a small part of the total cancer burden. The war will eventually turn more decisively against cancer when greater knowledge of biological processes and environmental factors help provide clearer targets and more effective weapons. This chapter is a general survey of the battleground.

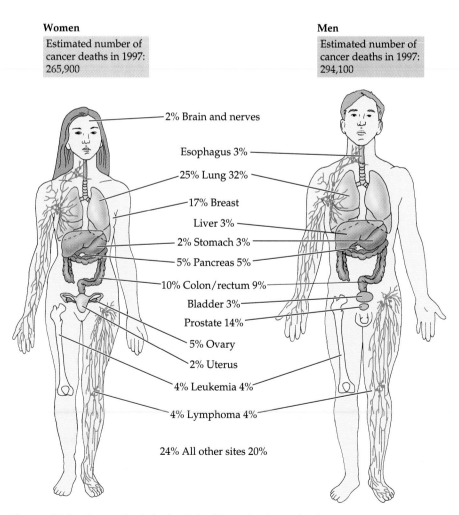

Women

Estimated number of cancer deaths in 1997: 265,900

Men

Estimated number of cancer deaths in 1997: 294,100

2% Brain and nerves

Esophagus 3%

25% Lung 32%

17% Breast

Liver 3%

2% Stomach 3%

5% Pancreas 5%

10% Colon/rectum 9%

Bladder 3%

Prostate 14%

5% Ovary

2% Uterus

4% Leukemia 4%

4% Lymphoma 4%

24% All other sites 20%

Figure 17.1 Cancer deaths in the United Sates for the ten leading sites in each sex (1997 estimates). (Data from Parker et al. 1997.)

following fact. Rather than being transmitted in germ cells, the genetic damage is usually confined to a person's *somatic* cells. The somatic mutations that lead to cancer, like germinal mutations that lead to typical inherited disorders, may be induced by environmental agents such as radiation and chemicals. The resulting damage to the DNA of somatic cells causes the orderly succession of mitotic divisions to go awry. Three classes of genes are involved in cancer: *tumor suppresser genes*, *proto-oncogenes*, and *DNA repair genes*. We discuss the first two in this chapter. (DNA repair syndromes were discussed in Chapter 7.) Researchers have begun to unravel how the proteins encoded by these genes regulate—or fail to regulate—cell growth. Through new insights into the genetic control of the cell division cycle, researchers hope to eventually provide much better methods for preventing, diagnosing, and treating cancer.

Unlike normal cells injured by infectious diseases, cancerous cells are quite healthy by the usual standards—so much so that

TABLE 17.2 Some cancers and the tissues that are affected

Cancer	Affected organ or tissue
Adenocarcinoma	Glandular structures of many organs
Angiosarcoma	Blood vessels
Astrocytoma	Supporting cells of brain
Carcinoma	Epithelial tissues: skin and linings of organs
Chondrosarcoma	Cartilage
Glioma	Supporting cells of nervous system
Hepatoma	Liver
Leukemia	White blood cells in bone marrow
Lymphoma	White blood cells in spleen and lymph nodes
Melanoma	Melanin-forming cells of skin
Meningioma	Covering of brain and spinal cord
Myeloma	Bone marrow
Nephroblastoma	Kidney
Neuroblastoma	Embryonic cells of nervous tissue
Osteosarcoma	Bone (often the femur)
Retinoblastoma	Retina of eye
Sarcoma	Connective tissues: bone, cartilage, muscle
Schwannoma	Sheath of nerve fibers

they interfere with the well-being of neighboring tissue. A mass of cells that multiplies when it should not is called a *tumor* or *neoplasm* (a term meaning "new form"). A *benign tumor* remains localized in the place of origin, often separated from the surrounding cells by a layer of connective tissue. (Common warts, for example, are benign skin tumors occurring primarily in children and young adults.) Although compact and slow growing, benign tumors may eventually contain billions of cells, become quite large, and, if internal, cause damage by pressing against a vital organ. They do not, however, spread to other sites or recur following surgical removal. In benign tumor cells, the chromosomal makeup is usually completely normal.

The cells of a **malignant tumor**, or **cancer**,* however, have the following characteristics:

• They do not respond to the mechanisms that normally restrict cell numbers to those required for growth or replacement. Like the growth of benign tumors, the proliferation of cancer cells may be slow, but it continues unabated.

• Cancer cells become disorganized and may revert to a more primitive form, losing their specialized structures and functions. Pathologists look for these characteristic changes under the microscope to distinguish benign from malignant tumors.

• Cancer cells are invasive, damaging adjacent tissues and often producing internal bleeding. They may

Cancer in Latin means "crab." The ancient physician Galen thought that the distended veins in a breast cancer looked like the extended legs of a crab. The crab reference was used even earlier by the "father of medicine," Hippocrates (460–370 B.C.). He thought that invasive growths of a cancer appeared to grasp surrounding tissues.

also be shed from the primary site and circulate through the blood and lymph to other locations in the body, where they initiate secondary cancers. The spreading to distant sites, called **metastasis**, poses a severe problem in treatment. Deaths of cancer patients often result from the destruction of vital organs at secondary sites or from overwhelming infections due to general debilitation and damage to the immune system.

• Cancer cells almost always have abnormal karyotypes. These include translocations, inversions, deletions, isochromosomes, monosomies, and extra chromosomes—the latter sometimes far beyond the normal diploid number. Specific chromosomal aberrations are sometimes associated with particular cancers.

Tumor cells pass through stages as they go from relatively benign to highly malignant. This process of **tumor progression** may follow the order of characteristics given in the preceding list; that is, cells first divide without proper control, then become disordered and less distinctive, and then invade surrounding tissue and metastasize. These steps may be accompanied by increasing chromosomal aneuploidy or by changes in genes that code for proteins that regulate growth and division.

The cells in a tumor—perhaps billions of them—are a *clone* of cells; that is, they all descend from one ancestral body cell that over time accumulates a critical load of mutations. The normal alleles of these genes keep the cell cycle under control in complex and redundant ways. Thus, when one gene mutates (or is deleted), there are other normal genes in the cell that provide backup, preventing faulty cell proliferation. For example, when certain control systems go awry, the cells are programmed to die rather than divide. (See the discussion of apoptosis in Chapter 15.) As additional mutations accumulate in the same line of cells, however, the backup plans also go awry, and the ability of the cells to manage timely cell division erodes. A cell population evolves with greater and greater genetic damage and fewer and fewer intact systems at its disposal to prevent runaway growth (Figure 17.2). Eventually, mutations overwhelm what once might have seemed a fail-safe arrangement, and the cell takes the path of unmistakable cancerous growth (Cavenee and White 1995; Weinberg 1996a).

In our discussion of specific cancers, we emphasize the following points:

1. Most mutations predisposing to cancer are confined to somatic cells and so are not transmitted from generation to generation; that is, most cancers—about 80%—occur in a hit-or-miss *noninherited* fashion and are called *sporadic*.

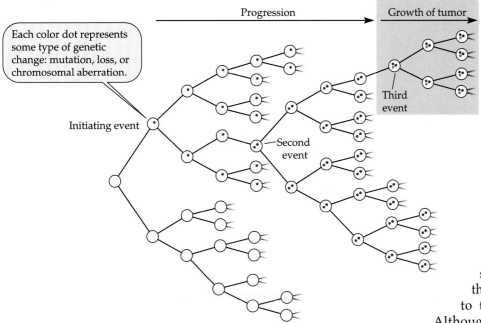

Figure 17.2 The cellular pathway by which accumulation of genetic damage leads to cancerous growth. Disturbing orderly cell divisions in some way, each genetic change acts in either a dominant or recessive fashion to promote runaway proliferation. In the figure, three changes are assumed to produce cancerous growth, but the actual number of such accumulated changes may vary.

2. Some mutations predisposing to cancer are transmitted in eggs and sperm; that is, some cancers—about 20%—are *familial* or *inherited* cancers.

3. The sporadic and familial cases sometimes result from the same mutant alleles. For that reason, intensive study of rare familial cancers can help identify the genetic changes responsible for the more frequent sporadic cancers.

Mutations of Tumor Suppressor Genes

Some of the mutations that prod cells along the path toward uncontrolled growth are recessive and some are dominant. We discuss first those cancer-predisposing mutations that act *within cells* in a recessive fashion. The normal dominant alleles of such genes are called **tumor suppressor genes**, and the nomenclature is summarized as follows:

Normal tumor suppressor gene (dominant) ——Mutation or loss (inactivation)——▸ Abnormal mutant or missing allele (recessive)

The presence of one normal allele is adequate for the control of cell division. Thus, both copies of the tumor suppressor gene must be nonfunctional or missing (as

a result of chromosomal deletion) for cancerous growth to begin. The two nonfunctional alleles may be identical (true homozygosity) or they may be different (e.g., one allele may have a point mutation while the other may have a deletion). The latter situation is often called *compound heterozygosity*.

About 20 human tumor suppressor genes have been isolated and cloned (Table 17.3). The first such gene to be studied in detail was the one that predisposes young children to the eye tumor *retinoblastoma*. Although retinoblastoma was historically important in understanding cancer, we discuss first the most commonly mutated tumor suppressor, the *p53* gene. Although no one gene is mutated in all human cancers, mutations of *p53* have been detected (as one of several mutational steps) in about half of them, especially those of the lung and colon/rectum. The *p53* gene has been shown to be a mutagenic target in the cancerous cells of people who smoke. More than 3,000 researchers worldwide are publishing paper after paper on *p53* and its protein product, which was chosen by *Science* magazine as the 1993 Molecule of the Year (Koshland 1993; Levine 1995; Marx 1993; Rawls 1997).

Li-Fraumeni Syndrome and Gene *p53*

In 1969, two researchers at the National Cancer Institute, Frederick Li and Joseph Fraumeni, described a syndrome in which individuals are extremely susceptible to not just one type of cancer but to a great many: cancers of the breast, brain, bone, white blood cells, as well as other sarcomas and carcinomas. The cancers usually develop in children and young adults, and if the victims survive a first bout with one type of cancer, they often are afflicted with a second type. The devastating *Li-Fraumeni syndrome* runs in families in an unmistakable autosomal dominant fashion (Figure 17.3). The disease is very rare, only about 100 families being known worldwide.

In a different line of investigation in 1979, cancer investigators David Lane of the University of Dundee (Scotland) and Arnold Levine of Princeton University discovered a tumor suppressor gene located in band p13 of chromosome 17. It was called *p53* because it coded for a protein with an approximate molecular weight of 53,000. Mutations of the *p53* gene (when

TABLE 17.3 Some tumor suppressor genes

Gene	Chromosome	Inherited cancer	Sporadic cancers
APC	5	Familial adenomatous polyposis	Carcinomas of colon/rectum, pancreas, and stomach
BRCA1	17	Breast, ovary	(None known)
BRCA2	13	Breast	(None known)
DCC	18	(None known)	Colorectal carcinoma
NF1	17	Neurofibromatosis type 1	Colon carcinoma; astrocytoma
NF2	22	Neurofibromatosis type 2	Schwannoma; meningioma
p16 or MTS1	9	Melanoma	Melanoma; brain tumors; leukemias; carcinomas of bladder, breast, kidney, lung, and ovary; sarcomas
p53	17	Li-Fraumeni syndrome	Carcinomas of bladder, breast, colon/rectum, esophagus, liver, lung, and ovary; brain tumors; osteosarcoma; lymphomas and leukemias
PTEN or MMAC1	10	Cowden syndrome	Gliomas (brain tumors); carcinomas of breast, kidney, and prostate
RB1	13	Retinoblastoma	Retinoblastoma; carcinomas of bladder, breast, esophagus, and lung; osteosarcoma
VHL	3	von Hippel-Lindau syndrome	Carcinoma of kidney
WT1	11	Wilms tumor	Wilms tumor (nephroblastoma)

Source: Adapted from Cooper 1995.

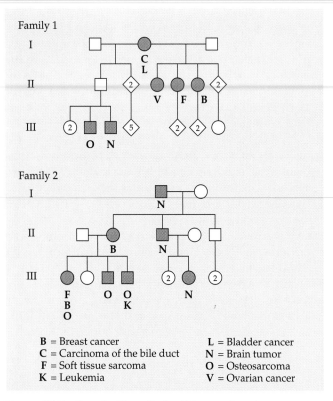

B = Breast cancer
C = Carcinoma of the bile duct
F = Soft tissue sarcoma
K = Leukemia
L = Bladder cancer
N = Brain tumor
O = Osteosarcoma
V = Ovarian cancer

Figure 17.3 Two families with the Li-Fraumeni syndrome, showing autosomal dominant inheritance (Malkin et al. 1990). As noted on the pedigrees, affected individuals had a great variety of cancers. In family 1, the age of onset of the cancers was from 16 to 53; in family 2, from age 2 to 38.

they became homozygous in a somatic cell) were later shown to be associated with many sporadic (noninherited) cancers of the lung, breast, colon, esophagus, bladder, brain, bone, and white blood cells. Since these sporadic cancers matched some of the common ones in Li-Fraumeni patients, *p53* became a so-called *candidate gene*, one that could possibly account for the cancers in Li-Fraumeni pedigrees.

With this clue, researchers showed that patients with the Li-Fraumeni syndrome do indeed have a single germ line mutation of the *p53* gene. To demonstrate this, they extracted DNA from normal and cancerous cells of patients with Li-Fraumeni syndrome, amplified the *p53* gene by the polymerase chain reaction, and sequenced the amplified product to identify any mutations. Affected individuals in family 1 (Figure 17.3), for example, were shown to have in all their somatic cells a particular base substitution mutation in one copy of the *p53* gene. That mutation is referred to as the "germ line" mutation, as it was already present in the germ line of the parent from whom it was inherited, and it is similarly present in the patient's germ line. The other *p53* allele was missing altogether in tumorous tissues as a consequence of deletion from somatic cells. In family 2, the germ line mutation was a different base substitution.

Thus, the *p53* mutations are present in both cancer-prone families with the Li-Fraumeni syndrome and in many sporadic cancers. In the *familial* cancer families, the *p53* mutation on one chromosome 17 is transmit-

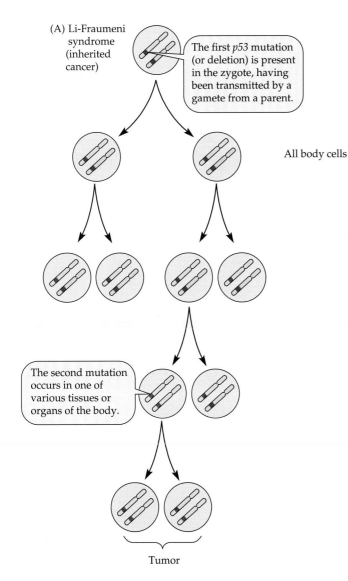

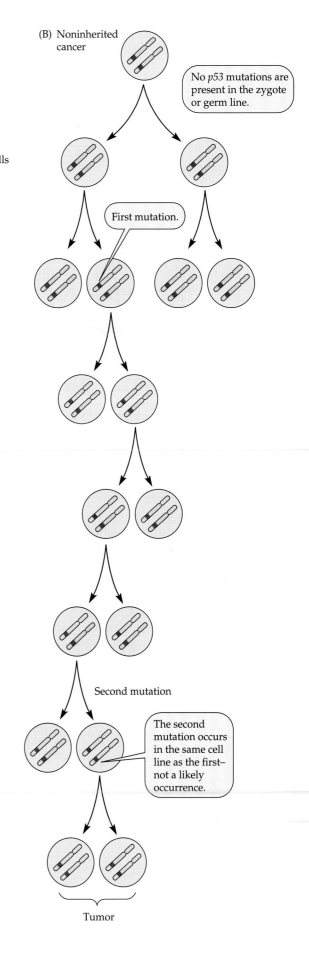

Figure 17.4 The two mutational events (red bars) in inherited and noninherited cancer involving the tumor suppressor gene *p53*. (A) Li-Fraumeni syndrome. (B) A noninherited form of cancer.

ted to a family member in an egg or sperm and copied to all cells of his or her body. The existence of this first mutation predisposes to cancer. A particular cancer actually develops when a second *p53* mutation occurs later in life in one or another tissue of the body (Figure 17.4A). In the *sporadic* cancer cases, both *p53* alleles are normal in the zygote and both normal alleles have to be inactivated for cancer to develop (Figure 17.4B). The low probability that both alleles will have mutational events is the reason for the sporadic nature of these *p53*-derived cancers.

The use of the terms *recessive* and *dominant* here is confusing but instructive. The Li-Fraumeni syndrome is transmitted between generations in a *dominant* fashion, but the expression of the *p53* gene is *recessive* at the cell level. This oddity is because the germ line mu-

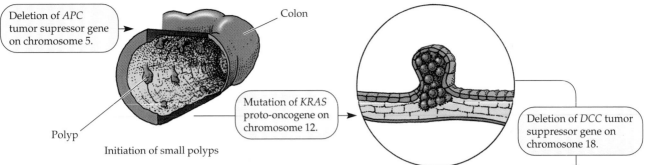

Growth of benign polyps

Growth of a larger tumor

Carcinoma

Blood vessels

Metastasis

tation is inherited by half the offspring of an affected parent. Such individuals develop cancer with high probability (about 95%), producing a pattern of dominant inheritance (but with an occasional skip due to incomplete penetrance). The germ line mutation (in heterozygous state) is reproduced in all the cells of the offspring's body, both somatic and germ line. The person develops cancer, however, only through a somatic mutation of the *other* allele. Both alleles of *p53* must be inactivated for cancer to develop, indicative of recessive expression.

In line with the general notion that several genetic changes are needed for the development of cancer, the *p53* mutation is not the only one required for the development of cancer. The stepwise progression in noninherited colon cancers has been studied in detail by Bert Vogelstein and colleagues at the Johns Hopkins Oncology Center in Baltimore and by others (Radetsky 1991). As illustrated in Figure 17.5, mutations of proto-oncogenes (discussed shortly) and tumor suppressor genes can be correlated with the clinical progression of tissue changes leading to malignant colon tumors. A half dozen genetic mutations have been discovered that are needed to produce colon cancer.

The first such change leading to noninherited colon cancer is a small deletion in chromosome 5. This site also appears to harbor a mutation that leads to an inherited form of colon cancer called *familial adenomatous polyposis*. In this dominantly inherited disease, which strikes about 1 in 5,000 persons, thousands of tiny blebs grow in the lining of the colon. These benign polyps may progress to malignant growths if the colon is not removed. Thus, like other types of cancer due to mutations of tumor suppressor genes, colon cancer exists in more common sporadic forms (accounting for perhaps 80% of colon cancers) as well as less common inherited forms.

Figure 17.5 Stepwise changes in the development of sporadic cancers of the colon. Three of the genetic changes involve tumor suppressor genes and one involves a mutation to an oncogene. The *p53* mutation occurs near the end of the progression. *APC*, adenomatous polyposis coli; *DCC*, deleted in colon cancer.

How Normal p53 Works. The product of the normal *p53* allele is an average-sized protein with 393 amino acids, but it plays a vital role in the cell cycle. Acting as a **transcription factor**, it binds to the regulatory region of a half dozen genes, causing them to be transcribed into RNA (i.e., turned on) and subsequently translated into protein. One of the genes that the p53 protein turns on encodes another protein called p21, which in turn binds to the cell cycle control proteins called **cyclins*** and inhibits their functioning. This last mechanism brings about a normal pause of the cell cycle just before DNA is synthesized at the S phase. Thus, through the p53 → p21 → cyclins pathway, the normal *p53* gene suppresses cell division, but at appropriate times and places.

Cells usually have a low concentration of p53 protein, which does not inhibit cell division. But in cells whose DNA is damaged (in regions other than *p53*), the p53 protein is found in increased amounts, leading to the chain of events outlined in the preceding paragraph and summarized in Figure 17.6. Several different types of DNA injury can activate the normal *p53* gene, including breaks in DNA due to harmful radiation or chemicals. The ensuing arrest of the cell cycle at the major G_1 → S checkpoint ensures that the damaged DNA is not replicated until it can be repaired. (Replication of unrepaired DNA would of course increase the frequency of mutations in subsequent cell generations.) When the damage is corrected, the levels of p53 and p21 proteins fall, the cell cycle proteins function, and mitosis continues. If there are no normal alleles at both *p53* loci, however, the check for DNA damage is omitted and mutations are transmitted to subsequent cell generations, increasing the risk of cancer. Hundreds of different *p53* mutations are known.

In combination with certain other mutations, the normal *p53* allele sometimes acts in a different way to suppress cancer. The alternative pathway is *apoptosis*, or programmed cell death, the orderly process whereby DNA degradation and cell shrinkage leads to elimination of the cell. (As noted in Chapter 15, apoptosis is a normal developmental process occurring, for example, in fashioning distinct fingers and toes free of the webbing present in early embryos.) Whether a cell line does not divide or simply dies, it is prevented from accumulating the mutations that may progress to cancer. The decision to stop cell division or to self-destruct is not well understood, although when DNA damage is greater, apoptosis is more likely to be triggered.

In summary, the *p53* gene is a major gatekeeper for cell growth and division. Through a chain of events

---

*A family of cyclin molecules helps regulate mitosis. The cyclins gradually increase in concentration during interphase and early mitosis and are destroyed during anaphase.

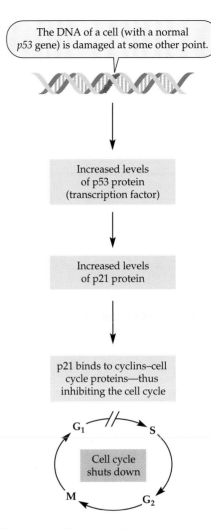

Figure 17.6 How p53 protein shuts down the replication of damaged DNA, which might otherwise contribute to the development of cancer.

occurring when DNA is damaged, the normal p53 protein suppresses division or kills the cell. The abnormal p53 protein, on the other hand, allows damaged DNA to be transmitted to subsequent cell generations, leading possibly to cancer. Another gatekeeper of the mitotic cycle that regulates the G_1 → S checkpoint is the protein product of the retinoblastoma gene (*RB1*), which we discuss next.

Retinoblastoma and Gene *RB1*

A cancer of the retina of one or both eyes, *retinoblastoma* occurs with a frequency of about 1 in 20,000 persons. This malignant tumor develops primarily in children younger than 3, because cancerous changes almost always start in *dividing* cells, and retinal cells stop dividing early in life. (The term *blast* means "immature cells.") Studies of cultured retinoblastoma cells suggest that the cancers develop specifically in the color-perceiving cone cells.

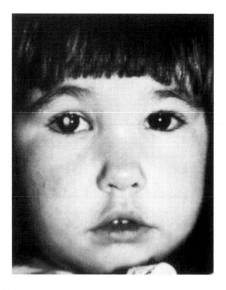

Figure 17.7 Light reflected off the surface of an eye tumor in a young girl with retinoblastoma. (Courtesy of B. L. Gallie, The Hospital for Sick Children, Toronto.)

A key symptom of the disease is the glassy appearance of the pupil of the eye in which the tumor is growing (Figure 17.7). If untreated, the cancer expands along the optic nerve to the brain; it may also metastasize to bone, liver, or other organs, eventually causing death. If the eye tumors are detected early enough, however, and the affected eye or eyes are removed by surgery or treated by cryotherapy, chemotherapy, and radiation, patients may live to adulthood, sometimes with partial vision (Angier 1987; Korf 1996).

Retinoblastoma was one of the first cancers shown to have a familial basis. It has a unusual pattern of occurrence: About 40% of cases run in families, while 60% of cases occur sporadically and are not inherited. Researchers usually find that inherited cases involve tumors in both eyes, whereas noninherited cases involve just one eye. The reasons for these patterns are similar to the reasons for familial and sporadic cancers induced by mutations of the *p53* gene.

Some of the *heritable* cases of retinoblastoma are accompanied by a small visible deletion in the long arm of one chromosome 13. These deletions vary in size, but they all include band q14, the site of a now isolated and cloned gene called *RB1*. Persons who inherit either the q14 deletion or a point mutation of *RB1* have it on just one chromosome 13 in all cells—cells that are not yet cancerous. Alfred Knudson of the Fox Chase Cancer Center in Philadelphia suggested in the 1970s that a second mutation in a retinal cell at the corresponding locus on the other chromosome 13 is the precipitating factor in the initiation of eye tumors. Thus, the expression of retinoblastoma at the cellular level is a *recessive* trait, requiring abnormal states at both of the relevant q14 sites on chromosome 13. Knudson's idea that two steps are needed to produce retinoblastoma—the so-called *two-hit hypothesis*—has been a model for explaining the general multistage development of cancer. Although in retinoblastoma the two hits are allelic, in many cancers (as noted for colon cancer in Figure 17.5) more than two hits are needed at different loci throughout the genome.

The likelihood of the second somatic mutation in inherited retinoblastoma is sufficiently high that several tumors often develop, thus explaining why these cases usually affect both eyes. Note that loss of function of both q14 sites on chromosome 13 can occur either via two mutations at the nucleotide level, two deletions, or one of each. In addition, somatic nondisjunction resulting in the loss of an entire chromosome 13 (or other cytological events) can produce the second "mutation" needed for malignancy. Although the expression of the retinoblastoma gene is *recessive* at the level of somatic cells, the disease is transmitted between generations in a *dominant* fashion, analogous to Li-Fraumeni syndrome illustrated in Figure 17.4A.

In persons with noninherited retinoblastoma, both chromosomes 13 in the zygote were normal at the relevant q14 locus (similar to Figure 17.4B). The mutational or cytological changes of the *RB1* loci must be present on *both* chromosomes 13 in the *same* retinal cell for retinoblastoma to develop. Such persons typically have a tumor in only one eye—a tumor that develops later in childhood, as expected by the very low probability of two rare events occurring in a single cell.

The DNA region including the *RB1* gene has been identified, cloned, and sequenced. It is a large gene with 27 exons, and it encodes a protein called RB with 928 amino acids (see Table 9.1). An important finding was that the number of phosphate groups attached to the normal RB protein varies in a set pattern during the cell cycle. There are relatively few during the G_1 period; in fact, the cell cycle will not proceed past the major $G_1 \rightarrow S$ checkpoint when the RB protein is in this unphosphorylated state. Thus, the normal RB protein acts as a major brake on cell division (Figure 17.8). When phosphate groups are added to RB, it releases a transcription factor that eventually turns on genes whose protein products are needed for DNA synthesis.

The *RB1* gene appears to be transcribed in cells throughout the body, and it is not clear why the mutant allele affects the retinal cone cells in particular. That other cells can also be targeted is suggested by the increased risk of osteosarcomas (bone cancers) among the survivors of retinoblastoma. In addition, some persons who do not get retinoblastoma but develop certain forms of breast, lung, and other cancers have structural abnormalities of the retinoblastoma gene. Thus, the *RB1* gene may play a key role in sus-

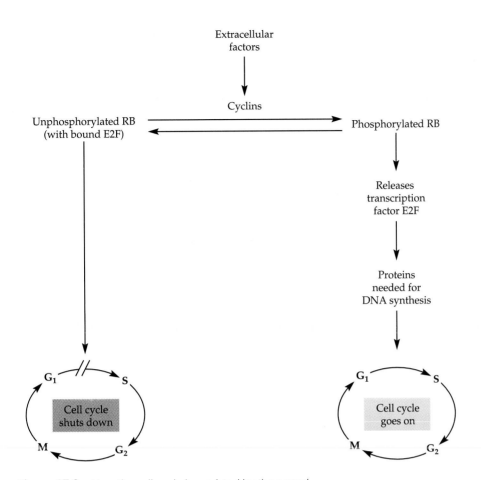

Figure 17.8 How the cell cycle is regulated by the normal retinoblastoma (RB) protein. When the protein is unphosphorylated (i.e., has few phosphate groups attached) it binds the transcription factor E2F and suppresses the cell cycle (left side). When the protein is phosphorylated (i.e., has many phosphates), it releases E2F, which turns on the genes that encode proteins needed for DNA synthesis (right side).

ceptibility to several forms of cancer, but other mutational changes are also required.

Another childhood cancer, called *Wilms tumor*, develops in the embryonic cells of one or both kidneys. Appearing in about 1 in 10,000 children under the age of 7, Wilms tumor is curable in 80% of cases by surgery, radiotherapy, and chemotherapy. Its pattern of occurrence and the types of genetic changes resemble those found in patients with retinoblastoma. For example, a noninherited form usually affects one kidney; a much less common inherited form due to a dominant germ line gene *WT1* may affect both kidneys. In a few patients with inherited Wilms tumor, a visible deletion occurs in all somatic cells in band p13 of one chromosome 11.* As with retinoblastoma, the

development of tumors is usually contingent on the occurrence of the genetic abnormality (deletion or point mutation) at both *WT1* loci on homologous chromosomes (Coppes et al. 1994). (Some cases of Wilms tumor are associated with genes other than *WT1*, but we will not discuss these complications here.)

The *WT1* Wilms tumor gene has been cloned, and the WT1 protein product is a transcription factor. Actually, there are several very similar protein products, because the *WT1* gene is transcribed in different ways due to alternative splicing sites for two of its exons. These curious so-called *isoforms* of the WT1 protein are not well understood. Unlike the protein products of *p53* and *RB1*, which are expressed in all cells, the WT1 proteins are expressed only in the kidney, the gonads, and a few other organs and tissues. The WT1 proteins belong to a class called **zinc-finger proteins**, in which zinc atoms bind to the protein and bend the molecule sharply into a finger shape (Rhodes and Klug 1993). The Wilms tumor proteins have four such fingers that interlock with the helical turns of DNA molecules (Figure 17.9), binding specifically to the sequence CGCCCCCGC. Interestingly, the WT1 protein may also form a physical complex with the p53 protein, modulating the effects of both transcription factors and making more complex their interactions with the cell cycle.

In summary, tumor suppressor genes such as *RB1*, *WT1*, and *p53* in their unmutated dominant forms act as brakes on cell division. Mutations of these genes to recessive alleles or their loss promote cancer in somatic cells only when they become homozygous and release the brake. Two complications are seen: (1) The mutations may sometimes occur in the germ line, predisposing several family members to a particular cancer; more usually, however, the mutations occur only in somatic tissue, producing noninherited, sporadic cases of the cancer. (2) In the familial cases, the inheritance from one generation to the next follows a dominant pattern; at the cell level, however, these cancers develop in a recessive fashion, that is, only when the cancer-promoting allele is homozygous.

*When the deletion is present, additional effects are often seen: absence of the irises of the eyes (aniridia), abnormalities of the genital and urinary systems, and mental retardation. It is presumed that the p13 deletion on chromosome 11 includes additional genes affecting these other aspects of the phenotype.

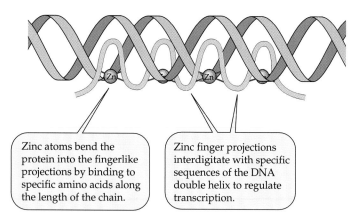

Zinc atoms bend the protein into the fingerlike projections by binding to specific amino acids along the length of the chain.

Zinc finger projections interdigitate with specific sequences of the DNA double helix to regulate transcription.

Figure 17.9   The product of the Wilms tumor gene is a zinc-finger protein with four loops.

Breast Cancer and Gene *BRCA1*

Because of scientific, social, political, and personal concerns, the race to clone a breast cancer susceptibility gene became an exciting and well publicized story. The race covered the period from 1990, when a breast cancer gene dubbed *BRCA1* was mapped to an approximate position on chromosome 17, to 1994, when *BRCA1* was isolated and cloned.

A key player in mapping the gene was Mary-Claire King, a pioneering mathematical geneticist then at the University of California, Berkeley, and now at the University of Washington, Seattle. Her dedicated 15-year quest began because she believed that some cases of breast cancer could be traced to a single gene. She also thought that breast cancer could not be prevented, but might be detected early enough by genetic

tests to allow effective therapy. A major obstacle to identifying a germ line gene that might influence the course of breast cancer was that many genes and many environmental factors (ionizing radiation, dietary components, chemical toxins) were possible causes, any one of which would likely account for just a small fraction of cases. It was known that about 90% of breast cancers were sporadic. They were sufficiently common that clustering in families could easily be due to chance events rather than being an indication of heredity. Nor could it be expected that a breast cancer gene is always completely penetrant, so inherited breast cancer might look sporadic. King called the situation an "epidemiological nightmare" (Roberts 1993). Furthermore, unlike the retinoblastoma or Wilms tumor genes, there were no visible chromosomal deletions to suggest possible locations. The mapping of a breast cancer gene was also begun before the advent of powerful molecular techniques such as DNA markers, and most investigators gave up.

Doggedly, King's team pushed on. Along the way, the 1985 invention of the polymerase chain reaction, which could amplify tiny amounts of DNA, allowed King to more fully analyze blood samples that were stored in her laboratory—samples taken from members of extended breast cancer families. She laboriously tested for genetic linkage of breast cancer to each of several hundred randomly selected DNA markers by the techniques described in Chapter 9. King found that band q21 of chromosome 17 (a region spanning many millions of bases) seemed to be significant in some breast cancer families but not in others (Figure 17.10). Good evidence of linkage came after the recognition by King's group in 1990 that a germ line gene predisposing to breast cancer would likely lead to cases with *early onset*—before about age 45 (see Figure 9.4). This pattern, along with bilateral expression (cancer in both breasts), is similar to the action of other tumor suppressor genes, such as *RB1* and *WT1*. Evidence that convinced the last skeptic still had to wait until French cancer researcher Gilbert Lenoir published similar results on early-onset cases involving *both* breast and ovarian cancer, a combination characteristic of the *BRCA1* gene.

After 1990, with an approximate location known, a dozen laboratories in the United States, Canada, Britain, France, and Japan feverishly competed

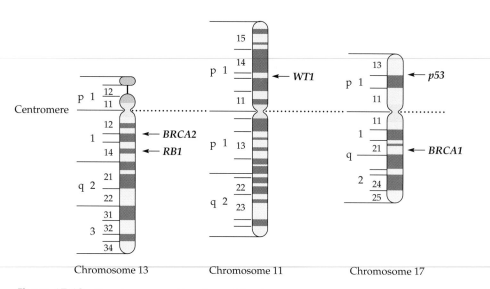

Figure 17.10   The chromosomal locations of five tumor suppressor genes: *p53*; the retinoblastoma gene, *RB1*; the Wilms tumor gene, *WT1*; the breast cancer genes *BRCA1* and *BRCA2*.

to isolate and clone *BRCA1* (Angier 1994). That so many groups jumped so quickly into this work testifies to the significance of King's mapping strike. The gene turned out to be well hidden in a difficult chromosomal region containing hundreds of genes. In the positional cloning contest, many egos and careers were on the line, and while the researchers were sometimes cooperative, they were sometimes secretive, especially as they got nearer the goal. The sentimental favorite to win the race was Mary-Claire King, who was collaborating with Francis Collins, the preeminent cloner of the cystic fibrosis gene and other genes. The victory did not go to King and Collins, however, but to Mark Skolnick of the University of Utah, a genetic epidemiologist, genealogist, and cofounder of Myriad Genetics, a biotechnology company devoted to isolating genes related to major health problems (Davies and White 1996). Skolnick had access to exceptionally informative Mormon family records, which he himself had computerized in the 1970s, crosslinking them to the official registry of cancer cases covering the whole state of Utah. Skolnick was also a coauthor of the influential 1980 article that drew attention to the remarkable potential of RFLPs (and later other DNA markers) for gene mapping (Botstein et al. 1980).

Many researchers were disappointed not to be on the winner's stand, but they agreed that the close-knit work of Skolnick and his many collaborators was outstanding (Nowak 1994). Skolnick's group had used several novel methods for pulling the gene out from the region delineated by mapping studies. They cleanly showed that dominant germ line mutations were present in family members affected with early-onset breast/ovarian cancer and usually absent in those family members without the cancer. In cancerous tissue (breast or ovary) of pedigree members, both copies of *BRCA1* were mutated or, more usually, one was mutated and one was missing. These characteristics are similar to other tumor suppressor genes. Skolnick's plans to patent the gene and some dilemmas associated with testing for breast cancer genes are described in Chapter 9.

The *BRCA1* gene turns out to be a large one, spanning about 100,000 bases in 24 exons (see Table 9.1). Its protein product of 1,863 amino acids seems in part to be a tumor suppressor, which binds to DNA through a zinc-finger region near one end. Another part of the BRCA1 protein is involved in the repair of double-stranded breaks in DNA (Marx 1997). Researchers have discovered more than 100 different mutations of *BRCA1* spread out along the gene (Figure 17.11). Testing for them in breast cancer families is tedious, involving some indirect methods, but often requiring the direct sequencing of splicing regions or long stretches of exons. About 60% of these mutations are frameshifts, 20% are nonsense (chain termination) mutations, 10% are missense mutations, and 10% involve mutations in splice sites. The frequency of women in the general population who carry these germ line mutations is estimated to be about 1 in 500 (more or less depending on the study). Some ethnic groups have distinctive germ line mutations of *BRCA1*. For example, a strikingly high 1% of Ashkenazi Jews (those of eastern European descent) carry a particular two-base deletion in the second exon of *BRCA1*.

From your knowledge of the tumor suppressors *p53*, *RB1*, and *WT1*, you will understand the excitement that accompanied the cloning of the familial *BRCA1* gene. As with the others, it was expected that the much more frequent sporadic cases of breast cancer would be associated with mutations of *BRCA1*. Thus, researchers and clinicians counted on knowledge of the newly cloned gene to advance the diagnosis and perhaps the treatment of virtually *all* cases of

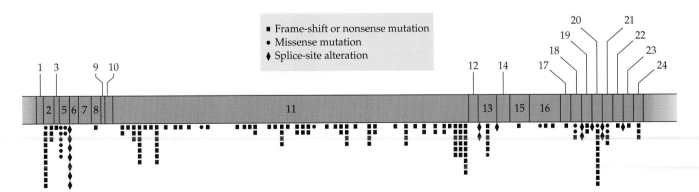

Figure 17.11 Germ line mutations spread out along the *BRCA1* gene. This diagram summarizes the various mutations found in all families that have been studied. Typically, only one of the sites is mutated for any mutant *BRCA1* allele. Numbers refer to exons of the gene. (After BIC 1997 and Collins 1996.)

breast cancer, not just the 3–5% due to germ line mutations of *BRCA1*.

Soon after cloning *BRCA1*, another 3–5% of breast cancer was shown to be due to germ line mutations of a different newly cloned gene, *BRCA2*, on chromosome 13 (Marx 1996). The *BRCA2* gene is not associated with ovarian cancer, but causes a breast cancer risk in men. A hypothetical *BRCA3* gene remains to be discovered, because about 30% of heritable breast cancers do not involve either *BRCA1* or *BRCA2* (Szabo and King 1997). In women who possess germ line mutations of *BRCA1*, breast cancer appears with about 85% probability, and ovarian cancer with about 50% probability, after the second "hit" occurs. In women who possess *BRCA2*, breast cancer appears with about 85% probability. (These probabilities are likely too high, because they are estimated from pedigrees having multiple patients.) Disappointingly, however, mutations of *BRCA1* and *BRCA2* do not usually appear in sporadic cases of breast cancer. Except to say the obvious, that other genes must be involved (*p53* is one of them), it is not known why. Some of these other mutations are also of the type of cancer-predisposing genes described in the next section.

Mutations of Proto-Oncogenes

Prior to the work on tumor suppressor genes, whose *recessive* mutations could lead to cancer, researchers recognized a class of genes whose *dominant* mutations could lead to cancer. The dominant mutant alleles were called **oncogenes** (Greek *onco*, "mass"). The normal recessive alleles were accordingly labeled **proto-oncogenes** (Greek *proto*, "first" or "early"):

Normal
proto-oncogene ——Mutation——→ Abnormal
(recessive)   (activation)   oncogene
                                 (dominant)

Oncogenes were first found in chickens and mice, although many identical or similar genes were later shown to be present in human cells too.

An important line of experimentation was carried out by Robert Weinberg (1983) and colleagues at the Massachusetts Institute of Technology. These researchers began with DNA extracted from cells of a human bladder carcinoma. Using this DNA, they were able to *transform* a special line of mouse fibroblast cells growing in tissue culture. In cancer research, **transformation** is the change of cultured cells from normal to malignant. During the transformation process, short sections of the human DNA had entered the mouse cells and had become inserted into the mouse DNA, although with low frequency (Figure 17.12). The transformed mouse cells were identified by their abnormal growth behavior: Rather than remaining in a flat layer one cell thick on the surface of the nutrient medium, they grew into a mound, or *focus*, of cells.

Weinberg was able to identify and clone the piece of human DNA—the oncogene—responsible for the transformation. This oncogene hybridized strongly with a DNA sequence in *normal* human cells—the presumed proto-oncogene. Further research showed that the nucleotide sequence of the bladder carcinoma oncogene is similar to that of a rodent oncogene called *ras* (for *rat* sarcoma) associated with a virus (specifically a *retrovirus*; see later). In humans, oncogenic mutations of three members of the *RAS** family are found in perhaps one-third of all human tumors, especially carcinomas of the bladder, pancreas, lung, and colon (see the second step in Figure 17.5). But there is no association of the human *RAS* genes with viruses.

*Gene symbols are usually lower case for mice, upper case for humans.

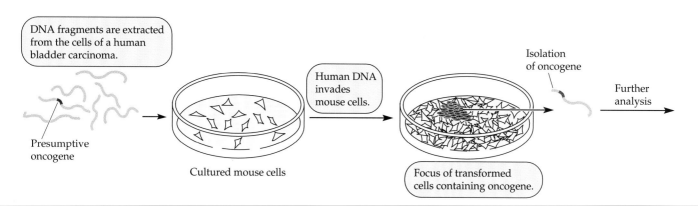

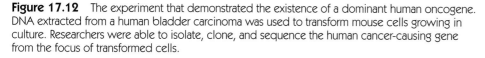

DNA fragments are extracted from the cells of a human bladder carcinoma.

Presumptive oncogene

Cultured mouse cells

Human DNA invades mouse cells.

Focus of transformed cells containing oncogene.

Isolation of oncogene

Further analysis

Figure 17.12 The experiment that demonstrated the existence of a dominant human oncogene. DNA extracted from a human bladder carcinoma was used to transform mouse cells growing in culture. Researchers were able to isolate, clone, and sequence the human cancer-causing gene from the focus of transformed cells.

By sequencing both the bladder carcinoma oncogene and its corresponding proto-oncogene, Weinberg and his colleagues (along with other investigative teams) were able to pinpoint the mutational event: a G in the *RAS* proto-oncogene became a T in the *RAS* oncogene. The encoded amino acid is glycine in the proto-oncogene product (the normal RAS protein) and valine in the oncogenic product (the altered RAS protein). Not all oncogenes result from simple base substitution mutations, however. Many are due to chromosomal changes (especially translocations in which the oncogene is at a breakpoint) or amplification of chromosomal pieces containing the oncogene.

About 70 oncogenes and their corresponding proto-oncogenes have been located in the genomes of humans and other animals by procedures involving DNA transformation, by studies of oncogenic retroviruses, and by other means (Table 17.4). As emphasized before, a single genetic change by itself is probably never sufficient to cause a cell to become cancerous. Several mutational events—some dominant, some recessive—are required for malignancy. Each misstep in the progression may foster additional abnormalities characteristic of cancer: cell proliferation, invasion of adjacent tissue, metastasis, and establishment of new tumorous sites.

The Proteins Encoded by Oncogenes

Often using the mouse as a convenient experimental model (Box 17B), researchers now have a good understanding of the functions of the protein products of many proto-oncogenes and their cancer-causing mutations. The encoded proteins act in information pathways that extend from the cell surface to nuclear genes—specifically those genes that control growth and division. This line of chemical communication from outside a cell across the plasma membrane, cytoplasm, and nuclear membrane is called **signal transduction** (Krontiris 1995, Saltiel 1995). When one or more of these chemical signals goes awry because of an oncogenic mutation, the coordination necessary for orderly cell division is disrupted, and uncontrolled growth of cells may result.

Beginning the process of signal transduction are proteins called **growth factors** that bind to receptor molecules on the surface of cells (step 1 in Figure 17.13). Growth factors encoded by normal proto-oncogenes are delivered to receptors at cell surfaces in a regulated manner, stimulating cell growth only at appropriate times. The oncogene-encoded protein may be produced continually or at inappropriate times or places. Other proto-oncogenes code for the **receptor proteins** (step 2 in Figure 17.13). The oncogenic form of a receptor may mislead the cell into thinking that the growth factor is bound to the receptor when it is not—similar to a switch being stuck in the "ON" position.

In the cell cytoplasm, the information pathway involves a number of **relay molecules** (step 3). For example, the signal from the receptor proteins may first pass to other molecules attached to the inner surface of the plasma membrane. Some of these are called *G proteins*, so called because they interact with the energy molecule GTP (guanosine triphosphate) rather than the more prevalent ATP (adenosine triphosphate) (Linder and Gilman 1992). The RAS proteins referred to earlier are similar to G proteins (Figure 17.14). In their active form, they pass a signal to additional relay molecules that traverse the cytoplasm. Although the

TABLE 17.4 Some human oncogenes

Oncogene	Chromosome	Some cancers implicated	Activating mechanism
ABL1	9	Chronic myelogenous leukemia; acute lymphocytic leukemia	Translocation
BCL2	18	B cell lymphoma	Translocation
ERBB2	17	Breast and ovarian carcinomas	Amplification
FOS	14	Osteosarcomas	Chromosomal rearrangements
GLI	12	Glioblastoma	Amplification
GNAI2	3	Adrenal cortical and ovarian carcinomas	Point mutation
HOX11	10	Acute T cell leukemia	Translocation
HRAS	11	Thyroid and bladder carcinomas	Point mutation
KRAS	12	Colon, lung, pancreas, and thyroid carcinomas	Point mutation
LYL1	19	Acute T cell leukemia	Translocation
MYC	8	Burkitt lymphoma; breast and lung carcinomas	Translocation or amplification
MYCL1	1	Lung carcinoma	Amplification
MYCN	2	Neuroblastoma, lung carcinoma	Amplification
NRAS	1	Acute myelogenous and lymphocytic leukemias; thyroid carcinoma	Point mutation
RET	10	Thyroid carcinoma	Chromosomal rearrangement

Source: From Cooper 1997 and OMIM 1997. The latter has 119 entries with the word *oncogene* in the title.

BOX 17B   *Tampering with Mouse Genes*

Laboratory strains of mice that have a foreign gene stably incorporated into their germ line DNA are called **transgenic mice**. The grafted transgene can be from another mouse strain, from an animal other than a mouse, or from laboratory tinkering. Examples of transgenic farm animals that secrete useful human drugs in their milk were discussed in Chapter 8. An example of transgenic mice is the strain carrying the human *HPRT* gene mentioned in Chapter 16. Studies of mice with transplanted *oncogenes* have been useful in exploring the multistep process leading to cancer. Transgenic mice that carry the *ras* oncogene, for example, develop cancer about 40% of the time. Those carrying the *myc* oncogene develop cancer about 10% of the time. But those carrying both oncogenes develop cancer 100% of the time.

To produce a transgenic mouse carrying an oncogene, researchers first clone in a bacterial plasmid the DNA sequences constituting the oncogene along with a promoter (Shuldiner 1996). A solution with this DNA is then injected via a thin glass needle (and skillful hands) into a pronucleus of a fertilized mouse egg, as shown in the photograph. The eggs that survive the microinjection are then surgically implanted into foster mothers. Some embryos that come to term will have incorporated the oncogene into their DNA at random sites in some of their somatic cells. In about 10% of births, the oncogene is also incorporated into the germ line and can thus be propagated by appropriate matings. As expected, the transplanted oncogene is dominant to the mouse's own proto-oncogenic alleles, causing tumors in a reproducible fashion. Several dozen strains of mice, each transgenic for a different oncogene, have been made and studied.

Researchers have also created mice that carry a specific recessive gene that is *homozygously inactivated*. Called

knockout mice, they are obtained through a more involved and amazing process (Capecchi 1994; Majzoub and Muglia 1996). (The University of Utah's pioneering investigator, Mario Capecchi, was refused research funds in 1980 because reviewers thought that his proposed goals were beyond belief.) The essence of the procedure (simplified here) involves a DNA construct that looks like this (in the case of the mouse *p53* gene):

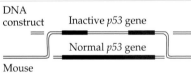

←Mouse *p53* gene→

| Long left homologous region | Insert of *neo* resistance gene | Long right homologous region |

The ends are homologous to portions of the mouse chromosome located to the left and right of its *p53* gene. In the middle of the *p53* gene (bold line), the researchers have inserted a gene (*neo*) that makes a cell resistant to the antibiotic neomycin. Of course, sticking this gene into the middle of *p53* inactivates *p53* at the same time. Early embryonic mouse cells (so-called *embryonic stem cells* or *ES cells*) growing in culture are treated with this DNA. Some ES cells take up the DNA construct after a brief electric shock. The DNA can then find and synapse with the corresponding region of the mouse chromosomes by virtue of the long homologous regions at the left and right ends. In about 1% of cells, double crossing over at the ends causes the inactivated gene to trade places with a *p53* gene in a normal mouse chromosome:

DNA construct   Inactive *p53* gene

Normal *p53* gene

Mouse chromosome

Somatic cells (in addition to cells in the first meiotic division) seem per-

fectly capable of this crossing over. The cells in which this exchange actually occurs can be selected because they will be resistant to neomycin, which kills the other cells in which the homologous replacement failed to occur. The short chromosome segment with the normal *p53* gene is lost from the cell.

The cells carrying the inactive *p53* gene (and *neo*) are then put into very young mouse embryos. Although the developing animals are chimeric—having some normal cells and some *p53*-inactivated ones—inbreeding the chimeric mice can achieve a line that is homozygous (in the germ line) for *p53*-inactivated genes.

It turns out that knockout mouse embryos having no functional *p53* genes make it to birth perfectly well, so it is inferred that *p53* is not needed for normal development. But as expected, by 6 months of age, they develop different kinds of cancers, similar to those found in patients with the Li-Fraumeni syndrome. Many hundreds of different knockout mouse strains have now been developed to study the action of a wide variety of genes with oncogenic, developmental, immunological, metabolic, or behavioral functions.

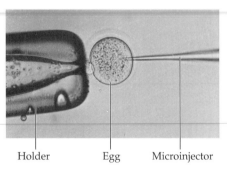

| Holder | Egg | Microinjector |

The egg is being held on the blunted tip of a glass micropipet (left) by slight suction. DNA solution will be microinjected into the egg from the glass needle on the right. (From Gordon and Ruddle 1983.)

RAS proteins encoded by the normal proto-oncogene switch back and forth appropriately between active and inactive forms, the RAS proteins encoded by the oncogenic alleles are stuck in the active form. Thus, they tell the cell to divide inappropriately. There may be several relay molecules in one signal transduction pathway, which can branch into separate tracks or intersect with other signaling systems.

The cascade of events continues into the nucleus. A number of oncogenic proteins have been shown to act in cell nuclei as **transcription factors** that turn genes on or off (step 4 in Figure 17.13). An example is the

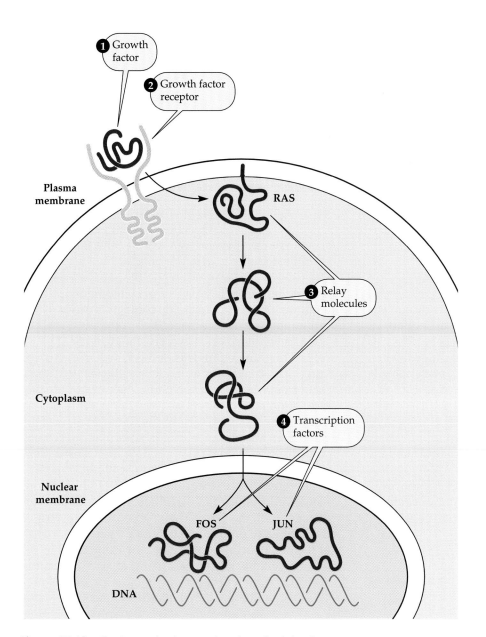

Figure 17.13 The four main elements (numbered) of signal transduction. The bold arrows trace the signaling pathway caused by successive protein phosphorylations.

The transfer alters the three-dimensional shape of the phosphorylated protein, increasing its activity. Signal transduction consists in some measure of one protein kinase activating a second, which activates a third, and so on.

In summary, carcinogenesis is at least partly precipitated by mutations from recessive proto-oncogenes to dominant oncogenes that alter signaling pathways. The action of a single dominant oncogene is to abnormally accelerate cell division. Although oncologists have an increasingly better understanding of the cascade of signals, these gains have not generally been translated into effective treatments for cancer.

Retroviruses

Because viruses have been key agents in understanding the genetic basis of malignancy, we discuss them briefly, although they have been implicated in only a few forms of human cancer. Viruses are, however, responsible for a wide range of human diseases other than cancer: AIDS, yellow fever, mumps, chicken pox, measles, rubella, polio, hepatitis, shingles, warts, mononucleosis, influenza, and common colds.

Many viruses consist of little more than a molecule of nucleic acid—DNA or RNA—surrounded by a protein coat. Recognition of certain coat proteins by receptors on the cells of their hosts determines what species, and what cell types, a virus can invade. Then by reproducing themselves inside the cells of a host organism, the viruses may disrupt the cell's normal metabolism and lead in some cases to cell death. Although not seen until the invention of the electron microscope, the presence of viruses was suspected early in the twentieth century when they were called *filterable agents*, particles so small that they could pass through the pores of filters developed by Pasteur to stop the passage of bacteria.

That some viruses can also be *carcinogenic*—that is, cancer-causing—was shown in 1911 by Peyton Rous while he was a young research assistant at Rockefeller

protein encoded by *JUN*, which interacts with specific DNA promoter sequences to regulate gene expression. The product of another proto-oncogene, *FOS*, binds with the *JUN* product to enhance its activity.

Many of the molecules in signal transduction pathways are **protein kinases** (Greek *kinase*, "to move"). The kinases act by adding phosphate groups to other proteins, a process called *phosphorylation*. (One specific type is a *tyrosine kinase*, in which the phosphate is added to tyrosine side chains of recipient proteins.) Usually, the phosphate comes from the energy molecule ATP (adenosine *tri*phosphate), which loses a phosphate to become ADP (adenosine *di*phosphate).

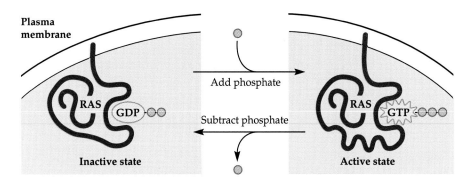

Figure 17.14 Regulation of RAS protein. RAS binds the guanine nucleotide as either GDP (with two phosphate groups; the inactive state) or GTP (three phosphates; the active state). In the active state, the RAS protein binds to another signaling protein.

Institute. Rous found a filterable agent in a ground-up chicken sarcoma that produced the same kind of tumor when injected into healthy chickens. Despite careful, well-designed experiments, his work was generally disregarded because it did not jibe with prevailing views on the causation of cancer. Not until the 1950s was it generally recognized that viruses could transmit some forms of cancer. In 1966, Rous, at age 86, was awarded a Nobel prize for his pioneering work started over a half century earlier (Dulbecco 1976).

The *Rous sarcoma virus* is one of a family of viruses called **retroviruses**, which have been implicated in many types of animal cancers. A retroviral genome consists of a short single-stranded molecule of RNA that codes for only about a half dozen proteins. One of the proteins is the enzyme **reverse transcriptase.** Immediately after entry of the retrovirus into a cell, this enzyme is used to reverse transcribe the RNA genome into a DNA copy (Temin 1972) (Figure 17.15). The DNA is inserted as a stable **provirus** at a random site in a host chromosome. During successive cell divisions of the host cells, the provirus is replicated along with the host DNA. The proviral DNA can also be transcribed (in the usual forward direction) into viral RNA and translated into viral proteins, which are packaged into complete viruses. These can escape and infect other cells.

Some retroviruses also carry an oncogene. The insertion of a single provirus of the Rous sarcoma virus into cultured cells, for example, is sufficient to change them from normal to malignant. The cancer-inducing gene of Rous sarcoma virus is called *src* (for *sarcoma*). The *src* oncogene present in the Rous sarcoma virus came from a proto-oncogene in a normal chicken cell some time in the past. This is how the transfer happens: When the provirus of an invading, nononcogenic retrovirus is inserted into the DNA of an infected cell (Figure 17.15), it may sit next to a proto-oncogene of the host. Then, when the provirus is transcribed to produce new virus particles, part or all of the neighboring proto-oncogene is included in the transcript. But the process is imperfect. Point mutations, deletions, or recombination events produce an altered allele—an oncogene—now part of the new retroviruses. The presence of proto-oncogenes in many different bird species suggest that they are essential to the regular metabolism of all avian cells.

Researchers have found scores of proto-oncogenes in

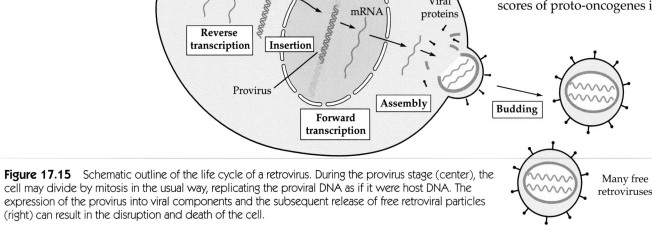

Figure 17.15 Schematic outline of the life cycle of a retrovirus. During the provirus stage (center), the cell may divide by mitosis in the usual way, replicating the proviral DNA as if it were host DNA. The expression of the provirus into viral components and the subsequent release of free retroviral particles (right) can result in the disruption and death of the cell.

birds and rodents that have been picked up by retroviruses to become viral oncogenes. Remarkably, almost all of the animal proto-oncogenes have homologous loci in the *human* genome, a few of which have already been mentioned (*RAS, FOS, JUN*) or will be in the next section (*ABL, MYC*). The DNA base sequence of a particular proto-oncogene is very similar whether taken from a fruit fly, bird, mouse, or human. The chromosomal sites of the human genes have been determined by somatic cell fusion or by in situ hybridization, methods described in Chapter 9. Note, however, that the human proto-oncogenes and their oncogenic mutations are not usually associated with viruses.

Chromosomal Changes

For many years, cancerous tissues have been known to harbor chromosomes that are abnormal—in structure or in number. The karyotypes of the cancer cells often become more and more abnormal in apparently haphazard ways over the course of the disease, a trend that usually signals an increasingly poor prognosis for patients. But from the time of inception of the malignancy, some *specific* aberrant chromosomes are consistently present in some *specific* cancers (Heim and Mitelman 1995; Rabbitts 1994). Thus, a structural change in a chromosome seems, in itself, to contribute to cancer induction. This discovery was first made in 1960 at the University of Pennsylvania in Philadelphia and focused on the role of the so-called *Philadelphia chromosome* in leukemia. This aberration has been shown to affect the activity of the proto-oncogene located at one of the breakpoints of a reciprocal translocation.

Proto-Oncogenes at Translocation Breakpoints

Leukemias are cancers of immature white blood cells. Due in some measure to ionizing radiation, leukemia syndromes occur with increased incidence among patients who receive large doses of therapeutic radiation, among radiologists not sufficiently protected from X-radiation, and among the people exposed to the atomic bomb explosions. In *myelogenous leukemia*, the cancer proliferates from stem cells (myeloblasts) in bone marrow. In *lymphocytic leukemia*, the cancer proliferates from stem cells (lymphoblasts) in lymph glands that mature into B and T immune cells (Chapter 18). Additional leukemias arise as a result of the abnormal maturation of other types of white blood cells.

In the bone marrow, the massive buildup of leukemic cells crowds out mature red blood cells (producing anemia) and platelets (producing internal bleeding). In addition, the liver, spleen, and other vital organs are damaged by large numbers of infiltrating cancer cells. Advance of the disease can either be slow (*chronic*) or rapid (*acute*), the latter type leading to death in a few weeks or months if untreated. A frequent cause of death in acute leukemias is uncontrolled infection due to the deficiency of mature white blood cells.

Chronic myelogenous leukemia (*CML*) affects mainly people of middle and older ages and is invariably fatal. It is easily recognized because the leukemic cells usually have a distinctive chromosomal marker, the so-called **Philadelphia chromosome**, which looks like a shortened chromosome 22 and was once thought to result from a simple deletion. The introduction of banding techniques, however, revealed that end pieces of the long arms of chromosomes 9 and 22—unequal in size—had switched places through a reciprocal translocation (Figure 17.16).

The regions around the breakpoints of chromosomes 9 and 22 have been cloned and sequenced. The breakpoint in chromosome 9 occurs within the cellular proto-oncogene *ABL*, which normally codes for a tyrosine kinase that binds to DNA. The breakpoint in chromosome 22 occurs within a gene that is called the *breakage cluster region* (*BCR*), which normally codes for a serine kinase. The fused *BCR-ABL* gene in the

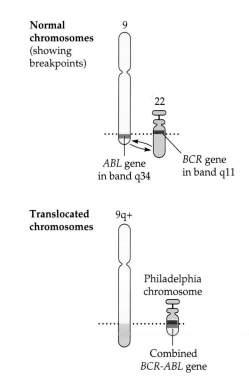

Figure 17.16 The reciprocal translocation between chromosomes 9 and 22 that produces the Philadelphia chromosome characteristic of chronic myelogenous leukemia. Breakage and reunion join together the *ABL* proto-oncogene on chromosome 9 with a so-called *breakage cluster region* (*BCR*) on chromosome 22.

Philadelphia chromosome is transcribed and translated into a joined protein, which retains kinase activity but which clearly acts in an abnormal fashion. In some patients, there is also a protein product from the joined genes on the reciprocal translocated chromosome, labeled 9q+ in Figure 17.16.

Another translocation with a breakpoint near a different proto-oncogene, *MYC* on chromosome 8, is a consistent feature of *Burkitt lymphoma* (*BL*). Occurring commonly in parts of central Africa (about 1 case per 10,000 children per year) and rarely in other regions of the world, BL is a malignancy of B lymphocytes (the antibody-producing cells). In a wide belt of equatorial Africa, it is the most frequent cancer among children, the average age at onset being about 7 years. First described in 1958 by Denis Burkitt, a British surgeon working at Makerere University in Uganda, the solid tumors typically affect two areas: the bones of the jaws and the organs in the abdomen. (Unlike leukemia, the bone marrow is not usually involved.) Extremely fast growing and aggressive, the tumors can reach prodigious size in a matter of days and obstruct neighboring organs (Figure 17.17). The kidneys may also be damaged by the excessive elimination of uric acid, a by-product of the breakdown of DNA, which results from the rapid growth and death of cancer cells. Fortunately, treatment with anticancer drugs often gives dramatic improvement and leads to long-term survival in about half the cases.

The malignant immune cells of BL patients have a translocation involving the *MYC* proto-oncogene at band q24 on chromosome 8. The other participant in the translocation in most cases is chromosome 14, or less commonly chromosome 22 or 2 (Figure 17.18). The breakpoints in the latter three chromosomes are all within immunoglobulin genes (*IGH*, *IGK*, *IGL*) that code for parts of antibody molecules.

The juxtaposed genes of *MYC* on chromosome 8 and one of the antibody genes have also been analyzed in detail at the nucleotide level. Unlike the *BCR-ABL* protein product, however, the normal *MYC* DNA-binding protein usually remains intact and unattached to any other polypeptide. As a result of its new neighbor, however, *MYC* becomes *deregulated*, being transcribed and translated at inappropriate times or at increased rates. This deregulation is thought to be one of the steps required to produce the malignant transformation, but it is probably not the only step. In the next chapter, we explain why the unusual antibody genes that are expressed in antibody-manufacturing cells might be involved in translocations; these genes *normally* undergo recombinations in the process of encoding antibodies. It is not known, however, why *MYC* (rather than some other oncogene or noncancer locus) is involved in these translocations.

Other Specific Chromosomal Abnormalities

In addition to CML and Burkitt lymphoma, many other cancers of blood cells are associated with chromosomal abnormalities involving proto-oncogenes, although usually with much less consistency. The most common rearrangements are translocations, inversions, and deletions (Cooper 1995; Mitelman et al. 1997). About half the cases result in a fusion protein like that seen in CML, and the other half result in aberrant expression like that seen in Burkitt lymphoma.

The same aberrations are sometimes seen in different cancers. For example, the Philadelphia chromosome characteristic of chronic myelogenous leukemia is also seen in the cancer cells of some other leukemias. Even more common than rearrangements are changes in chromosome numbers, but these changes are extremely variable from patient to patient even with the same type of cancer.

Conversely, several different aberrations may lead to the same cancer. In these cases, the prognosis for patients and their response to treatment may vary

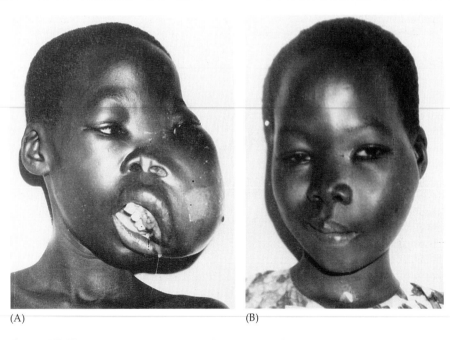

(A)   (B)

Figure 17.17 (A) A massive jaw tumor characteristic of Burkitt lymphoma in a 9-year-old Ugandan girl. (B) The same child 3 1/2 weeks after two injections of an anticancer drug, cyclophosphamide. (From Owor and Olweny 1978.)

Chromosome:	8	14	22	2
Breakpoint band:	q24	q32	q11	p11
Breakpoint gene:	*MYC*	*IGH*	*IGL*	*IGK*
Approximate percentage of cases:	100	89–90	5–15	5

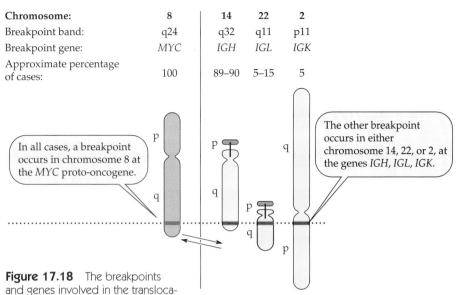

In all cases, a breakpoint occurs in chromosome 8 at the *MYC* proto-oncogene.

The other breakpoint occurs in either chromosome 14, 22, or 2, at the genes *IGH, IGL, IGK.*

Figure 17.18 The breakpoints and genes involved in the translocations producing Burkitt lymphoma (shown prior to the rearrangements). *IGH*, immunoglobulin heavy chain; *IGL*, lambda light chain; *IGK*, kappa light chain. These three genes code for the polypeptides of antibody molecules (see Chapter 18).

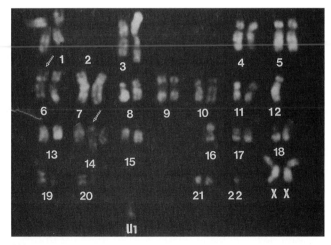

(A)

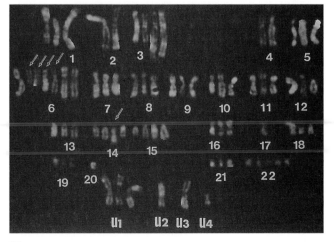

(B)

depending on the specific chromosomal defects. For example, Jorge Yunis of Hahnemann Medical School in Philadelphia found 17 different types of aberrations among 99 patients with *acute nonlymphocytic leukemia.* Those carrying an inversion of chromosome 16 had the best prognosis, many of them surviving many years. Those with trisomy 8 had intermediate survival—about one year. A number of patients with complex chromosomal defects usually lived only two or three months. The hope is that knowledge of the specific aberrations in a given patient can lead to more individualized therapy.

The leukemias and lymphomas whose chromosomal aspects have been so intensively studied represent only about 8% of all human cancers. Less is known about the cytogenetic aspects of the other 92%, including the common solid tumors of the breast, lung, colon, prostate, and pancreas. The cells of most solid tumors are difficult to grow in culture. In addition, because they divide much more slowly than the cells of blood cell malignancies, there may be at any given time relatively few cells in the mitotic divisions needed to obtain karyotypes. Another problem is that solid tumors of the same type have a range of chromosomal defects, sometimes even within the same tumor (Figure 17.19 and also Figure 2.10B). Dozens of different secondary changes, often characterized by a great many extra chromosomes, may obscure a primary karyotypic event that led directly to the cancerous transformation.

The idea that secondary chromosomal changes in cancer cells can lead to greater malignancy has been confirmed by the discovery of repetitive elements that contain an oncogene. In some lung cancers and in some cancers of neural tissue (*neuroblastomas*), researchers have found the *MYC* proto-oncogene repeated many times—a phenomenon sometimes called *gene amplification.* The increase in copy number of the

Figure 17.19 Two cells that are derived from the same ovarian cancer but show different karyotypes. (A) A cell showing a 6;14 translocation (arrows) and a small chromosome of unknown origin (U1), with a total of 45 chromosomes. (B) A cell showing the same 6;14 translocation, several chromosomes of unknown origin (U1–U4), and a total of 77 chromosomes. The translocation is thought to be the primary cancer-causing cytogenetic change. (From Sandberg 1988.)

MYC proto-oncogene often occurs in later stages of the cancers, probably contributing to the progression to more highly malignant forms.

We note finally that the chromosomal changes discussed here occur in the *cancerous* cells but *not* in the normal cells of the body and *not* in the germ line, where they might be included in eggs or sperm.

Environmental Factors

Although this chapter has been devoted to the *genetics* of cancer, we emphasize again that the disease is *not* usually transmitted from parents to offspring in the manner of typical heritable traits due to either single genes or the combined effects of several genes. All cancers are probably genetic, however, in that causative mutations are transmitted from one *cell* generation to the next during the development and maintenance of a person's organs and tissues. *Thus, the genetics of cancer is played out primarily in somatic rather than germinal cells.*

The environmental agents that bring about cancer-causing mutations in somatic cells are called **carcinogens**, and most of these apparently act by causing mutations. In fact, about 90% of carcinogens as revealed by animal testing are mutagens, as demonstrated by the Ames test (Chapter 7). Some of these mutations produce abnormal alleles of tumor suppressor genes, proto-oncogenes, or DNA repair genes. DNA repair genes encode enzymes that are needed to mend damage to any genes, including those involved in cell growth and replication. Most of the DNA repair syndromes listed in Table 7.2 include multiple types of cancer among their symptoms.

Identifying environmental causes of cancer in humans is unusually difficult, because years or even decades may pass between the action of carcinogens and the clinical detection of tumors. A clear illustration of this time lag involves the sex hormone diethyl-stilbestrol (DES). First administered to women during the 1940s, when it was thought to help maintain pregnancy, DES was later found to be ineffective, even detrimental, for this purpose. Among the teenage daughters resulting from these pregnancies, about 1 in 1,000 developed vaginal cancer, a type of malignancy that is otherwise extremely rare in young women. The clinical effect was clearly delayed for nearly two decades, because these cancers were initiated prior to birth. Note that if DES had instead induced a more common type of cancer, such as lung or breast cancer, its carcinogenic effect might never have been detected. This is because the few DES-induced cases would have been lost among the large number of cases associated with other causes.

Concentrations of cancer cases that occur in limited areas are often investigated, but rarely is a specific environmental cause identified. The occurrence of thyroid cancer in children exposed to fallout from the Chernobyl accident, however, is well documented (Chapter 7). But most community clustering probably happens by chance. Random statistical fluctuations alone can account for places where a given type of cancer occurs more frequently than average. Another factor in clustering is called "epidemiologic gerrymandering," in which, after cancer cases are identified, a geographical region is outlined that includes the maximum number in the smallest area (Kase 1996).

Nevertheless, most investigators—epidemiologists and others—think that about 80% of all cancers are caused by various environmental factors, including life-style considerations (Table 17.5). At the top of the list are tobacco and diet (Cohen 1987; Trichopoulos et al. 1996). Cigarette smoke contains about 50 known carcinogens and by itself accounts for perhaps 30% of all cancer deaths. Smoking is the greatest preventable cause of cancer.* *Secondhand smoke* is also associated with a small increase in lung cancer (about 3,500 deaths per year), as well as a large increase in heart disease (50,000 deaths per year) (Grady 1997). Even light smoking by parents in the home may add to the likelihood that children will acquire lung cancer later in life.

Dietary factors are thought to be responsible for another 30% of cancer deaths, although there is more uncertainty about this figure than for tobacco smoke. Still, high consumption of satu-

---

TABLE 17.5 Cancer deaths attributable to different environmental factors

Factor	Cancer deaths, best estimate (%)	Range of acceptable estimates (%)
Tobacco	30	25–35
Diet	30	10–70
Reproductive and sexual behavior	7	1–13
Occupational exposures	4	2–8
Alcohol	3	2–4
Radiation (mostly UV)	2	1–3
Pollution	2	1–5
Medicines and medical procedures	1	1–2
Industrial products	<1	1–2
Food additives	<1	−5–2
Other and unknown	20	?

Source: Doll and Peto (1987); Varmus and Weinberg (1993).

*One of the rugged models who played the "Marlboro Man" in billboard ads, rodeo rider Wayne McLaren, died at age 51 after fighting lung cancer for several years (Anonymous 1992).

rated fat and red meat are strongly linked to cancers of the colon and rectum. In addition, food plants manufacture their own toxic chemicals to ward off insects and other attackers. A number of these substances are either weak animal carcinogens or are converted to weak carcinogens by storage or preparation of foods or by metabolic events in the body. Bruce Ames of the University of California, Berkeley, states that we eat 10,000 times more of nature's pesticides than of human-made pesticides, and we generally need not worry about either group (Brody 1994, 1996). In addition, coffee, with or without caffeine, as well as artificial sweeteners in moderate amounts, are not human carcinogens. Some foods, such as fruits and vegetables and some dietary components (including vitamins C and E, β-carotene, and selenium), seem to act as protective agents *against* cancer. The health consequences of eating this or that type of food—for example, fat and fiber—are continuing areas of research.

Natural processes of reproductive biology are also linked to cancer and may account for about 7% of cases. For women, an early onset and late cessation of menstruation contribute to increased frequency of breast cancer. This may be due to extending the period that a women is exposed to her own estrogens, which are produced in quantity during each menstrual cycle and which stimulate breast tissue development. Several investigators have suggested that estrogen-like compounds (called *xenoestrogens*) used in herbicides, pesticides (DDT, for one), and some plastics may cause breast cancer (Davis and Bradlow 1995). Having children only later in life and having fewer children also contribute to increased breast cancer incidence.

Radiation and chemicals in the workplace and in the general environment are responsible for just a few percent of cancers. We hear a lot about some of the possible carcinogenic agents in these categories: radon, power lines, asbestos, and compounds used in the manufacture of pesticides, plastics, dyes, and paints. Pollution, food additives, alcohol, and medical drugs and procedures also appear to be responsible for only a few percent of cancers.

Certainly, we need to know more about both the environmental and genetic factors that contribute to the genesis of cancerous cells. Although the common cancers—those of the lung, colon/rectum, breast, and prostate—are generally random in their occurrence, there are subsets of each type of cancer that appear in increased frequency within families. The relationships between cancers that are substantially sporadic and cancers of the same types that are familial need to be worked out. This research involves proto-oncogenes, tumor suppressor genes, DNA repair genes, and the functions of their protein products. Many initially unconnected lines of investigation funded by taxes, charitable foundations, and business investments are now converging toward a fuller understanding. It is hoped that the new knowledge—especially at the molecular level—will eventually open up ways to alleviate the great suffering that accompanies cancer.

Summary

1. Cancer rates have remained high despite substantial research efforts. The largest killers are cancers of the lung, breast, prostate, and colon/rectum. The mutations responsible for about 80% of cancers are limited to somatic cells.

2. Cancers usually stem from single cells that accumulate multiple mutations, progressing through several stages as they become more malignant. Cancer cells reproduce without restraint, lose their normal form and function, invade nearby tissue, and spread to remote sites.

3. Tumor suppressor genes, whose pattern of inheritance appears dominant at the pedigree level, but recessive at the cellular level, normally keep the cell cycle under proper control. Examples are *p53*, *RB1*, *WT1*, and *BRCA1*. The normal alleles are transcription factors that bind to genes that synthesize proteins (e.g., cyclins) that regulate the cell cycle. The absence of normal alleles of tumor suppressor genes allows the cell cycle to proceed inappropriately.

4. Mutations of *p53* are present in families with the Li-Fraumeni syndrome. A mutation of *p53* is inherited in an egg or sperm, and cancers develop in somatic cells that develop a second allelic mutation. Thus, although Li-Fraumeni is transmitted through the germ line in a dominant manner, it is expressed at the cell level in a recessive manner. About half of noninherited cancers also involve *p53* mutations that become homozygous in somatic cells.

5. Retinoblastoma (eye) and Wilms tumor (kidney) are cancers whose patterns of occurrence—both familial and sporadic—are similar to cancers due to *p53* mutations. Knudson's 1970 two-hit hypothesis for the initiation of these cancers at the cell level became the model for the multiple mutation theory of cancer.

6. The competition was intense to map and clone the gene *BRCA1*, which predisposes to early-onset breast and ovarian cancers. Surprisingly, neither *BRCA1* nor *BRCA2* are commonly involved in sporadic breast cancers.

7. Recessively acting proto-oncogenes also keep the cell cycle under proper control. Mutant alleles, the dominant oncogenes, have been detected by using DNA taken from human cancer cells to transform mouse cells. Oncogenes were first discovered in RNA-containing retroviruses that infect birds and rodents. Retroviruses use reverse transcriptase to copy their genetic information from RNA to DNA.

8. The products of proto-oncogenes (e.g., RAS, JUN, FOS proteins) are involved in chemical signal transduction pathways from growth factor receptor molecules on the cell surface to transcription factors in the nucleus. Many proto-oncogenic proteins are kinases that alter the activity of other proteins by phosphorylation.

9. Cancer research has been aided by the use of transgenic and knockout mice. The former contain a gene of foreign origin (e.g., a human oncogene), and the later are homozygously missing an active gene (e.g., *p53*).

10. Cancerous cells usually have abnormal chromosomal makeups. In chronic myelogenous leukemia, a translocation with one breakpoint at the *ABL* proto-oncogene produces the Philadelphia chromosome. The protein product of *ABL* is thereby altered.

11. In Burkitt lymphoma, a translocation has one breakpoint at the *MYC* proto-oncogene and the other breakpoint at one of three different antibody genes. The protein product of *MYC* is not altered, but it is expressed at the wrong times or in the wrong amounts.

12. Researchers sometimes find the same chromosomal aberrations in different cancers and different aberrations in the same cancer. The progression of some solid cancers is accompanied by secondary chromosomal aberrations or by the occurrence of gene amplification.

13. Carcinogens are environmental agents that initiate and promote malignant changes, and they often act by causing a series of mutations. Agents in tobacco smoke (including secondhand smoke) and in diet are major factors in cancer causation, but the clinical effects of a carcinogen may be delayed for years or decades.

Key Terms

cancer	provirus
carcinogen	receptor protein
carcinoma	relay molecule
cyclins	retrovirus
growth factor	reverse transcriptase
knockout mouse	sarcoma
leukemia	signal transduction
lymphoma	transcription factor
malignant tumor	transformation
metastasis	transgenic mouse
oncogene	tumor progression
Philadelphia chromosome	tumor suppressor gene
protein kinase	zinc-finger protein
proto-oncogene	

Questions

1. A tumor the size of a marble, about 1 cubic centimeter, may contain 10^9 cells. Starting from a single cell, how many cell generations are required to produce this tumor? (*Note:* Such a lump is about the size of a breast tumor that a woman can find by self-examination. Mammography can detect smaller tumors.)

2. Some uterine tumors consist of as many as 10^{11} (100 billion) cells. In women heterozygous for a particular X-linked gene, researchers have discovered that every cell of such a tumor has the same active X-linked allele. Explain this observation in terms of the clonal nature of tumors and the Lyon hypothesis.

3. Identify an example of incomplete penetrance in the pedigrees in Figure 17.3.

4. Mutant alleles of the tumor suppressor genes discussed in this chapter can sometimes be transmitted in germ cells, leading to hereditary patterns of cancer occurrence. In other cases, the mutant alleles may be induced only in the somatic cells of individuals and not passed on in eggs or sperm. Discuss with regard to *p53*.

5. Distinguish between dominant inheritance and recessive expression for retinoblastoma.

6. The mapping of *BRCA1* was made difficult by confusing some sporadic cases for familial cases. Why the confusion? How was the issue resolved (at least in part)?

7. The mapping of *BRCA1* was also made difficult by confusing heritable cases due to mutations of *BRCA1* with those cases due to mutations of *BRCA2* and perhaps other genes. How were these separated (at least in part)?

8. A woman who inherits a mutant *BRCA1* allele does not necessarily get breast or ovarian cancer. Explain.

9. Although cancer researchers generally agree that the path to malignancy is a *multistep* process, Weinberg and his colleagues were able to transform mouse cells (that had long been maintained in culture) in *one* step, as outlined in Figure 17.12. Can you suggest an explanation for the apparent discrepancy?

10. The proto-oncogene *ERBB* encodes a cell surface receptor for a growth factor. When the growth factor outside the cell joins to the receptor, it signals the cell to divide. Speculate on how a mutation in the *ERBB* proto-oncogene might lead to malignancy.

11. Chronic myelogenous leukemia (CML) and Burkitt lymphoma (BL) both involve translocations that have a proto-oncogene at a breakpoint. But the consequences for the protein product of the proto-oncogene are somewhat different. Explain.

12. Most cases of lung cancer occur in people who smoke. A high-fat diet increases the risk of breast and colon cancer. Reconcile these facts with the idea that cancers are *genetic* diseases.

Further Reading

Excellent general discussions of cancer for nonscience students and laypersons are by Weinberg (1996b) and Varmus and Weinberg (1993). A marvelous history of the people and the science involved in mapping and cloning *BRCA1* is by Davies and White (1996). Chapters on cancer are included in many textbooks of cell and molecular biology, good ones (not too technical) being by Cooper (1997), Korf (1996), and Micklos and Freyer (1990). Cooper (1995) has also written a more detailed account of cancer genetics. More thorough information can be found in the large collections edited by Rimoin et al. (1997) and by Scriver et al. (1995).

The whole issue of *Scientific American* for September 1996 is devoted to cancer, including articles on its biology, causes, prevention, detection, and therapies. Other recent *Scientific American* articles include Perera (1996) on cancer detection, Leffell and Brash (1996) on skin cancer, Cavenee and White (1995) on multiple mutations in cancer cells, Davis and Bradlow (1995) on estrogen-like molecules in the environment, and Capecchi (1994) on making knockout mice. For browsing on the topic of cancer, see the Library of Congress classifications RC 254-282, where a variety of books, journals, and review series can be found.

CHAPTER 18

The Genetic Basis for Immunity

In 1971, a Texas family anxiously awaited the birth of another baby. They had already lost one son to an invariably fatal recessive disorder called *X-linked severe combined immunodeficiency disease* (*XSCID*). XSCID patients have no immune defense against microbes, so they easily succumb to pneumonia, influenza, bronchitis, meningitis, and fungal diseases. After delivery by cesarean section, baby David was quickly placed in a sterile chamber, for he had inherited the abnormal allele. Doctors believed that a cure for David's XSCID was at most a few years away, so the germ-free isolation was considered a temporary situation. This view turned out to be overly optimistic.

As David grew, so did his marvelously devised chambers and life-support machinery at Texas Children's Hospital in Houston (Figure 18.1). He was surrounded and loved by his parents, sister, friends, physicians, and dozens of devoted hospital staff. He had a positive attitude, a sharp mind, and a fine sense of humor. Although his days were filled with activities made as

Figure 18.1 David, a patient with X-linked severe combined immunodeficiency disease (XSCID), in his bubble chamber at Texas Children's Hospital in Houston around 1980. (Courtesy of Baylor College of Medicine.)

normal as possible, his outlook on the world originated almost entirely from two multichambered, antiseptic, plastic bubbles, one in the hospital and one at home. Tests showed that David's perceptions of space were distorted; he was unable, for example, to imagine a four-sided building. We will never fully understand how he felt about the limits on his freedom, although he seemed more comfortable with his confinement than those around him did. In any event, the case of the "bubble boy" and the increasingly technological approaches to his treatment received tremendous attention in the media (Carol Ann with Demaret 1984; Rennie 1985). At the age of 12, he received a bone marrow transplant from his sister, but the transplant failed, and he finally succumbed to his immune deficiency.

Immune systems help protect us against threats to our health from pathogens. *Immunology* is the study of these protective mechanisms, and *immunogenetics* deals with the genes that are involved. Beginning with Louis Pasteur, Robert Koch, and others in the late 1800s, a golden age of immunology brought under control many infectious diseases (yellow fever, diphtheria, whooping cough, tetanus) that had ravaged Western countries. In the 1970s, the deadly smallpox virus, which left disfiguring pockmarks on its surviving victims, was wiped out completely. Today's researchers hope to arrange a similar fate for a new threat to our health, the human immunodeficiency virus (HIV), but the time scale for conquering the acquired immune deficiency syndrome (AIDS) that HIV causes is unclear.

Starting in the 1950s, the growth of molecular biology ushered in a second golden age of immunology that is providing a deeper understanding of the immune system (Nossal 1993). Researchers have shown that the mechanisms of immunity consist of mar-

velously intricate webs of interactions among and between different cells and molecules. These processes are of practical importance in the prevention, diagnosis, and treatment of both infectious diseases (such as AIDS) and noninfectious diseases (such as cancer). But the immune system can function improperly or fail to function at all (as in XSCID), as well as causing serious problems in blood transfusions and organ transplants. In trying to understand such wide-ranging situations, immunologists have discovered unusual elements of gene structure and novel patterns of gene expression. These several aspects of immune processes are discussed in this chapter.

The Immune Response

The immune system (1) recognizes substances as either native to one's own body or foreign (i.e., self or nonself); (2) reacts specifically against any foreign substance and destroys it; and (3) remembers a particular foreign substance and responds to it more strongly upon a later exposure. **Antigens** are the substances that provoke the immune system. They are often protein molecules (or more usually parts of protein molecules) embedded in the surface of bacteria, viruses, or other small particles such as pollen grains. Other antigens are parts of glycoproteins (carbohydrate and protein), lipoproteins (fat and protein), or nucleoproteins (nucleic acid and protein). As a consequence of sequence differences in genes encoding particular antigens, they may differ in fine molecular details. These distinctions contribute to the large number of blood types and tissue types.

Components of the Immune System

Blood makes up about 8% of body weight (5 quarts in a 125-pound person). The plasma—the liquid, somewhat yellowish portion of blood—transports nutrients, hormones, clotting factors, waste materials, and other substances to and from various parts of the body. Suspended in the plasma are red blood cells, white blood cells, and platelets (which initiate the clotting process). The vital immunological properties—recognition, reaction, and memory—depend on several types of *white blood cells* that slip from the blood through capillary walls and wander the spaces between tissue cells. These spaces are filled with a fluid called **lymph**, which is derived largely from blood plasma. The lymph drains into inconspicuous networks of lymphatic vessels that run more or less parallel to, but separate from, the network of veins. The largest lymph vessel empties into a main vein near the heart, completing the joint blood-lymph circulation.

Associated with this *lymphatic system* are the following small organs: several dozen *lymph nodes* spaced along major lymph vessels; the *spleen*, lying behind the stomach; and the *thymus gland*, lying in the chest cavity between the lungs (Figure 18.2). These organs are rich in small, roundish white blood cells called **lymphocytes**, which respond specifically to foreign antigens. In addition, as lymph seeps through the finely divided spaces within the nodes, large, irregularly shaped white blood cells called **macrophages** engulf and digest foreign substances. (Macrophages are said to be *phagocytic*—literally, "to eat cells.") Lymph nodes in the neck may become swollen and painful during colds and other infections. The tonsils and adenoids in the throat are somewhat similar to lymph nodes in structure and function.

The thymus, which is particularly important in the immune response, consists of a spongy mass of lymphocytes. It is fully developed at birth but gradually regresses during adolescence, remaining very small throughout adult life. Experiments have shown that newborn mice whose thymus glands are removed grow fairly normally but cannot respond immunologically to some types of antigens. Mice that are homozygous for a certain autosomal recessive allele develop no thymus at all. Called *nude* because they are hairless, these mice will die unless they are protected from infection in a germ-free environment.

Three major types of small lymphocytes originate in the bone marrow and contribute to immune responses against a specific antigen: **B cells** (or B lymphocytes) and two types of T cells (or T lymphocytes), **killer T cells** and **helper T cells** (Figure 18.3). Both B and T cells are found throughout the blood, lymph, and lymphatic organs, each site being rich in one type or the other. The B and T cells can be distinguished by certain properties of their membranes and by the strengths of their responses to various foreign antigens. Human B cells respond particularly strongly to foreign antigens on the surface of invading bacteria by differentiating into **plasma cells**, which in turn make **antibodies**, protein molecules that recognize the foreign substances.* (Antibodies will be described more

Plasma cells should not be confused with *blood plasma*. Plasma cells are white blood cells that manufacture and secrete antibodies. They are suspended in blood plasma, and the antibodies are dissolved in blood plasma. Nor should *antibody* be confused with *antibiotic*, a drug (e.g., penicillin) that nonspecifically attacks invading bacteria.

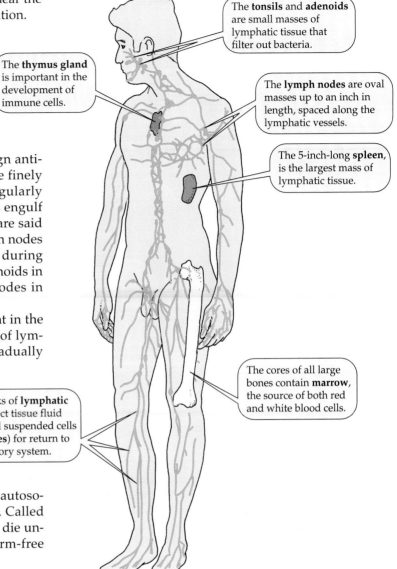

The **tonsils** and **adenoids** are small masses of lymphatic tissue that filter out bacteria.

The **thymus gland** is important in the development of immune cells.

The **lymph nodes** are oval masses up to an inch in length, spaced along the lymphatic vessels.

The 5-inch-long **spleen**, is the largest mass of lymphatic tissue.

The networks of **lymphatic vessels** collect tissue fluid (**lymph**) and suspended cells (**lymphocytes**) for return to the circulatrory system.

The cores of all large bones contain **marrow**, the source of both red and white blood cells.

Figure 18.2 Components of the lymphatic system.

fully in the next section.) T cells react particularly strongly to virus-infected cells, cancerous cells, and tissue grafts. Killer T cells do not make antibodies, but they respond to specific antigens, bringing about the destruction of their carriers by direct cell-to-cell contact. Helper T cells secrete a large variety of chemical messengers called **lymphokines** (for "lymphocyte movement") that regulate the activity of killer T cells, B cells, and macrophages. The critical role played by helper T cells is brought into focus by the disease AIDS, in which helper T cells (and other immune cells with the so-called CD4 receptor molecule) are invaded and destroyed by the HIV retrovirus. AIDS patients are susceptible to a variety of infections and malignancies as a consequence of a seriously weakened immune system (Greene 1993).

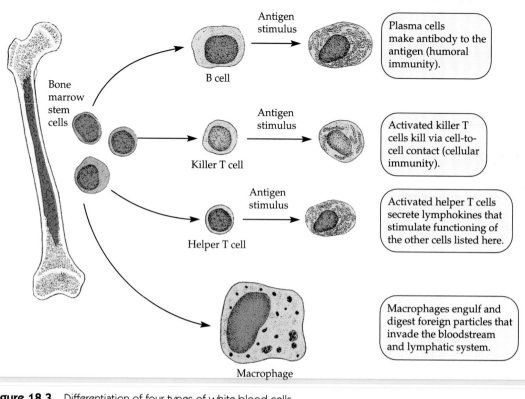

Figure 18.3 Differentiation of four types of white blood cells involved in the immune response. The B cells and two types of T cells respond specifically to the presence of a particular foreign antigen. The macrophages are not specific in their functions.

Although T cells originate in the bone marrow, they migrate to the thymus during development. There they somehow "learn" what is self and what is not self by associating with the cells and molecules about them. These processes are not well understood. (How, for example, does the immune system learn about substances like brain tissue, which probably never reaches the thymus, or substances like placental or milk proteins, which only appear later in life?) Only about 5% of bone marrow lymphocytes that enter the embryonic thymus are selected for and released to the circulation as T cells that react to foreign antigens. These T cells are said to be *self-tolerant* of the body's own repertoire of antigens, and this phenomenon of nonresponsiveness to self antigens is called immunological tolerance, or simply **tolerance**. (The breakdown of tolerance leads to autoimmune diseases, which we discuss later.) It is thought that the other 95% of incipient T cells would (if released) react to self antigens, and they are eliminated in the thymus by apoptosis.

Thus, self-recognition is not genetically determined, but seems to be learned by the immune system during developmental stages. In fact, injecting foreign antigens into a newborn mouse can trick its immune system into later tolerating the foreign antigens as if they were self antigens. These important ideas about toler-

ance were the focus of Nobel prizes awarded in 1960. Recently, however, the idea of self has been called into question. For example, depending on how and at what concentration an antigen is presented, it is possible to induce tolerance in an adult mouse or evoke an immune attack in a neonate. To some immunologists, the immune system does not, in fact, distinguish between self and nonself at all. Instead, they claim, it distinguishes between safe situations and dangerous events that cause tissue injury (Johnson 1996; Pennisi 1996).

The B Cell Response

By the 1950s, it was known that plasma cells congregate in large numbers at the sites of bacterial infection. Rich in the organelles of protein synthesis, these cells make antibodies against antigenic sites on bacterial surfaces. By culturing individual plasma cells, scientists showed that one plasma cell makes a single type of antibody active against only one specific antigen. This result was itself surprising, but the most intriguing question about the immune response was how it could manage to identify so many different foreign antigens. Over the course of a lifetime, a human being must be exposed (via food, air, skin contact, bites, etc.) to millions of different nonself materials, and the immune system responds specifically to each one.

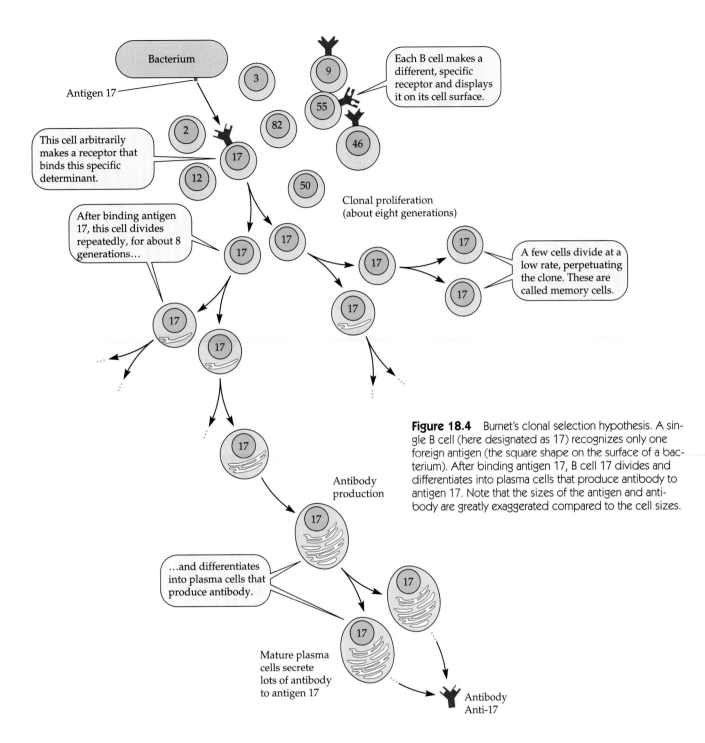

Figure 18.4 Burnet's clonal selection hypothesis. A single B cell (here designated as 17) recognizes only one foreign antigen (the square shape on the surface of a bacterium). After binding antigen 17, B cell 17 divides and differentiates into plasma cells that produce antibody to antigen 17. Note that the sizes of the antigen and antibody are greatly exaggerated compared to the cell sizes.

Indeed, researchers have shown that plasma cells can make antibodies to synthetic chemicals that no species has ever encountered over the course of evolution! The seemingly unlimited individual challenges that the immune system meets and overcomes is its most extraordinary property.

One model of the immune response that has been supported by much evidence was suggested by Nobel prize winners Niels Jerne, director of the Basel Institute of Immunology, and F. Macfarlane Burnet, director of Melbourne's Institute of Medical Research.

According to Burnet's **clonal selection theory**, each lymphocyte is *preprogrammed* to recognize and respond to just one antigen, but there is no advance "knowledge" of whether it will ever be called on to perform (Ada and Nossal 1987). The idea is illustrated in Figure 18.4, which shows a small sample of B cells. The preprogrammed antigen specificity of each cell is indicated by a number. When antigen 17 is present, for example, it interacts with the surface molecules on one or a few lymphocytes specific for 17, stimulating them to divide and differentiate over a period of several

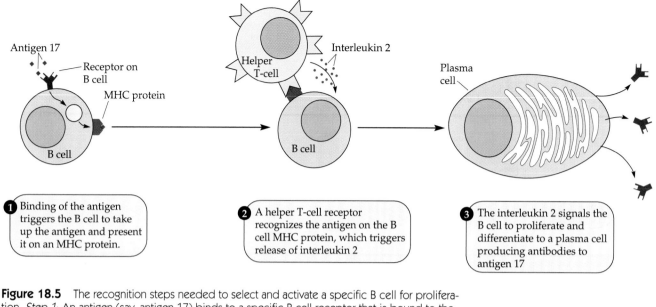

① Binding of the antigen triggers the B cell to take up the antigen and present it on an MHC protein.

② A helper T-cell receptor recognizes the antigen on the B cell MHC protein, which triggers release of interleukin 2

③ The interleukin 2 signals the B cell to proliferate and differentiate to a plasma cell producing antibodies to antigen 17

Figure 18.5 The recognition steps needed to select and activate a specific B cell for proliferation. *Step 1*: An antigen (say, antigen 17) binds to a specific B cell receptor that is bound to the cell surface. (Note that this receptor is shaped exactly like the antibody molecule later secreted by the cell.) The antigen is then drawn into the cell, processed, and passed to another membrane protein (MHC). *Steps 2 and 3*: The receptor of a helper T cell binds to the antigen-MHC complex. The helper T cell then secretes a lymphokine (interleukin-2) that stimulates the B cell to divide and differentiate into a plasma cell.

days to form a clone of several hundred identical plasma cells. These mature plasma cells secrete large amounts of antibodies, Y-shaped protein molecules with two pockets at the upper tips of the Y. The pockets, called the **combining sites**, have a shape complementary to the shape of the initiating antigen.

Recognition of antigen 17 by the B cells that are pre-programmed to manufacture anti-17 antibody begins when antigen 17 joins to antigen-binding molecules embedded in the surface of the B cells (step 1 of Figure 18.5). On any given B cell, all the antigen-binding molecules (receptors) are the same and virtually identical to the Y-shaped antibody molecules that the mature plasma cell will eventually secrete in large amounts. At this stage, however, the B cell is not secreting any antibody into the extracellular space; it is just displaying its name tag so that it can correctly engage with its antigen.

Another event is needed before the B cell can proliferate and mature: Antigen 17 must be processed inside the cell and a portion of it—a peptide made up of a dozen or more amino acids—moved back to the surface (Engelhard 1994). The processed peptide is displayed there in a narrow groove formed by another protein (step 2 in Figure 18.5). This other protein is a self molecule, synthesized within the B cell from genes called **MHC**, an abbreviation for **major histocompatibility complex**. This intimate association of a foreign peptide with a self protein is then recognized by a helper T cell, which had been previously stimulated

by the same antigen (see next section, "The T Cell Response"). The helper T cell then secretes a lymphokine called *interleukin-2* that stimulates the B cell to multiply and differentiate (step 3).*

The daunting complexity of this arrangement for antigen processing and eventual antibody production is thought to foster a fail-safe strategy. It helps ensure that the immune system counters only substances that are foreign to oneself by displaying them in the context of molecules known to be specific to oneself. The MHC proteins that present the foreign peptide are extremely polymorphic (as we explain in a later section), with thousands of combinations, so that each person is highly individualistic. Experiments in mice suggest that no one else's MHC proteins could work for you. The recognition step (Figure 18.5, step 2) is said to be **MHC-restricted**. Nobel prizes awarded in 1996 for work done decades earlier acknowledge this crucial role of MHC proteins (Gura 1996).

A week or so after initial contact with bacterial antigen 17, the responding plasma cells begin to secrete millions of antibody molecules per plasma cell into the fluid portion of blood and lymph (see Figure 18.4). The combining sites of the antibodies recognize and join with antigen 17 on the surface of an invading bacterium. When the tips of antibody 17 are bound to

---

*The immune deficiency disease that affected David, the "bubble boy," is due to a mutation in a gene that encodes part of the receptor for interleukin-2 (see later).

antigen 17, the stem of the Y-shaped antibody molecule sticks out from the surface and can react with receptor sites on the surface of macrophages. This reaction stimulates the phagocytic action of the macrophages, and the bacterium is engulfed and destroyed.

Alternatively, the antigen-antibody complex that forms on some cell surfaces invites other plasma proteins, called *complement*, to join up. Complement proteins injure the cell membranes of the bacteria, allowing fluid to enter. The bacteria swell and break open, a process called *lysis*. Thus, we see that antibody does not itself destroy the invading bacteria. Rather, antibody molecules mark cells for destruction by other agents.

The antibody-producing plasma cells of a selected clone do not live long after the removal of the stimulating antigen. Other members of the same clone that have not fully differentiated, however—called **memory cells**—can survive for years. Once primed by a given antigen, they will initiate the so-called **secondary immune response** when they meet the same antigen again. This response is characterized by a quicker, stronger production of antibody, primarily because more plasma cells are produced. The booster shots given in an immunization program have the effect of renewing the supply of memory cells.

The T Cell Response

Like B cells, T cells of both types also respond by clonal selection, multiplying and differentiating after they are specifically stimulated. Showing the same high degree of specificity, T lymphocytes have on their surfaces recognition molecules, called *T cell receptors*, which make the necessary fine distinctions between self antigens and various foreign antigens (Marrack and Kappler 1986). Although the T cell receptor molecules differ from the B cell receptors, the two systems share some amino acid sequences and are presumed to be products of genes that have evolved from a common primordial gene. Recall that the B cell receptors are membrane-bound forms of the antibody molecules that the B cells will eventually secrete when they are mature. The T cell receptor molecules, by contrast, are never secreted. Instead, T cells perform their functions by cell-to-cell contact.

The antigens that stimulate specific T cells to proliferate into a clone do not join directly to the T cell receptors. Rather, a foreign protein antigen is first phagocytized by macrophages and processed in a manner already described for the processing of antigens by B cells. A difference is that a B cell only internalizes and processes the one antigen recognized by its receptor. A macrophage will engulf and process any antigen, including self antigens present on cell debris from normal metabolic activity. A portion of the phagocytized antigen is displayed on the macrophage surface in a groove of its MHC protein. Only then is the antigen in a form that the receptors on T cells can recognize. (As you could predict, self antigens that may be displayed by macrophages are ignored.) When a killer T cell is so activated, it is ready to function by first attaching to the surface of a target cell through the stimulating antigen. The killing process involves proteins from the T cell that carve channels through the target cell's membrane, resulting in its lysis.

Antibodies

Now we consider antibodies, the end result of the B cell response. Collectively, antibody molecules of various types are called **immunoglobulins (Ig)**, and they constitute about 20% of all blood proteins. (Globular proteins have their polypeptide chains folded into more or less rounded shapes, rather than being long and fibrous like collagen.) Because antibodies, which are bound to the surface of B cells and later secreted, are so varied in their antigen-combining abilities, protein chemists long wondered how one differed from another. But the chemists were hampered by their inability to purify enough of any one antibody to analyze it chemically. This problem was solved by the discovery that persons suffering from *multiple myeloma*, a cancer of bone marrow, had in their plasma large amounts of a *single* antibody species. But each myeloma patient had a different antibody. Accounting for as much as 95% of the immunoglobulins in a given patient, the unusual antibody presence resulted from the malignant proliferation of one antibody-producing plasma cell.

Structure

The multiple myeloma researchers were able to show that the major form of antibody, *IgG* (immunoglobulin G, or gamma globulin), consists of four polypeptide chains connected to each other at several places (Figure 18.6). Each of the two identical **heavy (H) chains** consists of four regions of about 110 amino acids each. The regions, called *domains*, have similar amino acids. They are depicted by the elliptical outlines in Figure 18.6. Domains similar to those in the heavy chains also occur twice in each of the two identical **light (L) chains** that are present in the two branches of the Y-shaped molecule. The attachment of the branches to the stem of the Y is somewhat flexible.

Comparison of the amino acid sequences of light chains from different antibodies reveals a remarkable pattern: Within the domains at the top of the Y, about 25% of the amino acids in one antibody differ from those in another; thus, this region is called the **variable (V) domain**. In the lower half of the light chain, the amino acid sequences are almost exactly the same

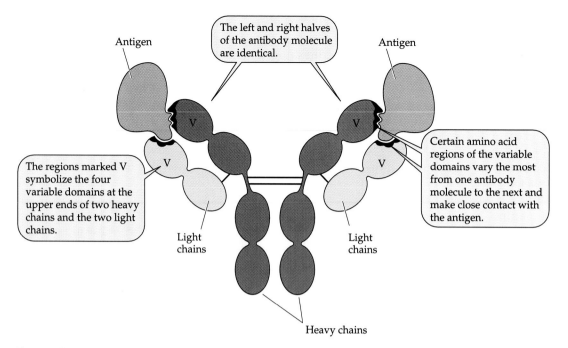

Antigen

The left and right halves of the antibody molecule are identical.

Antigen

The regions marked V symbolize the four variable domains at the upper ends of two heavy chains and the two light chains.

Certain amino acid regions of the variable domains vary the most from one antibody molecule to the next and make close contact with the antigen.

Light chains

Light chains

Heavy chains

Figure 18.6 The Y-shaped IgG antibody molecule. The left and right halves are identical. Each heavy chain is 446 amino acids long and each light chain, 214. The eight constant domains (those not marked V) do not vary much. The complex folding of the polypeptides within each chain is not shown.

from one antibody to another; thus, this region is called the **constant (C) domain**. When the heavy chains from different antibodies are examined, the domains at the top of the Y are also found to be variable; the remaining three domains of a heavy chain are constant. Interestingly, most of the variation in amino acids within variable domains is restricted to a few subregions totaling 20 to 30 amino acids. These so-called **hypervariable regions**—three in each light chain and three in each heavy chain—form the lining of the antigen-binding cavity (Figure 18.7). Antibody specificity arises from the three-dimensional shape of the cavity and from the particular chemistry of the hypervariable amino acids. The surfaces of the antigen and the heavy and light chains conform like parts of a jigsaw puzzle, but the fit may be either snug or loose.

IgG, whose structure we have been describing, constitutes about 80% of the immunoglobulin molecules. Different immunoglobulin classes (G, A, M, E, in the order of their concentration in plasma) have been analyzed: Each has different amino acids in the *stem* region of the heavy chains, giving the four classes somewhat different functions. For example, IgG readily crosses the placenta and provides protection to newborns before they develop their own immunological

competency. The IgA class is found not only in plasma but also in body secretions, such as saliva, nasal mucus, sweat, and breast milk. IgM molecules have somewhat larger heavy chains than the other classes do, and five of the Y-shaped molecules may be joined together in a circle. Because each IgM antibody then has ten combining sites, IgM molecules bind more firmly to antigens. IgE is involved in allergic reactions.

It is important to note that antibody of any of the classes can have the same combining site, directed against, say, antigen 17 (in our previous example). That is, the four functional classes of antibodies, differing in the *stems* of the Y's, can have exactly the same

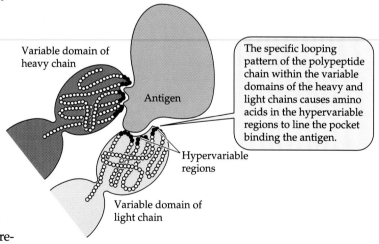

Variable domain of heavy chain

Antigen

The specific looping pattern of the polypeptide chain within the variable domains of the heavy and light chains causes amino acids in the hypervariable regions to line the pocket binding the antigen.

Hypervariable regions

Variable domain of light chain

Figure 18.7 One combining site of an antibody. Amino acids in hypervariable regions contain most of the differences between antibodies of different specificities. The actual folding of the polypeptides is more tortuous than indicated here.

antigen specificity in the *branches* of the Y's. A given B cell can synthesize antibodies of any of the classes, although not at the same time.

Antibody Diversity

As a group, the B cells of a person can make millions of different antibodies, each characterized by the distinctive array of amino acids lining the pockets where the antigens bind. Like all proteins, antibodies are encoded by genes. Yet we could not possibly have millions of antibody genes, each coding for the amino acid sequence of a different antibody, because millions of such genes are vastly more genes than we possess. The three-dimensional geometry of each combining site is formed, however, by two polypeptide chains, one heavy and one light. Thus, if there were 1,000 different H chains (from 1,000 genes) and 1,000 different L chains (from another 1,000 genes), then all possible combinations of the two proteins would generate $1,000 \times 1,000 = 1$ million different spatial patterns. Thus, 2,000 antibody genes might suffice.

Actually, research has shown that the enormous repertoire of antibody molecules is encoded by only about 200 genetic elements! The primary explanation of this number puzzle is simply stated: We inherit several separate sets of DNA pieces that are put together in various ways during the formation of individual B cells. Coding for the combining sites, the DNA pieces generate diversity by being shuffled around before being spliced together. By analogy, imagine outfitting somebody with a hat, shirt, and slacks (Figure 18.8). If the person has 10 different hats, 10 different shirts, and 10 different slacks (30 elements), you can arrange $10 \times 10 \times 10 = 1,000$ different costumes. With a *second* person and an additional 10 hats, 10 shirts, and 10 slacks, you can generate another 1,000 outfits. All together, from just 60 unique elements, you will be able to assemble a million different *pairs* of outfits! In a somewhat similar situation, the various DNA segments constituting the parts of antibody genes are combined together as B cells develop from undifferentiated stem cells.

The shuffling comes about by a process called **somatic recombination**, which has been studied by many researchers, including Nobel prize winner Susumu Tonegawa, then at the Basel Institute of Immunology. Tonegawa identified, mapped, and cloned the DNA elements in mice that undergo somatic recombination (Janeway 1993). He and other researchers have found, for example, that an extended region near the tip of the long arm of human chromosome 14 codes for the heavy chain. The top line of Figure 18.9 shows the general arrangement of heavy chain DNA elements as they are inherited. Note that there are four families of closely linked elements: *V* (variable), *D* (diversity), *J* (joining), and *C* (constant).

Figure 18.8 The large number of antibody "outfits" are assembled in a similar manner to items of clothing shown here, from several sets of DNA elements assembled in numerous combinations (see Figure 18.9).

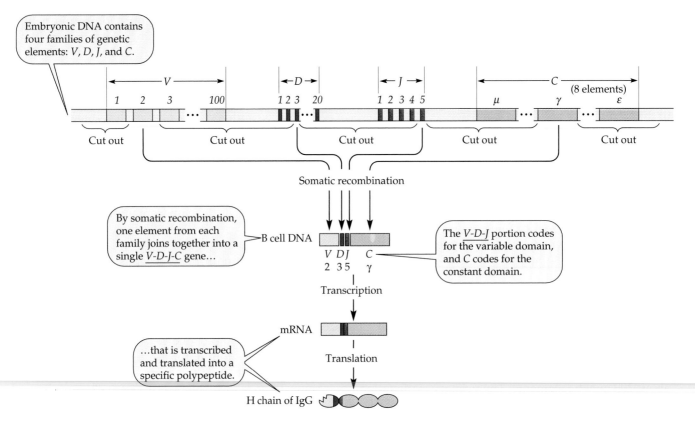

Figure 18.9   The DNA elements on human chromosome 14 that code for immunoglobulin heavy chains. The particular recombination events shown here bring together the second *V* element, the third *D* element, and the fifth *J* element. These join with a gamma *C* region to encode the heavy chain of an IgG molecule.

During the maturation of a particular B cell, intervening DNA regions are cut out, promoting the splicing together, by somatic recombination, of one member of each family to yield a unique **V-D-J-C gene** (second line of Figure 18.10). The gene is then transcribed and translated into a heavy chain.

The combined *V-D-J* portion of the heavy chain gene codes for the variable domain (including the hypervariable regions that characterize the antigen specificity); the *C* portion codes for all of the constant domains, which determine the class of the antibody: IgG, IgM, IgA, or IgE. The number of different *V-D-J* regions that a cell can construct is the product of the number of *V*, *D*, and *J* elements. For humans, the numbers are not known precisely, but there are approximately 100 *V*, 12 *D*, and 4 *J* elements. On the basis of these estimates, $100 \times 12 \times 4 = 4,800$ different heavy chain variable domains can be made (Nossal 1993).

Actually, the number of different heavy chain variable domain polypeptides is more than the 4,800 predicted by the combinatorial shuffling described here. The greater possibilities result from other unusual phenomena, including **somatic hypermutation**. In this process, base substitution mutations occur at a high rate in the combined *V-D-J* DNA during the differentiation of the selected B cell into a mature plasma cell, especially in the hypervariable regions. It is thought that some of these mutations lead to variant antibodies that fit even better with the antigen than the combining site configuration originally displayed on the surface of the selected B cell.

Note that in the case of immunoglobulins, several genes ("minigenes" might be a more accurate description) cooperate to encode one polypeptide, leading to a novel *two (or more) genes for one polypeptide* idea. Note also that we have described the generation of variability only for the H chains. To make an antibody, the H chain joins with an L chain from a somewhat similarly processed gene on chromosome 2 or 22 (two different L chain genes). The degree of light chain variation is less, because a *V* element joins directly with a *J* element; no family of *D* elements exists.

Antibody Engineering

When a virus or bacterium is injected into a test animal, it evokes the production of dozens or hundreds of different antibodies at low concentrations. These antibodies have different combining sites that recognize different aspects of the antigenic surface. All the antibody species, as well as other antibodies from prior

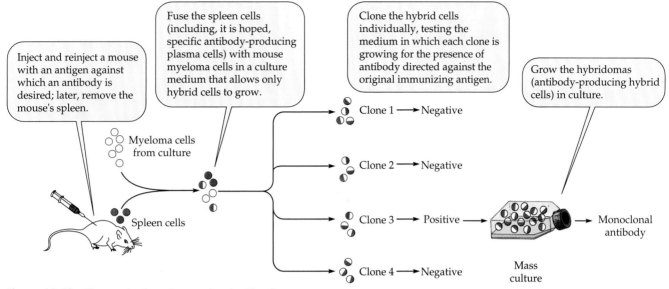

Figure 18.10 The production of monoclonal antibodies. Because each clone arises from a single hybrid cell, it produces a pure antibody.

immune provocations, mix together in the blood plasma. Only by very roundabout (and not always satisfactory) methods can technicians separate the antibodies from each other.

A special method of preparing large amounts of very pure antibodies has generated much interest because of its research, commercial, and medical possibilities. Each uniform product with identical combining sites is called a **monoclonal antibody**. It is made by a clone of cells derived by fusing together a single antibody-secreting *plasma* cell with a *cancer* cell. Although plasma cells by themselves do not live long in culture, cancer cells do, and the hybridization renders the fused cell immortal as well. The plasma cell–cancer cell fusion and the resultant clone is called a **hybridoma**.

The hybridoma technique was developed at Cambridge University by German researcher Georges Köhler and Argentina's César Milstein (1980). The basic procedure, a variant of the somatic cell fusion technique described in Chapter 9, is outlined in Figure 18.10. First, a mouse is injected with the antigen of interest to stimulate the proliferation of plasma cells, and the production of hybridomas follows the additional steps in Figure 18.10. Because each hybridoma clone arises from a single plasma cell, it produces a pure antibody, no matter how complex the original antigenic substance.

Dozens of pharmaceutical and biotechnology companies are developing monoclonal antibodies to use in a growing billion-dollar industry of commercial applications (Kuby 1997). For example, monoclonal antibodies can be used to purify interferon, a virus in-

hibitor, which exists in minute amounts mixed with other proteins. Monoclonals are also used to distinguish between cytotoxic and helper T cells and to type blood and tissue (see next sections). Therapeutic uses of monoclonal antibodies are also being developed. For example, a monoclonal antibody has been used to help prevent rejection of transplanted kidneys. And some cancer cells are vulnerable to immunological attacks, because they acquire tumor-associated (newly "foreign") antigens in the process of becoming malignant. Researchers are trying to bind monoclonal antibodies against these tumor antigens to drugs, toxins, and radioactive isotopes. Such cancer therapies might be better than current chemotherapies or radiation treatments, which kill all rapidly dividing cells, including many normal ones.

One of several problems in using these monoclonal antibodies for human immunotherapy, however, is that they are made by mouse cells, not human cells. As such, these molecules are themselves foreign to humans. If they are not rapidly destroyed when administered to humans (thus providing no therapy), they invite potentially serious immune reactions. For this reason, investigators have been trying to make monoclonal antibodies derived from a *human* plasma cell fused with a *human* myeloma cell. But this approach has not been very successful.

There are several more promising ways to produce useful monoclonal antibodies. In the transgenic mouse approach, sections of human <u>V-D-J-C</u> minigenes (stretching over several megabases) are incorporated into yeast artificial chromosomes (YACs). They are then transferred to a mouse (see Box 17B) whose own immunoglobulin genes are inactivated. Early results show that when such a transgenic mouse is challenged with an antigen, the *mouse* plasma cells respond by making perfectly good *human* antibodies (Neuberger and Brüggemann 1997). These antibodies can then be amplified by the monoclonal techniques described earlier.

Red Cell Antigens and Blood Transfusion

Blood transfusions were first attempted centuries ago, but interest in this form of treatment waxed and waned because patients exhibited an unfortunate tendency to die from it. The following graphic account by a French physician (Denis 1667) describes one patient's reaction to a second transfusion with calf's blood:

> We observ'd a plentiful sweat over all his face. His pulse varied extreamly at this instant, and he complain'd of great pains in his Kidneys, and that he was not well in his Stomach, and that he was ready to choak unless they gave him his liberty.

> Presently the Pipe was taken out that convey'd the blood into his Veins, and whilst we were closing the wound, he vomited... He made a great glass full of Urine, of a colour as black as if it had been mixed with the soot of Chimneys.

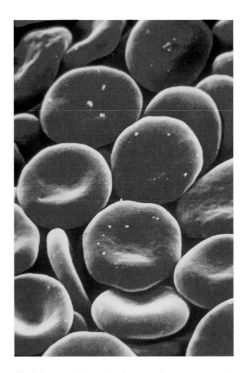

Figure 18.11 Red blood cells seen in a scanning electron microscope, which produces a three-dimensional appearance. The doughnut shape, with the hole partly filled, is clearly evident. (Copyright © David M. Phillips/Visuals Unlimited.)

The consequences of this second transfusion are typical of the hemolytic shock that may accompany an incompatible blood transfusion.

The discovery in 1900 of the ABO blood group system led again to the use of transfusions to treat wounded soldiers in World War I. Even then, the procedure sometimes produced unexpectedly disastrous results. Only after the discovery in 1940 of the Rh blood group system did transfusions become generally safe. Although ABO and Rh are the most important blood groups from a medical point of view, about 20 additional systems (not discussed here) make blood groups very useful in paternity testing, forensic medicine, and anthropology.

Red blood cells, or *erythrocytes*, make up about 45% of blood volume. Shaped like doughnuts with partly filled holes (Figure 18.11), they are seen to advance in single file through fine capillaries. Mature red blood cells have lost their nuclei, and their cytoplasm is packed with hemoglobin to the exclusion of almost everything else. Hemoglobin carries oxygen to the tissues (where it participates in energy production) and returns carbon dioxide (the waste product of these reactions) to the lungs. The average red blood cell circulates for about four months before it declines and is engulfed by macrophages.

On the surface of red blood cells are many types of proteins: receptors, enzymes, transport systems, and cell recognition molecules. Each biological species has some cell surface molecules that act as antigens for-eign to other species. (Hence, cross-species blood exchange usually produces severe immune attack.) Even within a given species, individuals may differ from one another in a set of red cell antigens inherited through corresponding alleles in the genotype. The alleles of one set, with their corresponding antigens, are referred to as a *blood group system*. For example, one set of antigens determines the ABO blood groups, another set determines Rh, and so on.

Some blood group genes have one very common allele and one very rare allele. Because almost all people would be homozygous for the common allele, these genes have limited general interest. About a dozen blood group genes, however, have sizable proportions of *two or more* alleles or phenotypes. Such genes are said to be **polymorphic** (*poly*, "many"; *morph*, "form"). For example, among Americans, the ABO blood types are distributed roughly as follows:

Blood type	Percent
A	40
B	7
AB	3
O	50

For a polymorphic gene, several people picked at random are not likely to have the same genotype, and a relatively high proportion of them will be heterozygous. The *ABO* and *Rh* genes are highly polymorphic.

The antigens present on a person's red blood cells determine his or her blood types. The presence of any antigen is revealed by a chemical reaction with a corresponding antibody. We have noted that the antigen-antibody reaction is a very specific one, so that antigen Z, say, will react only with the antibody anti-Z. The reaction is often done on a microscope slide or in a test tube. A saline drop with red blood cells is mixed with a saline drop containing a known antibody, say, anti-Z. Either the red blood cells clump, a process called **agglutination**, or nothing happens. Agglutination with anti-Z means that the corresponding antigen, Z, is present on the red blood cells; no agglutination means that the antigen Z is absent.

A microscope is rarely needed to see the clumps of red blood cells resulting from agglutination. The phe-nomenon is illustrated in Figure 18.12, showing a sample of red blood cells with antigen A. Two tests are shown, one with anti-A and another with anti-B. Because each antibody has *two* combining sites that can react with the antigen, adjacent red blood cells can be linked into a network of clustered cells large enough to be seen easily. Anti-B is unable to recognize and react with the A antigen because its combining sites are not complementary to the antigen.

The blood groups are thus defined in terms of dif-fering antigens on the surface of red blood cells. Their real functions and evolutionary histories are not clear (although *ABO* and *Rh* genes are similar in many pri-mates). It is important to recognize that the blood group antigens and the resultant phenotypes are gen-erally neutral attributes, not directly causing disease.

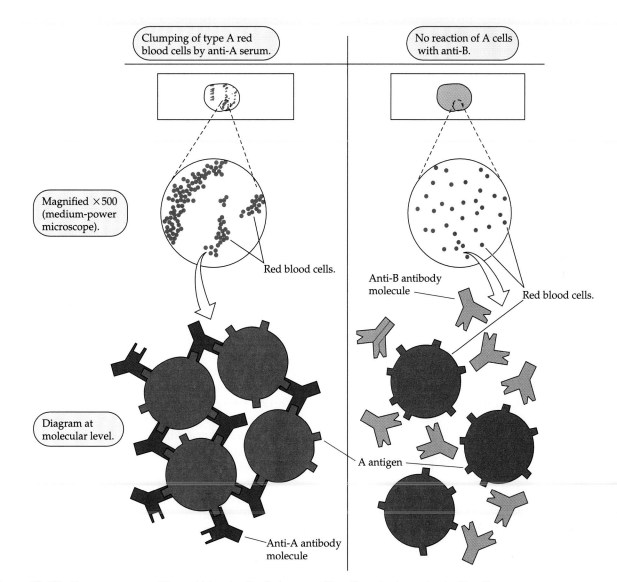

Figure 18.12 The appearance of the red blood cells of a person with antigen A when mixed with anti-A or anti-B. In the molecular view (bottom), the shapes of the A antigen and the antibodies are conjectur-al. The size of these molecules is much exaggerated compared to the size of the red blood cells.

The ABO Blood Group System

Already outlined in Chapter 4 to illustrate Mendel's second law, the ABO blood group system is described again in Table 18.1. The antibodies anti-A and anti-B are used to determine the presence or absence of the corresponding antigens on red blood cells. In the last column, we note that these antigens are controlled by the three alleles *A, B,* and *O* of the *ABO* gene (using shorthand notations for genotype). Note that the symbol A can refer to an antibody, an antigen, a blood type (phenotype), or an allele (genotype). The context in which the symbol is used makes clear which is meant.

Blood types AB and O correspond to genotypes *A/B* and *O/O*, respectively (Table 18.1). Blood types A and B can result either from the corresponding homozygote or from the heterozygous combination with the *O* allele. Thus, the *O* allele is recessive to *A* and to *B*, whereas the *A* and *B* alleles are *codominant* to each other. The specific genotype for phenotypes A and B can sometimes be inferred from the blood types of related individuals. For example, consider the mother, father, and daughter with the ABO blood types shown here:

The mother cannot be genotypically *A/A*, for she would then have had to transmit an *A* allele to her daughter (who lacks an *A* allele). Thus, the mother can only be *A/O*. The daughter cannot be *B/B*, because her mother does not have a *B* allele to transmit. The father, however, can be either *B/B* or *B/O*.

It turns out that the typing antibodies, anti-A and anti-B, can be isolated from the *normal* plasma of some persons, namely, those who *lack* the corresponding antigens. Thus, anti-A is found in the plasma of persons of blood type B or O, and anti-B is found in the plasma of persons of blood type A or O (Box 18A). It is

not known why these antibodies are present (but see question 5). For no other red cell antigens are any antibodies normally present in human plasma.

As you have probably figured out, the presence of anti-A and anti-B in some normal human plasma was the main stumbling block to successful transfusion. For example, type A red blood cells inadvertently given to a type B person will be attacked by the patient's anti-A antibodies.* This antigen-antibody reaction within the body results in the *hemolysis* (breaking open) of the red blood cells, the spilling of hemoglobin into plasma, and *anemia* (too few red blood cells to carry oxygen). Such sudden hemolytic anemia results in fatigue, breathlessness, heart fluttering, jaundice and kidney dysfunction (from breakdown of hemoglobin), low blood pressure, and perhaps death, since the transfusion reaction occurs in addition to the loss of blood that necessitated the transfusion. The prohibited combinations are indicated by ×'s in Table 18.2. The absence of an ×, however, does not guarantee a safe combination. There are two additional precautions in the choice of a donor:

1. The donor must be matched for Rh type, positive for positive, negative for negative (discussed shortly). It is especially important that Rh negative females not be given Rh positive blood.

2. After selecting donor blood by the ABO and Rh label, the donor and patient bloods should be tested for compatibility by *cross matching*. The patient's plasma is added to the donor cells to test for any significant antibody that is present for any reason. If positive reactions occur, another donor is chosen. Positive reactions may result because of a previous transfusion or pregnancy.

A person with type O blood is termed a *universal donor* because that person's blood has neither A nor B antigen to be attacked by recipient antibodies (if any). Some hospitals stock only type O blood to prevent any chance of a transfusion mistake. Blood bank researchers are trying to enzymatically convert stored type A and B blood to type O to increase its supply (Wu 1997). The use of the term *universal donor* is somewhat inappropriate, however. There are many reasons not related to ABO why a type O person may be an unsafe donor. Similarly, the term *universal recipient* for persons of type AB (because they have neither anti-A nor anti-B) may be a dangerous label, for they may have other antibodies that could react with red blood cells transfused to them.

TABLE 18.1   The ABO blood group system

Reaction of red bloods cells with[a]		Antigens on red blood cells	Phenotype (blood type)	Genotype
Anti-A	Anti-B			
+	0	A only	A	*A/A* or *A/O*
0	+	B only	B	*B/B* or *B/O*
+	+	Both A and B	AB	*A/B*
0	0	Neither A nor B	O	*O/O*

[a] + = agglutination; 0 = no reaction.

---

*The reverse reaction—the donor's antibodies attacking the patient's red blood cells—is usually harmless. This is because the donor's antibodies are diluted out when one or a few pints of blood are added to the several quarts of blood in the patient.

TABLE 18.2 Acceptable and prohibited blood transfusions

		Patient's blood type			
		A	B	AB	O
Donor's blood type	A	•	×	•	×
	B	×	•	•	×
	AB	×	×	•	×
	O	•	•	•	•

Dots indicate potentially acceptable transfusions with regard to the ABO blood group system; X's indicate transfusions that are prohibited because the donor's red blood cells will be attacked by the patient's ABO antibodies.

The *ABO* gene has now been isolated and cloned, and the chemical structures of the A and B antigens are known. They are short carbohydrate chains of six sugars that are attached to a carrier molecule, either protein or lipid (Figure 18.13). The carrier is embedded in the red blood cell membrane, and the carbohydrate chain extends out to form the business end of the molecule. That is, the antigenicity—what makes A react differently from B—is a property of the carbohydrate chain. Whereas the terminal sugar in antigen B is galactose, the sugar in antigen A is *N*-acetylgalactosamine. The small difference in these chemicals so alters the immunological specificity of the antigens that severe consequences ensue from a wrong transfusion. Such fine tuning of antigen-antibody reactions was first studied by Karl Landsteiner (Box 18A), who showed that small chemical groups attached to larger molecules were often the fundamental antigenic units.

The A and B antigens are both derived from the same chemical, the **H substance**, by the addition of the appropriate terminal sugars (see Figure 18.13). That is, the *ABO* gene codes for an enzyme that adds to the H substance either galactose (by the enzyme encoded by the *B* allele) or *N*-acetylgalactosamine (by the enzyme encoded by the *A* allele). Nucleotide sequencing has revealed that the *A* and *B* alleles differ by four bases, and the resultant enzymes differ by four amino acids. The *O* allele has a frameshift mutation (a single base deletion); its enzyme product is inactive, having no effect on the H substance. Although everyone has some H substance on their red blood cells, blood type O people have the most. Since the H substance can be an antigen in its own right, it is not quite correct to say that blood type O people lack ABO antigens on their red blood cells.

The Rh Blood Group System

Like the antigens of the ABO system, the antigens of the Rh (or rhesus) blood group system are found on the surface of human red blood cells. The system was so named because one path to its discovery began when Landsteiner and colleagues injected the blood of *rhesus* monkeys into rabbits. Unlike the ABO antigens, however, which differ in carbohydrate side groups, the Rh antigens are proteins directly encoded by alleles of the *Rh* genes. These proteins are apparently needed for stability of the red blood cell membrane, as evidenced by rare persons lacking all Rh antigens who have a lifelong mild to moderate hemolytic anemia. The Rh antigens are similar in structure to proteins that transport substances into and out of cells, but it is not known what the Rh proteins transport.

BOX 18A *The Simple Discovery of the ABO Blood Types*

Karl Landsteiner, of the University of Vienna, knew that mixing bloods of *different* species produced severe reactions (Dixon 1984). This knowledge made him wonder whether lesser reactions might occur between the bloods of individuals of the *same* species. In 1900, Landsteiner took blood from himself and from five of his colleagues and by centrifugation separated the sera* from the red cells. He then recombined the sera and cells in all possible pairings. He noted, for example, that the serum of Dr. Pletschnig strongly agglutinated the cells of Dr. Sturli, but did not affect his own.

*Serum (plural, sera) is blood plasma whose clotting elements are removed or inactivated.

Thus, the serum of Pletschnig seemed to have an antibody directed against an antigen present on one person's red blood cells but not on another's.

The following data are culled from Landsteiner's original paper (1901). (Note that + = agglutination and 0 = no reaction.)

Serum from	Red blood cells from		
	Ple	Stu	Lan
1. Ple	0	+	0
2. Stu	+	0	0
3. Lan	+	+	0

Interestingly, it is possible to conclude from this table that Landsteinder was blood type O, his serum thus containing both anti-A and anti-B. The reason is that the sera of Pletschnig and Sturli agglutinated each other's red cells (lines 1 and 2), so one was type A and the other B. Landsteiner's serum, however, agglutinated the red cells of both his colleagues (line 3).

Only the ABO blood groups could be discovered in the way that Landsteiner did his easy experiment, because only ABO antibodies occur in *normal* serum, that is, in the serum of persons who have not been exposed to foreign antigens. Also for this reason, the ABO antigens and antibodies of persons involved in transfusions must be very carefully checked to prevent an immediate and serious antigen-antibody reaction within the patient's bloodstream.

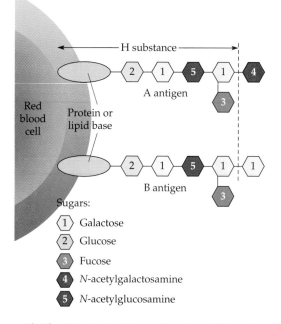

Sugars:
① Galactose
② Glucose
③ Fucose
④ N-acetylgalactosamine
⑤ N-acetylglucosamine

Figure 18.13 Chemical outlines of the A and B blood group antigens. The only difference between the two is the terminal sugar (far right) of the six-unit carbohydrate portion of the molecules. Enzymes encoded by the *ABO* gene add these terminal sugars to a common precursor called the H substance.

Researchers have recently determined that there are, in fact, *two* nearly duplicate *Rh* genes sitting next to each other near the tip of the short arm of chromosome 1.* One of them (called *RhCE*) encodes a variety of mRNA molecules as a result of base substitution mutations and alternative splicings of its exons. Because the resulting assortment of antigens is not of great medical significance, we will not discuss this gene. The other gene (called *RhD* or simply *D*) is apparently not subject to many mutations or alternative splicings. However, the *RhD* gene is easily deleted, which accounts directly for the medically relevant Rh positive (*D* gene present) and Rh negative (*D* gene absent) phenotypes.

Your Rh type is based on a laboratory test with the single typing antibody anti-D, which detects the D antigen encoded by the *D* allele:

Because red blood cells from both *D/D* and *D/d* carry the D antigen and are agglutinated equally well by anti-D, the *D* allele is dominant to *d*. (The lowercase *d* represents the absence of *RhD* gene.)

Upon its two independent discoveries in 1939 and 1940, the D antigen was immediately recognized as a very strong antigen, one that could be responsible for severe transfusion reactions. If blood with the D antigen is mistakenly given to a person who lacks it (Rh positive into Rh negative), the recipient responds by making the anti-D antibody. (No one normally possesses anti-D.) This antibody rarely produces a problem following a first Rh positive to negative transfusion, and it disappears after a time. However, the recipient's memory cells, sensitized to the D antigen, persist. Upon additional transfusions of blood containing the D antigen (should they also mistakenly be given), increasingly severe reactions can be expected as more and more anti-D antibodies are produced by the Rh negative person.

The D antigen is implicated in a disease called *hemolytic disease of the newborn* (HDN), whose occurrence has been drastically reduced since the 1970s. HDN results from an antigen-antibody reaction that occurs within the fetal blood circulation prior to and at the time of birth. The disease usually occurs when Rh negative women are carrying Rh positive fetuses.* The sequence of events leading to HDN is outlined in Figure 18.14.

During one or more Rh positive pregnancies, red blood cells with the D antigen leak into the Rh negative mother's circulation (Figure 18.14A). Although the bloodstreams of mother and fetus do not generally mix, parts of the placental structure separating the two circulatory systems are very thin. Tiny breaks may allow some fetal red blood cells to enter the mother's system. More significantly, at the time of birth, when the placenta separates from the uterine wall, larger amounts of fetal blood may enter the maternal circulation. In response to the D antigen on these fetal red blood cells, the woman is sensitized. But by the time the woman has built up any significant amount of anti-D, the first baby has already been born.

When the Rh negative woman carries a later Rh positive child, the leakage of just a small amount of fetal blood during pregnancy (Figure 18.14B) is sufficient to stimulate a lot of anti-D from memory cells. Then anti-D passes across the placenta and damages red blood cells of the fetus. The probability of HDN increases with subsequent Rh positive pregnancies.

Reaction of red cells with anti-D	Phenotype	Genotype	Frequency (%) in U.S.		
			Whites	Blacks	Natives
Agglutination	Rh positive	*D/D* or *D/d*	84	92	100
No reaction	Rh negative	*d/d*	16	8	0

*When the authors were graduate students, intense acrimony surrounded the question of whether the *Rh* locus was one gene with many different alleles or three closely linked genes, each with a few alleles. By the time molecular genetic researchers were able to clone the DNA segment and show that the truth lay halfway between, the original proponents of the disparate views were all dead.

*Technicians can determine the Rh type of the fetus using DNA from fetal cells obtained by amniocentesis or chorionic villus assay (Chapter 19). The DNA is amplified by PCR, and the *RhD* gene is shown to be present (Rh positive) or absent (Rh negative). The test is tricky, partly because the *RhD* gene is so similar to the *RhCE* gene (Simsek et al. 1995).

(A) **First pregnancy** of Rh negative mother with Rh positive fetus

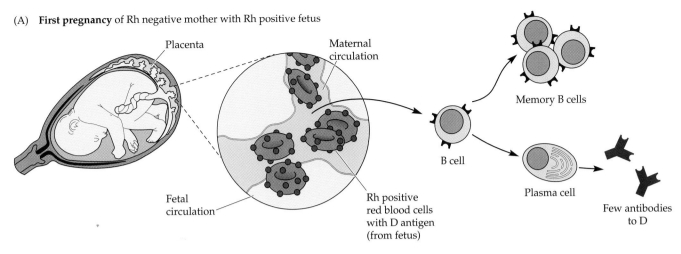

(B) **Later pregnancy** of Rh negative mother with Rh positive fetus

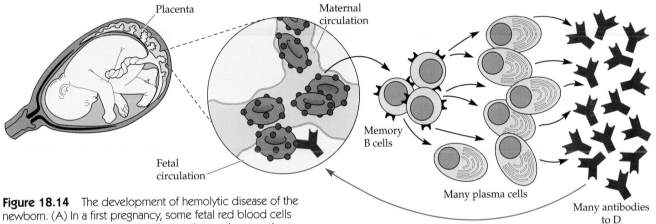

Figure 18.14 The development of hemolytic disease of the newborn. (A) In a first pregnancy, some fetal red blood cells with the D antigen leak into the Rh negative mother, mostly at childbirth. In response, the mother makes some anti-D and memory cells, but the baby is born before any significant damage is done. (B) In a later pregnancy, leakage of fetal red blood cells from an Rh positive fetus during pregnancy stimulates the mother's memory cells to quickly make much anti-D. Entering the fetus, the anti-D destroys the Rh positive fetal red blood cells.

Jaundice is one of the distinguishing symptoms. Another consequence is indicated by the medical term for HDN: *erythroblastosis fetalis*. When mature red blood cells are destroyed in large numbers, the body tries to compensate by releasing into the circulation immature ones, erythroblasts.

HDN is a potential problem whenever a fetus possesses a strong red cell antigen that is not present in the mother, a situation called **maternal-fetal incompatibility**.* In these cases, the allele for the relevant antigen is present in the father, who transmits it to the

*A notable exception is ABO incompatibility—for example, when the mother is type O and the fetus is type A. The mother does not even have to make anti-A, because she already has it. But the normally occurring anti-A of the mother is of a type that does not readily cross the placenta, and problems are rare, although they do occasionally arise.

fetus. If the father is heterozygous, however, not every fetus will be incompatible with its mother. Mothers who are incompatible with their fetuses for the D antigen account for most cases of HDN, although it can happen because of other red blood cell antigens too (including the C and E antigens encoded by the RhCE gene). Cases caused by the D antigen, however, are generally more severe, over 10% of affected fetuses being stillborn prior to full term.

Since the 1970s, couples starting their families have not needed to worry much about D-caused HDN, because the incidence of the disease has been reduced about 90%. The action required to prevent the disease is very simple: Within a few days of the birth of *every* Rh positive offspring to an Rh negative mother, the mother is given a dose of the anti-D antibody. It is important that the Rh negative woman be so injected starting with the *first* Rh positive fetus—whether it is born alive, stillborn, aborted, or miscarried. All pregnancies of Rh negative women married to Rh positive men must be monitored regardless of their outcome.

It might seem strange to administer the very antibody that is unwanted. But by injecting anti-D, the mother is prevented from making her own anti-D. Bear in mind that D-caused HDN occurs after the mother has built up anti-D memory cells as a consequence of a prior Rh positive pregnancy. Upon each subsequent exposure to the D antigen, her own immune mechanism responds more strongly through more and more memory cells. Most of the leakage of fetal red blood cells seems to occur at the time of birth, when the placenta separates from the uterine wall. Thus, the preventive shot of anti-D soon thereafter destroys the invading Rh positive fetal cells before their D antigen has an opportunity to stimulate the mother's immune system (Figure 18.15). Under these circumstances, the D antigen never evokes memory cells in the mother, and the injected anti-D breaks down after a time (Clarke 1968). Immunologically speaking, the mothers treated with anti-D enter their second pregnancy, and any succeeding pregnancy, as if it were their first. In other words, whenever the possibility of D sensitization occurs, it is prevented by injecting anti-D. The source of the therapeutic anti-D is usually the plasma of Rh negative men who volunteer to be injected with Rh positive blood under controlled conditions. Since this supply is not always adequate, researchers are trying to develop anti-D by monoclonal antibody techniques.

Tissue Antigens and Organ Transplantation

The Australian immunologist G. J. V. Nossal (1978) has written,

> There is perhaps something ghoulish in the thought that immediately after a person's death, a whole tribe of surgeons may descend on the body ready to remove kidneys, heart, liver, lungs, pancreas, and any other bits and pieces that they can get hold of. This block is largely in our minds. Most people in good health would agree that their own organs would be better employed, after their death, in helping to keep another person alive than in being burnt in a crematorium or buried in a coffin.

It is, of course, not the role of medicine to confer immortality by continually replacing worn-out parts. Rather, each of us hopes to live happily and, after a few score years, die peacefully with our affairs in order. In achieving these goals, the transplantation of organs has a limited (but much publicized) role. For an individual who might lead a productive life except for damage to one vital organ, a transplant operation is sometimes a last but logical recourse. The most common organ to be transplanted, and the one with the most hopeful prognosis over a long period of time, is a kidney. This and other transplant operations require more than a skilled surgeon working in a modern medical setting. The body's immune system must also be taken into account. If the new organ's foreignness is too great, reaction by the recipient's T cells will lead to the destruction of the intrusive tissue.

To improve the chances of a "take," two avenues are open. One is to repress the normal activity of the recipient's immune system. Drugs such as cyclosporin are routinely used to interfere with the normal working of T cells. But then the patient's weakened immune functions may lead to a fatal infection shortly after a transplant operation or to cancer later in the recovery period. The second method for thwarting the immune system is to choose a donor organ with a minimum degree of foreignness. This matching is done by tissue typing the recipient and potential donors for antigens important to the transplantation.

The HLA System and Tissue Typing

Knowledge of strong transplantation antigens was first obtained in the 1940s from grafts of skin between different inbred strains of mice. The fate of grafted skin depended on so-called **histocompatibility anti-**

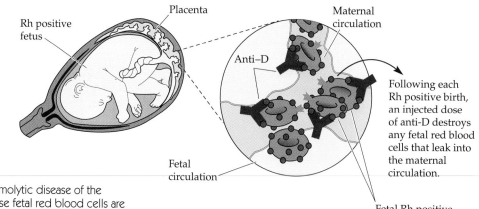

Figure 18.15   Protection against hemolytic disease of the newborn due to the D antigen. Because fetal red blood cells are destroyed, the mother is never sensitized to the D antigen. The injected anti-D disappears after a few months.

gens (*histo*, "tissue") on cell surfaces. Rejection occurred if tissue from the donor strain had histocompatibility antigens foreign to the recipient. (Perhaps a better term would be hist*oin*compatibility antigens.) Although many different genes and antigens were involved, researchers found that one series of closely linked genes coded for particularly strong cell surface antigens that led to quick rejection if not matched.

Beginning in the 1950s, work on human transplantation antigens and genes was done by several researchers. They found dozens of antigens on the surface of white blood cells and other cells of the body. The typing tests are somewhat more complicated than the simple agglutinations used to classify red cell antigens, and initially they gave inconsistent results. But now, pure preparations of monoclonal antibodies for typing white cell and tissue antigens have produced uniform results worldwide (Kuby 1997).

The strongest tissue antigens are encoded by six or more genes on human chromosome 6, spanning a region of over 3 million bases (Figure 18.16). The genes and antigens are known collectively as the *MHC system* (for major histocompatibility complex, a general name applying to any species) or the **HLA system** (for human leukocyte, or white blood cell, antigens). The gene products are cell surface polypeptides of two distinct classes, as indicated in the figure. Other less important, but still significant, histocompatibility loci are located at many sites elsewhere in the genome.

The MHC proteins were noted earlier in this chapter (see Figure 18.5). Recall that to be identified by the cells of one's immune system, a foreign antigen must be presented on cell surfaces in conjunction with one's particular MHC proteins. This recognition step helps ensure that immune cells react only to foreign antigens. Discriminating between self and nonself is, of course, the real function of the protein molecules encoded by the MHC genes, rather than tissue rejection. Organ transplants do not occur in nature. Had the antigen presentation function of MHC proteins been discovered before their transplant rejection side effects, the molecules would have been named differently: something like APP (for antigen presentation proteins).

With large numbers of fairly common alleles, the HLA genes are the most polymorphic protein-encoding loci known in humans. About 50 different alleles of the *B* gene, about 10 of the *C* gene, and about 25 of the *A* gene control the presence of roughly 85 different white cell antigens (detected by 85 corresponding HLA antibodies). In any given group of people, the frequencies of many of the alleles are substantial (Table 18.3). Differences between populations have been well documented by anthropologists and others. For example, in Table 18.3, note that the most frequent *B* allele among American whites is the one designated *44* (at 14%); among American blacks, it is allele *58* (at 10%); and among Mexicans, it is allele *35* (at 24%).

A single chromosome 6 will have all three genes, *HLA-B*, *-C*, and *-A*, lined up. The alleles of these genes will almost always be inherited together, because crossing over is uncommon within the short intervals involved. The number of possible combinations of *B*, *C*, and *A* alleles on one chromosome is about 50 × 10 × 25 = 12,500, since (in theory) any *B* allele can be associated with any *C* allele and with any *A* allele. A particular combination is called a **haplotype**; for example, a chromosome 6 carrying *B7*, *C4*,

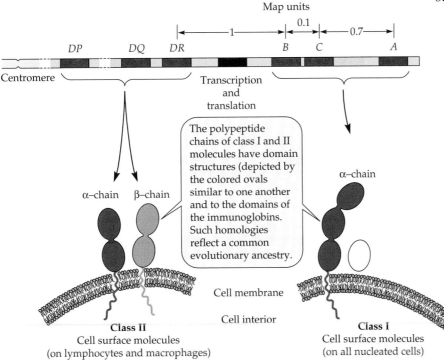

Figure 18.16 The genes and polypeptides of the HLA system. The genes on chromosome 6 are separated by map distances shown at the top. The polypeptides also have about 50 amino acids that continue the chains through the cell membrane into the cell interior. (The white chain on the class I molecules is not encoded by the HLA system.) This gene-rich region also includes (within the black bar) other immune-related systems (e.g., genes for complement) as well as the 21-hydroxylase genes involved in congenital adrenal hyperplasia (Chapter 16).

TABLE 18.3 The more common alleles of the *HLA-B*, *HLA-C*, and *HLA-A* genes

Allele	Percentage among		
	American whites	American blacks	Mexicans
B7	10	9	3
B8	9	3	3
B35	8	6	24
B44	14	7	19
B58	1	10	1
C2	5	12	4
C3	12	9	10
C4	10	16	22
A1	14	3	10
A2	28	15	25
A3	14	7	3
A23	3	10	1
A24	7	3	14
A25	2	0	11
A30	3	15	2

Source: Tiwari and Terasaki 1985.

and *A3* has the haplotype B7 C4 A3. For simplicity, we have omitted discussion of the several *D* loci, although polymorphisms for these genes are also very important in transplants.

The HLA genotype of a person consists of two haplotypes—one for each chromosome 6. Considering just the *B*, *C*, and *A* genes, we calculate that there are more than 78 million—(12,500)(12,501)/2—different HLA genotypes, and much, much more if we were to consider the dozens of alleles at the *D* loci as well. Because no one genotype is particularly common, no two unrelated people are likely to have exactly the same set of tissue antigens. This great diversity of haplotypes is one reason why organ transplants may fail when the donor tissue has not been specifically chosen to match the recipient.

Transplantation

The fewer the foreign HLA-encoded antigens present on a donor organ, the more likely it is to escape rejection by the host's T lymphocytes. By foreign, we mean HLA antigens on the donor tissue that are different from the host's own HLA antigens. In addition to the HLA system, transplants need to be matched for the ABO blood group system, because the strong A and B antigens appear on many cell types besides red blood cells.

The best donor for a *kidney transplant*, after an identical twin, is a sib with the same HLA and ABO antigens as the recipient. In more than 90% of these cases, the transplanted kidney is still functioning a year later; in more than 80% of cases, the kidney survives five years or more. The survival time of a grafted kidney is reduced when a sib donor has one or more HLA anti-

gens foreign to the recipient. In any event, the donor must be exceptionally well motivated, since the removal of an organ and subsequent existence with only one kidney entail some risk.

Kidneys from unrelated people can also be successfully transplanted, and tens of thousands of such operations have been performed since the 1960s. The donor is often an auto accident victim who is certified brain dead (the usual criterion for death), but whose other body systems can continue to function with external life-support machinery until organs are removed. Although not all transplant surgeons believe that the advantages of tissue typing outweigh possible technical and administrative delays, studies show that the long-term survival of cadaver grafts is related to the closeness of the HLA matching (Sanfilippo 1994).

Unlike solid organs, the cells of the *bone marrow* continually renew themselves from stem cells, and a portion may be removed from a living donor without long-term loss of function. The physician inserts a long needle into the upper part of the hip bone of the donor (who is under general anesthesia) to remove about 500 milliliters (1 pint) of the spongy marrow contained within. The cell suspension is filtered to remove chips of bone and clumps of fat and then slowly injected into a vein of the recipient. Such transplants are increasingly successful in treating patients with marrow defects, such as leukemia (Chapter 17), immunodeficiency diseases, and some types of anemia. Marrow transplants have also been tried on a number of patients suffering from inborn errors of metabolism (including thalassemia and sickle-cell disease) that are clinically expressed in the cells of bone marrow. Even more generalized genetic diseases may be treatable if the critical enzyme is transported from grafted marrow to tissues where it is needed.

Before a leukemia patient is given a bone marrow transplant, physicians try to kill 100% of the leukemic cells with drugs and radiation, forms of treatment that destroy *all* rapidly dividing cells (causing hair loss and nausea due to death of hair follicle cells and cells lining the intestines). Then they inject donor marrow cells, taken from a sib with the same HLA type, that may seed the patient's bone marrow cavities and eventually restore immune functions. However, several adverse consequences are possible: (1) recurrence of leukemia if all malignant cells were not destroyed, (2) serious microbial, viral, or fungal infections while the patient's immune system is knocked out, or (3) *graft-versus-host disease*. The latter reaction occurs when the lymphocytes normally present in the *donor* blood marrow mount a pervasive and painful immunological attack against antigens on the cells of the recipient—the reverse of the usual transplant complication. The frequent occurrence of graft-versus-host disease, even when donor marrow is from an HLA-identical sib, reveals the importance of histocompatibility antigens other than HLA. Graft-versus-host disease is fatal in about 25% of patients with acute leukemia receiving bone marrow.

The problem of sharply limited supply hampers all organ transplantation programs, and many patients die while waiting for their last chance to live. About 20,000 transplants are performed each year in the United States, but several times that number would be performed if the organs were available. For this reason, the development of artificial organs is being pursued, as is research devoted to **xenotransplantation**, the use of organs from animals (Greek xeno, "alien").

Recent attention has focused on pigs, some of whose organs have about the right size and physiology for transplanting to humans (Edwards 1996; Lanza et al. 1997). The use of pigs, which are regularly killed for food, does not engender as many emotional objections as the use of primates, such as chimpanzees and baboons. Furthermore, incorporating human genes into pigs has already been accomplished (Velander et al. 1997). (An appropriately transgenic pig organ might be more acceptable in a human host.) The pig donors will have to be free of viruses that could infect human cells. Although the possible cross-species transmission of viruses is a major consideration, the pig is a better choice in this regard than primates because of its greater evolutionary distance from us (Butler 1998; Murphy 1996). Researchers will also have to devise ways to overcome the quick (within minutes) and severe rejection—the so-called *hyperacute rejection*—that accompanies any organ transplanted across species lines. This and other problems will require extensive and careful investigation before xenotransplantation is proved feasible and safe.

Finally, we note the following transplantation-related puzzle: How does a pregnant female come to accept and nourish a fetus that usually has significant foreign histocompatibility alleles inherited from the father? The fetus is intimately "grafted" inside her uterus. If displayed to the mother's immune system after a surgical transplant, the same foreign antigens would elicit strong antibody responses leading to graft rejection. One interesting observation is that the fetal cells that contribute to the placenta and make contact with the maternal bloodstream do not in fact display major fetal HLA antigens. It is a mystery how this masking is achieved. The altered role of HLA antigens in pregnancies is further underscored by a few studies that show, surprisingly, that some HLA *mis*matches between mother and fetus are advantageous for a successful pregnancy (Ober 1995, 1998; Rodger and Drake 1987). Thus, the maternal-fetal union seems to have developed interesting but poorly understood modifications of immune mechanisms.

Immune System Disorders

The complexity of the immune system suggests that we should be subject to numerous immune disturbances, and we are. A person may lack fully functional B cells, T cells, or phagocytic cells. The complement system, which helps destroy invading bacteria, may not work. Cancers such as leukemia may develop in lymphocytes or other immune cells. Hyperactivity is the problem in allergies and in anaphylaxis, a severe and rapidly developing form of allergy (e.g., to a bee sting). Misdirected activity is seen in autoimmune disorders, in which antibodies attack one's own tissues. Our brief discussion includes a few immune deficiency and autoimmune diseases.

Immune Deficiencies

In about 50 mostly quite rare diseases, the immune system functions poorly. *Agammaglobulinemia* (*XLA*), for example, with a frequency of about 1 in 50,000 male births, is an X-linked, recessive disease (Figure 18.17). The encoded protein is a kinase involved in signal transduction through the cytoplasm, but its exact role is not understood. Some suggestions indicate that *V-D-J-C* rearrangements in B cell maturation do not occur properly. Affected male infants have no de-

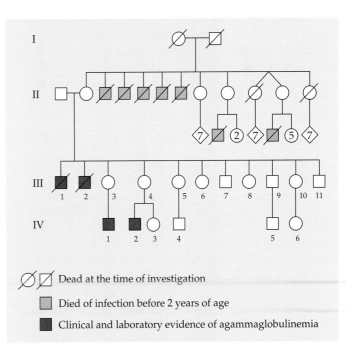

Figure 18.17 Pedigree illustrating X-linked inheritance of agammaglobulinemia. Affected males are hemizygous for a recessive mutant allele, and their mothers are heterozygous. (From Conley et al. 1986.)

tectable mature B cells (or subsequent antibody-secreting plasma cells) and virtually no immunoglobulin of any kind, except for their initial charge of maternally transmitted antibodies. (Recall that IgG, the major antibody class, readily crosses the placenta. A normal baby's ability to manufacture its own antibodies develops gradually after birth.) Infants remain well for about a year, but thereafter suffer from recurrent bacterial infections. Survival depends on periodic injections of gamma globulin that contain antibodies to common bacterial diseases. Because they possess normal T cells, however, the patients are no more susceptible to common viral diseases than other infants.

One interesting incident involves the initial description of XLA (the first explanation of *any* immunodeficiency disease) by Ogden Bruton at Walter Reed Army Hospital in 1952. Bruton ordered an analysis of the plasma proteins of a young boy suffering from recurrent infections. The hospital laboratory reported at first that their newly installed electrophoresis apparatus was not working properly because it revealed no gamma globulins in the boy's plasma (Manis and Schwartz 1996).

X-linked severe combined immunodeficiency disease (*XSCID*), about 1 in 100,000 male births, involves the malfunction of both B and T cells. Infants fail to thrive because of devastating bacterial, fungal, and viral infections, and they usually die in their first years. The gene defect involves the so-called gamma (γ) polypeptide chain, one of several polypeptides comprising the B cell receptor for the regulating molecule *interleukin-2*. The severity of the disease was a mystery for a while, because the complete absence of interleukin-2 results in a disease with far milder symptoms. Researchers have now found that the γ polypeptide chain is a constituent not only of the interleukin-2 receptor, but of receptors for other lymphokines (interleukin-4, -7, -9, and -15) present on many types of immune cells (Rennie 1993).

A few XSCID patients have been treated with a bone marrow transplant. Such was the case with David, the "bubble boy," described at the beginning of this chapter. At age 12, he received bone marrow from his older sister, who was the most closely (but still not perfectly) HLA-matched donor available. Prior to infusion into David, her marrow cells were treated with a monoclonal antibody against an antigen present on mature T cells. The object was to kill donor cells that might otherwise initiate graft-versus-host disease. Although most likely due to transplant complications, the exact cause of David's death four months after the operation is not clear.

A distinctive variant of SCID called *adenosine deaminase deficiency* is inherited as an autosomal recessive. This disease, in which DNA synthesis is abnormal, is discussed in Chapter 20, because two young girls with adenosine deaminase deficiency have become the first patients to be managed by the new procedure of *gene therapy*.

Acquired immune deficiency syndrome (*AIDS*) is due to the human immunodeficiency virus (HIV), which invades several cell types having the *CD4 receptor* on their surfaces, especially macrophages and helper T cells (Gallo and Montagnier 1988; Greene 1993). (Two variants of the virus, HIV-1 and HIV-2, are recognized.) Immune functions related to both T cells and B cells gradually deteriorate. Usually, no overt symptoms occur for up to a year after infection with HIV, although laboratories can detect anti-HIV antibodies against several different proteins in the viral coat long before clinical onset. For most patients, swollen lymph nodes are the first sign of disease, as the virus rapidly destroys T cells, which are at first replaced equally rapidly. Discovery of this frenzied (and previously unsuspected) activity in lymph nodes by David Ho of the Aaron Diamond AIDS Research Center in New York City helped earn him recognition as *Time* magazine 1996 Man-of-the-Year (Gorman 1996/1997).

Containing RNA rather than DNA, HIV is a retrovirus with about 10,000 ribonucleotide bases. It uses its own reverse transcriptase enzyme to copy the viral RNA into DNA, which is then inserted into the genome of an immune system cell (Figure 17.15). Once integrated, the provirus replicates along with the cell's own DNA. Later, the provirus DNA can be transcribed and translated, leading to the assembly and release of large numbers of new virus particles and the death of the cell.

So-called *opportunistic infections*—those caused by microbes that are commonly present but usually effectively controlled—take increasing advantage of a weakening immune system (Mills and Masur 1990). *Thrush*, a fungal growth in the mouth, is one of several diseases that afflict the mucous membranes and the skin. Other microorganisms include the parasite *Pneumocystis carinii*, which causes pneumonia. (In 1981, five surprising cases of this previously very rare type of pneumonia were the first indication of AIDS in the United States.) Fever, diarrhea, and weight loss weaken the body. Cancer, typically the otherwise rare Kaposi sarcoma, and brain damage are not uncommon. Although some of the disease symptoms can be prevented or eased with medical care, the hard reality is that many patients die within a dozen years after being infected with HIV.

The drug *AZT* (azidothymidine, or zidovudine) and several closely related drugs work by interfering with the reverse transcription step. The newer *protease inhibitor* drugs work by preventing the cutting of long viral proteins into shorter functional units that are assembled into new virus particles. The most effective therapies involve several drugs at once to overcome the high mutability of the virus that can produce resistance to any one drug (Culliton 1996). Although the drug combinations directed against different stages in the HIV life cycle are more potent than prior therapies, they are not free from serious side effects. In addition, perhaps one-third of patients fail to respond to the combination treatment from the outset or after a period of time (Stolberg 1997). The pills are also expensive, and dozens of them per day must be taken according to a strictly regulated protocol (Altman 1996; Cohen 1997). The next generation of therapies may be directed toward coreceptor molecules (in addition to CD4) to which the HIV virus must attach to gain entry to immune cells. Some men with defective coreceptors (due to homozygosity for the encoding gene) turn out to be completely healthy and immune to AIDS (Fauci 1996; O'Brien and Dean 1997).

AIDS is transmitted almost entirely by sexual acts or blood exchanges and also from mothers to their fetuses or nursing babies. It is prevalent among, but by no means limited to, homosexual men, intravenous drug users and their sexual partners, and recipients of blood products. Because of the mutability of the virus and the complexity of the disease processes, the development of protective vaccines has proved difficult. In addition to possible vaccines and new drugs, worldwide education and modification of transmission-related behaviors appear to be good strategies for containment (Francis and Chin 1987).

Autoimmune Disorders

As a group, people with autoimmune disorders develop antibodies—so-called *autoantibodies*—against many different normal components of the body. Present in about 5–7% of the population, the puzzling diseases are chronic, often developing in middle age, and are treatable to a limited extent with immunosuppressive, anti-inflammatory, or other drugs. Autoimmune diseases usually have a genetic component, but do not show simple Mendelian inheritance, a finding suggesting multiple causative genes and substantial environmental influences, the latter including exposure to particular microorganisms, foods, or drugs. Some environmental agents may contain molecules that mimic one's own molecules (Blakeslee 1996). Thus, normally produced antibodies against foreign antigens may also attack self antigens. Another explanation of autoimmune disorders was noted previously: T cells that could damage the body by binding to self antigens are normally removed by the thymus gland during early development. If this removal mechanism is imperfect, immune cells that damage one's own tissues could remain in the body (von Boehmer and Kisielow 1991). Curiously, autoimmune diseases often occur in people who carry particular alleles of the HLA system (Table 18.4).

In *systemic lupus erythematosus* (*SLE*, or simply *lupus*), for example, autoantibodies are found throughout the body. They are directed against a wide array of components of the cell nucleus (DNA, RNA, histones), cytoplasm (mitochondria, lysosomes, ribosomes), and elements of red blood cells and platelets. The disease affects women ten times more often than men, often striking those between 15 and 40 years of age (Rosenthal 1989). Periods of remission alternate with periods of suffering, including weakness, fever, seizures, psychoses, joint pains, and skin rashes. (It was once suggested that the cheek rash of SLE patients resembled the face of a wolf—thus the Latin term *lupus*.) Serious kidney damage results from deposits of antigen-antibody complexes that block fluid flow, contributing to a shortened life span. Genetic factors in SLE are indicated by much higher concordance in monozygotic than in dizygotic twins. SLE is weakly associated with the presence of HLA antigen DR3 for reasons that are not clear.

TABLE 18.4 Some autoimmune disorders and their HLA associations

Disorder	Target of autoantibodies	HLA antigen association	Relative risk[a]
Graves disease	Thyroid gland	B8, DR3	3
Juvenile diabetes	Insulin-making cells of the pancreas	DR3, DR4	20
Multiple sclerosis	Myelin sheaths around nerve cells	DR2	5
Myasthenia gravis	Receptors in nerve-muscle junctions	DR3	10
Rheumatoid arthritis	Joints	DR4	10
Systemic lupus erythematosus	Brain, joints, skin, kidneys, blood vessels, and other organs	DR3	5

[a]Likelihood of developing disease compared to the general population. For example, people with antigens DR3 and DR4 are 20 times more likely to have juvenile diabetes than others. (Risk figures from Kuby 1994.)

Juvenile diabetes (*type I, insulin-dependent diabetes mellitus*, or *IDDM*) affects about 700,000 people in the United States. It is second to asthma as the most serious chronic disease of children, the peak age of onset being 12. Although the beginning of clinical symptoms may be sudden, the disease develops only after about 80–90% of the insulin-producing (beta) cells scattered throughout the pancreas have been destroyed over a course of time by autoantibodies and T cells. Without adequate insulin, glucose in the blood is unable to move across cell membranes into cells, where it is used for energy production. The complications include blindness and kidney failure (Figure 10.6).

Histocompatibility antigens DR3 and DR4 are important contributors to the onset of juvenile diabetes in ways that are not well understood. In fact, 95% of diabetic individuals—but only 45% of the general population—have one or the other HLA antigen or both. It is puzzling that persons heterozygous for alleles DR3 and DR4 have the greatest risk of diabetes, and they develop the disease at the youngest ages (Raffel et al. 1997). Since about 1 in 20 children of a diabetic parent gets diabetes (versus 1 in 500 for the general population), several genes are probably involved. A major gene is closely linked to the HLA region on chromosome 6. However, a half dozen additional minor genes have been identified that contribute to the occurrence of juvenile diabetes, including on chromosome 11 the regulatory region that controls transcription of the insulin molecule itself. No one knows what genetic or environmental factors actually trigger the autoimmunity. Some researchers suggest the involvement of viruses that display antigenic proteins that happen to mimic proteins in insulin-producing cells of the pancreas (Atkinson and Maclaren 1990). Standard therapies for juvenile diabetes include careful control of diet and daily injections of insulin. Experimental therapies include transplantation of pancreatic beta cells taken from human cadavers or from pigs (Lacy 1995).

Summary

1. The immune system identifies foreign antigens, responds specifically to their shape and chemistry, and reacts more vigorously to them upon subsequent exposures.

2. Immune responsiveness relies on white blood cells (B cells, T cells, macrophages) found in the lymphatic and circulatory systems. The thymus gland removes T cells that would react to one's own tissues, providing tolerance to self antigens.

3. As an immune cell (B or T) matures, it develops receptors that recognize a single foreign antigen. When stimulated by that antigen (displayed in conjunction with a self antigen of the MHC system), the immune cell grows into a clone of immunologically active cells and a reserve of memory cells.

4. Antibodies are Y-shaped protein molecules. The specificity of the antigen-binding pockets at the tips of the Y is determined by the particular amino acids of the hypervariable regions.

5. The great diversity of antibodies that plasma cells can make results in part from the multitude of ways that *V, D,* and *J* "minigenes" recombine with each other during somatic differentiation of lymphocytes.

6. Although the many antibodies that are present in any person's plasma are a mixed lot, a monoclonal antibody preparation is pure and homogeneous. Monoclonals are produced by hybridomas, each made by fusing a specifically activated plasma cell with a cancer cell.

7. Red blood cells have antigens that determine the various blood group systems. Each set of antigens is controlled by a different—often polymorphic—blood group gene. Red blood cells carrying a given antigen are detected by agglutination with the corresponding antibody.

8. The ABO antigens are the most important in transfusions because anti-A and anti-B antibodies occur naturally in the plasma of certain persons. The A and B antigens differ by a single sugar residue. They are synthesized from the H substance (found in type O persons) via enzymes encoded by the *A* and *B* alleles, respectively.

9. Hemolytic disease of the newborn is due to maternal-fetal incompatibility for any strong antigen, but most cases result from the Rh (rhesus, or D) antigen. Nearly complete protection against Rh-caused hemolytic disease is afforded by injections of anti-D following every Rh positive pregnancy of Rh negative women.

10. The human histocompatibility antigens (HLA) are encoded by several highly polymorphic, closely linked genes—the major histocompatibility complex (MHC). The particular alleles of the linked HLA genes are inherited as a group called a haplotype.

11. Matching the HLA antigens of donor and host helps make transplants successful by minimizing immune rejection of donor cells. Bone marrow transplantation is useful for some types of leukemia and other diseases, but patients are subject to graft-versus-host disease. Xenotransplants (from animals to humans) are still experimental.

12. Immune deficiency diseases include inherited disorders such as agammaglobulinemia and XSCID. In AIDS, the HIV virus preferentially attacks helper T cells, leading to microbial diseases that take advantage of a weakened immune system.

13. In autoimmune diseases, such as systemic lupus erythematosus and juvenile diabetes, autoantibodies attack one's own tissues. These diseases have a mix of genetic and environmental causes.

Key Terms

agglutination	constant (C) domain
antibody	haplotype
antigen	heavy (H) chain
B cell	helper T cell
clonal selection theory	histocompatibility antigen
combining site	HLA system

Key Terms *(continued)*

H substance	memory cell
hybridoma	MHC restriction
hypervariable region	monoclonal antibody
immunoglobulin	plasma cell
killer T cell	polymorphic
light (L) chain	secondary immune response
lymph	somatic hypermutation
lymphocyte	somatic recombination
lymphokine	tolerance
macrophage	*V-D-J-C* gene
major histocompatibility	variable (V) domain
complex (MHC)	xenotransplantation
maternal-fetal incompatibility	

Questions

1. What is the cellular basis of immunological memory—that is, the stronger antibody response to a second exposure to an antigen?

2. Explain how patients with multiple myeloma contributed to the determination of antibody structure.

3. A protein-splitting enzyme called papain (from the papaya tree) cuts the antibody molecule into exactly three pieces. Two are identical to each other and bind antigen well; the third piece does not bind antigen at all. Where does papain split the antibody molecule? (See Figure 18.6.)

4. In 1945, a California woman accused Charlie Chaplin of fathering her child. The ABO blood types were: woman, A; Chaplin, O; child, B. When a jury decided that Chaplin was the father, an editorial writer commented that California had declared that black is white and up is down (*Boston Herald*, April 19, 1945). What do you think?

5. Substances that resemble the A, B, and H antigens are widespread in nature, and we undoubtedly eat some. Does this suggest where normally occurring anti-A and anti-B in human plasma might come from?

6. Within each pedigree symbol is recorded the reaction of each person's red blood cells with anti-A and anti-B of the ABO system and anti-D of the Rh system, in that order. Give the ABO and Rh *genotypes* of each person. (Note: + = agglutination, 0 = no reaction.)

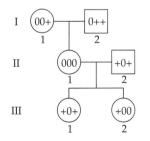

7. Considering the *ABO* and *Rh* loci together, what *phenotypes* would you expect among the offspring of the mating *A/B D/d* × *A/O D/d*? What is the expected proportion of each phenotype?

8. A baby is born with hemolytic disease of the newborn, but both the baby and its mother are Rh positive. Give a reasonable explanation for this occurrence.

9. Human volunteers can be used to produce the Rh antibody anti-D for typing and therapy. Who should or should not volunteer, and what is actually done?

10. In the mouse, there are approximately 150 *V* elements, 12 *D* elements, and 4 *J* elements coding the variable domain of the H chain of IgG. How many different heavy chain variable domains can a mouse make?

11. What are the HLA genotypes (for just the *B* and *A* genes) of the family members below? Indicate properly as haplotypes. (*Note:* + = antigen present, blank = antigen absent. *Hint:* Start with the first child, who is the only one homozygous at the *B* site.

	B antigens			A antigens			
	5	8	40	2	3	9	11
Father	+		+	+		+	
Mother	+	+			+		+
First child	+			+	+		
Second child	+	+		+			+
Third child		+	+			+	+

12. For the family in question 11, would the first or third child be a better kidney donor to the second child?

13. We stated that the *HLA* genes are the most polymorphic human genes. In previous chapters, however, we noted the hundreds of mutations of several disease-causing genes, such as those for breast cancer and cystic fibrosis. Wouldn't the latter be considered more polymorphic?

14. The cornea (the transparent front surface of the eyeball) has no blood or lymphatic vessels. Does this suggest why corneal transplants from unrelated donors are usually successful without using immunosuppressants?

15. In bone marrow transplants, patients are usually irradiated to destroy their immune cells, making way for sites to be populated by the donor's immune cells. Yet when David, the "bubble boy," received a bone marrow transplant from his sister, he was not irradiated. Why?

16. A normal father and mother have a son with XSCID. The mother is pregnant again. What is the probability that the newborn will have XSCID if prenatal diagnosis shows that the fetus is male (as in the case of David, the "bubble boy"? What if the fetus is female?

17. In conjunction with other modern medical practices, organ transplants have led to a reexamination of how we define the time of death. Discuss.

Further Reading

The immune system is a favorite subject for articles in *Scientific American* and other basic science magazines. *SA* collections include the September 1993 issue and edited books by Paul (1991) and Burnet (1976). Additional *SA* articles are by Lanza et al. (1997) on xenotransplantation, by Old (1996) on cancer immunotherapy, by Lacy (1995) on diabetes, and by Englehard (1994) on antigen processing. Excellent chapters on immunology are in the cell biology textbooks by Lodish et al. (1995) and by Alberts et al. (1994). More details on basic and clinical immunology can be found in the textbook by Kuby (1997).

PART 6

Practices
and Prospects

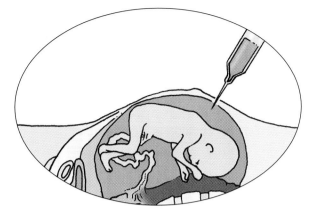

CHAPTER 19

Detecting Genetic Disease

Carol Terry's troubles seemed to begin suddenly at age 25 when she cut her wrists, although she said she didn't especially want to die. She had just wanted to take some action, but she hadn't known what action to take or why any action had been needed. Then, over the course of two years, she saw a series of psychiatrists who moved her in and out of mental hospitals, put her into group therapy sessions, and gave her antidepressants, tranquilizers, and electroshock treatments—all to no avail. Her hands became progressively more shaky and clumsy, her mouth became twisted, and she drooled. All this, the psychiatrists said, was the result of hysterical neurosis—her way of seeking attention and showing resentment toward her husband. After the divorce, her symptoms worsened: She had trouble swallowing and lost 25 pounds; also, her right leg went limp, her balance was unsteady, and her speech became badly slurred and squeaky. A series of physical tests revealed an enlarged and damaged liver, the result—it was thought—of all the drugs.

When her insurance ran out, Mrs. Terry entered a community mental health center as a welfare patient. In that unpromising situation, her frightening experience came to an end. David Reiser, a staff physician with a good memory, thought that her appearance and behavior were characteristic of a rare autosomal recessive disorder called *Wilson disease*, whose frequency is about 1 in 50,000 people. He had seen one case in medical school. In patients with Wilson disease, copper accumulates in the body in large amounts, for reasons that are not well understood. An essential trace element, copper is needed for the proper functioning of several enzymes. But excess copper gets deposited in the liver and brain, where it causes serious damage. It also forms golden-brown circles, called Kayser-Fleischer rings, that hide the outer portion of the iris of the eyes (Figure 19.1). The rings are definite marks of Wilson disease, and those in Mrs. Terry's eyes were easily visible.

People with Wilson disease can be helped if the disease is caught early enough. The drug penicillamine binds to copper and causes it to be excreted in the urine. With this treatment, Mrs. Terry's psychiatric symptoms quickly disappeared, her liver slowly returned to near normal, and the Kayser-Fleischer rings faded. She completed a university degree, went to work as an accountant, and remarried. The physical reminders of her brush with death were a slight limp, an occasional hesitation in speech, and a somewhat crooked smile (Roueché 1979).

Wilson disease was first described in 1912. At about the same time, Reginald Punnett (1911), the inventor of the checkerboard for solving genetics problems, wrote: "Increased knowledge of heredity means increased power of control over the living thing, and as we come to understand more and more the architecture of the plant or animal we realize what can and what cannot be done towards modification or improvement." Punnett was thinking of the application of Mendel's laws to farm production, but knowledge is power in human genetics as well.

Recent advances in molecular processes have expanded our understanding of our minds and bodies in regard to health and disease. Application of this knowledge to ourselves and to the world about us is viewed in different ways by different people. Pessimists predict that the new biotechnology will generate a host of problems: damage to the environment or to human health, intrusions into private affairs, unethical or irreligious experimentation on animals or human embryos, perversions of conventional methods of human reproduction, and corruption of human individuality. Optimists, meanwhile, expect genetic engineering to contribute greatly toward alleviating hunger and disease. And, especially in the last decade, researchers have developed powerful molecular methods for probing human disease—methods that may ease the burden of hereditary disorders (Reiss and Straughan 1996).

In this chapter, we examine further the implications of advances in human genetics for individuals, families, and society. We begin by surveying the growing field of genetic counseling, including prenatal diagnosis and genetic screening, areas of both promise and controversy. Although relatively few heritable illnesses can now be effectively treated, there is nothing inherent in genetic diseases that make them untreatable. Indeed, as many infectious diseases are being brought under control, disorders that are wholly or mostly genetic (e. g., cystic fibrosis, Duchenne muscular dystrophy, and Huntington disease) or only partly genetic (e.g., Alzheimer disease, breast cancer, and diabetes) have been receiving increasing attention. In Chapter 20, we will discuss treatments for genetic disease, including the limited use of gene therapy—a remarkable application of molecular genetic technology. We will also discuss the implications of the recent success in cloning mammals. Finally, we will ask if anything can (or should) be done to reduce the frequency of genetic conditions in future generations.

The problems inherent in many of the genetic technologies are out of the ordinary, and discussion of the technical, ethical, legal, and social issues has been widespread (Murray and Livny 1995). For each technique, consider the following questions:

• *Is it feasible?* Is a newly announced achievement now applicable only to microorganisms? To laboratory animals? How will it be tested for use in humans? Is the technique so complex, the cost so high, or the morality so debatable that the advancement would

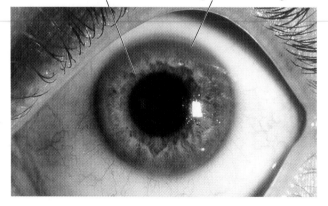

Iris   Kayser-Fleischer ring

Figure 19.1   A Kayser-Fleischer ring characteristic of a patient with Wilson disease. The golden-brown circle of copper hides the outer edge of the iris and is wider on the top and bottom than on the sides. It is deposited in the cornea, which is the transparent layer of the eyeball that covers the iris and the lens behind the iris. (Courtesy of Dr. I. H. Scheinberg.)

only be used by a handful of individuals? Is *my* family likely to benefit? Sensational pronouncements to the contrary, no human babies are likely to be decanted from flasks, as in Aldous Huxley's *Brave New World* (1932).

- *Who decides?* This question arises in several contexts. Who decides what type of research shall be supported by public funds, and who selects the investigators that receive support? Some decisions on the treatment of malformed newborns are painfully difficult. Who shall be the primary decision makers? Doctors? Nurses? Parents? Clergy? Judges? Administrators? Legislators? "Experts" of one sort or another? And just what are the criteria for judging "normal" and "healthy"?

- *Who pays?* In a world of limited resources, we must decide not only who receives an expensive treatment but also who pays the bill. Should scarce medical processes or facilities be available on a random basis? On the basis of one's ability to pay? On the basis of one's ability to mount a television appeal for funds? What procedures should or should not be covered by health insurance? Should insurability be influenced by genetic, or partly genetic, tendencies any more or less than conditions influenced by diet, smoking, or other aspects of life-style? Would expenditure of available funds on alternative projects produce greater social good?

- *Is it ethical?* Is a particular technology good or bad? Does it increase or diminish human freedom? Does it enhance our sense of human dignity, or does it dehumanize us? Although a genetic procedure that has been perfected may be acceptable, the morality of the experimentation on animals and humans to perfect the procedure may be questionable. Perhaps a fly, a mouse, or a sheep that develops badly can be discarded; but what will be done with a defective baby arising from experimentation?

Genetic Counseling

Every year, thousands of families are affected by the birth of an abnormal child. About 0.7% of newborns have a chromosomal abnormality with moderate to severe phenotypic effects. Another 0.5–1% suffer the consequences of single-gene defects (autosomal dominant, autosomal recessive, or X-linked), and about 2% have a malformation that may be due in part to heritable factors. Altogether, about 4% of newborns have a defect that is recognized at birth or within their first year. It has been estimated that about one-third of all children in pediatric hospitals are being treated for conditions that have a genetic component. Thus, there can be no doubt about the need for genetic education

and counseling services for the many families that require it.

The aim of **genetic counseling** is to convey medical and genetic facts to an affected or potentially affected family in a way that can be understood. Because genetic counseling is relatively new, few people know what to expect. This is also a consequence of the stigma that may still attach to genetic disease. Your friends may tell you all about their broken leg, breast lump, or other commonly discussed condition, but they may not share the upsetting episode at the genetic counseling center (Cohen 1994).

Genetic counseling requires professionals who are thoroughly grounded in both genetics and medicine. To this basic education* must be added patience, sensitivity, respect, and the ability to talk easily with people who are likely to be deeply troubled. Family members should be helped to make decisions they are comfortable with. Counselors must see that the families are provided with a full range of medical and social services while they adjust to their situation emotionally and intellectually. Illness and disability inevitably bring tremendous stresses to family life, and many broken marriages are found among the parents of children with genetic defects. Counselors must be prepared to deal with people who are experiencing denial, shock, anger, despair, and guilt and who may turn against the partner supposedly "at fault."

All this may require the joint efforts of a group of helpers: a clinician with genetic training, a geneticist with a mathematical background, laboratory personnel, plus a social worker, genetic counselor, or public health nurse skilled in working with families in their home environment. The first heredity clinic employing a team approach to genetic counseling was established at the University of Michigan in 1940. Today, several hundred centers worldwide—usually associated with medical schools—provide advice and guidance on genetic matters (Lynch et al. 1990).

The Procedure

The need for counseling often arises when a child is born with a possibly hereditary disorder and the parents or other family members are concerned with the well-being of future children. In addition, couples may be troubled by repeated miscarriages, or they may have heard about the greater risk of certain birth defects with increasing maternal age. People without any history of genetic disorder may also wonder whether they harbor a potentially harmful gene. This situation

---

*Several dozen training programs in the United States offer a master's degree in genetic counseling (Information and Education Committee 1996). Career information is available from the National Society of Genetic Counselors, 233 Canterbury Drive, Wallingford, PA 19086. (Internet http://members.aol.com/nsgcweb/nsgchrome.htm#marker)

might occur if a husband and wife are related to each other or if they belong to an ethnic group in which certain genetic diseases are more frequent than usual. In any event, the counseling process will often include the following steps.

Medical Diagnosis.   Genetic counseling requires precise diagnosis, because a number of conditions have multiple causes (Chapter 10). Sometimes these variants may be phenotypically indistinguishable from one another, but in other cases an experienced clinician may detect slight differences (such as hemophilia A versus hemophilia B). In addition to examining the propositus, the counselors may want to see other members of the family or possibly even family photographs. The clerical work involved in corresponding with relatives and searching for pertinent medical records and autopsy reports can be very time-consuming. Laboratory work often includes karyotyping; biochemical analyses of blood, urine, or cultured cells; or molecular analysis with restriction enzymes and DNA probes. Counselors need to be aware that the impact of a genetic illness is different from that of other diseases because a gene is innate—a part of one's self—and will not go away like an infection. A correct diagnosis is the cornerstone of useful genetic counseling, because a wrong diagnosis may have devastating consequences.

Pedigree Analysis.   Not only must a complete three- to four-generation family pedigree be obtained, but the reliability of the collected information must also be assessed. As pointed out in Chapter 5, some important clues—adoptions, illegitimate births, miscarriages, stillbirths, or mildly affected relatives—may be entirely missing. But even if the data are complete, more often than not there will be complicating factors of the types discussed in Chapter 10. Thus, deciding among various possible modes of inheritance may not be a simple matter.

Estimating Recurrence Risks.   How likely is it that a given condition will recur in a subsequent birth? The answer is straightforward in the case of simple Mendelian traits: 25%, for example, if both parents are known to be heterozygous for a completely penetrant, recessive allele.

With the advent of molecular techniques, more and more disease-causing

Mendelian genes can be analyzed prenatally using fetal cells obtained through amniocentesis or chorionic villus sampling (see later). Predictions of risk may then depend on the closeness of genetic linkage to a DNA marker. Direct detection of a disease-causing allele can also be done in some cases by using DNA probes complementary to one or another allele of the gene in question. In these cases, a tested fetus would either have the illness or not—with no uncertainty. The number of hereditary diseases that can be analyzed in this way is growing rapidly, and any list that is given will soon be out of date (Table 19.1).

If the trait in question is multifactorial, counselors cannot establish a risk figure based on rules of transmission. In these cases, they may make use of **empirical risk** figures that rely on the statistics of prior experience with the particular phenotype in question. For many congenital malformations, such as spina bifida, cleft lip, or clubfoot, the risk of recurrence of the particular malformation after the birth of one affected child to normal parents is about 2–5%. For certain

TABLE 19.1   **Some diseases whose causative genes have been cloned and whose presence can therefore be detected (at least experimentally) by a DNA probe**[a]

Disease	Gene symbol	Chromosome location
Achondroplasia	FGFR3	4
Agammaglobulinemia, type 1	XLA	X
Alzheimer disease, familial, type 3	PS1	14
Alzheimer disease, familial, type 4	PS2	1
Amyotrophic lateral sclerosis	SOD1	21
Ataxia telangiectasia	ATA	11
Bloom syndrome	BLM	15
Breast and ovarian cancer, early onset	BRCA1	17
Breast cancer, early onset	BRCA2	13
Cystic fibrosis	CFTR	7
Duchenne muscular dystrophy	DMD	X
Fragile X mental retardation	FMR1	X
Friedreich ataxia	FRDA	9
Hemochromatosis	HFE	6
Hemophilia A	HEMA	X
Hemophilia B	HEMB	X
Huntington disease	HD	4
Hypercholesterolemia, familial	LDLR	19
Lesch-Nyhan syndrome	HPRT	X
Marfan syndrome	FBN1	15
Myotonic dystrophy	DM	19
Neurofibromatosis, type 1	NF1	17
Phenylketonuria	PAH	12
Retinoblastoma	RB1	13
Sickle-cell disease	HBB	11
Tay-Sachs disease	HEXA	15
Thalassemia, α	HBA	16
Thalassemia, β	HBB	11
Wilms tumor	WT1	11
Wilson disease	WND	13

Source: Bassett et al. 1997 with additions.

[a]All of these diseases have been discussed elsewhere in this book.

traits, the geographical or racial background of a couple may be an important element of risk estimation. For Down syndrome, it is important to distinguish between a nondisjunctional cause and the presence of a translocation. Most cases of Down syndrome are due to nondisjunction, and the risk of recurrence is small. Less often, Down syndrome is due to a translocation in a parent, and the risk of recurrence is relatively high (Chapter 14). For some traits, little recurrence information is available, as is the case for certain chromosomal aberrations, imprinting, uniparental disomy (Chapter 10), mitochrondrial diseases, and trinucleotide repeat diseases, where the risk of change from premutation to full mutation is uncertain.

Whether clear-cut or not, the risk figures must be conveyed in such a manner that they can be understood. Repeating information in different ways may be useful. It should be emphasized that chance has no memory. The risk of recurrence for a simple Mendelian trait is the same for each birth, regardless of prior outcomes. For example, if unaffected carrier parents produce one child with a recessive disorder, this does *not* mean that the next three children will be normal. Rather, the risk of an affected child on each successive birth continues to be ¼. In any event, it is the family's *perception* of risk, not the actual risk, that is important. A good counselor will explore how a client feels about an unfavorable probability of ¼, which is the same as a favorable probability of ¾.

Options. When the genetic prognosis is unfavorable, a couple may refrain from childbearing or risk a defective birth. If abortion is an acceptable alternative, prenatal diagnosis (followed by possible termination of pregnancy) is now available in a growing number of cases. Alternatively, a couple may choose adoption or an appropriate reproductive technology using either a donor egg or donor sperm (Chapter 15).

One's perception of risk and willingness to accept it are highly subjective matters, depending on one's personality, experiences, moral convictions, and especially on the burden of care imposed by the condition in question. Some newborns with serious birth defects (whether genetic or not) may die within a few months. Others may require constant attention from their family throughout childhood and beyond or may require expensive medical treatment. We do not mean to minimize the grief of parents upon the loss of an infant, but we emphasize the difference between short-term and long-term care. A couple may decide to take a chance on a relatively mild disorder, or on a more serious disorder of short duration, or on one that can be ameliorated by treatment. But they may not be willing to accept the same percentage of risk on a disorder that imposes greater physical, financial, and emotional burdens on themselves and their family.

The humanistic aspects of counseling are extremely important (Walker 1997). Because a couple may need time to talk it out between themselves and with friends and relatives, the process should not be hurried. Decision making may be especially difficult if counseling occurs soon after the birth of a child with a severe defect. A couple may feel that they are betraying a first child when contemplating aborting a subsequent fetus with the same defect. The counselor must understand and help a mother who blames herself for her behavior during pregnancy (e.g., smoking or drinking) or a father who may deny that the child is his biological offspring. The setting of counseling sessions must also be considered; a hospital site may be intimidating, even if the counselor is skillful in personal interactions.

Follow-Up and Supportive Services. Because a counselor's spoken words may be misinterpreted or forgotten, the relevant medical and genetic information is often put in writing. Such a letter usually includes the names and addresses of specialists and appropriate social services, as well as written answers to the major questions raised by the clients. Families should be informed of useful new research results and be allowed to decide on the value of participating in experimental trials. Services can also be extended to relatives who may unknowingly be at risk, but giving unsolicited information raises a new set of ethical questions. Overall, the counselor must provide expert and caring assistance to persons whose self-esteem has been hurt and whose lives may have suddenly been complicated by the birth of a child with a genetic disorder.

Support groups for many of the various genetic diseases have sprouted up at the local, regional, national, and international levels. These organizations sometimes lobby on behalf of a particular disease or publish educational material for both laypersons and professionals. Their greatest value, however, may be in peer support, demonstrating to families affected by a genetic disease that they are not alone, that they can share their misfortunes, frustrations, and doubts. Over 300 of these groups belong to The Alliance of Genetic Support Groups, based near Washington, DC.*

Ethical Concerns

The goal of counseling is to enable couples under stress to plan their families according to their own values and with the maximum amount of information and support, but it is not always clear how this may best be done. Among counselors, opinions vary as to

---

*The Alliance of Genetic Support Groups, 35 Wisconsin Circle, Suite 440, Chevy Chase, MD 20815.
(Internet: http://medhlp.netusa.net/www/agsg.htm)

the degree of advice that should be given. At one extreme are those who specifically suggest a course of action based on the counselor's own values or experiences (as may occur in a doctor-patient relationship). Others may provide advice, if asked, but with the proviso that it is impossible for the counselor to fully appreciate all the variables that enter into an intensely personal situation. Most counselors, however, try to be nondirective—providing information, empathy, and support, but no specific advice about a course of action. In practice, however, complete nondirectiveness is difficult to achieve, especially when faced with the familiar and natural question, "What do you think I should do?" If a genetic counselor fails to answer forthrightly, clients may fear the worst or believe they are not getting their money's worth. Studies of the counselor-client relationship suggest that some counselors vary their style, becoming more directive with clients who are especially anxious about their situation or who are lower on the socioeconomic scale (Bernhardt 1997; Kessler 1997).

Another difference in outlook involves a couple's responsibilities to the human species as a whole. In most Western societies, medical personnel are trained to act primarily for the well-being of their patients, but individual reproductive decisions affect the welfare of future generations. Whatever our collective obligations to descendants are, however, the views of counseled couples, not those of the counselors, ought to prevail.

The question of withholding certain information also arises (Minogue et al. 1988). For example, in one case told to the authors by a genetic counselor, amniocentesis revealed a twin pregnancy with 46,XY and 47,XYY fetuses. The prognosis for double-Y males is uncertain, although most develop within the range of normal variation of XY males (Chapter 10). Although the existence of the two karyotypes was revealed to the prospective parents, they preferred not to be told which twin had which karyotype. Even if a counselor believes that withholding knowledge is in the best interest of the person being counseled, telling a lie might erode the trust necessary to the counseling relationship. (Also, counselors who withhold information may be subjected to lawsuits if the clients later learn of the secret information.) Generally, counselors tell only the truth—all of it, or virtually all of it.

The conflict between telling or withholding information is especially acute with regard to relatives who have not specifically sought counseling (Kolata 1994). Although many persons want to know such genetic information, some might not, especially if the "bad news" involves a serious late-onset condition such as Huntington disease. When no treatment is currently available, is a humanitarian purpose served by informing a person of a progressively debilitating neuro-

logical disease? If this person is still in childbearing years, does the counselor bear any obligation to help prevent the conception of additional affected persons? There are no pat answers to such questions.

Prenatal Diagnosis

The ability to detect prenatally virtually all major chromosomal aberrations and many gene-controlled biochemical defects has been a tremendous aid to genetic counseling. (See list in Table 19.1.) If a disorder is diagnosed early in pregnancy, the couple has the option of aborting the fetus and beginning again. Fortunately, in about 97% of prenatal analyses, the fetus is found *not* to have the disorder in question. This finding relieves anxiety for the parents, who can then anticipate a newborn no more at risk for abnormality than a random birth. Many couples at substantial risk of having a child with a serious defect would refrain from further pregnancies unless assured that the risk could be reduced. Thus, *because* of the availability of prenatal diagnosis, life is given to children who would otherwise never be conceived.

Procedures for Obtaining Fetal Cells

There are two major techniques for obtaining a sample of fetal cells for prenatal diagnosis (D'Alton and DeCherney 1993). Both are done on an outpatient basis with some discomfort for the woman. The physical discomfort, however, is typically not as great as the anxiety in waiting for the test results.

In **amniocentesis**, which has been performed since the 1960s, a physician removes a small amount of the amniotic fluid that surrounds the fetus. Suspended in this fluid are a few living cells that have sloughed off from the fetal skin, the lining of the respiratory or urinary systems, or the amnion. A disadvantage of amniocentesis is that these cells are not actively dividing. Thus, the cells must be put in culture medium and allowed to multiply for 8 to 12 days before technicians can do cytogenetic analyses. Further incubation for several weeks may be required before technicians can do many other analyses.

To obtain the cells, a long, thin needle is inserted through the woman's abdominal wall, uterus, and fetal membranes (Figure 19.2). This is often done at about the sixteenth week of pregnancy, when the 5-inch fetus floats in about a half pint (200 ml) of amniotic fluid. About 2 tablespoons (30 ml) of fluid are withdrawn. The position of the needle with respect to the fetus can be accurately monitored by ultrasound scanning, which makes major surface features of the fetus visible through reflected sound waves. The needle must not, of course, penetrate the fetus, the umbilical cord, or the placenta. When done by an experienced obstetrician, there is almost no risk to the

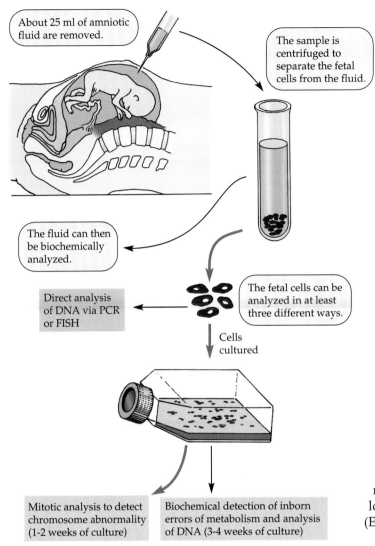

About 25 ml of amniotic fluid are removed.

The sample is centrifuged to separate the fetal cells from the fluid.

The fluid can then be biochemically analyzed.

Direct analysis of DNA via PCR or FISH

The fetal cells can be analyzed in at least three different ways.

Cells cultured

Mitotic analysis to detect chromosome abnormality (1-2 weeks of culture)

Biochemical detection of inborn errors of metabolism and analysis of DNA (3-4 weeks of culture)

Figure 19.2 Steps in the prenatal diagnosis of genetic disease by amniocentesis, which is done at about the sixteenth week of pregnancy. The blue arrows show the most common pathway of analysis—karyotyping to detect Down syndrome or some other chromosomal abnormality.

mother or fetus of needle injury that might cause bleeding or infection. There is, however, a risk of about 0.5% that the amniocentesis procedure will lead to premature onset of labor and loss of the baby.

Another method of obtaining cells for analysis is **chorionic villus sampling (CVS)**, a procedure that has been performed in the United States since about 1983. CVS is usually available only in specialized centers rather than being widely available in obstetricians' offices, as is the case for amniocentesis. In one variation of this process, the physician inserts a catheter guided by ultrasound imaging through the cervical canal (Figure 19.3). A few bits of fetal tissue, consisting of branched projections from the chorion (the chorionic villi), are sucked off from the developing placenta.

After removal, the fetal villi are separated from any attached maternal tissue with the aid of a low-power microscope.

One major advantage of CVS over amniocentesis is that several milligrams of fetal tissue are obtained. Because the tissue includes actively dividing cells, some chromosome studies and other tests can be done quickly—in a day or two. Another major advantage is that CVS is done much earlier in pregnancy—from the ninth to twelfth week of gestation, during which the fetus grows from about 2 to 3.5 inches and develops distinguishable external genitals. This earlier testing via CVS can avoid weeks of uneasiness. Moreover, the pregnancy can remain a private matter, and abortion, if chosen, can be performed during the first trimester, when it is medically safer for the mother.*

The safety of chorionic villus sampling continues to be evaluated. The procedure is technically more difficult to perform than amniocentesis and seems to result in slightly greater fetal loss according to some studies. The actual cause of the abortions following CVS is sometimes uncertain, because pregnancies that are viable at 10 weeks may miscarry in the *natural* course of events. Another concern is the risk of limb defects. These abnormalities have been noted in low frequency (but greater than the general population rate) in some but not all studies of newborns following CVS done prior to 10 weeks of gestation (Elias and Simpson 1997).

Isolating Fetal Cells from the Maternal Circulation.

To avoid any risk to the fetus or discomfort to the mother, researchers have been trying for years to obtain the relatively few fetal cells that regularly leak into the mother's bloodstream. As far back as the 1970s, researchers were able to isolate fetal *white blood cells* in the maternal circulation. Unfortunately, these cells can persist in the mother for many years, even decades, prohibiting their use in successive pregnancies.

Recent efforts have concentrated on fetal *erythroblasts*, immature red blood cells with a nucleus and a brief life span. From these cells, DNA can be amplified by the polymerase chain reaction and analyzed with probes that reveal a mutated allele. Researchers at the University of California, San Francisco, used a number of successive chemical-physical steps to identify 10 to 20 fetal erythroblasts among 10 million or so maternal

---

*Each trimester is approximately 13 weeks in duration. The fetus becomes viable (able to survive outside the womb) near the end of the second trimester, but it needs technological assistance to survive if born at that time.

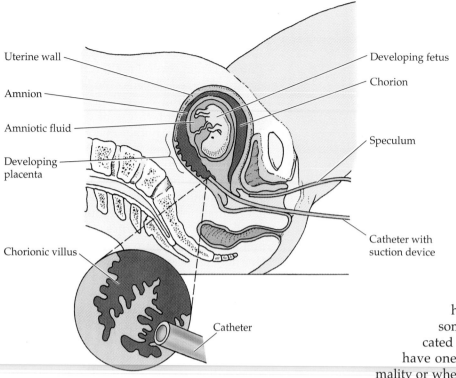

Figure 19.3 The transcervical method for obtaining fetal cells by chorionic villus sampling (CVS), which is done at the ninth to the twelfth week of pregnancy. The projecting villi that anchor the fetal sac to the uterine wall have been likened to a shaggy carpet.

cells in a small blood sample drawn from the mother's arm vein (Cheung et al. 1996; Williamson 1996). One step involved a monoclonal antibody (Chapter 18) specific for polypeptides (e.g., γ chains) present in fetal hemoglobin but not in adult hemoglobin (Chapter 6). The antibody was coupled with a dye, so that researchers could distinguish the fetal cells under a microscope.

Once identified, the nuclei of the fetal cells were scraped off the microscope slide with a microneedle and collected in a test tube for DNA analysis. The California researchers used this procedure with two pregnant women to correctly analyze the fetal genotype with regard to sickle-cell disease in one case and β-thalassemia in the other. (Both fetuses turned out to be normal.) Whether such complicated procedures can be used *routinely* for these and other single-gene disorders (e.g., cystic fibrosis, fragile X, Huntington disease, etc.) remains to be seen.

Indications and Results

Prenatal diagnosis using fetal cells or amniotic fluid is usually offered in three types of situations.

Increased Risk of a Chromosomal Abnormality. The major use of prenatal diagnosis is to detect Down syn-

drome or other chromosomal aberration (see bold arrows in Figure 19.2). A large percentage of all prenatal diagnosis—as much as 85% at some centers—is done in this situation. The risks of chromosomal abnormalities in fetuses increase gradually with maternal age, but more steeply when a pregnant woman is 35 years or older. At ages just above 35, the risk of Down syndrome in the fetus is roughly the same as the risk of miscarriage due to amniocentesis. Thus, routinely offering amniocentesis at 35 has become the standard of practice, but some younger women also have the test (Box 19A). Chromosomal analysis of fetal cells is also indicated when a couple (of any age) already have one child with a chromosomal abnormality or when one of the parents is a carrier of a translocation.

There is still no satisfactory explanation why advancing maternal age increases the likelihood of some aneuploid conditions in fetuses and subsequent newborns. We have previously noted that the frequency of Down syndrome increases gradually to about 1 in 100 births for 40-year-old women and becomes more frequent for yet older women (Chapter 13). Other chromosomal abnormalities that increase with maternal age include Klinefelter syndrome, trisomy 13, trisomy 18, and triple X, so that the total risk of chromosomal abnormality in the live-born offspring of a 40-year-old woman is about 1 in 65.

One example of prenatal chromosome analysis is shown in Figure 19.4. Here, fluorescence in situ hybridization, or FISH (Chapter 9), is used directly on nondividing interphase cells obtained by amniocentesis. This technique allows technicians to detect some chromosomal aberrations just one or two days after a visit to the doctor's office. Although it is not possible to see interphase chromosomes under a standard light microscope, the DNA probes hybridize to and light up regions on specific chromosomes. that have sequence homologues to the probe. Quick analysis by FISH is usually confirmed (about two weeks later) by standard karyotype analysis of mitotic chromosomes in cultured cells.

Increased Risk for a Single-Gene Disease. Prenatal diagnosis is offered to couples when a prior child is affected with a serious Mendelian disorder or when both parents are otherwise determined to be carriers

BOX 19A *The Age 35 Threshold*

Using age 35 as the starting point for offering amniocentesis as a routine procedure assumes a balance between two risks: having a child with Down syndrome (in the absence of prenatal diagnosis) and having a miscarriage of a normal child (induced by the procedure). At ages just above 35, the risk of both is about 1/200, but is this comparison between apples and oranges? Some couples may feel that the burden of caring for a child with Down syndrome is greater than the burden of miscarriage. For them, a threshold much lower than age 35 might be appropriate.

For women younger than 35, an analysis of certain chemicals in a pregnant mother's blood can help in making a decision about amniocentesis (or CVS). Laboratory personnel determine the concentrations in maternal blood of three chemicals: α-fetoprotein, estriol, and chorionic gonadotropin.* Compared with normal pregnancies, the values for women who are carrying a fetus with Down syndrome are on average reduced about 25% for the first two chemicals and increased

about 250% for the last one. It is not known why a Down syndrome pregnancy changes the "three-marker" concentrations in these ways.

A study by Haddow et al. (1992) examined 760 pregnant women (mostly under 35) who had significantly altered values of the three chemical markers and who underwent amniocentesis. Twenty of the pregnancies (about 3%) were found to involve fetuses with Down syndrome. In addition, seven other chromosomal disorders were found, including trisomy 13, Klinefelter and Turner syndromes, and the triple X karyotype. Through more extended survey techniques, the investigators showed that measuring the maternal blood markers in women

*The chemical α-fetoprotein is made in the fetal liver, but its function is unknown; it is also used as a marker for neural tube defects. Estriol is a breakdown product of the active form of estrogen. Chorionic gonadotropin, a major pregnancy hormone, is made in the placenta and acts to maintain pregnancy; it is also the basis of home pregnancy tests using urine samples.

younger than 35 detects most but not all cases of Down syndrome.

A logical extension of this policy is to use it also for women over 35—that is, to routinely offer amniocentesis *only* to women whose three-marker blood tests indicate a high risk of Down syndrome (Haddow et al. 1994). Amniocentesis would, of course, continue to be offered for any woman in special circumstances. Compared with the current practice of offering prenatal diagnosis to *all* women 35 and over, investigators calculate that this policy of fewer invasive tests would avoid the loss of about 1,400 fetuses per year due to miscarriages associated with amniocentesis or CVS (Pauker and Pauker 1994). On the other hand, routine testing for older women enables *all* their fetuses with Down syndrome to be identified, whereas limited testing would cause some fetuses with Down syndrome to be missed. It is hard to know what policy is in the best interest of individual couples and society as a whole.

---

and the disorder can be diagnosed in utero. Carrier status is often checked in members of ethnic groups that have unusually high frequencies of certain diseases: Tay-Sachs disease among Ashkenazi Jews and some French-Canadian groups; sickle-cell disease among some black African, Mediterranean, Arab, Indian, and Pakistani groups; and α- and β-thalassemia among some Mediterranean, Chinese, and southern Asian groups.

Although most inherited diseases in most populations are individually rare, medical geneticists are concerned with a great number of them. For about 1,000 clinical abnormalities, researchers have cloned and mapped the causative genes through variations in either the clinical phenotype or the gene itself (McKusick 1997). In general, this knowledge allows prenatal diagnosis on the basis of proteins or DNA in a cell or tissue sample obtained by amniocentesis or CVS. Some disorders that can (in theory) be detected through mutant DNA were presented in Table

19.1. Some disorders that can be detected in a fetus on the basis of the encoded protein are presented in Table 19.2. Prenatal diagnosis at the protein level can be made only if the encoding gene is expressed in the

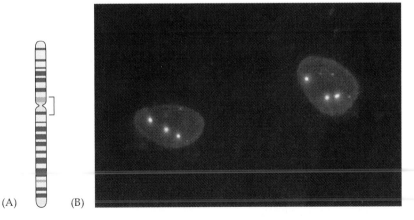

Figure 19.4 Quick prenatal diagnosis of the triple X karyotype by fluorescence in situ hybridization (FISH). Researchers used a fluorescent probe that was specific for DNA sequences in the centromere region of the X chromosome, indicated by the bar in A. They bathed nondividing, interphase amniotic cells with the probe. The three bright spots seen in each of the amniotic cells in B indicate three X chromosomes. (From Klinger et al. 1992.)

TABLE 19.2 Some inherited diseases that can be detected prenatally by analyzing for levels of the encoded protein[a]

Disease	Abnormal enzyme or other protein	Type of metabolism affected
Congenital adrenal hyperplasia	21-Hydroxylase	Hormone
Ehlers-Danlos syndrome	Collagen	Connective tissue
Fabry disease	α-Galactosidase	Fat
Galactosemia	Galactose-1-phosphate uridyltransferase	Sugar
Gaucher disease	β-Glucosidase	Fat
Hunter syndrome	Iduronate sulfatase	Complex carbohydrate
Hurler syndrome	Iduronidase	Complex carbohydrate
Hypercholesterolemia	LDL receptor	Cholesterol
Ichthyosis	Steroid sulfatase	Hormone
Lesch-Nyhan syndrome	Hypoxanthine-guanine phosphoribosyltransferase	Purines
Maple syrup urine disease	Branched-chain keto acid decarboxylase	Amino acid
Marfan syndrome	Fibrillin	Connective tissue
Porphyria (variegate)	Protoporphyrinogen oxidase	Heme
Severe combined immunodeficiency	Adenosine deaminase	DNA
Tay-Sachs disease	Hexosaminidase A	Fat
Wolman disease	Lipase	Fat

Source: Data is mostly from Besley 1992.
[a] Some of these may also be diagnosed by DNA analysis.

particular cells obtained by amniocentesis or CVS. (Many genes are only expressed in specialized tissues that are not obtained by CVS or amniocentesis. An example is the PKU gene, whose product, phenylalanine hydroxylase, is expressed only in liver cells.) Genes that are transcribed and translated in prenatal samples include those responsible for the basic housekeeping functions that occur in virtually all cell types. Tay-Sachs disease is an example of an inborn error of metabolism that can be detected prenatally by an enzyme assay (Chapter 16).

There are many experimental protocols for prenatal diagnosis based on fetal DNA (whether or not the gene is expressed and whether or not the protein product of the gene is even known). Some of these tests depend on linkage to DNA markers. Because now there are so many markers scattered throughout the entire genome, virtually no gene is without a DNA marker to which it can be linked more or less closely. Some analyses depend on *oligonucleotide probes* (Chapter 8), synthetic DNA segments of about 20 nucleotides. These short sequences can be made complementary to the portion of a gene that includes a single-base mutation; near the middle is the one base that

corresponds to either the normal allele or the mutant allele. Thus, the *normal* probe exactly matches the *normal* allele but has a one-base mismatch with the mutant allele; the *mutant* probe exactly matches the *mutant* allele but has a one-base mismatch with the normal allele. Under carefully controlled conditions of hybridization, each probe will bind only to its corresponding allele. For example, oligonucleotide probes have been used in the prenatal diagnosis of β-thalassemia in Sardinia. Here, the disease-causing change is a G-to-A mutation that encodes a stop triplet near the beginning of the gene. Bathed with the radioactive probes, the fetal DNA is hybridized by either the one probe, the other, or both probes in the case of heterozygotes (Figure 19.5).

Oligonucleotide probes can also be used in the prenatal diagnosis of cystic fibrosis. Recall that the most common mutation (accounting for about 70% of carriers in the United States) is a three-base deletion known as *ΔF508*. One oligonucleotide probe is made that binds only to the normal allele, and a second probe binds only to the *ΔF508* allele. These and a third probe, common to both alleles and which binds nearby, are used as primers for the polymerase chain reaction. The outcome is that DNA from a normal fetus generates an amplified DNA fragment exactly 63 bases long, and DNA from a CF fetus generates a DNA fragment 60 bases long. Heterozygotes yield fragments of both lengths. The fragments are then identified by gel electrophoresis, which separates them according to size.

Increased Risk for a Neural Tube Defect. In addition to possible chromosomal abnormalities or Mendelian disorders, **neural tube defects (NTDs)** constitute another indication for prenatal diagnosis. By the first month of embryonic development, the furrow of neural tissue that forms along the back closes upon itself to make a tubelike structure, the forerunner of the brain and spinal cord (Chapter 15). When the sides of the neural fold fail to close near the head of the embryo, the brain tissue remains exposed and becomes disorganized, resulting in *anencephaly* (literally, "no brain").

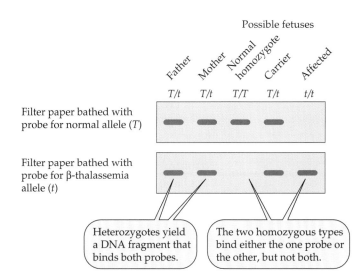

Possible fetuses

Father | Mother | Normal homozygote | Carrier | Affected

T/t | T/t | T/T | T/t | t/t

Filter paper bathed with probe for normal allele (*T*)

Filter paper bathed with probe for β-thalassemia allele (*t*)

Heterozygotes yield a DNA fragment that binds both probes.

The two homozygous types bind either the one probe or the other, but not both.

Figure 19.5 The use of radioactive oligonucleotide probes for prenatal diagnosis of β-thalassemia (genotype *t/t*) in Sardinia. One probe is specific for the normal allele; the other probe is specific for the thalassemia allele, which differs by a single base. Fetal DNA is cut with a restriction enzyme, then electrophoresed, Southern blotted onto filters, and bathed with one or the other probe.

Such fetuses are stillborn or survive only a few days. When the failure to close occurs lower down, the result is one or another form of *spina bifida*, often characterized by paralysis or weakness below the level of the spinal opening, incontinence of bowel and bladder, and hydrocephalus (abnormal accumulation of fluid in the brain, producing head enlargement). Some patients with less severe types of spina bifida may live many years with milder disabilities. In the United States, NTDs occur at a frequency of 1 or 2 per 1,000 births, but they are more frequent (up to 7 cases per 1,000) in some other countries, including Ireland, Scotland, and Wales. The incidence of NTDs has decreased in many countries over the past decade due, in part, to increased intake of folic acid. Diet supplementation with this vitamin has been shown to significantly decrease the occurrence and recurrence risks of NTDs.

Because the recurrence rate for couples with one child with an NTD is 2–5%, and because the prospect is often so bleak for affected children and their families, prenatal diagnosis is very useful. The determination of an NTD depends on the concentration of α-**fetoprotein**, a substance of unknown function that is present in fetal blood plasma. Small amounts are also present in the amniotic fluid of normal fetuses, but concentrations in the amniotic fluid of fetuses with NTDs are increased manyfold (especially in the more serious cases). It is thought that plasma with α-fetoprotein leaks into the amniotic cavity from the exposed neural tissue. Ultrasound scanning, done as an aid to safe amniocentesis, can also directly detect many of the deformities associated with neural tube defects.*

It is disheartening, of course, that it usually takes the birth of one child with an NTD to initiate the prenatal diagnosis that could prevent a second one. Screening of *all* fetuses might be desirable if amniocentesis carried no risk. It has been found that most (but not all) women carrying a fetus with an NTD have elevated levels of α-fetoprotein in their *own* circulation. Thus, amniocentesis and ultrasonography might be done on the group of pregnant women with elevated levels of serum α-fetoprotein. But this type of screening has presented some problems that are discussed later.

Sex Prediction and Preimplantation Diagnosis

When a woman is heterozygous for a disease-causing X-linked recessive gene (and her mate is hemizygous normal), half of her male fetuses are expected to inherit the disorder. When the disease is not diagnosable in utero, the birth of affected children can nevertheless be avoided by aborting all male fetuses. The dilemma posed by the abortion of some normal fetuses (50% on the average) to avert the birth of abnormal children (50%) is not easy to resolve. However, several accurate methods are available for the requisite sex determination of the fetus. These include traditional karyotyping of fetal cells obtained by CVS or amniocentesis, which reliably reveals the Y chromosome. A quicker method is to analyze fetal DNA from CVS or amniocentesis using the polymerase chain reaction. The primers that are used specifically hybridize to repetitive sequences found only on the Y chromosome.

Still in the experimental stage is a way to determine the sex of very early embryos prior to their implantation in the uterine wall—that is, before pregnancy is established. This process of **preimplantation diagnosis** could be useful to a couple who wish to avoid a male pregnancy and who are already making use of in vitro fertilization (see Box 20C). Its major advantage is that it sidesteps the difficult decision to terminate pregnancy that may follow amniocentesis or CVS months after pregnancy begins (Wertz 1996).

Using a microscope and fine micropipets, a skilled researcher teases out one cell from an eight-cell embryo (Figure 19.6B). Then the polymerase chain reaction (PCR) or fluorescence in situ hybridization (FISH) is used with the minute amount of DNA (a few picograms) to identify sequences specific to the Y

*Although ultrasound scanning has many beneficial purposes, using it routinely on fetuses for the sole purpose of detecting fetal anomalies is controversial. Although assumed to be safe, it may not be done because of its cost and possible inaccuracy. Especially to an untrained eye, the images may appear shadowy and indistinct.

(A)

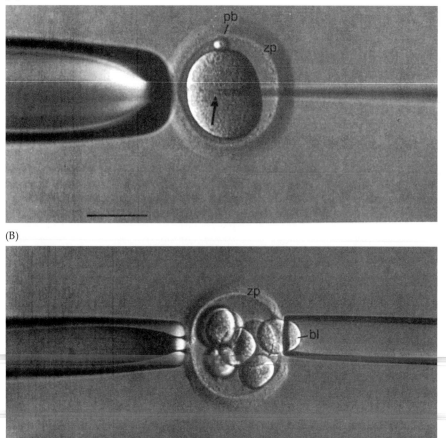

(B)

Figure 19.6 Micromanipulations for fertilization of a human egg and removal of an embryonic cell for genetic analysis. (A) A single sperm, barely visible at the tip of the arrow, is injected into an egg with a thin micropipet. The egg is held by gentle suction on the blunt end of a holding pipet. (B) After 45 hours, a single cell (a blastomere) is sucked away from the resulting eight-cell embryo. bl, blastomere; pb, polar body; zp, zona pellucida. Scale bar = 100 μm. (From Schimmel and Crumm 1994.)

chromosome. Meanwhile, the rest of the embryo, now with seven cells, continues what appears to be regular development in its nurturing culture medium. Only if the embryo is female would it be transferred to the woman's uterus.

This technically difficult and costly procedure is not practical on a large scale, in part because in vitro fertilization, even under the best circumstances, is not very successful—a low "take-home-baby" rate (Simpson and Carson 1992). Problems are also presented by the unneeded male embryos, which may be frozen to avoid the ethical questions posed by a more permanent disposal. Keep in mind, too, that half of the male embryos would not have inherited the abnormal X-linked gene from carrier mothers.

Preimplantation diagnosis that uses a single cell sucked from an eight-celled embryo also allows researchers to test for several specific disease alleles (in addition to the sex of the embryo). Preliminary investigations have involved several X-linked and autosomal genes, including those for Duchenne muscular dystrophy, cystic fibrosis, and Marfan syndrome (Table 19.3). Over 100 normal babies worldwide have been born following preimplantation diagnosis for sex or for a specific disease.

A major problem for any clinical application related to genetic disease is misdiagnosis stemming from PCR amplification of one allele of a heterozygous cell but not the other. In *one* cell, there is, of course, exactly *one*

TABLE 19.3 Genetic diseases that have been diagnosed in single cells from early embryos (prior to implantation)

Disease	Inheritance pattern	Gene symbol	Type of mutation detected
Duchenne muscular dystrophy	X-linked recessive	*DMD*	Large deletion
Fragile X	X-linked dominant	*FMR1*	Trinucleotide repeat expansion
Cystic fibrosis	Autosomal recessive	*CFTR*	Three-base deletion
Tay-Sachs disease	Autosomal recessive	*HEXA*	Four-base insertion
Lesch-Nyhan syndrome	Autosomal recessive	*HPRT*	Base substitution
β-Thalassemia	Autosomal recessive	*HBB*	Base substitution
Marfan syndrome	Autosomal dominant	*FBN1*	Base substitution
Myotonic dystrophy	Autosomal dominant	*DM*	Trinucleotide repeat expansion
Huntington disease	Autosomal dominant	*HD*	Trinucleotide repeat expansion

Source: Handyside and Delhanty 1997.

copy of each allele that could, and should, be amplified. But PCR may fail to amplify every template sequence marked by appropriate primer sites. When there are just two templates available for amplification (rather than the usual thousands), missing one of them can have major consequences. Consider, for example, an embryo that is heterozygous for the dominantly inherited Marfan syndrome (*M/m*). The affected embryo could be mistakenly declared unaffected if PCR amplified the one normal *m* allele but not the one disease-causing *M* allele. (See also question 5.) To get around this problem, researchers trying to diagnose Marfan syndrome from a single cell have first reverse transcribed the *multiple* molecules of mRNA transcribed from both Marfan alleles (Eldadah et al. 1995). One might not expect the Marfan gene, which codes for the elastic protein fibrillin, to be expressed in an eight-celled embryo, but it is. Whether other genes are expressed so early in development must be determined on a case-by-case basis.

Genetic Screening

Genetic screening is the search of populations for persons having a particular genotype or karyotype that can cause serious disease. Screening programs cover mainly people *without* a clear family history of the condition in question. Such screening—of everybody in a geographical area, of a certain age, or from a particular ethnic group—is done for relatively few diseases. But individual testing of any person *with* a family history of a genetic defect usually makes good sense.

Screening serves two general purposes. The first is to detect a serious disease—or predisposition to disease—before the onset of debilitating symptoms. This form of screening, typically on a population of newborns (or even prenatally), is consistent with an increasing emphasis on preventive medicine (Erbe and Levy 1997). It is similar in intent to *non*genetic screening—for example, periodic mammography, blood pressure checks, or cholesterol determinations. The second general purpose of genetic screening is to identify unaffected carriers (usually as adults) in order to counsel them about the risk of producing affected children, thereby helping them to make more informed reproductive decisions (McGinniss and Kaback 1997).

Although few would disagree with these goals, it may be hard to decide whether screening should be mandatory or voluntary. The health of its citizens is a proper governmental concern, but genetic diseases do not pose a danger to others in the same way that contagious diseases such as tuberculosis and AIDS do. Screening programs initiated with the best of intentions may cause ill will among the people who were supposed to be helped. Thus, the planning and execution of screening programs require careful attention to the following concerns.

1. *The test itself.* To be useful for screening large populations, a diagnostic test should be convenient, inexpensive, safe, and reliable. **False negatives** (persons who have the genotype but are not detected) and **false positives** (persons who are said to have the genotype but do not) must be minimized. State legislatures that often mandate screening programs are also concerned with cost-effectiveness: The money spent in testing and services should be less than the money that state agencies would spend on dealing with affected persons. This is likely to be true: The cost of long-term care for institutionalized patients with burdensome genetic conditions is often very great in measurable resources and beyond calculation in terms of human suffering.

2. *Treatment.* Unless the disease in question can be effectively treated, there may be little point in identifying it in a potentially affected person. Yet in the case of severe diseases with late onset, decisions about marriage and parenthood might be altered by early knowledge of the disease.

3. *Counseling.* Expert counseling services must be made available to guide affected families through difficult times. Considerable psychological harm—low self-esteem and feelings of worthlessness—can result when a person is told that he or she is a carrier of a recessive genetic disease.

4. *Safeguards.* Strict confidentiality of medical records is necessary, especially with the expansion and centralization of computerized records. Although many people fear the loss of health insurance due to genetic or other factors, no insurance company currently requires any kind of genetic testing (ASHG Ad Hoc Committee 1995). They may, however, have access to a doctor's medical data when applicants routinely give permission for release of all pertinent information. Insurance companies argue that genetic factors that influence illness or death should be treated in the same way as other kinds of information (Box 19B). As of this writing, no federal laws apply to the use of genetic information by insurance companies, except in limited circumstances (e.g., changing jobs). In about a dozen states,* however, comprehensive legislation bans state-regulated health insurers (but not life insurers) from denying insurance or setting higher rates for persons who are at genetic risk to disease (Holmes 1996/1997; Preston 1996). As genetic tests become more available, more predictive, and less expensive, the laws governing insurance operations will cer-

---

*California, Colorado, Georgia, Minnesota, New Hampshire, New Jersey, Ohio, Oregon, North Carolina, Virginia, and Wisconsin, as of October 1997.

BOX 19B *When Individuals Have More Information Than Insurance Companies*

The expansion of genetic testing raises many complicated questions for patients, doctors, insurance companies, lawmakers, and people seeking insurance. We present here one problem that has been raised by the insurance industry.

People buy health, life, and other forms of insurance to provide protection against a catastrophe. They may not be able to prevent a serious illness or an untimely death, but they can buy a financial cushion to soften the blow. The premiums they pay are calculated on the expected outcomes for large numbers of people with similar risks. Private insurers try to evaluate these risks, and they charge higher premiums for groups of people at higher risk, a process called *underwriting*. Thus, persons with chronic lung disease, high blood pressure, obesity, cardiovascular disease, diabetes, and so on, pay more for their life insurance policies than do persons in better health, who pay standard premiums.

But people shopping for insurance can be expected to act in their own best interest. Sometimes, individuals have more information about their risks than they share with insurance companies. For example, is estimated that several percent of applicants for life insurance fail to reveal tobacco use. Such imbalances allow dishonest individuals at higher risk to pay the same premiums as those at lower risk. Insurance companies may, of course, raise all premiums to protect themselves against such situations.

With regard to genetic conditions, which individuals may or may not reveal, some would argue that charging higher premiums is unfair. They argue that genetic conditions are not the individual's fault and should therefore not be the basis of differential treatment. Insurance companies do not buy this argument. First of all, nobody doubts the relevance of genetics to health and longevity. Insurers argue that predictive information—genetic or nongenetic—is essential to conducting their business. They maintain, for example, that persons prone to a particular cancer for genetic reasons should be treated the same as persons prone to the same cancer for nongenetic reasons. The cause is not relevant, only the level of risk. Moreover, "To give preference to genetic factors would be tantamount to a declaration that all other applicants were responsible for their risk status and disease processes, which is clearly not the case" (Pokorski 1997).

A related and important consideration is familiar to readers of this book: There is often no clear distinction between genetic and environmental causations of disease. Possession of a mutant allele at the *BRCA1* or *BRCA2* locus does not predict breast cancer for sure, but only a probability of developing breast cancer. Similarly, predictive tests for Alzheimer and other diseases are inexact. Thus, even if blame *should* attach to certain lifestyle choices and blamelessness to mutant genes, it is often difficult to assign a cause of illness.

In coming years, it is expected that many more genetic tests with varying degrees of predictive usefulness will be developed, making the potential problem of withholding genetic information more troublesome to insurance companies and their applicants. Insurers believe that customers who pay at standard rates will be troubled because their premiums will increasingly have to be raised to cover the growing number of people who may hide genetic risks. Furthermore, those people who have the most to hide genetically may seek the most insurance. (With regard to health insurance, we note that none of these problems apply to Canadians or the citizens of other countries covered by universal national health plans.)

Still, some people seeking insurance will certainly ask why they should be penalized in situations where the prognostic value of a particular genotype is not clear-cut. Can applicants and claimants be confident that insurers understand increasingly complicated genetic data and will play fair with them without endless hassles (Zallen 1997)? How many people actually avoid getting useful genetic tests because they fear that the results might be used in a discriminatory way? Revealing genetic test results to insurers may also violate the privacy of close relatives without their knowledge or consent. Good-faith discussions and negotiations between insurance companies, consumers, lawmakers, and regulatory agencies are in order:

"The economic aspects of health care in this country are driven by a system of private insurers, and technology is giving insurers an arsenal with which, unless regulated, they can justify excluding individuals from the insurance system. ... The overriding problem is how to assure that all citizens have affordable access to health and health-related insurance. Any solution ought to be a balanced one which has a societal consensus." (Holmes 1996/1997)

tainly change. Discrimination in employment is also worrisome; but perhaps as a result of the Americans with Disabilities Act of 1990 and efforts of the federal Equal Employment Opportunity Commission, there have been few publicized examples of genetic discrimination in the workplace.

Screening for Early Detection of Genetic Disease

The diagnosis of phenylketonuria (PKU) in newborns was the first large-scale screening program undertaken by government agencies and is still the most widely used. By the late 1950s, it was clear that the most severe abnormalities of PKU could be controlled by diet, but only if treatment was begun in the first few weeks of life. With the development of the so-called **Guthrie test**, the routine screening of newborns prior to their leaving the hospital became practical. In this procedure, a technician pricks a newborn's heel and prepares a small filter paper disk of dried blood. It is placed on top of an agar surface inoculated with a bacterial strain and an inhibitor that keeps the bacteria from reproducing *unless* phenylalanine diffuses from the disk into the

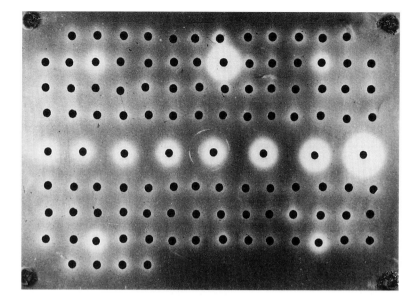

Figure 19.7 A Guthrie test plate for phenylketonuria screening. Filter paper disks of dried blood from 100 babies have been placed on top of bacteria that only grow if phenylalanine is present. The disks in the middle row are controls that contain increasing concentrations of phenylalanine; they support increasingly larger halos of bacterial growth. The test disk near the top center with a wide halo is from an infant with phenylketonuria. (From Levy 1973.)

nutrient medium. After incubation (along with control disks having known amounts of phenylalanine), a halo of bacterial growth around a test disk means that excess phenylalanine is present, and this is usually diagnostic of PKU (Figure 19.7). Confirmatory tests using more refined techniques are performed to rule out false positives. For example, oligonucleotide probes for the several base substitution mutations within the PKU gene can detect heterozygotes as well as homozygotes, but these methods are currently too expensive for a general newborn screening program.

In 1962, Massachusetts became the first state to require that newborn children be screened for PKU;* today, all states and several dozen foreign countries have PKU legislation. Tens of millions of newborns have been screened, and thousands of affected infants have been detected. Unfortunately, few states provide for financial aid for counseling services or for treatment of children found to have PKU. The PKU screening program has proved to be cost-effective: Tens of thousands of dollars are spent to detect one case of PKU (out of about 10,000 births on the average), but considerably more would be spent on institutionaliza-

*Massachusetts tests for eight genetic diseases: PKU, sickle-cell disease, galactosemia, maple syrup urine disease, congenital adrenal hyperplasia, congenital hypothyroidism, homocystinuria, and biotinidase deficiency. It also tests for congenital toxoplasmosis, caused by a protozoan. (Erbe and Levy 1997.)

tion of affected individuals in the absence of testing.

Some states make multiple use of the newborn blood sample that is taken to test for PKU. For example, the autosomal recessive disorder, *galactosemia*, with an incidence of about 1 in 75,000 births, can be diagnosed from dried newborn blood by a direct assay for the missing enzyme (galactose-1-phosphate uridyl transferase). Without treatment, these newborns suffer from severe vomiting, diarrhea, liver disease, jaundice, cataracts, and often death from *E. coli* infection. These symptoms would of course bring the infant to medical attention (without newborn screening), but not necessarily to a correct diagnosis. The progress of the disease can usually be halted if the infant is quickly identified and placed on a diet that includes specially formulated milk substitutes and eliminates all other sources of the sugar galactose, which these babies cannot metabolize.

Neural Tube Defects. Screening can also be done during fetal life. As we noted in Box 19A, chemical markers in maternal serum help detect a Down syndrome pregnancy. In a similar way, general prenatal screening for neural tube defects (NTDs) such as spina bifida can be done by testing the blood of pregnant women for α-fetoprotein. Recall that higher than normal levels of α-fetoprotein are present in the amniotic fluid of fetuses with spina bifida and anencephaly. The α-fetoprotein in the amniotic fluid diffuses into the maternal plasma, so about 80% of fetuses with open NTDs cause elevated levels in their mother's blood. Testing programs can screen *all* pregnant women in a defined population rather than just those mothers who have had a prior child with an NTD. The justification for general screening is that most newborns with NTDs occur in families with no prior history. In England and Wales, the prevalence of spina bifida dropped from about 1 in 500 births in 1970 to about 1 in 5,000 births in 1990 (Wald and Kennard 1997). Most of the decrease was due to prenatal screening and selective abortion, although part of the decrease was undoubtedly due to a diet richer in folic acid, which protects against the occurrence of NTDs.

These screening programs must be carefully managed to minimize uncertainties for the pregnant women. Their blood is tested for α-fetoprotein at the sixteenth to eighteenth week of pregnancy, but high levels do not necessarily—or even usually—mean a defective fetus. There are many benign reasons for an elevated level of maternal α-fetoprotein. A retest for α-fetoprotein plus ultrasonography and amniocentesis are needed to yield a more definitive diagnosis, but these additional procedures take two to four weeks. Since about 90% of women with elevated serum α-fetoprotein deliver *perfectly normal* babies, expert counseling and emotional support must be provided to offset the initially upsetting diagnosis of a possible defect.

Breast Cancer.   Recall from Chapter 9 that testing adult women for mutations of the *BRCA1* and *BRCA2* genes has moved from research settings to at least four commercial laboratories. It might be advantageous to be tested if you have several close family members with early-onset breast cancer, but we noted a host of complicated questions that should be carefully considered. In any case, testing should be accompanied by skillful genetic counseling, which the commercial companies may or may not provide (Brower 1997). Mutations of each of these genes account for only about 3–5% of breast cancer cases (Chapter 17). Thus, testing negatively leaves your risk of breast cancer at about the same level as that for women in the general population. Possessing a mutant allele of *BRCA1* (over 100 different mutations have been detected) leads to breast cancer in about 85% of cases and to ovarian cancer in about 50% of cases. Possessing a mutant allele of *BRCA2* also leads to breast cancer in about 85% of cases, but not to ovarian cancer. (Some researchers think that these figures are somewhat too high.) It is too early to know whether intervention can substantially alter these risks, but monitoring for early detection might be useful. It is also not known with certainty whether different mutant alleles carry different probabilities of cancer onset or different degrees of disease severity.

While testing may make sense within a family setting, screening general populations is not recommended at this time (ASHG Ad Hoc Committee 1994). The reasons deal with the uncertainties regarding breast cancer onset as well as the inability to forestall disease, problems of ensuring quality control in testing facilities, and the difficulties of providing adequate educational materials and psychological support (Hubbard and Lewontin 1996). The insurance problem accompanying any type of testing is discussed in Box 19B.

Screening Adults for Carrier Status

It is important to detect the carriers of recessive diseases so that counseling information can be provided to couples when both are heterozygous. For many autosomally inherited inborn errors of metabolism, heterozygotes show only half the level of enzyme activity exhibited by normal homozygotes. Two heterozygote screening programs were initiated in the 1970s, one (for sickle-cell disease) marked by confusion in the be-

ginning and one (for Tay-Sachs disease) fairly successful from the start. We look at these projects briefly and then at a recent screening proposal (for cystic fibrosis) based on DNA analysis.

Sickle-Cell Disease.   Recall that homozygotes for the sickle-cell allele (about 1 in 500 black Americans) suffer from severe anemia and painful crises. The symptoms are variable, however, and although many affected persons die young, other homozygotes live long, active lives. Heterozygotes (about 1 in 10 black Americans) have normal phenotypes. When used properly, a simple slide or solubility test can clearly separate persons with at least one sickle-cell allele from homozygous normals (see Figure 7.4), and electrophoretic analysis can easily and reliably distinguish all three genotypes.

The history of screening for sickle-cell phenotypes is intertwined with politics. The civil rights movement of the 1960s, combined with the appreciable frequency of the sickle-cell allele among blacks and the availability of good screening tests, led to an assortment of individual state laws to try to deal with what was seen as a neglected disease. These laws were conceived in good faith and were often passed without debate, having been sponsored by black politicians and supported enthusiastically by community leaders. Later experience showed, however, that some of the state laws were hastily drawn and poorly planned. For example, rather than being voluntary, some of the statutes required the testing of either prospective marriage partners (who could use the information) or schoolchildren (who could not). In addition, some programs failed to adequately inform the public that heterozygotes (with the so-called sickle-cell trait) are not sick. As a consequence, many heterozygotes were wrongly denied educational and employment opportunities (Vines 1994).

The National Sickle Cell Anemia Control Act of 1972 addressed the varied problems of the then-existing state laws by providing federal funds to states and private groups that planned voluntary programs incorporating substantial community representation, effective education, counseling services, and strict safeguards. Thus, testing for sickle-cell carriers is now well planned, comprehensive, and available to anyone who requests it.

The value of screening *newborn* populations for sickle-cell disease has received attention because early identification can reduce mortality by about 15%. Infants with sickle-cell disease may die quickly from overwhelming *Pneumococcus* infections, which often progress from the onset of fever to death in less than 12 hours. If diagnosed early, however, the frequency and severity of infections can be much reduced by daily administration of oral penicillin. In Chapter 7, we discussed the recently discovered treatment of sickle-cell disease with hydroxyurea.

Tay-Sachs Disease. The opportunity for screening young adults for the autosomal recessive allele causing Tay-Sachs disease (TSD) was unusually favorable. Tay-Sachs disease is an untreatable disorder with clinical onset toward the end of the first year and progressing to an invariably fatal outcome by about 4 years of age. The disease is extremely rare in almost all populations, but it has a moderate frequency among French Canadians in Quebec and an even higher frequency (about 1 in 4,000 births) in Ashkenazi Jews of eastern and central European descent. Furthermore, the initial screening test for heterozygosity (an assay for hexosaminidase A in serum) is inexpensive, convenient, reliable, and discriminating. When both partners are confirmed as carriers, the option of amniocentesis for all pregnancies and selective abortion of affected fetuses can be offered. Thus, couples who are tested can assure themselves of a family without fear of the emotional trauma imposed by the slow, inevitable death of a TSD child.

After a year of careful planning and organization, a pilot program was undertaken in Jewish communities in Baltimore and Washington, DC. Information about genetics in general and Tay-Sachs in particular was widely spread through newspapers, radio, television, and special pamphlets. About 7,000 persons volunteered for testing in the initial 1971–1972 period. Although some couples at first expressed shock, anger, or anxiety on learning that one or the other of them was a carrier, these reactions usually abated with further counseling. Nearly everyone was glad to have been tested, the implications were freely discussed with friends and relatives, and in the end, the carrier state was not generally regarded as a stigma. Nevertheless, the feelings of those couples identified as heterozygotes, and the feelings of their (perhaps untested) relatives, cannot be fully known.

The role of personal conscience is particularly acute in some Orthodox Jewish communities, where abortion is prohibited and where the revelation of a TSD allele in one person may adversely affect the marriageability of a whole kindred. Some ultra-Orthodox New York Jews have participated since 1983 in an interesting program in which confidentiality is assured by the traditional matchmaker. A person does not learn of his or her carrier status unless matched with another carrier, in which case the families can say that the pair

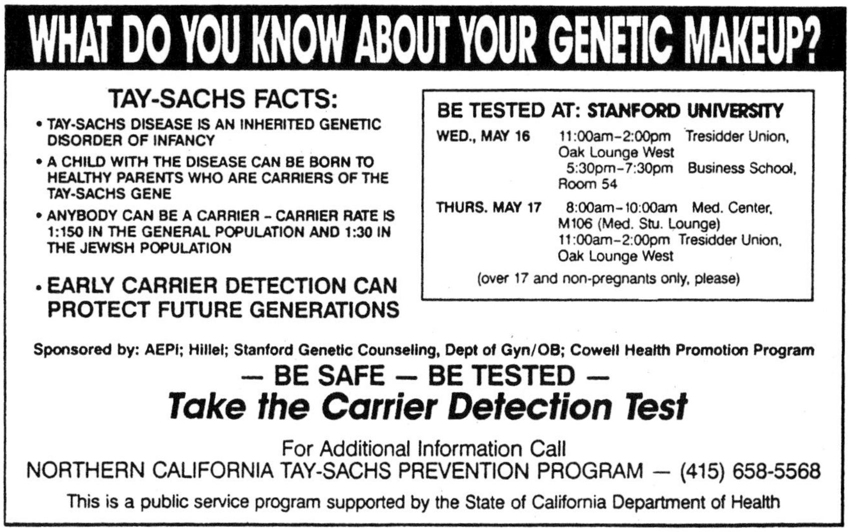

Figure 19.8 Advertising a Tay-Sachs prevention program in California. As part of a public education campaign, this poster promoted the screening program for carriers of Tay-Sachs disease. (Courtesy Alpha Epsilon Pi.)

broke up for other reasons. Although the program is now well established, it met with stubborn resistance when first proposed (Kolata 1993). Traditional Jews said that if God wanted them to have a Tay-Sachs child, they would have a Tay-Sachs child. But proponents argued that choosing a mate on the basis of religious devotion, community status, and some money was not left to God—"So why leave Tay-Sachs to God? He has enough to do." In fact, the program has been expanded (not without criticism) to include testing for other high-frequency (but not so serious) diseases.

Jewish communities throughout North America and elsewhere (South America, Europe, Israel, South Africa, Australia) have started TSD screening programs (Figure 19.8). About a million young Jewish adults have been voluntarily tested, with a carrier detection rate of about 1 in 30. About 1,000 couples without a prior history of affected children have been identified as *both* being carriers, and their pregnancies have been monitored. These efforts have contributed to a decline of over 90% in the incidence of TSD among Jewish populations in North America (Kaback et al. 1993; Mitchell et al. 1996).

Cystic Fibrosis. Testing for cystic fibrosis (CF) carriers among people *with a family history* of the disease makes good sense. But controversy surrounds screening proposals to detect CF heterozygotes in the *general* U.S. white population (Grody et al. 1997; Wilfond and Fost 1990). Carriers are present in a substantial frequency, about 1 in 25 people.

A large part of the problem stems from the multitude of different mutations that lead to the disease

TABLE 19.4 Nineteen different cystic fibrosis mutations detected in a French Celtic population in western Brittany

Code name of mutation	Frequency (%)	Amino acid change	Nucleotide change
ΔF508	81	One amino acid deletion	Three-base deletion
1078 delT	5	Many changes (frameshift)	One-base deletion
G551D	4	One amino acid substitution	Base substitution
1717–1 G to A	1	Many changes (splicing error)	Base substitution at end of intron
W846X	1	Shortened polypeptide	Base substitution to stop triplet
G91R	1	One amino acid change	Base substitution
1221 delCT	1	Many changes (frameshift)	Two-base deletion
G542X	1	Shortened polypeptide	Base substitution to stop triplet
4005+1 G to A	1	Many changes (splicing error)	Base substitution at end of intron
Ten other mutations	Each <1		

Source: Férec et al. 1992.

(Chapter 5). We have noted that the most common mutation (*ΔF508*) is present in about 70% of carriers. But among the remaining 30%, there are over 600 mutations that can also produce CF when homozygous or when present in compound heterozygotes with different CF mutations (Table 19.4). Screening tests often use oligonucleotide probes and the polymerase chain reaction on DNA from cells from the inside of the cheek. It is practical to screen for *ΔF508* and a limited number of other specific mutant alleles; a test that screens for about two dozen mutations would detect about 85–90% of carriers (in many populations of northern European descent) and cost about $150 to $200 per person. Mass screening for *all* the mutations known to be present in a particular population is usually impractical.

There are bound to be many *false negatives*—people who are told they are not carriers, but who nevertheless possess one of the untested CF mutations (Figure 19.9). In a mating between two false negatives, one-quarter of the children will unexpectedly be affected with CF. What level of carrier detection justifies a screening program for the general population? The question has economic, social, and personal ramifications and is not easily answered. The current level (85–90% of carriers detected) seems to some investigators to be on the borderline. Note that the detection of 85% of carriers means that only 72%—that is $(0.85)^2$—of carrier × carrier marriages will be discovered.

Another problem involves the perceived need for CF screening. CF patients have variable symptoms, some being more seriously affected than others. In a typical case of CF, the patient suffers chronic infections and debilitation, frequent hospitalizations, and a shortened life span, with accompanying psychological and financial tolls on the family. However, modern medical treatments continue to improve the quality and length of life. The stated average life span of 26 years is for a cohort of people born decades ago, not for people born more recently, when treatment is more effective. (We do not know the average life span of recently born cases, because these individuals have not yet lived out their lives.) Some argue that screening for heterozygotes and prenatal diagnosis for fetuses at risk are not needed at all. Although many patients suffer for extended periods of time, advocates for disabled persons (among others) note that some CF patients live productive and independent lives, even into middle age and beyond. Thus,

It's like this, Mrs. Cameron. The results are negative, but that doesn't mean not positive, exactly. Nor is it not negative, we wouldn't want a double negative there, would we ...

Figure 19.9 Explaining concepts like probability and false negatives is not easy. (Copyright © 1997 Donna Barstow. Reprinted by permission of the artist.)

they conclude that all people with CF deserve the chance for life.

Cystic fibrosis screening has been done on a limited basis, and population testing will undoubtedly increase as the percentage of detectable mutations rises and the cost falls. To offer screening to all Americans, however, would be a massive undertaking requiring an unprecedented educational effort. Experienced genetic counselors would be especially in demand to deal with people in uncertain situations—people who must confront the possible stigma of carrier status and who must deal with slippery probabilities involving the concept of false negatives. It is unclear where the expertise and resources would come from, and the American Society of Human Genetics does not believe that mass population screening for CF carriers is justified at this time (ASHG Ad Hoc Committee 1992). On the other hand, an advisory panel established by the National Institutes of Health recently recommended that DNA testing for CF be offered, but not promoted, to couples of childbearing age regardless of their family history (Milot 1997).

Summary

1. Inherent in the various procedures for genetic testing and counseling are some troublesome technical, ethical, legal, and social problems.

2. Genetic counselors determine relevant medical and genetic facts for a family confronting genetic disease. They must convey this information, as well as available options, in a sympathetic and understandable way.

3. Counselors should try to be nondirective and sensitive to the emotional needs of affected families. They must recognize that one's perception of risk and willingness to accept a burden depend on unique personalities, experiences, and moral convictions.

4. Prenatal diagnosis usually relies on analyses of fetal cells obtained by chorionic villus sampling or amniocentesis. The former is done at about the tenth week of pregnancy, and the latter (perhaps a bit more safely) at about the sixteenth week. If the fetus is found to be abnormal, abortion is an option.

5. To avoid any risk to the fetus, researchers are seeking ways to diagnose fetal conditions by isolating fetal cells from the maternal circulation.

6. Prenatal diagnosis is offered whenever there is increased risk of chromosome abnormalities. Analysis may utilize fetal DNA: fluorescence in situ hybridization, linkage to DNA markers, or oligonucleotide probes specific for mutant alleles. If a gene is expressed in fetal cells, researchers can examine the relevant protein. Diagnosis of neural tube defects is aided by determining α-fetoprotein concentrations in amniotic fluid and maternal blood.

7. Fetal sex can be easily determined by karyotyping, by Y chromosome probes, or (experimentally) by preimplantation diagnosis, in which one cell is excised from an eight-cell embryo and examined. The DNA of this cell can be amplified (with difficulty) by PCR.

8. Genetic screening is the systematic search of populations to detect a genetic abnormality prior to overt symptoms or to determine heterozygosity for recessive disorders. Screening programs should be carefully and sympathetically planned to ensure acceptability in the community. Problems of privacy and insurability for any kind of genetic testing are particularly troublesome.

9. The most widely used screening program tests newborns for phenylketonuria. Prenatal screening for neural tube defects has been undertaken in some populations by examining levels of α-fetoprotein in maternal blood. Testing for mutant alleles of two breast cancer genes is available.

10. Carrier screening can be done among black populations to reveal the sickle-cell allele and among Jewish populations to reveal mutant Tay-Sachs alleles. Screening for cystic fibrosis carriers has many problems, including the occurrence of false negatives due to the inability to test for all of the hundreds of mutant alleles.

Key Terms

amniocentesis	genetic counseling
chorionic villus sampling (CVS)	genetic screening
empirical risk	Guthrie test
false negative	neural tube defects (NTDs)
false positive	preimplantation diagnosis
α-fetoprotein	

Questions

1. In parts of Africa, Asia, and Latin America, diarrheal diseases, respiratory infections, tuberculosis, malaria, hepatitis, AIDS, and a half dozen other diseases kill tens of millions of people annually. Are the concerns raised in this chapter equally applicable to these populations? What funding should be given to research on tropical diseases versus genetic diseases?

2. In a so-called wrongful life case, an infant with Tay-Sachs disease sought damages for negligence when a medical laboratory failed to find that her parents were both carriers. Had the laboratory made a correct diagnosis, there would have been no child to endure the suffering, because it would have been aborted or not conceived. Would you award damages to the Tay-Sachs family in this case? (Courts in some states have recognized wrongful life actions, while others have rejected them.)

3. Can there be any point in amniocentesis or chorionic villus sampling if the parents do not agree beforehand to abort a seriously abnormal fetus?

4. How do you feel about using prenatal diagnosis for the sole purpose of determining fetal sex in advance of birth? Assuming that you were using in vitro fertilization to achieve pregnancy, how do you feel about using preimplantation diagnosis for sex determination?

5. Consider preimplantation diagnosis for cystic fibrosis, using PCR on the DNA from an embryonic cell. In what situation would researchers mistakenly discard a phenotypically normal embryo from heterozygous parents? Recall that PCR may not amplify both of the target sequences, each present exactly once in the single cell.

6. One reason that chorionic villus sampling is a bit less accurate than amniocentesis is that the CVS cell sample comes from fetal tissue that will form the placenta rather than from the embryo itself. Why should this matter, since all the tissue comes from one zygote?

7. Why is chorionic villus sampling not useful for the prenatal diagnosis of neural tube defects?

8. One aspect of genetic screening that has received attention is whether it should be mandatory or voluntary. Do you think that the widespread compulsory screening for phenylketonuria (in which you were probably included) represents an unwarranted government intrusion into personal lives, or is it a legitimate concern of a state's public health service? What about other screening programs?

9. Screening all newborns for cystic fibrosis or Duchenne muscular dystrophy is not likely to improve the lives of those who test positive, because nothing can be done until symptoms strike. Nevertheless, revealing the abnormal genotype earlier through screening might be useful to the parents. Explain.

In the following questions, assume a population in which the frequency of CF carriers is 1 in 20, or 5%. Assume also that CF screening detects 70% of carriers.

10. Show that slightly more than half of carrier × carrier marriages would be *missed*.

11. Assume that one parent is shown to be a carrier, and the other is not tested. What is the probability that this marriage has a carrier × carrier status?

12. Assume that one parent is shown to be a carrier, and the other parent is tested and said to be free of disease-causing CF alleles. What is the probability that this marriage has a carrier × carrier status?

Further Reading

The comprehensive reference work (3,000 pages) on medical genetics by Rimoin et al. (1997) includes several dozen introductory chapters on basic principles and practices. Chapter topics include genetic screening, prenatal diagnosis, ethical considerations, and legal issues. The chapter by Walker (1997) on genetic counseling is especially thoughtful. Another detailed text on genetic screening and prenatal diagnosis is by Brock et al. (1992). Although meant for professionals, some of the material in these books could be helpful to lay readers. Meant for general readers is Zallen's (1997) stimulating and useful book, subtitled "A Consumer's Guide to DNA Testing for Genetic Disorders." Wertz (1996) has written a thorough article on the social and ethical issues concerning preimplantation diagnosis. The long article by Holmes (1996/1997) considers the legal and legislative tangles presented by genetic testing and insurability. Finally, the article by Roueché (1979) on Carol Terry's experience with Wilson disease is gripping.

CHAPTER 20

Altering Genetic Traits

"I am the mother of five children, three of whom have sickle cell anemia," wrote Ola Mae Huntley in 1984. After explaining that genetic counseling was not offered in the 1960s, when her children were born, she continued:

> After our second child was diagnosed as sickle-cell-anemic, a well-meaning but uninformed doctor assured me that, since we already had one normal child and one with sickle cell anemia, our future children would be normal, purely on the basis of numerical chance. …

> Our sickle-cell-anemic children are now young adults. Only rarely can we all be home together as a family. Usually one or two, sometimes all three, are in hospitals being treated for acute disease crises or for the debilitating effects of the disease. This is now a way of life, or I should say, a way of existence. …

> In my twenty-five years as a mother very little has changed. The current state of medical art is to treat the symptoms, but little has been done about treating the cause. (Huntley 1984)

Huntley wrote of the mental anguish brought about by poor self-esteem and little hope for a better life and likened the

437

relief of physical pain to putting a small bandage on a large wound. Today, better treatments are available partly through daily administration of *hydroxyurea*. This drug, however, is not useful for all adults, is not approved for children at all, and is still being tested for long-term safety (Schechter and Rodgers 1995; also see Chapter 7). Huntley's 1984 wish list contained items that are still pertinent, including genetic engineering research that would lead to an effective cure or a way to prevent the disease in the first place. Although her specific interest was sickle-cell disease, Huntley noted that her concerns were relevant to other severe genetic diseases as well.

The treatment possibilities for heritable disorders—therapies with a wide range of benefits—are discussed in this chapter. We consider the potential for *antisense therapy*, that is, using a stretch of nucleotides to block the action of a disease-causing gene by binding specifically to its messenger RNA. We also look at attempts at *gene therapy*, whereby a new gene is added to somatic cells of the body to correct the consequences of a defective gene. In theory, gene therapy could also correct a gene in eggs or sperm, thus extending healing effects to future generations. Such direct attempts to improve the genetic endowment of our descendants is a form of *eugenics* and is a topic of current debate. Eugenic issues prompt the additional and novel question of whether we should try to leave replicas of ourselves. That it might be possible to create more persons with copies of one's own nuclear genes (i.e., delayed identical twins) arises from the recent *cloning* of a sheep and other animals. Thus, we also include a history and discussion of cloning.

Treating Genetic Disease

In principle, genetically influenced conditions *can* be treated. Effective therapy usually requires an understanding of the cascade of causes and effects. In only a few percent of the thousands of genetic diseases, however, can anything like a normal phenotype be restored, allowing treated individuals to follow ordinary life-styles and attain regular life spans. In one study, the treatment success in 1983 for 65 randomly chosen hereditary metabolic diseases was compared with the treatment success in 1993 for the same diseases (Treacy et al. 1995). During those ten years, about one-third of the causative genes were mapped and several were cloned, so a lot of basic knowledge was brought to light. Yet, progress in effective treatment of the diseases was modest. In both 1983 and 1993, the same 12% of diseases were considered to respond completely to treatment. Those that responded partially to treatment increased from 40% to 57% over the decade.

The shortcomings arise, in part, because the products of mutant genes often interfere with fundamental processes within cells. Therefore, it may be difficult to intervene in a substantive way without significant side effects. In addition, most genetic diseases are individually uncommon, and money for research may be limited. Yet even some of the better known genetic or chromosomal disorders with a long history of funded research (e.g., achondroplasia, albinism, alkaptonuria, Down syndrome, fragile X syndrome, Huntington disease, muscular dystrophy, Tay-Sachs disease) remain largely untreatable. Although the results have been generally disappointing, we can hope that recent developments in molecular genetics will open new avenues for intervention.

It would be best, of course, if genetic diseases could be prevented from occurring in the first place, which would make treatment unnecessary. At a basic level, this means reducing the frequency of mutations to abnormal alleles by cutting exposure to mutagenic agents. As noted in Chapter 7, however, some mutations are due to the instability of DNA itself, and the threat of germ line mutagenicity from generally existing levels of chemicals and radiation is already probably small. Without greater knowledge of the causes of germ line mutations, it is unclear what social policies would be sufficiently practical to achieve further reductions. Once new mutations have occurred, however, and have been transmitted to later generations (usually without any notice), prevention becomes a matter of genetic testing and counseling. These may include prenatal diagnosis, selective abortion, and alternative reproductive technologies.

Some Current Practices

To improve everyday functions, many of us alter our private environment by using corrective devices that act at a level far removed from the primary effect of any mutant gene. The list would include eyeglasses, hearing aids, artificial limbs, wheelchairs, and similar devices. In extreme cases, a few children with immune deficiencies of genetic origin have been confined to environments that seal out infectious agents (Chapter 18). Such external management of the physical environment may involve ingenious biomedical engineering, but it can be cumbersome and of limited value.

Treatments that alter the body rather than the environment are extremely varied (Desnick 1991; Friedmann 1991). Diet modifications may restrict the intake of a toxic substance or supplement a deficient metabolite. Therapy may involve drugs, transplantation, other surgery, or, much closer to the gene, replacement of a polypeptide that the mutant allele is unable to supply. Research at the level of the gene itself is in experimental and trial stages. Some tactics that can help persons with genetically influenced diseases are listed in Table 20.1 and discussed here.

TABLE 20.1 Some mostly genetic diseases that are treatable

Means of treatment	Disease
SURGICAL REPAIR AND REMOVAL	
Surgical repair	Cleft lip and cleft palate
Removal of spleen	Hereditary spherocytosis
Removal of colon	Familial polyposis coli
DRUG TREATMENT OF A MAJOR SYMPTOM	
Beta blocker (to reduce pressure on aorta)	Marfan syndrome
Hydroxyurea (to inhibit sickling)	Sickle-cell disease
DIETARY RESTRICTION OF A PRECURSOR	
Phenylalanine	Phenylketonuria
Galactose	Galactosemia
Leucine, isoleucine, valine	Maple syrup urine disease
Fava beans	Favism (G6PD deficiency)
DEPLETION OF AN EXCESSIVE SUBSTANCE	
Copper (by penicillamine)	Wilson disease
Cholesterol (by bile binders)	Familial hypercholesterolemia
Iron (by bloodletting)	Hemochromatosis
Uric acid (by several drugs)	Gout
REPLACEMENT OF A MISSING GENE PRODUCT	
Insulin	Juvenile-onset diabetes
Growth hormone	Pituitary dwarfism
Clotting factor VIII	Hemophilia A
Glucocerebrosidase (enzyme)	Gaucher disease
Adenosine deaminase (ADA)	ADA deficiency
ORGAN AND TISSUE TRANSPLANTATION	
Bone marrow	Severe combined immunodeficiency
Bone marrow	β-thalassemia
Bone marrow	Lysosomal storage diseases
Liver	α_1-antitrypsin deficiency
ANTISENSE THERAPY?	
Blocking synthesis of cell adhesion protein	Crohn disease
GENE THERAPY?	
Adding ADA gene to T cells	ADA deficiency

Surgical Repair and Removal.

Alteration of the phenotype by surgery can improve appearance or basic health. The disfigurement, speech difficulties, and swallowing problems of *cleft lip* and *cleft palate* (occurring by themselves or accompanying other genetic and chromosomal syndromes) can be restored to near normal by skillful surgery. The extra digits of *polydactyly* can also be easily removed. In the autosomal dominant disorder *spherocytosis*, the red blood cells assume rounded shapes because of increased permeability of the cell membrane. Although the spherical cells carry oxygen reasonably well, their fragility leads to chronic anemia and often to gallstones at an early age. The fragile cells are entrapped in the spleen, whose phagocytic cells normally remove aging red blood cells from the circulation. Removal of the spleen in patients with spherocytosis apparently allows the red blood cells to survive longer and relieves most of the disease symptoms.

Treating a Major Symptom with Drugs.

Rather than surgical repair, some genetic diseases can be managed by drugs that influence a major symptom. In Marfan syndrome, for example, the aorta (the main artery leaving the heart) is weakened by abnormal forms of the elastic protein fibrillin (Chapter 9). The weakening may not be apparent until the aorta ruptures and quickly leads to death. The hidden damage to the aorta (especially where it joins the heart) is accelerated by the rhythmic pumping of blood hard against its walls. *Beta blockers* can be used to reduce the activity of the heart and therefore alleviate some of the pressure on the aorta. Beta blockers are drugs that hinder nerve receptors in the heart and have long been used as the primary medication for many patients with heart disease or high blood pressure as well as Marfan syndrome. In advanced cases, the aorta and aortic valve can be surgically replaced by synthetic materials. These types of medical management have greatly improved the outlook and increased the life span of Marfan patients.

We have already mentioned the *hydroxyurea* treatment of sickle-cell disease. This drug (for unknown reasons) increases the production of fetal hemoglobin, thus reducing the sickling of red blood cells due to abnormal β chains of adult hemoglobin.

In the next three categories of treatment listed in Table 20.1, researchers often make use of knowledge of the relevant gene-controlled metabolism, discussed in Chapter 16 and summarized briefly as:

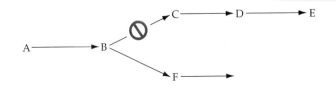

Recall that a metabolic block—here in the conversion of B to C and represented by the stop symbol—is due to a defective enzyme. The blocked chemical reaction can result in the accumulation of the precursor (beginning) substance, B, to toxic levels. Damage to the body might also result from the buildup of the substances F and G in the alternative pathway. Lack of the vital end product, E, is yet another way for damage to arise as the result of the metabolic block.

Dietary Restriction of a Precursor.

The diet therapy for *phenylketonuria* (*PKU*) detailed in Chapters 1 and 16 remains the prototype for this method of treatment. The recommended diet drastically lowers the intake of the amino acid phenylalanine, which is a component of proteins. In patients with PKU, an enzyme deficiency reduces the normal conversion of phenylalanine to tyrosine, so that phenylalanine builds up in the body to toxic levels. The low-phenylalanine diet prevents this buildup. Several dozen genetic diseases are treated by diet restrictions or diet exclusions, with varying degrees of success. In these cases, the end products that are not synthesized because of the blockage may be available from food or from alternative metabolic pathways.

Depletion of an Excessive Substance.

The rare recessive *Wilson disease* (Chapter 19) results from an error in copper metabolism. In affected persons, the abnormal deposition of copper causes liver and brain damage; if untreated, it leads to death after years of suffering. In some persons, the symptoms are primarily psychological: grossly inappropriate social behavior and bizarre personality changes, which can be mistaken for schizophrenia or manic depression. The age of onset varies from 6 to 50. Although the chain of causation of the symptoms is obscure, drugs that bind copper (so-called chelating agents) and lead to its excretion in the urine can provide dramatic improvement. One such drug is penicillamine (which is derived from penicillin but has no antibiotic activity). Early diagnosis and treatment can completely prevent the serious consequences of Wilson disease.

Recall from Chapter 5 that heterozygotes for *familial hypercholesterolemia* (having a frequency of about 1 in 500 persons) develop atherosclerosis and may have heart attacks beginning in their 30s. The symptoms are due to excessive LDL (low-density lipoprotein), which is the major cholesterol transport system in plasma. Elevated cholesterol levels result from a deficiency of cell receptors (especially in the liver) that take up LDL from the blood. Reductions in serum cholesterol levels are possible by an unusual route involving the bile acids. These substances, rich in cholesterol products, are made in the liver and transported to the intestines, where they help digest fats. Normally, the bile acids are reabsorbed in the intestines and used again, but drugs are available that cause them to be excreted in the stool instead. This action forces the liver cells to take up more cholesterol from the serum to manufacture the missing bile acids. Blood cholesterol levels are thereby reduced by about 15–30%.

Replacement of a Missing Gene Product.

Diabetes, pituitary dwarfism, hemophilia, and *Gaucher disease* can be fairly well treated with insulin, human growth hormone, clotting factor VIII, and an enzyme called glucocerebrosidase, respectively. Recombinant DNA techniques have been used to make these four therapeutic proteins (Table 8.2). Recall that their manufacture involves inserting the cloned human gene into the DNA of cells growing in culture, from which researchers isolate the gene product. Because proteins are broken down by digestive enzymes, oral administration is not usually possible. Thus, intravenous injections are required, often over a lifetime. Although annoying and painful, such treatments allow many patients to lead productive lives.

Attempts to replace defective or missing enzymes (rather than nonenzymatic proteins) have generally not been applied at the clinical level. Such treatment might be very helpful to people with albinism, Tay-Sachs disease, phenylketonuria, or hundreds of other enzyme defects. Two major difficulties stand in the way: the short time that the injected enzyme persists in the body and the inability to deliver the enzyme to the site where it is needed while protecting it from degradation.

A notable exception to these problems is the successful enzyme treatment of persons with Gaucher disease, which is due to recessive alleles of a chromosome 1 gene. Several common mutations and over 50 uncommon mutations are known to code for a defective form of glucocerebrosidase (Beutler 1991). This lysosomal enzyme normally breaks down a lipid called *glucocerebroside*. Recall that these basic facts are similar to those for Tay-Sachs disease, in which a defective form of the enzyme hexosaminidase A can no longer break down a different (but very similar) fatty substance (ganglioside G_{M2}; see Chapter 16). For reasons that are not understood, there is a further similarity: Both Gaucher and Tay-Sachs are prevalent in Ashkenazi Jews, but not in most other ethnic groups (Diamond 1994). Also, in both cases, the accumulation of the lipid injures cells, but not the same cells. The damage in Tay-Sachs disease, always fatal in early childhood, is primarily in nerve cells of the brain, while the damage in Gaucher disease is primarily in the macrophages of the immune system.

The symptoms of Gaucher disease are extremely variable. Depending on definitions, the disease is mild in about 50% of all cases, moderate in 40%, and serious in about 10% (see question 1). The lipid-engorged macrophages cluster in and cause swelling (sometimes massive) of the liver and spleen. This typically occurs in children and adolescents, often the first sign of disease. The abnormal macrophages also accumulate in the bone marrow, replacing normal cells. Acute bone pain, bone deterioration, severe anemia due to low levels of hemoglobin, and internal bleeding (easy bruising, frequent nosebleeds) are additional symptoms (Figure 20.1).

Early attempts to treat Gaucher disease with infusions (slow injections into a vein) of the natural enzyme (extracted from human placentas) were hopeful but limited, because not enough of the enzyme could be supplied* and not enough of the infused enzyme found its way into macrophages where it was needed. In the early 1990s, researchers chemically modified the natural enzyme, resulting in a form that was much more readily taken up by specific receptors on macrophages. In 1994, the Genzyme Corporation of Cambridge, Massachusetts used recombinant DNA meth-

*About 20,000 placentas were needed to produce a year's supply for one patient.

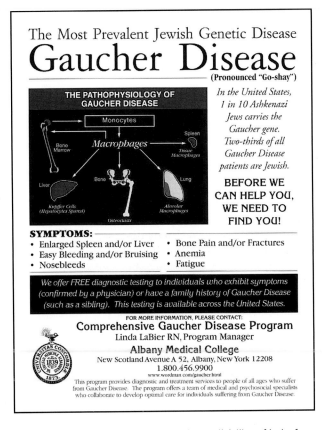

Figure 20.1 A poster advertising the availability of help for patients with Gaucher disease. The poster emphasizes free testing for persons with mild cases, which may not yet have been diagnosed. (Courtesy of Albany Medical College.)

ods to produce a macrophage-targeting form dubbed *Cerezyme*, promising to solve the supply problem. But in any form, the enzyme is very expensive. The company says that the average cost to patients is about $150,000 per year, a cost that must be borne for life (Leary 1995). Needless to say, there is controversy about pricing between producers, doctors, insurers, and regulators. Recent experimental trials are designed to determine the lowest dosages that are clinically effective. Unfortunately, response to the enzyme is varied and unpredictable, so each Gaucher patient must be individually monitored. But effective and safe treatment can be achieved in most cases, eventually reducing the size of enlarged organs, restoring hemoglobin levels, and easing bone pain and bone loss.

A novel method of delivery of the therapeutic agent has been used in the treatment of the rare autosomal recessive disease *adenosine deaminase (ADA) deficiency*. The normal ADA enzyme involves the metabolism of purines, and its deficiency affects white blood cells in particular. The functions of T cells are completely impaired, and antibody production by B cells is also reduced. As a result, ADA-deficient children succumb to overwhelming viral, bacterial, and fungal infections of the skin, airways, and digestive tract. In the new therapy, investigators inject children with the ADA enzyme isolated from cows. But first they coat the enzyme with the chemical polyethylene glycol, or PEG. Apparently, the life of the coated enzyme is significantly prolonged, and the dozen or so patients treated with PEG-ADA have usually shown some improvement in their immunological functions. ADA-deficient patients have also been treated by bone marrow transplants and by gene therapy (discussed shortly).

Organ and Tissue Transplantation. An interesting medical procedure for transferring the correct genetic information into patients who have a hereditary disease is to transplant an organ or tissue from a normal individual. Such a graft may provide a missing enzyme within the patient's body *on a continuing basis*. Within the graft, the enzyme might operate on circulating substrates. Or the enzyme might be secreted by the cells of the transplant to work elsewhere in the body. Most research in this area has been with the transplantation of bone marrow.

We noted in Chapter 18 that David, the "bubble boy" with *severe combined immunodeficiency disease (SCID)*, received a bone marrow transplant from his sister. Although he did not long survive the transplant, over 100 other patients with SCID (including the ADA deficiency type) have improved after receiving new bone marrow. Matching for all HLA antigens is not absolutely required in transplants to infants with SCID, since they lack the functioning B cells and T cells that mount immunological attacks on the transplanted tissue. However, the reverse reaction, graft-versus-host disease, remains a problem in mismatched transplants. Recent techniques to rid the donor marrow of mature T cells have sometimes prevented this problem.

The use of bone marrow transplants to treat *thalassemia* and other hemoglobin diseases is more difficult because immunosuppressive agents must be used to prevent the rejection of the transplanted bone marrow cells. Nevertheless, in one study, about three-fourths of patients survived entirely free of the disease for at least one to three years. About one-fourth died from graft-versus-host disease or other complications of marrow transplantation. Frequent blood transfusions provide alternative management of thalassemia. Multiple transfusions, however, lead to iron overload, requiring additional long-term treatment with iron-binding chemicals (chelating agents) that remove the potentially toxic excess iron from the body. Because these more conservative techniques can provide almost all affected children with 15 to 20 years of life, it is not always clear what therapy is to be preferred.

The impressive variety of therapeutic approaches outlined here should remove any pessimism over the possibility of treatment for genetic diseases. Still, many of the maneuvers require sophisticated medical or technical expertise and significant expense. Furthermore, some are halfway measures that cause significant side effects or discomfort and must be continued for a lifetime. It is clear that knowing the cause of a disease in profound detail does not guarantee effective therapy.

Antisense Therapy

An interesting and straightforward plan for the treatment of some diseases was formulated in the 1980s and is now in experimental and trial stages. The idea is to prevent messenger RNA from carrying out its essential function of translation. Blocking the synthesis of a harmful protein can be brought about by short nucleotide segments of about 20 bases (called **oligonucleotides** or simply **oligos**) that are complementary to a portion of the mRNA. Complementarity leads to binding, which invites a normal cell enzyme (ribonuclease) to break down the mRNA (Figure 20.2). Because the mRNA is often called the sense molecule, this treatment approach is called **antisense therapy** (Askari and McDonnell 1996; Cohen and Hogan 1994). Note that this tech-

nique is applicable in theory to diseases caused by a harmful protein, but not to diseases caused by a missing or ineffectual protein.

The first antisense oligos did not work very well during experiments with cells in culture. Perhaps because sulfur atoms were included in the nucleotide backbone to prevent its degradation by cell enzymes, these oligos bonded to components that had nothing to do with their logical antisense targets. They also produced some immunological side effects in experimental animals. Researchers have now designed oligos that minimize these problems.

Clinical trials of antisense therapy for *Crohn disease* have been encouraging. This disorder results in chronic inflammation of the bowel, leading to abdominal pain, diarrhea, fever, and general weakness. Although the disease usually harms the small and large intestines, it can affect any part of the gastrointestinal tract, from the mouth to the anus. It is not inherited in a simple Mendelian fashion, but involves several unknown genes and environmental factors. The antisense treatment attacks the mRNA that encodes a cell adhesion protein involved in the inflammatory process. For a small number of patients, the antisense therapy has lessened the major symptoms (Roush 1997). Antisense therapy is being investigated for a wide variety of genetic and nongenetic diseases,

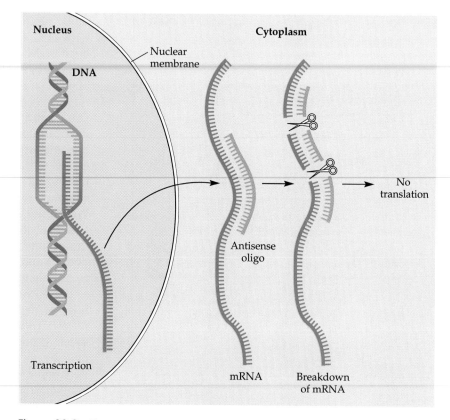

Figure 20.2 The strategy of antisense therapy. Complementary oligonucleotides bind to a portion of mRNA, preventing its translation.

including cancers, cardiovascular conditions, thalassemia, AIDS, and complicating infections associated with AIDS.

Gene Therapy

Investigators and patients alike look forward to the day when it will be possible to correct the effects of abnormal alleles by supplying the normal alleles themselves. In such **gene therapy**, the hope is that the transferred genes will cure the basic defect rather than merely alleviating its damage. We use the word *cure* to mean a limited course of treatment that restores a person's well-being, just as penicillin cures pneumonia or an appendectomy cures appendicitis. In such procedures, treatment is stopped once health is regained.

We consider here what is called **somatic gene therapy**, in which the body cells, but not the germ cells, of the patient receive the corrected gene. Of course, if the germ cells were also changed from mutant to normal (with exact replacement of the mutant gene by the normal one at its usual locus), the cure could extend through future generations. However, objections have been raised to this **germinal gene therapy** because of unsavory overtones—the possibility of unwarranted or unwise tampering with another generation's genes.* Therefore, investigations of human gene therapy have been limited to the correction of disease within individuals. This is, of course, the traditional aim of medicine, and there is general consensus that human somatic gene therapy is a rational option. Every proposal must pass through many levels of government review and scientific scrutiny and is examined for safety, therapeutic, and ethical issues (Friedmann 1996, 1997; Verma and Somia 1997).

ADA Deficiency.   Initially, it was thought that sickle-cell disease or some other hemoglobin disease would be the best candidate for gene therapy trials. It is easy to get bone marrow, including red cell progenitors, from a patient and return it to the patient after adding, say, a normal β-globin gene. A major problem, however, is gene regulation. Hemoglobin is composed of two α and two β chains (Figure 6.19), and the two polypeptides are produced in exactly equal amounts in red blood cells. An excess of either chain leads to cell damage. Since the addition of a new gene to a human chromosome is generally a hit-or-miss process, the proper genetic controls that regulate gene expression may not be in place. It would be better, therefore, to choose an *enzyme* disease where a small amount of new, normal enzyme would likely be advantageous and an extra large amount would not be harmful.

ADA deficiency seemed to be a good choice, since the disease is fatal unless corrected in some way. The PEG-ADA treatment (the cow-derived enzyme coated with polyethylene glycol) had been shown to increase the cellular level of ADA enzyme a small amount, and patients showed some improvement in immunological functions. Furthermore, it was found that cultured white blood cells deficient in ADA grew and divided slowly, whereas the ADA-corrected cells multiplied faster. Thus, when injected into patients, the latter might have a competitive advantage.

Getting sufficient amounts of the normal gene to use in laboratory manipulations was not a stumbling block. Hundreds of human genes have been isolated and characterized by the cloning and recombinant DNA techniques described in Chapters 8 and 9. The exact nucleotide sequence of the gene can be combined with whatever promoters or enhancers are required to achieve transcription and translation in the proper cells.

The major obstacle in gene therapy is getting the normal DNA into patients' cells so that it is stably integrated and working. These twin problems of *gene delivery* and *gene expression* are uppermost in all gene therapy trials. The most direct technique would be to simply inject the cloned DNA into recipient cells (in vitro or in vivo), where it might integrate into chromosomal DNA. But microinjection is usually not feasible for somatic human gene therapy; not nearly enough cells can be injected to be effective. Rather, in the most common approach, researchers package the correct genetic material into a **retrovirus**. Many retroviruses infect target cells with nearly 100% efficiency. Then, in the normal course of their life cycle, they cause a DNA copy of their genetic material (i.e., the provirus) to be inserted into the chromosomal DNA of the host cell (as shown in Figure 17.15). The most frequently used retrovirus for human gene therapy is the Moloney murine (mouse) leukemia virus. Before using it, researchers remove most of the genetic material of the virus to render it incapable of further reproduction. They sustitute suitable human RNA roughly equal in length to the deleted viral DNA (Figure 20.3). This substitution does not affect the ability of the virus to infect human cells and insert a DNA copy of its genetic material into the cells' DNA. Thus, the virus acts as a **vector**, or vehicle, for the delivery of a new gene to human cells.

After years of preparation, the first human trial of gene therapy was begun in 1990 at the National Institutes of Health. It involved a 4-year-old girl named Ashanthi DeSilva (and in 1991 another young girl) with ADA deficiency. Their gene therapy involved removing a blood sample, isolating the T cells, infecting them with the derived retrovirus, and returning the treated T cells to the patient (Figure 20.4). This

---

*Germinal gene therapy works fine in mice. Several genetic diseases have been corrected in the descendants of affected mice whose gametes were injected with normal genes.

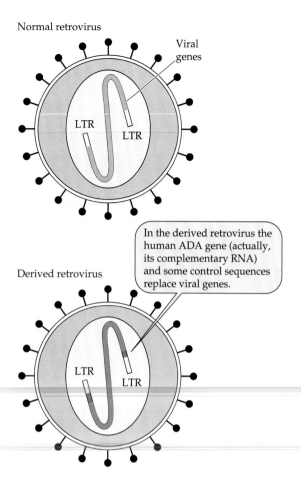

Normal retrovirus

Viral genes

LTR

LTR

Derived retrovirus

In the derived retrovirus the human ADA gene (actually, its complementary RNA) and some control sequences replace viral genes.

LTR

LTR

Figure 20.3 Schematic diagram of a normal retrovirus compared with a derived retrovirus used for gene therapy of ADA deficiency. LTR stands for long terminal repeat, part of the normal viral genome required for insertion of the provirus into the host chromosome. (See also Figure 17.15). The structures other than the genetic material are various types of proteins.

process is called the **ex vivo** ("out of body") **technique.** The girls received injections of their gene-corrected T cells at one- or two-month intervals for the first year, but less often thereafter. The injections were needed periodically because the T cells have a limited life span in the body. For the girls the treatment was simple, in contrast to the painstaking research (and extensive ballyhoo) surrounding the advent of gene therapy. Raj DeSilva, the father of Ashanthi, has said:

> The gene therapy treatment itself was a nonevent from Ashanthi's point of view. No overnight hospital stay was required and the treatment no more painful than the PEG-ADA shot she received every week. A simple infusion of cells from a plastic bag hanging from a metal stand not unlike a blood transfusion lasting less than an hour constituted the treatment for that day. (Committee on Science, Space, and Technology 1994)

Unfortunately, it is not entirely clear that the gene therapy worked as intended. The PEG-ADA treat-

ments were continued throughout the gene therapy (and continue to this day), as it would be unethical to stop even a minimally effective therapy for a potentially fatal disease unless a better treatment is available. Ashanthi has not received gene therapy injections now for several years. She makes about 25% of the normal amount of ADA in her T cells and lives a fairly normal life. The other girl has also been helped, although perhaps somewhat less (Culver 1996a; Marshall 1995a). For these girls and a handful of additional patients whose ADA symptoms have been alleviated, the limited but proven role of PEG-ADA obscures the possibly greater role of gene therapy. Thus, the evidence for a therapeutic benefit of gene therapy is suggestive but a little ambiguous. No gene therapy trial to date has had an unequivocally beneficial effect, and critics raise many unanswered questions of basic gene function and disease physiology (Marshall 1995b).

Additional Comments about the Technology.   One problem with ADA deficiency treatment is the need for recurrent injections of gene-corrected T cells. To address this problem, scientists would prefer to correct and administer, not the mature T cells, but the **stem cells** from which these lymphocytes are derived. The primordial stem cells exist in the bone marrow and serve as a source of both red and white blood cells throughout a person's life (Figure 20.5). When a primordial stem cell divides, it produces one stem cell like itself and one cell that differentiates in succeeding divisions into mature blood cells. Using marrow stem cells in ADA gene therapy thus has the advantage that it may have to be given just once, achieving a real cure. The problem is that out of a million bone marrow cells, only a dozen are actually stem cells, and they are hard to identify and collect. Whether they can be effectively gene-corrected in a permanent fashion and whether they will work in patients are still open questions.

A major worry is that insertion of the retrovirus-borne gene into the T cell appears to be at random places in the genome rather than at the locus of the abnormal gene. There is the risk of interruption of some normal gene activities or death of the cell. Researchers also worry that a cellular oncogene may be activated or that infectious viruses may be generated (by recombinational processes) and spread to other cells. Investigators are currently addressing these possible dangers.

Hundreds of protocols for gene therapy involving thousands of patients worldwide are now in progress. A few of these gene therapy procedures are listed in Table 20.2. A common method for getting the normal gene inside human cells is the ex vivo use of retroviruses, as with ADA deficiency. For cystic fibrosis tri-

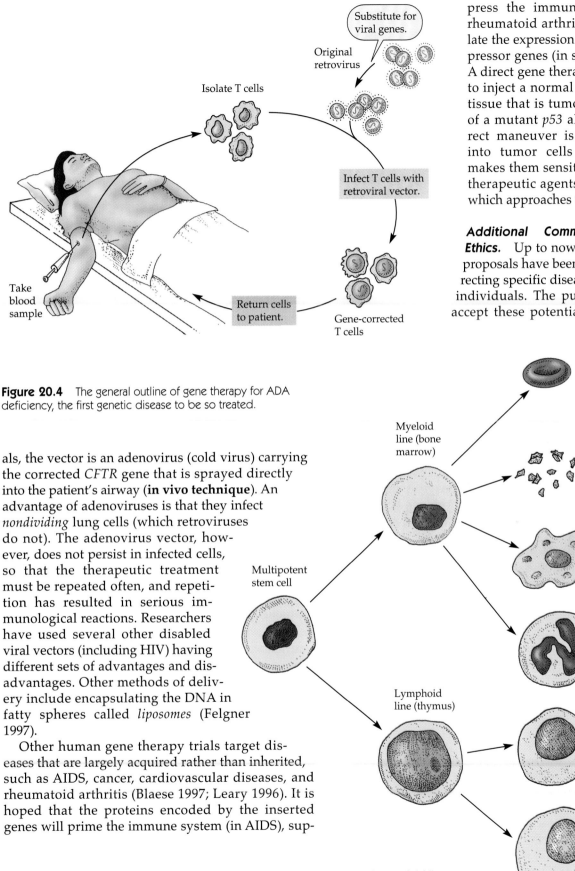

press the immune system (in rheumatoid arthritis), or stimulate the expression of tumor suppressor genes (in some cancers). A direct gene therapy strategy is to inject a normal *p53* gene into tissue that is tumorous because of a mutant *p53* allele. An indirect maneuver is to introduce into tumor cells a gene that makes them sensitive to chemotherapeutic agents. It is unclear which approaches will work.

Additional Comments About Ethics. Up to now, gene therapy proposals have been aimed at correcting specific diseases in specific individuals. The public seems to accept these potential "high-tech"

Figure 20.4 The general outline of gene therapy for ADA deficiency, the first genetic disease to be so treated.

als, the vector is an adenovirus (cold virus) carrying the corrected *CFTR* gene that is sprayed directly into the patient's airway (**in vivo technique**). An advantage of adenoviruses is that they infect *nondividing* lung cells (which retroviruses do not). The adenovirus vector, however, does not persist in infected cells, so that the therapeutic treatment must be repeated often, and repetition has resulted in serious immunological reactions. Researchers have used several other disabled viral vectors (including HIV) having different sets of advantages and disadvantages. Other methods of delivery include encapsulating the DNA in fatty spheres called *liposomes* (Felgner 1997).

Other human gene therapy trials target diseases that are largely acquired rather than inherited, such as AIDS, cancer, cardiovascular diseases, and rheumatoid arthritis (Blaese 1997; Leary 1996). It is hoped that the proteins encoded by the inserted genes will prime the immune system (in AIDS), sup-

Figure 20.5 The multipotent bone marrow stem cell and its derivatives. The myeloid line continues to differentiate in the bone marrow, while the lymphoid line travels to the thymus gland.

TABLE 20.2 Some experimental human gene therapy trials[a]

Disease	General technique[b]	Description of therapy
GENETIC DISEASES		
ADA deficiency	Ex	T cells removed from patient, gene-corrected for *ADA* gene by retrovirus vector, and returned. Some trials use stem cells
Familial hyper-cholesterolemia	Ex	Liver cells removed from patient, gene-corrected for LDL receptor gene by retrovirus vector, and returned
Gaucher disease	Ex	Marrow stem cells removed from patient, corrected for glucocerebrosidase gene by retrovirus, and returned
Cystic fibrosis	In	Adenovirus (cold virus) carrying normal *CFTR* gene introduced directly into airways of patient by an aerosol spray
NONGENETIC DISEASES		
Rheumatoid arthritis	Ex	Synovial (joint lining) cells removed, treated with a gene in retrovirus that encodes a protein that blocks inflammation, and injected into diseased knuckles
Cardiovascular disease	In	A growth factor gene (in a bacterial plasmid) that stimulates new blood vessel growth is delivered to the site of a blocked leg vessel by coating the plasmids onto an angioplasty balloon
Skin cancer	In	Melanomas injected with a gene in a plasmid that produces a protein to stimulate the immune system to attack the cancer
Lung cancer	In	Tumor sites injected with normal *p53* gene carried by retrovirus

[a]For more detail, see Culver 1996a.
[b]Ex = ex vivo technique; In = in vivo delivery.

applications as a reasonable part of medical practice. Two possible uses of gene therapy, however, involve areas of ethical concern. One deals with *germinal gene therapy*—modifying the genes of future generations. A second deals with *enhancement*—improving on characteristics that are generally viewed as being within a normal range.

With regard to the genes of future generations, one worry is that manipulation of the germ line could harm rather than help. Our knowledge of the inner workings and interactions of cells is still fragmentary, and many biological examples could be given of well-intentioned interventions producing unexpected side effects. While we all wish for physically and mentally healthy offspring, it is unclear to what extent we should go to pursue that end. Opinion varies. What steps should be taken to foster fetuses that are normal? Is it right to abort fetuses that are not normal? What is normal?

Using gene therapy to try to alter the height of a very short pituitary dwarf would probably not be controversial. Many people would argue, however, that gene therapy to make a somewhat shorter-than-average child taller would be wrong, promoting discriminatory attitudes and leading to greater inequalities. But at the borderline between abnormal and normal, the question is muddy. For the undersized youngster whose life is made miserable by continual taunts and bullying, would a new growth-promoting

gene be enhancement or therapy? The quandary applies to intelligence as well. A number of single genes, when mutated, can lead to forms of profound mental retardation, and these might reasonably be targets for gene correction. But what about a person whose measured IQ is, say, 80? (At present, no specific genes are known that affect intelligence within the normal range, so this form of gene therapy remains unavailable.)

What will be the impact of gene therapy in the future? Just a few years ago, the whole idea seemed revolutionary, even bizarre. Now, it is much less so. But will gene therapy ever affect you, your family, or your descendants? If the procedures remain highly specialized, cumbersome, expensive, and applicable only to rare inherited diseases, the impact of gene therapy will remain limited. But millions of people could be affected if researchers develop vectors that deliver genes in a simple way to alleviate common disorders such as cancer, cardiovascular disease, or arthritis.

Dealing with Normal Traits

We include here two somewhat offbeat topics not related to genetic disease: predetermining a child's sex and producing a clone of an adult animal like yourself. The former is possible by several means. The latter is not now possible for humans, but has been accomplished for farm and laboratory animals. We discuss the techniques and their repercussions.

Choosing Your Child's Sex

Fetal sex has always been a matter of scientific interest and speculation. Suppose, for example, that samples of pure X-bearing sperm or pure Y-bearing sperm were available for artificial insemination. Or suppose that more commonplace techniques (perhaps condoms with special filtering mechanisms) could predetermine a child's sex with considerable success. Do you think that using such technology is unethical? What proportion of people might avail themselves of the opportunity? Would the ability to easily choose a son or a daughter at each birth alter the composition of families or a population's sex ratio? What societal changes might ensue? In one study of the consequences of sex preselection, over 7,000 married American women were asked their preferences for the sex of their next child. Although many more women wanted the firstborn to be a boy rather than a girl, the desire for a balanced family was very strong (Figure 20.6). Thus, if the firstborn was male, over 80% of women who expressed a preference wanted the second to be *female*—that is, 72.3/(72.3 + 15.0) from Figure 20.6. For a third child, boys and girls were about equally desired, although some other studies suggest a preference for males. The major consequences of sex selection would certainly mean more firstborn children being male. Holmes (1985) summarizes many studies that show the greater achievements of the eldest child compared with those of later siblings and notes the benefits that would increasingly accrue to the male sex. A large male excess might be expected in countries that have social or religious traditions of male inheritance. In some cultures, females are perceived as family burdens because of poor employment opportunities or the need for large marriage dowries.

Postconception Techniques. It is now technically possible, of course, for a couple to choose the sex of their children with virtually 100% success. The required abortions of the "wrong" sex, however, carry some small risk and are morally offensive to many persons. Fetal cells obtained by amniocentesis or chorionic villus sampling can be reliably karyotyped for sex chromosomes or sexed with DNA probes specific for the Y chromosome. Abortion of the "wrong" sex can be avoided by sexing eight-cell embryos obtained from in vitro fertilization and transferring to the woman only embryos of the desired sex (Chapter 19). Currently, however, such preimplantation diagnosis is severely limited by the need for special technical expertise.

It may be that the need for an invasive technique to obtain fetal cells can be avoided: Some research centers claim that the sex of a fetus can be determined by examining the DNA from the few fetal cells that circulate in the blood of the pregnant woman. Successful amplification of Y-specific sequences by the polymerase chain reaction signifies a male. As noted in Chapter 19, however, some types of fetal cells can persist in the maternal circulation for years, so that fetal cells from a prior pregnancy may be mistaken for cells of the current pregnancy.

Although these types of procedures can be useful in the prenatal diagnosis of genetic diseases, only a fraction of American physicians would agree to perform

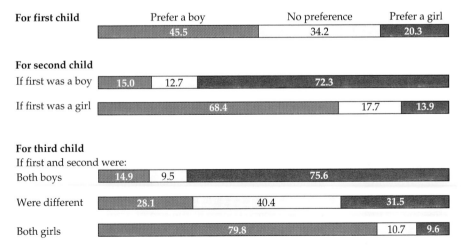

Figure 20.6 The preference of American married women for the sex of their next child. The figures above the bars express the percentages of women preferring a boy (blue bars), expressing no preference (yellow bars), or preferring a girl (red bars). (Data from Pebley and Westoff 1982.)

BOX 20A *Should Parents Be Able to Choose a Child's Sex for Personal Reasons?*

YES

*B*eing a desired sex furthers the happiness and well-being of both parents and child. What about a couple with, say, four sons who desperately want a daughter to balance their family? Being wanted is important. Since relatively few people will use the techniques, the change in the population sex ratio will be insignificant.

•Medical resources are often used for matters not related to health. Cosmetic surgery, for example, is commonplace and can improve a patient's mental outlook. Furthermore, techniques of sperm selection and preimplantation diagnosis will reduce reliance on abortion to achieve a desired sex.

•It is an issue of civil liberties, pure and simple. No one is obligated. Furthermore, it is not a doctor's job to judge the motives of his or her patients.

NO

•Sex is not a genetic disorder and should not merit the use of scarce medical resources.

•Successful parenting requires the acceptance of a child's individuality— its strengths, weaknesses, and sex.

Choosing a child's sex (to the point of abortion) is at odds with these values. When the technique fails, no one knows the possible adverse psychological consequences on a wrong-sex child.

•It is sexism, pure and simple. The preference for boys, especially for the firstborn, will benefit further the male sex and skew the sex ratio. Males will also benefit by being preferentially born into well-to-do families—that is, those who can afford the technologies.

---

prenatal diagnosis for the sole purpose of choosing the sex of a child. Nevertheless, they might agree to abort all male fetuses that are carried by women heterozygous for hemophilia or other serious recessive X-linked disorders, even though half the aborted fetuses are expected to be normal. In the United States, abortion before about 24 weeks is currently legal for any reason. Parents need to ask whether it is *ethical* to abort a fetus just to fulfill their desires for a boy or a girl (Box 20A) (Wertz and Fletcher 1989).

Preconception Techniques. In an area of abundant scientific and popular publications, one fact stands out about planning a child's sex prior to conception: Even a useless technique produces the desired outcome about half the time! The ancient who advised tying off the left testicle to produce a son could easily have become a seer on the basis of small samples. Over the centuries, hundreds of theories of sex determination were constructed, all of which worked in about 50% of the trials. More modern proposals would alter the behavior of the two sperm types in vivo or separate X-bearing from Y-bearing sperm in vitro for artificial insemination (Levin 1987; Reubinoff and Schenker 1996).

Many investigators have looked at the timing of intercourse relative to ovulation. One theory from the 1960s holds that Y-bearing sperm swim faster but are shorter-lived than X-bearing sperm. If this conjecture were true, then intercourse at the time of ovulation would favor boys because the faster Y-bearing sperm would find a waiting egg. (Intercourse that stopped some days before ovulation, on the other hand, would favor the birth of girls. In this case, sperm would be present in the oviduct waiting for an egg to be ovu-

lated, and the X-bearing sperm supposedly have greater staying power.) Another theory predicts just the opposite, that high levels of maternal gonadotropins at conception favor female offspring. Since these hormones peak as the egg matures, intercourse at the time of ovulation would favor girls. Neither theory has solid evidence to support it. In fact, there seems to be little change in sex ratio one way or the other when insemination occurs at the time of ovulation (Zarutskie et al. 1989).

Over the years, many researchers have tried to enrich sperm samples (from humans or farm animals) for X-bearing or Y-bearing sperm. Basic work was slow until recent times, because the only sure way of knowing whether the separation had been achieved was to use the putatively X- or Y-enriched samples in artificial insemination and then wait nine months (for both cows and people). Several methods are now available, however, that quickly identify whether sperm cells carry an X or a Y chromosome. In a FISH technique, unique DNA sequences on the X or on the Y can be specifically stained (but the procedure kills the cells, so they could not be used for insemination). Meanwhile, many investigators have tried to separate X-bearing from Y-bearing sperm using swimming tubes, centrifugation, electrophoresis, cell-sorting cytometry, or immunological techniques.

In the so-called *albumin gradient* method, experimenters place sperm at the top of a column containing progressively more viscous solutions of albumin toward the bottom. It is said that sperm cells that swim downward and accumulate in the most viscous layer are predominantly Y-bearing, perhaps because they are smaller or wriggle faster (but these properties are speculative). The putative Y-enriched samples are then used for artificial insemination. Does the method work? The latest journal article we have found from the promoters of the albumin gradient technique notes 72% male births among 1,034 couples desiring a male child (Beernink et al. 1993). This is certainly a statistically significant result. The study, however, did not have an appropriate control group, which might consist of couples going through the same artificial insemination routine but with sperm from the top or middle of the albumin gradients. Nor have independent groups confirmed the usefulness of the technique. Furthermore, we again point out the obvious: If 10 couples desiring a male child use no special sex preselection technique, about 5 of them succeed; with the albumin gradient technique in this study, about 7 succeed. Perhaps the fairly modest increase accounts for the relatively small number of people worldwide (about 2,000 at 65 different clinics) who have availed themselves of the technology over two decades (Steinbacher and Ericsson 1994).

Another technique for partial X and Y sperm separation is called *flow cytometry*. Currently being developed for farm animals by researchers at the U.S. Department of Agriculture in Beltsville, Maryland, its safety for humans is a matter of concern. Live sperm are treated with a fluorescent marker that attaches to DNA. Since X-bearing sperm have 2–3% more DNA than Y-bearing sperm, more of the marker attaches to X-bearing sperm. The sperm are then passed in single file through a thin tube and illuminated with ultraviolet light, which causes the sperm heads to glow, the X-bearing sperm more than the Y-bearing sperm. Each sperm is then deflected according to its brightness into one or another collecting tube. Hundreds of thousands of sperm can be sorted per hour. The accuracy of separation is about 82% for human X sperm and 75% for human Y sperm (but higher for cows, which have a greater DNA difference between X and Y sperm).

An efficient way to use such X- or Y-enriched sperm samples (from either farm animals or humans) would be to place the sample directly into the recipient's vagina (i.e., artificial insemination). Because this procedure requires millions of sperm, and this many are not usually available, the union of egg and sperm has usually been achieved through in vitro fertilization. For humans, the X-enriched fraction has been used at least once for the birth of a seemingly normal baby girl (Gillis 1995). In this case, technicians checked one of the cells of eight-cell embryos prior to transfer to the mother's uterus to make sure of the embryonic sex. A major concern is that the fluorescent dye or the ultraviolet light, both possible mutagens, might damage sperm DNA. Several hundred farm animals, however, have been born after sperm sorting without any obvious evidence of abnormalities (Reubinoff and Schenker 1996).

Cloning Yourself

An adult organism begins as a fertilized egg, which divides again and again, grows in size and complexity, and differentiates into specialized tissues and organs. This oversimplified statement barely hints at the intricately coordinated timetables of development, some of which have been deciphered by modern molecular research, but most of which remain a mystery (Chapter 15). One basic question of embryology has been whether the genome of a differentiated cell (e.g., skin or muscle or liver) is **totipotent**—that is, whether it retains the ability to produce a whole organism. If so—as it was suggested in the 1930s—why not take the nucleus of, say, a cell scraped from the inside of your cheek and transfer it to an egg cell whose own chromosomes had been removed? Would not the genes of the transferred nucleus direct normal development of the reconstituted zygote and result in a much younger twin, a clone of yourself?

The word **clone** has several related meanings. The items in a clone are usually exact genetic copies of each other (e.g., the members of a bacterial colony all derived from one cell by asexual reproduction). Identical twins are whole-body clones of each other. In animal husbandry, researchers produce clones by artificially splitting a young embryo into two, three, or more parts. In Chapter 8, we dealt with clones of molecules obtained by the polymerase chain reaction or by recombinant DNA methods. In the case of transferring a nucleus from one organism to an enucleated egg, the use of the word *clone* is slightly different, because the nucleus is put into a new cytoplasmic environment:

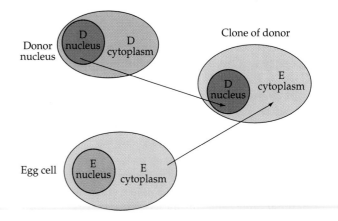

To whatever extent the cytoplasm is important, the clone may develop somewhat differently from the organism that supplied the donor nucleus. Among other cytoplasmic components, recall that several dozen genes constituting the mitochondrial DNA code for a variety of proteins used in energy production. Thus, although we and many others use the word *clone* in this context, *nuclear transfer* would be a more accurate description of the work described in this section.

Twenty years ago, researchers were optimistic that the cloning of a mammal by nuclear transfer might be possible. Some nuclear transfers leading to adult development had already been done with frogs and toads, whose eggs are a thousand times larger than those of many mammals and therefore much easier to handle (Mange and Mange 1980). In addition, these eggs are normally fertilized externally, in pond water, so no womb was needed to complete development. This work was pioneered in the 1950s by Robert Briggs and Thomas King at the Institute for Cancer Research in Philadelphia. They discovered that when the nuclei from early embryos were transferred to frog eggs whose own nuclei had been removed, a high percentage of the renucleated eggs developed into normal tadpoles. But when they transferred nuclei from later and later embryonic stages, fewer and fewer normal tadpoles developed. Thus, as development proceeds, the cell nuclei seem to become irreversibly differentiated so that they can no longer support the total development of an organism from scratch.

Research with mammalian eggs was more difficult, partly because of size: A rabbit egg has a diameter of about 0.1 mm (0.004 inch), requiring fine glass micropipets and steady hands for sucking out the egg nucleus and injecting the transferred nucleus. Bitter controversy surrounded the purported cloning of mice in 1981 by a skillful researcher, Karl Illmensee of the University of Geneva. This dispute—did he or didn't he?—muddied the field of cloning research and is still unresolved (Kolata 1998). Most experimental results of the 1980s were just plain disappointing: Although cloning small laboratory mammals, as well as cows and sheep, was possible using the transferred nuclei of very early embryonic cells (blastocysts), the nuclei of somewhat later cells simply would not work. Cloning research waned, and the belief grew stronger that the DNA of even slightly differentiated cells became modified in ways that prevent them from being totipotent. Only a few agricultural research centers pursued their cloning work into the 1990s.

Enter Dolly. One such place was the Roslin Institute near Edinburgh, Scotland, a government-financed cen-

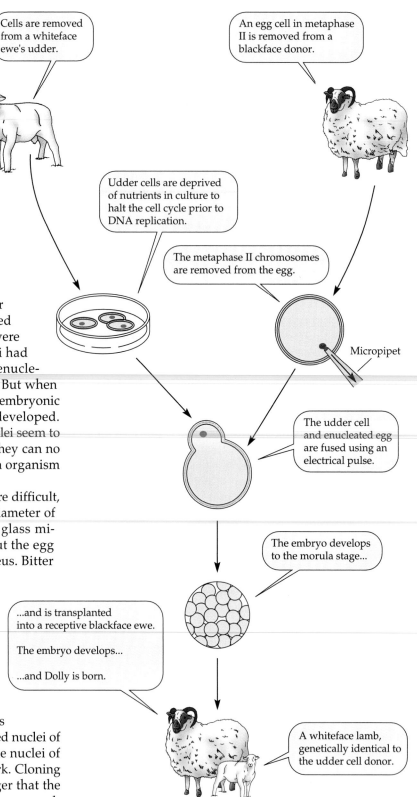

Figure 20.7 The procedure used by Roslin Institute scientists to clone a sheep.

ter for animal biotechnology. The researchers' interest in cloning continues to be related to their primary interest in the development of transgenic animals. They have transferred human genes to farm animals in such a way that the animals manufacture and incorporate the protein product in their milk. In Chapter 8, we noted that the production of any one useful transgenic animal is rare and unpredictable. But rapidly amplifying one valuable animal into a herd might be accomplished if successful cloning techniques were at hand.

Ian Wilmut and Keith Campbell, two researchers at the Roslin Institute, thought that the success of nuclear transfer with frogs and toads meant that success could be achieved with mammals too, if only the right conditions could be discovered. They suspected that the failure of previous cloning attempts by transferring nuclei from adult cells to egg cells might be due to disparities in the cell cycles between donor and recipient cells (see Figure 3.1). Indeed, the Roslin researchers noted that the transferred nuclei they customarily worked with were usually in S (DNA synthesis) or G_2 (preparing to divide), while the eggs were arrested in metaphase II of meiosis, awaiting sperm penetration. To possibly increase compatibility, Wilmut and his colleagues cut down on the concentration of nutrients in the culture medium surrounding the cells that contributed the nuclei. The cultured cells responded to their 5-day "starvation" diet by going into a resting stage (G_0 in Figure 3.1) similar, it was thought, to the metaphase II stage of the enucleated eggs.

These efforts to coordinate the cell cycles resulted in partial success of the cloning experiments (Specter and Kolata 1997; Stewart 1997; Wilmut et al. 1997). Wilmut and his colleagues isolated 277 sheep eggs (from ewes treated with fertility drugs to increase ovulation). As soon as feasible, they removed the small portion of the eggs near the surface that contained the metaphase II chromosomes (see Figure 3.6A). To these enucleated eggs they fused (by electrical pulses) cells that were taken from the udder of a donor animal, grown in culture, and induced to enter a quiescent G_0 stage (Figure 20.7).* Of the reconstituted eggs, 29 (12%) developed to a blastocyst or morula stage (Chapter 15) and were transferred to unrelated foster mothers. Of the transferred embryos, just one (0.04% of the original 277) resulted in a pregnancy and birth of a healthy lamb. That was Dolly (Figure 20.8). DNA fingerprinting and microsatellite analysis (Figure 20.9) has confirmed that Dolly has exactly the same nuclear DNA as her "nucleus" mother (of a whiteface breed), different from

*The donor animal had lived on a nearby farm. At age 6 *while pregnant*, a sample of her udder tissue was collected for purposes not related to the cloning experiments. These cells were conveniently available in Wilmut's freezer. At the time of her cloning, the donor was dead—perhaps slaughtered for meat—although records of her fate are apparently lacking.

her "egg" or "gestational" mothers (blackface breed) (Signer et al. 1998; Solter 1998).

Dolly is not an exact copy of the animal that donated the nucleus. Although there is the obvious influence of the nuclear genes of the donor, there is also the influence of the foster mother's womb and the influence of the cytoplasmic factors, including the mitochondrial genes of the egg cell. A human clone produced this way would also live in a culturally different time frame, with a new set of environmental influences modifying the development of the infant, child, teenager, and adult. The mitochondria of Dolly were in fact not tested. Because a *whole* udder cell was fused with the enucleated egg, Dolly's mitochondria could be all from the donor udder cell, all from the recipient egg, or more likely a mixture. For Dolly, the previous diagram of nuclear transfer should be modified to look like this:

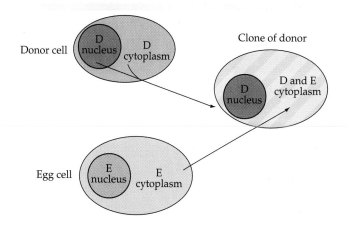

The methods of the Roslin Institute, or similar methods, have been applied to other species. Researchers at the Universities of Massachusetts and Wisconsin working with cattle have produced pregnancies from cloned adult cells (Goldberg and Kolata 1998; Kolata 1997a). Although these pregnancies apparently did not come to term, recent news reports suggest that successful cloning of cows from adult cells has been accomplished in Japan (Travis 1998). More remarkable is the successful cloning of dozens of mice from the nuclei of adult cells by researchers at the University of Hawaii (Wakayama et al. 1998). In these experiments, donor nuclei (not the whole cells as in the Dolly work) were obtained from the so-called cumulus cells that surround the egg after it is ovulated from the ovary (see Figure 3.9). These cells are naturally in the G_0 stage, so no culturing was needed to make them compatible with the enucleated egg cells into which they were injected. The Hawaiian researchers were even able to use cloned animals to derive a subsequent generation of cloned animals. In addition, they were much more successful than the Roslin Institute workers: Once transferred to foster mothers, 2–3% of the cloned

Figure 20.8 The cloned lamb Dolly (a Finn Dorset whiteface breed) and the foster mother (a Scottish blackface breed) who carried her to term. (Courtesy of Roslin Institute.)

embryos developed to term. There is no particular reason why human cloning would not work, but the safety issue—exemplified by the large proportion of sheep, cow, and mouse embryos that die—looms large.

In addition to the currently unacceptable safety risks for humans, fundamental ethical issues are being aired. Foremost are the mysteries of individuality and personal growth, which arise in part from chance combinations of different ancestries. It would be different for a clone:

> To become a new edition, indeed only a new imprint of a parent, could undermine these expectations and aspirations and produce a breed of passive creatures waiting for the familiar ancestral scenario to unfold, a breed for whom praise, blame, wonder, and fulfillment would have lost their meaning. (Freund 1972)

In this country, the President's National Bioethics Advisory Committee recommends legislation to ban nuclear transfer for the purpose of creating a child (Marshall 1997; Shapiro 1997). The committee's objections to cloning include possible psychological harm to the offspring because appropriate and caring family relationships might suffer from the strange circum-

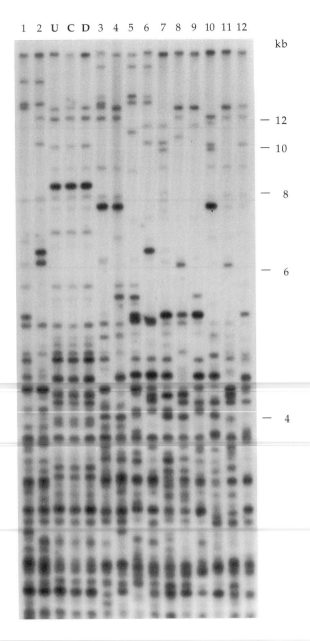

Figure 20.9 The DNA fingerprint of Dolly's blood (lane D) compared with udder cells from the nucleus donor (lane U) and cultured cells from this udder (lane C) that were used to make Dolly. The other lanes, labeled 1–12, represent controls–different sheep from Dolly's whiteface flock, including a mother-offspring pair (lanes 2 and 1, respectively). That Dolly is an authentic clone is shown by the identity of the bands in lanes U, C, and D, which are substantially different from the banding patterns in all other lanes. This DNA fingerprint was prepared using several multilocus probes to detect variation in different VNTR (or minisatellite) loci. The process is explained in Chapter 8. (From Signer et al. 1998.)

stances. (Such harm is speculative, however, and some argue that an individual would enjoy advantages based on being cloned.)

It is doubtful, in any case, that laws will prevent approaches to cloning and, if reliable and safe, actual attempts at cloning in humans. Some support already exists for the possibility of using cloning for infertile couples. If a husband fails to produce sperm, for example, his wife's eggs are sometimes fertilized in vitro with another man's sperm, a procedure many find acceptable. Using the infertile husband's somatic cells in the way described in this section might also be an acceptable option. Nuclear transfer using the wife's somatic nuclei and the cytoplasm of donor eggs might be justifiable if the wife suffered from a mitochondrial disease. Supporters of this approach point out that human clones have always existed in the form of identical twins (identical for nucleus *and* cytoplasm *and* womb), and that the twins, their families, and society have adjusted to the occurrence. In addition, basic research using cloning technology might aid in understanding early differentiation and other aspects of human embryology—perhaps leading to as yet unimagined medical breakthroughs. There will almost certainly be areas where cloning-related research will provide acceptable benefits to society (Kolata 1998).

Changing Frequencies Over Time

In 1883, Francis Galton (Box 11A) coined the word **eugenics** (literally, "well born") to mean the genetic improvement of the human species over time. Eugenic goals included the elimination of genetic disease and the enhancement of desirable traits. Galton wrote (1905):

> What Nature does blindly, slowly, and ruthlessly, man may do providently, quickly, and kindly. As it lies within his power, so it becomes his duty to work in that direction; just as it is his duty to succour neighbors who suffer misfortune. The improvement of our stock seems to me one of the highest objects that we can reasonably attempt.

In 1908, Galton organized the Eugenics Society in London to investigate human heredity and carry out social action programs. Similar societies were established at about the same time in the United States and Germany. Although some people who joined these groups were motivated by altruism, others were concerned with race, class, and privilege. They believed that society was degenerating, and they sought ways to keep their own positions of influence. Based, as it was, on classifications of superiority and inferiority, eugenics appealed to both idealists and cranks, and to moderates in between. One American genetic researcher of the early 1900s was Charles Davenport, director of the Eugenics Record Office at Cold Spring Harbor, Long Island, in New York. Davenport, along with others of his time, tried overly hard to fit virtually every human trait into a Mendelian frame, including poverty, moral degeneracy, insanity, and feeble-mindedness—as well as a number of bona fide genetic diseases. A small part of their work, including Davenport's analysis of skin color, described in Chapter 11, contributed to the early objective study of human genetics. But much of eugenics "research" was scientifically shoddy, influenced by racism and class consciousness. In fact, the systematic development of human genetics was actually set back because some serious researchers were reluctant to enter a field marked by racial and ethnic prejudice (Kevles 1985).

Although he advocated only voluntary actions, Galton's eugenic aspirations and plans were themselves heavily tinged with nationalism. He noted that a "high human breed" was especially important for the English because they colonized the world and planted the seeds of future millions of the human race. In the United States, eugenicists influenced many aspects of social policy, including enactment of discriminatory immigration laws in the 1920s after their lengthy "scientific" testimony before Congress. Adolf Hitler carried eugenic notions to cruel and bizarre extremes, starting with sterilization programs that championed Aryan elitism (Allen 1996). The horrors of the Holocaust soon followed.

Besides overestimating the role of hereditary factors in human behavioral traits, Galton naively believed that eugenic proposals would be easily and voluntarily accepted. Few persons, however, give more than passing thoughts to the quality of genetic material that is committed to future generations. Some are even offended by the very idea that methods used daily to improve agricultural plants and animals might be relevant to humans. Still, certain eugenic ideals are involved in traditional medical practices and new reproductive technologies that help individuals and families improve their prospects for health and happiness. In this section, we look at social action schemes that seek to prevent the spread of "disadvantageous" alleles (negative eugenics) and encourage the transmission of "advantageous" ones (positive eugenics).

Negative Eugenics

Lessening the incidence of hereditary disorders through the prevention of childbearing—that is, **negative eugenics**—has a particularly unsavory history related to the enforced sterilization of persons considered "unfit" or likely to have "socially inadequate" offspring. Most infamous was Hitler's policy of "racial hygiene."

In the United States, the first compulsory sterilization law was enacted by Indiana in 1907. Although this and other early statutes were declared unconstitutional, a more carefully drawn law in Virginia was upheld by the U.S. Supreme Court in the 1927 decision *Buck* v. *Bell*. The case involved an institutionalized patient, Carrie Buck, who was declared feebleminded, as

BOX 20B *The Kallikaks*

*T*he Kallikak Family (Goddard 1912) was a much publicized book that purported to be a scientific study of heritable mental defects among the descendants of one New Jersey man. We are told that Martin Kallikak, a young Minuteman, strayed from the path of virtue with a "nameless feeble-minded girl." Martin neglected mother and child, but through this line of descent he became the progenitor of 480 persons, among whom were 143 feebleminded souls, 36 illegitimate children, 33 sexually immoral persons, 24 alcoholics, 8 madams, 3 epileptics, and 3 criminals.

Martin later "straightened up and married a respectable girl of good family." On this "control" side, he became the proud ancestor of 496 persons, all with good credentials: "doctors, lawyers, judges, educators, traders, and landholders." From this dual family history, it was concluded that good alleles were perpetuated on one side and bad alleles on the other. In fact, the invented name Kallikak comes from the Greek words meaning "beauty" (*kallos*) and "bad" (*kakos*).

If it is to believed at all, the book is a moralistic tract that almost completely ignores the role of environmental factors and even the role of Martin himself. Far from proving anything about heredity, the Kallikak yarn could constitute a plea for upgrading the social conditions that perpetuate poverty. Gould (1996), for example, made the interesting discovery that the facial features in the book's photographs of persons on the bad side of the family were crudely altered to make them look sinister or stupid. Curiously, the Kallikak author criticizes an earlier analysis of crime, pauperism, and disease in the *Jukes* family as being inconclusive because of unknowable interactions of heredity and environment. In fact, as Rafter (1988) points out, many investigators of "bad-gened" degenerate clans simply "found" that which they assumed they would find.

---

were her mother and Carrie's 7-month-old daughter. The trio was found to be genetically defective by a eugenics "expert" who never examined them and who misrepresented the daughter. She died at age 8 of an infection, but her schoolteachers considered her very bright. Furthermore, Buck's ineffective defense lawyer was in collusion with judicial and legislative proponents of the new Virginia sterilization law, so the case was a sham (Lombardo 1985). Yet the patriotic rhetoric of Justice Oliver Wendell Holmes in *Buck* v. *Bell* (1927) rings out:

> We have seen more than once that the public welfare may call upon the best citizens for their lives. It would be strange if it could not call upon those who already sap the strength of the State for these lesser sacrifices, often not felt to be such by those concerned, in order to prevent our being swamped with incompetence. It is better for all the world, if instead of waiting to execute degenerate offspring for crime, or to let them starve for their imbecility, society can prevent those who are manifestly unfit from continuing their kind. The principle that sustains compulsory vaccination is broad enough to cover cutting the Fallopian tubes. Three generations of imbeciles are enough.

Not only was the case factually incorrect, but the judgment has been questioned on legal grounds: Are vaccinations and sterilizations equivalent? In addition, the genetic suppositions of the decision are groundless, for mental retardation is not a single entity. Some cases are due to the presence of a major gene with differing modes of inheritance, some to the chance combinations arising from polygenic inheritance, and some to various chromosomal aberrations. Other types of mental deficiency are entirely due to environmental damage occurring prenatally, to birth trauma, or to infectious diseases. If the various types of people

with mental problems did not reproduce, the subsequent decrease in frequency would vary and would usually be slight. Thus, it is silly to suggest that Carrie Buck and her "kind," whatever that is, would populate the world with criminals and imbeciles. *Buck* v. *Bell* has been soundly censured by judges and others, and today's readers are likely to be embarrassed by the insensitive and flippant words of Justice Holmes, but he undoubtedly reflected the public opinion of the time (Box 20B).

However misguided past practices have been, involuntary sterilization can be of value in specific cases to reduce or prevent unhappiness and suffering. Still, persons with mental retardation are entitled to the full array of constitutional rights that others enjoy. Exercising those rights, however, raises complex legal issues. After proper procedural safeguards, courts have occasionally decided that sterilization may be in their best interests, allowing them, for example, to live outside an institution without fear of becoming a parent. Fewer than 20 states retain statutes allowing for the eugenic sterilization of institutionalized retarded persons after due process of law, and their implementation is minimal. Tens of thousands of involuntary operations were performed, however, between 1930 and 1960 (Reilly 1985).

More recently, many normal adults have chosen voluntary sterilization (by tubal ligation in females and vasectomy in males) as the surest contraceptive. Although primarily employed to prevent the birth of a child of *any* phenotype, rather than as eugenic measures, voluntary sterilization would be appropriate for persons carrying alleles for serious genetic diseases. But as discussed in Chapter 12, the resultant change in frequency of recessive disease genes is very slight over the course of a moderate number of generations. This is true even when all affected persons do not reproduce. More effective programs to decrease the frequency of recessive diseases depend on the detection and counseling of *carriers* prior to the birth of any affected children. The screening program for Tay-Sachs disease has, in fact, dramatically lowered the occurrence of the disease among Jewish groups (but not the frequency of the mutant alleles) (Chapter 19).

Positive Eugenics

Rather than advocating the sterilization of the unfit (however defined), Galton stressed the need for increased propagation of men of talent and genius, as well as those with superior health, moral strength, and high "civic worth." **Positive eugenics** could be accomplished, he felt, by educational programs that would influence popular opinion and by wealthy persons taking an interest in and befriending poor but promising lads. He specifically recommended that exceptionally worthy young couples be provided convenient housing at low rentals.

Although most American eugenicists were obsessed with negative eugenics, the intellectual heir to the positive idealism of Galton was Hermann Muller (a 1946 Nobel laureate for mutation studies). Through his 1935 book *Out of the Night: A Biologist's View of the Future*, Muller hoped to stimulate interest in positive eugenics and other aspects of social reform. He was an early advocate of publicizing birth control methods, legalizing and regulating abortion, and reducing the burden of bearing and rearing children. He recommended child care facilities for women who wanted to work outside the home or further their education. Muller proselytized for positive eugenics, primarily through "germinal choice," that is, artificial insemination with semen freely selected from sperm banks to which eminent men of known identity made contributions. Muller (1961) recommended that the sperm be frozen for decades before use in order to better view the individual worth of donors and reduce the danger of hasty choices based on fads and fashions of the day. It was hoped that these methods would enhance basic values distinctive to humans, among which Muller included intelligence, curiosity, creativity, "genuineness and warmth of fellow feeling," and "joy in life and in achievement."

Although there has been no stampede to follow Muller's proposals, many sperm banks have been established in the United States (many of which are listed in telephone Yellow Pages). Most of these facilities, however, are not intended to fulfill the eugenic purposes envisioned by Muller. For example, some men deposit sperm prior to vasectomy so they can still father a child (Box 20C). One sperm bank with eugenic goals is the Repository for Germinal Choice, established in 1981 in southern California by Robert Graham, the wealthy inventor of plastic eyeglass and contact lenses. The bank accepts sperm from only a few men, initially just Nobel prize winners. When only three prize recipients donated sperm samples, and when no women chose them, the criteria were loosened to include other eminent scientists and exceptional athletes, including an Olympic gold medalist. Many people (including Muller's heirs) have ridiculed or castigated these blatantly elitist and narrow criteria. Nevertheless, this bank has "fathered" 218 children worldwide up to Graham's death in 1997 (Hotz 1997).

The net eugenic effect of these births on the human species is minuscule—and would no doubt remain so even if the number of such births was much, much larger. Because of meiotic segregation, polygenic inheritance, and especially environmental effects, the specific characteristics of a parent are rarely passed on intact to a child. For example, using Babe Ruth as a sperm donor would only slightly increase the probability of producing a slugger. (Indeed, none of Ruth's 15 descendants has shown any particular talent for baseball.) We wish to emphasize that phenotypic variability is influenced by differences in both genes and environment. Rather than trying to control the genes with which a person is born, society would be better served by providing the enriching environments that foster optimal expressions of the genotypes that currently exist. The benefits of a multitude of public and private programs for betterment—socioeconomic, health related, educational, cultural—have never been fully realized.

Still, there is some eugenic potential (depending in part on the heritabilities of the traits in question) in currently feasible and safe technologies that are voluntarily undertaken with minimal, if any, government oversight. Technologies such as genetic testing and reproductive assists can contribute in humane ways to the universal desire of parents for healthy, normal children. Because of its offensive history, however, the term eugenics should perhaps be discarded. "But the judicious use of genetic knowledge for the alleviation of human suffering and increase in the well-being of future generations is a noble ideal, whatever it is called" (Crow 1988).

Conclusion

How can we eliminate the pseudoscience that supported—and in some places still supports—bigotry in any form? Perhaps it would help if society and its leaders knew something about the methods of science, the nature of scientific evidence, and the motivations of scientists. Public debate is important to protect individual rights and to counter prejudice and injustice. Yet in many countries around the world—including

BOX 20C   *Different Ways to Get Pregnant*

Although not universally accepted on religious or moral grounds, diverse techniques, some very high-tech, have been developed to overcome infertility in both sexes (Stolberg 1997a). We give a few details.

Artificial insemination, dating back to the nineteenth century, is simple and relatively inexpensive. Perhaps 10,000 children per year in the United States are now conceived in this way, using fresh or frozen donor semen. Employed primarily when a man is infertile because of a low sperm count or poor sperm motility, it has also been used by couples when a fertile husband could transmit a deleterious allele. Unmarried women, too, have used artificial insemination. Usually a doctor selects an anonymous sperm bank donor, matched as closely as possible to the husband for obvious physical characteristics. Near the estimated day of ovulation, the donor's semen is deposited close to the woman's cervix using a syringe. The procedure is repeated for as many months as necessary to achieve fertilization. Miscarriages and birth anomalies among the resulting pregnancies appear to be no more frequent than among pregnancies achieved through sexual intercourse.

The donor and the "preadopting" couple remain unknown to each other, and a few doctors deliberately muddle the matter by mixing semen from several donors. The sperm is usually frozen and then thawed just before use. Tests for genetic diseases are not generally done unless the donor interview reveals a familial disorder and a test is available.

A California sperm bank (one of the largest and most reputable) is currently being sued because a resultant child developed polycystic kidney disease, a common, often serious disorder due to an autosomal dominant gene (Marquis 1997). The family blames the bank for allowing someone with a suspicious family history to donate sperm, although the natural overall rate of birth defects is in the range of 2–5%. As of this writing, many questions are unanswered. Did the anonymous donor give partial or incorrect information to the bank? Was the bank as careful as it should have been? What is the fate of other children sired by the donor? He supplied 320 vials of sperm over a period of four years (at $35 per

donation), and the bank sells two vials (for $85 apiece) for each insemination.

In vitro fertilization and embryo transfer (IVF-ET) has resulted in thousands of births. Over 100 centers in the United States now perform IVF-ET. The process requires several outpatient hospital visits for the prospective mother, numerous other examinations and tests, skilled medical and technical personnel, and a specialized laboratory.

The prospective mother is treated with fertility drugs to induce her oocytes to begin to mature. At the appropriate time, the ovaries are viewed with a laparoscope (a thin, lighted instrument with a magnifying lens and a collecting tube) inserted into the abdomen. The contents of the mature follicles on the surface of the woman's ovaries are sucked into a collecting chamber. The eggs are put into culture medium and bathed with fresh sperm from the husband. If fertilization occurs, the resultant zygote cleaves to form two, four, and eight cells over several days (see figure). The developing embryos are transferred to the female's uterus at the four- or eight-cell stage by means of a fine tube inserted through the cervical canal. The practice of using three or four embryos has resulted in many multiple births—about 20–25%, compared to just 1% in the general population. The multiple births often result in premature deliveries, low birth weights, stillbirths, infant mortality, and subsequent health problems at higher rates than those in the general population.

Although there is much variation among clinics, the overall success rate for IVF-ET seems disappointingly low to some people. For each month that an attempt is made, only about 20% of starts eventually end in a live birth (Stolberg 1997b). The major stumbling block seems to be the failure of the embryo to implant after transfer to the recipient's uterus, an increasing problem the older the woman. In addition to the emotional roller coaster, the financial costs are high—about $10,000 per cycle for as many monthly cycles as the process is attempted. But the overall low rate of success should be compared with the natural situation. For normally fertile young couples, it is estimated that only 30% of eggs exposed to sperm through sexual

intercourse produce a viable offspring.

There are now techniques for freezing and storing young embryos. This practice reduces the need to repeat ovulation stimulation and laparoscopy to retrieve additional eggs. Subsequent menstrual cycles occur without drug treatment (because no more eggs need be collected), and this may improve the chances of achieving pregnancy. Freezing also reduces the risk of multiple births, because the physician can transfer just one or two fresh embryos, saving the remainder for successive attempts. Some fertility clinics sell extra frozen embryos. The going rate at a major New York hospital is $2,750 (Kolata 1997b).

Surrogate motherhood is a simple procedure by which a husband and wife buy the services of another woman (typically for $30,000), who is artificially inseminated with sperm from the husband and is legally obligated to turn over the resulting newborn to the contracting couple. This unusual social situation has provoked strong public reactions and a sensational court case when the surrogate mother changed her mind and decided to keep the baby rather than accept the fee and relinquish the child. In one twist on the technique, the surrogate mother need not be related to her fetus. Instead, the child is conceived by IVF-ET using sperm from the husband, eggs from the wife, and the womb of the surrogate.

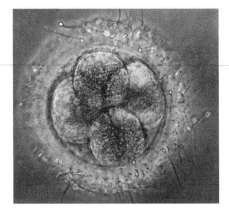

A four-cell human embryo derived from an egg recovered directly from a woman's ovary and fertilized in vitro. (Courtesy R. G. Edwards, Cambridge University.)

democratic ones—racism and other forms of discrimination are almost daily affronts.

Over the years, the positive impact of genetics on the fields of medicine, public health, and agriculture has been impressive. Increased knowledge of genes and increased skill in applying genetic technologies promise to contribute further to humankind. Scientists should, of course, make every effort to consider the ramifications of their own work and convey to the public at large what it is they hope to accomplish. But a precise accounting of future benefits and risks from genetic or any other scientific research is not possible. Scientists can predict neither the outcome of specific experiments nor the many ways their findings might be used. For example, the biochemists who discovered restriction enzymes could hardly have imagined all the applications to gene mapping, prenatal diagnosis, genetic screening, gene therapy—and who knows what to come!

As we learn more about our genetic heritage, it is also to be hoped that people will come to view themselves and *all* the world's fellow creatures in a clearer and more sympathetic light. Human genetics, a science of differences, emphasizes the uniqueness of each individual:

> It is astonishing that man, a species which displays such a range of variety and which lives in a world populated by tens of thousands of other species, should be so conformist and so intolerant of diversity. If knowledge can help to counter this limitation, the study of the origins of human individual differences is very important. Some understanding of the genetic determinants of behavior—their biological qualities, their extent, and their distribution—might … give people an enhanced sense of their uniqueness, as well as acceptance, perhaps even tolerance, of their kinship with others. It might lead, above all, to a more charitable view of those foibles, frailties, peculiarities, and eccentricities which cause many people, their virtues notwithstanding, to be set apart. The headlines of the newspapers on any day proclaim the urgency of the need for this understanding. (Barton Childs et al. 1976)

Summary

1. Genetic diseases are treatable in principle, although only a relatively few are really effectively managed. Physicians must generally understand the chain of events extending from mutant gene to altered phenotype, but this knowledge does not guarantee a remedy.

2. Some treatments rely on drugs to alter a major symptom of the disease or to deplete a metabolite in excess. Diet modifications can be used to restrict an otherwise toxic substance.

3. Replacing a missing gene-encoded protein or transplanting an organ or bone marrow may also be effective. Antisense therapy to block the translation of mRNA to form a harmful protein is being investigated.

4. In theory, gene therapy provides a cure by supplying a gene whose product alters the consequences of genetic or nongenetic disease. The major problems are getting the gene into the appropriate somatic cells and expressed at the right levels. Altered retroviruses are the most used vectors for these purposes.

5. The first approved gene therapy trial was for a fatal immune disorder, adenosine deaminase deficiency. The benefits of the therapy seem tangible but are still being debated. Many other gene therapy protocols have now begun.

6. Choosing the sex of a child by using abortion is controversial. Many investigators have sought a method for choosing the sex of a fetus before conception. Methods for selecting X-bearing or Y-bearing sperm for artificial insemination have been promoted, although the issue of safety and the actual degree of success remain unsettled.

7. Until recently, cloning an adult organism has seemed impossible due to the irreversible differentiation of DNA during development. The cloning of the sheep Dolly was accomplished, however, by inducing a state of quiescence in the adult nucleus before transferring it to an enucleated egg.

8. It remains to be seen whether or not human cloning is possible. Even if possible, many ethical objections have been raised.

9. Eugenics means improving the human species over generations through human genetics research and social action. Despite altruistic aims, the research has often been poor and the actions bigoted.

10. Negative eugenics has sometimes meant enforced sterilization to curtail the transmission of presumed undesirable traits. Some proposals for positive eugenics are based on the voluntary use of artificial insemination with banked sperm of selected men.

11. The study of human genetics should lead to a greater appreciation of the uniqueness of each individual.

Key Terms

antisense therapy	positive eugenics
clone	retrovirus
ex vivo technique	somatic gene therapy
germinal gene therapy	stem cell
in vivo technique	totipotent
negative eugenics	vector
oligonucleotide	

Questions

1. As with many other genes that have been cloned and sequenced, many different mutations have been discovered in the gene for Gaucher disease. Could this help explain the great variability of disease symptoms?

2. The cloned sheep Dolly had three mothers. Explain.

3. The following questions refer to the DNA fingerprint in Figure 20.9.

(a) Note that the ewe (lane 2) has bands that are not present in her offspring (lane 1), and vice versa. Explain.
(b) Note also that some bands seem to be present in all sheep, straight across the autoradiograph (or approximately so). Explain.

4. It has been pointed out that Dolly may grow old more quickly than "regular" newborns. Explain.

The following questions are meant to stimulate discussion, and we do not provide answers.

5. An objection to human cloning is that it might be used to create a group of "second-class" citizens—even slaves. Comment.

6. Ethical considerations aside, gene therapy is not likely to raise the intelligence of persons already within a normal range, but it might someday be used to overcome some specific causes of severe mental retardation. Why is this so?

7. Do you approve or disapprove of the following processes for your own family? For other families? What conditions would you impose on their use?

    Abortion to avoid the birth of a boy or a girl
    Artificial insemination or in vitro fertilization to overcome infertility
    Artificial insemination or in vitro fertilization with samples of X- or Y-enriched sperm
    Sterilization of sexually mature Down syndrome patients
    Banking of sperm from distinguished and identified men
    Surrogate motherhood
    Cloning an adult by nuclear transfer

8. Is there a law in your state that allows sterilization of some people without their consent? Is the law utilized?

9. Should people with certain genetic diseases be prevented from marrying? If so, which diseases?

10. If enriching environments and opportunities were available to everybody on an equal basis, would the apparent differences between people disappear?

11. Are Muller's criteria for selecting sperm for a bank of frozen samples the same as yours? Comment on the heritability of these characteristics.

12. Some sperm banks may sell the same donor's sperm 50 or more times. Is there anything wrong with this?

13. If a democratic government decided that some line of research (such as human cloning) should be stopped, what could the government do?

14. Does the expression "playing God" seem appropriate in reference to genetic engineering research? Why or why not?

Further Reading

Although not meant for laypeople, the books by Desnick (1991) and Friedmann (1991) provide complete accounts of treatments for genetic diseases. More up to date (but just as technical) information can be found by consulting the indices in the encyclopedic volumes edited by Rimoin et al. (1997) and Scriver et al. (1995). An article by Friedmann (1997) on gene therapy introduces four other articles in the same issue of *Scientific American*. See Culver (1996b) for a nice poster and Library of Congress classification RB 155.8 for books on gene therapy. Kolata (1998) is a well-written account of the history and implications of the cloned lamb Dolly. Gould (1996) and Kevles (1985) are the most cited accounts of the excesses of the eugenics movement, but Crow (1988) points out some biases in Kevles book. Bajema (1976) reprints many original papers relating to the history of eugenics, and Rafter's (1988) large collection of turn-of-the-century studies of purportedly degenerate clans makes fascinating reading.

Appendix 1: Some Common Units of Measurement

Quantity	Unit	Abbreviation	Equivalents
Length	kilometer	km	10^3 m, 0.621 mile
	meter	m	39.4 in, 3.28 ft, 1.09 yd
	centimeter	cm	10^{-2} m (10^2 cm = 1 m), 0.394 in
	millimeter	mm	10^{-3} m (10^3 mm = 1 m)
	micrometer	μm	10^{-6} m (10^6 μm = 1 m)
	nanometer	nm	10^{-9} m (10^9 nm = 1 m)
	Angstrom unit	Å	10^{-10} m (10^{10} Å = 1 m)
	mile	mi	1.61 km
	yard	yd	0.914 m
	foot	ft	0.305 m, 30.5 cm
	inch	in	2.54 cm
Mass	kilogram	kg	10^3 g, 2.20 lb
	gram	g	0.0353 oz
	milligram	mg	10^{-3} g (10^3 mg = 1 g)
	microgram	μg	10^{-6} g (10^6 μg = 1 g)
	nanogram	ng	10^{-9} g (10^9 ng = 1 g)
	picogram	pg	10^{-12} (10^{12} pg = 1 g)
	ton	ton	2000 lb, 907 kg
	pound	lb	454 g, 0.454 kg
	ounce	oz	28.3 g
Volume (liquid)	liter	l	1.06 qt
	milliliter	ml	10^{-3} l (103 ml = 1 l)
	microliter	μl	10^{-6} l (106 μl = 1 l)
	quart	qt	0.946 l
	pint	pt	0.473 l
	tablespoon	tbsp	15 ml
	teaspoon	tsp	5 ml
Temperature	Fahrenheit	°F	0°C = 32°F
	Celsius (centigrade)	°C	20°C = 68°F
			100°C = 212°F

Appendix 2: Some Basic Chemistry

Here we present a few chemical concepts that underlie our discussions of molecules and chemical reactions throughout the book.

Atoms and Molecules

The materials around us are all composed of tiny particles called **atoms**. One simple model likens the atom to an ultraminiature solar system, with the positively charged atomic nucleus corresponding to the sun and the negatively charged **electrons** corresponding to the orbiting planets. Just as our solar system is mostly space, so are atoms mostly space.

The smallest atom is that of the element *hydrogen*, symbolized by H. It has just one electron spinning about an atomic nucleus containing one **proton**. These two particles carry exactly the same electric charge, but the electron is negative and the proton is positive. The *mass* of a proton, however, is almost 9,000 times greater than that of an electron. The number of protons (or electrons) is called the *atomic number*, which is the same for all atoms of a particular element. The number and arrangement of the electrons about the atomic nucleus largely account for the chemical reactions of an element.

The four most common elements in living matter are (starting with the most abundant) hydrogen (H), oxygen (O), carbon (C), and nitrogen (N). These and other elements join together to form **molecules**. The simplest molecules consist of just a few atoms of the same kind (e.g., molecular oxygen, O_2) or of different kinds (e.g., water, H_2O; carbon dioxide, CO_2; or ethyl alcohol, C_2H_6O). The subscripts give the number of atoms of each element that are present in the molecule.

The atoms in any molecule are held together by **chemical bonds**. Because energy input is often required for the formation of chemical bonds, a molecule, such as the sugar glucose, contains more energy than the sum of its separated atoms. For this reason, the bonds within a molecule represent *stored chemical energy*. This energy gets into our food because green plants convert radiant energy from the sun into the chemical energy of sugars and other molecules. Animals, in turn, extract energy from plant molecules for use in their own bodies. This is why animals need plants, and plants need the sun, in order to survive.

The breakdown of bonds in energy-rich compounds within our body releases some of the sun's energy that was put there when the bonds were originally formed. Complex networks of reactions occurring in all cells release this bond energy in *small steps*, often storing it temporarily in intermediate compounds for future use in other reactions. What remain afterward are simpler, energy-poor molecules.

The most common type of bond in living organisms is the **covalent bond**. This strong and stable bond occurs when two atoms *share* electrons between them. In a molecule of water, for example, one *shared pair* of electrons connects oxygen to each hydrogen. Each connection, a covalent bond, is represented in a structural formula by a short line, as in H—O—H. Hydrogen is always connected by one covalent bond to other atoms; oxygen is connected by two covalent bonds. Nitrogen atoms always form three covalent bonds, and carbon atoms four. Two covalent bonds may be directed toward one atom, as in carbon dioxide (Figure A.1). This arrangement is called a *double bond*.

Another type of connection, much weaker than a covalent bond, is also exemplified by water molecules. The oxygen atom in H_2O attracts the shared electrons more strongly than the hydrogen atoms do, so this part of the molecule carries a partial negative charge. Conversely, because the shared electrons are rather far

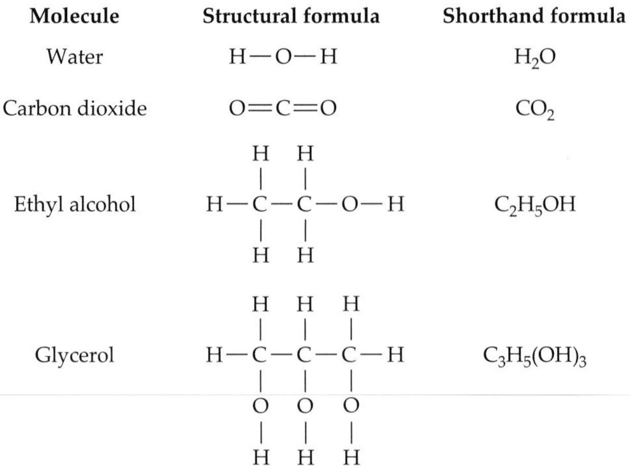

Molecule	Structural formula	Shorthand formula
Water	H—O—H	H_2O
Carbon dioxide	O=C=O	CO_2
Ethyl alcohol	H—C—C—O—H	C_2H_5OH
Glycerol	H—C—C—C—H	$C_3H_5(OH)_3$

Figure A.1   Covalent bonding of some simple substances.

from the hydrogen nuclei, the hydrogen atoms carry a partial positive charge. Although the molecule as a whole is neutral, with no net charge, there is a separation of charge within the molecule. Such molecules are called *polar molecules*. With both positive and negative parts, they tend to attract other polar molecules. In liquid water, for example, the hydrogens tend to orient toward the oxygens of *other* water molecules. The weak attractions *between* molecules of water are called **hydrogen bonds**. Very little energy is needed to make or break such bonds.

Although hydrogen bonds have only about 1/20 the strength of covalent bonds, they are important in forming the structure of many biological molecules. For example, the helical formation of DNA is stabilized by hydrogen bonding between different parts of the molecule, and proteins owe their three-dimensional conformation (a critically important property) in part to hydrogen bonding. In these cases, some of the hydrogen bonds link a partial positive charge on hydrogen with a partial negative charge on a nitrogen atom.

A third type of bond occurs when an acceptor atom attracts an electron so strongly that the electron's connection with a donor atom is severed completely. In this way, the donor atom loses an electron, becoming a *positively charged ion*; the acceptor atom becomes a *negatively charged ion*. The oppositely charged ions then attract each other to form an **ionic bond**. In table salt, for example, the chlorine atom has pulled an electron away from the sodium atom. Ionic bonds are important in the chemistry of many salts, acids and bases.

Organic Molecules

At one time, only the molecules present in living things were called organic. Now, the definition is broader: All molecules that contain carbon atoms are called **organic molecules**. Here we discuss some carbon compounds that are important in the structure and function of living things.

Hydrocarbons contain only hydrogen and carbon. These compounds might have been the first organic molecules on earth, predating the appearance of anything that could be called living. The simplest hydrocarbons, constituents of natural gas, are methane, ethane, and propane:

Methane	Ethane	Propane
H	H H	H H H
|	| |	| | |
H—C—H	H—C—C—H	H—C—C—C—H
|	| |	| | |
H	H H	H H H

Note that each carbon atom is involved in four covalent bonds, and each hydrogen is involved in one.

Other hydrocarbons have longer chains of carbon atoms, sometimes branched, sometimes in circles of six carbon atoms. Compounds with double bonds between adjacent carbon atoms are *unsaturated*, containing fewer hydrogen atoms than they would if these bonds were single (and the freed electrons were shared with hydrogen instead). Hydrocarbons play a key role in our lives. Examples include synthetic and natural rubber, petroleum, gasoline, fuel oil, asphalt, plastics, and many other petrochemicals.

Many classes of organic compounds have a hydrocarbon skeleton to which is attached a small group of atoms that enter readily into reactions with other compounds. Such attachments, replacing one or more hydrogens, are called *functional groups*. A set of compounds with the same functional group engages in similar reactions. About a dozen functional groups are important in living systems. Of these, we mention three:

1. *Alcohols* have a *hydroxyl* (—O—H) group that replaces one of the hydrogens in a hydrocarbon.

2. Organic acids have a carboxyl group: $-C\diagup\!\!\!\!\!\overset{\displaystyle O}{\underset{\displaystyle O-H}{}}$

3. Amines have an amino group: $-N\diagdown\!\!\!\!\overset{\displaystyle H}{\underset{\displaystyle H}{}}$

Many important organic compounds possess more than one functional group. **Amino acids**, for example, have both a carboxyl group and an amino group. Many organic molecules are very large, being composed of long chains of similar subunits. Four types of large molecules are extremely common in living organisms: proteins, nucleic acids, lipids, and carbohydrates.

Proteins are chains of dozens to hundreds of *amino acids*. The carboxyl group of one amino acid reacts with the amino group of the next one to make the links. Protein structure is coded by genes. The average mammalian cells produces over 10,000 different proteins; most of these are *enzymes*, which catalyze all the chemical reactions within an organism. Other proteins include *hemoglobin*, the oxygen-transporting protein of red blood cells; *antibodies*, the immunity-providing protein of blood plasma and other body fluids; *myosin*, the contractile protein of muscles; *collagens*, the fibrous protein of bones and tendons; and *keratin*, the structural protein of hair and skin.

Nucleic acids consist of long chains of repetitive subunits called *nucleotides*, whose structure and function are discussed in Chapter 6. *Deoxyribonucleic acid* (*DNA*), the genetic material of all organisms except certain viruses, contains four different nucleotides. A slightly different set of four nucleotides is found in the *ribonucleic acid* (*RNA*) molecules that aid in the gene-decoding process. DNA and RNA are *informational* molecules: The information contained in a sequence of DNA nucleotides is translated by RNA intermediates into a sequence of amino acids—that is, into proteins.

Lipids are fatty substances that do not dissolve in water. One type consists of three long-chain organic acids attached, one on one, to the three hydroxyl groups of a compound called glycerol (Figure A.1). These *triglycerides*, or *neutral fats*, pack a caloric wallop. In *phospholipids* and *glycolipids*, the third hydroxyl group of glycerol is replaced, respectively, by a phosphate group and by a short chain of sugars. Both of these lipid types are major components of membranes. Another kind of lipid, structurally quite different, is the group of *steroids* containing several six-member rings of carbon atoms. The steroids include the *sex hormones* (*estrogen* and *testosterone*) as well as *cholesterol*, *cortisone*, and *vitamin D*.

Carbohydrates include simple sugars and long chains of these sugars. The breakdown of the simple six-carbon sugar *glucose* is the primary pathway yielding the chemical energy that drives other cellular reactions. In animal cells, excess glucose may be temporarily stored as *glycogen*, which links together hundreds of glucose subunits. In plant cells, the corresponding storage carbohydrate is *starch*. The major component of plant cell walls, *cellulose*, is also composed of long chains of glucose subunits, but these are attached to each other differently than in the storage compounds. About half of the carbon atoms in all living things are components of cellulose. We cannot utilize cellulose as food because we lack the enzymes necessary to break the bonds between its glucose subunits. Cows manage the trick by harboring in their largest stomach compartment bacteria that do the job for them.

Biological Reactions

Many chemical reactions can be generalized this way:

Reactant(s) ⟶ Product(s) + Energy

In any chemical reaction, energy is conserved; that is, when the chemical bonds joining the atoms of the reactants contain more energy than in the rearranged bonds of the products, this excess energy appears on the product side of the reaction. Part of this extra energy may be picked up as chemical bond energy in other compounds. For example, consider the breakdown of glucose, which occurs in all cells. Although it involves several dozen separate small steps, the overall reaction is:

Glucose + Oxygen ⟶ Carbon dioxide + Water + Energy

The chemical energy within the glucose molecule and oxygen is much greater than the energy of the corresponding amounts of carbon dioxide and water. Most of this excess energy appears as heat or is simply unavailable for any kind of useful work in the cell. About

one-third of the energy, however, is transferred to the chemical bond energy of a molecule called **adenosine triphosphate (ATP)**. The formation of ATP is *coupled* with each of several of the small steps in the breakdown of glucose. That is, generation of ATP occurs simultaneously with the glucose reactions. This is diagrammed as follows:

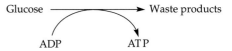

As indicated, the ATP is made from *ADP*, or *adenosine diphosphate*. These two molecules differ by just a single phosphate group (which is present in solution in the cell). ATP is a *high-energy compound* that plays a crucial role in the networks of cellular reactions. As an *energy transfer molecule*, it picks up energy from an energy-yielding reaction and delivers it to a different, energy-requiring reaction. Without ATP and other molecules with similar functions, *metabolism* (the totality of chemical reactions in cells) would be extremely slow and cumbersome.

Most chemical reactions in the body are *energy-requiring* and may be generalized as follows:

Reactant(s) + Energy ⟶ Product(s)

All cells must synthesize thousands of biochemicals from simpler substances, including DNA and proteins and many of the necessary lipids and carbohydrates. Almost all of these reactions need energy, which is usually supplied by ATP. When ATP loses a phosphate group, becoming ADP, some of the energy that was put into the ATP molecule in its formation becomes available to drive other reactions. Thus, the energy transfer circle is completed:

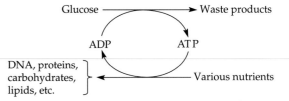

The reactions are far from 100% efficient, however. Thus, the energy content of the final products—DNA, proteins, and so on—is only a small fraction of the energy available in glucose.

Enzymes

Most chemical reactions, regardless of whether they produce energy or need it, will not proceed unless given an initial nudge. For example, paper burns fiercely, yielding large amounts of energy, but only if it is first brought to its kindling temperature. In living things (with fairly constant, low temperatures) the activation of reactions is accomplished by **enzymes**.

These are large protein molecules that *catalyze* (speed up) a reaction by bringing the reactants together in one spot and also by straining the molecular bonds that get broken and remade. Without enzymes, random molecular movements might bring the reactants together, but so infrequently as to be negligible.

Each enzyme is highly specific, catalyzing only one reaction (or sometimes a few reactions of similar type). For example, the enzyme *hexokinase* catalyzes the first of many small steps in the breakdown of glucose. This is a coupled reaction involving ATP. Hexokinase brings together glucose and ATP and causes a phosphate group from ATP to be transferred to the number 6 carbon atom of glucose:

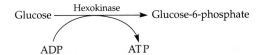

Glucose-6-phosphate has more energy than does glucose, the difference (as well as the phosphate group itself) being supplied by the high-energy molecule ATP. Although the total breakdown of glucose to carbon dioxide and water ultimately yields considerable energy, this particular starting step requires it.

Enzymes are able to catalyze a reaction partly because of their three-dimensional shape. The folding of the long chain of amino acids leaves an inpocketing of the surface, into which the reactants fit. Hydrogen bonding also helps to anchor the reactants to the enzyme and to orient them to each other. After the reaction, the products are released, but the enzyme is unchanged in the process. Because they are recycled, enzymes are required in only small amounts. They do eventually wear out, however, so new enzymes must be synthesized as needed.

Biochemical Pathways

A sequence of several dozen steps is required for the complete breakdown of glucose. Some steps involve just the splitting or the making of one bond or the transfer of a functional group from one molecule to another. So, too, are other molecules constructed or degraded, each small reaction step balanced for energy and catalyzed by a specific enzyme. Figure A.2 shows how the amino acid *phenylalanine* is converted, in five steps, into the hormone *adrenaline*. Note that each step involves the addition or removal of a single functional group.

Within just one cell, thousands of enzymes, involved in many dozens of metabolic pathways, are at

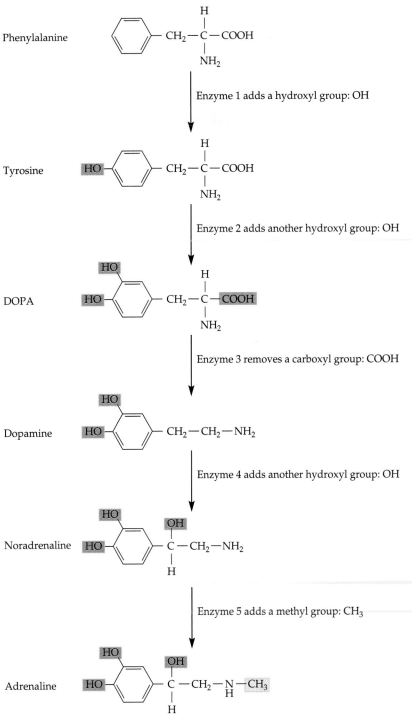

Figure A.2   An outline of the synthesis of the hormone adrenaline from the amino acid phenylalanine, which is present in food. (See also Figure 16.7.)

work simultaneously. These reactions are dependent on each other, so that an intermediate compound in one reaction may be the starting compound of another. For example, the intermediate compound DOPA in Figure A.2 is the precursor of the melanin pigments found in skin, eyes, and hair. The next two compounds, dopamine and noradrenaline, are themselves neurotransmitters that chemically transfer an electrical impulse from one nerve cell to the next. The biochemical pathways are thus linked, forming networks of diverging and converging patterns.

All digestion products from the foods we eat, as well as many other substances coming into the cell, are gradually degraded to extract the energy needed for doing the work of the cell. During the first stage,

which requires no oxygen, certain enzymes partially break down small molecules such as sugars, amino acids, fatty acids, and glycerol. This process yields a chemical called pyruvate and some energy-rich ATP molecules. But much more energy is released during the second stage, in which pyruvate is gradually degraded through the *citric acid cycle* to carbon dioxide and water (Figure A.3).

In normal cells, each substance appears in the proper concentrations for the correct functioning of the body and its parts. And because the structure of each enzyme depends on one gene (or a few), DNA provides the blueprint for all cellular activity. These networks of chemical reactions help determine the final appearance of an organism.

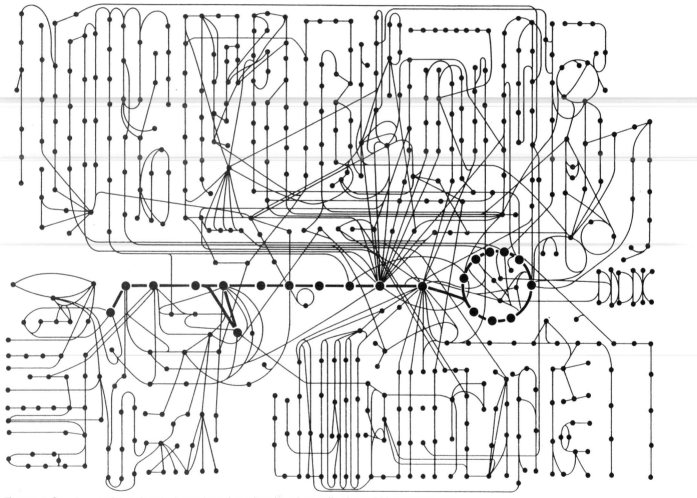

Figure A.3   Some of the chemical reactions that take place in a cell. About 500 common conversions of small molecules are diagrammed here, with each type of molecule represented by a dot. The breakdown of glucose and the citric acid cycle, key segments of this metabolic complex, appear in bold. (Adapted from Alberts et al. 1994.)

Appendix 3: MIM and OMIM Numbers for Some Genetic Conditions

MIM refers to *Mendelian Inheritance in Man*, currently in its twelfth edition (McKusick 1997); OMIM is the online version (http://www3.ncbi.nlm.nih.gov/Omim/). For over 35 years this database has provided continually updated summaries of research and relevant citations for all provisional and confirmed genetic conditions. The text is difficult to wade through, even for professionals. The printed version contains less arduous introductory material with many tables, gene maps, statistics, historical information on gene discovery, and other pertinent matter. The online version provides up-to-date point-and-click links to considerable additional data.

The **first digit** of the 6-digit MIM/OMIM number indicates the mode of inheritance of the condition:

    1 = Autosomal dominant locus
    2 = Autosomal recessive locus
    3 = X-linked locus
    4 = Y-linked locus
    5 = Mitochondrial locus
    6 = Autosomal locus entered in the database after May 15, 1994

In the following list, the name of the condition is followed by modifying terms (e.g., Alzheimer disease, familial, type 1; Cancer, breast, early onset). A condition with several common, distinct names (e.g., Christmas disease, factor IX deficiency, hemophilia B) is listed under each name. For the the many diseases known to be influenced by several genes, only one is usually given. We list here the diseases that are covered in the text, although some conditions that appear only in a table are omitted. Some genes have several symbols, but only one is given.

Genetic condition	Gene symbol	MIM and OMIM number
ABO blood group	ABO	110300
Acatalasia	CAT	115500
Achondroplasia	FGFR3	100800
Achoo syndrome	(none)	100820
Acromegaly	SMTN	102200
Adenomatous polyposis, familial	APC	175100
Adenosine deaminase (ADA) deficiency	ADA	102700
Adrenal hyperplasia, congenital	CYP21	201910
Adrenal hypoplasia congenita	DAX1	300200
Adrenoleukodystrophy	ALD	300100
Agammaglobulinemia, type 1	XLA	300300
Albinism, Chediak-Higashi syndrome	CHS1	214500
Albinism, Hermansky-Pudlak syndrome	HPS	203300
Albinism, ocular, type 1	OA1	300500

Genetic condition	Gene symbol	MIM and OMIM number
Albinism, oculocutaneous, brown phenotype	OCA3	203290
Albinism, oculocutaneous, tyrosinase-negative	OCA1	203100
Albinism, oculocutaneous, tyrosinase-positive	OCA2	203200
Albinism, oculocutaneous, yellow variant	OCA1	203100
Alcoholism	DRD2(?)	126450
Alkaptonuria	AKU	203500
Alzheimer disease	AD	104300
Amylase, salivary	AMY1	104700
Amyotrophic lateral sclerosis	ALS	105400
Androgen insensitivity	AIS	300068
Angelman syndrome	ANCR	105830
Angioneurotic edema, hereditary	HANE	106100
Anhidrotic ectodermal dysplasia	EDA	305100

Genetic condition	Gene symbol	MIM and OMIM number	Genetic condition	Gene symbol	MIM and OMIM number
Aniridia	PAX6	106210	Celiac disease	CD	212750
Antitrypsin deficiency, α-1	AAT	107400	Charcot-Marie-Tooth disease	CMT	118200
Apert syndrome	FGFR2	101200	Christmas disease	HEMB	306900
Ataxia telangiectasia	AT	208900	Chronic granulomatous disease	CGD	306400
Baldness, male pattern	MPB	109200	Cleft chin	(none)	119000
Basal cell nevus syndrome	PTCH	109400	Cleft lip and palate	EEC	129900
Becker muscular dystrophy	DMD	310200	Cockayne syndrome	CKN1	216400
Beckwith-Wiedemann syndrome	BWS	130650	Color blindness, green-shift	CBD	303800
Blood group, ABO	ABO	110300	Color blindness, red-shift	CBP	303900
Blood group, MN	MN	111300	Cowden syndrome	PTEN	158350
Blood group, Rh, CcEd antigens	RHCE	111700	Craniosynostosis, nonsyndromic	FGFR3	602849
Blood group, Rh, D antigen	RHD	111680	Craniosynostosis, type I	CRS1	123100
Bloom syndrome	BLM	210900	Cretinism, goitrous	FIDD	228355
Burkitt lymphoma	BL	113970	Creutzfeldt-Jakob disease	PRNP	123400
Campomelic dysplasia	SOX9	114290	Crohn disease	IBD1	266600
Camptodactyly syndrome	CAP	208250	Crouzon syndrome	FGFR2	123500
Cancer, adrenal, ovarian	GNAI2	139360	Cystic fibrosis	CFTR	219700
Cancer, bladder, thyroid	HRAS	190020	Dentatorubral-pallidoluysian atrophy	DRPLA	125370
Cancer, breast and ovarian	ERBB2	164870	Dentinogenesis imperfecta	DGI1	125490
Cancer, breast and ovarian, early onset	BRCA1	113705	Deuteranomaly	CBD	303800
Cancer, breast, early onset	BRCA2	600185	Diabetes insipidus	ND1	304800
Cancer, Burkitt lymphoma, breast, lung	MYC	190080	Diabetes mellitus, adult onset	NIDDM	125853
Cancer, colon, lung, pancreas, thyroid	KRAS	190070	Diabetes mellitus, juvenile onset	IDDM	222100
			DiGeorge syndrome	DGS	188400
Cancer, colorectal	DCC	120470	Disaccharide intolerance I	SI	222900
Cancer, colorectal	NRAS	164790	Distichiasis	(none)	126300
Cancer, glioblastoma	GLM	137800	Dosage-sensitive sex reversal	DAX1	300018
Cancer, leukemia, acute lymphoblastic	LALL	247640	Down syndrome	(none)	190685
			Duchenne muscular dystrophy	DMD	310200
Cancer, leukemia, acute T-cell lymphocytic	HOX11	186770	Dyslexia	DYX1	127700
Cancer, leukemia, chronic myelogenous	ABL1	189980	Ear malformation	(none)	128600
			Ear pits	(none)	128700
Cancer, leukemia, lymphoid	LYL1	151440	Ear wax, wet type	(none)	117800
Cancer, lung	MYCL1	164850	Ehlers-Danlos syndrome, type I	EDSI	130000
Cancer, lung, neuroblastoma	MYCN	164840	Elliptocytosis-1	EL1	130500
Cancer, lymphoma, B-cell	BCL2	151430	Ellis-van Creveld syndrome	EvC	225500
Cancer, melanoma	MTS1	154280	Emphysema	PI	107400
Cancer, multiple myeloma, familial	AL	254500	Epicanthus	(none)	131500
Cancer, osteosarcoma	FOS	164810	Epidermolysis bullosa	COL7A1	131750
Cancer, thyroid	RET	164761	Extra nipples	(none)	163700
Cancer, thyroid, ovary, colorectal	NRAS	164790	Fabry disease	GLA	301500
Cat-cry syndrome	(none)	123450	Factor IX deficiency	HEMB	306900
			Factor VIII deficiency	HEMA	306700
Cataracts	FTL	134790	Fanconi anemia, A	FACA	227650

Genetic condition	Gene symbol	MIM and OMIM number
Favism	*G6PD*	305900
Fibrodysplasia ossificans progressiva	*BMP4*	135100
Fish-odor syndrome	*FMO3*	602079
Fragile X syndrome	*FMR1*	309550
Friedreich ataxia	*FRDA*	229300
G6PD deficiency	*G6PD*	305900
Galactosemia	*GALT*	230400
Gaucher disease, type 1	*GBA*	230800
Glycogen storage disease	*G6PT*	232200
Graves disease	*GDA*	139080
Hairy palms / soles	(none)	139650
Hapsburg jaw	(none)	176700
Hemochromatosis	*HFE*	235200
Hemoglobin α	*HBA1, HBA2*	141800 /850
Hemoglobin β	*HBB*	141900
Hemophilia A	*HEMA*	306700
Hemophilia B	*HEMB*	306900
Histone H4	*H4F2*	142750
HLA histocompatibility type, A locus	*HLA-A*	142800
HLA histocompatibility type, B locus	*HLA-B*	142830
HLA histocompatibility type, DR locus	*HLA-DRA*	142860
Hodgkin disease	(none)	236000
Holoprosencephaly, type 3	*SHH*	142945
Homeobox genes (example, *HOXA7*)	*HOXA7*	142950
Homeobox genes (example, *PAX3*)	*PAX3*	193500
Homocystinuria	*CBS*	236200
Homosexuality, male	*HMS1*	306995
Hunter syndrome	*MPS2*	309900
Huntington disease	*HD*	143100
Hurler syndrome	*MPS1*	252800
Hydrometrocolpos	*HMCS*	236700
Hydroxylase deficiency, 21-	*CYP21*	201910
Hypercholesterolemia, familial	*LDLR*	143890
Hyperphenylalaninemia	*PAH*	261600
Hyperthermia, malignant	*RYR1*	145600
Hypochondroplasia	*FGFR3*	146000
Hypogonadotropic hypogonadism	(none)	215470
Hypoparathyroidism	*HYPX*	307700
Hypophosphatemia	*HYP*	307800
Hypothyroidism	*TRH*	275120
Ichthyosis	*TGM1*	242300
Isoniazid sensitivity	*NAT2*	243400
Jackson-Weiss syndrome	*FGFR2*	123150

Genetic condition	Gene symbol	MIM and OMIM number
Kallman syndrome	*KAL1*	308700
Kennedy disease	*SBMA*	313200
Leber optic neuropathy	*LHON*	535000
Lesch-Nyhan syndrome	*HPRT*	308000
Li-Fraumeni syndrome	*LFS*	151623
Lou Gehrig disease	*ALS*	105400
Machado-Joseph disease	*MJD*	109150
Major histocompatibility complex, A locus	*HLA-A*	142800
Major histocompatibility complex, B locus	*HLA-B*	142830
Major histocompatibility complex, DR locus	*HLA-DR*	142860
Mandibulofacial dysostosis	*TCOF1*	154500
Maple syrup urine disease	*MSUD*	248600
Marfan syndrome	*FBN1*	154700
Melanocyte stimulating hormone receptor	*MC1R*	155555
Methylmalonic acidemia	*MUT*	251000
Michelin tire baby syndrome	(none)	156610
Midphalangeal hair	(none)	157200
MN blood group	*MN*	111300
Multiple sclerosis	*MS1?*	126200
Myasthenia gravis	*MG1*	254210
Myotonic dystrophy	*DM*	160900
Nail-patella syndrome	*NSP1*	161200
Neurofibromatosis, type 1	*NF1*	162200
Neurofibromatosis, type 2	*NF2*	101000
Nevus flammeus	(none)	163100
Ocular albinism	*OA1*	300500
Odd-shaped teeth	(none)	187000
Ornithine transcarbamylase deficiency	*OTC*	311250
Osteogenesis imperfecta, type I	*COL1A2*	166200
Pfeiffer syndrome	*FGFR1*	101600
Phenylketonuria	*PAH*	261600
Phenylthiocarbamide (PTC) tasting	*PTC*	171200
Piebald trait	*PBT*	172800
Pituitary dwarfism, type I	*GH1*	262400
Platelet fibrinogen receptor disorder (Grinkov)	*ITGB3*	173470
Polycystic kidney disease, type 2	*PKD2*	173910
Polydactyly	*PAPA*	174200
Polyposis coli, familial	*APC*	175100
Poor metabolizer phenotype	*CYP2D6*	124030
Porphobilinogen deaminase deficiency	*AIP*	176000

Genetic condition	Gene symbol	MIM and OMIM number
Porphyria cutanea tarda	PCT	176100
Porphyria, acute intermittent	AIP	176000
Porphyria, variegate	PPOX	176200
Prader-Willi syndrome	PWS	176270
Primaquine sensitivity	G6PD	305900
Prion protein	PRNP	176640
Progeria	HGPS	176670
Prognathism, mandibular	(none)	176700
Pseudocholinesterase deficiency	BCHE	177400
Psoriasis vulgaris	PSORS2?	177900
Pyruvate kinase-1 deficiency	PK1	266200
Reductase deficiency, 5-α-	SRD5A2	264600
Retinitis pigmentosa-1	RP1	180100
Retinoblastoma	RB1	180200
Rh blood group, CcEd antigens	RHCE	111700
Rh blood group, D antigen	RHD	111680
Rheumatoid arthritis	HLADR?	180300
Sandhoff disease	HEXB	268800
Severe combined immunodeficiency disease	SCIDX	308380
Sex determining region of the Y	SRY	480000
Sexual orientation, male	HMS1	306995
Sickle-cell disease	HBB	141900
Spherocytosis, hereditary	HS	182900
Spinal and bulbar muscular atrophy	SBMA	313200
Spino-cerebellar ataxia	(none)	117200
Succinylcholine sensitivity phenotype	CHE1	177400
Synpolydactyly	HOXD	186000
Systemic lupus erythematosus	SLE	152700
Tay-Sachs disease	HEXA	272800
Testicular feminization	AIS	300068

Genetic condition	Gene symbol	MIM and OMIM number
Testis determining factor	TDFX	306100
Thalassemia α	HBA1	141800
	HBA2	141850
Thalassemia β	HBB	141900
Thanatophoric dysplasia	FGFR3	187600
Transfer RNA (example: mitochondria, proline)	MTTP	590075
Treacher Collins syndrome	TCOF1	154500
Trembling chin	GSM1	190100
Trimethylaminuria	FMO3	602079
Trisomy 21	(none)	190685
Tuberous sclerosis	TSC1	191100
Tune deafness	(none)	191200
Uncombable hair	(none)	191480
Ventricular hypertrophy	CMH1	192600
Vitamin D-resistant rickets	HYP	307800
von Hipple-Lindau syndrome	VHL	193300
von Recklinghausen disease	NF1	162200
Von Willebrand disease	VWD	193400
Waardenburg syndrome	PAX3	193500
WAGR syndrome	(none)	194072
Werner syndrome	WRN	277700
Williams syndrome	WBS	194050
Wilms tumor	WT1	194070
Wilson syndrome	WND	277900
Wolf-Hirschhorn syndrome	WHS	194190
Wolman disease	LIPA	278000
X-inactivation-specific transcript	XIST	314670
X-linked mental retardation	MRX20	300047
Xeroderma pigmentosum I	XPA	278700
Zellweger syndrome	ZWS1	214100

Answers to Questions

Chapter 2

2. GCCTACATTGGGA (Note: For reasons given in Chapter 6, this chain of bases should actually be written in the opposite direction, but we will ignore that technicality here.)

3. (a) 47,XXY   (b) 45,X

4. (a) 47,XX,+21   (b) 47,XY,+13   (c) 47,XX,+18

5.

	New order	Detectable in unbanded spreads?	Detectable in banded spreads?
(a)	*a b c ♦ d*	no	yes
(b)	*a b d ♦ c*	no	no
(c)	*e f ♦ h g*	yes	yes
(d)	*e f g h ♦*	no	no
(e)	*j ♦ l k m*	no	yes
(f)	*j l k ♦ m*	yes	yes

Chapter 3

1. *Embryo/fetus until birth:* Oogonia become primary oocytes and enter a very long meiosis I. This period takes roughly 3 to 7 months. *Birth until beginning of oocyte maturation:* Arrest in meiosis I. This period takes roughly 12 to 50 years. *Oocyte maturation until ovulation:* end of meiosis I, metaphase I, anaphase I, telophase I, cytokinesis (to produce secondary oocyte and first polar body), prophase II, metaphase II. This period takes about 1 month. *Fertilization until first cleavage division:* anaphase II, telophase II, cytokinesis (to produce egg and second polar body). This period takes 24 to 36 hours.

2. At metaphase the chromosome arms extend out to the sides of each centromere, but at anaphase they are pulled behind the centromere.

3. (a) Mitosis   (b) First meiotic division   (c) Either

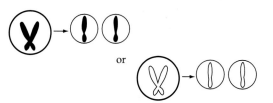

4. One convenient way to systematize such combinations of sets of things is by setting up a branching diagram like the one below. This method works not only for homologous chromosomes but also for gene pairs or any other combinatorial data.

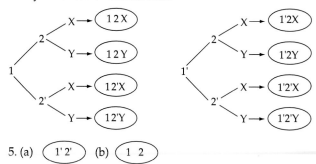

5. (a) 1' 2'   (b) 1 2

6. 46 chromosomes, 92 chromatids

7.

	Chromosomes	Chromatids
(a)	46	—
(b)	46	92
(c)	23	46
(d)	23	—
(e)	23	—

8. $3(23) = 69$; $4(23) = 92$

9. $(64/2) + (62/2) = 32 + 31 = 63$

Chapter 4

1. C/c ♂ × c/c ♀ produces $1/2$ normal (C/c) and $1/2$ albino (c/c).

2. Presence of the albino child shows that she is heterozygous (C/c). No other genotype is possible, barring mutation of C to c in a C/C woman (or adoption of the child).

3. Most likely she is homozygous normal (C/C), in which case only normally pigmented children are expected. She could, however, be heterozygous (C/c), in which case half of her children are expected to be albino. The probability is low (but definitely not 0) that a heterozygote produces no albinos among several children. For ten children, this probability is $(3/4)^{10} = 0.056$. (Note: This situation exactly parallels Mendel's method for progeny testing his tall F_2 plants (see page 51), running about a 5% chance of mislabeling a heterozygote as a homozygote.)

4. (a) $(3/4)^3 (1/4)^2 = 27/1024$

   (b) $(3/4)^5$

   (c) $1 - (3/4)^5$

   (d) One-fourth. Each birth is considered separately and is not influenced by the outcomes of previous births.

5. (a) $B/b \times B/B$ ⎫
      $B/b \times B/b$ ⎬ 50% of progeny expected to be B/b
      $B/b \times b/b$ ⎭

   (b) $B/B \times b/b$ 100% of progeny expected to be B/b

6. (a) $B/B \times B/B$ (b) $B/b \times B/B$ (c) $B/b \times B/b$
      $B/B \times b/b$ $B/b \times b/b$
      $b/b \times b/b$

7. $2/3\ Q/q$, $1/3\ q/q$

8. $1/4\ AB$, $1/4\ ab$, $1/4\ Ab$, $1/4\ aB$

9. $1/8$ each of the gametes carrying the following sets of genes (with only superscripts used for the Hb gene):

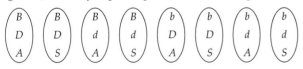

   Can you see how the sets of three follow from a tree diagram?

10. Sixteen. The 8 kinds above, with Xg^a or Xg^b added to each.

11. $9/16$ normal, type B; $3/16$ affected, type B; $3/16$ normal, type O; $1/16$ affected, type O. This is the Mendelian 9:3:3:1 dihybrid ratio for two independent genes with dominance.

12. (a) $P(B/b\ P/p\ S/s) = (1/2)^3 = 1/8$

    (b) P(all three dominant traits) $= (3/4)^3 = 27/64$

    (c) $P(b/b\ p/p\ s/s) = (1/4)^3 = 1/64$

13. Four are homozygous (A/A, S/S, E/E, C/C) and six are heterozygous (A/S, A/E, A/C, S/E, S/C, and E/C).

14. (a) $1/2\ A/A$, $1/2\ A/S$

    (b) $1/4\ A/A$, $1/4\ A/S$, $1/4\ A/C$, $1/4\ C/S$

    (c) $1/4\ A/C$, $1/4\ A/S$, $1/4\ E/C$, $1/4\ E/S$

15. (a) $1/2\ g^1/(Y)$ = green-shifted male; $1/2\ G/g^1$ = normal (carrier) female

    (b) $1/4\ G/(Y)$ = normal male; $1/4\ g^2/(Y)$ = missing-green-pigment male; $1/4\ G/G$ = normal female; $1/4\ G/g^2$ = normal (carrier) female

    (c) $1/4\ G/(Y)$ = normal male; $1/4\ g^1/(Y)$ = green-shifted male; $1/4\ G/g^2$ = normal (carrier) female; $1/4\ g^1/g^2$ = green-shifted female

    (d) Parents: g^1/g^2 female × $G/(Y)$. Offspring are as follows: $1/4\ g^1/(Y)$ = green-shifted male; $1/4\ g^2/(Y)$ = missing-green-pigment male; $1/4\ G/g^1 + 1/4\ G/g^2 = 1/2$ normal (carrier) female

Chapter 5

1. The terms *homozygous* and *heterozygous* imply the existence of two alleles (same or different) in an individual. X-linked genes in males are present only once.

2. (b) It appears that everybody produces and excretes a rather musky-smelling intermediate product (methanethiol) of asparagus metabolism, but not everybody can smell it (Lison et al. 1980). The ability to smell this substance is probably an autosomal dominant trait.

3. Autosomal recessive. Note that the parents of the affected individual are first cousins.

4. (a) Parents: Gl/gl. Child: gl/gl.

   (b) $1/4$. No; the same probabilities exist independently for each child, regardless of the outcome of previous pregnancies.

   (c) $1/300$; that is, the chance of marrying a heterozygote ($1/150$) times the chance of then producing an affected child ($1/2$).

5. (a) Queen Victoria; her children, Alice of Hesse and Beatrice; her grandchildren, Irene, Alix, Alice of Athlone, and Victoria Queen of Spain.

   (b) 10 hemophiliacs to 10 normal.

   (c) No, as far as descent from Victoria is concerned. The father of Elizabeth II, being unaffected with hemophilia, must have inherited (and passed on to her) an X chromosome lacking Victoria's hemophilia allele.

6. Autosomal dominant. If this were an X-linked dominant: In family 1, male II-1 should not be affected; in family 2, males II-3 and II-6 would be severely affected (as is male III-3); and in family 4, female III-8 would be affected moderately (like II-1 and II-6) rather than severely. The condition represented here is familial hypercholesterolemia.

7. (a)

	Among daughters	Among sons
G/G ♂ × $G/(Y)$ ♂ →	all G/G	all $G/(Y)$
G/G ♂ × $g/(Y)$ ♂ →	all G/g	all $G/(Y)$
G/g ♂ × $G/(Y)$ ♂ →	1/2 G/G, 1/2 G/g	1/2 $G/(Y)$, 1/2 $g/(Y)$
G/g ♂ × $g/(Y)$ ♂ →	1/2 G/g, 1/2 g/g	1/2 $G/(Y)$, 1/2 $g/(Y)$
g/g ♂ × $G/(Y)$ ♂ →	all G/g	all $g/(Y)$
g/g ♂ × $g/(Y)$ ♂ →	all g/g	all $g/(Y)$

(b) All but the first and last matings listed in part (a). All the daughters will be carriers when the mother is homozygous and the father is hemizygous for the other allele (second and fifth mating).

(c) Those in which the mother is heterozygous. For sons, the genotype of the father is irrelevant with regard to X-linked genes.

(d) g/g ♂ × $G/(Y)$ ♀ → all daughters normal, all sons green-shifted

(e) G/g ♂ × $g/(Y)$ ♀

8. X-linked recessive. The gene has recently been cloned.

9. Daughters. He transmits a gene-rich X chromosome to daughters but a gene-poor Y chromosome to sons; as a result, daughters carry about 5% more of his genes than do sons.

10. A female receives no sex chromosomes from her father's father, whereas a male receives no sex chromosomes from his father's mother.

11. Does not skip generations; expressed only in females and transmitted only through females; affected females produce only half as many sons as daughters, but all sons are normal. All but the first feature distinguish this pattern from an autosomal dominant. This mode of inheritance is also rare, but one example is oral-facial-digital (OFD) syndrome type I. Affected females have a cleft jaw and tongue, other malformations of the skull, face, and hands, and mental retardation.

12. (a) Autosomal dominant or X-linked dominant.

(b) In addition to (a), could also be autosomal recessive or X-linked recessive if the mother were heterozygous. These ambiguities point up the importance of having extensive and reliable family data and also knowing something about the frequency of the trait.

13. (a) The rare allele would have had to be present in spouses I-2 and III-2. Because these persons are genetically unrelated to I-1 and III-1, it is highly unlikely that a rare allele would occur this way.

(b) The rare allele would have had to occur in I-2.

(c) Individual III-1 would have had to exhibit the trait.

(d) Individuals I-2, II-1, III-1, III-3, and III-4 would have had to exhibit the trait.

(e) III-2 and III-6 should exhibit the trait.

(f) There are two simple possibilities. It could be a common X-linked recessive trait [affected = a/a or $a/(Y)$],

with I-2 as a carrier; III-1, III-3, and III-4 would also be carriers who inherited the mutant allele from their fathers. It could also be a common autosomal recessive trait (affected = a/a), with I-2, III-1, and III-2 (as well as III-3, III-4, and III-6) being carriers.

14. Yes, there could be a mutation in one of the nuclear genes that encodes a mitochondrial protein. In fact, Bourgeron et al. (1995) describe such a situation, in which two sibs (whose parents were first cousins) had Leigh syndrome associated with a nuclear gene mutation.

15. These pedigrees, describing hypophosphatemic rickets, are consistent with either X-linked dominant or autosomal dominant inheritance (Holm, Huang, and Kunkel 1997). The former may be more likely, however, because there is no male-to-male transmission of the trait and all affected males have an affected mother. But the numbers are small, so these two factors may be chance occurrences.

16. In this situation, FH homozygotes result only from the matings of FH heterozygotes, from which 1/4 of the offspring are expected to be homozygotes. So $1/500 \times 1/500 \times 1/4 = 1/1,000,000$.

Chapter 6

1. P is present in DNA but not in protein; S is present in virtually all proteins but not in DNA. These differences make possible some elegant experiments in which either DNA or protein is radioactively labeled.

2. Chargaff's rules only apply to double-stranded nucleic acids. The DNA molecules in some viruses and most RNAs are single-stranded.

3. 1/4 intermediate, 3/4 light; that is, of the eight double helices, two would be intermediate and six would be light.

4.

	(a)	(b)	(c)
A	33%	21%	27%
T	21%	33%	27%
G	18%	28%	23%
C	28%	18%	23%

5. mRNA: G G G U U U
   DNA: C C C A A A

6. Three pieces having amino acids 1–22, 23–29, and 30 only.

7. The β polypeptide has 146 amino acids and so is coded for by (146)(3) = 438 bases. 438/1,606 = 27.3% (see Figure 6.17).

8. The number of nucleotides required to code for the five polypeptides is (5)(438) = 2,190 bases. 2,190/50,000 = 4.4%.

9. Only mutation combinations (b) and (e) produce reading frame shifts.

10. Ala–Gly–Cys–Lys–Asn–Phe–Phe–Trp–Lys–Thr–Phe––Thr–Ser–Cys

11. (a) ...Lys–Lys–Lys–Lys...

(b) ...Ser–Ser–Ser–Ser... and ...His–His–His–His... and ...Ile–Ile–Ile–Ile...

(c) ...Phe–Leu–Pro–Ser... (this foursome repeated in tandem)

12. ACC: Trp   CAC: Val   CCA: Gly (still)

GCC: Arg   CGC: Ala   CCG: Gly (still)

TCC: Arg   CTC: Glu   CCT: Gly (still)

13. While the absence of effective telomerase would limit the growth of cancer cells after a number of generations, the same treatment might also stop the replenishment of necessary blood cells.

Chapter 7

1. The mutant allele encodes an altered receptor protein on the surface of melanocytes. The altered protein apparently no longer binds the hormone that stimulates the production of brown-black melanins. In the absence of dark melanins, the reddish melanins are expressed.

2. C was changed to G.

3. Germinal mutations occur in the cells leading to sperm or eggs, while somatic mutations occur in other cells. If a mutation occurred early in development, before specific cells were set aside for the germ line, a mutation could affect both somatic and germinal tissues.

4. The mutation could occur outside the amino acid coding regions in a completely neutral site, such as in an intron or in the extensive DNA regions between genes. Alternatively, it could change an amino acid triplet to a triplet coding for the same amino acid.

5. (a) The resultant polypeptide would be too short, since translation would stop at the newly created stop codon.

(b) The resultant polypeptide would be too long, since translation would continue through the normal stop codon and into a region that is normally not translated at all.

6. Of the 64 possible triplets, three are stop triplets. If the bases were arranged randomly, then a stop triplet would occur, on the average, once in about every 21 triplets (64/3).

7. Either G to A or T to A.

8. Three adjacent bases are deleted from the gene.

9. In the case of Huntington disease, the polyglutamine stretches might bind more or less strongly to a critical protein in the brain. But for the most part, nobody knows.

10. In theory, misalignment and unequal crossing over at meiosis could account for the increase in the number of trinucleotide repeats in successive generations. One problem is that trinucleotide repeats that have been studied tend to increase rather than decrease, whereas unequal crossing over should result in a decrease as often an an increase (see Figure 7.12).

11. Enzymes from mammalian liver are sometimes added to the growth medium for whatever effect they might have in metabolizing the tested chemicals.

12. Although radiation can cause mutations, this particular man has no case. He transmitted a *Y chromosome* to his son; the mutant hemophilia gene in question was inherited on an *X chromosome* from the mother.

13. Scientists estimate that it would take 400 rem of chronic radiation to yield the observed rate of spontaneous mutations in humans. The average exposure of a human is much less than this amount (about 10 rem).

14. The α chains are a major component of hemoglobin both pre- and postnatally, but β chains are synthesized in quantity only after birth.

15. (a) Nicks the sugar-phosphate backbone.

(b) Helps open up the double helix to achieve single strands.

(c) Forms a new strand of DNA complementary to a template strand.

(d) Seals the sugar-phosphate backbone.

16. (a) Fragile X syndrome; about 1 in 1,250 males (about 1 in 2,500 females); due to an expanded trinucleotide repeat.

(b) Charcot-Marie-Tooth disease; about 1 in 2,500; due to a duplicated gene brought about by unequal crossing over.

(c) Myotonic dystrophy; about 1 in 8,000; due to an expanded trinucleotide repeat.

Chapter 8

1. 20 cycles. That is, 2^{20} (1,048,576) just exceeds 1 million. At 5 minutes per cycle, this will take (in theory) 1 hour and 40 minutes. In practice, it will take longer, since later PCR cycles require more time.

2. Reading the base sequence of one strand to the right is the same as reading the other strand to the left. Almost all restriction enzymes recognize palindromes. Other examples are *Sau*3A (GATC), *Bam*HI (GGATCC), *Hind*III (AAGCTT), and *Not*I (GCGGCCGC).

3. Treat the recombinant plasmid with *Eco*RI. This treatment will yield DNA segments of two sizes: 1 kb for the human gene and something else (one hopes) for the rest of the plasmid. The segments can be physically separated by electrophoresis.

4. $10,000 = (3 \times 10^9 \text{ total bases})/(3 \times 10^5 \text{ bases per YAC})$

5. *Eco*RI lane: bands at 3, 7, and 12 kb

*Hind*III lane: bands at 9 and 13 kb

"Both" lane: bands at 3, 6, and 7 kb

6. Animal rights activists have objected, for example, to injections of recombinant bovine growth hormone on dairy herds to increase milk yields because it is bad for cows (possibly increasing udder infections, for example). While not taking a position on that issue, we defend the humane use of laboratory animals in medical research as indispensable to wide-ranging efforts to understand, treat, and eliminate many diseases. Research centers are subject to comprehensive policies, accreditation standards, and laws meant to ensure appropriate and respectful treatment of their animals.

Consider also the vastly larger number of animals that are used for human food, clothing, and work.

7. TGCA

 TGCAA

 TGCAATGGCA

 TGCAATGGCACA

 TGCAATGGCACAGTGTA

 TGCAATGGCACAGTGTAGCCCTA

 TGCAATGGCACAGTGTAGCCCTAGA

8. AGCTGTTTCC

9. 90%. $(300 \times 10^6$ bases sequenced$)/(3 \times 10^9$ bases total$)$ = 0.1 = 10% will have been sequenced. As of March 1998, about 90 million bases (about 3%) had been sequenced.

10. Approximately $2(0.03)(0.02) = 0.0012$; that is, about 1 in 800 people would be expected to have the indicated genotype

11. If independent, these probabilities can be multiplied together: $3 \times 8 \times 2 \times 7 \times 12 \times 10^{-12}$ = approximately 4×10^{-9} = 1 in 250 million (the population of the U.S). The report of the National Research Council (1996) suggests that such an estimate might be wrong by a factor of 10 in either direction, so the true value is estimated to lie between 1 in 25 million and 1 in 2.5 billion.

12. We don't know. If the probability is one in such a large number that it approaches the world population, can we regard the fingerprint as unique?

Chapter 9

1. Look at the phenotype of the sons. Each son gets one of the four chromatids as their X chromosome, and they get no corresponding genes from their father (who instead contributes a Y chromosome). They should be either green blind with hemophilia (from chromatid 1), green blind with normal blood clotting (chromatid 2, a recombinant type), normal color vision with hemophilia (chromatid 3, a recombinant type), or normal color vision with normal blood clotting (chromatid 4). In practice, about 3% of sons are recombinant, indicating 3 map units between the two loci.

2. To distinguish a recombinant from a parental assortment. If either gene is homozygous, crossing over produces no new assortment.

3.
Gamete type (%)				
AB	**ab**	**Ab**	**aB**	
(a) 25	25	25	25	Independent assortment
(b) 50	50	0	0	Linked as one unit
(c) 46	46	4	4	A total of 8% recombination
(d) 25	25	25	25	The maximum of 50% recombination

4. $30,000 \times 10^5/3700 = 8 \times 10^5$ = 800,000 bases per map unit

5. Simple laboratory tests based on DNA (from, say, easily obtained white blood cells) can document variation (= polymorphism) at specific genetic sites in pedigree members. Such variation is due to mutations in restriction sites (RFLPs) or changes in copy number (VNTRs and microsatellites). This variation is widespread.

Thus, DNA markers provide sites around which "real" genes can be mapped by old-fashioned genetic linkage analysis.

6. 4,096 bases, or about 4 kb. This is the denominator of the probability of cutting any specified 6-base sequence, which is $(1/4)^6$.

7. In DNA fingerprinting, the fragment-length variation depends on different numbers of repeated segments between two unmutated restriction recognition sites. In mapping using RFLPs, the restriction recognition sites themselves have mutations (so that different alleles may be cut or not cut).

8. Chromosome 1 is about 290 map units long; chromosome 21 is about 60 map units long.

9. Females, 360 map units; males, 220 map units

10. This is an artifact of map construction. The only way to know that there is more chromosome beyond a mapped site is to find another mapped site more distally, in which case the new site marks the end.

11. Chromosome 11. This chromosome is the only one present in all three "+" lines and absent in all three "−" lines.

12. No, but mouse genes could be assigned to mouse chromosomes.

13. No; genes on the same chromosome that are more than about 50 map units apart (without intervening marker genes) will appear to be unlinked. See also question 3.

14. Amino acids 1 through 1,755 are normal. Amino acids 1,756 through 1,828 (the next 73) are largely wrong because of the frameshift at codon 1,756 and the stop at 1,829. (Amino acids 1,829 through 1,863 (the next 35) are missing because of the stop at codon 1,829.)

15. Sequencing of the alleles of both genes in affected members might be necessary. This procedure is generally done in research laboratories, and not in commercial testing laboratories, although the latter might detect one of the several common mutations using specific DNA probes.

Chapter 10

1. It is easier to calculate first the probability of a normal child, and then get the probability of a deaf child by subtraction:

 P(normal child) = $P(D/–\ E/–)$ = $(3/4)(3/4)$ = $9/16$

 P(deaf child) = $1 – P$(normal child) = $1 – 9/16 = 7/16$

2. All sons become bald (B_1/B_2) and all daughters are non-bald (B_1/B_2).

3. Yes, except for female II-5, who should have received the B allele from her father and been nonbald. As noted in the text, however, environmental factors may have contributed to her baldness.

4. Use a checkerboard or the gene × gene method to get:

 $9/16\ B/–\ C/–$ = black

 $3/16\ b/b\ C/–$ = brown

$$\left.\begin{array}{l} 3/16 \; B/-\; c/c \\ 1/16 \; b/b \; c/c \end{array}\right\} = 4/16 \text{ albino}$$

5. Use a gene × gene method to get:

1/2 –/– c/c –/–	= albino
1/8 B/b C/c D/d	= black
1/8 B/b C/c d/d	= gray
1/8 b/b C/c D/d	= brown
1/8 b/b C/c d/d	= light chocolate

6. $C/c \times c^b/c$

7. The parents were homozygous for different autosomal recessives leading to albinism (or deafness). For example, assume that c/c is albino at one locus and that w/w is albino at another locus. (The c is allelic to its normal dominant allele C; likewise, w is allelic to W.) One parent could be c/c W/W, and the other parent C/C w/w. All their children would then be double heterozygous, C/c W/w, and phenotypically normal.

8. (a) $1/2 \times 1/2 = 1/4$ (b) $1 \times 1 = 1$; all affected

(c) A/A B/B; A/A B/b; A/A b/b; A/a B/B; A/a b/b; a/a B/B; a/a B/b; and a/a/ b/b.

9. False. "The mean contribution from each chromosome is 50%, but because the process is generally random, a normal female may vary considerably from the mean." (Belmont 1996).

10. As illustrated in Figure 10.11, genotypically identical twin females may differ in the phenotypic expression of heterozygous X-linked genes. Because of the random inactivation of X chromosomes, it is unlikely that mosaic patches in twin females would be identical in size and distribution. Identical twin 46,XY males, however, must express all the same genes on their single X chromosome.

11. Each tumor must have arisen from a single cell (either A or B) rather than from a group of cells. The cell that gave rise to a tumor had *either* an active A allele *or* an active B allele of the G6PD gene. This active allele remained the active allele in all descendants of that original tumor cell.

12. All phenotypically normal carrier daughters, and all phenotypically normal sons.

13. In Figure 10.13, note that males with a premutation will have all unaffected daughters who carry a premutation; in fact, several cases have been reported of males with FXS who fathered normal carrier daughters. Sons, who receive their father's Y chromosome, should be normal with regard to this trait.

14. During early development, the X-inactivation patterns of the cells destined to give rise to the brain probably differed between the twin embryos: in one, the X with the full mutation remained active in many or most brain cells, but in the other, the X with the full mutation was inactivated.

15. (a) No children are expected to be affected, although half are expected to receive allele D.

(b) Half of the children are expected to receive and express allele D.

Chapter 11

1. All the events in their lives that were different—perhaps very different—are not specified. Did either one like something other than math, or dislike something other than spelling, during all their school years? Did they both drive only blue Chevrolets during their lives? Did either one vacation elsewhere than on the same stretch of Florida beach? We are not told.

2. Probably the genetic variability in maze-running ability was exhausted. After seven generations of selection, virtually all the alleles that tended toward maze-brightness or maze-dullness were already concentrated in one group or the other.

3. (a) Higher heritability when genetically heterozygous.

(b) Higher heritability when environmentally uniform.

4.

G/G H/H i/i	G/G h/h I/I	g/g H/H I/I
G/G H/h I/i	G/g H/H I/i	G/g H/h I/I

5. (a) Hazel × blue-green: G/g H/h I/i × G/g H/h i/i, for example. (b) Light brown.

6. (a) True. In addition, no progeny can be darker than medium. (b) False. The child can be medium when the other parent is light, and the child can be dark when the other parent is medium.

7. The paternal contribution to the unusual pair is similar to that for DZ twins, but the maternal contribution is similar to that for MZ twins. Note that in the absence of crossing over, the second polar body is genetically identical to the egg.

8. The range of differences—up to 24 points—shows that environmental factors are important in IQ variation. Being a twin reduces IQ an average of about 5 points below that for single births (this is seen in other studies also).

9. The sibs of the severely retarded would generally have IQ scores in the normal range. They have only a 25% chance of being homozygous for the postulated abnormal recessive, and a much smaller chance of abnormality through repetition of a possible developmental error. On the other hand, the sibs of the moderately retarded would tend to have relative low IQ scores because of multiple gene inheritance of the same additive alleles, or perhaps because of similar environmental factors.

10. Within both countries, of IQ is 100%. There would almost certainly be an average difference in IQ scores between persons from A and B, caused in part by the obvious environmental factors. There is nothing in this situation as stated that rules out genetic factors contributing to the difference, however.

11. The most difficult calculation is for the gay man's aunt (second female in generation II):

P(she got the red X) = (1/2)(1) + (1/2)(1/2) = 3/4

In this expression, the four probabilities in parentheses are, in order: 1/2 = the left-hand case in generation I; 1 = the probability that the aunt got the red X in *this* case; 1/2 = the right-hand case in generation I; 1/2 = the probability that the aunt got the red X in *this* case. For the uncle (the last person in generation II), the similarly calculated probability that he got the red X = (1/2)(0) + (1/2)(1/2) = 1/4. For the cousin (the last person in generation III), the probability that he got the red X is (1/2)(3/8) —that is, half the likelihood that his mother had it.

12. For the male cousins, the calculation of penetrance is 8/37.5 = 21%; and for the uncles, 7/25 = 28%.

Chapter 12

1. freq(A) = p = 0.1 + ½(0.4) = 0.3

   freq(a) = q = 0.5 + ½(0.4) = 0.7

2. freq(A/A) = p^2 = 0.3^2 = 0.09

   freq(A/a) = $2pq$ = 2(0.3)(0.7) = 0.42

   freq(a/a) = q^2 = 0.7^2 = 0.49

3.

	Phenotype		
Population	**M**	**MN**	**N**
Bushmen	0.36	0.48	0.16
Amish	0.25	0.50	0.25
Polynesians	0.16	0.48	0.36

4. (a) freq(M) = p = (82 + 38)/200 = 0.6; freq(N) = q = (42 + 38)/200 = 0.4. (*Note:* Hardy-Weinberg law not assumed.)

   (b) freq(type M) = p^2 = 0.36; freq(type MN) = $2pq$ = 0.48; freq(type N) = q^2 = 0.16.

5. p = freq(A) = (1,210 + 390)/2,000 = 0.80; q = 1 − p = 0.20.

Genotype	Expected frequency	Expected Number
A/A	p^2 = 0.64	640
A/S	$2pq$ = 0.32	320
S/A	q^2 = 0.04	40
Total	= 1.0	1,000

6. Assuming the Hardy-Weinberg law, q = freq (c) = $\sqrt{1/10,000}$ = 1/100; p = freq (C) = 1 − q = 99/100. Fraction of population that is heterozygous = $2pq$ = 199/10,000 = 2%.

7. $(1/10)^2$ = 1/100 = 1%

8. freq(type A) = freq($A/A + A/O$) = $(0.3)^2$ + 2(0.3)(0.6) = 0.45

   freq(type A) = freq($B/B + B/O$) = $(0.1)^2$ + 2(0.1)(0.6) = 0.13

   freq(type AB) = freq(A/B) = 2(0.3)(0.1) = 0.06

   freq(type O) = freq(O/O) = $(0.6)^2$ = 0.36

   Sum = 1.00

9.

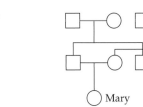

10. (a) 10^{-6}

    (b) 10^{-3}(1/16) + 10^{-6}(15/16) = 63.4 × 10–6

    (c) 63.4

11. Sperm cells from older fathers are more likely to carry a mutated gene because the precursor cells from which they are derived continue to divide and therefore continue to mutate.

12. The calculated rate would be too high because some affected children would be tallied as new mutations when, in reality, an "old" mutant gene, unexpressed in the parent, was transmitted to them.

13. The first consequence might be a sharp reduction in the appearance of isolated (so-called sporadic) cases of dominant diseases. All such patients would have an affected parent (except for instances of incomplete penetrance).

14. As the allele becomes rarer, the proportion of alleles present in heterozygotes, where they are masked, becomes greater.

15. Because there are so many more heterozygotes than homozygous recessives.

16. (1) Selection acts against the detrimental sickle-cell homozygote (without favoring the heterozygote, as happens in malarial regions of Africa). (2) Some admixture with white populations has occurred.

17. When a molecule of insulin is formed, the C segment is cut out; thus, its amino acids might be expected to be less constrained—freer to change without affecting critical functions. Mutational changes would therefore be more likely to be found here than in the A or B sections.

18. Mitochondrial DNA accumulates mutations at a faster rate than nuclear DNA. Therefore, it is a more sensitive indicator of evolutionary changes.

Chapter 13

1. (a) Chromosome loss at meiosis I

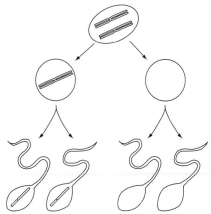

2 normal:2 nullisomic sperm

(b) Chromosome loss at meiosis II

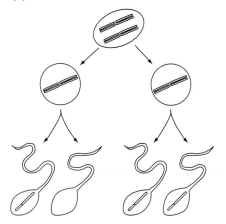

3 normal:1 nullisomic sperm

(Some cytologists see chromosome loss as a form of nondisjunction; others distinguish between the two.)

2. The zygote will be euploid, although both chromosomes 21 came from the mother.

3. Autosomal diploidy is required for normal embryonic development, and monosomy is lethal. One reason for the latter may be that any recessive lethal present on the hemizygous chromosome is expressed in the early embryo. Such embryos may be spontaneously aborted before a pregnancy is recognized.

4.

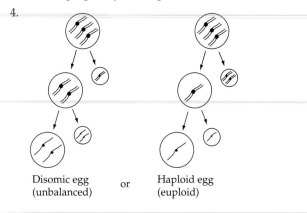

Disomic egg (unbalanced)   or   Haploid egg (euploid)

Of the two dozen or so cases in which females with Down syndrome have given birth, about half of their children were normal and half had Down syndrome.

5. Theoretically, if the split occurred at the two-cell stage and then mitotic nondisjunction occurred in one of these cells, the result would be one normal twin and one with Down syndrome. (Actually, identical twinning probably does not take place at the two-cell stage.)

6. Spontaneous abortions of trisomy 21 embryos and fetuses are probably more likely to happen in pregnant women with poorer prenatal care.

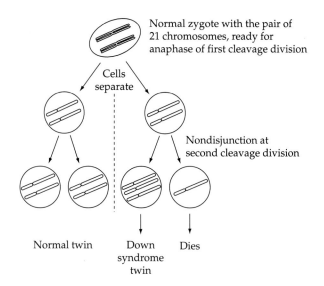

Normal zygote with the pair of 21 chromosomes, ready for anaphase of first cleavage division

Cells separate

Nondisjunction at second cleavage division

Normal twin   Down syndrome twin   Dies

7. Females with Turner syndrome will express color blindness whenever the gene occurs on their single X chromosome, as is true for 46,XY males. Males with Klinefelter syndrome, however, need two doses of the mutant gene to be color-blind, as is the situation with 46,XX females. Thus the frequencies of X-linked genes in females with Turner syndrome and males with Klinefelter syndrome should parallel the frequencies usually found, respectively, among males and females in the general population.

8. The allele for color-blindness must have been present in the heterozygous mother (G/g), and the father was hemizygous normal, ($G/(Y)$). The females with Turner syndrome probably resulted from fertilization of a g-bearing egg by an X- or Y-bearing sperm, and the male-derived sex chromosome was later lost. The males with Klinefelter syndrome must have resulted from fertilization of a g/g egg by a Y-bearing sperm.

9. The twins were derived from a single 46,XY zygote. If the presumptive embryo split in half and a cell in one of the two halves lost a Y chromosome, then one twin would be a normal male and the other a female with Turner syndrome.

10. Nondisjunction in females in division(s):

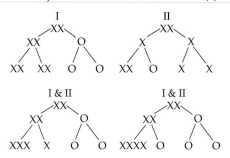

Nondisjunction in males in division(s):

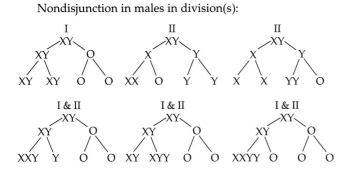

11. The first three examples require only one nondisjunctional event. The other scenarios, requiring two nondisjunctional events in one parent, are extremely unlikely.

		Nondisjunctional events		
Egg	**Sperm**	**In mother**	**In father**	**Total**
XX	Y	Meiosis I	—	1
XX	Y	Meiosis II	—	1
X	XY	—	Meiosis I	1
XX	Y	Meiosis I	Meiosis I & II	3
XX	Y	Meiosis II	Meiosis I & II	3
X	XY	Meiosis I & II	Meiosis I	3
O	XXY	Meiosis I	Meiosis I & II	3
O	XXY	Meiosis II	Meiosis I & II	3

12. Not all the loci are inactivated on the one inactive X chromosome possessed by 47,XXY males. Thus, unlike completely hemizygous 46,XY males, they will have abnormal double doses of all those X-linked loci (at least 20) that escape inactivation.

13. Mitotic loss of one X chromosome from a cell that was originally XXY, or nondisjunction of an X chromosome during mitotic division of an XY cell (in any division after the first cleavage) to give an XXY cell + a Y cell that dies. These two possibilities are diagrammed as follows:

14. Endocrinology clinics would attract those males with Klinefelter syndrome who seek treatment for failure to develop secondary sexual characteristics (i.e. undergo normal puberty); these cases would be more seriously affected. Subfertility clinics would attract married males with Klinefelter syndrome—that is, those who are masculine enough to find mates but who later discover that they are infertile.

15. *Advantages:* (1) We can learn more about physical and behavioral phenotypes—including the study of XYY individuals with no abnormalities—as well as the frequencies of such individuals in the general population. (2) If physical or behavioral problems do turn up, early identification and counseling might help the subjects and their parents. (3) Some genetic conditions are treatable, but it is impossible to develop treatments without first studying the phenotypes.

Disadvantages: (1) Publicity about the 47,XYY karyotype makes it difficult to do prospective studies: subjects might be stigmatized or treated differently by their family, and develop problems as a self-fulfilling prophecy. (2) There is no point in studying XYY individuals, when no treatment for possibly abnormal traits is available.

Chapter 14

1. Both (a) and (b) will show a looping out of one chromosome:

2. Neither (a) nor (b) require any unusual pairing configurations, since the homologues match:

3. (a) Half the gametes would contain the (21q21q) translocation chromosome (giving rise to progeny with Down syndrome), and half would contain no chromosome 21 at all (causing lethal monosomy 21). (b) No. (Hypothetically, if there were uniparental disomy for the other parent's 21, a normal offspring could result. But this has never been found.)

4. Because both chromosomes 15 were stuck together, all her eggs contained either two attached chromosomes 15 or no chromosome 15. Fertilized by normal sperm, all the resultant zygotes were therefore either trisomic or monosomic for a medium-sized chromosome and thus inviable. One abortus was in fact recovered and found to be trisomic for chromosome 15.

5. By an inversion that includes the centromere:

The reverse of this phenomenon is probably what accounts for the few metacentric Y chromosomes that have been reported in human males.

6. A triploid could arise (1) if a haploid sperm fused with a diploid egg. The diploid egg could result from failure of either the first or second meiotic division, or by retention of the second polar body; (2) if two sperm fertilized a haploid egg; and (3) theoretically, if nondisjunction in meiosis I or meiosis II of spermatogenesis gave rise to a diploid sperm, which fertilized a haploid egg. But diploid sperm have never been found to have fertilizing capacity in humans.

7. Only if the two aneuploidies were complementary, so that the duplicated chromosomes or chromosome segments present in one were balanced by deletions of the same genetic material in the other, and vice versa (i.e., fertilization involving the pair of gametes shown in the orange panels of Figure 14.11). This event is highly unlikely, even if the two parents were related and heterozygous for the same translocation.

8. One possibility starts with a normal XX zygote. Before the first cleavage division, both X chromosomes replicate, and the two chromatids of one X segregate normally. But the centromere of the other replicated X misdivides to form an isochromosome of the long arm, with loss of the replicated short arm. Thus the dividing zygote will contain one isochromosome plus two copies of the normal X. Following mitosis, each daughter cell contains one normal X chromosome, but the isochromosome will be present in only one of the two cells. Thus a mosaic results, in which one cell line has 46 chromosomes (including the isochromosome of Xq) and the other cell line is 45,X.

9. The son must have inherited both his father's X and Y chromosomes, due to nondisjunction in Meiosis I of spermatogenesis in the father. The XY sperm must have fertilized a no-X egg that resulted from maternal nondisjunction. Replication of the affected father's X chromosome in the zygote resulted in uniparental isodisomy for that chromosome.

10. If crossing over occurred before nondisjunction, part(s) of the recombinant disomic chromosome pair inherited from one parent would be identical (i.e., derived from just one member of that parental pair), and the other part(s) of the disomic chromosome pair would be different (i.e., derived from both members of that parental pair).

11. Especially for individuals exhibiting rare recessive disorders whose chromosomal locations are totally unknown, the occurrence of uniparental disomy can—when combined with linkage tests for heterozygous molecular markers present in the parents—help to locate the gene on a particular chromosome.

12. The severely affected child was homozygous for the mutant chromosome, having inherited a *CMT1* duplication from each parent.

13. The syndrome is associated with a supernumerary chromosome consisting of a duplicated chromosome part. Thus, affected individuals have four rather than the normal two copies of the duplicated region.

14. (a) $\overline{A\ 1\ B\ C\ D\ E\ F}$ Balanced, noninverted

   $\overline{A\ 2\ B\ C\ d}$ $\overline{e\ b\ 4\ a}$ Unbalanced, carrying fragments of the broken dicentric

   $\overline{f\ c\ d\ e\ b\ 3\ a}$ Balanced, inverted

   (b) $A\ B\ C\ D \blacklozenge E\ F$ Balanced, noninverted

   $A\ B\ C\ d \blacklozenge e\ b\ a$ Unbalanced

   $F\ E \blacklozenge D\ c\ f$ Unbalanced

   $a\ b\ e \blacklozenge d\ c\ f$ Balanced, inverted

   (c) $W\ X\ Y\ Z\ A\ B\ C\ D \blacklozenge E\ F$ Balanced, noninverted

   $W\ X\ Y\ Z\ A\ B\ C\ d \blacklozenge e\ b\ a\ z\ y\ x\ w$ Unbalanced

   $F\ E \blacklozenge D\ c\ f$ Unbalanced

   $w\ x\ y\ z\ a\ b\ e \blacklozenge d\ c\ f$ Balanced, inverted

   Here the unbalanced gametes are even more unbalanced (by too many or too few loci outside the inversion) than those in (b), and zygotes formed by union with normal gametes would be less viable than those in (b).

15. Chances are that this small supernumerary chromosome arose during meiosis in an ancestor. Through a process of chromosome breakage and reunion, the short arms of two different acrocentric chromosomes got joined together to form a Robertsonian translocation (see Figure 14.10A). Presumably, this supernumerary chromosome was included in a gamete with 23 normal chromosomes and became part of a normal zygote.

Chapter 15

1. Either pattern eventually causes an unequal apportionment of certain substances among the various cells in a developing embryo. And since maternal gene products govern the expression of zygotic genes involved in axis formation in the developing embryo, different embryonic cells will express different arrays of zygotic genes and thus give rise to different cell fates.

2. It appears that there is a fair amount of functional redundancy built into the process of limb development, whereby the loss of an individual Hox locus can be compensated for by the remaining Hox loci.

3. Ellis et al. (1991) suggest several reasons why cell death can be advantageous. (1) Certain cells or structures seem to have no function, probably because they are evolutionary vestiges. In some cases they are totally lost, but in other cases they can be re-sculpted by apoptosis to a more useful form. (2) Some cells are not needed after they complete their transient functions. Examples in mammals are the degeneration of Müllerian ducts in male embryos and Wolffian ducts in female embryos, or (in female adults) the death and menstrual loss of certain cells lining the uterus. (3) Some cells are normally produced in excess. In many organisms, for example, many more embryonic nerve cells are generated than needed. (4) Some cells develop abnormally, and their death benefits the organism. (5) Some

cells are harmful to the organism and should be eliminated. An example (discussed in Chapter 18) is the destruction of certain white blood cells that might attack an organism's own tissues.

4. What criteria should be used in defining sex? Although nearly all males are indeed 46,XY and females 46,XX, exceptions do exist—and it cannot be assumed that chromosomal sex always leads to a given pattern of adult sexual development and behavior. Testicular feminization accounts for most cases of 46,XY females who are barred from competition, but confers no athletic advantage over 46,XX women. And some male pseudohermaphrodites brought up as females (with appropriate surgery and hormone treatment from birth) would also be unfairly excluded. For discussions of sex testing for Olympic athletes, see Ljungqvist and Simpson (1992), Ferguson-Smith et al. (1992), de la Chapelle (1986), and Turnbull (1988).

5. Absence or inactivity of the Müllerian-inhibiting substance or its receptor.

6. All human embryos tend to develop as females unless deflected from this path by the action of testosterone at several critical steps. "If any one of your sexual systems fail, you can coast down the female road, but not down the male" (Money and Ehrhardt 1972).

7. The simplest possibility is X-linked recessive inheritance, but it could also be an autosomal dominant gene expressed only in males.

Chapter 16

1. The intermediate may be metabolized via a different pathway.

2. This allele is more common among the Irish and their descendants, many of whom live in Massachusetts. In contrast to this, so few PKU cases were found in Washington, D.C. (most of whose citizens are black) that the screening program there was discontinued.

3. The error might occur during the process of polymerization of dopaquinone, blocking the formation of black and brown pigments (left fork at bottom of Figure 16.6).

4. Random inactivation of one X chromosome in each cell during early development would lead to a variety of phenotypes—some with a preponderance of cells in which the mutant X-linked gene is expressed and others with a majority of cells expressing the normal allele.

5. It makes little sense. Milk can induce serious illness among the children and great discomfort in the adults; such experiences may lead recipients of this aid to reject other, more useful, food products.

6. The "cross-correction" studies showed that the two mutants are nonallelic, because each excreted a normal substance that was deficient in the other (in this case, two different acid hydrolase enzymes). If they had been allelic, neither could have provided the normal substance missing in the other. Complementation also

occurs in cell hybrids formed from two phenotypically abnormal parental strains. If the hybrid cells show a normal phenotype, the two mutants must involve different genes.

7. The Lyon effect may not be observed in all tissues. Nobody is sure why, but there are two main hypotheses: (1) Random inactivation occurs in the precursors of the blood cells, but those expressing the mutant allele are at a disadvantage and become overgrown by their normal counterparts; (2) X chromosome inactivation is nonrandom (favoring the chromosome with the normal allele) in the blood cell precursors but random in other tissues.

8. Metabolites in minimal medium $\xrightarrow{8}$ ant $\xrightarrow{2}$ IGP $\xrightarrow{3}$ indole $\xrightarrow{6}$ trp

9. **Advantages:** Identify those people who are unable to metabolize standard dosages of the drug before they suffer a possible drug reaction; we might also save some of the large amounts of money that are currently spent on hospitalization and/or treatment of people who suffer drug reactions.

   Disadvantages: Perhaps stigmatize individuals, in the eyes of insurers and employers, by making it appear that they have a special condition that renders them "abnormal" in some way.

10. People with ultrarapid alleles, because they may inactivate some drugs more quickly than do people with the "standard" alleles, might not get the full effect of the drug therapy. In order to get the expected beneficial drug response, they may need higher-than-standard dosages or more frequent administration of the medication.

11. Kalow and Grant (1995) note that people with pharmacogenetic mutations might metabolize carcinogenic substances in ways that alter processes of tumor initiation or progression. Second, their altered metabolisms could influence the effectiveness of chemotherapy for existing tumors. These authors also mention, however, that there are conflicting data on the possible association between metabolizer phenotypes and cancer.

Chapter 17

1. About 30. Each generation doubles the number of cells, so we need to know the power of 2 that gives 10^9. Use the to-the-power key on a hand-held calculator, or the approximation, $2^{10} \approx 10^3$.

2. That all 100 billion members of the tumor have the same active X chromosome suggests, first of all, that the tumor started from a single cell. The X chromosome that was active in this original cell remained the active one throughout all cell divisions, as postulated by the Lyon hypothesis.

3. In family 1, the first person in generation II seems to have transmitted the *p53* gene that he received from his mother to two children, although he was cancer-free.

4. When inherited in the germ line, *p53* mutations produce a variety of cancers in many members of the same family, as shown in the pedigrees of the Li-Fraumeni syndrome in Figure 17.3. In other people whose body cells become homozygous for *p53* mutations, these same cancers of the breast, brain, bone, and so on, can occur. But these cases occur sporadically, that is, in unrelated individuals.

5. Pedigrees of the disease show the standard characteristics of dominant inheritance; that is, about 50% of children of an affected parent get the disease. But in persons carrying the dominant gene, cells become cancerous only if they become homozygous for the genetic defect.

6. Because breast cancer is relatively frequent, several noninherited cases might occur in the same family, thus mixing up "mappable" breast cancer with "unmappable" breast cancer. The issue was resolved by concentrating on early-onset, often bilateral, breast cancers, which are more often heritable cases.

7. *BRCA1* mutations can lead to both breast and ovarian cancer, whereas *BRCA2* mutations do not produce ovarian cancer.

8. In addition to the germ line mutation of *BRCA1*, a second allelic mutation in breast or ovarian tissue (and perhaps still other mutations) is needed to produce cancer.

9. The cultured mouse cells may have already undergone most of the changes on the pathway to malignancy, so only the one final step remained. In fact, the mouse cell line commonly used for such experiments is known to be unusual in several respects.

10. The oncogenic form of the receptor might behave as if it were *always* joined to the growth factor. The cell is thus fooled by its malfunctioning receptor, being stimulated to continually divide whether growth factor is present or not.

11. In CML, the proto-oncogene fuses with a gene at the other breakpoint to produce a "hybrid" protein product. In BL, the proto-oncogene remains intact, but its new neighbors cause it to be expressed abnormally.

12. Environmental factors like smoking and a fatty diet act by inducing mutations in somatic cells.

Chapter 18

1. During proliferation of a specific, stimulated B cell, some cells of the clone do not fully mature to produce antibody, but remain for a long time as memory cells for further quick proliferation.

2. The large amount of a single antibody species in each patient provided a nearly pure substance for investigation. Antibodies from different patients led to the analysis of different variable domains.

3. Somewhere along the heavy chains between the bonds holding the heavy chains together and the bonds connecting the heavy and light chains.

4. The columnist had a point. (Note that blood grouping can only exclude paternity; any man of type B or AB may have been the father.)

5. Anti-A and anti-B may be antibodies like any others that are produced in response to foreign antigens. For a person of type A, substance B, if taken up somehow by the circulatory system, would stimulate the production of anti-B.

6. I-1: *O/O D/d*   II-1: *O/O d/d*   III-1: *A/O D/d*
   I-2: *B/O D/d*   II-2: *A/– D/d*   III-2: *A/O d/d*

7. A, Rh pos: 3/8; A, Rh neg: 1/8
   B, Rh pos: 3/16; B, Rh neg: 1/16
   AB, Rh pos: 3/16; AB, Rh neg: 1/16

8. Maternal-fetal incompatibility for an antigen other than Rh. This phenomenon is known to occur sometimes with ABO.

9. Rh negative *males* could volunteer to be immunized with Rh positive boood. Rh negative *females* certainly should not volunteer, because that would sensitize them to subsequent Rh positive pregnancies.

10. $150 \times 12 \times 4 = 7{,}200$.

11.

Subject	Haplotypes
Father	5 2 / 40 9
Mother	5 3 / 8 11
First child	5 2 / 5 3
Second child	5 2 / 8 11
Third child	40 9 / 8 11

12. The first child would be better, possessing only one antigen (A3) foreign to the recipient.

13. No. A gene is said to be polymorphic if it has several *common* alleles. For the disease-causing genes, the normal allele is the only common one; all the others are rare. Polymorphic genes such as those for blood groups or tissue antigens almost always code for neutral attributes.

14. T cells, the agents of immunological rejection, do not reach the site of the graft.

15. He had no effective immune system to attack any transplant.

16. If male, $P = 1/2$. If female, $P = 0$.

17. The sooner a vital organ can be removed from a dead person, the more likely its success as a transplant. Therefore, it is important to define the moment of death.

Chapter 19

3. Prenatal diagnosis may be undertaken in order to prepare parents for a possibly adverse outcome, without any consideration given to abortion.

5. If the embryo were heterozygous, *C/c*, and the PCR process amplified the single *c* allele but not the *C* allele, the embryo would mistakenly appear to have cystic fibrosis.

6. Mosaicism is possible; that is, mutations or chromosomal aberrations could occur in the embryo itself that do not occur in cell lines giving rise to extraembryonic membranes, or vice versa.

7. The detection of neural tube defects is based on the amniotic *fluid*, which is not obtained by CVS.

9. They might decide to curtail additional pregnancies, might choose alternative reproductive options, or they might ask that future pregnancies be monitored.

10. Carrier × carrier marriages that would be detected = (0.70)(0.70) = 0.49 = 49%.

    Missed marriages = 100 − 49 = 51%.

11. 5% = 0.05, the probability that the untested parent is a carrier.

12. The probability is the product of 0.05 (the frequency of carriers in the population) and 0.3 (the frequency of nondetection). This equals 0.015 = 1.5%.

Chapter 20

1. Definitely yes. In Ashkenazi Jews, the most common mutation produces an amino acid change, and homozygotes have mild, if any, symptoms. The next most common mutation is a frameshift near the promoter, and homozygotes die as embryos. Many persons are compound heterozygotes with varying symptoms.

2. The ewe that supplied the nucleus; the ewe that supplied the egg cytoplasm; the ewe that was pregnant with Dolly.

3. (a) Some alleles in the mother that are revealed by the fingerprinting probes were not transmitted to her offspring, and some in the offspring were inherited not from the mother, but from the father.

   (b) Such a band represents an allele of a VNTR locus that is present (at least heterozygously) in all members of Dolly's flock (or at least in all that were tested). Perhaps darker bands represent homozygotes.

4. Dolly's nuclear chromosomes came from a somatic cell of a 6-year-old ewe and may have accumulated somatic mutations and lost telomeric regions (among other possible changes). Also, some of Dolly's cytoplasmic elements may be from the 6-year-old donor cell.

Glossary

We briefly define here many of the terms, concepts, and processes associated with the study of genetics. For more detail, or for information about specific disorders, genes, or gene products, consult the index and text of this book or a dictionary of genetics (King and Stansfield 1996; Rieger, Michaelis, and Green 1991).

acentric A chromosome or chromosome fragment that lacks a centromere.

acrocentric A chromosome with a centromere near one end and thus having one very short arm and one long arm.

age of onset The age at which phenotypic consequences first appear in an individual with a particular genotype.

agglutination The clumping together of cells (or viruses) in the presence of a specific immune serum.

allele An alternative form of a gene at a particular locus.

allele frequency The proportion of a particular allele in a particular population.

alleles identical by descent Two alleles that have originated from one particular allele in a common ancestor.

alpha-fetoprotein A substance of unknown function that is present in fetal blood plasma. Its concentration in amniotic fluid is increased when the fetus has a neural tube defect, but decreased when a fetus has Down syndrome.

Alu sequence A short (about 300 base pairs), repeated DNA sequence that is dispersed among structural genes and makes up about 5% of human DNA. An example of repetitive DNA.

Ames test A rapid, simple test (developed by B. Ames) that uses bacteria to screen chemical compounds for mutagenicity.

amino acid A small organic molecule with a carboxyl group and an amino group, which links to other amino acids to form polypeptide chains.

amniocentesis A method of obtaining fetal cells for prenatal diagnosis (to detect chromosomal abnormalities and some biochemical disorders), usually done about the sixteenth week of pregnancy and involving the removal of amniotic fluid from the sac surrounding the fetus.

anaphase A stage of mitosis or meiosis, during which chromatids of chromosomes move to opposite poles of the spindle.

aneuploid Having too many or too few chromosomes or chromosome segments, compared to the normal genotype.

antibody A protein (immunoglobulin) produced by plasma cells in response to a particular foreign substance (antigen) and capable of binding specifically to that antigen.

anticipation The decreased age of onset and/or increased severity of an inherited disorder as it passes from one generation to the next.

anticodon In a transfer RNA molecule, a nucleotide triplet whose base sequence is complementary to that of a particular messenger RNA codon, thereby allowing recognition and binding to the appropriate codon during the process of translation (protein synthesis).

antigen A foreign substance that stimulates the production of a particular antibody.

antisense therapy Blocking the action of a gene by using an oligonucleotide complementary to its RNA product.

apoptosis See programmed cell death.

artificial selection The development of special breeds of organisms through the selective mating of individuals of particular phenotypes.

ascertainment bias A deviation in the observed proportion (compared to the expected proportion) of affected offspring, occurring when sampled families are not randomly chosen—for example, when two heterozygous parents with *no* affected children are excluded from a study.

autoimmune disorder A disease that results when affected individuals make antibodies against their own cells or tissues.

autoradiograph An image produced on a photographic film that was placed in close contact with an electrophoresis gel or tissue section, and which shows the positions of radioactive molecules in the gel or tissue.

autosome Any nuclear chromosome that is not a sex chromosome.

B cell A lymphocyte (white blood cell) that matures in the bone marrow and responds strongly to bacterial antigens by differentiating into antibody-producing plasma cells.

bacterial artificial chromosome (BAC) A stable cloning vector that can contain an insert of about 150 kb.

Barr body The sex chromatin; a dark-staining body (discovered by M. Barr) representing a condensed, inactivated X chromosome found in the nuclei of somatic cells in female mammals.

base pair A pair of hydrogen-bonded DNA bases (one purine and one pyrimidine) located between the two backbones of a DNA double helix.

base substitution A type of mutation in which a single DNA base is replaced by a different base.

beta-amyloid A protein present in amyloid plaques in the brains of people with Alzheimer disease; a degradation product of the larger amyloid precursor protein.

bioinformatics The storage, indexing, and retrieval of nucleotide or amino acid sequences of an organism, including methods to compare the sequences of different species.

biometry   The application of statistical methods to biological traits.

bivalent   A pair of homologous chromosomes synapsed during prophase of meiosis I.

blastocyst   An early stage of animal development; a hollow sphere of cells containing the inner cell mass, which will become the embryo.

blood group   A blood type defined by genetically determined antigens present on the surface of red blood cells; alleles of a gene coding for such a set of antigens.

bottleneck effect   A change in gene frequency that results when a population is greatly reduced in size but then expands again with an altered gene pool.

cancer   A pathological condition in which cells grow inappropriately without restraint, losing their normal form and function, invading nearby tissue, and spreading to remote sites.

candidate gene   A gene in a chromosome region where a disease phenotype has been mapped. The gene is known to code for a protein that is plausibly associated with the disease phenotype.

carcinogen   A substance that causes cancer.

carcinoma   A solid cancer derived from epithelial tissues (skin, linings of organs, glands, breasts, and nerves).

carrier   An individual who is heterozygous for a recessive mutant allele.

cDNA library   Complementary (or copy) DNA: A collection of DNA molecules made by cloning segments of the DNA molecules obtained by reverse transcribing the mRNA in a cell. Such cloned DNA has no introns or spacer DNA.

cell adhesion molecule (CAM)   A cell surface protein in animals that binds cells to other cells or to substrates such as the extracellular matrix.

cell cycle   The sequence of events from one eukaryotic cell division to the next, including all stages of interphase (G_1, S, G_2) and mitosis (prophase, prometaphase, metaphase, anaphase, telophase).

centriole   A tiny, self-reproducing structure located just outside of the nuclear membrane in animal cells; it replicates during interphase, and daughter centrioles migrate to opposite spindle poles during nuclear division.

centromere   The indented region of a metaphase chromosome that divides it into two arms; also, the region of spindle fiber attachment during mitosis or meiosis.

chain termination (stop) codon   A codon (UAA, UAG, or UGA) that does not specify any amino acid; instead, it stops the translation process and allows the translated polypeptide to be released from the ribosome.

chain termination mutation   A mutation which changes a triplet that encodes an amino acid into a stop triplet.

checkpoint   A point in the cell cycle at which cell surveillance systems check to see if conditions are suitable for cell division to continue on to the following stage.

chiasma   (plural, chiasmata) The physical manifestation of crossing over; a cross-shaped connection between synapsed nonsister chromatids that is seen during prophase I of meiosis.

chorionic villus sampling   A method of obtaining fetal cells for prenatal diagnosis (to detect chromosomal abnormalities and some biochemical disorders), usually done about the tenth week of pregnancy, in which bits of chorionic tissue are removed from the developing placenta.

chromatid   One of the two daughter replicas (strands) of a duplicated chromosome, which separate during anaphase of mitosis or meiosis II.

chromatin   The complexed substances (DNA, histones, and nonhistone proteins) that make up eukaryotic chromosomes; the structural unit of chromatin is the nucleosome.

chromosome   A self-reproducing structure in eukaryotic nuclei that consists of DNA and proteins and contains a set of linked genes. Its appearance varies during different stages of the cell cycle.

chromosome arms   The two segments of a chromosome, called p (short) and q (long), that are defined by the position of the centromere.

chromosome banding   Staining techniques (such as G-banding, Q-banding, R-banding, or C-banding) that give rise to a unique pattern of lateral bands along the length of each chromosome and thus provide a means of identification and analysis.

chromosome loss   The disappearance of a chromosome during mitosis or meiosis, often because of lagging on the spindle during anaphase.

chromosome painting   The use of molecular probes, fluorescent dyes, special microscopes and computer technology to label and visualize chromosomes, resulting in preparations with brilliantly colored whole chromosomes or chromosome parts. See also fluorescent in situ hybridization (FISH).

cleavage division   Mitotic cell division of the zygote and early embryonic cells without any intervening periods of cell growth.

clonal selection theory   A model for explaining how the immune system can respond so specifically to vast numbers of antigens. It proposes that a lymphocyte is preprogrammed to recognize and respond to just one antigen, whether or not that antigen ever appears.

clone   A group of genetically identical cells descended by repeated mitoses from one ancestral cell. In molecular biology a clone means multiple replicas of a particular piece of DNA produced by recombinant DNA methods.

cloning   Making copies of a gene by using the polymerase chain reaction or inserting it into a bacterium. Making copies of a whole organism by transferring a nucleus of the animal to be cloned to an enucleated egg.

codominant alleles   Alleles (for example, those of blood group genes) that are both expressed in a heterozygote.

codon   A group of three nucleotides in messenger RNA that specifies one particular amino acid.

coefficient of inbreeding   The probability that two alleles in one person are identical by descent from precisely the same allele in an ancestor of both parents.

collagen   A fibrous protein found in connective tissues (bone, cartilage, skin, and tendons) and the extracellular matrix.

combining site   The pocket, at the upper tip of a Y-shaped antibody molecule, whose shape is complementary to the shape of an antigen.

complementary bases   Nucleotide bases that can pair, purine to pyrimidine (i.e., adenine with thymine or uracil, and guanine with cytosine), by hydrogen bonding.

complementary DNA library   See cDNA library.

compound heterozygote   A heterozygote having two different mutant alleles at the same locus, and whose phenotype mimics a homozygote.

concordant twins A pair of twins who both exhibit a particular trait.

consanguineous mating Mating between genetically related individuals.

constant (C) domains The portions of the heavy and light chains of antibodies that are the same in molecules having different antigenic specificities. See also variable (V) domain.

contig A set of cloned DNA segments that span a region of interest. The DNA segments can be individually sequenced and arranged in proper order by their overlapping ends.

contiguous gene syndrome A syndrome that involves a number of phenotypically unrelated features, resulting from the deletion or duplication of several closely linked genes in a chromosomal region.

cosmid A cloning vector derived from the bacteriophage lambda.

coupling phase The situation in a double heterozygote when two linked but nonallelic recessive mutants (*ab*) are present on one chromosome and their two dominant alleles (*AB*) are present on the homologous chromosome. See also repulsion phase.

CpG island A region of DNA rich in the dinucleotide CG. Such regions are often present near the promoters of genes.

crossing over The exchange of chromosome parts between the chromatids of synapsed homologues during prophase I of meiosis.

cyclins Proteins that regulate the cell cycle by activating protein kinases.

cytokinesis Division of the cytoplasm (as opposed to nuclear division), which occurs during telophase of mitosis or meiosis.

cytoplasm The protoplasm that surrounds the nucleus of a eukaryotic cell.

Darwinian fitness A measure of the average number of offspring left by an individual or by a class of individuals.

degeneracy of code The occurrence of several codons that specify the same amino acid.

deletion Loss of a chromosome (or DNA) segment of any length.

deletion mapping Gene mapping based on the principle that the absence of a chromosome segment is correlated with reduced expression, nonexpression, or abnormal expression of genes on that segment.

denaturation Separation of the two strands of DNA in a double helix by breaking of the hydrogen bonds between base pairs.

deoxyribose The sugar found in the sugar-phosphate backbones of DNA.

dicentric A chromosome or chromatid having two centromeres rather than one.

digenic inheritance A mode of inheritance involving two loci, in which only the double heterozygote is affected.

diploid Having two representatives of every chromosome, as occurs in most somatic cells of higher organisms. The diploid (2*n*) number in humans is 46.

discordant twins Twin pairs in which only one member exhibits a given trait.

disomic Having two representatives of a particular chromosome.

dizygotic (DZ) twins Nonidentical (fraternal) twins, which arise from two different zygotes (i.e., two separate eggs fertilized by two separate sperm).

DNA Deoxyribonucleic acid, the genetic material; a double helix consisting of two deoxyribose-phosphate backbones joined together by hydrogen-bonded purine-pyrimidine base pairs.

DNA fingerprinting A method for identifying individuals by using the unique electrophoretic banding patterns generated by DNA probes for highly polymorphic repeated sequences.

DNA helicase An enzyme that helps open up the double helix to expose single strands for replication.

DNA ligase An enzyme that seals together fragments of newly replicated DNA or repairs nicks in one strand of a DNA double helix.

DNA marker A variable site on a DNA molecule, which can be used as a "signpost" for mapping or detecting closely linked genes. DNA markers include RFLPs (restriction fragment length polymorphisms) and VNTRs (variable number of tandem repeats).

DNA polymerase An enzyme that forms new DNA by linking together a string of deoxyribonucleotides, using single-stranded DNA as a template.

DNA repair syndrome A genetic disease in which the cause is a faulty enzyme normally used in the repair or replication of DNA.

dominant allele The allele or phenotype that is expressed in a heterozygote.

dopamine A neurotransmitter in the brain.

dopamine receptor A protein receptor molecule that binds dopamine, thus continuing a nerve impulse across a synapse.

dosage compensation The equalization of expression of X-linked genes in the two sexes, despite a 2:1 difference in their normal dosage.

double heterozygote An organism or individual that is heterozygous at two loci.

double-blind study A study in which neither the researchers nor the subjects know whether a given individual belongs to the control group or to the experimental group until after all the data are recorded.

doubling dose The amount of ionizing radiation needed to double the rate at which mutations occur spontaneously in a given species.

drift See genetic drift.

duplication An extra segment of chromosome or DNA, resulting in excess dosage of the genes present on that segment.

dynamic mutations Mutations involving trinucleotide (triplet) repeats that can increase or decrease in size from one generation to the next. They may cause considerable variability in phenotype and age of onset of the associated disorder between generations. See also triplet repeat.

ecogenetics The study of genetic predisposition to the toxic effects of chemicals.

EcoRI A restriction enzyme that recognizes a sequence of six bases (GAATTC/CTTAAG) and cuts double-stranded DNA whenever this sequence occurs.

egg The female gamete, a haploid product of meiotic cell division.

electrophoresis A method for separating a mixture of molecules (in filter paper or in a gel placed in an electric field) according to their electric charge, size, and shape.

ELSI Ethical, Legal, and Social Implications discussion group; a component of the Human Genome Project.

embedded gene A gene found inside (often within an intron) of another gene.

empirical risk A risk figure based on the statistics of prior experience, rather than on Mendel's laws, genetic linkage, or other model.

enzyme A protein molecule that, in small amounts, speeds up the rate of a specific biochemical reaction without itself being used up in the reaction.

epigenetics The study of mechanisms that can influence a phenotype without changing the genotype; gene silencing through heritable but potentially reversible changes in chromatin structure and/or DNA methylation.

epistasis The interaction of nonallelic loci, whereby one gene alters or masks the expression of a different gene.

equatorial plate The midregion of the spindle, where chromosomes congregate during metaphase of mitosis or meiosis.

euchromatin The chromatin that is relatively uncondensed during interphase, and which stains differently from the more tightly condensed heterochromatin; it contains nearly all of the known genes.

eugenics Attempted improvement of the genetic endowment of human populations by encouraging the matings of people with supposedly beneficial genes (positive eugenics) and discouraging the matings of people with supposedly harmful genes (negative eugenics).

eukaryote A cell (or organism made of such cells) that contains a membrane-bounded nucleus with chromosomes, and which undergoes mitosis and/or meiosis.

euploid Having a chromosome number that is an exact multiple of the haploid number characteristic of that species.

evolution Changes in gene frequency in a population, over time, that may produce better-adapted organisms and new species.

ex vivo technique A method of gene therapy in which body cells (often white blood cells) are removed, treated to insert an altered gene, and returned to a patient. See also in vivo technique.

excision repair A cut-and-patch process for repairing damaged DNA molecules; enzymatic removal of an abnormal piece of DNA, and resynthesis of that piece by using its intact complement as a template.

exon An expressed coding sequence; the part of a split gene whose complementary mRNA sequences remain after the primary transcript is processed to remove the introns.

exon trapping A method to see if a piece of DNA contains a gene as opposed to mere "junk" DNA. If the DNA contains an exon, as a real gene would, it lengthens an mRNA construct. Otherwise it does not.

expressed sequence tags (ESTs) Short (100–200 bases) sequenced stretches within genes. ESTs have been used as markers for the genes in cDNA obtained from a particular cell type.

F_1 generation The first generation of offspring from a mating between two parental individuals or inbred lines.

false negative An error of testing in which an affected individual is said to be unaffected.

false positive An error of testing in which an unaffected individual is said to be affected.

FISH See fluorescent in situ hybridization.

fitness See Darwinian fitness.

flanking markers DNA markers (RFLPs, VNTRs, etc.) that are mapped on opposite sides of a gene of interest.

fluorescent in situ hybridization (FISH) A cytological technique for locating specific gene loci in a metaphase chromosome spread by using fluorescent DNA probes that hybridize to and "light up" the appropriate chromosomal region. See also chromosome painting.

founder effect A change in allele frequencies that occurs when the founders of a new population are not genetically representative of the original population from which they came.

fragile site A heritable constriction or gap that occurs at a specific chromosomal site, and which may cause chromosome breakage.

frameshift mutation A reading frame error caused by the deletion or addition of one or more nucleotide bases (but not multiples of three).

full mutation A trinucleotide repeat region within a gene that is sufficient to cause an abnormal phenotype. It is derived from an unstable premutation in a prior generation. See also dynamic mutation.

functional cloning Identifying and cloning a gene starting from knowledge of the protein product of the gene. See also positional cloning.

G_0 phase A period of withdrawal from the eukaryotic cell cycle, often entered by differentiated cells.

G_1 phase In the eukaryotic cell cycle, the initial interphase time gap between cell "birth" (at the end of mitosis) and the beginning of DNA synthesis. Its duration is the most variable of all the cell cycle stages.

G_2 phase In the eukaryotic cell cycle, the interphase time gap between the end of DNA synthesis and the beginning of mitosis (or meiosis).

gain-of-function mutation A mutation that changes a gene to an allele (usually a dominant allele) that actively causes something to go wrong, rather than not doing something right.

gamete Egg or sperm.

gene A hereditary unit or segment of DNA that occupies a specific site on a chromosome and contains genetic information that is replicated, transcribed into messenger, transfer, or ribosomal RNA, and (if a structural gene) translated to form a polypeptide.

gene expression Transcription and translation of a gene, resulting in the synthesis of a protein.

gene family A group of genes coding for similar polypeptides and sharing a common evolutionary history.

gene flow The introduction of new alleles into a population, owing to the migration of individuals.

gene mapping Determining the position of genes along a chromosome by either genetic crossing over between them or by physical means.

gene pool All the genes present in a population of organisms.

gene splicing Attaching a piece of DNA from one species (say, human) to another (bacterium) for the purpose of cloning the human DNA.

gene therapy Treatment of a genetic disorder by inserting normal alleles into abnormal cells.

genetic code   The 64 triplet nucleotide bases in DNA and mRNA that specify the 20 amino acids found in polypeptides, as well as the chain termination signals.

genetic counseling   A process of clinical diagnosis, risk assessment, explanation of options, and provision of social services for families who are affected or potentially affected with genetic disorders.

genetic drift   Random and haphazard changes in allele frequencies, occurring especially in small populations.

genetic equilibrium   The maintenance of constant allele frequencies in a population over many generations.

genetic heterogeneity   Multiple genetic causes of the same, or nearly the same, phenotype.

genetic linkage map   A chromosome map based on the crossover distances between successive genes. See also physical map.

genetic screening   The systematic search of the general population for persons having a particular genotype (or karyotype) that might cause later trouble for them or for their children.

genome   The sum total of all the genes (or DNA sequences) within a single gamete or within a cell's mitochondria.

genomic library   A collection of DNA molecules made by cloning segments of the total DNA (including introns and spacer DNA) in a cell.

genotype   The precise allelic makeup of an organism or cell.

genotype frequency   The proportion of a particular genotype among all genotypes in a population.

germ cell   A gamete (egg or sperm).

germ line   Specialized cells (as opposed to somatic cells) that give rise to gametes.

germinal gene therapy   Altering a gene of future generations by changing the genes in a gamete. This form of gene therapy has unsavory overtones. See also somatic gene therapy.

germinal mosaicism   A mixture of cells with different chromosome numbers or structural features or abnormalities among gonadal cells destined to give rise to gametes; it can lead to aneuploidy among the offspring of phenotypically normal parents.

germinal mutation   A mutation in a cell that produces gametes (as opposed to a somatic cell mutation).

glycoprotein   A protein combined with a carbohydrate.

growth factor   A protein that binds to a cell surface receptor, stimulating cell growth at appropriate times.

Guthrie test   A test (devised by R. Guthrie) that screens newborn babies for phenylketonuria, using bacteria that grow only when excess phenylalanine is present in a dried sample of a baby's blood.

H substance   The short sequence of sugars common to all the antigens determining the ABO blood types.

haploid   The chromosome number (n) of a normal eukaryotic gamete, including one representative of each chromosome type.

haploid autosomal length (HAL)   The total amount of chromosomal material present in a haploid set of autosomes.

haplotype   A unique array of closely linked alleles, such as the HLA complex, that is usually inherited as a unit.

Hardy-Weinberg Law   A formula that predicts constant gene and genotype frequencies within a large, randomly mating population where there is no selection, mutation, or migration.

heavy (H) chains   The two identical chains that form the stem of antibody molecules and a portion of the combining sites that bind an antigen. See also light (L) chains.

helper T cell   A white blood cell that responds to a specific antigen and helps in the maturation of both B cells and killer T cells.

hemizygous   Present in a single dose—either in a haploid organism or gamete, or in a male who has just one X chromosome.

heritability, broad-sense   The proportion of phenotypic variation that is due to the genotypic differences among members of a population.

heritability, narrow-sense   The proportion of variation in a population that is due to just the additive portion of the genetic differences among the members.

hermaphrodite   An intersex; an individual who has both male and female sex organs.

heterochromatin   Highly condensed, darkly staining, late replicating regions of chromatin that contain repetitive DNA but few known genes.

heterozygous   Having two different alleles at the same gene locus (or at more than one locus) on homologous chromosomes.

high-resolution banding   Special techniques for staining prophase and prometaphase chromosomes, which can produce up to 2,000 bands.

histocompatibility antigen   A genetically encoded tissue or cell surface antigen, which can influence whether a transplanted organ will be accepted or rejected.

histones   Small DNA-binding proteins found in nucleosomes of chromatin, where they regulate chromosome compaction and gene activity.

HLA system   The major complex of cell surface histocompatibility molecules, the *h*uman *l*eukocyte *a*ntigens, and the cluster of highly polymorphic genes that encode them.

homeobox   A sequence of about 180 nucleotide base pairs found within many developmental genes; it encodes a DNA-binding protein (homeodomain) that acts as a gene regulator.

homeotic genes   Developmental genes that regulate pattern formation within body segments. Mutants may have extra, misplaced body parts.

homologues   Chromosomes that carry the same genes, pair during meiosis, and look alike in metaphase nuclei.

homozygous   Having the same alleles at a particular gene locus.

Human Genome Project   An international organization of research teams set up to map and sequence all the DNA in the human genome.

hybridoma   A fused cell (or its clone), derived from an antibody-secreting B cell and a cancer cell, that produces just one kind of antibody and can be indefinitely maintained in tissue culture.

hydrogen bond   A chemical attraction, weaker than a covalent bond, between polar organic molecules that share a hydrogen atom; found in the bases linking two strands of a DNA double helix, as well as the amino acids in a polypeptide chain.

hypervariable region   Within the variable domains of an antibody molecule, amino acid sequences that form the lining of the antigen-binding cavity.

immunoglobulin   An antibody molecule typically consisting of four polypeptide chains and forming pockets for binding an antigen.

imprinting   The differential expression of genes, depending on whether they are inherited from the male or female parent. An imprinted allele is not expressed.

in vivo technique   A method of gene therapy in which an altered gene is introduced into a patient by inserting it in a vector and injecting or spraying the vector into the body. See also ex vivo technique.

inactive X hypothesis   The Lyon hypothesis of dosage compensation. In female mammalian cells, one or the other X chromosome is randomly inactivated during early development, so that females heterozygous for X-linked genes show mosaic expression for these loci, and gene expression is equalized between the sexes.

inborn error of metabolism   A metabolic disorder caused by the mutation of an enzyme-producing gene and the consequent block of a specific step in a biochemical pathway.

inbreeding   Mating between closely related individuals.

incestuous mating   A mating between close relatives.

incomplete dominance   The situation in which a heterozygote has a phenotype intermediate between those of the two homozygotes.

incomplete penetrance   Nonexpression of a gene that usually gives rise to a characteristic phenotype.

independent assortment   The random and independent distribution of each pair of chromosomes (or each pair of unlinked alleles) during gamete formation, as stated by Mendel's second law.

induced mutation   A mutation whose cause is known to be exposure to a mutagen like ionizing radiation or certain chemicals.

insertion mutation   A mutant allele brought about by the insertion of DNA sequences from elsewhere in the genome.

insertional translocation   A three-break chromosomal abnormality in which a deleted chromosomal fragment moves from its normal location to a new site on the same or a different (nonhomologous) chromosome.

interphase   A period of growth and metabolism, comprising the G_1, S, and G_2 phases of the cell cycle, that occurs between cell divisions.

intron   A noncoding intervening sequence within a gene; a nucleotide sequence that does not code for amino acids, and whose complementary mRNA sequence is excised during mRNA processing.

inversion   A chromosome abnormality formed when two breaks occur in one chromosome, and the intervening segment is inverted before the broken ends rejoin.

ionizing radiation   Radiation of sufficient energy (e.g., X-rays) to knock an electron out of an atom.

isochromosome   A chromosome abnormality that occurs when a centromere appears to split crosswise rather than lengthwise during mitosis or meiosis; the result is a metacentric chromosome with two identical arms.

karyotype   A photomicrograph of stained metaphase chromosomes, which may be arranged by size and chromosome number, representing an individual or species.

killer T cell   A white blood cell that responds to a specific antigen on foreign cells, bringing about their destruction by direct cell-to-cell contact.

kinetochore   A proteinaceous structure, formed during late prophase at the centromere region of a mammalian chromatid, to which spindle fibers attach.

knockout mouse   A mouse in which a gene is homozygously inactivated. Such a mouse is obtained after replacing the normal gene with an abnormal one by genetic recombination.

lethal mutation   A mutation that causes premature death by disrupting processes essential to life; it can be dominant or recessive and take effect at any time, from early embryonic stages through adulthood.

leukemia   A type of cancer involving overproduction of white blood cells manufactured in the bone marrow.

light (L) chains   The two identical chains that form a portion of the arms of an antibody molecule, including the combining sites that bind an antigen. See also heavy (H) chains.

LINE element   A long interspersed repetitive DNA element present in human DNA. See also SINE element.

linkage group   A group of genes found on the same chromosome.

linked genes   Two or more genes located on the same chromosome and showing substantially less than 50% recombination.

locus   (plural, loci) The place on a chromosome where a particular gene is found.

looped domain   A segment of chromatin that attaches at either end to the central scaffold and contains a specific group of genes regulated as an independent unit.

loss of imprinting   A process by which an imprinting pattern gets reversed, as is sometimes observed in cancerous cells.

lymph   The fluid in lymphatic vessels and which also bathes the spaces between tissue cells, derived largely from blood plasma and containing various types of white blood cells.

lymphocyte   A white blood cell, either a B cell or a T cell, that responds to foreign antigens in one of several ways.

lymphokine   A glycoprotein, secreted by T cells in response to foreign antigens, that affects various lymphocytes in the immune reaction.

lymphoma   A type of cancer affecting cells of the spleen and lymph nodes.

lyonization   X chromosome inactivation, as described by M. Lyon's inactive X hypothesis.

lysosome   A tiny, membrane-bounded cytoplasmic structure containing enzymes that digest large molecules no longer needed in the cell.

macrophage   An irregularly shaped white blood cell that engulfs and digests foreign substances.

major histocompatibility complex   The HLA complex, a closely linked cluster of genes on human chromosome 6 that code for histocompatibility antigens and control activities of immune cells.

malignant tumor   See cancer.

map unit   A unit of genetic distance on a chromosome, equivalent to 1% recombination between two loci.

marker chromosome   A chromosome with a distinctive cytological feature that is found in several generations of a family.

maternal age effect   The increased (or decreased) incidence of a particular disorder among newborns, based on the mother's age; an example is the greatly increased probability of Down syndrome births from women above age 30.

maternal effect genes   Maternal genes that affect early embryonic development in offspring by producing gradients of gene products in the egg cytoplasm.

maternal-fetal incompatibility   The situation when a fetus possesses a strong red blood cell antigen not present in the mother. It can result in hemolytic disease of the newborn.

mating frequency   The proportion of a particular mating among all matings in a population.

meiosis   A type of cell (nuclear) division that occurs in sexually reproducing organisms, whereby diploid ($2n$) cells give rise to haploid (n) gametes; one chromosome replication is followed by two nuclear divisions, called meiosis I and meiosis II.

meiosis I   The first of two successive cell divisions that will produce haploid gametes from diploid oocytes or spermatocytes. During meiosis I, homologous chromosomes pair, recombine, and then (each still consisting of two chromatids) separate into two daughter cells.

meiosis II   The second of two successive cell divisions that will produce haploid gametes from diploid oocytes or spermatocytes. During meiosis II, the two chromatids that make up each chromosome separate into haploid daughter cells.

melanocyte   A cell that produces the melanin pigment found in skin, hair, and other tissues.

memory cells   Cells proliferating from an antigen-stimulated B cell that do not fully differentiate into antibody-producing cells, but can react quickly to a subsequent exposure to the same antigen.

messenger RNA (mRNA)   Ribonucleic acid containing genetic information that is transcribed from DNA and (after processing to remove introns) translated to produce polypeptides.

metabolic block   A block in one step of a biochemical pathway, which is caused by an absent or defective enzyme and which may result in an abnormal phenotype.

metabolism   The sum of all chemical and physical changes that take place in a cell or organism, allowing it to function and grow.

metacentric chromosome   A chromosome with a centromere at or near its midpoint and two arms of about equal length.

metaphase   The stage of cell (nuclear) division during which the chromosomes are most tightly condensed and lined up on the equatorial plate of the spindle.

metastasis   The spreading of cancer cells from one area to distant sites.

methylation   The addition of methyl (i.e., CH_3) side groups to cytosine bases of DNA in gene control regions.

MHC restriction   The manner in which foreign antigens are presented to T cells, i.e., in combination with a self protein encoded by the major histocompatibility complex.

microchromosome   Artificial human chromosomes, synthesized from centromeric and telomeric DNA plus some genomic DNA, which might be used as vectors for human gene therapy.

microdeletion   A chromosomal deletion so small that it is detectable only by fluorescent in situ hybridization (FISH) or other molecular techniques.

microduplication   A chromosomal duplication so small that it is detectable only by fluorescent in situ hybridization (FISH) or other molecular techniques.

microsatellite   A site along DNA with a variable number of repetitions of a very short sequence (2 to 5 nucleotides). See also minisatellite.

migration   The movement of individuals from one place to another, possibly resulting in altered allele frequencies between populations.

minisatellite   A site along DNA with a variable number of repetitions of a fairly long sequence (dozens to hundreds of nucleotides). See also microsatellite.

missense mutation   A base substitution mutation that causes the wrong amino acid to be encoded.

mitochondrial gene   A gene, present on the DNA of mitochondria, that is transmitted in a strictly maternal pattern from mothers to all of their offspring; mitochondrial genes exhibit a high mutation rate.

mitochondrion   A self-reproducing, maternally inherited cytoplasmic structure in which the process of cell respiration yields energy in the form of ATP molecules. The DNA found in mitochondria evolved from DNA of aerobic bacteria and contains a slightly different genetic code.

mitosis   Cell (nuclear) division that gives rise to two daughter cells with identical chromosomes and genotypes.

modifying gene   A gene that alters the degree of expression of another gene at a different locus.

monoclonal antibody   A pure and uniform antibody type produced by a clone of hybridoma cells.

monohybrid cross   A cross in which the parents differ from each other in just one characteristic or trait.

monosomic   Having just one representative of a particular chromosome, as occurs normally in a gamete or abnormally in an otherwise diploid cell or individual.

monozygotic (MZ) twins   Two genetically identical sibs who are derived from one zygote that split into two during very early development.

mosaic   An individual who has two or more genetically different cell lines.

mRNA processing   The changes occurring between transcription and translation, that is, conversion of an original transcript to mature mRNA through the removal of introns and other modifications.

MSH receptor protein   A receptor on the surface of melanocytes that binds melanin stimulating hormone.

multilocus probe   A radioactive probe (or set of probes) whose sequence(s) can bind to the alleles of several loci in the genome. See also single-locus probe.

multiple alleles   The existence of more than two alternative forms of a gene in a population.

multiple gene inheritance   The type of inheritance in which many genes, each with a small effect, influence the expression of a quantitative trait. See also quantitative trait locus.

multiregional theory   The idea that *Homo sapiens* evolved separately from a previous species in Africa, Europe, Asia, and Indonesia over the last million years or so. Opposed to the out-of-Africa theory.

mutagen   A substance that produces mutations above the rate at which they occur spontaneously.

mutation   A heritable change in a gene, typically a deleterious change.

natural selection   See selection.

nested gene   See embedded gene.

neural tube defect An abnormal phenotype brought about by the failure of the embryonic furrow of neural tissue to close completely. Normally, the neural tube forms the brain and spinal cord.

neurotransmitter A small molecule that is released by one nerve cell and binds to receptor molecules of the next nerve cell in a neural network.

nondisjunction The failure of two homologous chromosomes or sister chromatids to separate (disjoin) during anaphase of meiosis or mitosis, giving rise to daughter cells with either a missing chromosome or an extra chromosome.

nonhistone proteins A heterogeneous mix of chromosomal proteins that are not histones.

nuclear envelope The two-layered membrane that surrounds the nucleus.

nucleolus A dark-staining nuclear structure, formed by a nucleolus organizer region of a chromosome and full of ribosomal RNA.

nucleolus organizer region The region of a chromosomes containing genes that produce ribosomal RNA and thus give rise to a nucleolus.

nucleosome A beadlike complex, consisting of a histone core wrapped by DNA and capped by H1 histone, that forms a repeating structural unit of chromatin.

nucleotide A deoxyribose or ribose sugar attached to a phosphate and to a purine or pyrimidine base, and which forms the basic building block of nucleic acids.

nucleus The membrane-bounded structure in eukaryotic cells that contains chromosomes and is surrounded by cytoplasm.

nullisomic A cell or organism that has no representative of a particular chromosome.

oligonucleotide A sequence of nucleotides, often synthesized in the laboratory.

oligonucleotide primer A short sequence of bases, complementary to a particular DNA sequence, that is used as a primer in the polymerase chain reaction.

oncogene A gene that can convert normal cells into cancer cells by causing uncontrolled cell division.

oogenesis In sexually reproducing organisms, the process by which a diploid primary oocyte undergoes meiosis to form one haploid egg and one or more nonfunctional polar bodies.

oogonium A diploid ovarian cell that divides to form more oogonia or enlarges to form a primary oocyte.

open reading frame (ORF) A coding sequence of DNA nucleotide triplets that does not contain any stop codons.

origin of replication The specific site, within each looped domain of chromatin, at which the unwinding of nucleosomes and replication of DNA starts and then proceeds in both directions during the S (synthesis) period of the cell cycle.

out-of-Africa theory The idea that *Homo sapiens* evolved from a previous species in Africa and from there spread over the earth in the last 100,000 years or so. Opposed to the multiregional theory.

p and q Algebraic symbols that represent the frequencies of alleles. See also Hardy-Weinberg law.

paralogues Genes that have arisen by duplication and evolutionary divergence, such as the highly conserved *Hox* (homeotic) developmental genes in animals.

parental gamete With respect to two heterozygous genes, a nonrecombinant gamete formed by an individual. It contains the same alleles of genes as those in one of the gametes that joined to form the individual. See also recombinant gamete.

parental generation (P) The first generation of an experimental cross, which may consist of two pure-breeding lines.

pattern formation genes Homeotic genes that act during early development by specifying the structures unique to each embryonic region.

pedigree A diagram of two or more generations of a family's lineage, often showing the expression of genetic traits among its members.

phage artificial chromosome (PAC) A cloning vector that can contain inserts of about 100 kb suitable for contigs.

pharmacogenetics The study of genetic variation leading to differences in the ways that individuals metabolize drugs, foods, and other substances.

phenocopy An environmentally induced mimic of a genetic condition in an individual who lacks the usual causative gene.

phenotype The observed attribute(s) of a cell or individual, brought about by the interaction of genotype and environment.

phenotype frequency The proportion of a particular phenotype among all phenotypes in a population.

Philadelphia chromosome A reciprocal translocation between human chromosomes 9 and 22 that is found in bone marrow cells of most people with chronic myelogenous leukemia.

physical map A chromosome map based on physical landmarks as opposed to a genetic map based on crossover distances. The landmarks can be stainable bands, break points, deletions, overlapping contigs, or the sequence of bases. See also genetic linkage map.

plasma The clear, straw-colored fluid in which blood cells are suspended; it contains specialized blood proteins, salts, nutrients, hormones, intermediate metabolites, and many other substances.

plasma cell An antibody-secreting cell derived from a B cell in response to an antigenic stimulation.

plasma membrane The two-layered membrane that surrounds a cell.

plasmid A tiny, circular molecule of nonchromosomal DNA found in bacteria and used as a vector for transferring genes from one cell to another.

pleiotropy The situation in which a single gene has multiple and seemingly unrelated phenotypic effects, owing to a cascade of reactions stemming from the original gene product.

polar body A tiny nonfunctional cell, containing a nucleus but very little cytoplasm. It is formed during the first or second meiotic cell division during oogenesis.

polygenic inheritance Quantitative inheritance; the determination of a particular trait by several genes, each with a small effect.

polymerase chain reaction (PCR) A technique for quickly amplifying a particular DNA sequence, starting with just a tiny amount of double-stranded DNA.

polymorphism The presence in a population of two or more relatively common alleles of a particular gene (or forms of a chromosome)—the more alleles, the greater the polymorphism.

polypeptide   An unbranched string of amino acids linked together by peptide bonds that are formed during the process of gene translation.

polyploid   Having three or more haploid sets of chromosomes.

population   A collection of individuals, often a breeding group important in evolutionary processes.

positional cloning   Identifying and cloning a gene starting from knowledge of the chromosomal map position of the gene. See also functional cloning.

preimplantation diagnosis   A method of determining certain genetic traits of the embryo prior to its implantation in the uterine wall on the basis of one cell extracted from an eight-celled embryo.

premutation   An expansion in length of a trinucleotide repeat region within a gene to just under the number of repeats needed to give an abnormal phenotype. Such an allele is sensitive to further expansion to a full mutation. See also dynamic mutation.

primary oocyte   A diploid ovarian cell, which is formed from the enlargement of an oogonium and which undergoes meiosis I.

primary spermatocyte   A diploid testicular cell, which is formed from the enlargement of a spermatogonium and which undergoes meiosis I.

probability   The frequency or chance or likelihood of some event, expressed as a fraction or decimal between 0 (if the event never occurs) and 1 (if the event always occurs).

proband   The clinically affected individual through whom a pedigree is discovered. See also propositus.

probe   A substance (such as a radioactively labeled single-stranded DNA or RNA nucleotide sequence or a monoclonal antibody) that is used to identify or isolate a gene or a gene product.

programmed cell death   Apoptosis; an orderly process whereby certain cells are eliminated at certain stages of development, as controlled by specific cell death genes.

prokaryote   A cell or organism (such as bacteria or blue-green algae) that lacks a membrane-bounded nucleus and complex chromosomes and does not undergo mitosis or meiosis.

prometaphase   The stage of nuclear division, occurring just before metaphase, when the nuclear envelope breaks down, kinetochores form at the centromere regions, and chromosomes become attached to spindle fibers.

promoter   A specific sequence of DNA nucleotide bases to which RNA polymerase binds to initiate gene transcription.

pronucleus   The haploid egg or sperm nucleus in a fertilized egg.

prophase   The first stage of mitosis or meiosis, during which chromosomes condense, the nucleoli disappear, and the spindle forms.

propositus (female, *proposita*)   The clinically affected individual through whom a human pedigree is discovered. See proband.

prospective study   Research that involves the selection of a group of individuals who are then periodically observed or tested to see what happens to them in the future.

protein   A gene product; a molecule consisting of one or more polypeptide chains, each one an unbranched string of amino acids.

protein kinase   The type of molecule involved in intracellular signal transduction that acts by phosphorylating other molecules.

proto-oncogene   A cellular gene that normally regulates cell division, but which can mutate to a dominant oncogene that causes cancer.

provirus   The DNA copy of the RNA of a retrovirus when it is inserted into the chromosome of a host cell.

pseudoautosomal region   A tiny region of homology at the tip of the X and Y chromosome short arms, which normally pairs and crosses over during meiosis in males; genes within this region, being present in duplicate in males, are inherited as though they were autosomal rather than sex-linked.

pseudogene   A length of DNA that is similar in nucleotide sequence to a normal gene (at a different locus) but is not expressed, having one or more mutations that render it nonfunctional.

pseudohermaphrodite   An individual who has either testes or ovaries, but whose external genitals are the opposite of the gonadal sex, or ambiguous, or abnormal in some other way.

purine (A, G)   The larger of two types of nucleotide bases found in DNA and RNA; includes adenine (A) and guanine (G).

pyrimidine (C, T, U)   The smaller of two types of nucleotide bases found in DNA and RNA; includes cytosine (C), thymine (T), and (replacing thymine in RNA only) uracil (U).

quantitative trait loci (QTL)   Genes that influence the expression of a quantitative trait. Typically there are many genes, each with a small effect. See also multiple gene inheritance.

quantitative variation   The situation in which a trait (such as height, weight, or behavioral activity) varies continuously along a range of values, rather than being observable as discrete, nonoverlapping phenotypic groups; involves the interaction of the environment with several or many genes.

race   A genetically or geographically distinct subgroup of a species.

radioactive probe   A single-stranded, radioactive sequence of about 20 or more nucleotides, which is used to bind to and thus identify a complementary sequence on a Southern blot.

random mating   Mating, within a population, that occurs without regard to an individual's phenotype or genotype.

reading frame   Successive groups of three in a sequence of nucleotide bases. Depending on the starting position, there are three possible reading frames for a particular sequence.

receptor protein   A cell surface protein that binds signalling molecules (e.g., growth factor) and transmits the signal to other molecules within the cell.

recessive allele   An allele that is expressed when homozygous but not expressed when heterozygous.

reciprocal matings   Matings in which the sexes representing the two differing parental genotypes or phenotypes are reversed.

reciprocal translocation   A two-break chromosomal abnormality consisting of an interchange of chromosomal pieces between two nonhomologous chromosomes, each translocated chromosome having one centromere.

recombinant DNA   A hybrid DNA molecule formed by the fusion of pieces of DNA derived from different species.

recombinant gamete   With respect to two heterozygous genes, a gamete formed by an individual that contains a new allelic array of genes (compared to the assortment in the gametes that formed the individual). See also parental gamete.

recombination   The reassortment of parental genes to form new combinations of alleles in the offspring of a multiple heterozygote, as a result of independent assortment or crossing over.

relay molecules   Molecules within a cell that transmit a signal from receptor molecules near the cell surface to transcription factors in the nucleus.

repetitive DNA   Highly repeated or moderately repeated multiples of DNA nucleotide sequences, making up a major fraction of eukaryotic DNA and found in heterochromatin, ribosomal RNA genes, and transfer RNA genes.

repulsion phase   The configuration of two linked and heterozygous genes, when the recessive genes a and b are on opposite homologues (aB/Ab). See also coupling phase.

restriction enzyme   A bacterial enzyme that can cut across a DNA molecule. Each enzyme recognizes a unique target sequence of bases, often producing staggered cut ends that can join to complementary sequences of any foreign DNA—thus being an important tool of molecular biology.

restriction fragment length polymorphism (RFLP)   Variation in the lengths of DNA fragments cut by a specific restriction enzyme, due to mutations in the target sequences.

restriction point (R)   The main checkpoint of the mammalian cell cycle, occurring late in the G_1 period, which determines whether or not a cell will continue through the division cycle.

retinoic acid   A derivative of vitamin A that (when combined with intracellular retinoic acid receptors) acts as a transcription factor, with particular specificity for development of the anterior-posterior axis and of limbs.

retrospective study   Research that involves the selection of a group of individuals who are then interviewed and/or tested to find out what happened to them in the past.

retrovirus   An RNA-containing virus that can enter a cell, use reverse transcriptase to reproduce a DNA copy of itself, and integrate into the host's genome. If the retrovirus carries an oncogene, the host cell can be changed into a cancer cell.

reverse transcriptase   A retrovirus polymerase enzyme for "backward" transcription, that is, making DNA from an RNA template.

ribosomal RNA (rRNA)   A type of RNA that becomes part of the ribosomes in the cytoplasm.

ribosome   A cytoplasmic structure composed of proteins and ribosomal RNA, on which protein synthesis (messenger RNA translation) occurs.

ring chromosome   A ring-shaped chromosome resulting from the loss of both chromosome ends and the joining together of the broken ends of the remaining central segment.

RNA   Ribonucleic acid, a single-stranded molecule transcribed from DNA, containing the sugar ribose and the bases adenine, uracil, guanine and cytosine. The three types of RNA, all involved in protein synthesis, are messenger RNA, ribosomal RNA, and transfer RNA.

RNA polymerase   An enzyme that forms RNA by linking together a string of ribonucleotides in a sequence complementary to that of a single-stranded DNA template.

RNA splicing   The processing of messenger RNA by the removal of all intron sequences from the primary transcript and the joining together of all the exons.

Robertsonian translocation   Translocations between and among acrocentric chromosomes, in which the long arms fuse to form a single chromosome; the short arms are usually lost.

S period   The interphase period of the eukaryotic cell cycle, between G_1 and G_2, during which DNA and histones are synthesized.

sarcoma   A solid cancer derived from bone, cartilage, muscle, or fat.

scaffold   The central structure of a chromosome, composed of nonhistone proteins, to which looped domains of DNA are attached.

secondary constriction   A thin strand of chromatin, present on some chromosomes, connecting a knoblike tip to the rest of the chromosome.

secondary immune response   The stronger, quicker response to a second exposure to an antigen, dependent on the presence of memory cells formed on the previous exposure.

secondary oocyte   A diploid female cell that undergoes meiosis II to form an egg and a second polar body.

secondary spermatocyte   A diploid male cell that undergoes meiosis II to form two spermatids.

segmentation genes   Genes that act in very early development to divide an embryo into a series of primitive segments, which are later acted on by the homeotic genes.

segregation   The separation of the two alleles at any given locus into different daughter cells during meiosis.

selection   The differential reproduction of the various genotypes in a population over a period of time.

sequence tagged site (STS)   Short (100–200 bases), randomly selected stretches of cloned DNA that are sequenced in order to mark the longer segments in which they occur.

sex chromatin   A Barr body; a dark-staining body that represents a condensed, inactivated X chromosome in somatic cell nuclei of female mammals.

sex chromosomes   The pair of dissimilar chromosomes (X and Y) found in organisms with separate sexes; chromosomes that are not autosomes.

sex reversal   A normal or abnormal or environmentally induced or experimentally induced change in the sex of an individual at any stage of development.

sex-influenced inheritance   The differential expression of autosomally inherited traits (such as baldness) in the two sexes.

sex-limited trait   A trait (such as milk production) affecting structures or processes that exist in only one sex.

sexual reproduction   The alternation of a diploid generation of cells (the body cells) with a haploid generation of cells (the eggs and sperm) produced by meiosis.

short tandem repeat (STR)   A locus with repetitive units used in DNA fingerprinting, each unit being 3, 4, or 5 nucleotides long. Similar to a VNTR locus, but with a shorter repeat unit.

signal transduction   The line of communication from outside a cell to the cell nucleus involving growth factors, receptors, relay molecules, and transcription factors.

signaling molecule   A molecule involved in intercellular communication; its binding by a specific target cell receptor initiates a response in that cell.

SINE element   A short interspersed repetitive DNA element (like *Alu*) that is present in human DNA. See also LINE element.

single-copy DNA   The approximately 3% of nonrepetitive human DNA that codes for polypeptides.

single-locus probe   A radioactive probe whose sequence binds to the alleles of just one locus in the genome. See also multilocus probe.

sister chromatids   The two identical replicas (strands) of a replicated chromosome. See chromatid.

sister-chromatid exchange   Abnormal crossing over between two chromatids of one chromosome in somatic cells during mitosis.

somatic cell   Any cell in a eukaryote that is not in the germ cell lineage.

somatic cell fusion   The fusion of cells from two different species or tissues to form a hybrid cell, used for gene mapping and other genetic studies.

somatic gene therapy   Introduction of a normal gene in somatic cells to replace or supplement an abnormal gene, generally considered to be an extension of current medical practice. See also germinal gene therapy.

somatic hypermutation   The high rate of mutation in the combined *V-D-J* DNA during the differentiation of a B cell into an antibody-secreting plasma cell.

somatic mutation   A mutation occurring in a somatic cell and thus not transmitted to offspring.

somatic recombination   The shuffling of linked genetic elements in a B cell to form a gene that encodes one of the polypeptide chains of an antibody.

Southern blotting   A technique of molecular biology, invented by E. M. Southern, in which pieces of single-stranded, electrophoretically separated DNA are transferred from a gel to a nitrocellulose filter and then treated with a radioactive probe to identify the segment of interest.

species   A group of organisms whose members are capable of interbreeding with one another but incapable of producing viable and fertile offspring with members of other species.

sperm   A haploid male gamete.

spermatid   One of four haploid products of male meiosis, which undergo cytoplasmic changes to become sperm.

spermatogenesis   In sexually reproducing organisms, the process by which a diploid primary spermatocyte undergoes meiosis to form four haploid spermatids which become functional sperm.

spermatogonium   A diploid cell in testes that divides mitotically to form more spermatogonia or that enlarges to form a primary spermatocyte.

spindle   A structure made up of spindle fibers, formed during prophase of mitosis or meiosis, on which chromosomes congregate during metaphase and move to opposite poles during anaphase.

spindle microtubule   One of numerous microtubular filaments that arise during prophase of cell division and collectively form the spindle.

spontaneous mutation   A mutation that occurs haphazardly, irrespective of known environmental causes.

SRY gene   A gene on the Y chromosome of mammals that encodes the testis-determining factor and initiates the development of maleness.

stem cell   A cell in the bone marrow that produces the precursors of various types of differentiated blood cells as well as more stem cells.

sticky (cohesive) end   The single-stranded tip to an otherwise double-stranded DNA molecule. Such tips tend to bind to complementary single-stranded tips.

stop codon   See chain termination codon.

structural gene   A unique sequence gene, which is transcribed and translated to form a polypeptide product.

submetacentric chromosome   A chromosome whose centromere is located off-center, creating chromosome arms of somewhat unequal length.

superior heterozygote   The situation that exists when the fitness of a heterozygous genotype is greater than that of either homozygote.

supernumerary chromosome   An extra structurally abnormal chromosome (also called a marker or accessory chromosome) that may or may not be associated with phenotypic effects. They are highly variable in size and composition, and some cause mental or physical abnormalities.

suppressor mutation   A mutation that reduces the effect of another mutant gene to a normal or near-normal phenotype, or prevents the expression of another gene.

synapsis   The pairing of homologous chromosomes during prophase I of meiosis.

synaptonemal complex   A proteinaceous structure that forms during prophase I of meiosis and that binds homologous chromosomes together during synapsis.

T cell   A white blood cell that is produced in the bone marrow, matures in the thymus gland, and is involved in the immune response.

Taq DNA polymerase   A DNA replicating enzyme obtained from a bacterium (*Thermus aquaticus*) that lives in hot springs. The enzyme is stable at near-boiling temperatures.

telomerase   An enzyme that extends the tips (telomeres) of chromosomes with their repetitive TTAGGG units.

telomere   The tip of a chromosome, containing repeated DNA sequences.

telophase   A stage of mitosis or meiosis during which the spindle disappears, daughter chromosomes decondense, the nuclear envelope re-forms, and nucleoli appear; the cytoplasm splits too.

template   A mold or pattern; a sequence of nucleotides from which a complementary DNA or RNA strand is made.

template strand   The strand of DNA that is transcribed into complementary RNA.

testcross   An experimental mating between an individual with a dominant phenotype and one with a recessive phenotype for the same trait(s), to determine whether the individual with the dominant phenotype is homozygous or heterozygous.

testis-determining factor (TDF)   Product of the Y-linked *SRY* gene, which initiates the development of maleness in mammals.

testosterone   a male sex hormone produced in the testes and adrenal glands.

tetrad   The four-chromatid structure that makes up a pair of synapsed homologous chromosomes during prophase I and metaphase I of meiosis.

tetraploid   A cell or organism with four haploid sets of chromosomes.

threshold   With regard to radiation, a dose below which there is no risk. There is probably no threshold for the induction of mutations.

threshold trait   A polygenic trait (such as cleft lip) that is either present or absent, rather than varying continuously.

thymine dimer   A type of ultraviolet-induced mutation in which two adjacent thymine nucleotides on one DNA strand join to form a double base that distorts the DNA molecule and interrupts both DNA replication and transcription.

tolerance   The nonresponsiveness of a person's immune system to his or her own molecules.

totipotent cell   A cell that retains the ability to produce a whole organism.

transcription   The formation of a strand of RNA from a DNA template by complementary base pairing, catalyzed by RNA polymerase.

transcription factor   A protein that binds to the regulatory region of a gene, causing the gene to be transcribed.

transcription termination sequence   A DNA sequence that causes RNA polymerase to move off the DNA strand, thus stopping transcription.

transfer RNA (tRNA)   A type of RNA that acts as an adapter during protein synthesis by binding to specific amino acids and fitting them to appropriate codons of messenger RNA.

transformation   The conversion by an oncogene of a normal eukaryotic cell in culture to a cancer cell.

transgene   A foreign gene present in the cells of a different species.

transgenic organism   An organism that has a foreign gene transferred and stably incorporated into its germ line.

translation   The process of protein formation, occurring in a ribosome, whereby the information contained in a sequence of nucleotide bases in messenger RNA directs the synthesis of a sequence of amino acids in a polypeptide.

translocation   A chromosome aberration formed by the exchange of parts between nonhomologous chromosomes.

transmitter male   Within fragile X syndrome pedigrees, a phenotypically unaffected male who has the abnormal X-linked *FMR1* allele but does not express it or show the fragile X site in any of his cells.

transposon   A transposable element or jumping gene; a piece of DNA that can move from one place to another within the genome.

trinucleotide repeat   See triplet repeat.

triplet repeat   The repetition of three bases in tandem, involved in several genetic diseases (e.g., CAG in Huntington disease).

triploid   A cell or organism with three haploid sets of chromosomes.

trisomic   A diploid cell or organism having one too many chromosomes, that is, three (rather than two) representatives of one chromosome.

tumor progression   The stages through which cancer cells pass. Typically, they go from uncontrolled division to disordered growth to invasiveness to metastasis.

tumor suppressor gene   A gene whose normal, dominant allele checks cell division, but whose mutant allele or deletion releases the cell from growth controls.

unbalanced gamete/zygote   A gamete or zygote containing duplications or deletions of chromosomal material, resulting in more than or fewer than the normal number of genes. Phenotypic effects (which may include lethality) vary according to the size and position of the chromosomal abnormality.

unequal crossing over   A crossover that occurs between misaligned copies of duplicated loci or repetitive DNA sequences, resulting in recombinant chromatids of unequal size.

uniparental disomy   Homozygosity for a particular chromosome or chromosome segment, due to the inheritance of two chromosomes (rather than one) from one parent.

V-D-J-C gene   The gene that codes for the heavy chain of antibodies. It is assembled from separate *V*, *D*, *J*, and *C* "genelets" by somatic recombination.

variable (V) domains   The portions of the heavy and light chains of antibodies having different amino acids, thereby producing different antigenic specificities. See also constant (C) domain.

variable expressivity   Differences in the observed effects of a given allele or genotype in different individuals.

variable number of tandem repeats (VNTR)   Short DNA sequences that are tandemly repeated variable numbers of times at many defined regions throughout the genome. This variation forms the basis of DNA fingerprinting procedures.

vector   A plasmid, retrovirus, bacterial, or yeast artificial chromosome that is used to transfer a DNA segment from one cell or species to another.

X chromosome   The sex chromosome present in two copies in female mammals (and in many other female animals).

X-linked gene   A gene located on the X chromosome.

xenotransplantation   The transplantation of organs between species.

Y chromosome   The sex chromosome present in one copy in male mammals (and in many other male animals). In mammals it carries the male-determining *SRY* gene.

Y-linked gene   A gene located on the Y chromosome, which is transmitted from a male to all of his sons.

yeast artificial chromosome (YAC)   A synthetic chromosome that can accept a large piece of foreign DNA, thus acting as a vector for making gene libraries and for transferring genes from one species to another.

zinc-finger protein   A protein molecule that is bent into fingerlike projections in regions that bind to the grooves of DNA.

zona pellucida   A dense, jellylike layer secreted by the oocyte, that surrounds the oocyte and later the egg. It protects the egg and early embryo and prevents the entry of foreign sperm.

zoo blot   A Southern blot that uses a cloned human DNA segment as a probe to see if its sequence is present in the DNAs of other species.

zygote   A fertilized egg; a diploid cell formed by the union of haploid male and female gametes.

Bibliography

The numbers in [brackets] at the end of each entry refer to the chapters in which the citation is found.

To save space, we have cited papers with six or more authors as the first author et al. Also, we use the following abbreviations for frequently cited works:

Journals:
AJHG: *The American Journal of Human Genetics*
NEJM: *The New England Journal of Medicine*
NYT: *The New York Times*
PNAS: *Proceedings of the National Academy of Science, U.S.A.*
SA: *Scientific American*
TIG: *Trends in Genetics*

Books:
Rimoin et al.: RIMOIN, D. L., J. M. CONNOR and R. E. PYERITZ (eds.). 1997. *Emery and Rimoin's Principles and Practice of Medical Genetics*, 3rd ed. 2 volumes. Churchill Livingstone, New York.

Scriver et al: SCRIVER, C. R., A. L. BEAUDET, W. S. SLY and D. VALLE (eds.). 1995. *The Metabolic and Molecular Bases of Inherited Disease*, 7th ed. 3 volumes. McGraw-Hill, New York.

ADA, G. L. and G. NOSSAL. 1987. The clonal-selection theory. *SA* 257 (Aug.): 62–69. [18]

ADAMS, M. D. et al. 1995. Initial assessment of human gene diversity and expression patterns based upon 83 million nucleotides of cDNA sequence. *Nature* 377 Supplement: 3–174. [9]

AKAM, M., P. HOLLAND, P. INGHAM AND G. WRAY (eds.). 1994. *Development 1994 Supplement: The Evolution of Developmental Mechanisms.* The Company of Biologists Limited, Cambridge. [15]

ALBERTS, B. et al. 1994. *Molecular Biology of the Cell*, 3rd ed. Garland, New York. [2,3,6,15,18]

ALLEN, G. E. 1996. Science misapplied: The eugenics age revisited. *Technology Review* 99 (Aug./Sep.): 23–31. [20]

ALTMAN, L. K. 1993. Surprise discovery about split genes wins Nobel prize. *NYT* Oct. 12, p. C3. [6]

ALTMAN, L. K. 1996. A discovery energizes AIDS researchers: Preliminary steps are under way to explore a genetic mutation. *NYT* Aug. 10, pp. N7, L12. [2]

ALTMAN, L. K. 1996. New AIDS therapies arise, but who can afford the bill? *NYT* Feb. 6, pp. A1ff. [18]

ANASTASI, A. 1988. *Psychological Testing*, 6th ed. Macmillan, New York. [11]

ANDERSON, C. 1992. NIH, under fire, freezes grant for conference on genetics and crime. *Nature* 358: 357. [11]

ANGELL, R. 1997. First-meiotic-division nondisjunction in human oocytes. *AJHG* 61: 23–32. [13]

ANGIER, N. 1987. Light cast on a darkling gene. *Discover* 8 (March): 85–96. [17]

ANGIER, N. 1994. Cause of cystic fibrosis is traced to the Stone Age. *NYT* June 1, p. B9. [5]

ANGIER, N. 1994. Keys emerge to mystery of "junk" DNA. *NYT* June 28, pp. C1 ff. [6]

ANGIER, N. 1994. Family of errant genes is found to be related to variety of skeletal ills. *NYT* Nov. 1, pp. C1ff. [12]

ANGIER, N. 1994. Family of errant genes is found to be related to variety of skeletal ills. *NYT* Nov. 1, p. C1. [15]

ANGIER, N. 1994. Vexing pursuit of breast cancer gene. *NYT* Jul 12, pp. C1ff. [17]

ANGIER, N. 1995. Gene hunters pursue elusive and complex traits of mind. *NYT* Oct. 31, p. C1. [11]

ANONYMOUS 1992. (Obituaries) Wayne McLaren, 51, played "Marlboro Man"; of cancer. *Boston Globe* July 24, p. 25. Wayne McLaren, 51, Rodeo Rider and Model. *NYT* July 25, p. 11. [17]

ANONYMOUS. 1995. Biotechnology and genetics. *The Economist* 334 (Feb. 25): insert 1–18 following p. 56. [8]

ANONYMOUS. 1996. Romanovs find closure in DNA. *Nature Genetics* 12: 339–340. [5]

ANONYMOUS. 1996. Facing our fears. *Consumer Reports* 61 (Dec.): 50–53. [7]

ANTONARAKIS, S. E. 1993. Human chromosome 21: Genome mapping and exploration, circa 1993. *TIG* 9: 142–148. [13]

APPLE, B. 1994. Coffee and health. *Consumer Reports* 59 (Oct.): 650–651. [7]

ARVEY, R. D. et al. 1994. Mainstream science on intelligence. *The Wall Street Journal*, Dec. 13, p. A18. [11]

ASHG AD HOC COMMITTEE ON BREAST AND OVARIAN CANCER SCREENING. 1994. Statement of The American Society of Human Genetics on genetic testing for breast and ovarian cancer predisposition. *AJHG* 55 (Nov.): i–iv. Also available on the Internet: http://www.faseb.org/genetics/ashg/policy/pol–11.htm [19]

ASHG AD HOC COMMITTEE ON CYSTIC FIBROSIS CARRIER SCREENING. 1992. Statement of The American Society of Human Genetics on cystic fibrosis carrier screening. *AJHG* 51: 1443–1444. Also available on the Internet: http://www.faseb.org/genetics/ashg/policy/pol–10.htm [19]

ASHG AD HOC COMMITTEE ON GENETIC TESTING/INSURANCE ISSUES. 1995. Background statement: Genetic testing and insurance. *AJHG* 56: 327–331. Also available on the Internet: http://www.faseb.org/genetics/ashg/policy/pol–12.htm [19]

ASHTON, G. C. 1980. Mismatches in genetic markers in a large family study. *AJHG* 32: 601–613. [5]

ASKARI, F. K. and W. M. McDONNELL. 1996. Antisense-oligonucleotide therapy. *NEJM* 334: 316–318. [20]

ATKINSON, M. A. and N. K. MACLAREN. 1990. What causes diabetes? *SA* 263 (July): 62–63, 66–71. [10,18]

BACA, M. and L. ZAMBONI. 1967. The fine structure of human follicular oocytes. *Journal of Ultrastructural Research* 19: 354–381. [3]

BACETTI, B. 1984. The human spermatozoon. In *Ultrastructure of Reproduction*, J. Van Blerkom and P. M. Motta (eds.), pp. 110–126. Martinus Nijhoff, Boston. [3]

BAILAR, J. C., III and H. L. GORNIK. 1997. Cancer undefeated. *NEJM* 336: 1569–1574. [17]

BAIRD, P. A. and B. McGILLIVRAY. 1982. Children of incest. *Journal of Pediatrics* 101: 854–857. [12]

BAJEMA, C. J. (ed.). 1976. *Eugenics, Then and Now. Benchmark Papers in Genetics*, Vol. 5. Dowden, Hutchinson & Ross, Stroudsburg, PA. [20]

BALLABIO, A. and H. Y. ZOGHBI. 1997. Kallman syndrome. In Rimoin et al., pp. 4549–4557. [15]

BALLANTYNE, J. 1997. Mass disaster genetics. *Nature Genetics* 15: 329–330. [8]

BALTER, M. 1996. Children become the first victims of fallout. *Science* 272: 357–360. (This issue includes many other articles and sidebars on the Chernobyl accident.) [7]

BANFI, S. et al. 1996. Identification and mapping of human cDNAs homologous to *Drosophila* mutant genes through EST database searching. *Nature Genetics* 13: 167–174. [15]

BARINAGA, M. 1996. An intriguing new lead on Huntington's disease. *Science* 271: 1233–1234. [7]

BARLOW, D. P. 1995. Gametic imprinting in mammals. *Science* 270: 1610–1613. [10]

BARR, P. J. and L. D. TOMEI. 1994. Apoptosis and its role in human disease. *Bio/Technology* 12: 487–493. [15]

BARSH, G. S. 1995. Pigmentation, pleiotropy, and genetic pathways in humans. *AJHG* 57: 743–747. [16]

BARTECCHI, C. E., T. D. MACKENZIE and R. W. SCHRIER. 1994. The human costs of tobacco use. (First of two parts.) *NEJM* 330: 907–912. [7]

BARTECCHI, C. E., T. D. MACKENZIE and R. W. SCHRIER. 1995. The global tobacco epidemic. *SA* 272 (May): 44–51. [7]

BARTOLOMEI, M. S. and S. M. TILGHMAN. 1997. Genomic imprinting in mammals. *Annual Review of Genetics* 31: 493–525. [12]

BARTON, J. C. and L. F. BERTOLI. 1996. Hemochromatosis: The genetic disorder of the twenty-first century. *Nature Medicine* 2: 394–395. [5]

BASERGA, R. 1981. The cell cycle. *NEJM* 304: 453–459. [3]

BASERGA, R. 1985. *The Biology of Cell Reproduction*. Harvard University Press, Cambridge, MA. [3]

BASSETT, D. E., JR. et al. 1997. Genome cross-referencing and *XREFdb*: Implications for the identification and analysis of genes mutated in human disease. *Nature Genetics* 15: 339–344. [9,19]

BAUER, H. H. 1992. *Scientific Literacy and the Myth of the Scientific Method*. University of Illinois Press, Urbana. [1]

BEADLE, G. W. 1980. The ancestry of corn. *SA* 242 (January): 112–119. [1]

BEADLE, G. W. and E. L. TATUM. 1941. Genetic control of biochemical reactions in *Neurospora*. *PNAS* 27: 499–506. [16]

BEARDSLEY, T. 1994. Big-time biology. *SA* 271 (Nov.): 90–97. [8]

BEARDSLEY, T. 1996. Vital data. *SA* 274 (March): 100–105. [9]

BEARN, A. G. 1993. *Archibald Garrod and the Individuality of Man*. Clarendon Press, Oxford. [16]

BEARN, A. G. 1996. Oswald T. Avery and the Copley Medal of the Royal Society. *Perspectives in Biology and Medicine* 39: 550–554. [6]

BEAUDET, A. L. et al. 1989. Genetics and biochemistry of variant human phenotypes. In *The Metabolic Basis of Inherited Disease*, 6th ed., C. R. Scriver, A. L. Beaudet, W. S. Sly and D. Valle (eds.), pp. 3–53. McGraw-Hill, New York. [10]

BECKER, R. C. 1996. Antiplatelet therapy. *Science and Medicine* 3 (July/August): 12–21. [5]

BEERNINK, F. J., W. P. DMOWSKI, and R. J. ERICSSON. 1993. Sex preselection through albumin separation of sperm. *Fertility and Sterility* 59: 382–386. [20]

BEIR, V. Committee on the Biological Effects of Ionizing Radiation. 1990. *Health Effects of Exposure to Low Levels of Ionizing Radiation*. National Academy Press, Washington, D. C. [7]

BELL, J. and J. B. S. HALDANE. 1937. The linkage between the genes for colour-blindness and haemophilia in man. *Royal Society of London, Proceedings, Series B* 123: 119–150. [9]

BELLUS, G. A. et al. 1995. Achondroplasia is defined by recurrent *G380R* mutations of *FGFR3*. *AJHG* 56: 368–373. [12]

BELLUS, G. A. et al. 1995. A recurrent mutation in the tyrosine kinase domain of fibroblast growth factor receptor 3 causes hypochondroplasia. *Nature Genetics* 10: 357–359. [15]

BELLUS, G. A. et al. 1996. Identical mutations in three different fibroblast growth factor receptor genes in autosomal dominant craniosynostosis syndromes. *Nature Genetics* 14: 174–176. [15]

BELMONT, J. W. 1996. Genetic control of X inactivation and processes leading to X-inactivation skewing. *AJHG* 58: 1101–1108. [10]

BENNETT, W. R. JR. 1995. Electromagnetic fields and power lines. *Science and Medicine* 2 (July/August): 68–77. [7]

BERG, J. M., H. KARLINSKY and A. J. HOLLAND. 1994. *Alzheimer's Disease, Down Syndrome, and Their Relationship*. Oxford University Press, New York. [13]

BERG, P. and M. F. SINGER. 1995. The recombinant DNA controversy: Twenty years later. *PNAS* 92: 9011–9014. [8]

BERNHARDT, B. A. 1997. Empirical evidence that genetic counseling is directive: Where do we go from here? *AJHG* 60: 17–20. [19]

BERRA, T. M. 1990. *Evolution and the Myth of Creationism: A Basic Guide to the Facts in the Evolution Debate*. Stanford University Press, Stanford, CA. [12]

BESLEY, G. T .N. 1992. Enzyme analysis. In *Prenatal Diagnosis and Screening*, D. J. H. Brock, C. H. Rodeck and M. A. Ferguson-Smith (eds.), pp. 127–145. Churchill Livingstone, Edinburgh. [19]

BEUTLER, E. 1991. Gaucher's disease. *NEJM* 325: 1354–1360. [20]

BIC 1997. Breast Cancer Information Core of the National Human Genome Research Institute. http://www.nhgri.nih.gov/Intramural_research/Lab_transfer/Bic [17]

BICKEL, H. 1987. Early diagnosis and treatment of inborn errors of metabolism. *Enzyme* 38: 14–26. [16]

BISHOP, J. et al. 1996. Aneuploidy in germ cells: Etiologies and risk factors. *Environmental and Molecular Mutagenesis* 28: 159–166. [13]

BISHOP, J., C. A. HUETHER, C. TORFS, F. LOREY and J. DEDDENS. 1997. Epidemiological study of Down syndrome in a racially diverse California population, 1989–1991. *American Journal of Epidemiology* 145: 134–147. [13]

BITTLES, A. H. and J. V. NEEL. 1994. The costs of human inbreeding and their implications for variations at the DNA level. *Nature Genetics* 8: 117–121. [12]

BITTLES, A. H., W. M. MASON, J. GREENE and N. A. RAO. 1991. Reproductive behavior and health in consanguineous marriages. *Science* 252: 789–794. [12]

BLAESE, R. M. 1997. Gene therapy for cancer. *SA* 276: 111–115. [20]

BLAKESLEE, S. 1994. Researchers find gene that may link dyslexia with immune disorders. *NYT*, Oct. 18: C3. [11]

BLAKESLEE, S. 1995. Newfound brain protein may be "smoking gun" in Huntington's. *NYT* Nov. 14, p. C3. [7]

BLAKESLEE, S. 1996. Genetic questions are sending judges back to classroom. *NYT* July 9, p C1. [8]

BLAKESLEE, S. 1996. Researchers track down a gene that may govern spatial abilities. *NYT*, July 23, p. C3. [11]

BLAKESLEE, S. 1996. Virus's similarity to body's proteins may explain autoimmune diseases. *NYT* Dec. 31, pp. C1ff. [18]

BLAKESLEE, S. 1997. Cause of brain cells' death in seven diseases is discovered. *NYT* Aug. 8, pp. A1ff. [7]

BLEIER, R. 1984. *Science and Gender*. Pergamon, New York, pp. 80–114. [15]

BODMER, W. F. 1986. Human genetics: The molecular challenge. *CSHS* 51: 1–13. [1]

BODMER, W. F. and L. L. CAVALLI-SFORZA. 1970. Intelligence and race. *SA* 223 (Oct.): 19–29. [11]

BODMER, W. F. and L. L. CAVALLI-SFORZA. 1976. *Genetics, Evolution, and Man*. W. H. Freeman and Company, San Francisco. [11]

BOFFEY, P. M. 1996. Are the mad cows bad cows? *NYT* March 23, p. A20. [17]

BOGUSKI, M. S. 1995. Hunting for genes in computer data bases. *NEJM* 333: 645–647. [8]

BOHMAN, M. 1978. Some genetic aspects of alcoholism and criminality; a population of adoptees. *Archives of General Psychiatry* 35: 269–276. [11]

BOIS, E. et al. 1978. Cluster of cystic fibrosis cases in a limited area of Brittany (France). *Clinical Genetics* 14: 73–76. [5]

BONGIOVANNI, A. M., W. R. EBERLEIN, A. S. GOLDMAN and M. NEW. 1967. Disorders of adrenal steroid biogenesis. *Recent Progress in Hormone Research* 23: 375–449. [16]

BOTSTEIN, D., R. L. WHITE, M. SKOLNICK and R. W. DAVIS. 1980. Construction of a genetic linkage map in man using restriction fragment length polymorphisms. *AJHG* 32: 314–331. [9,17]

BOUCHARD, T. J., JR. and M. McGUE. 1981. Familial studies of intelligence: A review. *Science* 212: 1055–1059. [11]

BOUCHARD, T. J., JR., D. T. LYKKEN, M. McGUE, N. L. SEGAL, and A. TELLEGEN. 1990. Sources of human psychological differences: The Minnesota study of twins reared apart. *Science* 250: 223–228. [11]

BOURGERON, T. et al. 1995. Mutation of a nuclear succinate dehydrogenase gene results in mitochondrial respiratory chain deficiency. *Nature Genetics* 11: 144–149. [5]

BOWEN, P. et al. 1965. Hereditary male pseudo-hermaphroditism with hypogonadism, hypospadias and gynecomastia (Reifenstein's syndrome). *Annals of Internal Medicine* 62: 252–270. [15]

BOWER, B. 1996. Gene tied to excitable personality. *Science News* 149: 4. [11]

BOYER, S. H., IV (ed.). 1963. *Papers on Human Genetics*. Prentice-Hall, Englewood Cliffs, NJ. [18]

BROAD, W. J. 1995. Cancer fear is unfounded, physicists say. *NYT* May 14, p. 14. [7]

BROCK, D. J. H., C. H. RODECK and M. A. FERGUSON-SMITH (eds.). 1992. *Prenatal Diagnosis and Screening*. Churchill Livingstone, Edinburgh. [14,19]

BRODY, J. E. 1994. Scientist at work: Bruce N. Ames: Strong views on origins of cancer. *NYT* July 5, pp. C1ff. [17]

BRODY, J. E. 1996. Chemicals in food? A panel of experts finds little danger. *NYT* Feb. 16, pp. A1ff. [17]

BRODY, J. E. 1997. The potential dangers of iron overload in the rich American diet. *NYT* March 5, p. C8. [5]

BROOK, J. D. 1994. Positional cloning. *SA Science & Medicine* 1 (Nov./Dec.): 48–57. [9]

BROWN, J. L. and E. POLLITT. 1996. Malnutrition, poverty, and intellectual development. *SA* 274 (Feb.): 38–43. [11]

BROWN, M. S. and J. L. GOLDSTEIN. 1984. How LDL receptors influence cholesterol and atherosclerosis. *SA* 251(Nov.): 58–66. [5]

BROWN, M. S. and J. L. GOLDSTEIN. 1986. A receptor-mediated pathway for cholesterol homeostasis. Nobel lecture, 9 Dec. 1985. *Les Prix Nobel 1985*. Almqvist & Wiksell International, Stockholm. Reprinted in *Science* 232: 34–47. [5]

BUCK v. BELL. 1927. *United States Reports: Cases Adjudged in the Supreme Court* 274: 200–208. [20]

BULMER, M. G. 1970. *The Biology of Twinning in Man*. Oxford University Press, London. [11]

BURNET, F. M. (ed.). 1976. *Immunology: Readings from Scientific American*. W. H. Freeman and Company, San Francisco. [18]

BUTLER, D. 1998. Last chance to stop and think on risks of xenotransplants. *Nature* 391: 320–324. [18]

CAMPION, E. W. 1997. Power lines, cancer, and fear. *NEJM* 337: 44–46. [7]

CANN, R. L. 1997. Phylogenetic estimation in humans and neck riddles. *AJHG* 60: 755–757. [12]

CAPECCHI, M. R. 1994. Targeted gene replacement. *SA* 270: 52–59. [17]

CAROL ANN with K. DEMARET. 1984. David's story. *People Weekly* 122 (Oct. 29): 120–141 and 122 (Nov. 5): 107–127. [18]

CAROTHERS, E. E. 1913. The Mendelian ratio in relation to certain Orthopteran chromosomes. *Journal of Morphology* 24: 487–511. [4]

CASKEY, C. T. 1986. Summary: A milestone in human genetics. *CSHS* 51: 1115–1119. [1]

CASKEY, C. T., R. G. WORTON and J. D. WATSON. 1991. ASHG Human Genome Committee Report. The human genome project: Implications for human genetics (an exchange of letters). *AJHG* 49: 687–691. [9]

CAVALLI-SFORZA, L. L. 1991. Genes, peoples and languages. *SA* 265 (Nov.): 104–110. [12]

CAVALLI-SFORZA, L. L. 1998. The DNA revolution in population genetics. *TIG* 14: 60–65. [12]

CAVALLI-SFORZA, L. L., P. MENOZZI and A. PIAZZA. 1994. *The History and Geography of Human Genes*. Princeton University Press, Princeton NJ. [12]

CAVENEE, W. K. and R. L. WHITE. 1995. The genetic basis of cancer. *SA* 272 (March): 72–79. [17]

CHAMBON, P. 1981. Split genes. *SA* 244 (May): 60–71. [6]

CHANDLEY, A. C. 1988. Meiosis in man. *TIG* 4: 79–84. [3]

CHANDLEY, A. C. 1997. Infertility. In Rimoin et al., pp. 667–675. [14]

CHANDRASEKHAR, S. 1990. Science and scientific attitudes. *Nature* 344: 286–286. [1]

CHEN, E. 1979. Twins reared apart: A living lab. *New York Times Magazine*, Dec. 9, pp. 112ff. [11]

CHETTOUH, Z. et al. 1995. Molecular mapping of 21 features associated with partial monosomy 21: Involvement of the *APP-SOD1* region. *AJHG* 57: 62–71. [14]

CHEUNG, M., J. D. GOLDBERG and Y. W. KAN. 1996. Prenatal diagnosis of sickle cell anaemia and thalassemia by analysis of fetal cells in maternal blood. *Nature Genetics* 14: 264–268. [19]

CHILDS, B., J. M. FINUCCI, M. S. PRESTON, and A. E. PULVER. 1976. Human behavior genetics. *Advances in Human Genetics* 7: 57–97. [20]

CLARKE, C. A. 1968. The prevention of "Rhesus" babies. *SA* 219 (Nov.): 46–52. Reprinted in Burnet (1976). [18]

CLARKE, J. T. 1996. *A Clinical Guide to Inherited Metabolic Diseases*. Cambridge University Press, Cambridge. [16]

CLAYTON-SMITH, J. and D. DONNAI. 1997. Human malformations. In Rimoin et al., pp. 383–394. [15]

COHEN, J. 1997. AIDS: Advances painted in shades of gray at a D.C. conference. *Science* 275: 615–616. [18]

COHEN, J. 1997. How many genes are there? *Science* 275: 769. [9]

COHEN, J. S. and M. E. HOGAN. 1994. The new genetic medicines. *SA* 271: 76–82. [20]

COHEN, L. A. 1987. Diet and cancer. *SA* 257 (Nov.): 42–48. [17]

COHEN, M. M., JR., R. J. GORLIN and F. C. FRASER. 1997. Craniofacial disorders. In Rimoin et al., pp. 1121–1147. [15]

COHEN, R. 1994. About men. Waiting. *New York Times Magazine* Jan. 16, p. 16. [19]

COLD SPRING HARBOR SYMPOSIA. 1986. *Molecular Biology of Homo sapiens*. Vol. 51 of *CSHS*. Cold Spring Harbor Laboratory, New York. [8]

COLLINS, F. S. 1996. *BRCA1*: Lots of mutations, lots of dilemmas. *NEJM* 334: 186–188. [17]

COLLINS, F. and D. GALAS. 1993. A new five-year plan for the U.S. Human Genome Project. *Science* 262: 43–49. [9]

COLLINS, M. 1925. *Colour-Blindness: With a Comparison of Different Methods of Testing Colour-Blindness*. Harcourt Brace, New York. [4]

COMMITTEE ON SCIENCE AND CREATIONISM. 1984. *Science and Creationism: A View from the National Academy of Science*. National Academy Press, Washington, D. C. [12]

COMMITTEE ON SCIENCE, SPACE, AND TECHNOLOGY. 1994. Gene therapy: status, prospects for the future and government policy implications. (Hearing before the Committee of the U.S. House of Representatives.) Publication 88–241, U.S. Government Printing Office, Washington, DC. [20]

CONLEY, M. E. et al. 1986. Expression of the gene defect in X-linked agammaglobulinemia. *NEJM* 315: 564–567. [18]

CONNOR, J. M. and M. A. FERGUSON-SMITH. 1993. *Essential Medical Genetics*, 4th ed. Blackwell Scientific Publications, Oxford. [10]

CONNORS, E., T. LUNDREGAN, N. MILLER, and T. McEWEN. 1996. *Convicted by Juries, Exonerated by Science: Case Studies in the Use of DNA Evidence to Establish Innocence After Trial*. U.S. National Institute of Justice, Washington, DC. [8]

COOK-DEEGAN, R. 1994. *The Gene Wars: Science, Politics, and the Human Genome Project*. Norton, New York. [9]

COOPER, G. M. 1995. *Oncogenes*, 2nd ed. Jones and Bartlett, Boston. [17]

COOPER, G. M. 1997. *The Cell: A Molecular Approach*. Sinauer Associates, Sunderland MA and ASM Press, Washington DC. [2,3,17]

COPPES, M. J., D. A. HABER and P. E. GRUNDY. 1994. Genetic events in the development of Wilms' tumor. *NEJM* 331: 586–590. [17]

CORWIN, H. O. and J. B. JENKINS (eds.). 1976. *Conceptual Foundations of Genetics: Selected Readings*. Houghton Mifflin, Boston. [4]

COX, D. R. and R. M. MYERS. 1996. A map to the future. *Nature Genetics* 12: 117–118. [9]

COX, D. W. 1995. α_1-Antitrypsin deficiency. In Scriver et al., pp. 4125–4158. [16]

CRICK, F. 1988. *What Mad Pursuit*. Basic Books, New York. [6]

CRONISTER, A. 1996. Genetic counseling. In *Fragile X Syndrome*, R. J. Hagerman and A. Cronister (eds.), pp. 251–282. Johns Hopkins University Press, Baltimore. [10]

CROW, J. F. 1983. *Genetics Notes: An Introduction to Genetics*, 8th ed. Burgess, Minneapolis. [4]

CROW, J. F. 1987. Population genetics history: A personal view. *Annual Review of Genetics* 21: 1–22. [12]

CROW, J. F. 1988. A diamond anniversary: The first chromosome map. *Genetics* 118: 1–3. [9]

CROW, J. F. 1988. Eighty years ago: The beginnings of population genetics. *Genetics* 119: 473–476. [12]

CROW, J. F. 1988. Eugenics: Must it be a dirty word? *Contemporary Psychology* 33: 10–12. [20]

CROW, J. F. 1993. Francis Galton: Count and measure, measure and count. *Genetics* 135: 1–4. [11]

CULLITON, B. J. 1996. AIDS—a treatable disease at last. *Nature Medicine* 2: 253. [18]

CULOTTA, E. 1995. Asian hominids grow older. *Science* 270: 1116–1117. [12]

CULVER, K. W. 1996a. *Gene Therapy: A Primer for Physicians*, 2nd ed. Liebert, Larchmont, NY. [20]

CULVER, K. W. 1996b. Gene therapy (Poster). *TIG* 12 (supplement to June issue). [20]

CUNNINGHAM, C. 1996. *Understanding Down Syndrome: An Introduction for Parents*. Brookline Books, Cambridge, Massachusetts. [13]

CUNNINGHAM, P. 1991. The genetics of thoroughbred horses. *SA* 264 (May): 92–98. [1]

D'ALTON, M. E. and DeCHERNEY, A. H. 1993. Prenatal diagnosis. *NEJM* 328: 114–120. [19]

DARNELL, J. E., JR. 1985. RNA. *SA* 253 (Oct.): 68–78. [6]

DARWIN, C. 1859. *On the Origin of Species by Means of Natural Selection.* John Murray, London. A facsimile of the first edition has been reprinted by Harvard University Press (1964). The 6th edition (1872) is also often cited. [12]

DARWIN, C. 1875. *The Variation of Plants and Animals under Domestication,* 2nd ed., p. 319. John Murray, London. [5]

DARWIN, C. 1969. *The Autobiography of Charles Darwin, 1809–1882.* (With original omissions restored; edited with appendix and notes by his grand-daughter, Nora Barlow) W. W. Norton, New York. [12]

DAUSSET, J. and H. CANN. 1994. Our genetic patrimony. *Science* 264: 1991. [9]

DAVIES, K. and M. WHITE. 1996. *Breakthrough: The Race to Find the Breast Cancer Gene.* Wiley, New York. [5,17]

DAVIS, A. P., D. P. WITTE, H. M. HSIEH-LI, S. S. POTTER and M. R. CAPECCHI. 1995. Absence of radius and ulna in mice lacking *hoxa-11* and *hoxd-11*. *Science* 375: 791–795. [15]

DAVIS, D. L. and H. L. BRADLOW. 1995. Can environmental estrogens cause breast cancer? *SA* 273 (Oct.): 166–172. [17]

DAVIS, G. T. and A. J. BRUWER. 1991. Two book reviews: (1) *Health Effects of Exposure to Low Levels of Ionizing Radiation: BEIR V* by the Committee on the Biological Effects of Ionizing Radiation, National Academy Press, Washington, D. C. 1990; (2) *Radiation-Induced Cancer from Low-Dose Exposure: An Independent Analysis* by John W. Gofman, Committee for Nuclear Responsibility Book Division, San Francisco. 1990. *NEJM* 324: 497–499. [7]

DAWKINS, R. 1986. *The Blind Watchmaker.* W. W. Norton, New York. [12]

DAY, S. 1990. Uveal tract. In *Pediatric Ophthalmology,* D. Taylor (ed.), pp. 276–298. Blackwell Scientific, Boston. [15]

DE BOULLE, K. et al. 1993. A point mutation in the *FMR-1* gene associated with fragile X mental retardation. *Nature Genetics* 3: 31–35. [10]

DE LA CHAPELLE, A. 1986. The use and misuse of sex chromatin screening for gender identification of female athletes. *JAMA* 256: 1920–1923. [15]

DE ROBERTIS, E. M., G. OLIVER and C. V. E. WRIGHT. 1990. Homeobox genes and the vertebrate body plan. *SA* 263 (July): 46–52. [15]

DEAN, G. 1957. Pursuit of a disease. *SA* 196 (March): 133–142. [16]

DEFORD, F. 1986. *Alex: The Life of a Child.* Signet, New American Library, New York. [5]

DENIS, J. 1667. An extract of a letter … touching a late cure of an inveterate phrensy by the transfusion of blood. *The Royal Society of London, Philosophical Transactions* 2: 617–623. [18]

DESMOND, E. W. 1987. Out in the open: Changing attitudes and new research give fresh hope to alcoholics. *Time* 130 (Nov. 30): 80–90. [11]

DESNICK, R. J. (ed.). 1991. *Treatment of Genetic Diseases.* Churchill Livingstone, New York. [20]

DIAMOND, J. M. 1991. Curse and blessing of the ghetto. *Discover* 12: 60–65. [16]

DIAMOND, J. M. 1994. Jewish lysosomes. *Nature* 368: 291–292. [20]

DIAMOND, M. and K. SIGMUNDSON. 1997. Sex reassignment at birth: Long-term review and clinical implications. *Archives of Pediatrics and Adolescent Medicine* 151: 298–304. [15]

DIB, C. et al. 1996. A comprehensive genetic map of the human genome based on 5,264 microsatellites. *Nature* 380: 152–154 and extended reprint (paper copy available on request or on the internet at http:\ \www.genethon.fr). [9]

DICKMANN, Z., T. H. CHEWE, W. A. BONNEY, JR. and R. W. NOYES. 1965. The human egg in the pronuclear stage. *The Anatomical Record* 152: 293–302. [3]

DISTECHE, C. M. 1995. Escape from X inactivation in human and mouse. *TIG* 11: 17–22. [10]

DISTECHE, C. M. 1997. The great escape. *ASHG* 60: 1312–1315. [10]

DIXON, B. 1984. Of different bloods. (20 discoveries that shaped our lives). *Science 84* (Nov.): 65–67. [18]

DOBZHANSKY, THEODOSIUS. 1937. *Genetics and the Origin of Species.* Columbia University Press, New York. [12]

DOLL, R. and PETO, R. 1987. Epidemiology of cancer. In *Oxford Textbook of Medicine,* 2nd ed. Vol. 1, D. J. Weatherall, J. G. G. Ledingham and D. A. Warrell (eds.), pp. 4.95–4.123. Oxford Medical Publishers, Oxford. [17]

DOLPHIN, C. T., A. JANMOHAMED, R. L. SMITH, E. A. SHEPHARD and I. R. PHILLIPS. 1997. Missense mutation in flavin-containing mono-oxygenase 3 gene, *FMO3,* underlies fish-odour syndrome. *Nature Genetics* 17: 491–494. [16]

DONAHUE, R. P., W. B. BIAS, J. H. RENWICK and V. A. McKUSICK. 1968. Probable assignment of the Duffy blood group locus to chromosome 1 in man. *PNAS* 61: 949–955. [4]

DONOHOUE, P. A., K. PARKER and C. J. MIGEON. 1995. Congenital adrenal hyperplasia. In Scriver et al., pp. 2929–2966. [16]

DOOLITTLE, R. F. 1985. Proteins. *SA* 253 (Oct.): 88–99. [6]

DOWN, J. L. H. 1866. Observations on an ethnic classification of idiots. *London Hospital Clinical Records and Reports* 3: 259–262. [13]

DRLICA, K. 1992. *Understanding DNA and Gene Cloning: A Guide for the Curious,* 2nd ed. Wiley, New York. [8]

DRLICA, K. 1994. *Double-Edged Sword: The Promises and Risks of the Genetic Revolution.* Addison-Wesley, Reading, MA. [8]

DUCHENNE, G. B. A. 1868. Recherches sur la paralysie musculaire pseudo-hypertrophique ou paralysie myosclerosique. *Archives Generales de Medicine* 11: 5–25. [5]

DUKE, R. C., D. M. OJCIUS and J. D.-E YOUNG. 1996. Cell suicide in health and disease. *SA* 275 (December): 80–87. [15]

DULBECCO, R. 1976. Francis Peyton Rous. *National Academy of Sciences, Biographical Memoirs* 48: 274–306. [17]

DUNN, L. C. 1962. Cross currents in the history of human genetics. *AJHG* 14: 1–13. [1]

DUPRAW, E. J. 1970. *DNA and Chromosomes.* Holt, Rinehart & Winston, New York. [2]

DVORÁK, M, J. TESARIK, L. PILKA and P. TRÁVNÍK. 1982. Fine structure of human two-cell ova fertilized and cleaved in vitro. *Fertility and Sterility* 37: 661–667. [3]

DYKENS, E. M., R. M. HODAPP and J. F. LECKMAN. 1994. *Behavior and Development in Fragile X Syndrome.* SAGE, Thousand Oaks, CA. [10]

EDWARDS v AGUILLARD. 1987. *United States Reports: Cases Adjudged in the Supreme Court* 482: 578–640. [12]

EDWARDS, C. Q. and J. P. KUSHNER. 1993. Screening for hemochromatosis. *NEJM* 328: 1616–1620. [5]

EDWARDS, S. A. 1996. Pork liver, anyone? *Technology Review* 99 (July): 42–49. [18]

EICHENLAUB-RITTER, U. 1996. Parental age-related aneuploidy in human germ cells and offspring: A story of past and present. *Environmental and Molecular Mutagenesis* 28: 211–236. [13,14]

EICHER, E. M. and L. L. WASHBURN. 1986. Genetic control of primary sex determination in mice. *Annual Review of Genetics* 20: 327–360. [15]

EISENSMITH, R. C. et al. 1992. Multiple origins for phenylketonuria in Europe. *AJHG* 51: 1355–1365. [16]

EISENSTEIN, B. I. 1990. The polymerase chain reaction: A new method of using molecular genetics for medical diagnosis. *NEJM* 322: 178–183. [8]

ELDADAH, Z. A., J. A. GRIFO and H. C. DIETZ. 1995. Marfan syndrome as a paradigm for transcript-targeted preimplantation diagnosis of heterozygous mutation. *Nature Medicine* 1: 798–803. [19]

ELIAS, S. and J. L. SIMPSON. 1997. Prenatal diagnosis In Rimoin et al., pp. 563–580. [19]

ELLIS, R. E., J. YUAN and H. R. HORVITZ. 1991. Mechanisms and functions of cell death. *Annual Review of Cell Biology* 7: 663–698. [15]

EMERY, A. E. H. 1994. *Muscular Dystrophy: The Facts.* Oxford University Press, New York. [5]

EMERY, A. E. H. and M. L. H. EMERY. 1995. *The History of a Genetic Disease: Duchenne Muscular Dystrophy or Meryon's Disease.* Royal Society of Medicine, London. [5]

ENGELHARD, V. H. 1994. How cells process antigens. *SA* 271 (Aug.): 54–61. [18]

EPEL, D. 1977. The program of fertilization. *SA* 237 (Nov.): 128–138. [3]

EPEL, D. 1980. Fertilization. *Endeavor* (new series) 4: 26–31. [3]

EPHRUSSI, B. and M. C. WEISS. 1969. Hybrid somatic cells. *SA* 220 (April): 26–34. [9]

EPSTEIN, C. J. (ed.). 1991. *The Morphogenesis of Down Syndrome.* Wiley-Liss, New York. [13]

EPSTEIN, C. J. (ed.). 1993. *The Phenotypic Mapping of Down Syndrome and Other Aneuploid Conditions.* Wiley-Liss, New York. [13]

EPSTEIN, C. J. 1986. *The Consequences of Chromosome Imbalance: Principles, Mechanisms, and Models.* Cambridge University Press, New York. [13,14]

EPSTEIN, C. J. 1988. Mechanisms of the effects of aneuploidy in mammals. *Annual Review of Genetics* 22: 51–75. [13,14]

EPSTEIN, C. J. 1993. Seven momentous years. *AJHG* 53: 1163–1166. [1]

EPSTEIN, C. J. 1995. Down syndrome (trisomy 21). In Scriver et al., pp. 749–794. [13]

EPSTEIN, C. J. 1997. 1996 ASHG Presidential Address: Toward the Twenty-First Century. *AJHG* 60: 1–9. [1]

EPSTEIN, C. J., T. HASSOLD, I. T. LOTT, L. NADEL and D. PATTERSON (eds.). 1995. *Etiology and Pathogenesis of Down Syndrome.* Progress in Clinical and Biological Research, Vol 393. Wiley-Liss, New York. [13]

ERBE, R. W. and H. L. LEVY. 1997. Neonatal screening. In Rimoin et al., pp. 581–593. [19]

ERNST, M. et al. 1996. Presynaptic dopaminergic deficits in Lesch-Nyhan disease. *NEJM* 334: 1568–1572 [16]

ESKENAZI, B. 1993. Caffeine during pregnancy: Grounds for concern? *JAMA* 270: 2973–2974. [7]

ESTABROOKS, L. L. et al. 1995. Preliminary phenotypic map of chromosome 4p16 based on 4p deletions. *American Journal of Medical Genetics* 57: 581–586. [14]

ESTIVILL, X. 1996. Complexity in a monogenic disease. *Nature Genetics* 12: 348–350. [5]

EVANS, D. A. P. 1997. Pharmacogenetics. In Rimoin et al., pp. 455–477. [16]

EVANS, T. 1997. *Porphyria.* New Horizon Press, Far Hills, NJ. [16]

EWART, A. K. et al. 1993. Hemizygosity at the elastin locus in a developmental disorder, Williams syndrome. *Nature Genetics* 5: 11–16. [14]

FACKELMANN, K. 1995. Drug wards off sickle-cell attacks. *Science News* 147: 68. [7]

FACKELMANN, K. 1997. Rusty organs: researchers identify the gene for iron-overload disease. *Science News* 151: 46–47. [5]

FAUCI, A. S. 1996. Resistance to HIV-1 infection: It's in the genes. *Nature Medicine* 2: 966–967. [18]

FAUSTO-STERLING, A. 1995. Animal models for the development of human sexuality: A critical evaluation. *Journal of Homosexuality* 28: 217–236. [15]

FEDER, J. N. et al. 1996. A novel MHC class I-like gene is mutated in patients with hereditary haemochromatosis. *Nature Genetics* 13: 399–408. [5]

FEINBERG, A. P. 1993. Genomic imprinting and gene activation in cancer. *Nature Genetics* 4: 110–113. [10]

FEINBERG, A. P. 1995. A domain of abnormal imprinting in human cancer. In *Genomic Imprinting: Causes and Consequences*, R. Ohlsson, K. Hall and M. Ritzen (eds.), pp. 273–292. Cambridge University Press, Cambridge. [10]

FELGNER, P. L. 1997. Nonviral strategies for gene therapy. *SA* 276: 102–106. [20]

FELSENFELD, G. 1985. DNA. *SA* 253 (Oct.): 58–67. [6]

FÉREC, C. et al. 1992. Detection of over 98% cystic fibrosis mutations in a Celtic population. *Nature Genetics* 1: 188–191. [19]

FERGUSON-SMITH, M. A. et al. 1992. Olympic row over sex testing. *Nature* 355: 10. [15]

FERNANDES, J., J. M. SAUDUBRAY and G. VAN DAN BERGHE (eds.). 1995. *Inborn Metabolic Diseases: Diagnosis and Treatment.* Springer-Verlag, Berlin. [16]

FERNÁNDEZ-CAÑÓN, J. et al. 1996. The molecular basis of alkaptonuria. *Nature Genetics* 14: 19–24. [16]

FERRIS, T. 1990. The space telescope: A sign of intelligent life. *NYT* April 29, Section 4, pp.1ff. [1]

FIELDS, C., M. D. ADAMS, O. WHITE and J. C. VENTER. 1994. How many genes in the human genome? *Nature Genetics* 7: 345–346. [9,10]

FIGUERA, L. E., M. PANDOLFO, P. W. DUNNE, J. M. CANTU and P. I. PATEL. 1995. Mapping of the congenital generalized hypertrichosis locus to chromosome Xq24-q27.1. *Nature Genetics* 10: 202–207. [5]

FINN, R. 1991. Different minds. *Discover* 12 (June): 55–58. [11]

FINN, R. 1997. As Alzheimer's studies progress, debate on cause persists. *The Scientist* 11 (Jan. 20): 14, 17. [13]

FISCHEL-GHODSIAN, N. and R. E. FALK. 1997. Deafness. In Rimoin et al., pp. 1149–1170. [10]

FISCHMAN, J. 1996. Evidence mounts for our African origins—and alternatives. *Science* 271: 1364. [12]

FISHER, J. M., J. F. HARVEY, N. E. MORTON and P. A. JACOBS. 1995. Trisomy 18: Studies of the parent and cell division of origin and the effect of aberrant recombination on nondisjunction. *AJHG* 56: 669–675. [13]

FLEMMING, W. 1879. Contributions to the knowledge of the cell and its life phenomena. *Archiv fur Mikroskopische Anatomie* 16: 302–406. (Abridged translation reprinted in Gabriel and Fogel 1955.) [2]

FLETCHER, R. 1991. *Science, Ideology, and the Media: The Cyril Burt Scandal.* Transaction Publishers, New Brunswick, NJ. [11]

FØLLING, A. and K. CLOSS. 1938. Occurrence of 1-phenylalanine in urine and blood in phenylpyruvic imbecility. *Zeitschrift fur Physiol. Chem.* 254: 115–116. [1]

FRANCIS, D. P. and J. CHIN. 1987. The prevention of acquired immunodeficiency syndrome in the United States. *JAMA* 257: 1357–1366. [18]

FRANCKE, U. 1981. High-resolution ideograms of trypsin-Giemsa banded human chromosomes. *Cytogenetics and Cell Genetics* 31: 24–32. [2]

FRANCKE, U. et al. 1985. Minor Xp21 chromosome deletion in a male associated with expression of Duchenne muscular dystrophy, chronic granulomatous disease, retinitis pigmentosa, and McLeod syndrome. *AJHG* 37: 250–267. [14]

FRANGISKAKIS, J. M. et al. 1996. *LIM-kinase1* hemizygosity implicated in impaired visuospatial constructive cognition. *Cell* 86: 59–69. [14]

FRANKLIN, D. 1989. What a child is given. *The New York Times Magazine*, Sept. 3, pp. 36ff. [11]

FREIFELDER, D. 1978. *The DNA Molecule: Structure and Properties. Original Papers, Analyses, and Problems.* W. H. Freeman, San Francisco. [6]

FREUND, P. A. 1972. Genetic engineering: By whom and for what? Unpublished speech at Worcester Polytechnic Institute, May 2. [20]

FRIEDMAN, R. C. and J. I. DOWNEY. 1994. Homosexuality. *NEJM* 331: 923–930. [11]

FRIEDMANN, T. (ed.). 1991. *Therapy for Genetic Disease.* Oxford University Press, Oxford. [20]

FRIEDMANN, T. 1996. Human gene therapy—an immature genie, but certainly out of the bottle. *Nature Medicine* 2: 144–147. [20]

FRIEDMANN, T. 1997. Overcoming the obstacles to gene therapy. *SA* 276: 96–101. [20]

FUJITA, T. 1975. Abnormal spermatozoa and infertility (man). In *Scanning Electron Microscopical Atlas of Mammalian Reproduction*, E. S. E. Hafez (ed.), pp. 82–87. Springer-Verlag, New York. [3]

FUTUYMA, D. J. 1995. *Science on Trial: The Case for Evolution.* Sinauer Associates, Sunderland MA. [12]

GAJDUSEK, D. C. 1977. Unconventional viruses and the origin and disappearance of kuru. *Science* 197: 943–960. [10]

GALINAT, W. C. 1977. The origin of corn. In *Corn and Corn Improvement*, G. F. Sprague (ed.).

pp. 1–47. American Society of Agronomy, Madison, Wisconsin. [1]

GALLO, R. C. and L. MONTAGNIER. 1988. AIDS in 1988. *SA* 259 (Oct.): 40–48. (The entire issue is devoted to AIDS.) [6,18]

GALTON, F. 1905. Eugenics: Its definition, scope and aims. In *Sociological Papers*, pp 45–50. Macmillan & Company, London. Reprinted in Bajema 1976). [20]

GALTON, F. 1909. *Memories of My Life*, 3rd ed. Methuen & Co., London. [11]

GARATTINI, S. (ed.). 1993. *Caffeine, Coffee, and Health.* Raven Press, New York. [7]

GARBER, A. P., R. SCHRECK and D. E. CARLSON. 1997. Fetal loss. In *Emery and Rimoin's Principles and Practice of Medical Genetics*, 3rd ed., D. L. Rimoin, J. M. Connor and R. E. Pyeritz (eds.), pp. 677–686. [14]

GARDNER, R. J. M. and G. R. SUTHERLAND. 1996. *Chromosome Abnormalities and Genetic Counseling.* Oxford University Press, New York. [10,13,14]

GARROD, A. 1909. *Inborn Errors of Metabolism.* Frowde, Hodder and Stoughton, London. Reprinted in 1963 with a supplement by H. Harris. Oxford University Press, London. [16]

GEDDA, L. 1961. *Twins in History and Science.* Charles C. Thomas, Springfield, IL. [11]

GEJMAN, P. V. et al. 1994. No structural mutation in the dopamine D_2 receptor gene in alcoholism or schizophrenia. *JAMA* 271: 204–208. [11]

GERHART, J. and M. KIRSCHNER. 1997. *Cells, Embryos, and Evolution.* Blackwell Science, Malden, MA. [15]

GIBBONS, A. 1997. Ideas on human origins evolve at anthropology meeting. *Science* 276: 535–536. [12]

GILBERT, S. F. 1997. *Developmental Biology*, 5th ed. Sinauer Associates, Sunderland, MA. [15]

GILLIS, A. M. 1994. Turning off the X chromosome. *BioScience* 44: 128–132. [10]

GILLIS, A. M. 1995. Finding the right sperm for the job. *BioScience* 45: 525–526. [20]

GLADER, B. E. and L. NAUMOVSKI. 1997. Other hereditary red blood cell disorders. In Rimoin et al., pp. 1627–1649. [16]

GLAUSIUSZ, J. 1995. Hidden benefits. *Discover* (March): 30–31. [5]

GLICK, B. R. and J. J. PASTERNAK. 1994. *Molecular Biotechnology: Principles and Applications of Recombinant DNA.* ASM Press, Washington D.C. [9]

GODDARD, H. H. 1912. *The Kallikak Family: A Study in the Heredity of Feeble-Mindedness.* Macmillan, New York. [20]

GOFFEAU, A. et al. 1996. Life with 6000 genes. *Science* 274: 546,563–567. [9]

GOLDBERG, C. 1998. DNA databanks giving police a powerful weapon and critics. *NYT* Feb. 19, pp. A1ff. [8]

GOLDBERG, C. and G. KOLATA. 1998. Scientists announce births of cows cloned in new way. *NYT* Jan. 21, p. A14. [20]

GOLDMAN, D. 1996. High anxiety. *Science* 274: 1483. [11]

GOLDMAN, M. 1996. Cancer risk of low-level exposure. *Science* 271: 1821–1822. [7]

GOLDSTEIN, J. L., H. H. HOBBS and M. S. BROWN. 1995. Familial hypercholesterolemia. In *The Metabolic and Molecular Basis of Inherited Disease*, 7th ed. C. R. Scriver, A. L. Beaudet, W. S.

Sly and D. Valle (eds.), pp. 1981–2030. McGraw-Hill, New York. [5]

GOLEMAN, D. 1988. An emerging theory on Black's I.Q. scores. *New York Times Education Life* (Section 12), Apr. 10: 22–24. [11]

GOODFELLOW, P. 1995. A big book of the human genome. *Nature* 377: 285–286. [9]

GOODFELLOW, P. N. and R. LOVELL-BADGE. 1993. *SRY* and sex determination in mammals. *Annual Review of Genetics* 27: 71–92. [15]

GOODWIN, D. W. 1991. The genetics of alcoholism. In *Genes, Brain, and Behavior*, P. R. McHugh and V. A. McKusick (eds.), pp 219–226. Raven Press, New York. [11]

GORDEEVA, E., with E. M. SWIFT. 1996. *My Sergei: A Love Story*. Warner Books, New York. [5]

GORDON, J. W. and F. H. RUDDLE. 1983. Gene transfer into mouse embryos: Production of transgenic mice by pronuclear injection. *Methods in Enzymology* 101: 411–433. [17]

GORMAN, C. 1996/97. Man of the year: The disease detective. *Time* 148 (Dec. 30/Jan. 6): 56–64. (See also other articles in this issue.) [18]

GOULD, S. J. 1986. Reflections from an interior world. *Nature* 320: 647–648. [1]

GOULD, S. J. 1995. *Dinosaur in a Haystack: Reflections in Natural History*. Harmony Books, New York. [12]

GOULD, S. J. 1996. *The Mismeasure of Man*, revised ed. W. W. Norton, New York. [11, 20]

GRADY, D. 1996. Tracing a genetic disease to bits of traveling DNA. *NYT* Mar. 3, p. C1. [7]

GRADY, D. 1997. Study finds secondhand smoke doubles risk of heart disease. *NYT* May 20, pp. A1ff. [17]

GRANDJEAN, P. (ed.). 1991. *Ecogenetics: Genetic Predisposition to the Toxic Effects of Chemicals*. Routledge, Chapman, & Hall/World Health Organization, New York. [16]

GRAVEL, R. A. et al. 1995. The GM$_2$ gangliosidoses. In Scriver et al., pp. 2839–2879. [16]

GREEN, E. D., D. R. COX and R. M. MYERS. 1995. The human genome project and its impact on the study of human disease. In Scriver et al., pp. 401–436. [9]

GREENBERG, D. S. 1986. What ever happened to the war on cancer? *Discover* 7 (March): 47–64. [17]

GREENBERG, D. S. 1991. A sober anniversary of the "War on Cancer." *The Lancet* 338: 1582–1583. [17]

GREENE, W. C. 1993. AIDS and the immune system. *SA* 269 (Sep.): 99–105. [18]

GREENSPAN, R. J. 1995. Understanding the genetic construction of behavior. *SA* 272: 72–78. [11]

GREIDER, C. W. and E. H. BLACKBURN. 1996. Telomeres, telomerase, and cancer. *SA* 274 (Feb.): 92–97. [6]

GRIFFIN, D. K. 1994. Fluorescent *in situ* hybridization for the diagnosis of genetic disease at postnatal, prenatal and preimplantation stages. *International Review of Cytology* 153: 1–40. [9]

GRIFFIN, J. E., M. J. McPHAUL, D. W. RUSSELL and J. D. WILSON. 1995. The androgen resistance syndromes: Steroid 5-α-reductase 2 deficiency, testicular feminization and related disorders. In Scriver et al., pp. 2967–2998. [15]

GRIMALDI, D. A. 1996. Captured in amber. *SA* 274 (April): 84–91. [12]

GROBSTEIN, C. 1979. External human fertilization. *SA* 240 (June): 57–67. [3]

GROCE, N. E. 1985. *Everyone Here Spoke Sign Language: Hereditary Deafness on Martha's Vineyard*. Harvard University Press, Cambridge, MA. [10]

GRODY, W. W. et al. 1997. PCR-based screening for cystic fibrosis carrier mutations in an ethnically diverse pregnant population. *AJHG* 60: 935–947. [19]

GRUNSTEIN, M. 1992. Histones as regulators of genes. *SA* 267 (Oct.): 68–74B. [2]

GRUSS, P. and C. WALTHER. 1992. *Pax* in development. *Cell* 69: 719–722. [15]

GUMBINER, B. M. 1996. Cell adhesion: The molecular basis of tissue architecture and morphogenesis. *Cell* 84: 345–357. [15]

GURA, T. 1996. Ten get the call to Stockholm. *Science* 274: 345. [18]

GUSELLA, J. F. et al. 1983. A polymorphic DNA marker genetically linked to Huntington's disease. *Nature* 306: 234–238. [9]

GUTMANN, D. H. and F. S. COLLINS. 1995. von Recklinghausen neurofibromatosis. In *Molecular and Metabolic Basis of Inherited Disease*, 7th ed., C. R. Scriver, A. L. Beaudet, W. S. Sly and D. Valle (eds.), pp. 677–696. McGraw-Hill, New York. [10]

HABER, D. A. 1995. Clinical implications of basic research: Telomeres, cancer, and immortality. *NEJM* 332: 955–956. [6]

HADDOW, J. E. et al. 1992. Prenatal screening for Down's syndrome with use of maternal serum markers. *NEJM* 327: 588–593. [19]

HADDOW, J. E. et al. 1994. Reducing the need for amniocentesis in women 35 years of age or older with serum markers for screening. *NEJM* 330: 1114–1118. [19]

HAFEZ, E. S. E. and P. KENEMANS (eds.). 1982. *Atlas of Human Reproduction by Scanning Electron Microscopy*. MIT Press, Boston. [3]

HAGERMAN, R. J. 1996. Physical and behavioral phenotype. In *Fragile X Syndrome*, R. J. Hagerman and A. Cronister (eds.), pp. 3–87. Johns Hopkins University Press, Baltimore. [10]

HAGERMAN, R. J. and A. CRONISTER (eds.). 1996. *Fragile X Syndrome*. Johns Hopkins University Press, Baltimore. [10]

HALL, J. G. 1990. Genomic imprinting: Review and relevance to human diseases. *AJHG* 46: 857–873. [10]

HALL, J. G. 1992. Genomic imprinting and its clinical implications. *NEJM* 326: 827–829. [10]

HALL, J. G., and E. LOPEZ-RANGEL. 1997. Twins and twinning. In Rimoin et al., pp. 395–404. [11]

HALL, J. M. et al. 1990. Linkage of early-onset familial breast cancer to chromosome 17q21. *Science* 250: 1685–1689. [9]

HALLIDAY, J. L., L. F. WATSON, J. LUMLEY, D. M. DANKS and L. J. SHEFFIELD. 1995. New estimates of Down syndrome risks at chorionic villus sampling, amniocentesis and livebirth in women of advanced maternal age from a uniquely defined population. *Prenatal Diagnosis* 15: 455–465. [13]

HAMOSH, A. and M. COREY, for the Cystic Fibrosis Genotype-Phenotype Consortium. 1993. Correlation between genotype and phenotype in patients with cystic fibrosis. *NEJM* 329: 1308–1313. [5]

HANDYSIDE, A. H. and J. D. A. DELHANTY. 1997. Preimplantation genetic diagnosis: Strategies and surprises. *TIG* 13: 270–275. [19]

HANSON, I. and V. VAN HEYNINGEN. 1995. *Pax6*: More than meets the eye. *TIG* 11: 268–272. [15]

HARDY, G. H. 1908. Mendelian proportions in a mixed population. *Science* 28: 49–50. (Reprinted in Peters 1959, Morris 1971, and Jameson 1977.)] [12]

HARPER, P. S. 1990. Myotonic dystrophy and related disorders. In *Emery and Rimoin's Principles and Practice of Medical Genetics*, 2nd ed., D. L. Rimoin, J. M. Connor and R. E. Pyeritz (eds.), pp. 579–597. Churchill Livingstone, Edinburgh. [7]

HARRIS, A. and M. SUPER. 1995. *Cystic Fibrosis: The Facts*, 3rd ed. Oxford University Press, New York. [5]

HARRIS, H. 1980. *The Principles of Human Biochemical Genetics*, 3rd ed. Elsevier/North-Holland Biomedical Press, Amsterdam. [16]

HART, S. 1994. The drama of cellular death. *BioScience* 44: 451–455. [15]

HARTL, D. L. 1988. *A Primer of Population Genetics*, 2nd ed. Sinauer Associates, Sunderland, MA. [12]

HARTWELL, L. 1995. Introduction to cell cycle controls. In *Cell Cycle Control*, C. Hutchison and D. M. Glover (eds.), pp. 1–15. Oxford University Press, New York. [3]

HASELTINE, W. A. 1997. Discovering genes for new medicines. *SA* 276: 92–97. [9]

HASSOLD, T., D. PETTAY, A. ROBINSON and I. UCHIDA. 1992. Molecular studies of parental origin and mosaicism in 45,X conceptuses. *Human Genetics* 89: 647–652. [13]

HASSOLD, T. et al. 1996. Human aneuploidy: Incidence, origin and etiology. *Environental and Molecular Mutagenesis* 28: 167–175. [13,14]

HAYDEN, M. R. and B. KREMER. 1995. Huntington disease. In Scriver et al., pp. 4483–4510. [7]

HEARD, E., P. CLERC and P. AVNER. 1997. X-chromosome inactivation in mammals. *Annual Review of Genetics* 31: 571–610. [10]

HEARNSHAW, L. S. 1979. *Cyril Burt, Psychologist*. Cornell University Press, Ithaca, NY. [11]

HECHT, C. A. and E. B. HOOK. 1994. The imprecision in rates of Down syndrome by 1-year maternal age intervals: A critical analysis of rates used in biochemical screening. *Prenatal Diagnosis* 14: 729–738. [13]

HEIM, S. and F. MITELMAN. 1995. *Cancer Cytogenetics*, 2nd ed. Wiley-Liss, NY. (See Chapter 3, Overview) [17]

HENIKOFF, S. and M. A. MATZKE. 1997. Exploring and explaining epigenetic effects. *TIG* 13: 293–295. [10]

HERRICK, J. B. 1910. Peculiar elongated and sickle-shaped red blood corpuscles in a case of severe anemia. *Archives of Internal Medicine* 6: 517–521. [5]

HERRNSTEIN, R. J. and C. MURRAY. 1994. *The Bell Curve: Intelligence and Class Structure in American Life*. The Free Press, New York. [11]

HESLOP-HARRISON, J. S. and R. B. FLAVELL (eds.). 1993. *The Chromosome*. Bios, Oxford. [2]

HESLOP-HARRISON, J. S., A. R. LEITCH and T. SCHWARZACHER. 1993. The physical organization of interphase nuclei. In *The Chromosome*, J. W. Heslop-Harrison and R. B. Flavell (eds.), pp. 221–232. Bios, Oxford. [2]

HEUSSNER, R. C., JR. and M. E. SALMON. 1988. *Warning: The Media May Be Harmful to Your Health! A Consumer's Guide to Medical News and Advertising.* Andrews and McMeel, Kansas City, MO. [1]

HILEMAN, B. 1995. Views differ sharply over benefits, risks of agricultural biotechnology. *Chemical & Engineering News*, August 21: 8–17. (See also editorial by R. M. Baum on p. 3) [8]

HILTS, P. J. 1991. Gene-altered pigs produce key part of human blood. *NYT* June 26, p. 1. [6]

HILTS, P. J. 1997. Gene may be responsible for cancer cells' immortality. *NYT* Aug. 19, p. C3. [6]

HODGSON, J. 1995. Does less mean more for the genome projects? Large-scale sequencing requires smaller, more powerful processors. *Bio/Technology* 13: 231–233. [8]

HOFFMAN, M. 1991. How parents make their mark on genes. *Science* 252: 1250–1251. [10]

HOGAN, B. L. M. 1996. Bone morphogenetic proteins: multifunctional regulators of vertebrate development. *Genes and Development* 10: 1580–1594. [15]

HOGAN, K. 1997. To fire the train: A second malignant-hyperthermia gene. *AJHG* 60: 1303–1308. [16]

HOLDEN, C. 1987. The genetics of personality. *Science* 237: 598–601. [11]

HOLM, I. A., X. HUANG and L. M. KUNKEL. 1997. Mutational analysis of the *PEX* gene in patients with X–linked hypophosphatemic rickets. *AJHG* 60: 790–797. [5]

HOLMES, E. M. 1996/1997. Solving the insurance/genetic fair/unfair discrimination dilemma in light of the human genome project. *Kentucky Law Journal* 85 (3): 503–664. [19]

HOLMES, H. B. 1985. Sex preselection: Eugenics for everyone? In *Biomedical Ethics Reviews: 1985*, J. M. Humber and R. F. Almeder (eds.). Humana Press, Clifton, NJ. [20]

HOOK, E. B. 1992. Chromosome abnormalities: Prevalence, risks and recurrence. In *Prenatal Diagnosis and Screening*, D. J. H. Brock, C. H. Rodeck and M. A. Ferguson-Smith (eds.), pp. 351–392. Churchill Livingstone, Edinburgh. [14]

HORGAN, J. 1992. Profile: Francis H. C. Crick. *SA* 266 (Feb.): 32–33. [1]

HORGAN, J. 1994. Radon's risks. *SA* (Aug.) 271: 14–16. [7]

HORSTHEMKE, B., B. DITTRICH and K. BUITING. 1995. Parent-of-origin-specific DNA methylation and imprinting mutations on human chromosome 15. In *Genomic Imprinting: Causes and Consequences*, R. Ohlsson, K. Hall and M. Ritzen (eds.), pp. 295–308. Cambridge University Press, Cambridge. [10]

HOTZ, R. L. 1997. Robert Graham, founder of exclusive sperm bank, dies. *Los Angeles Times* Feb. 18, pp. A1ff. [20]

HOUSMAN, D. E. 1995. DNA on trial: The molecular basis of DNA fingerprinting. *NEJM* 332: 534–535. [8]

HOYER, L.W. 1994. Hemophilia. *NEJM* 330: 38–47. [7]

HOYLE, R. 1996. Delaney clause knocked out by surprise compromise act. *Nature Biotechnology* 14: 1056. [7]

HRUBEC, Z. and C. D. ROBINETTE. 1984. The study of human twins in medical research. *NEJM* 310: 435–441. [11]

HSU, T.C. 1979. *Human and Mammalian Cytogenetics: An Historical Perspective.* Springer-Verlag, New York. [2]

HUBBARD, R. and R. C. LEWONTIN. 1996. Pitfalls of genetic testing. *NEJM* 334: 1192–1193. [19]

HUNTER, R. H. F. 1995. *Sex Determination, Differentiation, and Intersexuality in Placental Mammals.* Cambridge University Press, Cambridge. [15]

HUNTINGTON'S DISEASE COLLABORATIVE RESEARCH GROUP. 1993. A novel gene containing a trinucleotide repeat that is expanded and unstable on Huntington's disease chromosomes. *Cell* 72: 971–983. [7]

HUNTINGTON, G. 1872. On chorea. *The Medical and Surgical Reporter* 26: 320–321. Reprinted in 1973 in *Advances in Neurology* 1: 33–35. [5]

HUNTLEY, O. M. 1984. A mother's perspective. *Hastings Center Report* 14(2): 14–15. [6,20]

HUXLEY, A. 1932. *Brave New World*. Harper & Row, New York. [19]

ILTIS, H. 1924. *Gregor Johann Mendel: Leben, Werk und Wirkung.* Julius Springer, Berlin. [English translation by E. Paul and C. Paul . 1932. *Life of Mendel*. Allen & Unwin, London.] [4]

IMPERATO-McGINLEY, J. and T. GAUTIER. 1986. Inherited 5-α-reductase deficiency in man. *TIG* 2: 130–133. [15]

INFORMATION AND EDUCATION COMMITTEE. 1996. *1996–1997 Guide to North American Graduate and Postgraduate Training Programs in Human Genetics.* American Society of Human Genetics, Bethesda, MD. [19]

INMAN, K. and N. RUDIN. 1997. *Introduction to Forensic DNA Analysis*. CRC Press, Boca Raton, FL. [8]

ISHIHARA, S. 1968. *Tests of Colour-Blindness*, 38-plate edition. Kanehara, Tokyo, Japan. [4]

ISSA, J.-P. J. and S. B. BAYLIN. 1996. Epigenetics and human disease. *Nature Medicine* 2: 281–282. [10]

JACOBS, P. A. 1982. The William Allen Memorial Award Address. Human population cytogenetics: The first twenty–five years. *AJHG* 34: 689–698. [1,2]

JACOBS, P. A. 1992. The chromosome complement of human gametes. *Oxford Reviews of Reproductive Biology* 14: 48–72. [14]

JACOBS, P. A. and T. J. HASSOLD. 1995. The origin of numerical chromosomal abnormalities. *Advances in Genetics* 33: 101–133. [13]

JACOBS, P. A. and J. A. STRONG. 1959. A case of human intersexuality having a possible XXY sex-determining mechanism. *Nature* 183: 302–303. [13]

JACOBS, P. A., M. BRUNTON and M. M. MELVILLE. 1965. Aggressive behavior, mental subnormality and the XYY male. *Nature* 208: 1351–1352. [13]

JACOBSON, M.D., M. WEIL and M. C. RAFF. 1997. Programmed cell death in animal development. *Cell* 88: 347–354. [15]

JAENISCH, R. 1997. DNA methylation and imprinting: Why bother? *TIG* 13: 323–329. [10]

JAENISCH, R., C. BEARD and E. LI. 1995. DNA methylation and mammalian development. In *Genomic Imprinting: Causes and Consequences*, R. Ohlsson, K. Hall and M. Ritzen (eds.), pp.118–126. Cambridge University Press, Cambridge. [10]

JAMESON, D. L. (ed.). 1977. *Evolutionary Genetics*. Benchmark Papers in Genetics, Vol. 8. Dowden, Hutchinson & Ross, Stroudsburg, Pennsylvania. [12]

JANEWAY, C. A. 1993. How the immune system recognizes invaders. *SA* 269 (Sep.): 73–79. [18]

JENKINS, T. 1996. The South African malady. *Nature Genetics* 13: 7–9. [16]

JENSEN, A. R. 1969. How much can we boost IQ and scholastic achievement? *Harvard Educational Review* 39 (1): 1–123. Reprinted in *Environment, Heredity, and Intelligence* (1969). Reprint Series Number 2 compiled from the Harvard Educational Review. [11]

JENSEN, A. R. 1981. *Straight Talk about Mental Tests*. The Free Press, New York. [11]

JERVIS, G. A. 1947. Phenylpyruvic oligophrenia deficiency of phenylalanine-oxidizing system. *Society for Experimental Biology and Medicine, Proceedings* 82: 514–515. [1]

JIMÉNEZ, R., A. SÁNCHEZ, M. BURGOS and R. D. DE LA GUARDIA. 1996. Puzzling out the genetics of mammalian sex determination. *TIG* 12: 164–166. [15]

JOHNS, D. R. 1995. Mitochondrial DNA and disease. *NEJM* 333: 638–644. [5]

JOHNSON, G. 1996. Findings pose challenge to immunology's central tenet. *NYT* March 26, pp. C1ff. [18]

JOYNSON, R. B. 1989. *The Burt Affair*. Routledge, London. [11]

JUDSON, H. F. 1979. *The Eighth Day of Creation: Makers of the Revolution in Biology*. Simon & Schuster, New York. [6]

JUDSON, H. F. 1996. *The Eighth Day of Creation: Makers of the Revolution in Biology*, expanded edition. Cold Spring Harbor Laboratory Press, Plainview, NY. [1]

KABACK, M. M. et al. 1993. Tay-Sachs disease. Carrier screening, prenatal diagnosis, and the molecular era: An international perspective. 1970–1993. *JAMA* 270: 2307–2315. [19]

KAHN, P. 1996. Coming to grips with genes and risk. *Science* 274: 496–498. [9]

KAHN, P. 1996. Gene hunters close in on elusive prey. *Science* 271: 1352–1354. [11]

KAHN, P. and A. GIBBONS. 1997. DNA from an extinct human. *Science* 277: 176–178. [12]

KALOW, W. and D. M. GRANT. 1995. Pharmacogenetics. In Scriver et al., pp. 293–326. [16]

KANDEL, E. R., J. H. SCHWARTZ and T. M. JESSELL. 1995. *Essentials of Neural Science and Behavior*. Appleton & Lange, Norwalk, CT. [15]

KAPPAS, A., S. SASSA, R. A. GALBRAITH and Y. NORDMANN. 1995. The porphyrias. In Scriver et al., pp. 2103–2159. [16]

KASE, L. M. 1996. Why community cancer clusters are often ignored. *SA* 275 (Sept.): 85–86. [17]

KASTAN, M. 1995. Ataxia-telangiectasia: Broad implications for a rare disorder. *NEJM* 333: 662–663. [7]

KAYAALP, E. et al. 1997. Human phenylalanine hydroxylase mutations and hyperphenylalaninemia phenotypes: a metanalysis of genotype-phenotype correlations. *AJHG* 61: 1309–1317. [16]

KEATS, B. 1997. Population Genetics. In Rimoin et al., pp. 347–357, Churchill and Livingstone, New York. [12]

KESSLER, D. A. and K. L. FEIDEN. 1995. Faster evaluation of vital drugs. *SA* 272 (March): 48–54. [8]

KESSLER, S. 1997. Genetic counseling is directive? Look again. *AJHG* 61: 466–467. [19]

KEVLES, D. J. 1984a. A secular faith. (Annals of eugenics, part I). *The New Yorker* 60 (Oct. 8): 51ff. [1]

KEVLES, D. J. 1984b. A secular faith (Annals of eugenics, part II). *The New Yorker* 60 (Oct. 15): 52ff. [1]

KEVLES, D. J. 1984c. A secular faith (Annals of eugenics, part III). *The New Yorker* 60 (Oct. 22): 32ff. [1]

KEVLES, D. J. 1984d. A secular faith (Annals of eugenics, part IV). *The New Yorker* 60 (Oct. 29): 51ff. [1]

KEVLES, D. J. 1985. *In the Name of Eugenics: Genetics and the Uses of Human Heredity.* Alfred A. Knopf, New York. [20]

KIMURA, D. 1992. Sex differences in the brain. *SA* 267 (Sept.): 118–125. [15]

KIMURA, M. 1979. The neutral theory of molecular evolution. *SA* 241 (Nov.): 98–126. [12]

KING, R. A., V. J. HEARING, D. J. CREEL and W. S. OETTING. 1995. Albinism. In Scriver et al., pp. 4353–4392. [16]

KING, R. A., W. S. OETTING, D. J. CREEL and V. HEARING. 1997. Abnormalities of pigmentation. In Rimoin et al., pp. 1171–1203. [16]

KING, R. C. and W. D. STANSFIELD. 1996. *A Dictionary of Genetics,* 5th ed. Oxford University Press, New York. [Glossary]

KINGSLEY, D. M. 1994. What do BMPs do in mammals? Clues from the mouse short-ear mutation. *TIG* 10: 16–21. [15]

KINGSTON, H. M. 1997. Genetic assessment and pedigree analysis. In Rimoin et al., pp. 505–520. [5]

KISHINO, T., M. LALANDE and J. WAGSTAFF. 1997. *UBE3A/E6-AP* mutations cause Angelman syndrome. *Nature Genetics* 15: 70–73. [10]

KLEIN, J. 1980. *Woody Guthrie: A Life.* Knopf, New York. [5]

KLINE, A. D. et al. 1993. Molecular analysis of the 18q– syndrome and correlation with phenotype. *AJHG* 52: 895–906. [14]

KLINE, A. H., J. B. SIDBURY, JR. and C. P. RICHTER. 1959. The occurrence of ectodermal dysplasia and corneal dysplasia in one family. *Journal of Pediatrics* 55: 355–366. [10]

KLING, J. 1996. Could transgenic supercrops one day breed superweeds? *Science* 274: 180–181. [8]

KLINGER, K. et al. 1992. Rapid detection of chromosomal aneuploides in uncultured amniocytes by using fluorescence in situ hybridization (FISH). *AJHG* 51: 55–65. [19]

KOLATA, G. 1993. Nightmare or the dream of a new era in genetics? *NYT* Dec. 7YT A1ff. [19]

KOLATA, G. 1994. Two chief rivals in the battle over DNA evidence now agree on its use. *NYT* Oct. 27, p. B14 [8]

KOLATA, G. 1994. Should children be told if genes predict illness? *NYT* Sep. 26YT A1ff. [19]

KOLATA, G. 1996a. Map of yeast genes promises clues to diseases of humans. *NYT* April 25, p. B11. [8]

KOLATA, G. 1996b. Biology's big project turns into challenge for computer experts. *NYT* June 11, pp. C1 ff. [8]

KOLATA, G. 1997a. Ten cloned cows soon to be born, company reports, duplicating a lamb experiment. *NYT* Aug. 8, p. A10. [20]

KOLATA, G. 1997b. Clinics selling embryos made for "adoption." *NYT* Nov. 23, pp. A1ff. [20]

KOLATA, G. 1998. Scientists see a mysterious similarity in a pair of deadly plagues. *NYT* May 26, pp. C1, C5. [2]

KOLATA, G. 1998. *Clone: The Road to Dolly and the Path Ahead.* William Morrow, New York. [20]

KONOTEY-AHULU, F. I. D. 1991. *The Sickle Cell Disease Patient.* Macmillan, Houndsmills, Hants. [5]

KORENBERG, J. R. 1993. Toward a molecular understanding of Down syndrome. In *The Phenotypic Mapping of Down Syndrome and Other Aneuploid Conditions,* C. J. Epstein (ed.), pp. 87–115. Wiley-Liss, New York. [13]

KORENBERG, J. R. 1995. Mental modelling. *Nature Genetics* 11: 109–111. [13]

KORENBERG, J. R. et al. 1994. Down syndrome phenotypes: the consequences of chromosome imbalance. *PNAS* 91: 4997–5001. [13]

KORF, B. R. 1996. *Human Genetics: A Problem-Based Approach.* Blackwell Science, Cambridge, MA. [17]

KORNBERG, T. B. and M. P. SCOTT. 1997. Regulation of *Drosophila* development by transcription factors and cell-cell signaling. In *Exploring Genetic Mechanisms,* M. Singer and P. Berg (eds.), pp. 459–513. University Science Books, Sausalito, CA. [15]

KOSHLAND, D. E. 1987. Nature, nurture, and behavior. *Science* 235: 1445. [11]

KOSHLAND, D. E. 1989. The molecule of the year. *Science* 246: 1541. [8]

KOSHLAND, D. E. 1993. Molecule of the year. *Science* 262: 1953. [17]

KOSHLAND, D. E. 1994. Molecule of the year: The DNA repair enzyme. *Science* 266: 1925. [7]

KREEGER, K. Y. 1995. First completed microbial genomes signal birth of new area of study. *The Scientist,* Nov. 27: 14–15. [9]

KRONTIRIS, T. G. 1995. Oncogenes. *NEJM* 333: 303–306. [17]

KRUSH, A. J. and K. A. EVANS. 1984. *Family Studies in Genetic Disorders.* Charles C. Thomas, Springfield, IL. [5]

KUBY, J. 1997. *Immunology,* 3rd ed. W. H. Freeman, New York.

LABIE, D., J. PAGNIER, H. WAJCMAN, M. E. FABRY and R. L. NAGEL 1986. The genetic origin of the variability of the phenotypic expression of the Hb_S gene. In *Genetic Variation and its Maintenance,* D. F. Roberts and G. F. De Stefano (eds.), pp. 149–155. Cambridge University Press, Cambridge. [12]

LACY, P. E. 1995. Treating diabetes with transplanted cells. *SA* 273: 50–58. [18]

LaFOLLETTE, M. C. 1990. *Making Science Our Own.* The University of Chicago Press, Chicago. [1]

LAHN, B. T. and D. C. PAGE. 1997. Functional coherence of the human Y chromosome. *Science* 278: 675–680. [15]

LAKE, J. A. 1981. The ribosome. *SA* 245 (Aug.): 84–97. [6]

LALANDE, M. 1996. Parental imprinting and human disease. *Annual Review of Genetics* 30: 173–195. [10]

LANDER, E. 1996. The new genomics: Global views of biology. *Science* 274: 536–538. [9]

LANDER, E. S. and B. BUDOWLE. 1994. DNA fingerprinting dispute laid to rest. *Nature* 371: 735–738. [8]

LANDSTEINER, K. 1901. Uber Agglutinationserscheinungen normalen men-

schlichen Blutes. *Wiener klinische Wochenschrift* 14: 1132–1134. An English translation is in Boyer (1963). [18]

LANZA, R. P., D. K. C. COOPER, and W. L. CHICK. 1997. Xenotransplantation. *SA* 277 (July): 54–59. [18]

LE BEAU, M. M. 1996. One FISH, two FISH, red FISH, blue FISH. *Nature Genetics* 12: 341–344. [2]

LEARY, W. E. 1995. A partial cure for drug bills for a $370,000-a-year ailment. *NYT* March 8, p. C12. [20]

LEARY, W. E. 1996. Arthritis researchers try gene therapy. *NYT* July 18, p. A16. [20]

LEDBETTER, D. H. and A. BALLABIO. 1995. Molecular cytogenetics of contiguous gene syndromes: mechanisms and consequences of gene dosage imbalance. In Scriver et al., pp. 811–839. [14]

LEDERBERG, J. 1986. Forty years of genetic recombination in bacteria: A fortieth anniversary reminiscence. *Nature* 324: 627–628. [1]

LEFFELL, D. J. and D. E. BRASH. 1996. Sunlight and skin cancer. *SA* 275 (July): 52–59. [17]

LEMAIRE, P. and L. KODJABACHIAN. 1996. The vertebrate organizer: structure and molecules. *TIG* 12: 525–531. [15]

LEMONICK, M. D. 1995. Cosmic close-ups: Stunning new photos from Hubble Space telescope put the mysteries of the universe into sharp focus. *Time* 146 (21; Nov. 20): 90–99. [1]

LENHOFF, H. M., P. P. WANG, F. GREENBERG and U. BELLUGI. 1997. Williams syndrome and the brain. *SA* 277 (Dec.): 68–73. [14]

LeVAY, S. and D. H. HAMER. 1994. Evidence for a biological influence in male sexuality. *SA* 270: 44–49. [11]

LEVIN, R. J. 1987. Human sex pre-selection. In *Oxford Reviews of Reproductive Biology,* Vol. 9, J. R. Clarke (ed.), pp. 161–191. Clarendon Press, Oxford. [20]

LEVINE, A. J. 1995. Tumor suppressor genes. *SA Science & Medicine* 2 (Jan./Feb.): 28–37. [17]

LEVINE, L. (ed.) 1971. *Papers on Genetics: A Book of Readings.* Mosby, St. Louis. [4]

LEVITAN, M. 1988. *Textbook of Human Genetics,* 3rd ed. Oxford University Press, New York. [4]

LEVY, H. 1973. Genetic screening. *Advances in Human Genetics* 4: 1–104. [19]

LEVY, H. 1996. *And the Blood Cried Out.* Basic Books, New York. [8]

LEWIN, B. 1994. *Genes V.* Oxford University Press, New York. [3]

LEWIN, B. 1997. *Genes VI.* Oxford University Press, New York. [2,6]

LEWIN, T. 1996. Move to patent cancer gene touches off storm of protest. *NYT,* May 5, p. A14. [9]

LEWITTER, F. 1996. Genetwork: Resources for human geneticists. *TIG* 12: 531–532. [1]

LEWONTIN, R. C., S. ROSE and L. J. KAMIN. 1984. *Not in Our Genes.* Pantheon Books, New York. [11]

LINDAHL, T. 1997. Facts and artifacts of ancient DNA. *Cell* 90: 1–3. [12]

LINDER, D. and S. M. GARTLER. 1965. Glucose-6-phosphate dehydrogenase mosaicism: Utilization as a cell marker in the study of leiomyomas. *Science* 150: 67–69. [10]

LINDER, M. E. and A. G. GILMAN. 1992. G proteins. *SA* 267: (July): 56–65. [17]

LISON, M., S. H. BLONDHEIM and R. N. MELMED. 1980. A polymorphism of the ability to smell urinary metabolites of asparagus. *British Medical Journal* 281: 1676–1678. [5]

LJUNGQVIST, A. and J. L. SIMPSON, for the International Amateur Athletic Federation Work Group on Gender Verification. 1992. Medical examination for health of all athletes replacing the need for gender verification in international sports: The International Amateur Athletic Federation plan. *JAMA* 267: 850–852. [15]

LODISH, H., D. et al. 1996. *Molecular Cell Biology*, 3rd ed. Scientific American Books/W. H. Freeman, New York. [2,3,5,6,18]

LOMBARDO, P. A. 1985. Three generations, no imbeciles: New light on *Buck v. Bell*. *New York University Law Review* 60: 30–62. [20]

LONGO, F. J. 1987. *Fertilization*. Chapman & Hall, London. [3]

LUND, J. C. 1860. Chorea St. Vitus Dance in Saetersdalen. *Report of Health and Medical Conditions in Norway in 1860*. (Norway's Official Statistics, 1862, C. No. 4), p. 137. Partial translation in Ørbeck 1959. [5]

LUPSKI, J. R. et al. 1991. DNA duplication associated with Charcot-Marie-Tooth disease type 1A. *Cell* 66: 219–232. [14]

LUZZATTO, L. and A. MEHTA. 1995. Glucose 6-phosphate dehydrogenase deficiency. In Scriver et al., pp. 3367–3398. [16]

LYMAN, F. L. 1963. Diets and recipes. In *Phenylketonuria*, F. L. Lyman (ed.). Charles C. Thomas, Springfield, IL. [1]

LYNCH, E. D. et al. 1997. Nonsyndromic deafness DFNA1 associated with mutation of a human homolog of the *Drosophila* gene *diaphanous*. *Science* 278: 1315–1318. [10]

LYNCH, H. T., R. H. HODEN and N. W. PAUL (eds.). 1990. *International Directory of Genetic Services*, 9th ed. March of Dimes Birth Defects Foundation, White Plains, NY. [19]

LYON, M. F. 1992. Some milestones in the history of X-chromosome inactivation. *Advances in Genetics* 26: 17–28. [10]

LYON, M. F. 1995. X chromosome inactivation and imprinting. In *Genomic Imprinting: Causes and Consequences*, R. Ohlsson, K. Hall and M. Ritzen (eds.), pp. 129–141. Cambridge University Press, Cambridge. [10]

MABUCHI, H. et al. 1978. Homozygous familial hypercholesterolemia in Japan. *American Journal of Medicine* 65: 290–297. [5]

MACALPINE, I. and R. HUNTER. 1969. Porphyria and King George III. *SA* 221 (July): 38–46. [11,16]

MACKENZIE, T. D., C. E. BARTECCHI and R. W. SCHRIER. 1994. The human costs of tobacco use. (Second of two parts.) *NEJM* 330: 975–980. [7]

MACKINTOSH, N. J. 1995. *Cyril Burt: Fraud or Framed?* Oxford University Press, Oxford. [11]

MAGA, E. A. and J. D. MURRAY. 1995. Mammary gland expression of transgenes and the potential for altering the properties of milk. *Bio/Technology* 13: 1452–1457. [8]

MAJZOUB, J. A. and L. J. MUGLIA. 1996. Knockout mice. *NEJM* 334: 904–907. [17]

MALKIN, D. et al. 1990. Germ line p53 mutations in a familial syndrome of breast cancer, sarcomas and other neoplasms. *Science* 250: 1233–1238. [17]

MANAK, J. R. and M. P. SCOTT. 1994. A class act: conservation of homeodomain protein functions. In *Development 1994 Supplement: The Evolution of Developmental Mechanisms*, M. Akam, P. Holland, P. Ingham and G. Wray (eds.), pp.

61–71. The Company of Biologists Limited, Cambridge. [15]

MANGE, A. P. and E. J. MANGE. 1980. *Genetics: Human Aspects*. Saunders, Philadelphia. [20]

MANGE, A. P. and E. J. MANGE. 1990. *Genetics: Human Aspects*, 2nd ed. Sinauer Associates, Sunderland, MA. [4,8,10,12,14]

MANIS, J. and R. S. SCHWARTZ. 1996. Agammaglobulinemia and insights into B-cell differentiation. *NEJM* 335: 1523–1525. [18]

MANN, C. C. 1994. Behavioral genetics in transition. *Science* 264: 1686–1689. [11]

MANUELIDIS, L. 1990. A view of interphase chromosomes. *Science* 250: 1533–1540. [3]

MARAN, S. P. 1991. Science is dandy, but promotion can be lucrative. *Smithsonian* 22 (May): 72–80. [1]

MARION, R. 1995. The girl who mewed. *Discover* 16 (Aug.): 38–40. [16]

MARQUIS, J. 1997. Gift of life, questions of liability. *Los Angeles Times*, Aug. 9, pp. A1ff. [20]

MARRACK, P. and J. KAPPLER. 1986. The *T* cell and its receptor. *SA* 254 (Feb.): 36–45. [18]

MARSHALL, A. 1996. Genetic tests forge ahead, despite scientific concerns. *Nature Biotechnology* 14: 1642–1643. [9]

MARSHALL, E. 1994. The company that genome researchers love to hate. *Science* 266: 1800–1802. [9]

MARSHALL, E. 1995a. Gene therapy's growing pains. *Science* 269: 1050–1055. [20]

MARSHALL, E. 1995b. Less hype, more biology needed for gene therapy. *Science* 270: 1751. [20]

MARSHALL, E. 1996. Academy's about-face on forensic DNA. *Science* 272: 803–804. [8]

MARSHALL, E. 1996. The Genome Program's conscience. *Science* 274: 488–490. [9]

MARSHALL, E. 1997. Clinton urges outlawing human cloning. *Science* 276: 1640. [20]

MARSHALL, E. and E. PENNISI. 1996. NIH launches the final push to sequence the genome. *Science* 272: 188–189. [9]

MARX, J. 1993. Learning how to suppress cancer. *Science* 261: 1385–1387. [17]

MARX, J. 1995. Mammalian sex determination: Snaring the genes that divide sexes for mammals. *Science* 269: 1824–1825. [15]

MARX, J. 1996. A second breast cancer gene is found. *Science* 271: 30–31. [17]

MARX, J. 1997. Possible function found for breast cancer genes. *Science* 276: 531–532. [17]

MASON, V. R. 1922. Sickle cell anemia. *JAMA* 79: 1318–1820. [5]

MASOOD, E. 1996. Swifter, higher, stronger: Pushing the envelope of performance. *Nature* 382: 12–13. [1]

MASSIE, R. K. 1985. *Nicholas and Alexandra*. Dell, New York. [5]

MASSIE, R. K. 1995. *The Romanovs: The Last Chapter*. Random House, New York. [5]

MATTSON, M. P. and K. FURUKAWA. 1997. Short precursor shortens memory. *Nature* 387: 457–458. [13]

MAWER, S. 1998. *Mendel's Dwarf*. Harmony Books, New York. [12]

MAZIA, D. 1987. The chromosome cycle and the centrosome cycle in the mitotic cycle. *International Review of Cytology* 100: 49–92. [3]

McCALL, K. and H. STELLER. 1997. Facing death in the fly: Genetic analysis of apoptosis in *Drosophila*. *TIG* 13: 222–226. [15]

McCLEARN, G. E. et al. 1997. Substantial genetic influence on cognitive abilities in twins 80 or more years old. *Science* 276: 1560–1563. [11]

McCONKEY, E. H. 1993. *Human Genetics: The Molecular Revolution*. Jones and Bartlett, Boston. [9]

McDONALD, J. D. and C. K. CHARLTON. 1997. Characterization of mutations at the mouse phenylalanine hydroxylase locus. *Genomics* 39: 402–405. [16]

McELFRESH, K. C., D. VINING-FORDE and I. BALAZS. 1993. DNA-based identity testing in forensic science. *BioScience* 433: 149–157. [8]

McGINNISS, M. J. AND M. M. KABACK. 1997. Carrier screening. In Rimoin et al., pp. 535–543. [19]

McGINNIS, W. and R. KRUMLAUF. 1992. Homeobox genes and axial patterning. *Cell* 68: 283–302. [15]

McGINNIS, W. and M. KUZIORA. 1994. The molecular architects of body design. *SA* 270 (Feb.): 58–66. [15]

McGOVERN, M. M., K. E. ANDERSON, K. H. ASTRIN and R. J. DESNICK. 1997. Inherited porphyrias. In Rimoin et al., pp. 2009–2036. [16]

McINERNEY, J. D. 1996. Why biological literacy matters: a review of commentaries related to *The Bell Curve: Intelligence and class structure in American Life*. *The Quarterly Review of Biology* 71: 81–96. [11]

McKUSICK, V. A. 1965. The royal hemophilia. *SA* 213 (Aug.): 88–95. [5]

McKUSICK, V. A. 1975. The growth and development of human genetics as a clinical discipline. *AJHG* 27: 261–273. [1]

McKUSICK, V. A. 1997. *Mendelian Inheritance in Man: A Catalog of Human Genes and Genetic Disorders*, 12th ed. Johns Hopkins University Press, Baltimore. (Internet: http://www3.ncbi.nlm.nih.gov/omim/) [5,7,15,16,19, Appendix 3]

McKUSICK, V. A. and D. L. RIMOIN. 1967. General Tom Thumb and other midgets. *SA* 217 (July): 102–110. [4]

McLEAN v ARKANSAS BOARD OF EDUCATION. 1982. Opinion of U. S. District Court Judge William R. Overton. Reprinted in *Science* 215: 934–943. [12]

MEIER, B. 1996. Infected hemophiliacs and tainted lawsuits. *NYT* June 11, p. D1. [7]

MEISSNER, P. N. et al. 1996. A *R59W* mutation in human protoporphyrinogen oxidase results in decreased enzyme activity and is prevalent in South Africans with variegate prophyria. *Nature Genetics* 13: 95–97. [16]

MENDEL, G. 1865. Versuche über pflanzenhybriden. *Verhandlungen des Naturforschenden Vereines in Brünn* 4: 3–47. (An English translation by W. Bateson is reprinted in Peters 1959 and in Jameson 1977. Another English translation, by E. R. Sherwood, is in Stern and Sherwood 1966.) [4]

MendelWeb, established by R. B. Blumberg, at internet address http://www.netspace.org/MendelWeb/ [4]

MERTENS, T. R. (ed.). 1975. *Human Genetics: Readings on the Implications of Genetic Engineering*. John Wiley & Sons, New York. [20]

MERYON, E. 1852. On granular and fatty degeneration of the voluntary muscles. *Medico-Chirurgical Transactions (London)* 35: 73–84. [5]

METZ, T. 1980. New Genentech issue trades wildly as investors seek latest high-flier. *The Wall Street Journal*, Oct. 15, p. 31. [8]

MICKLOS, D. A. and G. A. FREYER. 1990. *DNA Science: A First Course in Recombinant DNA Technology.* Cold Spring Harbor Laboratory Press, Cold Spring Harbor, New York. [17]

MIKI, Y. et al. 1994. A strong candidate for the breast and ovarian cancer susceptibility gene *BRCA1. Science* 266: 66–71. [9]

MILLER, O. L., JR. 1973. The visualization of genes in action. *SA* 228 (March): 34–42. [6]

MILLS, J. and H. MASUR. 1990. AIDS-related infections. *SA* 263 (Aug.): 50–57. [18]

MILOT, C. 1997. Panel backs widening net of genetic test. *Science News* 151 (April 26): 253. [19]

MILSTEIN, C. 1980. Monoclonal antibodies. *SA* 243 (Oct.): 66–74. [18]

MILUNSKY, A. 1991. *Heredity and Your Family's Health.* Johns Hopkins University Press, Baltimore. [5]

MINOGUE, B. P., R. TARASZEWSKI, S. ELIAS and G. J. ANNAS. 1988. The whole truth and nothing but the truth? *Hastings Center Report* 18 (Oct./Nov.): 34–36. [19]

MITCHELL, J. J., A. CAPUA, C. CLOW and C. R. SCRIVER. 1996. Twenty-year outcome analysis of genetic screening programs for Tay-Sachs and β-thalassemia disease carriers in high schools. *AJHG* 59: 793–798. [19]

MITCHELL, S. C. 1996. The fish-odor syndrome. *Perspectives in Biology and Medicine* 39: 515–526. [16]

MITELMAN, F., F. MERTENS and B. JOHANSSON. 1997. A breakpoint map of recurrent chromosomal rearrangements in human neoplasia. *Nature Genetics* 159 (Special Issue): 417–474 [17]

MITTWOCH, U. 1986. Males, females, and hermaphrodites. *Annals of Human Genetics* 50: 103–121. [15]

MITTWOCH, U. 1992. Sex determination and sex reversal: Genotype, phenotype, dogma, and semantics. *Human Genetics* 89: 467–479. [15]

MLOT, C. 1989. On the trail of transfer RNA identity. *BioScience* 39: 756–759 [6]

MONACO, A. P. et al. 1986. Isolation of candidate DNAs for portions of the Duchenne muscular dystrophy gene. *Nature* 323: 646–650. [9]

MONEY, J. and A. A. EHRHARDT. 1972. *Man & Woman, Boy & Girl: The Differentiation and Dimorphism of Gender Identity from Conception to Maturity.* The Johns Hopkins University Press, Baltimore. [15]

MOORE, K. L. and T. V. N. PERSAUD. 1998. *The Developing Human: Clinically Oriented Embryology,* 6th ed. Saunders, Philadelphia. [15]

MORELL, V. 1993. Huntington's gene finally found. *Science* 260: 28–30. [5]

MORGAN, T. H., A. H. STURTEVANT, H. J. MULLER and C. B. BRIDGES. 1915. *The Mechanism of Mendelian Heredity.* Henry Holt and Company, New York. [9]

MORRIS, J. M. 1953. The syndrome of testicular feminization in male pseudohermaphrodites. *American Journal of Obstetrics and Gynecology* 65: 1192–1211. [15]

MORRIS, L. N. (ed.). 1971. *Human Populations, Genetic Variation, and Evolution.* Chandler Publishing, San Francisco. [12]

MORTLOCK, D. P., L. C. POST and J.W. INNIS. 1996. The molecular basis of hypodactyly (Hd): A deletion in *Hoxa13* leads to arrest of digital arch formation. *Nature Genetics* 13: 284–289. [15]

MOTULSKY, A. G. 1960. Metabolic polymorphism and the role of infectious disease in

human evolution. *Human Biology* 32: 28–62. (Reprinted in Morris 1971.) [12,16]

MOTULSKY, A. G. 1978. Medical and human genetics 1977: Trends and directions. *AJHG* 30: 123–131. [1]

MOURANT, A. E., A. C. KOPEC and K. DOMANIEWSKA-SOBCZAK. 1976. *The Distribution of the Human Blood Groups and Other Polymorphisms,* 2nd ed. Oxford University Press, London. [12]

MOXON, E. R. and C. F. HIGGINS. 1997. A blueprint for life. *Nature* 389: 120–121. [9]

MOYZIS, R. K. 1991. The human telomere. *SA* 265 (Aug.): 48–55. [6]

MUELLER, R. F. and J. COOK. 1997. Mendelian genetics. In Rimoin et al., pp. 87–102. [10]

MUENKE, M. and U. SCHELL. 1995. Fibroblast growth factor receptor mutations in human skeletal disorders. *TIG* 11: 308–313. [2,15]

MULLER, H. J. 1935. *Out of the Night: A Biologist's View of the Future.* Vanguard Press, New York. [20]

MULLER, H. J. 1961. Human evolution by voluntary choice of germ plasm. *Science* 134: 643–649. Reprinted in Mertens (1975). [20]

MULLIS, K. B. 1990. The unusual origin of the polymerase chain reaction. *SA* 262 (April): 56–65. [1]

MULLIS, K. B. 1990. The unusual origin of the polymerase chain reaction. *SA* 262 (April): 56–65. [8]

MULLIS, K. B., F. FERRÉ, and R. A. GIBBS (eds.). 1994. *The Polymerase Chain Reaction.* Birkhäuser, Boston. [8]

MURAGAKI, Y., S. MUNDLOS, J. UPTON and J. R. OLSEN. 1996. Altered growth and branching patterns in synpolydactyly caused by mutations in *HOXD13. Science* 272: 548–551. [15]

MURPHY, F. A. 1996. The public health risk of animal organ and tissue transplantation into humans. *Science* 273: 746–747. [18]

MURRAY, A. and M. W. KIRSCHNER. 1991. What controls the cell cycle? *SA* 264 (March): 56–61. [3]

MURRAY, A. and T. HUNT. 1993. *The Cell Cycle: An Introduction.* Oxford University Press, New York. [3]

MURRAY, C. 1994. The real Bell Curve. *The Wall Street Journal,* Dec. 2: A12. [11]

MURRAY, T. H. and E. LIVNY. 1995. The human genome project: Ethical and social implications. *Bulletin of the Medical Library Association* 83: 14–21. [19]

MUSCATELLI, F. et al. 1994. Mutations in the *DAX-1* gene give rise to both X-linked adrenal hypoplasia congenita and hypogonadotropic hypogonadism. *Nature* 372: 672–676. [15]

NADEL, L. and C. J. EPSTEIN (eds.). 1992. *Down Syndrome and Alzheimer Disease.* Wiley-Liss, New York. [13]

NAGEL, R. L. 1984. The origin of the hemoglobin S gene: Clinical, genetic and anthropological consequences. *Einstein Quarterly Journal of Biology and Medicine* 2: 53–62. [12]

NATHANS, J. 1989. The genes for color vision. *SA* 260 (Feb.): 42–49. [4]

NATIONAL RESEARCH COUNCIL. 1992. *DNA Technology in Forensic Science.* National Academy Press, Washington, DC. [8]

NATIONAL RESEARCH COUNCIL. 1996. *Carcinogens and Anticarcinogens in the Human*

Diet. National Academy Press, Washington, DC. [7]

NATIONAL RESEARCH COUNCIL. 1996. *The Evaluation of Forensic Evidence.* National Academy Press, Washington, DC. [8]

NEBERT, D. W. 1997. Polymorphisms in drug-metabolizing enzymes: What is their clinical relevance and why do they exist? *AJHG* 60: 265–271. [16]

NEEL, J. V. 1994. *Physician to the Gene Pool: Genetic Lessons and Other Stories.* Wiley, New York. [1,12]

NEEL, J. V. 1995. New approaches to evaluating the genetic effects of the atomic bombs. *AJHG* 57: 1263–1266. [7]

NEEL, J. V. and W. J. SCHULL. 1954. *Human Heredity.* University of Chicago Press, Chicago. [1,10]

NEI, M. 1995. Genetic support for the out-of-Africa theory of human evolution. *PNAS* 92: 6720–6722. [12]

NELKIN, D. 1987. *Selling Science: How the Press Covers Science and Technology.* W. H. Freeman, New York. [1]

NELKIN, D. and M. S. LINDEE. 1995. *The DNA Mystique: The Gene as a Cultural Icon.* W. H. Freeman, New York. [1,11]

NELSON, K. and L. B. HOLMES. 1989. Malformations due to presumed spontaneous mutations in newborn infants. *NEJM* 320: 19–23. [12]

NEUBERGER, M. and M. BRÜGGEMANN. 1997. Mice perform a human repertoire. *Nature* 386: 25–26. [18]

NEUFELD, P. J. and N. COLMAN. 1990. When science takes the witness stand. *SA* 262 (May): 46–53. [8]

NEW, M. I. and L. S. LEVINE. 1973. Congenital adrenal hyperplasia. *Advances in Human Genetics* 4: 251–326. [16]

NEW, M. I., C. CRAWFORD and R. C. WILSON. 1997. Genetic disorders of the adrenal gland. In Rimoin et al., pp.1441–1476. [16]

NEWMAN, H. H., F. N. FREEMAN, and K. J. HOLZINGER. 1937. *Twins: A Study of Heredity and Environment.* The University of Chicago Press, Chicago. [11]

NOMURA, M. 1984. The control of ribosome synthesis. *SA* 250 (Jan.): 102–112. [6]

NORA, J. J. and F. C. FRASER. 1989. *Medical Genetics: Principles and Practice,* 3rd ed. Lea & Febiger, Philadelphia. [13]

NOSSAL, G. J. V. 1978. *Antibodies and Immunity,* 2nd ed. Basic Books, New York. [18]

NOSSAL, G. J. V. 1993. Life, death and the immune system. *SA* 269 (Sep.): 53–62. [18]

NOVICK, G. E., M. A. BATZER, P. L. DEININGER, and R. J. HERRERA. 1996. The mobile genetic element *Alu* in the human genome. *BioScience* 46: 32–41. [6]

NOWAK, R. 1994. Mining treasures from "junk DNA." *Science* 263: 608–610. [6]

NOWAK, R. 1994. Breast cancer gene offers surprises. *Science* 265: 1796–1799. [17]

NUSSBAUM, R. L. and D. H. LEDBETTER. 1995. The fragile X syndrome. In *Molecular and Metabolic Basis of Inherited Disease,* 7th ed., C. R. Scriver, A. L. Beaudet, W. S. Sly and D. Valle (eds.), pp. 795–810. McGraw-Hill, New York. [10]

NÜSSLEIN–VOLHARD, C. 1996. Gradients that organize embryo development. *SA* 275 (Aug.): 54–61. [15]

NYHAN, W. L. and D. F. WONG. 1996. New approaches to understanding Lesch-Nyhan disease. *NEJM* 334: 1602–1604. [16]

OBE, G. and A. T. NATARAJAN (eds.). 1994. *Chromosomal Alterations: Origin and Significance.* Springer-Verlag, Berlin. [13,14]

OBER, C. 1995. HLA and fertility. *AJHG* 57: 1242–1243. [18]

OBER, C. 1998. HLA and pregnancy: The paradox of the fetal allograft. *AJHG* 62: 1–5. [18]

O'BRIEN, S. J. and M. DEAN. 1997. In search of AIDS-resistance genes. *SA* 277 (Sep.): 44–51. [18]

OETTING, W. S., M. H. BRILLIANT and R. A. KING. 1996. The clinical spectrum of albinism in humans. *Molecular Medicine Today* 2: 330–335. [16]

OHLSSON, R., K. HALL and M. RITZEN (eds.). 1995. *Genomic Imprinting: Causes and Consequences.* Cambridge University Press, New York. [10]

OLD, L. J. 1996. Immunotherapy for cancer. *SA* 275 (Sep.): 136–143. [18]

OLSON, M. V. 1995. A time to sequence. *Science* 270: 394–396. [8]

OMIM. 1997. National Center for Biotechnology Information: Online Mendelian Inheritance in Man. http://www3.ncbi.nlm.nih.gov/omim/ [5,17]

ØRBECK, A. L. 1959. An early description of Huntington's chorea. *Medical History* 3: 165–168. [5]

OREL, V. 1996. *Gregor Mendel: The First Geneticist.* Oxford University Press, New York. [4]

OSBORN, D. 1916. Inheritance of baldness. *Journal of Heredity* 7: 347–355. [10]

OSTRANDER, E. and E. GINIGER. 1997. Semper fidelis: What man's best friend can teach us about human biology and disease. *AJHG* 61: 475–480. [1]

OVERZIER, C. (ed.). 1963. *Intersexuality.* Academic Press, New York. [15]

OWERBACH, D., G. I. BELL, W. J. RUTTER and T. B. SHOWS. 1980. The insulin gene is on chromosome 11 in humans. *Nature* 286: 82–84. [9]

OWOR, R. and C. OLWENY. 1978. Malignant neoplasms. In *Diseases of Children in the Subtropics and Tropics,* 3rd ed., D. B. Jelliffe and J. P. Stanfield (eds.), pp. 605–623. Edward Arnold, London. [17]

PÄÄBO, S. 1993. Ancient DNA. *SA* 269 (Nov.): 86–92. [12]

PALCA, J. 1994. The promise of a cure. *Discover* (June): 76–86. [5]

PARKER, S. L., T. TONG, S. BOLDEN and P. A. WINGO. 1997. Cancer statistics, 1997. *CA: A Cancer Journal for Clinicians* 47: 5–27. [17]

PARTINGTON, M. W. 1959. Mother and daughter showing the white forelock of Waardenburg's syndrome. *Archives of Disease in Childhood* 34: 154–157. [10]

PATTERSON, D. 1987. The causes of Down syndrome. *SA* 257 (Aug.): 52–60. [13]

PAUKER, S.P. and S. G. PAUKER. 1994. Prenatal diagnosis: Why is 35 a magic number? *NEJM* 330: 1151–1152. [19]

PAUL, W. E. (ed.). 1991. *Immunology: Recognition and Response. Readings from Scientific American Magazine.* W. H. Freeman, New York. [18]

PEBLEY, A. R. and C. F. WESTOFF. 1982. Women's sex preferences in the United States, 1970 to 1975. *Demography* 19: 177–189. [20]

PENA, S. D. J. and R. CHAKRABORTY. 1994. Paternity testing in the DNA era. *TIG* 10: 204–209. [8]

PENNINGTON, B. F. 1997. Using genetics to dissect cognition. *AJHG* 60: 13–16. [11]

PENNISI, E. 1996. Premature aging gene discovered. *Science* 272: 193–194. [7]

PENNISI, E. 1996. Teetering on the brink of danger. *Science* 271: 1665–1667. [18]

PENNISI, E. 1997. Laboratory workhorse decoded. *Science* 277: 1432–1434. [9]

PENNISI, E. 1997. The architecture of hearing. *Science* 278: 1223–1224. [10]

PENROSE, L. S. 1951. Measurement of pleiotropic effects in phenylketonuria. *Annals of Eugenics* 16: 134–141. [16]

PERERA, F. P. 1996. Uncovering new clues to cancer risk. *SA* 274 (May): 54–62. [17]

PETERS, J. A. (ed.). 1959. *Classic Papers in Genetics.* Prentice-Hall, Englewood Cliffs, NJ. [4,6,12]

PETERSON, K. and C. SAPIENZA. 1993. Imprinting the genome: Imprinted genes, imprinting genes and a hypothesis for their interaction. *Annual Review of Genetics* 27: 7–31. [10]

PIERCE, J. A. 1997. Hereditary pulmonary emphysema. In Rimoin et al., pp. 2727–2750. [16]

PLOMIN, R. 1990. The role of inheritance in behavior. *Science* 248: 183–188. [11]

PLOMIN, R. and G. E. McLEARN (eds.). 1993. *Nature, Nurture, and Psychology.* American Psychological Association, Washington, DC. [11]

PLOMIN, R. and J. C. DeFRIES. 1998. The genetics of cognitive abilities and disabilities. *SA* 278 (May): 62–69. [11]

PLOMIN, R., J. C. DeFRIES, G. E. McLEARN and M. RUTTER. 1997. *Behavior Genetics,* 3rd ed. W. H. Freeman, New York. [11]

POKORSKI, R. J. 1997. Insurance underwriting in the genetic era. *AJHG* 60: 205–216. [19]

POLLEN, D. 1993. *Hannah's Heirs: The Quest for the Genetic Origins of Alzheimer's Disease.* Oxford University Press, New York. [5,13]

PRESTON, J. 1996. Trenton votes to put strict limits on use of gene tests by insurers. *NYT* June 18, pp. A1ff. [19]

PRESTON, R. A. et al. 1994. A gene for Crouzon craniofacial dysostosis maps to the long arm of chromosome 10. *Nature Genetics* 7: 149–153. [15]

PUESCHEL, S. M. 1990. *A Parent's Guide to Down Syndrome: Toward a Brighter Future.* P. H. Brookes, Baltimore. [13]

PUNNETT, R. C. 1911. Mendelism. In *The Encyclopaedia Britannica,* 11th ed. The Encyclopaedia Britannica Company, New York. [19]

PUY, H. et al. 1997. Molecular epidemiology and diagnosis of PBG deaminase gene defects in acute intermittent porphyria. *AJHG* 60: 1373–1383. [16]

QUIRING, R., U. WALLDORF, U. KLOTER and W. J. GEHRING. 1994. Homology of the *eyeless* gene of *Drosophila* to the *small eye* gene in mice and *Aniridia* in humans. *Science* 265: 785–789. [15]

RABBITTS, T. H. 1994. Chromosomal translocations in human cancer. *Nature* 372: 143–149. [17]

RADETSKY, P. 1991. The roots of cancer. *Discover* 12 (May): 60–64. [17]

RADMAN, M. and R. WAGNER. 1988. The high fidelity of DNA duplication. *SA* 259 (Aug.): 40–46. [6]

RAFF, R. A. 1996. *The Shape of Life: Genes, Development and the Evolution of Animal Form.* University of Chicago Press, Chicago. [15]

RAFFEL, L. J., M. T. SCHEUNER, D. L. RIMOIN and J. I. ROTTER. 1997. Diabetes mellitus. In Rimoin et al., pp. 1401–1440. [10,18]

RAFTER, N. H. (ed.). 1988. *White Trash: The Eugenic Family Studies, 1877–1919.* Northeastern University Press, Boston. [20]

RAMSAY, M. 1996. Protein trafficking violations. *Nature Genetics* 14: 242–245. [16]

RAMSEY, B. W. 1996. Management of pulmonary disease in patients with cystic fibrosis. *NEJM* 335: 179–188. [5]

RAO, E. et al. 1997. Pseudoautosomal deletions encompassing a novel homeobox gene cause growth failure in idiopathic short stature and Turner syndrome. *Nature Genetics* 16: 54–63. [13]

RASHBASS, J. and L. WALSH (Eds.). 1996. *The Trends Guide to the Internet.* Supplement to the February 1996 Elsevier Trends Journals. [1]

RAWLS, R. L. 1997. Keeping cancer in check with p53. *Chemical & Engineering News* Feb. 17, pp 38–41. [17]

READ, A. P. 1995. *Pax* genes—Paired feet in three camps. *Nature Genetics* 9: 333–334. [15]

REDDY, P. G. 1987. Effects of consanguineous marriages on fertility among three endogamous groups of Andhra Pradesh. *Social Biology* 34: 68–77. [12]

REIK, W. 1989. Genomic imprinting and genetic disorders in man. *TIG* 5: 331–336. [10]

REIK, W. and E. R. MAHER. 1997. Imprinting in clusters: Lessons from Beckwith-Wiedemann syndrome. *TIG* 13: 330–334. [10]

REILLY, P. R. 1985. Eugenic sterilization in the United States. In *Genetics and the Law III,* A. Milunsky and G. J. Annas (eds.), pp. 227–241. Plenum Press, New York. [20]

REINER, W. 1997. To be male or female—that is the question. *Archives of Pediatrics and Adolescent Medicine* 151: 224–225. [15]

REISS, M. J. and M. STRAUGHAN. 1996. *Improving Nature? The Science and Ethics of Genetic Engineering.* Cambridge University Press, Cambridge. [19]

RENNIE, D. 1985. "Bubble boy." *JAMA* 253: 78–80. [18]

RENNIE, J. 1993. David's victory: Gene causing "bubble boy" illness is finally found. *SA* 268 (June): 34–35. [18]

REUBINOFF, B. E. and J. G. SCHENKER. 1996. New advances in sex preselection. *Fertility and Sterility* 66: 343–350. [20]

REVKIN, A. 1993. Hunting down Huntington's. *Discover* 14: 98–108. [5]

RHODES, D. and A. KLUG. 1993. Zinc fingers. *SA* 268 (Feb.): 56–65. [17]

RICCARDI, V. M. 1990. The phakomatoses. In *Emery and Rimoin's Principles and Practice of Medical Genetics,* 2nd ed., D. L. Rimoin, J. M. Connor and R. E. Pyeritz (eds.), pp. 435–445. Churchill Livingstone, Edinburgh. [7]

RICHARDS, F. M. 1991. The protein folding problem. *SA* 264 (Jan.): 54–63. [6]

RICHARDS, R. I. and G. R. SUTHERLAND. 1992. Fragile X syndrome: The molecular picture comes into focus. *TIG* 8: 249–255. [2,10]

RIEGER, R., A. MICHAELIS and M. M. GREEN. 1991. *Glossary of Genetics: Classical and Molecular.* Springer-Verlag, Berlin. [Glossary]

RIMOIN, D. L., J. M. CONNOR and R. E. PYERITZ (eds.). 1997. *Emery and Rimoin's Principles and Practice of Medical Genetics*, 3rd ed. 2 volumes. Churchill Livingstone, New York. [throughout text]

ROBERTS, L. 1993. Taking stock of the Genome Project. *Science* 262: 20–22. [9]

ROBERTS, L. 1993. Zeroing in on a breast cancer susceptibility gene. *Science* 259: 622–625. [17]

ROBINSON, A. and A. DE LA CHAPELLE. 1997. Sex chromosome abnormalities. In Rimoin et al., pp. 973–997. [13,14]

RODGER, J. C. and B. L. DRAKE. 1987. The enigma of the fetal graft. *American Scientist* 75: 51–57. [18]

ROESSLER, E. et al. 1996. Mutations in the human *sonic hedgehog* gene cause holoprosencephaly. *Nature Genetics* 14: 357–360. [15]

ROMEO, G. and V. A. McKUSICK. 1994. Phenotypic diversity, allelic series and modifier genes. *Nature Genetics* 7: 451–453. [10]

ROOT, M. 1988. Glow-in-the-dark biotechnology. *BioScience* 38: 745–747. [6]

ROSE, R. J. 1982. Separated twins: Data and their limits. (Book review.) *Science* 215: 959–960. [11]

ROSEN, C. M. 1987. The eerie world of reunited twins. *Discover* 8 (Sept.): 36–46. [11]

ROSENBERG, R. N. 1996. DNA triplet repeats and neurological disease. *NEJM* 335: 1222–1224. [7]

ROSENTHAL, E. 1989. The wolf at the door. *Discover* 10 (Feb.): 34–37. [18]

ROSENTHAL, N. 1995. Fine structure of a gene: DNA sequencing. *NEJM* 332: 589–591. [8]

ROSENTHAL, O. and R. H. PHILLIPS. 1996. *Coping with Color-Blindness.* Avery, Garden City Park, NY. [4]

ROSES, A. D. and M. A. PERICAK-VANCE. 1997. Alzheimer disease and other dementias. In Rimoin et al., pp. 1807–1825. [13]

ROSS, P. E. 1992. Eloquent remains. *SA* 266: 115–125. [12]

ROSSITER, B. J. F. and C. T. CASKEY. 1995. Hypoxanthine-guanine phosphoribosyltransferase deficiency: Lesch-Nyhan syndrome and gout. In Scriver et al., pp. 1679–1706. [16]

ROUECHÉ, B. 1979. Annals of medicine: live and let live. *The New Yorker* 55 (July 16): 82–87. [19]

ROUHI, A. M. 1996. Biotechnology steps up pace of rice research. *Chemical and Engineering News* 74 (Oct 7): 10–14. [8]

ROUSE, B. et al. 1997. Maternal Phenylketonuria Collaborative Study (MPKUCS) offspring: facial anomalies, malformations and early neurological sequelae. *American Journal of Medical Genetics* 69: 89–95. [16]

ROUSH, W. 1995. Conflict marks crime conference. *Science* 269: 1808–1809. [11]

ROUSH, W. 1996. Corn: A lot of change from a little DNA. *Science* 272: 1873. [1]

ROUSH, W. 1997. Antisense aims for a renaissance. *Science* 276: 1192–1193. [20]

ROWEN, L., G. MAHAIRAS and L. HOOD. 1997. Sequencing the human genome. *Science* 278: 605–607. [9]

RUDDLE, F. H. and R. S. KUCHERLAPATI. 1974. Hybrid cells and human genes. *SA* 231 (July): 36–44. [9]

RUSHTON, W. A. H. 1975. Visual pigments and color-blindness. *SA* 232 (March): 64–74. [4]

RUTHEN, R. 1993. Profile: Frederick Sanger. Revealing the hidden sequence. *SA* 269 (Oct.): 30–31. [8]

RUTLAND, P. et al. 1995. Identical mutations in the *FGFR2* gene cause both Pfeiffer and Crouzon syndrome phenotypes. *Nature Genetics* 9: 173–176. [15]

RUTLEDGE, S. L. and A. K. PERCY. 1997. Gangliosidoses and related lipid storage diseases. In Rimoin et al., pp. 2105–2130. [16]

SACHS, B. 1887. On arrested cerebral development, with special reference to its cortical pathology. *Journal of Nervous and Mental Disease* 14: 541–553. [16]

SACHSE, C., J. BROCKMÖLLER, S. BAUER and I. ROOTS. 1997. Cytochrome P450 2D6 variants in a Caucasian population: Allele frequencies and phenotypic consequences. *AJHG* 60: 284–295. [16]

SAENGER, P. 1996. Turner's syndrome. *NEJM* 335: 1749–1754. [13]

SAGAN, C. 1995. *The Demon-Haunted World: Science as a Candle in the Dark.* Random House, New York. [1]

SALTIEL, A. R. 1995. Signal transduction pathways as drug targets. *SA Science & Medicine* 2 (Nov./Dec.): 58–67. [17]

SANDBERG, A. A. 1988. Chromosomal lesions and solid tumors. *Hospital Practice* 23 (10): 93–106. [17]

SANDHOFF, K., E. CONZELMANN, E. F. NEUFELD, M. M. KABACK and K. SUZUKI. 1989. The GM$_2$ gangliosides. In *The Metabolic Basis of Inherited Disease*, 6th ed., C. R. Scriver, A. L. Beaudet, W. S. Sly and D. Valle (eds.), pp. 1807–1839. McGraw-Hill, New York. [16]

SANFILIPPO, F. 1994. HLA matching in renal transplantation. *NEJM* 331: 803–804. [18]

SAPIENZA, C. 1990. Parental imprinting of genes. *SA* 263 (Oct.): 52–60. [10]

SAPIENZA, C. and J. G. HALL. 1995. Genetic imprinting in human disease. In *Metabolic and Molecular Basis of Inherited Disease*, 7th ed., C. R. Scriver, A. L. Beaudet, W. S. Sly and D. Valle (eds.), pp. 437–458. McGraw-Hill, New York. [10]

SARGENT, T. D., C. D. MILLAR and D. M. LAMBERT. 1997. Industrial melanism and natural selection. *Evolutionary Biology*, in press. [12]

SATHANANTHAN, A. H., A. O. TROUNSON and C. WOOD. 1986. *Atlas of Fine Structure of Human Sperm Penetration, Eggs and Embryos Cultured In Vitro.* Praeger, New York. [3]

SAYRE, A. 1975. *Rosalind Franklin and DNA.* W. W. Norton, New York. [6]

SCARR, S. 1987. Three cheers for behavior genetics: Winning the war and losing our identity. *Behavior Genetics* 17: 219–228. [11]

SCARR, S. and L. CARTER-SALTZMAN. 1983. Genetics and intelligence. In *Behavior Genetics: Principles and Applications*, John L. Fuller and Edward C. Simmel (eds.), pp. 217–335. Lawrence Erlbaum Associates, Hillsdale, NJ. [11]

SCAZZOCCHIO, C. 1997. Alkaptonuria: From humans to moulds and back. *TIG* 13: 125–k127. [16]

SCHECHTER, A. N. and G. P. RODGERS. 1995. Sickle cell anemia: Basic research reaches the clinic. *NEJM* 332: 1372–1374. [7,20]

SCHIMMEL, T. W. and K. F. CRUMM 1994. Microsurgical fertilization and blastomere removal for genetic analysis. *NEJM* 331: 1200. [19]

SCHNIEKE, A. E. et al. 1997. Human factor IX transgenic sheep produced by transfer of nuclei from transfected fetal fibroblasts. *Science* 278: 2130–2133. [8]

SCHROTT, H. G., J. L. GOLDSTEIN, W. R. HAZZARD, M. M. McGOODWIN and A. G. MOTULSKY. 1972. Familial hypercholesterolemia in a large kindred: Evidence for a monogenic mechanism. *Annals of Internal Medicine* 76: 711–720. [5]

SCHULL, W. J. 1995. *Effects of Atomic Radiation; A Half-Century of Studies from Hiroshima and Nagasaki.* Wiley-Liss, New York. [7]

SCIENTIFIC AMERICAN. 1993. Special Issue: *Life, Death, and the Immune System.* 269 (Sep.): Entire issue. [18]

SCOTT, M. P. 1994. Intimations of a creature. *Cell* 79: 1121–1124. [15]

SCRIVER, C. R. 1996. Alkaptonuria: Such a long journey. *Nature Genetics* 14: 5–6. [16]

SCRIVER, C. R., R. C. EISENSMITH, S. L. C. WOO and S. KAUFMAN. 1994. The hyperphenylalaninemias. *Annual Review of Genetics* 28: 141–165. [16]

SCRIVER, C. R., A. L. BEAUDET, W. S. SLY and D. VALLE (eds.). 1995. *The Metabolic and Molecular Bases of Inherited Disease*, 7th ed. 3 volumes. McGraw-Hill, New York. (Also available in a CD-ROM version.) [throughout text]

SCRIVER, C. R., S. KAUFMAN, R. C. EISENSMITH and S. L. C. WOO. 1995. The hyperphenylalaninemias. In Scriver et al., pp. 1015–1075. [16]

SCULLEY, D. G., P. A. DAWSON, B. T. EMMERSON and R. B. GORDON. 1992. A review of the molecular basis of hypoxanthine-guanine phosphoribosyltransferase (HPRT) deficiency. *Human Genetics* 90: 195–207. [16]

SEACHRIST, L. 1995. Gene for rare disease gives cancer clues. *Science News* 148: 372. [7]

SEEGMILLER, J. E. 1997. Purine and pyrimidine metabolism. In Rimoin et al., pp. 1921–1943. [16]

SELIKOWITZ, M. 1997. *Down Syndrome: The Facts*, 2nd ed. Oxford University Press, New York. [13]

SELKOE, D. J. 1991. Amyloid protein and Alzheimer's disease. *SA* 265 (Nov.): 68–78. [13]

SERRAT, A. and A. G. de HERREROS. 1993. Determination of genetic sex by PCR amplification of Y-chromosome-specific sequences. *The Lancet* 1: 1593. [8]

SHAFER, A. J. 1995. Sex determination and its pathology in man. *Advances in Genetics* 33: 275–329. [15]

SHAKESPEARE, W. 1611. *The Tempest.* In *Complete Works of William Shakespeare: Library of Literary Classics.* 1990. Random House Value, Avenal, NJ. [16]

SHAPIRO, H. T. 1997. Ethical and policy issues of human cloning. *Science* 277: 195–196. (Full report at http://www.nih.gov/nbac/nbac.htm) [20]

SHAW, D. J. (ed.). 1995. *Molecular Genetics of Human Inherited Disease.* Wiley, New York. [15]

SHELL, E. R. 1998. The Hippocratic Wars. *The New York Times Magazine*, June 28, pp. 34–38. [1]

SHERMAN, S. 1996. Epidemiology. In *Fragile X Syndrome*, R. J. Hagerman and A. Cronister (eds.), pp. 165–192. Johns Hopkins University Press, Baltimore. [10]

SHERMAN, S. L. et al. 1997. Behavioral genetics '97: ASHG statement. Recent developments in human behavioral genetics, past accomplishments, and future directions. *AJHG* 60: 1265–1275. (See other articles following.) [11]

SHINE, I. and S. WROBEL. 1976. *Thomas Hunt Morgan, Pioneer of Genetics*. The University of Kentucky Press, Lexington. [9]

SHULDINER, A. R. 1996. Transgenic animals. *NEJM* 334: 653–655. [17]

SIGNER, E. N. et al. 1998. DNA fingerprinting Dolly. *Nature* 394: 329–330. [20]

SILVER, L. M. 1997. *Remaking Eden: Cloning and Beyond in a Brave New World*. Avon Books, New York. [20]

SIMPSON, J. L. 1997. Disorders of the gonads, genital tract and genitalia. In Rimoin et al., pp. 1477–1500. [15]

SIMPSON, J. L. and S. A. CARSON. 1992. Preimplantation genetic diagnosis. *NEJM* 327: 951–953. [19]

SIMSEK, S. et al. 1995. Rapid Rh *D* genotyping by polymerase chain reaction-based amplification of DNA. *Blood* 85: 2975–2980. [18]

SINCLAIR, A. H. et al. 1990. A gene from the human sex-determining region encodes a protein with homology to a conserved DNA-binding motif. *Nature* 346: 240–245. [15]

SINGER, M. and P. BERG (eds.). 1997. *Exploring Genetic Mechanisms*, University Science Books, Sausalito, CA. [15]

SKUSE, D. H. et al. 1997. Evidence from Turner's syndrome of an imprinted X-linked locus affecting cognitive function. *Nature* 387: 705–708. [13]

SLOAN, H. R. and D. S. FREDRICKSON. 1972. GM₂ gangliosidosis: Tay-Sachs disease. In *The Metabolic Basis of Inherited Disease*, 3rd ed., J. B. Stanbury et al. (eds.), pp. 615–638. McGraw-Hill, New York. [16]

SOLTER, D. 1998. Dolly *is* a clone—and no longer alone. *Nature* 394: 315–316. [20]

SPECTER, M. and G. KOLATA. 1997. After decades and many missteps, cloning success. *NYT* March 3, p. A1ff. [20]

SPENCE, J. E. et al. 1988. Uniparental disomy as a mechanism for human genetic disease. *AJHG* 42: 217–226. [14]

SPINNER, N. B. and B. S. EMANUEL. 1997. Deletions and other structural abnormalities of the autosomes. In Rimoin et al., pp. 999–1025. [14]

SRB, A. M., R. D. OWEN and R. S. EDGAR (eds.). 1970. *Facets of Genetics. Readings from Scientific American*. W. H. Freeman, San Francisco. [4]

STAPLES, B. 1996. Death by discrimination? *NYT*, Section 4, Nov. 4, p. 14. [11]

STAPLETON, P., A. WEITH, P. URBÁNEK, Z. KOZMIK and M. BUSSLINGER. 1993. Chromosomal localization of seven *PAX* genes and cloning of a novel family member, *PAX-9*. *Nature Genetics* 3: 292–298. [15]

STEEL, K. P. and S. D. M. BROWN. 1994. Genes and deafness. *TIG* 10: 428–435. [10]

STEEN, R. G. 1996. *DNA and Destiny: Nature and Nurture in Human Behavior*. Plenum Press, New York. [11]

STEINBACHER, R. A. and R. ERICSSON. 1994. Should parents be prohibited from choosing the sex of their child? *Health* 8 (March/April): 24. [20]

STEINBERG, D. 1998. NIH jumps into genetic variation studies. *The Scientist* 12: 1ff. [12]

STEITZ, J. A. 1988. "Snurps." *SA* 258 (June): 56–63. [6]

STELLER, H. 1995. Mechanisms and genes of cellular suicide. *Science* 267: 1445–1449. [15]

STEPHENS, F. E. and F. H. TYLER. 1951. Studies in disorders of muscle. V. The inheritance of childhood progressive muscular dystrophy in 33 kindreds. *AJHG* 3: 111–125. [5]

STERN, C. 1949. *Principles of Human Genetics*. W. H. Freeman, San Francisco. [7]

STERN, C. 1965. Mendel and human genetics. *Proceedings of the American Philosophical Society* 109: 216–226. [4]

STERN, C. and E. R. SHERWOOD (eds.). 1966. *The Origin of Genetics: A Mendel Source Book*. W. H. Freeman, San Francisco. [4]

STEWART, A. (ed.). 1990. The functional organization of chromosomes and the nucleus. Special issue. *TIG* 6: 377–437. [3]

STEWART, C. 1997. An udder way of making lambs. *Nature* 385: 769–780. [20]

STEWART, G. D., T. J. HASSOLD and D. M. KURNIT. 1988. Trisomy 21: Molecular and cytogenetic studies of nondisjunction. *Advances in Human Genetics* 17: 99–140. [13]

STOLBERG, S. G. 1997. Despite new AIDS drugs, many still lose the battle. *NYT* Aug. 22, pp. A1ff. [18]

STOLBERG, S. G. 1997a. For the infertile, a high-tech treadmill. *NYT* Dec. 14, pp. A1ff. [20]

STOLBERG, S. G. 1997b. U.S. publishes first guide to treatment of infertility. *NYT* Dec. 19, p. A20. [20]

STOLBERG, S. G. 1998. New cancer cases decreasing in U.S. as deaths do, too. *NYT* March 13, pp. A1ff. [17]

STONEKING, M. et al. 1995. Establishing the identity of Anna Anderson Manahan. *Nature Genetics* 9: 9–10. [5]

STRACHAN, T. and A. P. READ. 1996. *Human Molecular Genetics*. Wiley-Liss, New York. [9]

STRANGE, C. 1992. Cell cycle advances. *BioScience* 42: 252–256. [3]

STRICKBERGER, M. W. 1985. *Genetics*, 3rd ed. Macmillan, New York. [4]

STRICKBERGER, M. W. 1990. *Evolution*. Jones and Bartlett, Boston. [12]

STRINGER, C. B. 1990. The emergence of modern humans. *SA* 263 (Dec.): 98–104. [12]

STUART, E. T., C. KIOUSSI and P. GRUSS. 1994. Mammalian *PAX* genes. *Annual Review of Genetics* 28: 219–236. [15]

STURTEVANT, A. H. 1959. Thomas Hunt Morgan. *National Academy of Sciences, Biographical Memoirs* 33: 283–299. [9]

STURTEVANT, A. H. 1965. *A History of Genetics*. Harper & Row, New York. [1]

SUMMITT, R. L. 1973. Abnormalities of the autosomes. A *Pediatric Annals* reprint. Insight Publishing, New York. [13]

SUTHERLAND, G. R. and J. C. MULLEY. 1997. Fragile X syndrome and other causes of X-linked mental handicap. In Rimoin et al., pp. 1745–1766. [10]

SUTHERLAND, G. R. and R I. RICHARDS. 1994. Dynamic mutations. *American Scientist* 82 (March/April): 157–163. [7]

SUTHERLAND, G. R. and R. I. RICHARDS. 1994. DNA repeats: A treasury of human variation. *NEJM* 331: 191–193. [7]

SVERDRUP, A. 1922. Postaxial polydactylism in six generations of a Norwegian family. *Journal of Genetics* 12: 217–240. [10]

SWAIN, A., V. NARVAEZ, P. BURGOYNE, G. CAMERINO and R. LOVELL-BADGE. 1998. *Dax1* antagonizes *Sry* action in mammalian sex determination. *Nature* 391: 761–767. [15]

SYKES, B. 1991. The past comes alive. *Nature* 352: 381–382. [8]

SZABO, C. I. and M.-C. KING. 1997. Population genetics of *BRCA1* and *BRCA2*. *AJHG* 60: 1013–1020. [17]

TATTERSALL, I. 1997. Out of Africa again … and again? *SA* 276 (April): 60–67. [12]

TAUBES, G. 1995. Epidemiology faces its limits. *Science* 269: 164–169. [1]

TAY, W. 1881. Symmetrical changes in the region of the yellow spot in each eye of an infant. *Transactions of the Ophthalmological Society, U.K.* 1: 55–57. [16]

TELLEGEN, A. et al. 1988. Personality similarity in twins reared apart and together. *Journal of Personality and Social Psychology* 54: 1031–1039. [11]

TEMIN, H. M. 1972. RNA-directed DNA synthesis. *SA* 226 (Jan.): 24–33. [17]

TERRY, R. D. and M. WEISS. 1963. Studies in Tay-Sachs disease. II. Ultrastructure of the cerebrum. *Journal of Neuropathology and Experimental Neurology* 22: 18–55. [16]

THAYER, A. M. 1996. Biotech stocks strong in 1995 despite lackluster earnings. *Chemical and Engineering News*, March 11: 22–23. [8]

THERMAN, E. and M. SUSMAN. 1993. *Human Chromosomes: Structure, Behavior, and Effects*. Springer-Verlag, New York. [14]

THERMAN, E. and M. SUSMAN. 1995. *Human Chromosomes: Structure, Behavior, and Effects*, 3rd ed. Springer-Verlag, New York. [2,14]

THORNE, A. G. and M. H. WOLPOFF. 1992. The multiregional evolution of humans. *SA* 266 (April): 76–83. [12]

TISHKOFF, S. A., K. K. KIDD and N. RISCH. 1996. Interpretations of multiregional evolution (Response to letter). *Science* 274: 706–707. [12]

TIWARI, J. L. and P. I. TERASAKI. 1985. *HLA and Disease Associations*. Springer-Verlag, New York. [18]

TJIAN, R. 1995. Molecular machines that control genes. *SA* 272 (Feb.): 54–61. [6]

TOLMIE, J. L. 1997. Down syndrome and other autosomal trisomies. In Rimoin et al., pp. 925–971. [13]

TRAFFORD, A. (ed.). 1995. The prostate: An owners manual. *Washington Post* (Health: Special Issue), May 23: 1–28. [17]

TRAVIS, J. 1995. End games: Tips of chromosomes may contain secrets of cancer and aging. *Science News* 148: 362–365. [6]

TRAVIS, J. 1998. Cloned cows provide company for Dolly. *Science News* 154 (July 11): 21. [20]

TREACY, E., B. CHILDS, and C. R. SCRIVER. 1995. Response to treatment in hereditary metabolic disease: 1993 survey and 10-year comparison. *AJHG* 56: 359–367. [20]

TRENDS IN GENETICS. 1997. *Epigenetics: A Special Issue.* Vol. 13 (Aug.): 293–341. [10]

TRICHOPOULOS, D., F. P. LI and D. J. HUNTER. 1996. What causes cancer? *SA* 275 (Sept.): 80–88. [17]

TRYON, R. C. 1940. Genetic differences in maze-learning ability in rats. In *The Thirty-Ninth Yearbook of the National Society for the Study of Education,* G. M. Whipple (ed.), pp. 111–119 and 154–155. Public School Publishing Company, Bloomington, IL. [11]

TUDOR, A. 1989. Seeing the worst side of science. *Nature* 340: 589–592. [1]

TURNBULL, A. 1988. Woman enough for the games? *New Scientist* 119 (15 Sept.): 61–64. [15]

TURNER, G. 1996. Finding genes on the X chromosome by which *Homo* may have become *sapiens. AJHG* 58: 1109–1110. [10]

TURNER, G., T. WEBB, S. WAKE and H. ROBINSON. 1996. The prevalence of the fragile X syndrome. *American Journal of Medical Genetics* 64: 196–197. [10]

TURNER, H. H. 1938. A syndrome of infantilism, congenital webbed neck, and cubitus valgus. *Endocrinology* 23: 566–574. [13]

TYCKO, B. 1994. Genomic imprinting: mechanism and role in human pathology. *American Journal of Pathology* 144: 431–443. [10]

UPTON, A. C. 1982. The biological effects of low-level ionizing radiation. *SA* 246 (Feb.): 41–49. [7]

VALENTINE, G. H. 1986. *The Chromosomes and Their Disorders: An Introduction for Clinicians,* 4th ed. W. Heineman Medical Books, London. [13,14]

VALVERDE, P., E. HEALY, I. JACKSON, J. L. REES and A. J. THODY. 1995. Variants of the melanocyte-stimulating hormone receptor gene are associated with red hair and fair skin in humans. *Nature Genetics* 11: 328–330. [7]

VAN HEYNINGEN, V. 1996. Changing tack on the map. *Nature Genetics* 13: 134–137. [9]

VAN OMMEN, G. J. B. et al. 1986. A physical map of 4 million bp around the Duchenne muscular dystrophy gene on the human X chromosome. *Cell* 47: 499–504. [14]

VARMUS, H. and R. A. WEINBERG. 1993. *Genes and the Biology of Cancer.* Scientific American Library, New York. [17]

VELANDER, W. H., H. LUBON, and W. N. DROHAN. 1997. Transgenic livestock as drug factories. *SA* 276: 70–74. [8,18]

VENTER, J. C. et al. 1998. Shotgun sequencing of the human genome. *Science* 280: 1540–1542.

VERMA, I. M. and N. SOMIA. 1997. Gene therapy—promises, problems and prospects. *Nature* 389: 239–242. [20]

VIDAUD, D. et al. 1989. Father-to-son transmission of hemophilia A due to uniparental disomy. *AJHG* 45: A226. [14]

VIG, B. K. (ed.). 1993. *Chromosome Segregation and Aneuploidy.* NATO ASI Series H: Cell Biology, Vol. 72. Springer-Verlag, New York. [13,14]

VINES, G. 1994. Gene tests: The parents' dilemma. *New Scientist* 144 (Nov. 12): 40–44. [19]

VON BOEHMER, H. and P. KISIELOW. 1991. How the immune system learns about self. *SA* 265 (Oct.): 74–81. [18]

WACHTEL, S. S. (ed.). 1994. *Molecular Genetics of Sex Determination.* Academic Press, New York. [15]

WADE, N. 1997. Male chromosome is not a genetic wasteland, after all. *NYT* Oct. 28, p. C3. [5]

WADE, N. 1997. Scientists map a bacterium's genetic code. *NYT* Aug. 7, p. A30. [9]

WADE, N. 1997. To people the world, start with 500. *NYT* (Nov 11): C1ff. [12]

WADE, N. 1997a. How cells unwind tangled skein of life. *NYT* Oct. 21, pp. C1, C6. [2]

WADE, N. 1997b. Artificial human chromosome is new tool for gene therapy. *NYT* Apr. 1, p. C3. [2]

WADE, N. 1998. Cell rejuvenation may yield rush of medical advances. *NYT* Jan. 20, pp. C1ff. [6]

WADE, N. 1998. The struggle to decipher human genes. *NYT* March 10, pp. C1ff. [9]

WAGNER, R. P., M. P. MAGUIRE and R. L. STALLINGS. 1993. *Chromosomes: A Synthesis.* Wiley-Liss, New York. [2]

WAKAYAMA, T., A. C. F. PERRY, M. ZUCCOTTI, K. R. JOHNSON and R. YANAGIMACHI. 1998. Full-term development of mice from enucleated oocytes injected with cumulus cell nuclei. *Nature* 394: 369–374. [20]

WALD, N. J. and A. KENNARD. 1997. Prenatal screening for neural tube defects and Down syndrome. In Rimoin et al., pp. 545–562. [13,19]

WALKER, A. P. 1997. Genetic counseling. In Rimoin et al., pp. 595–618. [19]

WALLACE, A. R. 1855. On the law which has regulated the introduction of new species. *Annals and Magazine of Natural History,* Series 2, 16 (Sept): 184–196. (Reprinted in Wallace 1875) [12]

WALLACE, A. R. 1875. *Contributions to the Theory of Natural Selection: A Series of Essays.* Macmillan and Co., London. [12]

WALLACE, A. R. 1889. *Darwinism: An Exposition of the Theory of Natural Selection with Some of Its Applications.* Macmillan and Co., London. [12]

WALLACE, A. R. 1908. *My Life; A Record of Events and Opinions.* Chapman and Hall, London. [12]

WALLACE, D. C. 1995. Mitochondrial DNA variation in human evolution, degenerative disease, and aging. *AJHG* 57: 201–223. [5]

WALLACE, D. C. 1997. Mitochondrial DNA in aging and disease. *SA* 277 (Aug.): 40–47. [5]

WALLACE, I. and A. WALLACE. 1978. *The Two.* Simon & Schuster, New York. [11]

WALLACE, M. R. et al. 1991. A *de novo Alu* insertion results in neurofibromatosis type 1. *Nature* 353: 864–866. [7]

WALLIS, C. 1991. A puzzling plague. *Time* Jan 14, pp. 48–55. [17]

WAMBAUGH, J. 1989. *The Blooding.* William Morrow, New York. [8]

WARBURTON, D. 1987. Reproductive loss: How much is preventable? *NEJM* 316: 158–160 [14]

WARBURTON, D. and A. KINNEY. 1996. Chromosomal differences in susceptibility to meiotic aneuploidy. *Environmental and Molecular Mutagenesis* 28: 237–247. [13]

WARD, R. and C. STRINGER. 1997. A molecular handle on the Neandertals. *Nature* 388: 225–226. [12]

WARREN, S. T. 1996. The expanding world of trinucleotide repeats. *Science* 271: 1374–1375. [7]

WASSARMAN, P. M. 1988. Fertilization in mammals. *SA* 259 (Dec.): 78–84. [3]

WATSON, J. D. 1968. *The Double Helix.* Atheneum, New York. [6]

WATSON, J. D. and F. H. C. CRICK. 1953. Molecular structure of nucleic acids: A structure for deoxyribose nucleic acid. *Nature* 171: 737–738. Reprinted in Peters (1959) and Freifelder (1978). [6]

WATSON, J. D. and J. TOOZE. 1981. *The DNA Story: A Documentary History of Gene Cloning.* W. H. Freeman, New York. [8]

WATSON, J. D., N. H. HOPKINS, J. W. ROBERTS, J. A. STEITZ, and A. M. WEINER. 1987. *Molecular Biology of the Gene,* 4th ed. Vol. I: *General Principles.* Vol. II: *Specialized Aspects.* Benjamin/Cummings, Menlo Park, California. [6]

WEBB, T. 1994. Inv dup (15) supernumerary marker chromosomes. *Journal of Medical Genetics* 31: 585–594. [14]

WEBER, W. W. 1997. *Pharmacogenetics.* Oxford University Press, New York. [16]

WEIGALL, A. 1924. *The Life and Times of Cleopatra, Queen of Egypt.* G. P. Putnam's Sons, New York. [12]

WEINBERG, R. A. 1983. A molecular basis of cancer. *SA* 249 (Nov.): 126–142. [17]

WEINBERG, R. A. 1996a. How cancer arises. *SA* 275 (Sept.): 62–70. [17]

WEINBERG, R. A. 1996b. *Racing to the Beginning of the Road: The Search for the Origins of Cancer.* Harmony Books, New York. [17]

WEIR, B. S. 1995. DNA statistics in the Simpson matter. *Nature Genetics* 11: 365–368. [8]

WEIR, R. F., S. C. LAWRENCE and E. FALES (eds.). 1994. *Genes and Human Self-Knowledge: Historical and Philosophical Reflections on Modern Genetics.* University of Iowa Press, Iowa City. [1]

WEISSENBACH, J. 1995. Genetic and molecular analysis of the human Y chromosome. In *Molecular Genetics of Human Inherited Disease,* D. J. Shaw (ed.), pp. 155–174. Wiley, New York. [15]

WELSH, M. J. and A. E. SMITH. 1995. Cystic fibrosis. *SA* 273 (December): 52–59. [5]

WERNER, M. H., J. R. HUTH, A. M. GRONENBORN and G. M. CLORE. 1996. Molecular determinants of mammalian sex. *Trends in Biological Science* 21: 302–308. [15]

WERTH, B. 1991. How short is too short? *New York Times Magazine* June 16, pp. 14ff. [6]

WERTZ, D. C. 1996. Ethical and social issues in preimplantation genetic diagnosis. *The Genetic Resource* 10 (1): 17–26. [Published by the Massachusetts Department of Public Health] [19]

WERTZ, D. C. and J. C. FLETCHER. 1989. Fatal knowledge? Prenatal diagnosis and sex selection. *Hastings Center Report* 19 (May/June): 21–27. [20]

WEXLER, ALICE. 1995. *Mapping Fate: A Memoir of Family, Risk and Genetic Research.* Oxford University Press, New York. [1]

WHITE, R. and J. LALOUEL. 1988. Chromosome mapping with DNA markers. *SA* 258 (Feb.): 40–48. [9]

WILFOND, B. S. and N. FOST. 1990. The cystic fibrosis gene: Medical and social implications for heterozygote detection. *JAMA* 263: 2777–2783. [19]

WILFORD, J. N. 1995. In Kenya, fossils are found of earliest walking species. *NYT* Aug. 17, pp. A1,A8. [12]

WILFORD, J. N. 1996. Three human species coexisted on earth, new data suggest. *NYT* Dec 13, pp. A1,B14. [12]

WILKE, C. M. et al. 1994. Multicolor FISH mapping of YAC clones in 3p14 and identification of a YAC spanning both *FRA3B* and the t(3;8) associated with hereditary renal cell carcinoma. *Genomics* 22: 319–326. [9]

WILKIE, A. O. M. et al. 1995. Apert syndrome results from localized mutations of *FGFR2* and is allelic with Crouzon syndrome. *Nature Genetics* 9: 165–172. [15]

WILLARD, H. F. 1990. Centromeres of mammalian chromosomes. *TIG* 6: 410–416. [3]

WILLARD, H. F. 1995. The sex chromosomes and X-chromosome inactivation. In *The Metabolic and Molecular Bases of Inherited Disease*, 7th ed., C. R. Scriver, A. L. Beaudet, W. S. Sly and D. Valle (eds.), pp. 719–737. McGraw-Hill, New York. [10,13]

WILLARD, H. F. 1996. X-chromosome inactivation, *XIST* and pursuit of the X-inactivation center. *Cell* 86: 5–7. [10]

WILLIAMS, M. 1997. Little prisoner of the light. *NYT* May 14, p. A19. [7]

WILLIAMS, S. A., B. E. SLATKO, L. S. MORAN and S. M. DeSIMONE. 1986. Sequencing in the fast lane: A rapid protocol for [^{35}S] dATP dideoxy DNA sequencing. *BioTechniques* 4: 138–147. [8]

WILLIAMSON, B. 1996. Towards non-invasive prenatal diagnosis. *Nature Genetics* 14: 239–240. [19]

WILMUT, I., A. E. SCHNIEKE, J. McWHIR, A. J. KIND and K. H. S. CAMPBELL. 1997. Viable offspring derived from fetal and adult mammalian cells. *Nature* 385: 810–813. [20]

WILSON, A. C. and R. L. CANN. 1992. The recent African genesis of humans. *SA* 266 (April): 68–73. [12]

WILSON, E. B. 1925. *The Cell in Development and Heredity.* Macmillan, New York. [2]

WINN, M. 1990. New views of human intelligence. *New York Times Magazine* (Part 2, *Good Health Magazine*) April 29, pp. 16 ff. [11]

WINTERS, R. W., J. B. GRAHAM, T. F. WILLIAMS, V. C. McFALLS and C. H. BURNETT. 1957. A genetic study of familial hypophosphatemia and vitamin D-resistant rickets. *Transactions of the Association of American Physicians* 70: 234–242. [5]

WOLFFE, A. P. 1995. Genetic effects of DNA packaging. *SA Science & Medicine* 2 (Nov./Dec.): 68–77. [2]

WOLPERT, L. 1992. *The Triumph of the Embryo.* Oxford University Press, Oxford. [15]

WOLPERT, L. et al. 1998. *Principles of Development.* Current Biology, London. [15]

WRIGHT, L. 1995. Double mystery. *The New Yorker*, Aug. 7, pp. 45–62. [11]

WRIGHT, S. 1930. *The Genetical Theory of Natural Selection* (Book review). *Journal of Heredity* 21: 349–356. [12]

WU, C. 1997. Banking on blood conversion. *Science News* 151 (Jan. 11): 24–25. [18]

WYLLIE, A. 1998. An endonuclease at last. *Nature* 391: 20–21. [15]

YODER, J. A., C. P. WALSH and T. H. BESTOR. 1997. Cytosine methylation and the ecology of intragenomic parasites. *TIG* 13: 335–340. [10]

YOON, P. W. et al. 1996. Advanced maternal age and the risk of Down syndrome characterized by the meiotic stage of the chromosomal error: A population-based study. *AJHG* 58: 628–633. [13]

YUNIS, J. J. 1976. High resolution of human chromosomes. *Science* 191: 1268–1270. [2]

ZALLEN, D. T. 1997. *Does It Run in the Family? A Consumer's Guide to DNA Testing for Genetic Disorders.* Rutgers University Press, New Brunswick, NJ. [19]

ZARUTSKIE, P. W., C. H. MULLER, M. MAGONE, and M. R. SOULES. 1989. The clinical relevance of sex selection techniques. *Fertility and Sterility* 52: 891–905. [20]

ZHANG, Y.-H. et al. 1998. *DAX1* mutations map to putative structural domains in a deduced three-dimensional model. *AJHG* 62: 855–8864. [15]

ZIGAS, V. 1990. *Laughing Death: The Untold Story of Kuru.* Humana Press, Clifton, NJ. [10]

ZINN, A. R. 1997. Growing interest in Turner syndrome. *Nature Genetics* 16: 3–4. [13]

ZOGHBI, H. Y. and A. BALLABIO. 1995. Waardenburg syndrome. In Scriver et al., pp. 4575–4580. [15]

ZOURLAS, P. A. and H. W. JONES. 1965. Clinical, histologic and cytogenetic findings in male hermaphroditism. *Obstetrics and Gynecology* 25: 768–778. [15]

ZSCHOCKE, J., J. P. MALLORY, H. G. EIKEN and N. C. NEVIN. 1997. Phenylketonuria and the peoples of Northern Ireland. *Human Genetics* 100: 189–194. [16]

ZURER, P. 1994. DNA profiling fast becoming accepted tool for identification. *Chemical and Engineering News* 72 (October 10): 8–15. [8]

Index

About the Book

Editor: Andrew D. Sinauer

Project Editor: Carol J. Wigg

Copy Editor: Janet Greenblatt

Production Manager: Christopher Small

Electronic Book Production: Wendy Beck

Illustrations: J/B Woolsey Associates

Book and Cover Design: Jefferson Johnson

Cover Manufacturer: Henry N. Sawyer Company, Inc.

Book Manufacturer: The Courier Companies, Inc.